Soft-Switching Technology for Three-phase Power Electronics Converters

IEEE Press
445 Hoes Lane
Piscataway, NJ 08854

Soft-Switching Technology for Three-phase Power Electronics Converters

Dehong Xu, Rui Li, Ning He, Jinyi Deng, Yuying Wu

IEEE Press Series on Power and Energy Systems
Ganesh Kumar Venayagamoorthy, Series Editor

IEEE PRESS

WILEY

Published by John Wiley & Sons, Inc., Hoboken, New Jersey.
Published simultaneously in Canada.

Library of Congress Cataloging-in-Publication Data

Names: Xu, Dehong, 1961– author.
Title: Soft-switching technology for three-phase power electronics converters / Dehong Xu, Rui Li, Ning He, Jinyi Deng, Yuying Wu.
Description: First edition. | Hoboken, New Jersey : John Wiley & Sons, Inc., [2022] | Series: IEEE Press series on power and energy systems | Includes bibliographical references and index.
Identifiers: LCCN 2021048504 (print) | LCCN 2021048505 (ebook) | ISBN 9781119602514 (cloth) | ISBN 9781119602521 (adobe pdf) | ISBN 9781119602552 (epub)
Subjects: LCSH: Electric current converters. | Switching power supplies. | Electic currents, Alternating–Three-phase.
Classification: LCC TK7872.C8 X83 2022 (print) | LCC TK7872.C8 (ebook) | DDC 621.3815/322–dc23/eng/20211122
LC record available at https://lccn.loc.gov/2021048504
LC ebook record available at https://lccn.loc.gov/2021048505

Cover Design: Wiley

MIX
Paper from responsible sources
FSC® C013604

Contents

Preface

The three-phase power electronic converter is one of the most important power conversion building blocks in power electronics systems. It is widely used in renewable energy generation, power supply for data centers, drives for industry or transportations, etc. The converter is composed of power semiconductor devices that turn on or turn off in less than a microsecond. Power flow is controlled accurately, efficiently, and quickly by the converter due to the fast action of power semiconductor devices. However, the voltage and current of a device have an overlapping time during turn-on or turn-off transients, which results in switching loss, higher stress on the devices, and Electromagnetic Interference (EMI) noise. The switching loss of a converter is proportional to its operating switching frequency. Accordingly, switching frequency is restricted in order to achieve expected conversion efficiency. The lower switching frequency may result in bulky passive components, lower dynamics, higher audible noise, etc. Aiming to reduce or eliminate the voltage and current overlapping during the switching actions, soft-switching technology occurs. It is a way to shape the voltage and current of the power device during switching transients so that the overlapping of voltage and current on the power device during switching commutations is reduced. It not only reduces switching loss and EMI noise but also suppresses stress on the power devices. Soft-switching has been widely utilized in DC–DC converters in switching power supply and single-phase inverters for induction heating. However, there is almost no book to systematically investigate soft-switching technology of three-phase converters or inverters. As we know, applications of soft switching to three-phase converters/inverters have a large potential in the future in distributed power generation, data centers, industrial power supplies, Electric Vehicle (EV) charging stations, high-speed motor drives, etc. This book tries to give readers an overview of the progress of the soft-switching three-phase conversion.

This book emphasizes circuit analysis, pulse-width-modulation (PWM) control, and the design of zero-voltage-switching (ZVS) three-phase converters. First, this book gives an introduction to the fundamentals of soft-switching three-phase conversion. Requirements for three-phase power conversions such as conversion efficiency, power density, and impact of soft-switching technique are explained. It includes an overview of the progress of soft-switching technology for three-phase converters. It introduces applications of soft-switching three-phase power conversion to renewable energy generation, industry, transportation, etc. A ZVS space vector modulation (ZVS-SVM) is introduced, which is modified from SVM to adapt to the soft-switching operation of the three-phase converters. A generic soft-switching modulation method for three-phase converters, edge-aligned PWM (EA-PWM) is introduced. Then circuit analysis, soft-switching condition, and control of active-clamped three-phase converter with EA-PWM are provided. Second, the book will introduce applying soft-switching technology to three-phase rectifiers. Two types of soft-switching circuits are investigated. It includes circuit analysis, soft-switching condition derivation, circuit parameters design, and experiment results of the three-phase rectifier prototypes. Afterward, applying soft-switching technology to three-phase grid inverters are studied. Two types of soft-switching circuits are introduced. Similarly, it includes circuit analysis, soft-switching condition derivation, circuit parameters design, and experiment results of the three-phase inverter prototypes. Since resonant inductor is a critical component with respect to magnetic and copper loss, thermal characteristics, and size, the design of the resonant inductor is explained. In addition, the optimization of a grid inverter is investigated with regard to power density and efficiency. Finally, due to the increasing application of the wide-bandgap device to three-phase converters, this book investigates the impact of SiC devices on soft-switching three-phase converters with respect to efficiency and power density enhancement. Three examples of soft-switching converters with SiC devices are studied, which include soft-switching SiC three-phase grid inverter, soft-switching SiC single-phase grid inverter with active power decoupling, and soft-switching SiC three-phase back-to-back (BTB) converter. They are helpful for the readers to understand how the practical system works and can be a guide to build a test bench in a laboratory. This book may be helpful for readers who hope to have knowledge of circuit analysis, PWM control and design of ZVS three-phase converters, and deep understanding of the soft-swithcing technique of three-phase conversions. It is suitable for both undergraduate and graduate levels and may serve as a useful reference for academic researchers, engineers, managers, and other professionals in the industry. Most of the chapters include descriptions of fundamental and advanced concepts, supported by many illustrations.

Dr. Dehong Xu has designed the contents and writing plan of the book. He wrote Chapters 1, 2, and 3. Dr. Rui Li wrote Chapters 4, 5, and 6. Dr. Ning He wrote Chapters 7, 8, and 10. Mr. Jinyi Deng wrote Chapters 9 and 12. Mr. Yuying Wu wrote Chapter 11. Dr. Dehong Xu has helped revise all the chapters of the book.

The authors would like to acknowledge the contribution of colleagues and former graduate students of Zhejiang University, Dr. Gang Chen, Dr. Bo Feng, Dr. Chengrui Du, Dr. Keyan Shi, Dr. Yenan Chen, Dr. Min Chen, Dr. Changsheng Hu, Dr. Ke Ma, Mr. Zhiyuan Ma, Mr. Yingfeng Zhu, Ms. Yawen Li, Mr. An Zhao, Mr. Zhengyu Ye, and Mr. Ruizhe Wang, for developing the soft-switching technology for the three-phase converters. We acknowledge the tireless efforts and assistance of Wiley Press editorial staff.

Dehong Xu
Rui Li
Ning He
Jinyi Deng
Yuying Wu

Nomenclature

Subscripts

a	phase A
b	phase B
c	phase C
g	gate of switch
m	modulation
r	resonance
s	switching cycle

Superscripts

$\vec{x}^{ref}$	reference vector
x^{ref}	reference value

Variables

v	variable voltage
V	constant voltage
$\vec{V}$	voltage vector
i	variable current
I	constant current
$\vec{I}$	current vector
v_{dc}	instantaneous voltage of DC side
V_{dc}	DC component of voltage
i_{dc}	instantaneous current of DC side
I_{dc}	DC component of current
T_s	switching period
T_r	resonant period
f_s	switching frequency
f_r	resonant frequency

f_l	grid frequency
ω_r	resonant angular frequency
ω_l	grid angular frequency
φ_0	initial phase angle of grid
P	active power
Q	reactive power
p	instant power
PF	power factor
η	efficiency
d	duty cycle
ϕ	magnetic flux
ψ	magnetic flux linkage

Part 1

Fundamental of Soft-switching

1

Introduction

In this chapter, an overview of soft-switching technology for three-phase power electronics converters and its evolution are briefly introduced, and the challenges and trends in the soft-switching three-phase converters are discussed.

1.1 Requirement of Three-phase Power Conversions

Three-phase converters are widely used as grid connecting inverters for renewable energy systems, rectifiers, or inverters for uninterruptible power supply (UPS) for data centers, rectifiers for electrical vehicle (EV) fast charging stations, inverters for EV, inverters for industrial motor drives, etc. For these applications, power flow is quickly and accurately controlled because the power converter is composed of power semiconductor devices, which can be turn on or off within less than a microsecond. With the application of the converters, we can realize high efficiency power conversion between sources and loads or vice versa. In addition, if a converter system operates at high frequency, its volume or weight is reduced due to size reduction of passive components such as inductors, capacitor, electric motors, etc. The higher the switching frequency, the smaller are the passive components. Thus the power density, processing power per liter, of the converter is increased [1, 2]. In addition, the dynamics of converter systems are enhanced with increased switching frequency.

1.1.1 Three-phase Converters

Three-phase converters are used as either grid converters to connect the utility or inverters to drive a motor or supply high-quality alternating current (AC) power to the load as shown in Figure 1.1. When a grid converter is used to convert the utility AC voltage into direct current (DC) voltage, it is usually called a rectifier. When it converts DC voltage to the grid AC voltage, it is usually called an inverter. Actually, it is the same entity, but it has two names. It is sometimes confusing for new learners. In most applications such as battery storage systems, the grid converter is required to control power flow bidirectionally. It can operate in either rectifier mode or inverter mode according to the system requirement. When a three-phase converter is used to drive a motor or supply AC power to a load, it is usually called as an inverter since it converts the DC bus voltage into three-phase AC voltages. In the book, a general name "converter" is used, which covers names such as rectifier, inverter, bidirectional converter that swaps between the inverter mode and the rectifier mode according to operation requirement.

Soft-Switching Technology for Three-phase Power Electronics Converters, First Edition.
Dehong Xu, Rui Li, Ning He, Jinyi Deng, and Yuying Wu.

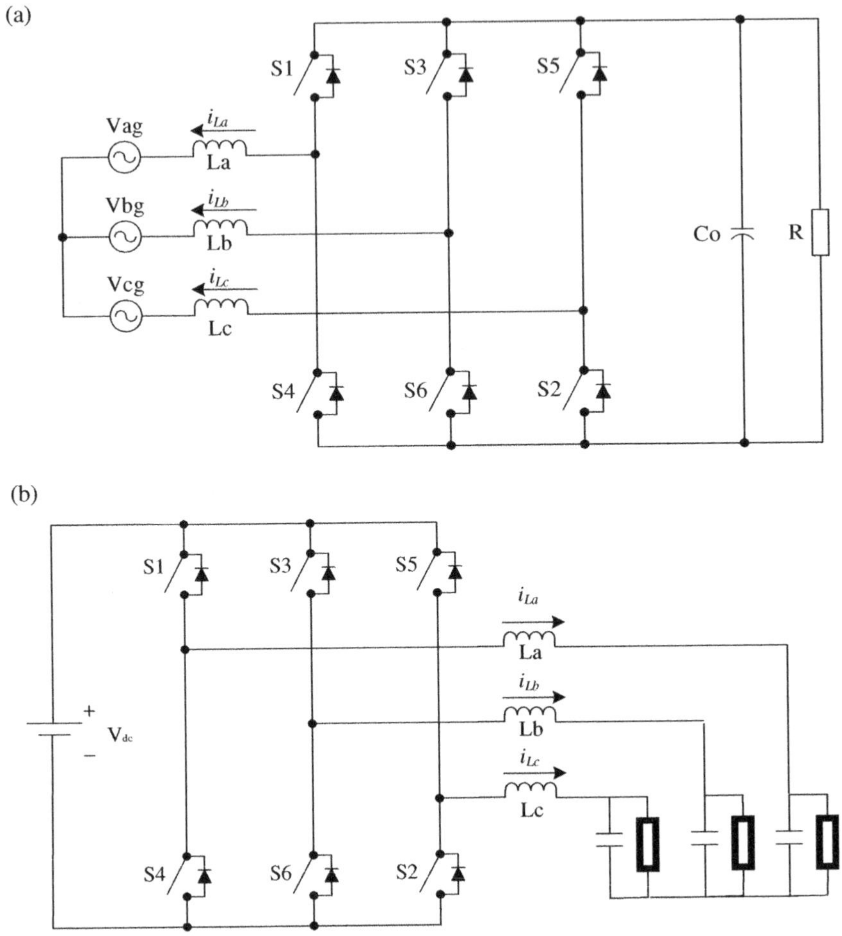

Figure 1.1 Three-phase converters: (a) grid converters; (b) inverter.

The three-phase converter is one of the most important power conversion building blocks in power electronic systems. It is widely used in various applications due to its distinct advantages as follows:

- It has the simplest converter topology that can realize DC/AC or AC/DC conversions with minimum number of power switches. It has lower cost.
- It has lower voltage stress on the power devices because the maximum voltage on them is capped by the DC bus. It has lower current stress on the device for AC loads/sources to operate at current continuous mode (CCM).
- There are well-established pulse-width-modulation (PWM) control methods such as sinusoidal PWM (SPWM), third harmonics injected SPWM, space vector modulation (SVM), etc.
- It has well-established system control methods for the converter under *abc* static frame, $\alpha\beta$ static frame, and *dq* synchronous rotating frame.

Because of these advantages, the three-phase converters have been used almost everywhere from low power to high power such as disk drives, inverters for pumps, inverters for EV, drive inverters for bullet trains, solar inverters, wind turbines, UPS, etc. For these applications, there is an ever increasing demand for the performance of the converter. In addition to basic functions such as AC to DC or DC to AC conversion and output power quality, following demands are critical for the converter.

First, it is expected that the converter has higher efficiency, which has become a more stringent requirement than ever before due to increasing public concern of impact of energy consumption on the environment. Besides, high efficiency can also bring economic benefits to the users. A Photovoltaic (PV) inverter with high efficiency can harvest more electricity and also improves the utilization of the PV panel. A UPS with high efficiency can save the operating cost of data centers. It also can reduce the footprint of the power supply due to lower energy loss. An EV power train with high efficiency increases the utilization of the kWh of the battery and extends mileage at the same time.

Second, we hope that the converter has small size. It is especially critical for moving vehicles such as electric vehicles, electric railway, boats, and airplanes or aerospace applications. Smaller size means less use of material, copper, iron, isolation material, etc., which cuts down the cost. For users, it can reduce space, which is expensive in many large cities. Usually, you may hear the word *power density*. It means the ratio of power capacity to the volume or weight of the converter. It represents the ability of a converter to process the power at a given size. For a given power capacity of a converter, the higher the power density, the smaller is the size of the converter.

Third, the converter is required to have good dynamics. The dynamics of a converter mainly depends on bandwidth of the close loop control systems. It is mainly constrained by the switching frequency of the converter. Generally speaking, power electronic converters have high dynamics since they use power semiconductor devices as the switch. In many applications such as ultra-high-speed pumps and compressors in industrial applications, power generation for aeronautics, EV, etc., it is required that the converter drives the motor to reach ultra-high speed from tens to hundreds of thousands rpm. In some applications such as moving vehicles, the size of the electric machines can be reduced by increasing their operating frequency. To increase the fundamental frequency of the converter, it is natural to increase switching frequency. What is the limit of the switching? We will discuss it in detail later.

Another demand is lower cost. The cost of a converter or a converter system should be optimized. A converter system is generally composed of power semiconductor devices and passive components. Size of the passive components depends on operating frequency. If we can increase the frequency, their size can be reduced.

Actually, switching frequency is a critical parameter for the converter. It plays an important role in efficiency, power density, dynamics, and cost reduction.

1.1.2 Switching Frequency vs. Conversion Efficiency and Power Density

In this section, we will discuss the effects of the switching frequency on the converter. First, an inverter used for UPS is taken as an example. The main circuit of UPS is back to back (BTB) converter as shown in Figure 1.2. It is composed of three-phase grid converter as the rectifier cascaded with a three-phase converter as the inverter to provide high-quality power to the load. To satisfy the load requirement, output voltage of UPS needs to be an almost sinusoidal waveform. Its output voltage quality is usually described by total harmonic distortion (THD). Filters are installed in load sides. In addition, to satisfy the grid standard, filters in utility side are also installed. Filter cost ranges from 15 to 30% of total material expenditure in the UPS (excluding the battery cost). Besides, it also occupies a large footprint. Size of the passive components depends on operating switching frequency. If we can increase the switching frequency, their size will be reduced.

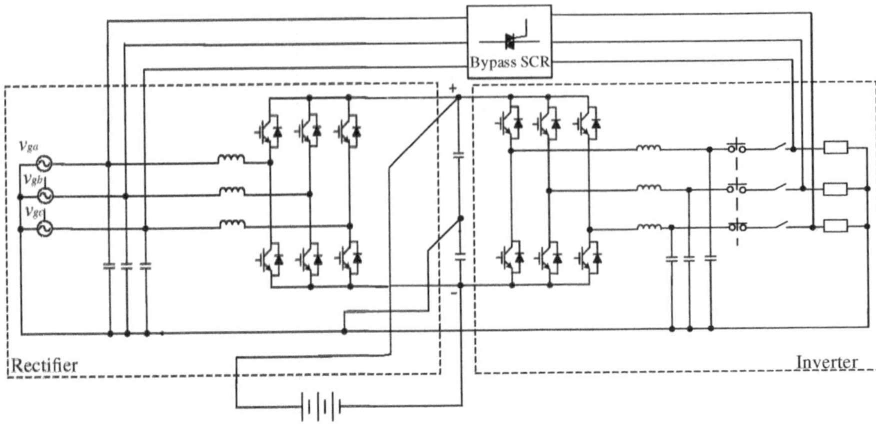

Figure 1.2 Circuit diagram of UPS.

It is assumed that a UPS equipment has rated power 100 kVA at load power factor PF = 0.8. Both the input grid phase voltage and output phase voltage are 220 Vrms. Internal DC bus voltage is 750 Vdc. THD of output voltages is less than 5%. Inductor-capacitor filters are used to improve output voltage waveform. The filter inductors are designed with amorphous core, and their maximum magnetic flux density is Bmax = 1.2 T.

According to the aforementioned assumption, filter inductance of the output converter is designed with different switching frequency as shown in Figure 1.3a. It is observed that the filter inductance decreases with an increase in the switching frequency. Similarly, the size of the filter inductor is also reduced with an increase in the switching frequency as shown in Figure 1.3b. The shaded bar shows the weight of the magnetic core of the filter. The blank bar shows the weight of the copper winding of the filter. The weight of the inductor is reduced about to one of the fifth by increasing switching frequency from 10 to 100 kHz. Figure 1.3c shows total loss of three output filter inductors vs. the switching frequency. The inductor loss also decreases with an increase in the switching frequency. It is because we use a smaller inductance when the switching frequency is higher. A smaller inductance means it has short length of winding so that copper loss is reduced. A smaller inductance needs a smaller core, whose loss depends on its magnetic core volume for a given maximum magnetic flux density. The smaller the core, the smaller the core loss. Thus, a small inductance results in both smaller copper loss and core loss. As a result, with an increase in the switching frequency, both the filter inductor size and loss decrease. Cost of the filter inductor is also cut down. It seems that it is better to design the UPS at higher switching frequency. Unfortunately, there is a switching frequency limit due to the loss of the power semiconductor devices in UPS. It will be discussed later.

Now an inverter used for power trains of electric vehicles is investigated. The main circuit of power train is shown in Figure 1.4. It is composed of battery, film capacitor, three-phase switch bridge and motor. Power density is critical for the power train. The film capacitor C_{dc} occupies a large footprint. To suppress voltage ripple on the battery to a certain value, following capacitance is required [3]:

$$C_{dc} = \frac{3\sqrt{3}I_0 \cos(\varphi)}{16 f_s \Delta V_{dc}} \tag{1.1}$$

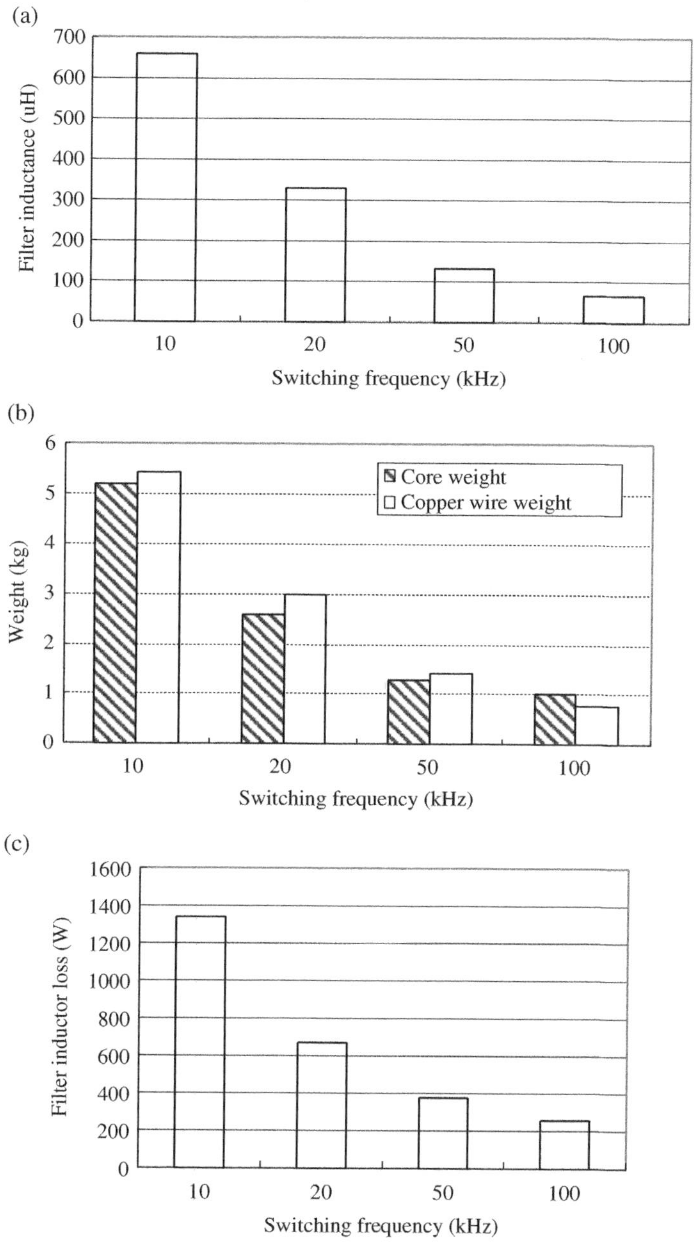

Figure 1.3 Filter inductance, weight and loss vs. switching frequency. (a) Filter inductance vs. fs. (b) Weight of each filter inductor vs. fs. (c) Total loss of three filter inductors vs. fs.

where I_o is output phase current (rms value), $\cos(\varphi)$ is load power factor, f_s is switching frequency, and ΔV_{dc} is maximum DC voltage ripple allowed on the battery. It is observed that the required DC bus capacitance is inversely proportional to switching frequency f_s.

It is assumed that DC bus voltage V_{dc} is 320 V, the inverter power is 120 kW, load current is 400 A, power factor of the motor is 0.93, and maximum DC voltage ripple allowed on the battery $\Delta V_{dc} = 32$ V.

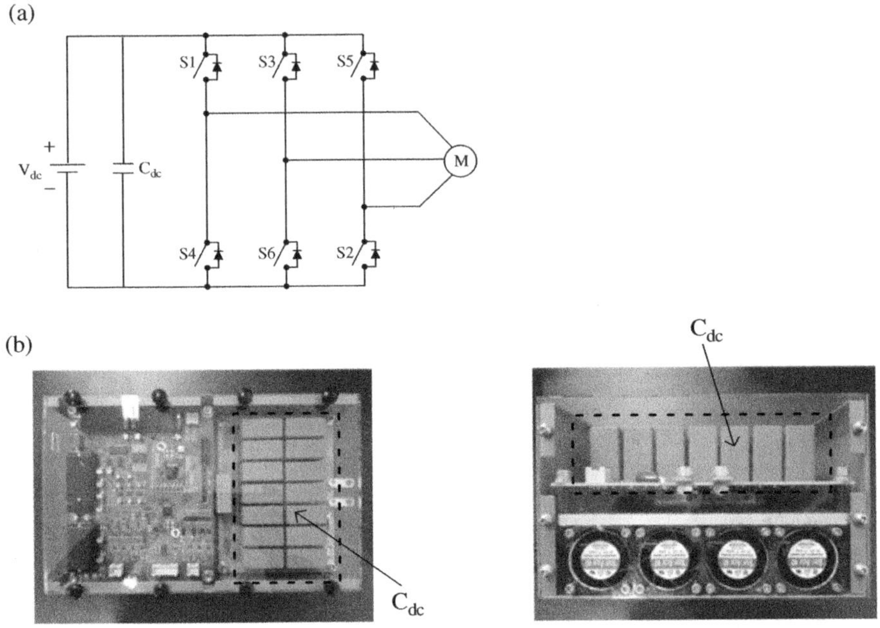

Figure 1.4 Power trains of electric vehicles: (a) circuit of power trains; (b) air-cooled 34 kW inverter for battery EV.

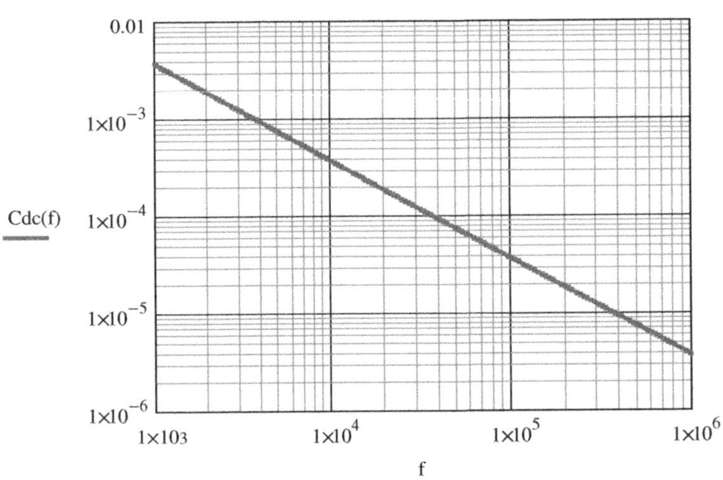

Figure 1.5 DC side capacitance vs. switching frequency.

The relationship between the DC side capacitance and the switching frequency is shown in Figure 1.5. The capacitance is reduced to one of the tenth if the switching is increased by ten times. Therefore, the DC side film can be reduced if we can increase the switching frequency. It is helpful to increase the power density of the power train of the EV. Similar to the discussion earlier, there is an upper limit to the switching frequency due to the loss of the power semiconductor devices in the power train.

1.1.3 Switching Frequency and Impact of Soft-switching Technology

As we mentioned earlier, if the inverter operates at higher switching frequency, the inverter will have smaller filter size and smaller loss on the filter in UPS and smaller DC-side film capacitor in EV power train. In addition, we can get better dynamics. In some applications of ultra-high-speed drives for the industry or aeronautics, inverters are required to operate at very high switching frequency to provide lower ripple fundamental frequency current to the motor/generators. It seems higher switching frequency operating has advantage. What is the maximum switching frequency the inverter can operate at?

Actually, the switching frequency is limited by loss of the power semiconductor devices in the inverter. Figure 1.6 shows loss of the power semiconductor switches of the inverter of 100 kVA UPS exampled in the last section. Typical Insulated Gate Bipolar Transistor (IGBT) devices (Si IGBT FF300R12KT) are used as the switches. The loss of the inverter power semiconductor devices is composed of conduction loss, turn-on loss, turn-off loss, and reverse recovery loss. Conduction loss is static loss, which does not change with the switching frequency shown as the black part of the bar at the bottom of the figure. The other three losses – turn-on loss, turn-off loss, and reverse recovery loss – are dynamic losses, which increase linearly with the switching frequency as shown in the figure. As a result the dynamic loss of the power device depends on the switching frequency. If we want to design the inverter with required efficiency, its maximum switching frequency should be restricted to constrain total power device loss to certain value. Dynamic loss has another commonly used name: switching loss. It means the loss is caused by the device switching actions, either turning on or turning off.

For the same inverter parameters, when SiC MOSFET (CAS300M12BM2) is used, total SiC MOSFET loss of the inverter vs. the switching frequency is shown in Figure 1.7. Since the recovery loss is smaller, it is ignored. The loss of the inverter power semiconductor devices is composed of conduction loss, turn-on loss, and turn-off loss. Similar to the IGBT inverter, the conduction loss is constant while the dynamic loss is proportional to the switching frequency. Although the power device loss of the SiC inverter loss is much smaller than that of the IGBT inverter, the dynamic loss is still the main factor to limit the upper switching frequency.

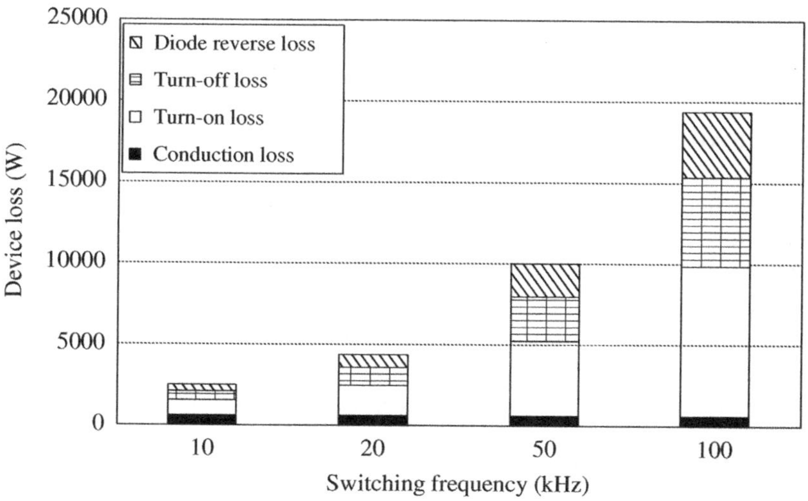

Figure 1.6 Power semiconductor loss of the inverter vs. switching frequency.

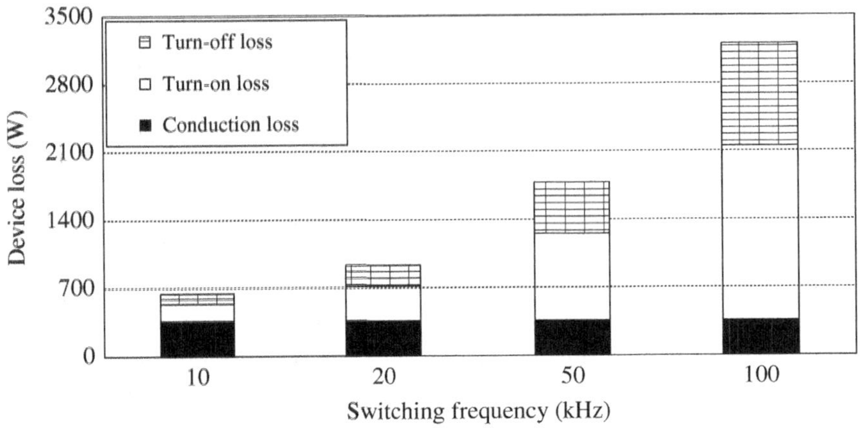

Figure 1.7 Total SiC MOSFET loss of the inverter vs. switching frequency.

As a result, conducting loss is almost constant while switching loss increases with switching frequency. Actually, the voltage and the current of the power device have an overlap time during the switching transient process, which causes switching losses. The average switching loss is proportional to the switching frequency. To get an expected conversion efficiency, the switching frequency of inverters needs to be restricted. This type of converter is called hard-switching converter. Hard-switching converters can operate only at lower switching frequency. Lower switching frequency operation in turn results in bulkier passive components and higher audible noise. The wall to prevent the switching frequency from increasing is due to the switching loss. Can we reduce the switch loss of the inverter so that it can operate at higher switching frequency? To realize this goal, a technology to shape voltage and current of the power device during switching transient process, known as soft-switching, occurs. Before a power device in a converter changes its status, either from on-state to off-state or from off-sate to on-state, the voltage across it or the current through it is set to zero with the help of the resonance between inductance and capacitance in the circuit. Soft-switching is able to reduce the switching loss of the power semiconductor devices so that the converter can operate at higher switching frequency.

1.2 Concept of Soft-switching Technique

A power device undergoes a switching transient process during its turn-on as shown in Figure 1.8. From time t_1 to t_2, both voltage u_{S1} across the device S_1 and current i_{S1} through it have high values during turn-on transient process. The voltage and current overlap on the device occurs during switch turn-on process, which results in turn-on loss. The turn-on loss E_{on} is equal to integral of multiplication of the voltage u_{S1} across the device S_1 and the current i_{S1} through it in the turn-on transient duration. Similarly, when the device turns off, it also undergoes a turn-off transient process. The voltage and current overlap on the device occurs in the switch turn-off process, which results in turn-off loss E_{off}.

To reduce the turn-on and turn-off loss of the power device, soft-switching techniques occur. The soft-switching technique is a way to shape voltage and current of the power device during switching transient process by changing converter topology and/or introducing a unique control. Thus the

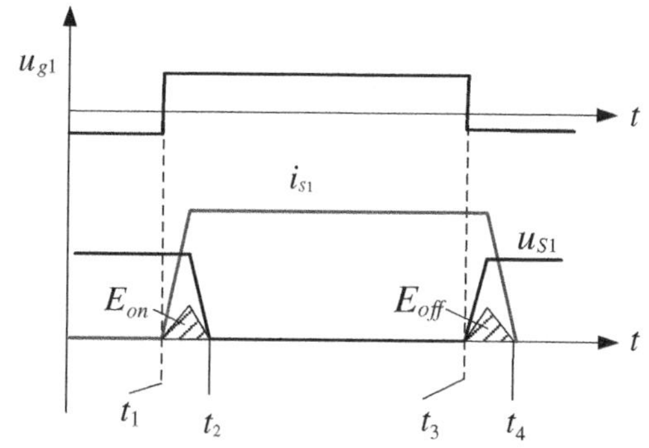

Figure 1.8 Typical switching waveforms of a power device.

overlapping of voltage and current on the power device during switching commutations is reduced. It not only reduces switching loss but also suppresses voltage stress on the power devices and electromagnetic interference (EMI) noise.

1.2.1 Soft-switching Types

Soft-switching techniques are realized with innovated converter topologies and/or by introducing a unique control. There are many soft-switching converter topologies and their control methods. Soft-switching techniques can be summarized as four types as follows.

Zero-Voltage-Switching Turn-on (ZVS-on): During a power semiconductor device turn-on process, voltage applied on the power device decreases almost to zero before the gate drive signal jumps to the high level for turning on the device. Thus the voltage and current overlapping of the device during turn-on transient process is eliminated. As shown in Figure 1.9, voltage u_{S1} on the device S_1 is set to zero before its gate drive signal u_{g1} goes to the high level. Typically, a diode is antiparalleled with device S_1. Once voltage u_{S1} on the device S_1 decreases to zero, the antiparalleled diode D_1 will conduct. It creates zero voltage turn-on condition for the device S_1. Thus the overlapping of voltage and current of the power device during turn-on process is got rid of. The integral of multiplication of u_{S1} and i_{S1} during turn-on process becomes zero. Thus turn-on loss of the device is avoided. ZVS-on is ideal turn-on process since it has no turn-on loss.

Zero-Current-Switching Turn-on (ZCS-on): During a power device turn-on process, its current gradually increases from zero value while its voltage quickly goes down. Thus the voltage and current overlapping of the device during turn-on transient process is reduced. As shown in Figure 1.10, when gate drive signal steps up, the current i_{S1} of the device gradually increases from zero value while the voltage quickly goes down. Typically, there is external inductor serially connected with the power device, which creates zero current turn-on condition for the device S_1. Thus the

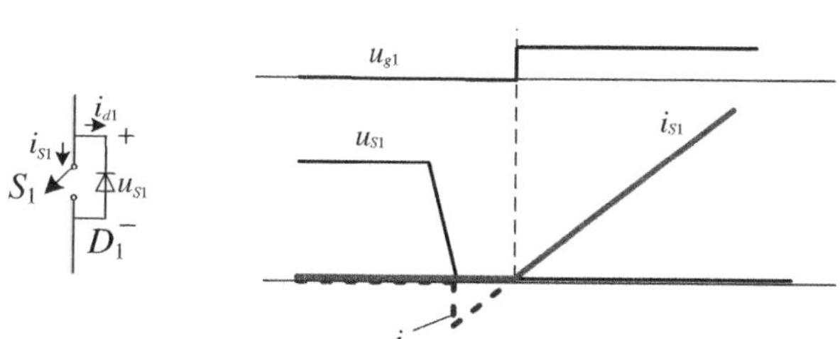

Figure 1.9 Zero-voltage-switching turn-on.

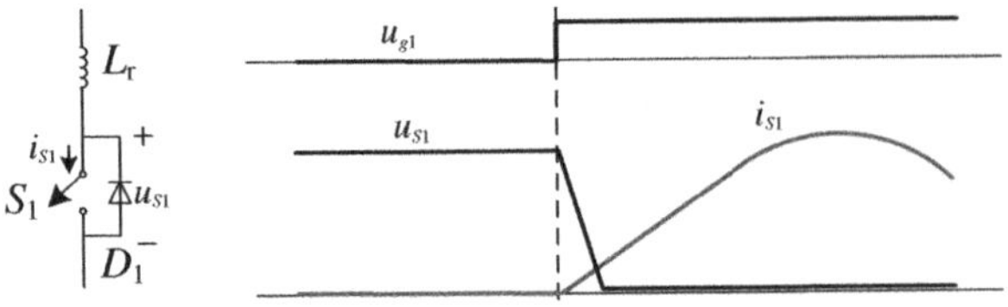

Figure 1.10 Zero-current-switching turn-on.

overlapping of voltage and current of the power device during turn-on process is reduced. The integral of multiplication of u_{S1} and i_{S1} during turn-on duration becomes smaller than that of the hard switch. Thus turn-on loss of the device is reduced. However, ZCS-on is not ideal turn-on process because turn-on loss still exists. It still has the overlapping of the voltage and current in turn-on transient process.

Zero-Voltage-Switching Turn-off (ZVS-off): During a power device turn-off process, its voltage gradually increases from zero value while its current quickly goes down. Thus the voltage and current overlapping of the device during turn-off transient process is reduced. As shown in Figure 1.11, when gate drive signal steps down for turning off the device, the voltage u_{S1} of the device gradually increases from zero while the current i_{S1} quickly goes down. Typically, there is capacitance paralleled with the power device, which suppresses the voltage increasing rate when the device turns off. Thus the overlapping of voltage and current of the power device during turn-off process is reduced. The integral of multiplication of u_{S1} and i_{S1} during turn-off process becomes smaller. Thus turn-off loss of the device is reduced. However, ZVS-off is also not ideal. There still exists turn-off loss.

Zero-Current-Switching Turn-off (ZCS-off): Current through a power device already decreases to zero before the gate drive signal steps down to the lower level for turning off the device. Thus the voltage and current overlapping of the device during turn-off transient process is eliminated. As shown in Figure 1.12, current i_{S1} through the power device S_1 is set to zero before its gate drive signal u_{g1} goes to the low level for turning off the device. Typically, there is an external inductor or L-C resonance branch serially connected with the power device S_1, which causes the current of the device S_1 decrease to zero automatically before the turn-off signal is applied to the gate drive.

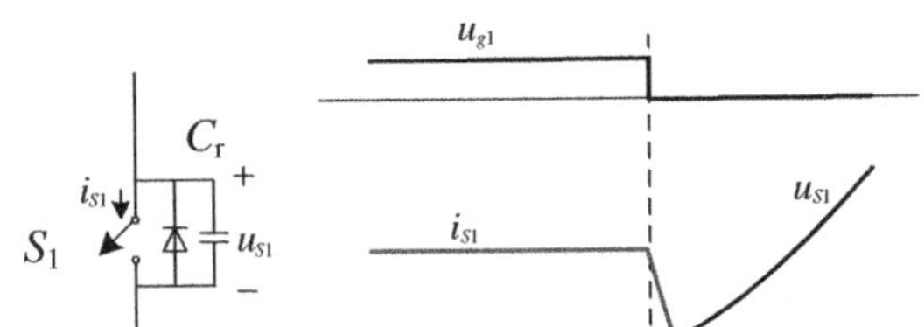

Figure 1.11 Zero-voltage-switching turn-off.

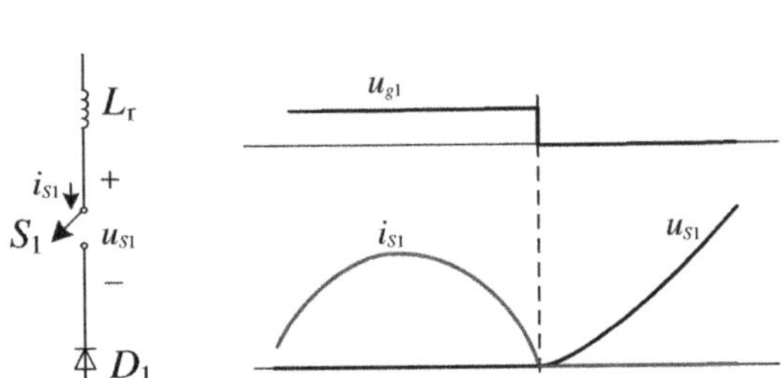

Figure 1.12 Zero-current-switching turn-off.

Thus the overlapping of voltage and current of the power device during turn-off process is eliminated. The integral of multiplication of u_{S1} and i_{S1} in turn-off duration becomes zero. Thus turn-off loss of the device is avoided. ZCS-off is ideal turn-off. There is no turn-off loss.

Among the four soft-switching techniques mentioned earlier, two methods, ZVS-on and ZCS-on, are used for turn-on loss reduction while other two, ZVS-off and ZCS-off, are used for turn-off loss reduction. ZVS-on and ZCS-off are ideal and can totally eliminate the switching loss. However, ZCS-on and ZVS-off are not ideal. They reduce switching loss, but the switching loss still exists.

1.2.2 Soft-switching Technique for Three-phase Converters

Soft-switching techniques for three-phase converters have been investigated by many predecessors. For three-phase applications, soft-switching converters can be divided into three classes: DC resonance converter, AC resonance converter, and soft-switching converters with triangular current mode (TCM) control [4].

In the DC-side resonance converters, an auxiliary resonant circuit is installed between DC input source and DC side of three-phase switch bridge of the converter. The fundamental philosophy of the DC-side resonance is to use an auxiliary resonant circuit to create zero-voltage duration at the DC side of the three-phase switch bridge at the desired switching instant. Thus all devices of the switch bridge are turned on or turned off when the voltage on them is equal to zero so that both turn-on loss and turn-off loss are significantly reduced. Besides, the DC-side resonance converter only needs one auxiliary resonant circuit regardless of the number of AC phases of the converter. This simple structure makes DC-side resonance attractive in multiphase converter applications. The resonant DC link (RDCL) converter [5] is milestone topology in evolution of soft-switching history. To reduce voltage stress on the devices, a revised version known as active clamped RDCL (ACRDCL) converter occurred [6, 7]. Both RDCL and ACRDCL converters are controlled with discrete pulse modulation (DPM). It is found that the soft-switching converters with DPM require higher switching frequency than that of the PWM converter for comparable current spectral performance. Many other topologies have been developed such as the quasi-resonant DC link (QRDCL) PWM inverter with PWM control [8–11]. They often use more complex auxiliary circuit. Zero-voltage-switching SVM (ZVS-SVM) for three-phase active clamping converters was proposed by Dehong Xu [13, 14]. The auxiliary power device only switches once in each switching cycle to realize ZVS for all the switches. It features fixed switching frequency and lower voltage stress of the power switch devices. The converter basically operates like PWM converter [15, 16]. Afterward it is generalized to edge-aligned PWM (EA-PWM) [17–19]. EA-PWM is suitable to three-phase converter, three-phase four-wire converter, three-phase four-wire BTB converter, etc.

The second class of the soft-switching converter is AC resonance converters. Auxiliary resonant circuits are installed in AC side of the switch bridge. Distinctive advantage of the AC-side resonance is that the auxiliary circuits are in shunt with the switch bridge and does not carry the load current. Thus the conduction loss in the auxiliary circuits is smaller. In addition, SPWM and SVM control can be applied because the converters basically operate as conventional PWM converters. Auxiliary resonant commutated pole (ARCP) converter is one of the earliest AC resonance converters [12, 20]. It achieves ZVS-on for main switches and ZCS-off for auxiliary switches. The inductor coupled zero-voltage transition (ZVT) inverter achieves ZVS-on for main switches and near-zero current turn off for auxiliary switches [21]. DC-side split capacitor voltage control needed for ARCP converter is avoided. The zero-current transition (ZCT) inverter achieves zero current switching for all of the main and auxiliary switches and their antiparallel diodes [22, 23]. It is suitable to converters with IGBT devices, which can reduce turn-off loss of IGBT due to its tail current. Other AC resonance circuits are developed [24–27]. The AC resonance converter has complex circuit because it generally needs three auxiliary resonant circuits. The number of power devices to be controlled are almost doubled in comparison to the original converter.

The third class of the soft-switching converter is known as the soft-switching converters with TCM. The concept comes from critical conduction mode (CRM) in DC-DC converter to achieve ZVS-on [28, 29]. With TCM control, AC-side filter inductor currents of three-phase converters are controlled as the triangle waveform in a switching period [30, 31]. TCM is originally introduced to single-phase totem-pole power factor correction (PFC) converters and then extended to three-phase converters. The distinct advantage of the soft-switching converter with TCM is that no auxiliary resonant circuit is needed. However, there are some drawbacks such as wider range of variable switching frequency, higher rms current, and turn-off current. Recently, there are many studies about TCM control. For readers interested in this research can read the related references on IEEE Xplore.

As mentioned earlier, the DC-side resonance converter only needs one auxiliary resonant circuit shared by three-phase legs of the converter. It has issues such as variable switching frequency, higher voltage stress on devices, etc. To overcome these drawbacks, the ZVS-SVM and EA-PWM for three-phase active clamping converters were proposed. The auxiliary power device only switches once in each switching cycle to realize ZVS for all the switches. It features fixed switching frequency and lower voltage stress of the power switch devices. The converter basically operates like the conventional PWM converter.

As an example of the soft-switching technique, the concept of three-phase active clamping converters with EA-PWM is briefly explained. With EA-PWM, the converter operates at the fixed switching frequency. An auxiliary circuit with auxiliary switch S_7 is installed in the DC side of the converter in Figure 1.13a. Once a high loss switch commutations in the switch bridge happens, the auxiliary switch S_7 will turn off, which starts a resonant process to create a slot with duration $\lambda_7 T_S$ on the DC bus voltage waveform v_{bus} where λ_7 is the turn-off duty ratio of the auxiliary switch S_7 and T_S is the switching period in Figure 1.13b. During $v_{bus} = 0$, the switches in the three-phase switch bridge make the commutation under zero voltage on their terminals, which is known as zero voltage switching. Thus the switching loss is reduced. With EA-PWM, all high loss commutations of the switch bridge in each switch cycle are synchronized. The auxiliary power device only switches once in a switching cycle to realize ZVS for all the switches. More detailed introduction will be given in Chapter 3.

1.3 Applications of Soft-switching to Three-phase Converters

1.3.1 Renewable Energy and Power Generation

The increasing applications of renewable energy and distributed power generation have become a new driving force for application of power electronics. Three-phase converters are playing more and more important role in the grid. More efficient and more reliable power converters are required. Soft-switching technique can be applied to converters for renewable energy integration, energy storage systems, and Flexible AC power transmission (FACTS) devices to increase power density and dynamics and reduce size of the equipment [32].

A single-phase PV inverter is widely used for residential PV applications as shown in Figure 1.14. The front boost converter is used to extend solar energy harvesting duration each day so that the inverter can feed power to grid with a wider PV panel output voltage. By introducing an auxiliary

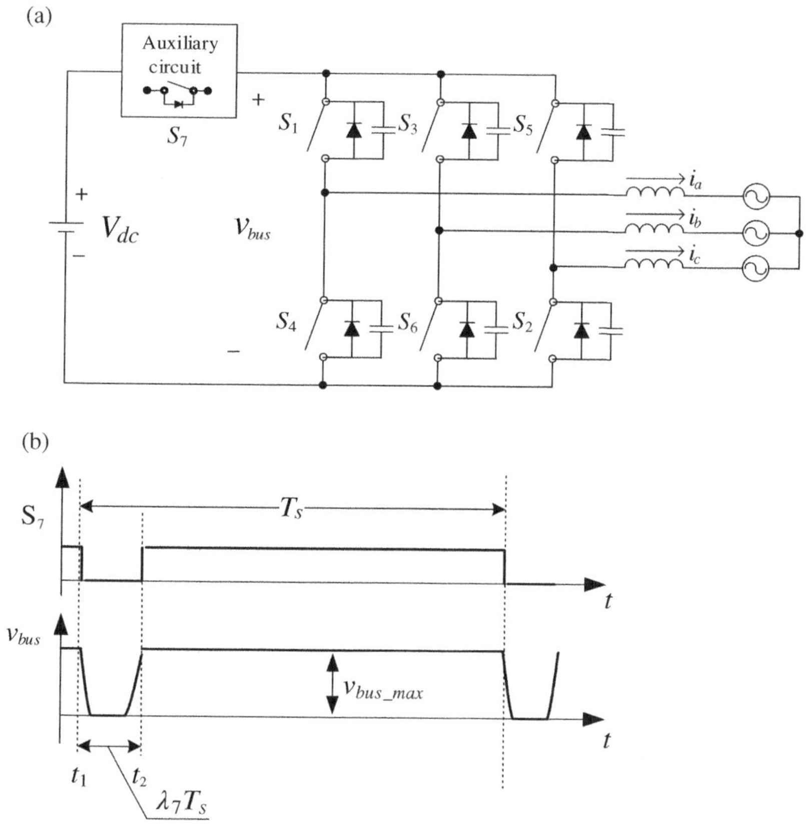

Figure 1.13 Active clamping converters: (a) circuit; (b) auxiliary switch state and DC bus voltage.

resonant circuit and EA-PWM control, both the boost stage and the H-bridge inverter stage can realize soft-switching operation. Due to the soft-switching, the switching frequency is increased so that passives such as inductors and capacitors may be reduced. The PV inverter becomes more compact.

Three-phase PV inverters are widely used in medium power or large power PV power generation systems. Six switch inverter circuits are commonly used due to their circuit simplicity. To reduce its switch loss, an auxiliary resonant circuit is introduced in the DC side of the converter as shown in Figure 1.15. The soft-switching SiC MOSFET grid inverter achieves a high efficiency of 98.6% at 300 kHz switching frequency, which is about three times of the original hard-switching counterpart with the same conversion efficiency [33]. Thus the inverter becomes more compact due to small passives. Besides, EMI noise caused by high dv/dt of SiC MOSFET is relieved due to the soft-switching.

Another circuit used in distributed PV generation systems is the string inverter [17]. The string inverter is basically composed of two conversion stages: the DC-DC and DC-AC stages. The DC-DC stage is adopted to extend the PV voltage operation region and harvest more solar energy as mentioned before. It usually has multiple DC-DC boost converters, as shown in Figure 1.16, which are connected in parallel to increase the maximum power point tracking (MPPT) efficiency and power capacity as well. By introducing the soft-switching technique, higher power conversion and higher power density can be obtained.

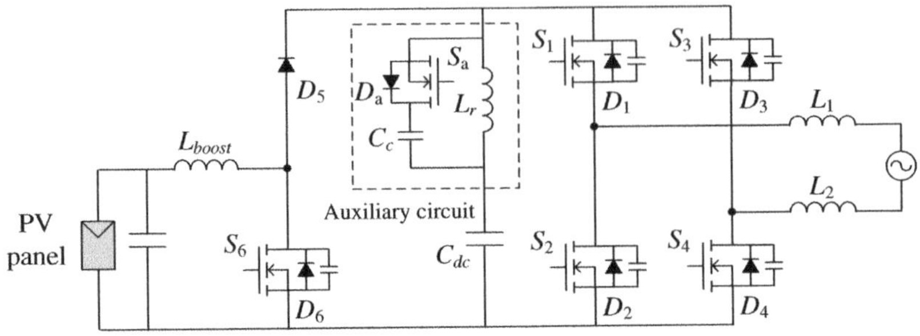

Figure 1.14 Single-phase PV inverter for residential applications.

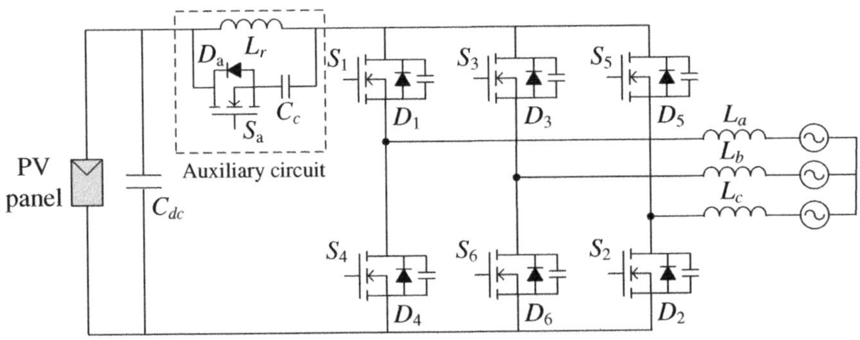

Figure 1.15 Three-phase ZVS PV inverter.

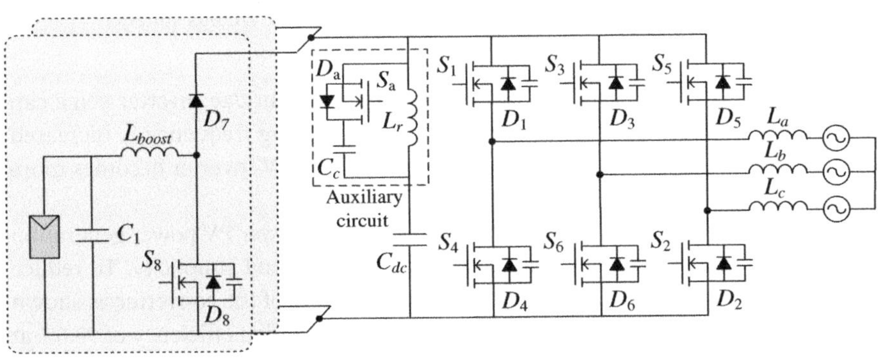

Figure 1.16 Two-stage three-phase ZVS inverter for PV system.

Similar to PV inverters, the soft-switching technique can also be applied to wind power systems. Two most popular wind power conversion systems (PCSs) are doubly fed induction generator (DFIG) and permanent magnet synchronous generator (PMSG). Both of them utilize a BTB converter to interface the grid side. The BTB converter for the PMSG system with the soft-switching technique is shown in Figure 1.17. In a typical wind power system, the entire power converter is packed in a cabinet and placed in a nacelle with limited space. The soft-switching BTB converter operates with higher switching frequency so that the volume and weight of the passive components can be significantly reduced. The reduced size and weight of the power converter can spare more

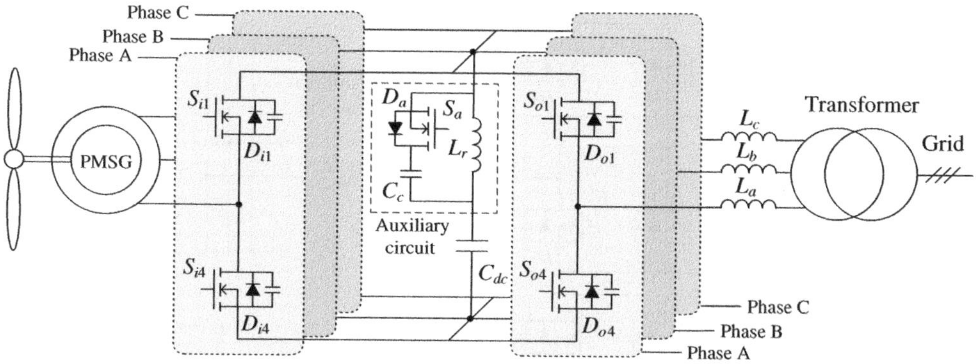

Figure 1.17 ZVS back-to-back converter for PMSG system.

room in the nacelle. Thus a step-up transformer can be accommodated in the nacelle to reduce the cable cost and energy losses.

1.3.2 Energy Storage Systems

Energy storage systems have become a key enabling technology for a robust, high efficiency, and cost-effective power grid. Grid level energy storage systems are used in frequency regulation, spinning reserve, peak load shaving, load leveling, and so on. Besides, energy storage systems are also introduced in distributed systems to stabilize the power output of renewable energy. The converter is the interface to connect the energy storage component with the grid. Energy storage systems require a bidirectional power flow control such as the battery energy storage system (BESS). The energy loss is also doubled during the whole energy utilization cycle by charging and discharging the energy storage component. Therefore the efficiency of the converter becomes more critical than that of the unidirectional converters. The soft-switching technique has a potential in the energy storage applications.

The PCS for the BESS can be divided into single-stage and double-stage structures. For the single-stage PCS, the battery voltage should not fluctuate widely during the discharging or charging process, which is typically related to the characteristics of the battery technology. The soft-switching technique can be applied as PV inverters, as shown in Figure 1.18. To extend the system power capacity and improve fault tolerance, a PCS structure with paralleled three-phase converters is adopted as shown in Figure 1.18. An auxiliary resonant circuit is installed in the common DC bus. Both converters can realize the ZVS operation with EA-PWM.

The double-stage PCS consists of a bidirectional DC/DC conversion stage and a DC/AC stage as shown in Figure 1.19. The DC/DC stage boosts the battery voltage to the suitable level so that the inverter stage can be directly interfaced to the grid. This type of PCS is suitable for maximum utilization of the battery stored energy, whose voltage has a wider variation during the entire SOC. With the ZVS auxiliary circuit, it can realize soft-switching for both DC/DC stage and DC/AC stage.

The double-stage interface has the advantage of a common DC bus line. Different energy storage units can be integrated into the system, which makes the system more expandable and fault tolerant. Figure 1.20 shows the diagram of the double-stage interface with two boost converters interfacing to the fuel cell and super-capacitor, respectively. All power devices can realize ZVS operation with only one auxiliary circuit in the middle DC link. Further extension to multiple energy sources system is shown in Figure 1.21 where various types of energy storage components are integrated by

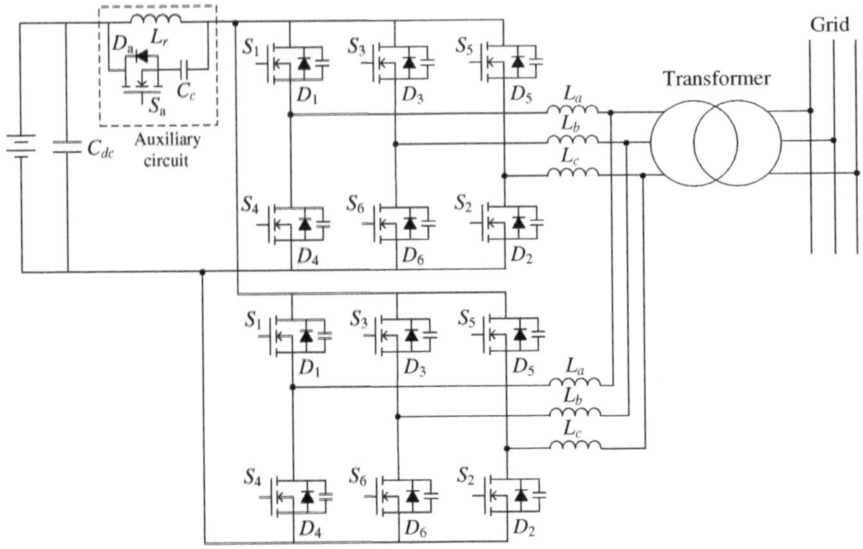

Figure 1.18 Paralleled three-phase ZVS inverter for BESS.

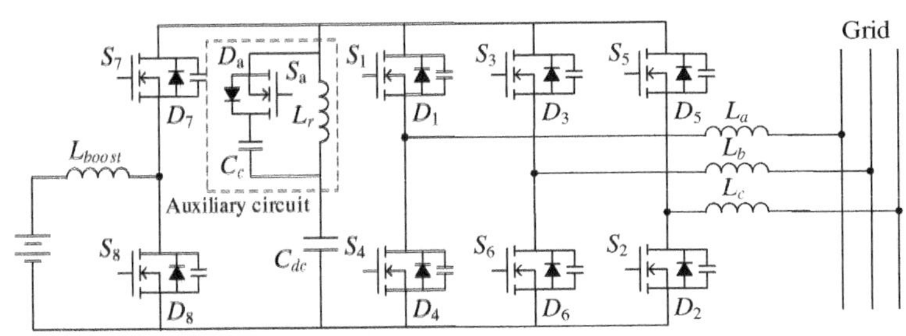

Figure 1.19 ZVS inverter with front boost stage for BESS.

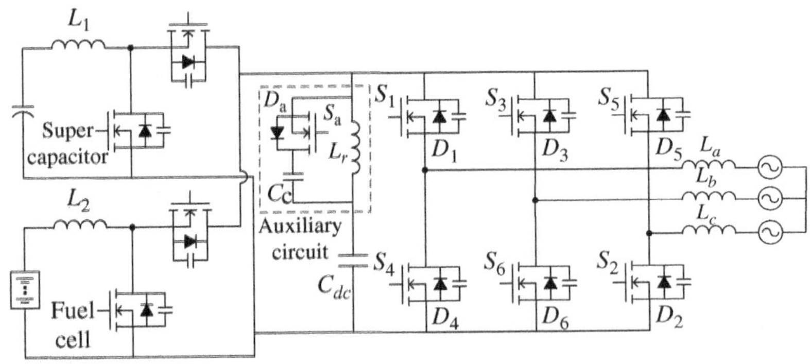

Figure 1.20 ZVS inverter with paralleled DC/DC converters for BESS.

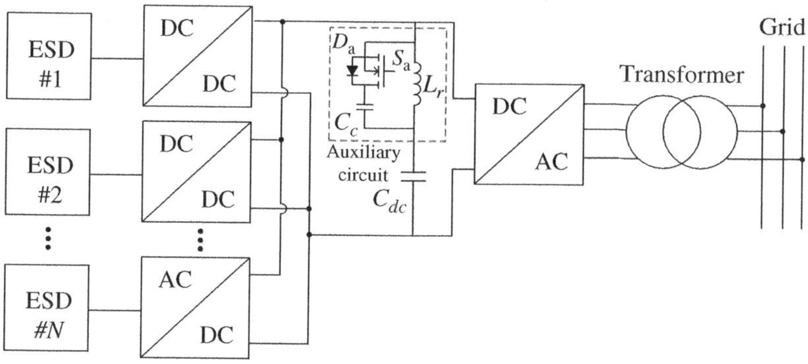

Figure 1.21 Multiple energy storage system with ZVS converters.

bidirectional DC/DC converter to the DC bus with only one auxiliary resonant circuit. All the converters connected to the DC bus can realize the soft-switching operation with only one auxiliary resonant circuit.

1.3.3 Distributed FACTS Devices

Distributed FACTS devices are used to improve reliability, power quality, and efficiency of the distributed systems. It includes static var compensators (STATCOMs), active power filter (APF), unified power flow controller (UPFC), unified power quality controller (UPQC), dynamic voltage restorer (DVR), etc.

A shunt APF is usually installed near the nonlinear load for the compensation of the harmonic current generated by the nonlinear load. The power converter can operate with higher switching frequency with soft-switching technique. Higher switching frequency means the APF has better dynamics and precisely cancel harmonic of the nonlinear load. Figure 1.22 shows soft-switching APF. The same topology can also be utilized in STATCOM.

DVR can stabilize voltage of the utility to loads. It also benefits from the application of the soft-switching technique with regard to fast control response and reduced system size. Figure 1.23 shows a DVR using the soft-switching technique.

The UPQC combines the functions of the shunt var compensation and series var compensation. It consists of two BTB connected converters shown in Figure 1.24. The one converter is connected in series with the grid while the other is connected in parallel with the grid. Each converter can generate reactive power at its own AC output terminal. By inserting the auxiliary circuit in the middle DC link, both converters can realize ZVS operation. The dynamics or power density of the UPQC are enhanced since the switching frequency is increased due to the soft-switching.

1.3.4 Uninterruptible Power Supply

Uninterruptible power supply (UPS) is widely used in data centers and manufacturing processes to provide continuous and higher quality power. UPS is composed of the rectifier, inverter, and neutral line control half-bridge. Totally there are seven switch legs. By installing an auxiliary resonant circuit in the DC side as shown in Figure 1.25, all switches in these seven switching legs can realize soft-switching. Thus the power density of the UPS can be enhanced [19].

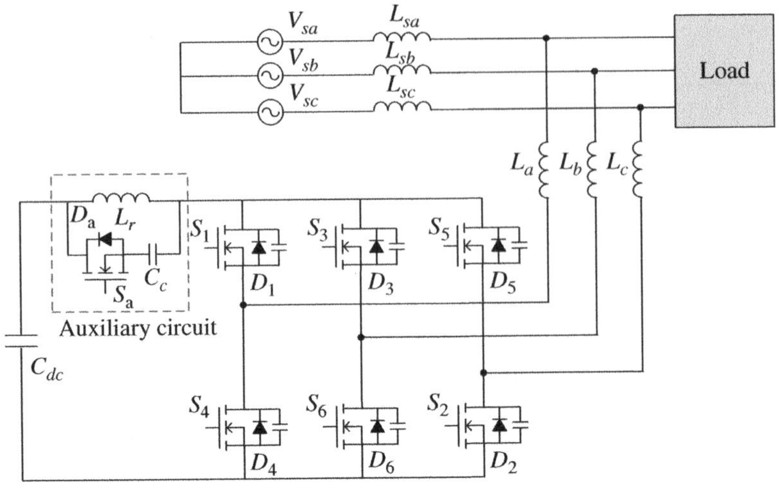

Figure 1.22 ZVS inverter for APF/STATCOM.

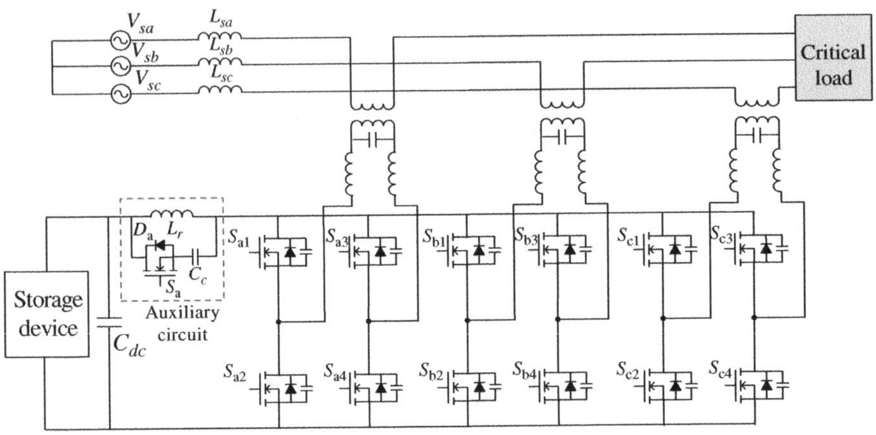

Figure 1.23 ZVS converter for DVR.

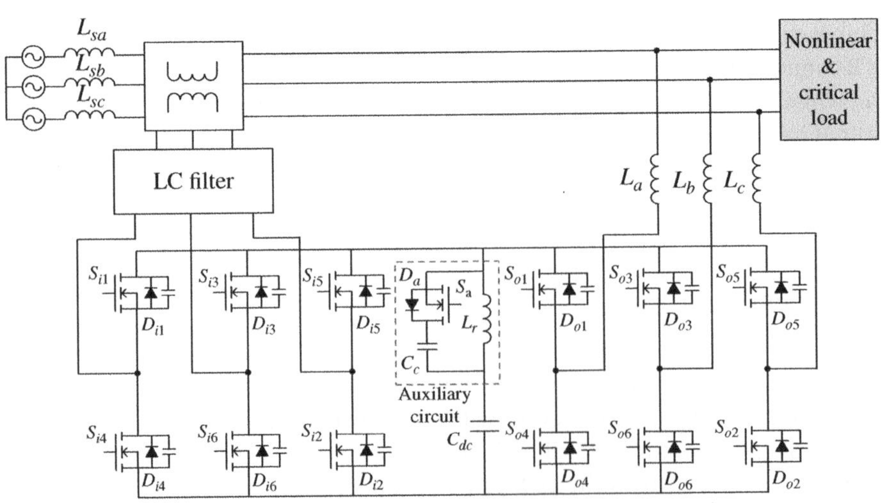

Figure 1.24 ZVS converter for UPQC. *Source:* Based on Shi et al. [19].

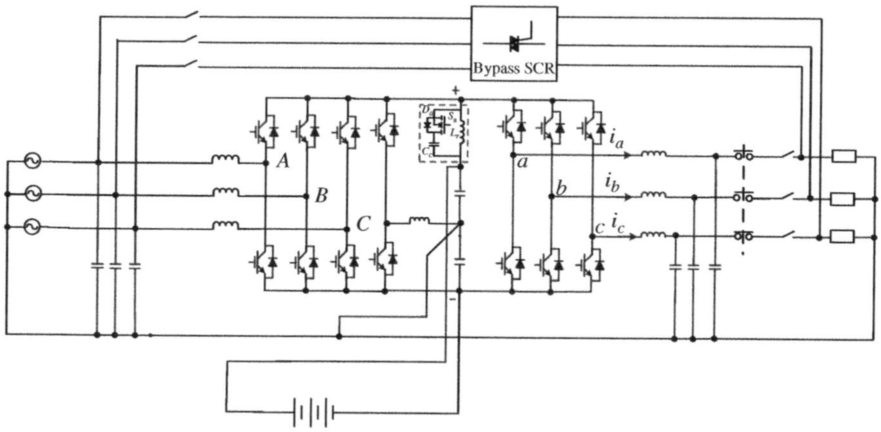

Figure 1.25 UPS with the auxiliary resonance circuit for soft-switching.

1.3.5 Motor Drives

High speed drives such as pumps or compressors are applied in the industry. The inverter of high speed drives needs to operate at very high switching frequency up to tens of kHz. Since switching loss is proportional to switching frequency, switching loss of the power semiconductor in a high speed drive is so high that the power devices have to operate at its derating state for the safety. Thus the power rating of power device is unable to be fully utilized. An auxiliary resonance circuit is introduced to the DC side of the inverter as shown in Figure 1.26. The switch loss of the inverter is significantly reduced due to soft-switching of the power devices in the inverter.

1.3.6 Fast EV Chargers

To reduce charging time of EV, power rating of EV chargers is becoming large and large. The EV charger shown in Figure 1.27 is composed of two power conversion stages. To increase power density of the charger, soft-switching technique is used. The first stage is three-phase rectifier where an auxiliary resonant circuit is introduced to realize soft-switching for the rectifier. For the second stage, double inductors and one capacitor (LLC) resonant DC/DC converter is used. It can realize soft-switching for all switches in the DC/DC converter. All switches in EV charger can realize soft-switching so that its power density is enhanced.

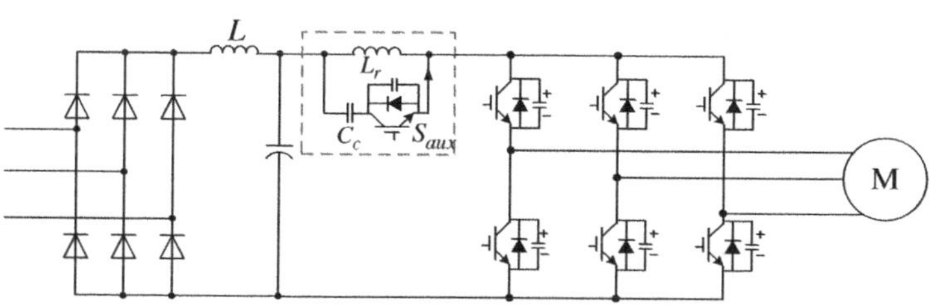

Figure 1.26 High speed drives with auxiliary resonance circuit.

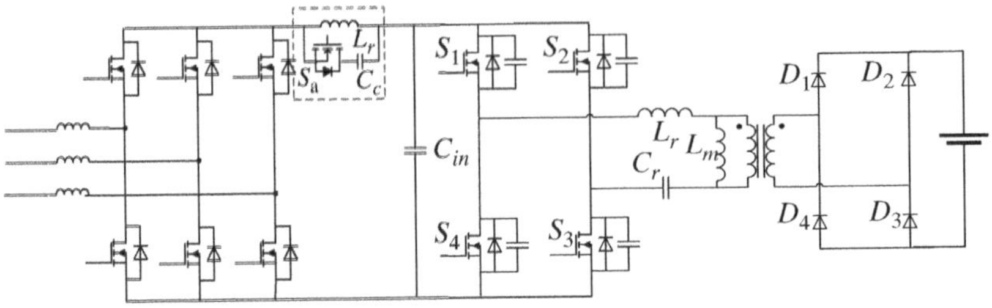

Figure 1.27 Soft-switching EV charger.

1.3.7 Power Supply

With development of big data, cloud computing, etc., it is more stringent than ever to require the data centers to have higher computing speed and density with less energy consumption. Compared with super junction Metal-oxide-semiconductor field effect transistors (MOSFETS), Gallium nitride high Electron mobility Transistor (GaN HEMT) shows significant improvement on the switching performance, but its switching frequency is still limited when hard switching is used. A ZVS totem-pole PFC circuit with fixed switching frequency is investigated for server power supply [34]. In Figure 1.28, the right leg with switch S_{2H} and S_{2L} operates at utility frequency and silicon MOSFET is used. The left leg with S_{1H} and S_{1L} operates at 500 kHz and GaN HEMT is chosen. With the auxiliary circuit, the GaN device can operate at ZVS condition. The turn-on loss of the GaN is eliminated and its turn-off loss is reduced.

1.4 The Topics of This Book

This book will focus on the soft-switching technology for three-phase converters or inverters and their applications. Aiming to reduce or even eliminate the voltage and current overlapping during the switching transient process, soft-switching techniques provide a solution for power converters to achieve high conversion efficiency with dramatic reduction of the switching losses. This book is divided into four parts:

Part 1(Chapters 1–3) will provide an introduction to fundamentals of soft-switching technology for three-phase conversion. Impacts of the soft-switching technique on three-phase converter performance such as conversion efficiency, power density, and EMI noise is explained. Applications of

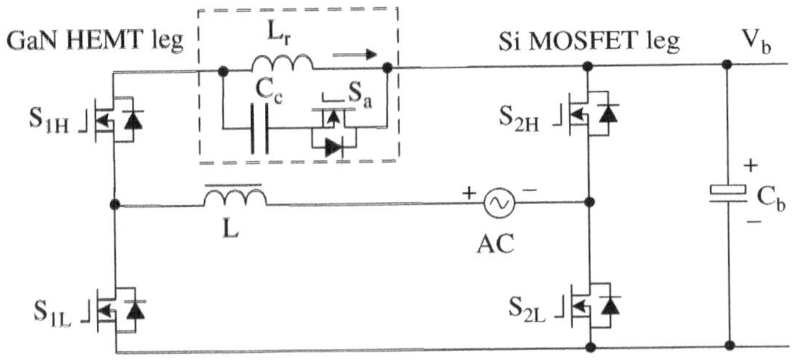

Figure 1.28 ZVS totem power-factor-correction circuit.

three-phase power converters in renewable energy, industry drives, power supplies, etc. are introduced. Development of soft-switching technology for three-phase converters is reviewed. A general soft-switching PWM method for three-phase converters, edge-aligned PWM (EA-PWM), is introduced.

Part 2(Chapters 4 and 5) will investigate applying soft-switching technology to three-phase rectifiers. Two types of soft-switching circuits are investigated. It includes circuit analysis, soft-switching condition derivation, and circuit parameters design. Then experimental result of the soft-switching rectifier prototypes are provided.

Part 3(Chapters 6–9) will aim at applying soft-switching technology to three-phase grid inverters. Two types of soft-switching circuits are investigated. It includes circuit analysis, soft-switching condition derivation, and circuit parameters design. Then experimental result of the soft-switching grid inverter prototypes are provided. Since the resonant inductor is a critical component with respect to its loss, size, and thermal design, design of the resonant inductor is introduced. In addition, optimization method for the grid inverter based on the loss model is provided.

Part 4(Chapters 10–12) will introduce the impact of SiC devices on soft-switching converters. Improvement of efficiency and power density by introducing SiC to soft-switching three-phase converter will be investigated. Converter circuit layout design and its effect are explained. Designs of single-phase grid inverter, a three-phase grid inverter, and a BTB converter with soft-switching technique are provided.

References

1 N. Mohan, T. M. Undeland, W. P. Robbins, *Power Electronics: Conversion, Applications, and Design*, New York: John Wiley &Son, 2003.

2 P. T. Krein, *Elements of Power Electronics*, Oxford: Oxford University Press, 2015

3 X. Pei, W. Zhou and Y. Kang, "Analysis and calculation of DC-link current and voltage ripples for three-phase inverter with unbalanced load," *IEEE Transactions on Power Electronics*, vol. **30**, no. 10, pp. 5401–5412, 2015.

4 M. D. Bellar, T. S. Wu, A. Tchamdjou, J. Mahdavi, and M. Ehsani, "A review of soft-switched DC-AC converters," *IEEE trans on Industrial Applications*, vol. **34**, no. 4, pp. 847–860, 1998.

5 D. M. Divan, "The resonant DC link converter – a new concept in static power conversion," *IEEE Transactions on Industry Applications*, vol. **25**, no. 2, pp. 317–325, 1989.

6 G. Venkataramanan, D. M. Divan, and T. M. Jahns, "Discrete pulse modulation strategies for high frequency inverter systems," *IEEE Transactions on Power Electronics*, vol. **8**, no. 3, pp. 279–287, 1993.

7 D. M. Divan and G. Skibinski, "Zero-switching-loss inverters for high-power applications," *IEEE Transactions on Industry Applications*, vol. **25**, no. 4, pp. 634–643, 1989.

8 J. He and N. Mohan, "Parallel resonant DC-link circuit – a novel zero switching loss topology with minimum voltage stresses," *IEEE Transactions on Power Electronics*, vol. **6**, no.4, pp. 687–694, 1991.

9 J. Cho, H. Kim, and G. Cho, "Novel soft-switching PWM converter using a parallel resonant DC-link," in *Proc. of IEEE Power Electronics Specialist Conference – PESC1991*, pp. 241–247, 1991.

10 M. Nakaoka, H. Yonemori, and K. Yurugi, "Zero-Voltage soft-switched PDM three-phase AC-DC active power converter operating at unity power factor and sinewave line current," in *Proc. of IEEE Power Electronics Specialist Conference – PESC '93*, pp. 787–794, 1993.

11 V. Agelidis, P. Ziogas, G. Joos, "An optimum modulation strategy for a novel "notch" commutated 3-Φ PWM inverter," *IEEE Transactions on Industry Applications*, vol. **30**, no. 1, pp 52–61, 1994.

12 R. W. DeDoncker and J. P. Lyons, "The auxiliary commutated resonant pole converter," in *Proc. of IEEE Industry Applications Society Annual Meeting*, pp. 1228–1235, 1990.

13 B. Feng and D. Xu, "Novel ZVS three-phase PFC Converters and zero-voltage-switching space vector modulation (ZVS-SVM) control," in *First International Conference on Power Electronics Systems and Applications (PESA)*, pp. 30–37, 2004.

14 D. Xu, R. Li, Z. Ma, C. Du, M. Chen, "A family of novel zero-voltage switching three-phase PWM converters topology for distributed generation", *Proc. of 8th International Conference on Power Electronics – ECCE Asia*, pp. 1–10, 2011

15 D. Xu, B. Feng, R. Li, "A Zero Voltage Switching SVM (ZVS–SVM) controlled three-phase boost rectifier," *IEEE Transactions on Power Electronics*, vol. **22**, no. 3, pp. 978–986, 2007.

16 R. Li, Z. Ma, D. Xu, "A ZVS grid-connected three-phase inverter," *IEEE Transactions on Power Electronics*, vol.**27**, no.8, p 3595–3604, 2012.

17 K. Shi, J. Deng, D. Xu, "A general pulse width modulation Method for Zero-Voltage-Switching Active-clamping Three-phase Power Converters: Edge Aligned Pulse Width Modulation (EA-PWM)," *IEEE Open Journal of Power Electronics*, vol. **1**, pp. 250–259, 2020.

18 N. He, Y. Zhu, A. Zhao and D. Xu, "Zero-voltage-switching sinusoidal pulse width modulation method for three-phase four-wire inverter," *IEEE Transactions on Power Electronics*, vol. **34**, no. 8, pp. 7192–7205, 2019.

19 K. Shi, A. Zhao, J. Deng and D. Xu, "Zero-voltage-switching SiC-MOSFET three-phase four-wire back-to-back converter," *IEEE Journal of Emerging and Selected Topics in Power Electronics*, vol. **7**, no. 2, pp. 722–735, 2019.

20 M. Lockwood, A. M. Fox, "A novel high power transistor inverter", in *Proceeding of 1983 IPEC-Tokyo*, pp. 637–648, 1983.

21 W. Dong, D. Peng, H. Yu, F. C. Lee, and J. Lai, "A simplified control scheme for zero voltage transition (ZVT) inverter using coupled inductors," in *Proc. of IEEE Annal Power Electronics Specialists Conference-PESC2000*, pp. 1221–1226, 2000.

22 H. Mao, F. C. Lee, X. Zhou, and D. Boroyevich, "Improved zero-current transition converters for high power applications," in *Proc. of IEEE Industry Applications Conference Thirty-First IAS Annual Meeting*, pp. 1145–1152, 1996.

23 Y. Li, F. C. Lee, J. Lai, and D. Boroyevich, "A Novel Three-phase Zero-Current-Transition and Quasi-Zero-Voltage-Transition (ZCT-QZVT) inverter/rectifier with reduced stresses on devices and components," in *Proc. of IEEE Applied Power Electronics Conference and Exposition (APEC)*, pp. 1030–1036, 2000.

24 V. Vlatkovic, D. Borojevic, F. C. Lee, C. Cuadros and S. Gataric, "A new zero-voltage-transition, three-phase rectifier/inverter," in *Proc. of IEEE Annual Power Electronics Specialists Conference-PESC'93*, pp. 868–873, 1993.

25 J. Lai, J. Zhang and H. Yu, "Source and load adaptive design for a high-power soft-switching inverter", *IEEE Transactions on Power Electronics*, vol. **21**, no. 6, pp.1667– 1675, 2006.

26 X. Yuan and I. Barbi, "Analysis, designing, and experimentation of a transformer-assisted PWM zero-voltage switching pole inverter," *IEEE Transactions on Power Electronics*, vol. **15**, no. 1, pp. 72–82, 2000.

27 N. Hoshi and K. Oguchi, "Control schemes for auxiliary switches of three-phase PWM resonant snubber inverters," *Electrical Engineering in Japan*, vol. **152**, no. 4, pp. 57–67, 2005.

28 Z. Liu, Z. Huang, F. C. Lee, and Q. Li, "Digital-based interleaving control for GaN-based MHz CRM totem-pole PFC," *IEEE Journal of Emerging and Selected Topics in Power Electronics*, vol. **4**, no. 3, pp. 808–814, 2016.

29 C. Marxgut, F. Krismer, D. Bortis, and J. W. Kolar, "Ultraflat interleaved triangular current mode (TCM) single-phase PFC rectifier," *IEEE Transactions on Power Electronics*, vol. **29**, no. 2, pp. 873–882, 2014.

30 Z. Huang, Z. Liu, F. C. Lee and Q. Li, "Critical-mode-based soft-switching modulation for high-frequency three-phase bidirectional AC–DC converters," *IEEE Transactions on Power Electronics*, vol. **34**, no. 4, pp. 3888–3898, 2019.

31 M. Haider, J. A. Anderson, S. Mirić, N. Nain, G. Zulauf, J. W. Kolar, D. Xu, G. Deboy, "Novel ZVS S-TCM modulation of three-phase AC/DC converters," *IEEE Open Journal of Power Electronics*, vol. **1**, pp. 529–543, 2020.

32 K. Shi, J. Deng, A. Zhao, D. Xu, "Soft-switching SiC power electronic conversion for distributed energy resources and storage applications," *Journal of Modern Power Systems and Clean Energy*, vol. **7**, no. 5, pp. 1340–1354, 2019.

33 N. He, M. Chen, J. Wu, N. Zhu, D. Xu, "20 kW Zero-Voltage-Switching SiC-MOSFET Grid Inverter with 300 kHz Switching Frequency", *IEEE Transactions on Power Electronics*, vol. **34**, no. 6, pp. 5175–5190, 2019.

34 K. Dong, Y. Guo, D. Shuailin, T. Hong, J. Zhang, D. Xu, "A GaN Totem-Pole PFC converter with zero-voltage-switching PWM control," in *Proceeding of 2021 IEEE Applied Power Electronics Conference and Exposition (APEC)*, pp. 247–251, 2021.

2

Basics of Soft-switching Three-phase Converters

Aiming to reduce or even eliminate the voltage and current overlap time during the switching actions, soft-switching techniques provide a solution for power converters to achieve high conversion efficiency with dramatic reduction of the switching losses. Given a power semiconductor device, soft-switching technology is also an effective way to increase switching frequency and power density. In this chapter, the basic knowledge of soft-switching three-phase converters is introduced.

2.1 Introduction

Three-phase converters are widely used as grid connecting inverters for renewable energy systems, rectifiers, and inverters of uninterruptible power supplies (UPS) for data centers, rectifiers of electrical vehicle (EV) fast charging stations, inverters of electrical vehicles, inverters of industrial motor drives, etc. The power flow of the converters is controlled by power semiconductor devices. To realize high energy conversion efficiency, the devices operate in a switching mode where they are required to turn on and turn off the load current in each switching cycle. Since the device is not an ideal switch and undergoes a transient process during either turn-on or turn-off, it is subjected to power loss and voltage or current stress. Actually, the voltage and the current of the power device have an overlap time during the switching transient process, which causes power losses. The average switching loss power is proportional to the switching frequency. In addition, these switching mode operations result in electric magnetic interference (EMI) to the environment due to high di/dt and dv/dt during turn-on or turn-off transient process of the devices. These drawbacks of the switching-mode operation of the converters are intensified with increase of switching frequency in order to reduce the converter size and weight and hence to increase the power density. Upper limit of the switching frequency is limited due to these drawbacks. This type of converter is called hard-switching converter. Accordingly, hard-switching converters are usually designed with lower switching frequency to limit the switching losses in order to achieve acceptable conversion efficiency. Lower switching frequency operation in turn results in bulkier passive components and may cause audible noise.

To overcome these shortcomings, a technology of shaping voltage and current of the power device during switching transient process, known as soft-switching, occurs. Before a power device in a converter changes its state, either from on-state to off-state or from off-sate to on-state, the voltage across it or the current through it is set to zero with the help of the resonance between

Soft-Switching Technology for Three-phase Power Electronics Converters, First Edition.
Dehong Xu, Rui Li, Ning He, Jinyi Deng, and Yuying Wu.

inductance and capacitance in the circuit. Therefore, the overlapping of the voltage and current on the device in either turn-on or turn-off transients is reduced. As a result, the switching loss is reduced.

2.2 Switching Characteristics of Three-phase Converters

Three-phase converters are used as either grid converters to connect the utility or inverters to drive a motor or supply high-quality alternating current (AC) power to the load as shown in Figure 2.1. As the grid converter, it converts grid AC into direct current (DC) bus voltage, which is usually called a rectifier, or converts DC bus voltage to the grid AC voltage, which is usually called an inverter. Actually, it is the same entity but has two names. In most cases, a grid converter operates in either rectifier mode or inverter mode according to power flow requirement. When three-phase converters are used to drive a motor or supply AC power to a load, it is usually called as an inverter since it converts the DC bus voltage into three-phase AC voltages. The three-phase converter is one of the most important power conversion building blocks in power electronic systems.

2.2.1 Control of Three-phase Converters

A converter is composed of three parts: DC bus, a switch bridge, filters, and three-phase loads (sources) as shown in Figure 2.2. The switch bridge comprises phase-a, phase-b, and phase-c switch legs.

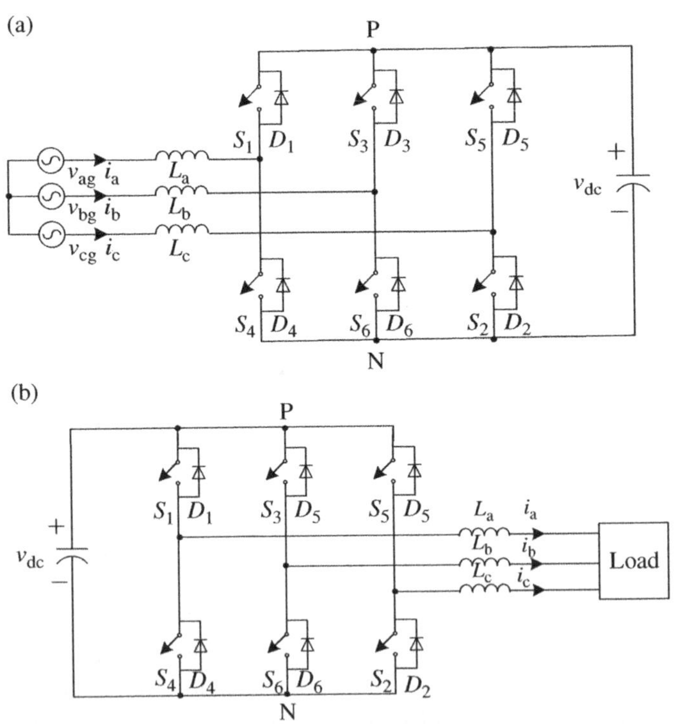

Figure 2.1 Three-phase converters: (a) grid converters; (b) inverters to supply passive loads.

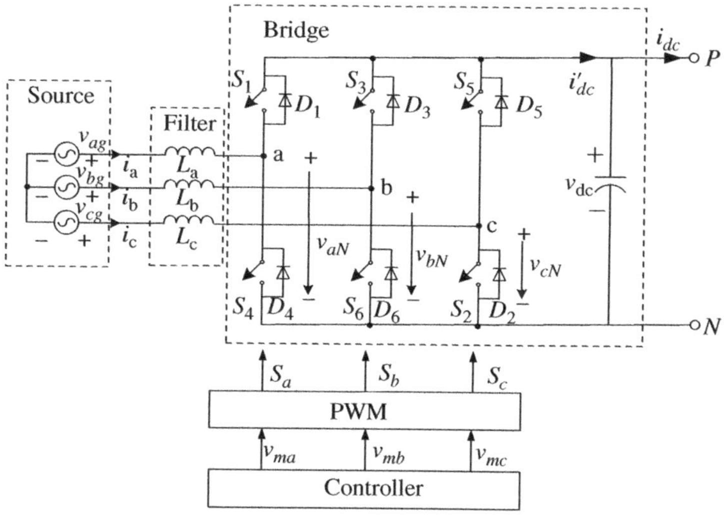

Figure 2.2 Control of three-phase converter.

It is used to convert DC-side voltage into pulse width modulation (PWM) waveform with the amplitude and frequency of its fundamental component adjustable [1, 2]. Switches S_1, S_3 and S_5 are connected between the positive rail *P* of the DC bus and phase-a, phase-b, and phase-c filters. Switches S_4, S_6 and S_2 are connected between the negative rail N of the DC bus and phase-a, phase-b, and phase-c filters. v_{dc} is the DC bus voltage, and i'_{dc} is the DC-side current of the switch bridge. The control diagram of a three-phase grid converter is also shown in Figure 2.2. Three reference voltages v_{ma}, v_{mb}, and v_{mc}, which are modulation signals of the PWM modulator, are generated by the higher level controller. They enter into the PWM modulator to produce PWM pulses to turn on or off the switches in the converter so that the expected three-phase midpoint voltages v_{aN}, v_{bN}, and v_{cN} are obtained. The output filter is used to filter switching frequency harmonics from output PWM voltage v_{aN}, v_{bN}, and v_{cN} so that there mainly exists the utility frequency current component, which exchanges between the converter and the grid. v_{ag}, v_{bg}, and v_{cg} are phase voltages of the three-phase sources.

PWM waveform is generated by comparing phase-a modulation signal v_{ma} with the triangle carrier waveform v_{car} as shown in Figure 2.3. $s_a(t)$ is called switching function of phase-a switch leg, which is equal to "1" when the modulation voltage is higher than the carrier, while it equals to "0" when the modulation voltage is lower than the carrier. Thus the switching function $s_a(t)$ is discontinuous in time domain. The pulse width of $s_a(t)$ is modulated by the modulation signal v_{ma}. The PWM generation shown in Figure 2.3 is called natural sampling PWM. If the modulation signal is replaced by a staircase waveform that is obtained from sampling the original modulation signal at the peaks or valleys instant of the carrier and hold the sampled value at the remaining time, it is called regular sampling PWM. In addition, modulation signals are classified into sinusoidal and sinusoidal with the third harmonic injection, which can be further divided into continuous PWM and discontinuous PWM. As we know, PWM can also be realized with space vector modulation (SVM), which is a popular modulation method in three-phase converters. It can be proved that SVM is equivalent to regular sampling PWM theoretically.

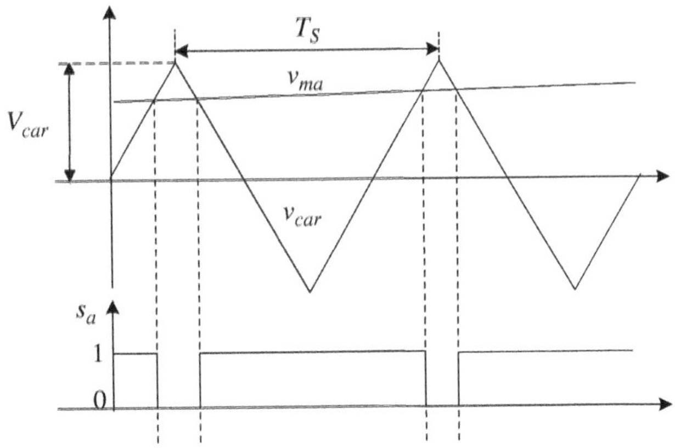

Figure 2.3 PWM signal generation.

It is easy to calculate the duty cycle of the switch leg according to the geometry relationship between the regular sample modulation waveform and the carrier. The duty cycle of phase-a switch leg is derived as

$$d_a = \frac{1}{2}\left(1 + \frac{v_{ma}}{V_{car}}\right) \tag{2.1}$$

where V_{car} is the peak value of the carrier v_{car}. The range of the modulation signal v_{ma} is $[-V_{car}, V_{car}]$. The range of the duty cycle d_a is [0, 1].

For sinusoidal pulse width modulation (SPWM), the modulation signal is expressed by

$$v_{ma} = V_{m\_\max} \sin\theta \tag{2.2}$$

where$V_{m_\max}$ is the amplitude of the sinusoidal modulation wave and is less than V_{car}.

By substituting Equation (2.2) into Equation (2.1), the duty cycle of phase-a switch leg is represented as

$$d_a = \frac{1}{2}(1 + m_a \sin\theta) \tag{2.3}$$

where $m_a = \dfrac{V_{m\_\max}}{V_{car}}$, which is known as amplitude modulation ratio. m_a ranges inside [0,1] for SPWM. If the third harmonic injected modulation is used, the modulation signal changes to

$$v_{ma} = V_{m\_\max}\left(\sin\theta + \frac{1}{6}\sin 3\theta\right) \tag{2.4}$$

The duty cycle becomes

$$d_a = \frac{1}{2}\left[1 + m_a\left(\sin\theta + \frac{1}{6}\sin 3\theta\right)\right] \tag{2.5}$$

With this choice, the highest value of m_a is $2/\sqrt{3} = 1.155$. However, the modulation signal still does not exceed the carrier amplitude.

Without losing generality, it is assumed the amplitude of the carrier is equal to 1, i.e. $V_{car} = 1$. $v_{ma} = m_a \sin\theta$ since $m_a = V_{m\_\max}$. Then Equation (2.1) is simplified as

$$d_a = \frac{1}{2}(1 + v_{ma}) \tag{2.6}$$

Similarly, we can derive duty cycles of phase-b and phase-c. The duty cycles for the three switch legs are summarized as follows:

$$\begin{aligned} d_a &= \frac{1}{2}(1 + v_{ma}) \\ d_b &= \frac{1}{2}(1 + v_{mb}) \\ d_c &= \frac{1}{2}(1 + v_{mc}) \end{aligned} \tag{2.7}$$

The three-phase duty cycles of the converter can be transformed to *dq0* rotating reference frame. Since there is no zero-sequence current component for the three-phase three-wire system, it can be briefly expressed with d-axis component and q-axis component as follows:

$$\begin{aligned} d_d &= \frac{1}{2}v_{md} \\ d_q &= \frac{1}{2}v_{mq} \end{aligned} \tag{2.8}$$

where d_d is the d-axis duty cycle and d_q is the q-axis duty cycle. v_{md} is the d-axis component of the modulation signal, and v_{mq} is the q-axis component of the modulation signal.

2.2.2 Switching Transient Process and Switching Loss

To investigate the switching transient process of the power devices in the three-phase converter controlled with PWM, we take one switch leg of the three-phase converter shown in Figure 2.4 as an example. Since the filter current changes not significantly in a switching cycle, it is assumed the filter current is constant in a switching cycle. In addition, the filter current i_a is assumed to be positive. Figure 2.5 shows PWM waveforms and voltage and current of the upper switch S_1. v_{ma} is the modulation signal of phase-a, and v_{car} is the carrier signal. PWM signal is derived by comparing the modulation signal v_{ma} with the carrier signal v_{car} as shown in Figure 2.5. When $v_{ma}(t) > v_{car}(t)$, phase leg upper switch S_1 turns on, while phase leg lower switch S_4 turns off. The midpoint voltage is $v_{aN} = V_{dc}$. When $v_{ma}(t) < v_{car}(t)$, phase leg upper switch S_1 turns off, while phase leg lower switch S_4 turns on. The midpoint voltage is $v_{aN} = 0$. The pulse width of the phase midpoint voltage v_{aN} is proportional to the modulation signal v_{ma}. The higher the modulation signal v_{ma}, the wider the pulse width of the midpoint voltage. If the modulation signal is sinusoidal waveform, then the pulse width of the midpoint voltage also varies with sinusoidal.

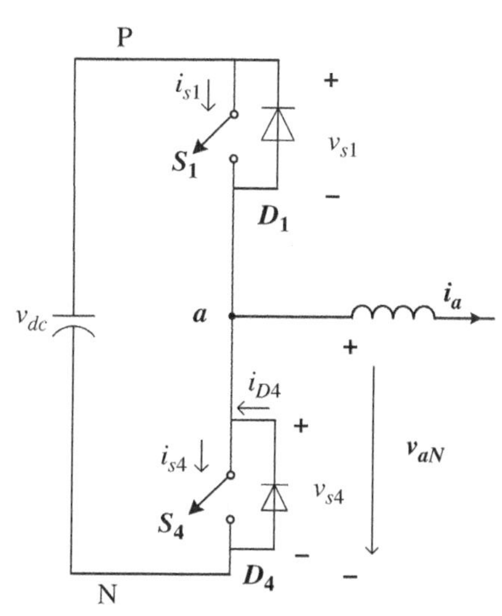

Figure 2.4 Switch leg of phase-a.

There are two kinds of switching commutations in a switch cycle T_s. They have different characteristics, which are explained as follows:

Switch commutation Type 1: It is a commutation from a switch to the antiparallel diode of its

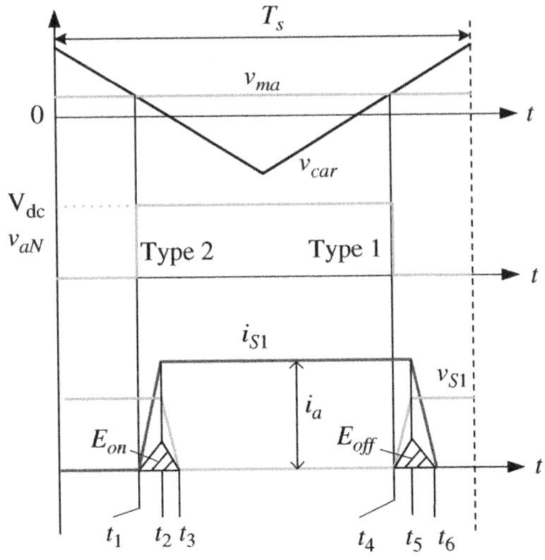

Figure 2.5 Key waveforms of phase-a switch leg in a switching cycle.

complementary switch in the same leg. Originally the filter current i_a flows through upper switch S_1 as shown in Figure 2.6a. At time t_4 in Figure 2.5, the switch S_1 is turned off. Since the power semiconductor device is not an ideal switch, the switch S_1 is unable to cut the load current suddenly. Firstly, the voltage v_{S1} across the switch S_1 will increase linearly, while the current i_{S1}through it keeps constant until v_{S1} increases to the DC bus voltage v_{dc} at time t_5. At t_5, the antiparallel diode D_4 of the lower switch S_4 is positively biased and starts to conduct as shown in Figure 2.6b. Then the upper switch S_1 begins to transfer its current to the lower diode D_4. The current through upper switch S_1 decreases linearly, while the current through lower diode D_4 increases as shown in Figure 2.5. At time t_6, the current through the upper switch S_1 decays to zero and turns off as shown in Figure 2.6c. The filter current totally commutates from switch S_1 to the antiparallel diode D_4 of its complementary switch S_4. As shown in Figure 2.5, the switch S_1 undergoes a switching transient process during its turn-off. From time t_4 to t_6, both voltage across it and current through it have high values, which will not happen at switch static states, either conducting state or blocking state. It is known as voltage and current overlap during switch transient process. The voltage and current

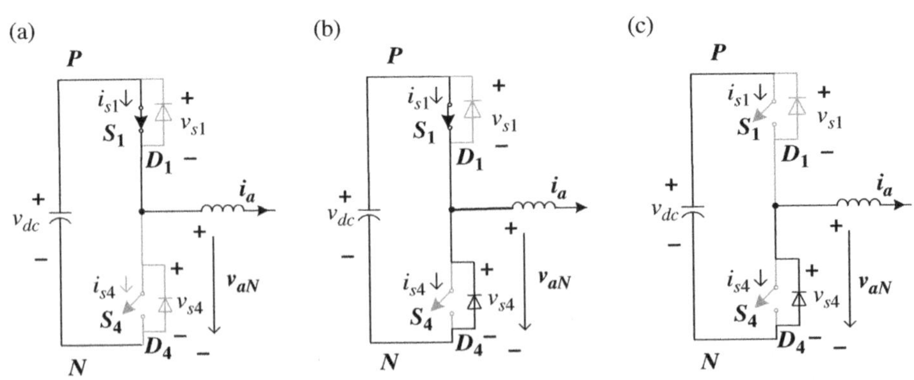

Figure 2.6 Commutation from upper switch S_1 to the antiparallel diode D_4 of lower switch.

overlap during the switching turn-off is the origin of turn-off loss. It can be estimated by integral of multiplication of voltage across the switch S_1 and its flowing current over the overlapping time as follows:

$$E_{off} = \int_{t_4}^{t_6} u_{S1} i_{S1} dt \tag{2.9}$$

In practice, E_{off} is measured with a digital oscilloscope. Figure 2.7 shows a turn-off transient waveforms of insulated gate bipolar transistor (IGBT) device in a phase leg. After sampling voltage and current traces of the switch device, the mathematical function is used to calculate the time integral of the product of both traces. Alternatively you may use Matlab to estimate the loss after exporting the measured waveform data from the oscilloscope.

If the switch S_1 is turned off with extra paralleled capacitor, the overlapping is reduced, and the turn-off loss is reduced. It is usually called turned off with zero voltage switching, or briefly ZVS turn-off. For wide bandgap (WBG) device, since turn-off loss is much smaller than turn-on loss, Type 1 commutation may not need to take special measure.

Switch commutation Type 2: It is a commutation from the antiparallel diode of a switch to its complementary switch in the same switch leg. Before the switching commutation, the filter current i_a flows through the antiparallel diode D_4 of the lower switch S_4 as shown in Figure 2.8a. If the

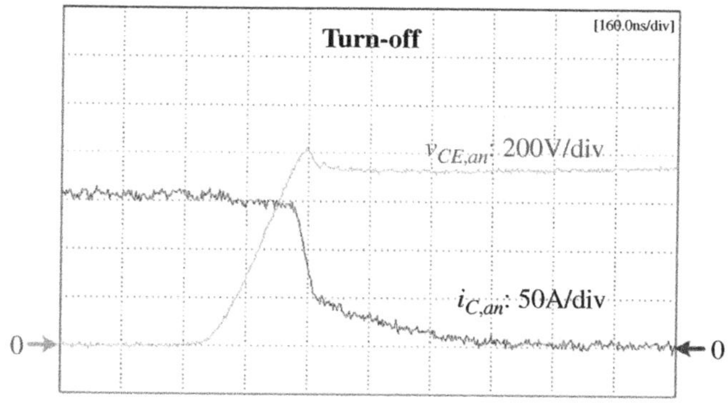

Figure 2.7 Turn-off transient waveforms of an IGBT device (1.7 kV/400 A module), test condition: v_{dc} = 750 V, load current i_a = 150 A. Time: 160ns/div.

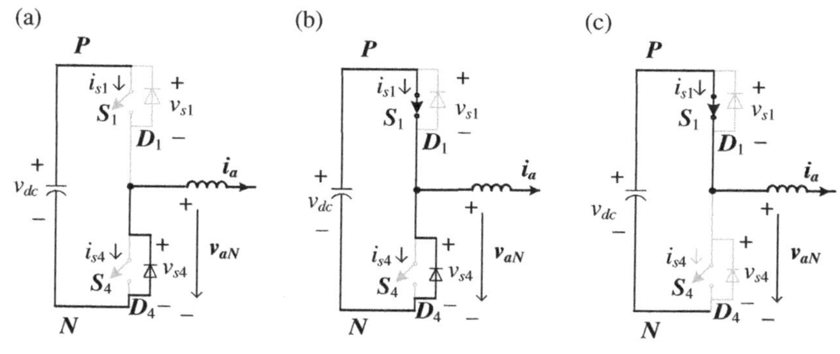

Figure 2.8 Commutation from lower diode D_4 to upper switch S_1.

upper switch S_1 is turned on, the filter current i_a will eventually commutate from the antiparallel diode D_4 of the lower switch S_4 to the upper switch S_1 as shown in Figure 2.8c. In Figure 2.5, at time t_1, the switch S_1 is turned on, its current through switch S_1 cannot jump to the load current suddenly. During t_1 to t_2 with the circuit shown in Figure 2.8b, the current i_{S1} of switch S_1 increases linearly from zero to the load current i_a, while the current of the antiparallel diode D_4 of the lower switch S_4 decreases. The voltage v_{S1} across the switch S_1 keeps constant as before until time t_2. At time t_2, the current through the diode D_4 is reduced to zero and turns off. The circuit is changed to what is shown in Figure 2.8c. From t_2 to t_3, the voltage v_{S1} on switch S_1 decreases linearly to zero, while the current i_{S1} of switch S_1 keeps constant. As shown in Figure 2.5, the switch S_1 undergoes a switching transient process during its turn-on. From time t_1 to t_3, both voltage v_{S1} across the switch S_1 and current i_{S1} through it have high values, which does not happen at the switch static state. The voltage and current overlap also occurs during switch turn-on process. The voltage and current overlap during the switching turn-on causes turn-on loss. It can also be estimated by integral of multiplication of voltage across the switch S_1 and its flowing current over the overlapping time as follows:

$$E_{on} = \int_{t_1}^{t_3} v_{S1} i_{S1} dt \tag{2.10}$$

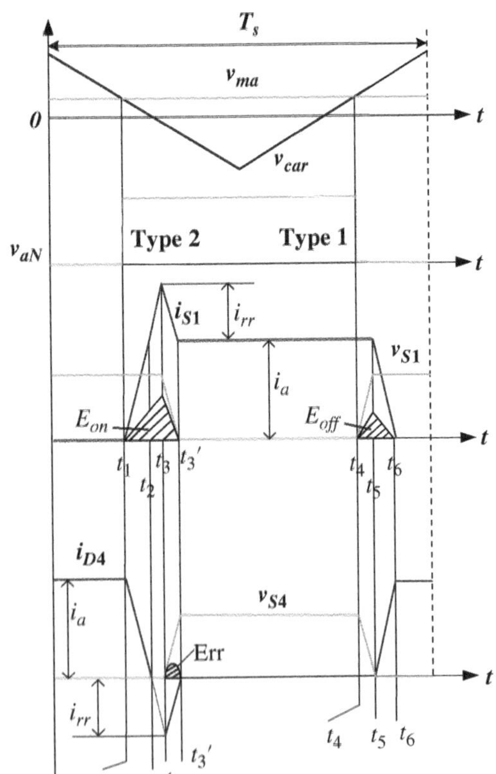

Figure 2.9 Key waveforms of phase-a switch leg in a switching cycle with consideration of diode reverse recovery.

2.2.3 Diode Turn-off and Reverse Recovery

When we turn on or turn off a diode, it also undergoes transient process as power semiconductor switch such as metal-oxide-semiconductor field effect transistors (MOSFETs) and IGBTs. When a diode turns on, it will undergo forward recovery process, while when it turns off, it undergoes reverse recovery process. Nowadays, forward recovery process can be ignored with improvement of the diode technology. However, the reverse recovery of the diode still needs to be paid attention to for high frequency applications except for the Schottky diode. Actually the diode reverse recovery process will affect the Type 2 switch commutation.

With consideration of the diode reverse recovery process, the switching waveforms in Figure 2.5 are changed as shown in Figure 2.9. In duration from t_1 to t_2, the current through the switch S_1 increases linearly from zero to the load current i_a, while the current of the antiparallel diode D_4 of the switch S_4 decreases. The waveform in duration t_1 to t_2 is the same as the earlier waveforms in Figure 2.5. From t_2, the current through diode D_4 reverses its direction and increases negatively until it reaches the negative peak value i_{rr} at t_3. Between time t_2 and t_3, the switch S_1 current keeps

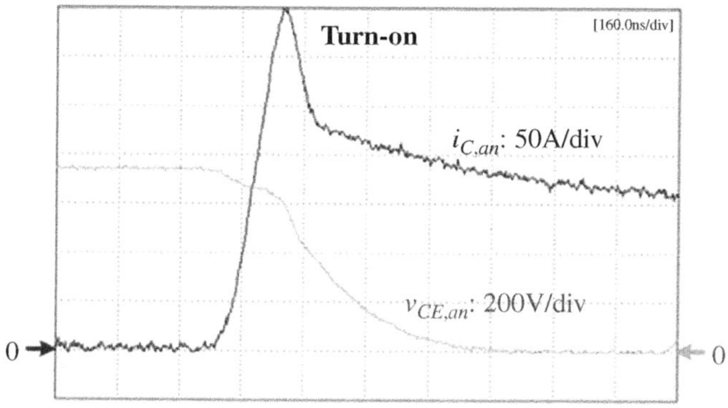

Figure 2.10 Turn-on transient waveforms of an IGBT device (1.7 kV/400 A module), test condition: v_{dc} = 750 V, load current i_a = 150 A. Time: 160ns/div.

increasing. Its peak value is equal to $i_a + i_{rr}$. The voltage across the switch S_1 keeps constant until time t_3, which is delayed from t_2. After time t_3, the diode D_4 current begins decaying, while its voltage v_{S4} begins to increase. At time $t_3^{'}$, the diode D_4 current decreases to zero and finally turns off. During reverse recovery process from t_2 to $t_3^{'}$, a reverse current with its peak of i_{rr} flows through the diode D_4, which adds a positive current spike into the current $i_{S1}$of the switch S_1. Due to the diode reverse recovery process when the switch S_1 turns on, the voltage and current overlap time on the switch S_1 is increased, and turn-on loss E_{on} is increased. Besides, there is additional loss on the turning-off diode D_4, which is due to the voltage $v_{S4}$and current i_{D4}overlap time on the diode D_4 between time t_3 and $t_3^{'}$. It is called diode reverse recovery loss. Figure 2.10 shows the turn-on transient waveforms of an IGBT device in a switch leg. A high current overshoot is added to the collect current i_c due to the diode reverse recovery of its complementary IGBT in the same leg. Similarly turn-off loss E_{rr} of the diode can be estimated by

$$E_{rr} = -\int_{t_3}^{t_3^{'}} v_{S4} i_{D4} dt \tag{2.11}$$

The diode reverse recovery process not only increases loss in both the turning-on switch and the turning-off diode but also results in high *di*/*dt* and high *dv*/*dt* in the conversion circuit, which may cause EMI noises. Sometimes it interfaces with the operation of gate drive circuit and other control circuits and even results in malfunction of other electronic equipment nearby. Many countries have set up standards to restrict the EMI noises.

IGBT device with antiparallel SiC Schottky barrier diode (SiC SBD), known as hybrid IGBT, has been designed to reduce the drawbacks of the conventional IGBT. Figure 2.11 shows the comparison of Si IGBT and hybrid IGBT with respect to the turn-on process. Extra current in IGBT is reduced from 118 to 44 A by replacing the antiparallel diode with SiC SBD during IGBT turn-on. However, special measure needs to be taken to suppress the switching ringing. The high frequency ringing at about 10 MHz in the hybrid IGBT current is due to the resonance between SBD parasitic capacitance and parasite inductance in the circuit.

2.2.4 Stray Inductance on Switching Process

As mentioned before, switching power loss happens due to nonideal characteristics of semiconductor switching power devices. For MOSFET or IGBT, there is voltage and current overlap on the

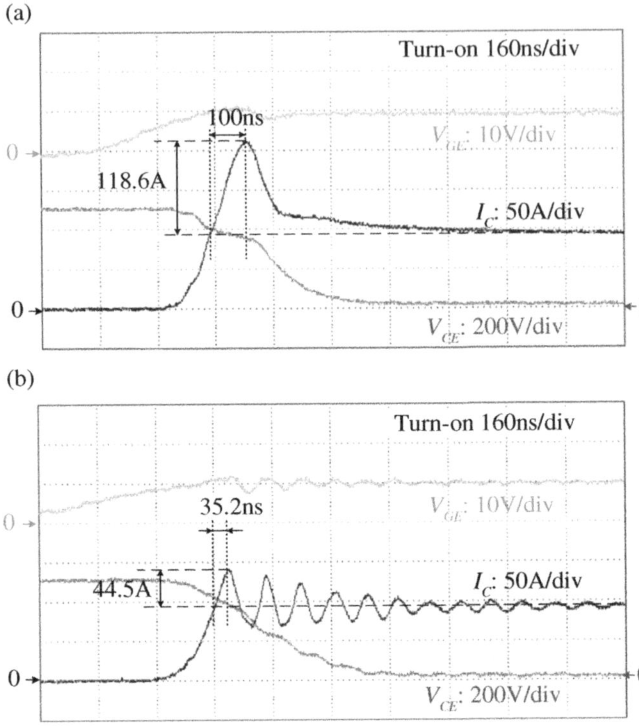

Figure 2.11 Turn-on waveforms: (a) Si IGBT device; (b) hybrid IGBT device. Test condition: v_{dc} = 500 V, i_a = 100 A. Device rating: 1.7 kV/400 A.

devices during switch commutations. For diode, there is a reverse recovery process when it is turned off. When we use devices to build a converter, conductor materials such as copper wire, print circuit board (PCB), bus bar, bonding wire, etc. are used to connect these devices together. These conductor materials have parasitic inductance and capacitance besides resistance. In addition, passive components such as capacitors, inductors, and transformers are used to build the converter. It is unavoidable that parasitic inductance or capacitance is introduced to the converter circuit. We still take one switch leg of the three-phase converter as the example. As shown in Figure 2.12a, two parasitic inductances L_{S1} and L_{S4} are introduced to upper and lower switches, respectively. They may be caused by circuit layout and bonding wires of the power semiconductor devices' package.

For simplicity, it is assumed that switches S_1 and S_4 are ideal and can turn on or turn off abruptly. C_{ds1} and C_{ds4} are the output capacitance of the switches S_1 and S_4, respectively. Originally the load current flows through upper switch S_1. At time t_4 in Figure 2.9, the switch S_1 is turned off. Then the load current is commutated to the diode D_4 of the lower switch S_4. The equivalent circuit is shown in Figure 2.12b. This is a second-order series resonant circuit. Once the transient process settles down, the voltage stress on the switch S_1 is equal to DC-side voltage v_{dc}. However, during the transient process of the second-order system, high voltage overshoot occurs on the switch S_1 due to lower damping in the circuit.

The resonant frequency is estimated by

$$\omega_0 = \frac{1}{\sqrt{(L_{s1} + L_{s4})C_{ds1}}} = \frac{1}{\sqrt{2L_{s1}C_{ds1}}} \tag{2.12}$$

where it is assumed that $L_{s1} = L_{s4}$ for simplicity.

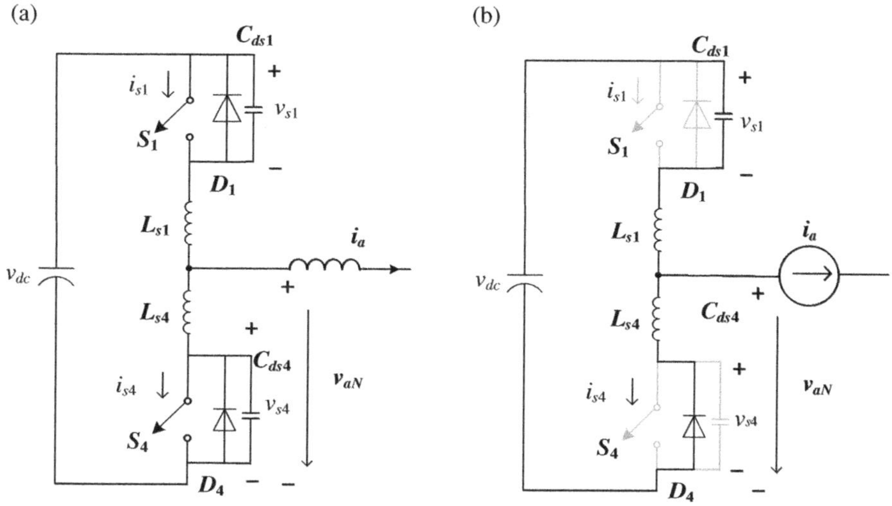

Figure 2.12 Effect of stray inductance: (a) a switch leg circuit; (b) second-order resonant circuit during commutation.

The peak voltage of the switch S_1 during its turn-off can be estimated by

$$V_{S1m} = v_{dc} + L_{S1}\frac{di_{S1}}{dt} + L_{S4}\frac{d(i_{S1} - i_a)}{dt} \tag{2.13}$$

It is assumed that the drain current i_{S1} of the switch S_1 falls linearly during turn-off. The load current i_a in one switching cycle is constant, i.e. $i_a = I_a$; then Equation (2.13) is represented as

$$V_{S1m} = v_{dc} + 2L_{S1}\frac{di_{S1}}{dt} = v_{dc} + 2L_{S1}\frac{I_a}{t_f} \tag{2.14}$$

where t_f is the switch S_1 current falling time during turn-off. In order to not jeopardize the power device, V_{S1m} must stay below the switch block voltage under all circumstances. The peak voltage V_{S1m} is composed of two terms in Equation (2.14). The first term is the DC bus voltage, which can be seen as a constant. The second term in the equation depends on both parasitic inductance L_{S1} and di/dt of the switch current, which causes the extra voltage stress on the switch. The larger the parasitic inductance, the higher the extra voltage stress.

For 1.2 kV/100 A SiC MOSFET half-bridge module, its current fall time t_f= 30ns when I_a = 100A, v_{dc} = 600V. Parasitic inductance L_{S1} = 15nH. The peak voltage of the switch is calculated by Equation (2.15):

$$V_{S1m} = v_{dc} + 2L_{S1}\frac{I_a}{t_f} = 600\text{V} + 2 \times 15 \times 10^{-9} \times \frac{100}{30 \times 10^{-9}}\text{V} = 700\text{V} \tag{2.15}$$

Total voltage stress of the switch is 700 V. Among it, 100 V extra voltage stress results from the parasitic inductance.

Figure 2.13 shows the turn-off waveforms of a SiC MOSFET. Beside extra voltage stress on device, there is a high frequency ringing, which causes EMI issues. Figure 2.14 shows the common-mode voltage envelopes comparison of Si IGBT and SiC MOSFET. SiC MOSFET has larger high frequency amplitudes.

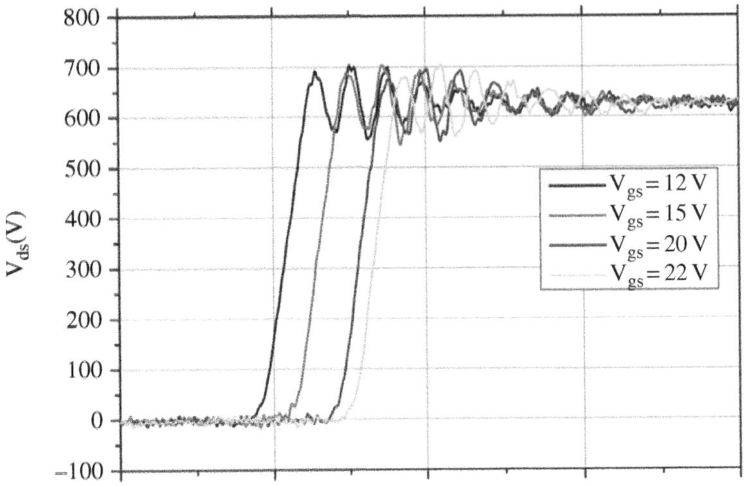

Figure 2.13 Turn-off waveforms of a SiC MOSFET with device rating: 1.2 kV/80 m Ohm, testing condition: V_{dc} = 600 V, I_a = 20 A. Time: 50ns/div.

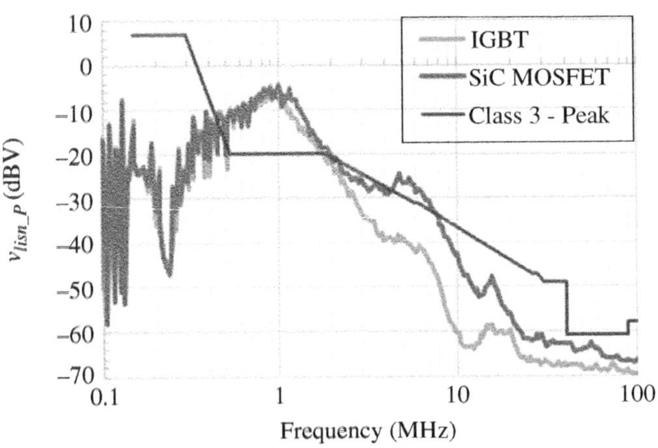

Figure 2.14 DC-side common-mode voltage envelopes comparison of Si IGBT inverter and SiC MOSFET inverter.

2.2.5 Snubber

Switch stress can be reduced by connecting simple dissipative snubber circuit. The basic principle of the turn-on snubber is to equivalently insert a series inductance at the turn-on path of the switch and thus limit the current-rising speed. The principle of the turn-off snubber is to use the parallel capacitor to divert the current during the turn-off, and limiting dv/dt. The energy stored in the snubber inductors and the snubber capacitors needs either to be circulated in the circuit or to be dissipated in the resistor. The circuit approach to circulate the snubbered energy is called the lossless snubber, which is usually complex and not commonly used. Generally the snubbered energy is dumped in the resistor. The snubber is typically composed of resistor and capacitor as shown in Figure 2.15. It can reduce voltage stress on the power semiconductor switches. In addition it

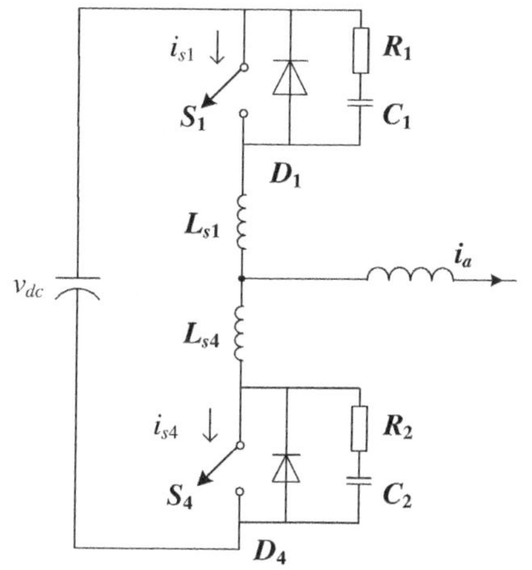

Figure 2.15 Dissipative snubber circuit for power semiconductor devices.

can also damp parasitic oscillations in the circuit due to switching turn-on or turn-off actions. Thus, EMI noise is reduced. However, the snubber cannot reduce the overall loss since it only shifts partial loss inside the switches to itself. Generally the overall loss of the circuit increases after introducing the snubber. Because of higher switching loss, the passive snubber approaches are deemed an impractical solution in higher switching frequency applications.

2.3 Classification of Soft-switching Three-phase Converters

Soft-switching is realized by either changing converter topology and/or introducing a unique control to reduce the overlapping of voltage and current of the semiconductor power switches during switching commutation. It can eliminate or reduce switching loss, problems of switching stress, and damping parasitic circuit oscillations. For soft-switching converters, either the switch voltage or switch current is set to zero when the switching transient happens [3]. There are four types of soft-switching for a switch as follows:

1) Zero-Voltage-Switching Turn-on (ZVS-on)
2) Zero-Current-Switching Turn-on (ZCS-on)
3) Zero-Voltage-Switching Turn-off (ZVS-off)
4) Zero-Current-Switching Turn-off (ZCS-off)

For three-phase applications, soft-switching converters can be divided into three classes: soft-switching converters with DC-side auxiliary resonance circuit, soft-switching converters with AC-side auxiliary resonance circuit, and soft-switching converters with triangle current mode (TCM) control, as shown in Figure 2.16. For the first and second classes, the auxiliary resonant circuit is introduced to the converter, which creates ZVS or ZCS conditions for all the switches of the converters. In the soft-switching converter with DC-side auxiliary resonance circuit, an auxiliary resonant circuit is installed in the DC side of the converter. In the soft-switching converter with AC-side auxiliary resonance circuit, an auxiliary resonant circuit is installed for each phase leg. Thus totally three auxiliary resonant circuits are needed. For the soft-switching converter with TCM control, soft-switching is realized by introducing a specific PWM control without using auxiliary circuit.

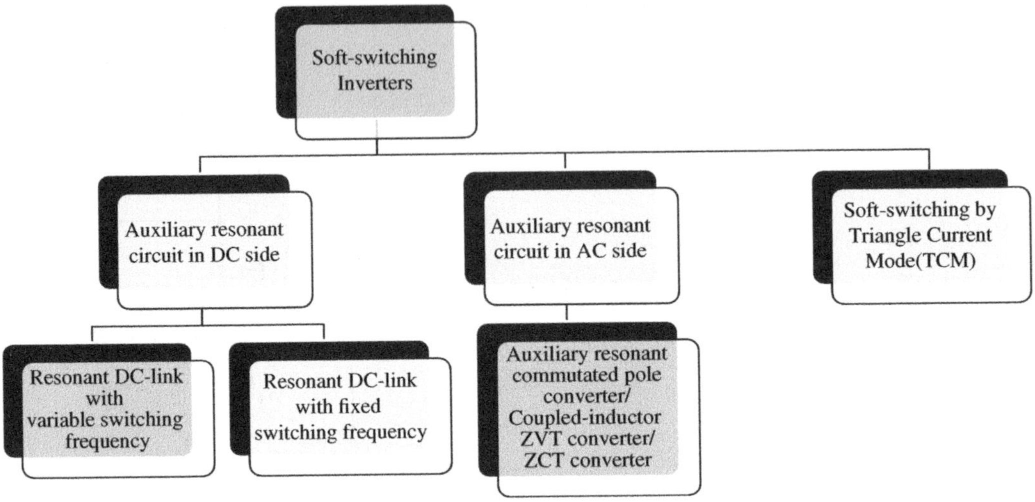

Figure 2.16 Classification of three-phase soft-switching converters.

2.4 DC-side Resonance Converters

Figure 2.17 shows the configuration of the DC-side resonance converters. An auxiliary resonant circuit is installed between the DC source and DC side of the switch bridge of the converter. The fundamental philosophy of the DC-side resonance is to use an auxiliary resonant circuit to create zero voltage duration at the DC side of the three-phase switch bridge at the desired switching instant. Then switching devices to commutate in the three switch legs can be switched under the zero voltage condition. The DC-side resonance only needs one auxiliary resonant circuit regardless of the number of AC phases of the converter. This simple structure makes DC-side resonance attractive in multiphase converter applications.

2.4.1 Resonant DC-link Converters

Resonant DC-link (RDCL) converter, shown in Figure 2.18, is a milestone in the evolution of soft-switching topologies for three-phase converters, which was invented by Deepak Divan [4]. The auxiliary resonant circuit installed in the DC side of the switch bridge consists of a resonant inductor L,

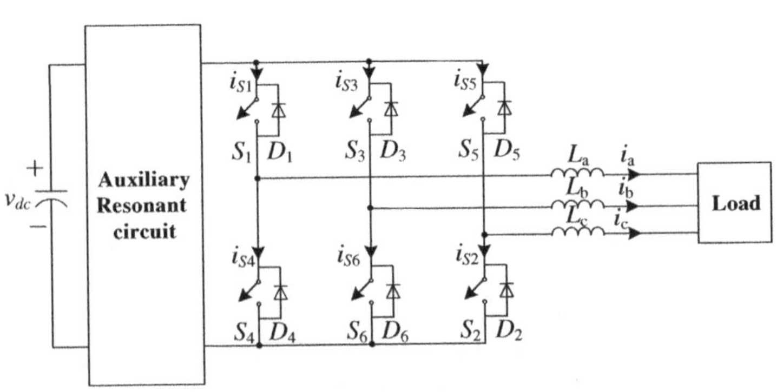

Figure 2.17 Configuration of DC-side resonance converters.

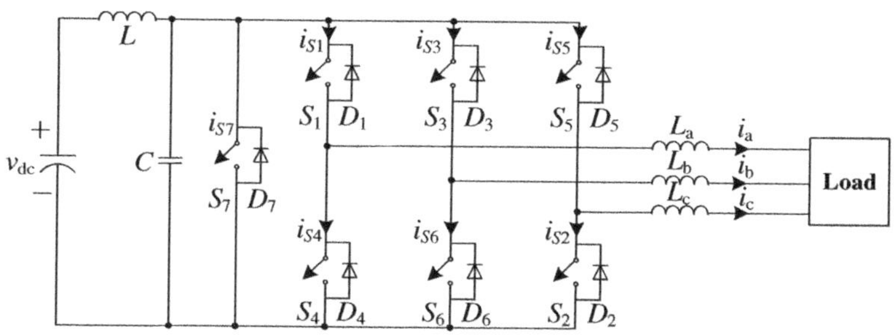

Figure 2.18 Circuit of RDCL converter. Source: Based on Divan [4].

Figure 2.19 Simplified circuit of RDCL converter.

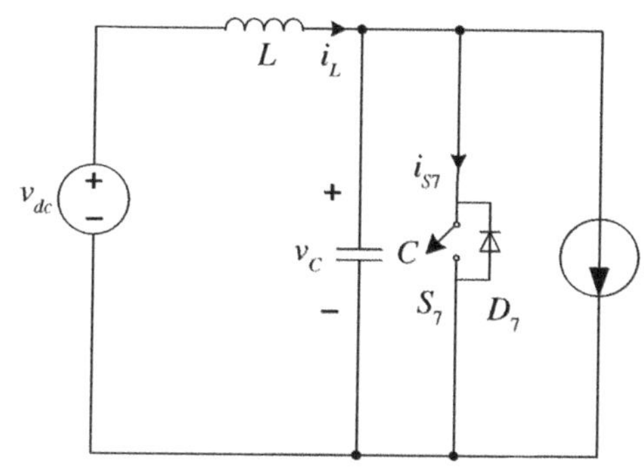

a resonant capacitor C, and a power semiconductor switch S_7. Different from the common voltage source converter, input power must pass through the resonant tank before flowing to the converter bridge. In this way, the voltage input to the inverter is no longer DC voltage but a series of resonant voltage pulses with high frequency. The DC-side voltage of the switch bridge oscillates periodically between its resonant peak and zero, thus generating a zero voltage time interval periodically. It creates a ZVS condition for the subsequent switch bridge.

In order to make the voltage v_c of the resonant capacitor C return to zero periodically, it is necessary to add energy to the resonant tank to compensate the loss in the circuit. The resonant inductor L is charged with enough energy before the resonance tank starts to oscillate. The switch device S_7 is used to control the charging of the resonant inductor L.

Since filter inductors L_a, L_b, and L_c of the converter are generally much larger than the resonant inductance L, DC-side current of the inverter can be approximately regarded as constant in a resonant period. Its value depends on the instantaneous value of the filter inductor currents and the state of the inverter bridges. Thus the inverter as a whole can be regarded a constant current source I_0 in a switching period as Figure 2.19.

The operation of the circuit can be divided into two stages as shown in Figure 2.20a. The first stage:$[t_0,t_1]$ is the pre-charge stage of the resonant inductor. In this stage, the switch S_7 is turned on, the resonant capacitor voltage v_c is clamped at zero, and the inductance current i_L increases linearly with the action of DC source voltage v_{dc}. The first stage ends when the current of the resonant inductor $i_L = I_{L1}$. The setting of the pre-charge current threshold I_{L1} depends on the load current I_0 and loss in the circuit.

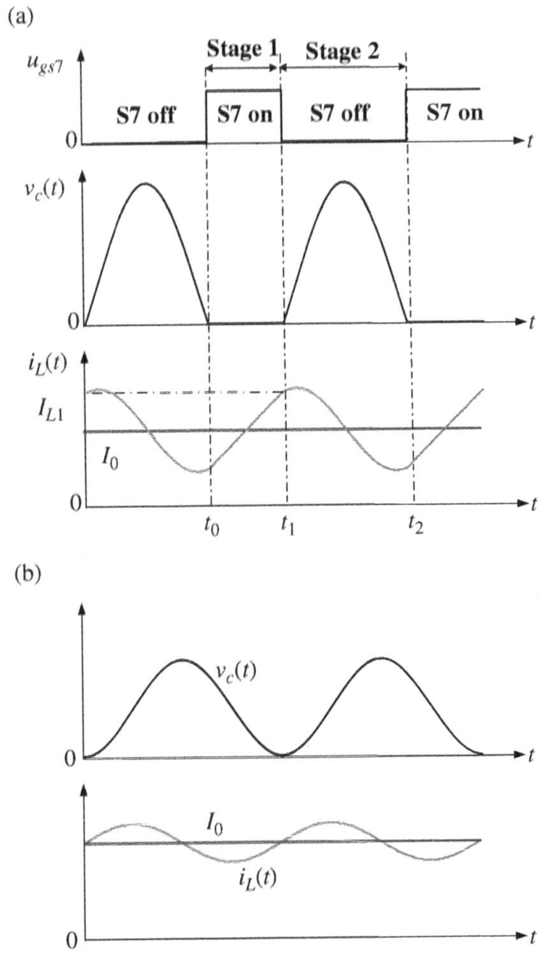

Figure 2.20 Waveforms of resonant tank voltage and current: (a) $I_{L0} > I_0$; (b) $I_{L0} = I_0$.

The second stage:$[t_1,t_2]$ is the resonance stage. At the beginning of this stage, switch S_7 is turned off. Then the circuit becomes LC resonant circuit. It is described by the following state equations:

$$\begin{cases} L\dfrac{di_L}{dt} + v_C = v_{dc} \\ C\dfrac{dv_C}{dt} = i_L - I_0 \end{cases} \tag{2.16}$$

Its initial condition is

$$\begin{cases} i_L(t_1) = I_{L1} \\ v_C(t_1) = 0 \end{cases} \tag{2.17}$$

By solving the differential Equation (2.16) with the initial conditions, the following solution is obtained:

$$\begin{cases} v_C(t) = v_{dc}[1 - \cos\omega_r(t-t_1)] + \omega_r L(I_{L1} - I_0)\sin\omega_r(t-t_1) \\ i_L(t) = I_0 + \dfrac{v_{dc}}{\omega_r L}\sin\omega_r(t-t_1) + (I_{L1} - I_0)\cos\omega_r(t-t_1) \end{cases} \tag{2.18}$$

where $\omega_r = \frac{1}{\sqrt{LC}}$.

The first equation in Equation (2.18) can be rewritten as

$$v_C(t) = v_{dc} - \sqrt{v_{dc}^2 + [\omega_r L(I_{L1} - I_0)]^2} \cos[\omega_r (t - t_1) + \phi] \tag{2.19}$$

where $\phi = \arctan\left[\frac{\omega_r L(I_{L1} - I_0)}{v_{dc}}\right]$.

To realize the ZVS turn-on of the main switches, the capacitor voltage $v_C(t)$ is required to resonance to zero. By observing Equation (2.19), $v_C(t)$ is composed of a constant term and a sinusoidal term. To make $v_C(t)$ negative, the amplitude of the sinusoidal term should be larger than the constant term. Thus, the following inequality needs to be satisfied:

$$v_{dc} < \sqrt{v_{dc}^2 + [\omega_r L(I_{L1} - I_0)]^2} \tag{2.20}$$

This inequality seems to be easily satisfied if the inductor $i_L(t_1) = I_{L1}$ at t_1 is a little larger than the load current I_0. Because there is power loss in practical circuit, the inductor needs to be charged to certain value I_{L1} in stage 1 so that the capacitor voltage $v_C(t)$ will resonance to zero in stage 2 for soft-switching of the switches.

If we assume that the pre-charge current threshold $I_{L1} = I_0$, the capacitance voltage and inductance current are simplified as

$$\begin{cases} v_C = v_{dc}(1 - \cos\omega_r t) \\ i_L = I_0 + \dfrac{v_{dc}}{\omega_r L}\sin\omega_r t \end{cases} \tag{2.21}$$

Equation (2.21) shows that when there is no loss in the resonant circuit, as long as the threshold value of the inductive pre-charge current is equal to the instant load current, i.e. $I_{L1} = I_0$, the capacitance voltage v_c will oscillate periodically between 0 and $2v_{dc}$, while the inductive current will become a DC biased sinusoidal waveform with the average value equals to I_0. The voltage and current in the circuit are shown in Figure 2.20b. Since there is no loss in the circuit, the resonant inductor current i_L returns to its initial value $I_{L1} = I_0$ at the end of the second stage. Thus, there is no need to pre-charge the resonant inductor in the first stage. As a result the duration of the first stage contracts to a point in the time axis.

Actually loss exists in the circuit. In order to make up for the loss, the threshold value of the inductive pre-charge current needs to be higher than the load current, i.e. $I_{L1} > I_0$. Then the voltage and current in the resonant tank changes as shown in Figure 2.20a. During the first stage, the resonant capacitor voltage is clamped to zero due to the conduction of the switch S_7, while the resonant inductor current linearly increases. This stage ends once the resonant inductor current reaches I_{L1}. During the second stage, the switch S_7 turns off, and the capacitor voltage will bounce to voltage higher than $2v_{dc}$ due to the higher initial inductor current. When the capacitor voltage returns to zero again, the second stage ends, and the switch S_7 is turned on again. Then it starts the first stage of a new switching cycle.

From the previously stated analysis, it is observed that the switching devices will bear stress more than twice the input voltage. In addition, the working period of the resonant circuit will vary with the instantaneous value of the DC-side current of the inverter. It means the switching frequency of this converter is variable. It not only complicates the control design but also affects the power quality of the inverter output.

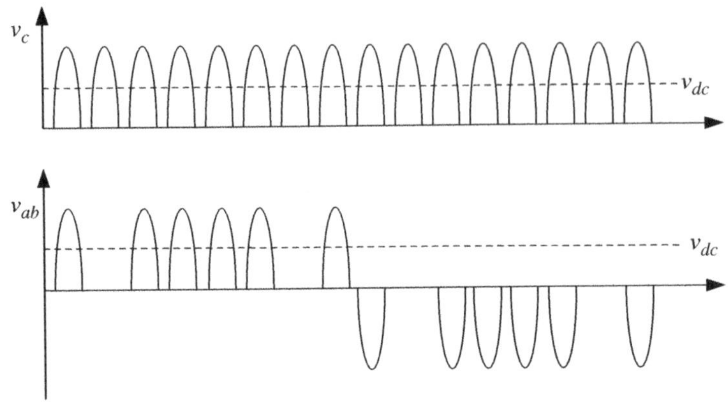

Figure 2.21 Generation of output waveforms of RDCL.

With the control of the resonant tank mentioned before, the resonant capacitor voltage becomes a series of discrete pulses. It is observed that the DC-side voltage of the inverter becomes a series of discrete pulses. To control the output of the inverter, the discrete pulses are distributed to inverter AC side through changing states of inverter switches. Figure 2.21 shows how to generate output line voltage v_{ab} of the inverter. If we turn on the upper switch of phase-a switch leg and the lower switch of phase-b switch leg at the same time, v_{ab} will output the resonant capacitor voltage v_c. If we turn on the lower switch of phase-a switch leg and the upper switch of phase-b switch leg at the same time, v_{ab} will output the reversed resonant capacitor voltage, i.e. $-v_c$. If we turn on either the lower switches or the higher switches of both phase-a switch leg and phase-b switch leg at the same time, the resonant capacitor voltage v_c is bypassed and v_{ab} outputs zero voltage. Thus we can integrate required v_{ab} waveform with a low frequency fundamental component by controlling states of phase-a and phase-b switch legs. We can improve v_{ab} waveform with more complicated control such as distributed pulse mode (DPM) control so that its switching period averaged value is more like a sinusoidal [5]. Similarly v_{bc} can also be generated. For three-phase line voltage, there exists the relationship $v_{ab} + v_{bc} + v_{ca} = 0$. Then v_{ca} can be automatically obtained after v_{ab} and $v_{bc}$are decided. Therefore, output fundamental component amplitude and the frequency of the inverter are able to be controlled.

With the control of the resonate tank mentioned before, the resonant capacitor voltage becomes a series of discrete pulses. Between two consecutive pulses is durations with zero voltage as shown in Figure 2.21. The state transition of the inverter switch bridge only happens when the resonant capacitor voltage equals to zero. Therefore, all switch devices of the inverter operate at ZVS switching condition. The switching loss of power semiconductor devices is reduced. It is possible to operate the inverter at higher switching frequency, which reduces the size of passive components such as inductor or capacitors and increases the power density.

From the previously mentioned analysis, it is observed that the switching devices bear a stress more than twice the input voltage. Power switch devices with higher blocking voltage are needed. Because switch devices with higher blocking voltage usually have relatively high conducting loss and switching loss than those of the lower blocking voltage ones, it not only increases cost but also deteriorates the conversion efficiency of the RDCL inverter. Finally, both efficiency and power density of the RDCL inverter may be unable to be enhanced.

If you carefully analyze the topology and working principle of RDCL inverter, you will find that the switch S_7 is actually redundant. Its function can be realized by switches of the three-phase converter. For example, the resonant capacitor voltage can be short-circuited by conducting both upper and lower switches of a switch leg or all three switch legs at the same time.

2.4.2 Active-clamped Resonant DC-link (ACRDCL) Converter

To reduce the voltage stress on the switch devices, an extra branch that consists of a switch S_c and a clamping capacitor C_c is paralleled to the resonant inductor L_r as shown in Figure 2.22. This modified circuit is known as active-clamped resonant DC-link (ACRDCL) inverter [6].

It is assumed that the clamping capacitor C_{cl} is so large that its voltage v_{cl} can be seen as constant. Whenever the resonant capacitor voltage is higher than the sum of DC source voltage and the clamping capacitor voltage $v_{dc} + v_{cl}$ in the second stage, the antiparallel diode D_c of the switch S_c will conduct automatically so that the resonant capacitor voltage v_c is clamped. Instead of diode D_c, the switch S_c with antiparallel diode D_c is used so that the current to the clamping capacitor is bidirectional and its voltage can be controlled. ACRDCL inverter has lower voltage stress, which is about 1.3–1.4 times of the DC bus voltage.

DPM is also used for output control. Since the inverter output is controlled by distributing the discrete pulses from the resonant capacitor, resolution of the output is limited by the pulse. DPM requires the DC-link resonating frequency to be several times higher than the switching frequency of the PWM converter to reach comparable current spectral performance. In addition, the switching frequency of both RDCL and ACRDCL inverters is variable and depends on instant inverter output current waveforms. Therefore, the pulses are not evenly spaced in time axis. It may cause undesirable subharmonics in the inverter output.

Predecessors tried to overcome the drawbacks of RDCL and ACRDCL converters by combining the soft-switching features of RDCL with characteristics of pulse width modulation converter [7–10]. One example is the quasi-resonant DC-link (QRDCL) PWM converter [7, 11]. The DC-side voltage resonates to zero only during the bridge switching transition. Therefore, the PWM control can be used, and power quality is guaranteed. Usually more auxiliary switches are used. It makes converter circuits more complex and may introduce extra loss.

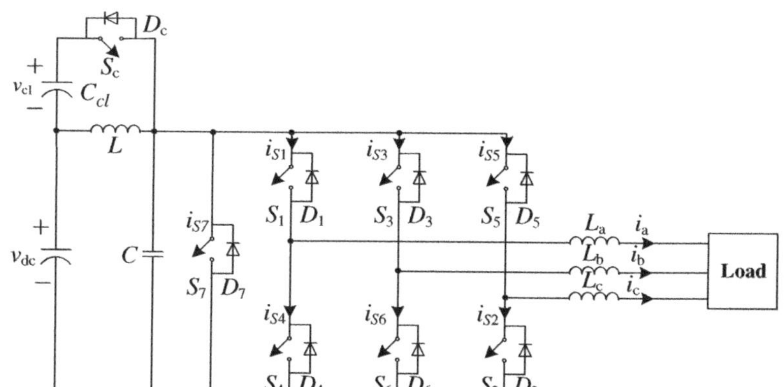

Figure 2.22 Circuit of active-clamped resonant DC-link (ACRDCL) inverter. *Source:* Based on Divan and Skibinski [6].

2.4.3 ZVS-SVM Active-clamping Three-phase Converter

2.4.3.1 Active-clamping DC–DC Converter

Similar to switch legs of the three-phase converter, DC–DC converters such as buck converter, boost converter, and buck–boost converter also have two types of commutations when their inductors operate at continuous current mode. We take boost converter as example, as shown in Figure 2.23a. Type 1 commutation is from switch S_1 to diode D_2. If there is capacitor paralleled with the switch, S_1 is turned off with ZVS so that turn-off loss is reduced. This commutation has smaller loss. However, Type 2 commutation from the diode D_2 to the switch S_1 has larger loss [12]. The equivalent circuits and key waveforms during Type 2 commutation are shown in Figure 2.23b and c. Before Type 2 commutation, the inductor current flows through the diode D_2, and power transfers from the source to the load. At t_0, the switch S_1 is turn-on, and its current cannot jump

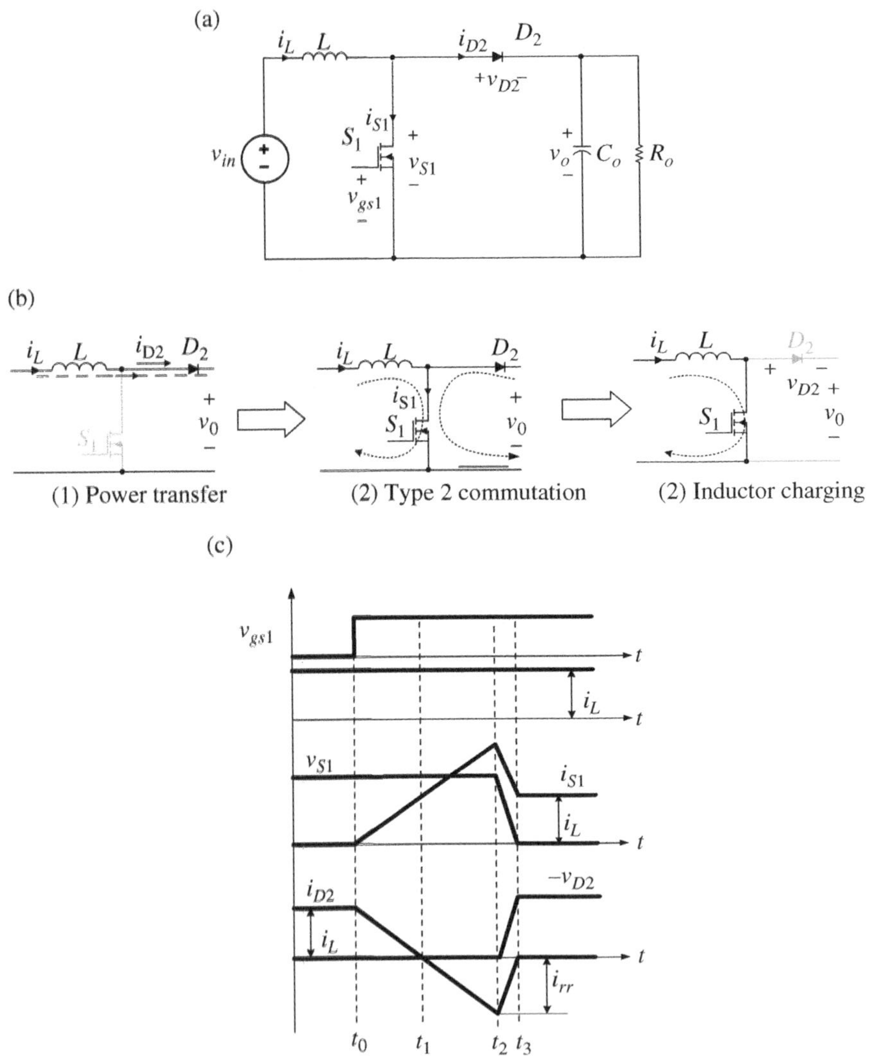

Figure 2.23 Boost converter and Type 2 commutation.

to the inductor current suddenly. During t_0 to t_1, the current i_{S1} of switch S_1 increases linearly from zero toward the inductor current value i_L, while the current of the diode D_2 decreases shown in Figure 2.23b. The voltage v_{S1} across the switch S_1 keeps constant. At time t_1, the current through the diode D_2 is reduced to zero and reverses its direction due to the diode reverse recovery process. It increases negatively until it reaches the negative peak value i_{rr} at t_2. Between time t_1 and t_2, the switch S_1 current keeps increasing. Its peak value is equal to $i_L + i_{rr}$. The voltage across the switch S_1 keeps constant until time t_2. After time t_2, the diode D_2 current begins decaying from its i_{rr}, while its voltage u_{D2} begins to increase while voltage on S_1 decreases. At time $t_3$when the diode D_2 current decreases to zero, D_2 recovers blocking voltage ability and finally turns off while the voltage on the switch S_1 becomes zero. During reverse recovery process from t_1 to t_3, a reverse current with its peak of i_{rr} flows through the diode D_2, which adds a positive current spike into the current $i_{S1}$of the switch S_1. Due to the diode reverse recovery process when the switch S_1 turns on, the voltage and current overlap time on the switch S_1 is increased, and turn-on loss E_{on} is increased. Besides, there is additional loss on the turning-off diode D_2, which is due to the voltage v_{D2} and current i_{D2} overlap time on the diode D_2 between time t_2 and t_3. It is known as diode reverse recovery loss E_{rr}.

For current continuous mode (CCM) boost converter, during Type 2 commutation, both voltage v_{S1} across the switch S_1 and current i_{S1} through it have high values, which does not happen at static state of the boost converter. Besides there is voltage and current overlap time on the diode D_2 due to the diode reverse recovery.

In order to suppress the loss due to Type 2 commutation, an extra inductor L_r is introduced and serially connected with the main switch S_1 as shown in Figure 2.24a. The inductor L_r controls di/dt rate of the diode reverse current when the diode D_2 turns off. To recycle the energy stored in L_r

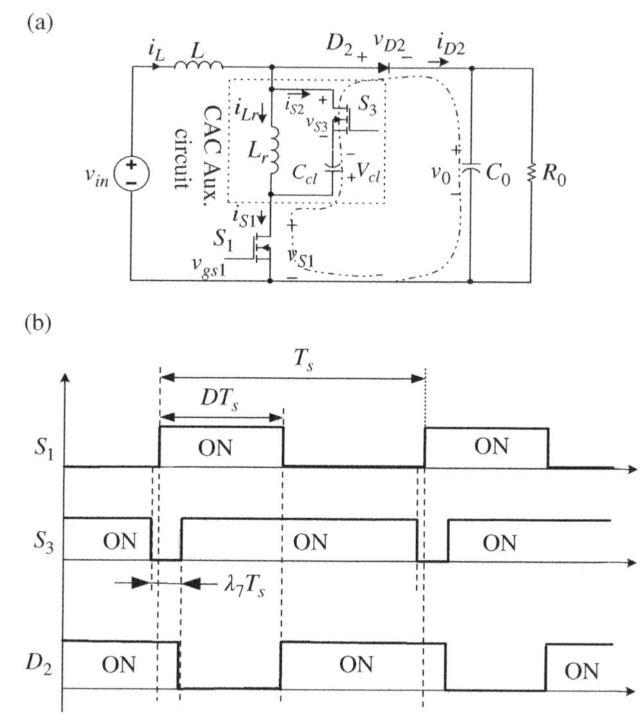

Figure 2.24 Compound active-clamping (CAC) boost converter: (a) circuit; (b) state of three devices.

when the switch S_1 is turned off, an active-clamping branch composed of an auxiliary switch S_3 and a clamping capacitance C_{cl} is introduced [13–15]. The auxiliary circuit inside dash-line box in the figure is called compound active-clamping (CAC) circuit. The boost converter with CAC circuit is called CAC boost converter.

In comparison to the conventional active-clamping boost converter, the main switch S_1 and the auxiliary switch S_3 of CAC boost converter do not operate in complementary mode [16, 17]. For the CAC converter, at any time there always exist two conducting devices among S_1, D_2, and S_3 so that the voltage across the one switch device, which is in off-state, is clamped. By applying Kirchhoff's voltage law (KVL) law to the loop shown in Figure 2.24a, the following equation is obtained:

$$v_{S1} + v_{S3} - v_{D2} = v_O + v_{cl} \tag{2.22}$$

Since there are always two devices conducting, the third device, which is in off-state, will block voltage in the loop. For example, if S_1 and S_3 are conducting, then D_2 is clamped with voltage $v_O + v_{cl}$. If S_1 and D_2 are conducting, then S_3 is clamped with voltage $v_O + v_{cl}$. If S_3 and D_2 are conducting, then S_1 is clamped with voltage $v_O + v_{cl}$. Since the terminal voltage of the turn-off device is clamped, parasitic oscillation in the circuit is eliminated. Therefore, the voltage stresses of all these three devices in the converter are lower.

Actually the auxiliary switch S_3 switches off only in a short interval in a switch period when the boost inductor current commutates from D_2 to S_1 as shown in Figure 2.24b. Besides turn-off of S_3 starts a resonance process to provide ZVS turn-on condition for the main switch S_1. In addition, both the main switch and the auxiliary switch can achieve ZVS.

To simplify the stage analysis in a switching period, the following assumptions are made:

- The converter has reached a steady state.
- All the power switches and diodes are ideal. The resonant inductor and capacitor are ideal and have linear characteristics.
- The current through boost inductor L is constant during a switch period and can be seen as a DC current source I_L. The output capacitor C_O is represented by a DC voltage source V_O.
- The capacitance of C_{cl} is so large that it can been seen as a DC voltage source V_{cl}.

With the previously mentioned assumption, the original circuit in Figure 2.24 is redrawn in Figure 2.25. A switching period can be divided into six stages. Circuit states at various stages

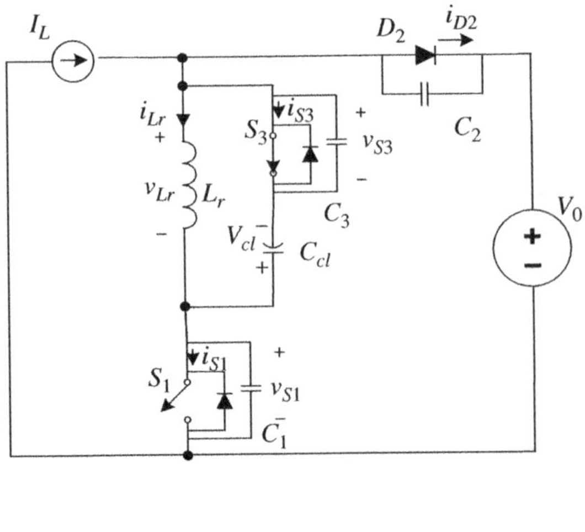

Figure 2.25 Equivalent circuit for stage analysis.

and key waveforms of the CAC converter in a switch period are shown in Figures 2.26 and 2.27, respectively.

Now let us discuss the operating stages of the converter in a switch period:

Stage 1 (t_0–t_1): *The first resonant stage.* Before t_0, the main switch S_1 is off and diode D_2 is conducting. The auxiliary switch S_3 is in conducting state, and the resonant inductor is clamped with V_{cl}. The circuit is transferring power from DC source to the load. At t_0, S_3 is turned off, which starts a resonant process. The equivalent circuit is shown in Figure 2.26a. C_1 and C_3 are paralleled,

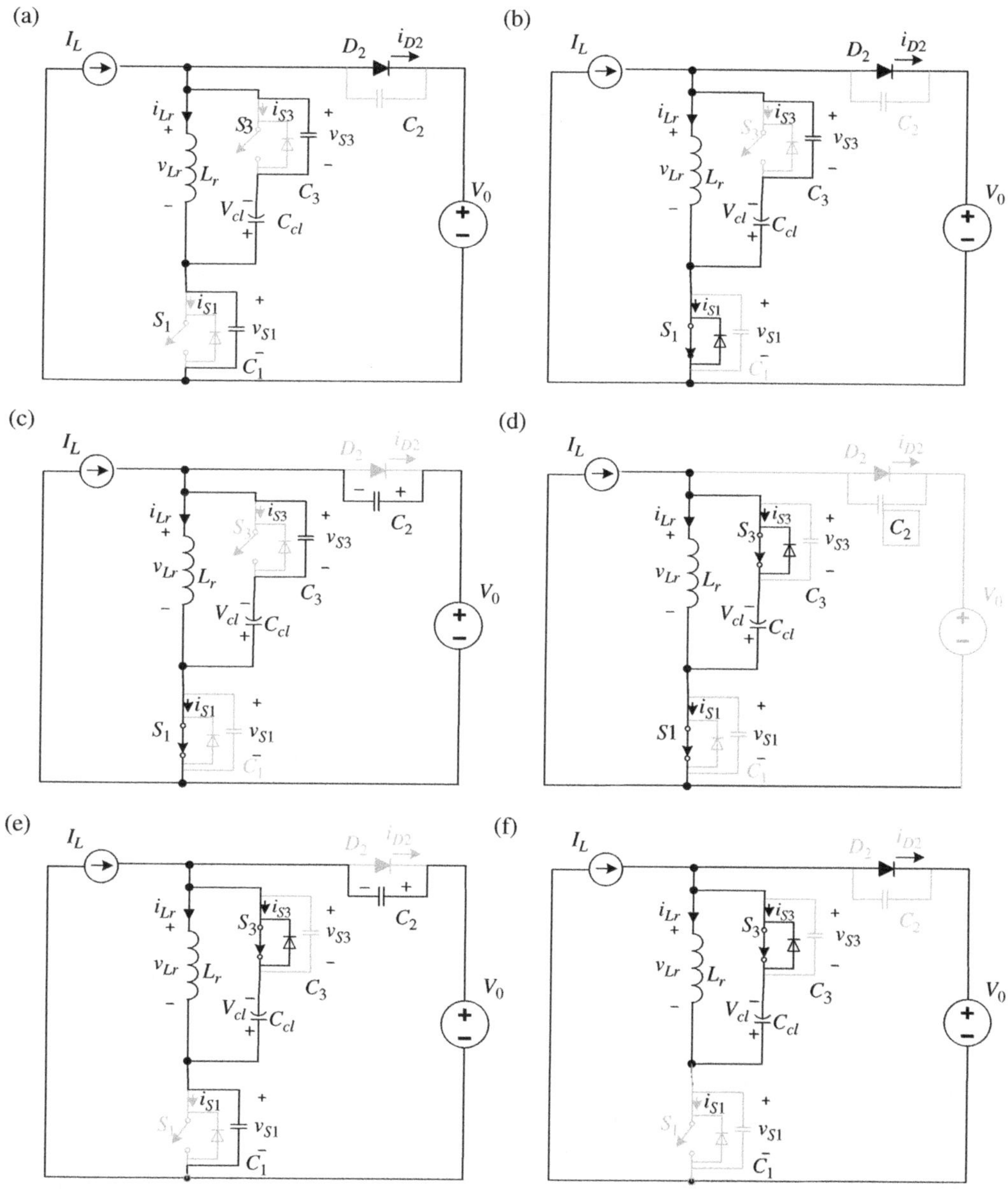

Figure 2.26 Circuit states at various stages: (a) Stage 1 (t_0–t_1); (b) Stage 2 (t_1–t_2); (c) Stage 3 (t_2–t_3); (d) Stage 4 (t_3–t_4); (e) Stage 5 (t_4–t_5); (f) Stage 6 (t_5–t_6).

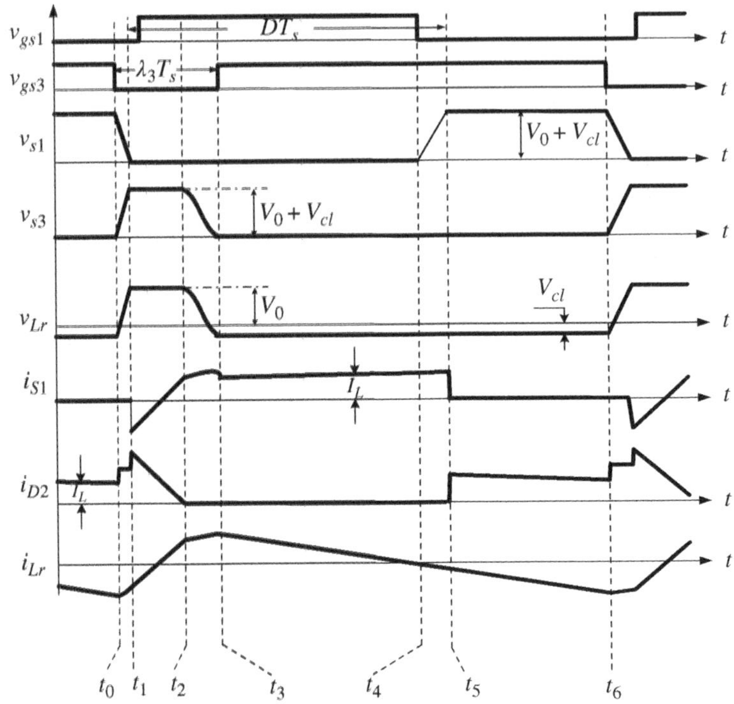

Figure 2.27 Steady-state waveforms of the converter.

which is resonance with resonant inductor L_r. Since the resonant inductor current i_{Lr} is negative, C_3 is in the charging state, while C_1 is in discharging state. Once voltage on C_1, i.e. voltage on S_1 v_{S1}, is resonance to zero, the antiparallel diode of S_1 will conduct. It creates ZVS turn-on condition for S_1 in next stage. At t_1, voltage on C_1 is resonance to zero, and this stage ends.

Stage 2 (t_1–t_2): Flywheel stage. At time t_1, the voltage across S_1 decreases to zero, and its antiparallel diode starts to conduct. The equivalent circuit is shown in Figure 2.26b. The voltage across S_3 is clamped at $V_O + V_{cl}$. Voltage V_O is applied on the resonant inductor L_r. Thus the resonant inductor current increases linearly with time, while the current through the diode D_2 decreases. Current decreasing rate of D_2 is decided by

$$\frac{di}{dt} = \frac{V_O}{L_r} \tag{2.23}$$

Thus we can adjust the di/dt by selecting the resonant inductance.

At the beginning of this stage, there is current flowing through S_1's antiparallel diode D_1. Then its current quickly decreases with the quick increasing of the resonant inductor current. Once D_1's current decays to zero, the current reverses direction and starts to flow through S_1.

To realize ZVS turn-on of S_1, it is required that the gate signal for S_1 is enabled during the conduction duration of S_1's antiparallel diode D_1.

At t_2 the current in D_2 decreases to zero, and this stage ends.

Stage 3 (t_2–t_3): The second resonant stage. At time t_2, D_2 becomes blocked and turns off due to its current decaying to zero. It starts the second resonant process. The equivalent circuit is shown in Figure 2.26c. C_2 and C_3 are paralleled, which is resonance with L_r. C_2 is in the charging state, while

C_3 is in discharging state. Once the voltage on C_3, i.e. the voltage across S_3, resonances to zero, this stage ends. The antiparallel diode of S_3 will conducts. It creates ZVS turn-on condition for S_3 in next stage.

At t_3, voltage on C_3 is resonance to zero, and this stage ends.

Stage 4 (t_3–t_4): Boost inductor charging stage. At t_3, the antiparallel diode D_3 of S_3 begins to conduct. The equivalent circuit is shown in Figure 2.26d. S_3 can be turned on with zero voltage condition. During this stage, actually resonant inductor is clamped by the clamping capacitor. The resonant inductor current decrease linearly with time as follows:

$$i_{Lr}(t) = i_{Lr}(t_3) - \frac{V_{Cc}}{L_r}(t - t_3) \tag{2.24}$$

D_2 is off and its voltage stress is $V_0 + V_{cl}$.

The boost inductor L is charging state in this stage.

The duration of this stage is decided by PWM duty cycle.

Stage 5 (t_4–t_5): Type 1 commutation. At t_4, S_1 is turned off with ZVS because of the existence of C_1 and C_2 in Figure 2.26e. C_1 is charging, while C_2 is discharging with action of the current source I_L. This stage ends when voltage on C_2 decays to zero, while D_2 starts to conduct. This stage is very short. The resonant inductor current still decreases linearly with time as last stage.

Stage 6 (t_5–t_6): Power transfer stage. At t_5, D_2 starts to conduct. The current source I_L flows through D_2 in Figure 2.26f. The voltage across S_1 is clamped at $V_0 + V_{cl}$. The resonant inductor current still decreases linearly with time as before.

The duration of this stage is decided by PWM scheme. At t_6,S_3 turns off, which starts a new switching period.

From the operation analysis of the converter, it can be found that all the off-state voltages across S_1, D_2, and S_3 are clamped to $V_0 + V_{cl}$.

When the converter is in steady state, voltage on the resonant inductor $v_{Lr}(t)$ satisfies the voltage second balance principle in a switching cycle.

$$\int_0^{T_s} v_{Lr}(t)dt = 0 \tag{2.25}$$

Except in the first and second resonant stages, the resonant inductor voltage v_{Lr} is either equal to V_0 or $-V_{cl}$ in a switching period as shown in Figure 2.27. By neglecting these two short resonant durations, Equation (2.25) is approximated as

$$\int_0^{T_s} v_{Lr}(t)dt \approx V_o\lambda_3 T_s - V_{Cc}(1 - \lambda_3)T_s = 0 \tag{2.26}$$

where $\lambda_3$is the turn-off duty cycle of the auxiliary switch S_3. Then the clamping capacitor voltageV_{cl} is derived:

$$V_{cl} = \frac{\lambda_3}{1 - \lambda_3} V_0 \tag{2.27}$$

Voltage stress of all the switches of the converter can be represented by

$$V_{s\max} = V_0 + V_{cl} = \frac{1}{1 - \lambda_3} V_0 \tag{2.28}$$

λ_3 is very short compared with the whole switching cycle and can be derived as

$$\lambda_3 = \frac{2L_r\left(\frac{V_{in}i_L}{V_0} + V_0\sqrt{\frac{C_2 + C_3}{L_r}}\right)}{V_0 T_s} \tag{2.29}$$

By introducing the CAC auxiliary circuit to the boost converter, we can realize ZVS switching for Type 2 commutation. It realizes ZVS condition for main switch and auxiliary switch, while the diode reverse recovery is suppressed effectively. Thus the switching loss is reduced.

If we made a small change of the CAC auxiliary circuit in the CAC boost converter in Figure 2.24a by swapping positions between L_r and S_3, a new auxiliary circuit is obtained as shown Figure 2.28. This auxiliary circuit is called minimum-voltage active-clamping (MVAC) auxiliary circuit. The boost converter is known as MVAC boost converter [18].

Compared with CAC boost converter, the position of L_r and S_3 is exchanged in the MVAC boost converter. The PWM scheme of MVAC converter is the same as that of CAC converter. At any time during operation, there are two devices conducting among S_1, D_2, and S_3 so that the voltage across the third device turned off is clamped. The feature of MVAC converter is that all the power devices, which include S_1, D_2, and S_3, have the minimum voltage stresses, which is equal to that of hard-switching boost converter as follows:

$$V_{s\max} = V_0 \tag{2.30}$$

CAC or MVAC auxiliary circuits can be applied to almost all types of DC–DC converter such as buck converter, boost converter, buck–boost converter, etc. to suppress switching loss of Type 2 commutation. CAC and MVAC DC/DC converter can realize ZVS condition for main switch and auxiliary switch, while the diode reverse recovery is suppressed. Figure 2.29 shows CAC and MVAC buck converters.

It is easy to extend the concept of active-clamping techniques to synchronous buck converter as shown Figure 2.30. First, CAC or MVAC auxiliary circuit is inserted in the circuit as active-clamping boost converters. Then the auxiliary circuit is moved to DC side of the converter. Similar to active-clamping boost or buck converters, all the Type 2 commutations in the synchronous buck converter can be eliminated by using the auxiliary circuit [19].

2.4.3.2 Active-clamping Three-phase Converter

CAC and MVAC auxiliary circuits suppress Type 2 commutations in DC–DC converter. It can realize ZVS turn-on for main switch and auxiliary switch. The three-phase converter is composed of three switch legs, and each leg can be seen as a synchronous buck converter. Thus the concept of active-clamping techniques can also be extended to three-phase converter as shown in Figure 2.31. One CAC or MVAC auxiliary circuit is shared by three switch legs.

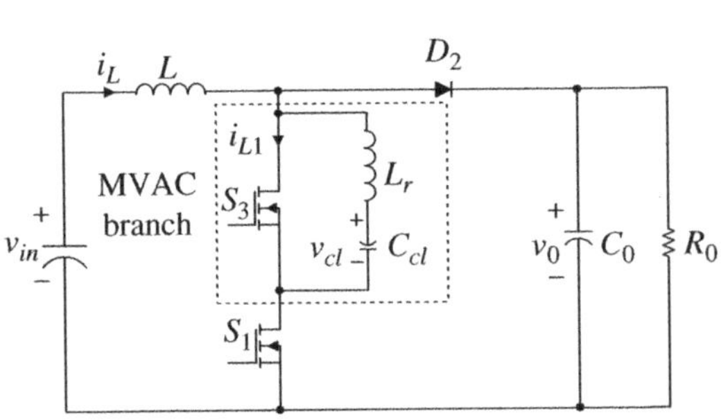

Figure 2.28 Minimum-voltage active-clamping (MVAC) boost converter.

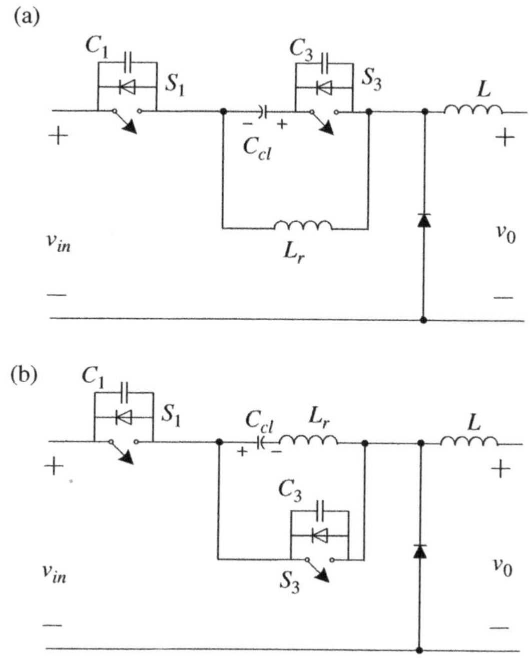

Figure 2.29 Active-clamping buck converters: (a) CAC buck converter; (b) MVAC buck converter.

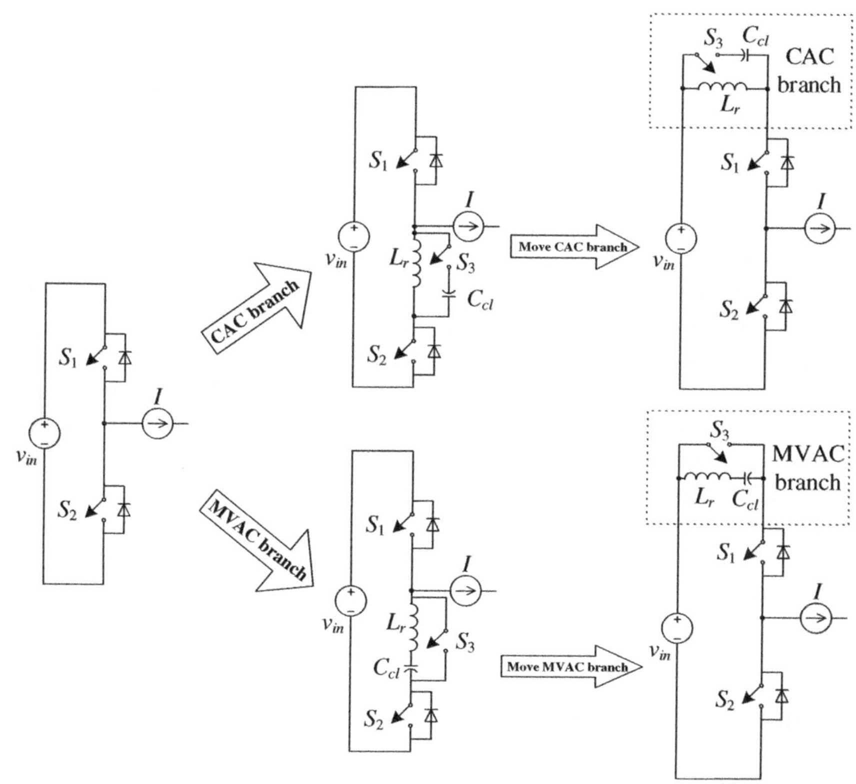

Figure 2.30 Active-clamping synchronous buck converters.

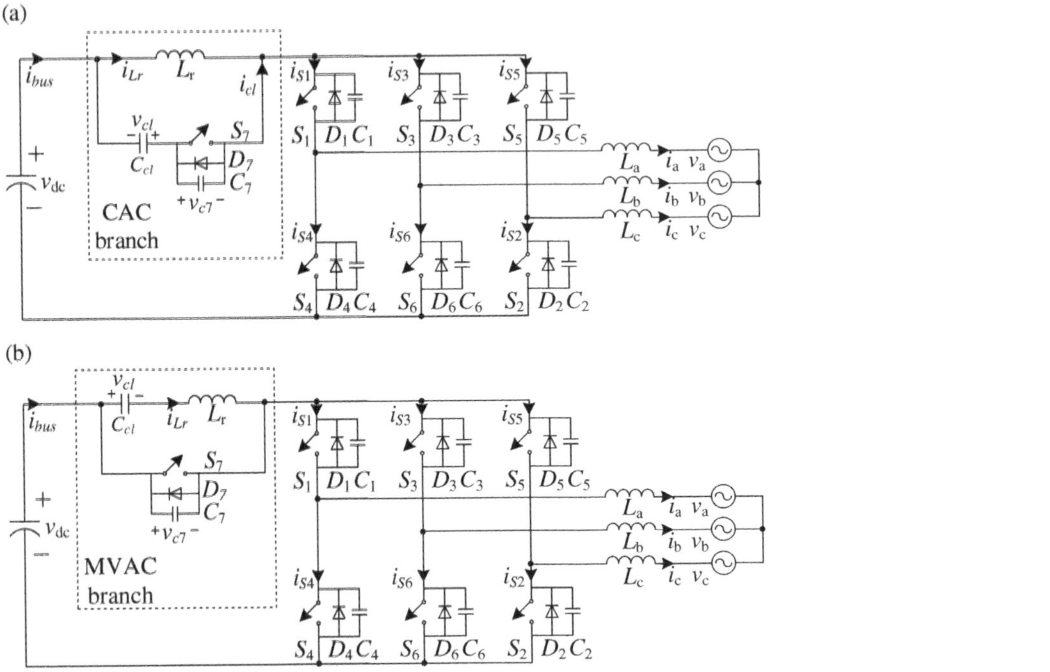

Figure 2.31 Active-clamping three-phase converter: (a) CAC three-phase converter; (b) MVAC three-phase converter.

Since there are three legs in the three-phase converter, normally the auxiliary circuit needs to be activated three times per switching period if the switches of the three legs are controlled asynchronously. To reduce the operating frequency of the auxiliary switch, a novel SVM is proposed to control the converter. Dehong Xu proposed ZVS-SVM scheme for actively clamped three-phase AC/DC rectifiers and inverters [19–24]. It retains advantages of ACRDCL while overcoming disadvantages of ACRDCL such as variable switching frequency and high voltage stress. More detailed introduction of ZVS-SVM scheme for active-clamping three-phase AC/DC converter will be given in later chapters.

A generic soft-switching PWM for active-clamping three-phase AC/DC converter, known as edge-aligned PWM (EA-PWM) [25], is invented and applied to three-phase inverter, three-phase four-wire inverter [26], and three-phase four-wire back-to-back converter [27]. In the aforementioned ZVS-SVM or EA-PWM actively clamped converter, the auxiliary switch only needs to switch once in every switching period to realize ZVS for all switches.

As a summary, the DC-side resonance converter only needs one auxiliary resonant circuit regardless of the number of AC phases of the converter. This simple structure makes DC-side resonance attractive in multiphase converter applications. However, since the auxiliary resonant circuit flows the current all the time, it results in certain conduction loss. A careful design is needed to suppress this extra loss in the auxiliary circuit.

2.5 AC-side Resonance Converters

Figure 2.32 shows the configuration of the AC-side resonance converter. Auxiliary resonant circuits are installed in AC side of the switch bridge instead of the DC side as its counterpart of DC-side resonance converters. Distinctive advantage of the AC-side resonance is that the auxiliary circuits

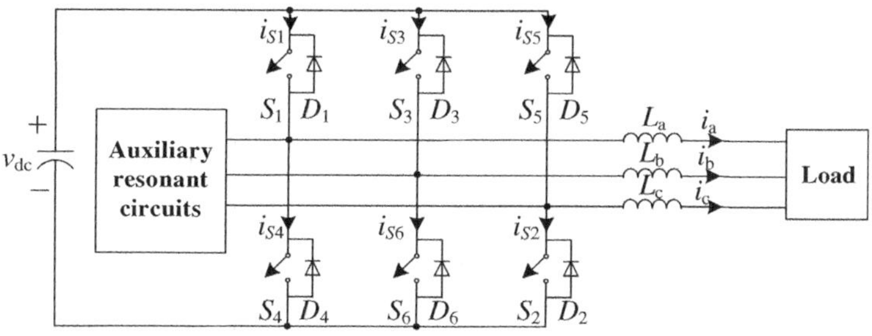

Figure 2.32 Configuration of AC-side resonance converters.

are in shunt with the switch bridge and conduct the current only in a small fraction of a switching period. They do not carry the load current throughout the inverter operation. Thus the conduction loss in the auxiliary circuits is smaller. In addition, SPWM and SVM control can be applied because the converter basically operates as the conventional converter in most of time.

Generally three auxiliary resonant circuits are needed for a three-phase converter. Each one helps realizing soft-switching of one switch leg. It means more components need to be added to the converter. Therefore, the resultant soft-switching converter circuit becomes complex. On the contrary, the DC-side resonance only needs one auxiliary resonant circuit regardless of the number of AC phases of the converter.

2.5.1 Auxiliary Resonant Commutated Pole Converter

Figure 2.33 shows the auxiliary resonant commutated pole (ARCP) converter, which was proposed by Rik DeDoncker et al. [28, 44]. Two large capacitors with the same capacitance are serially connected, which is paralleled to DC bus of the converter and create a middle point of the DC bus. Three auxiliary branches are located between the middle points of the DC bus and switch bridge of the converter. The middle point of the DC bus is shared by three auxiliary branches, whose another terminals are connected to the midpoint a, b, and c of the converter switch legs, respectively. Each auxiliary branch consists of a resonant inductor and a bidirectional switch. The bidirectional switch is usually implemented by two power semiconductor switches serially connected with the reverse direction. The auxiliary switches only need to block half the DC bus voltage. This topology achieves zero voltage turn-on for main switches and zero-current turn-off for auxiliary switch.

Here auxiliary resonant circuit for phase-a switch leg is taken as an example. The auxiliary resonate circuit is composed of the resonant inductor L_{r1}, a pair of auxiliary switches S_{x1} and S_{x2}, which constitute a bidirectional switch, and a pair of capacitors C_{r1} and C_{r4}, which are paralleled to upper

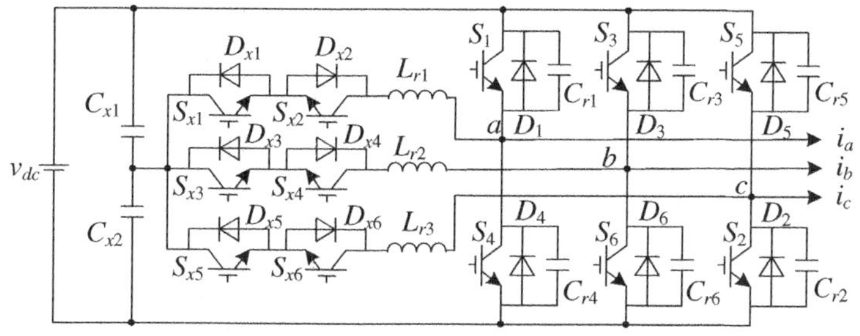

Figure 2.33 Auxiliary resonant commutated pole (ARCP) converter. *Source:* Based on Doncker and Lyons [28].

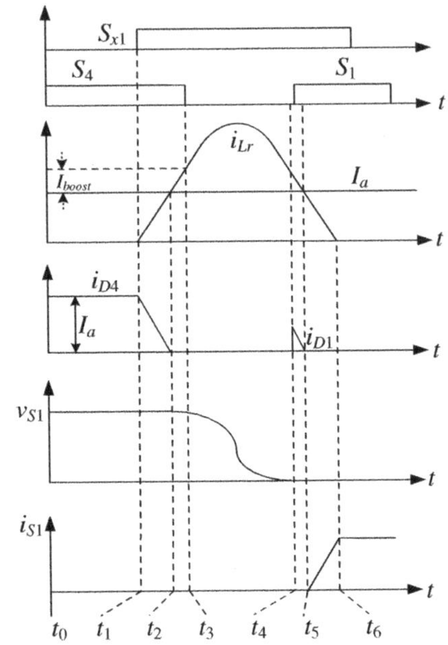

Figure 2.34 Commutation waveforms (from diode to switch) of ARCP converter.

switch S_1 and lower switch S_4, respectively. S_{x1} controls the resonant current to flow into phase-a switch leg, while S_{x2} enables the resonant current to flow out of phase-a switch leg.

Typically in more than 90% duration of a switching period, the auxiliary switches are in off-state, and have no current flowing through them. The ARCP inverter behaves like a PWM inverter. The auxiliary switches S_{x1} and S_{x2} conduct the current only during the current commutation of the phase-a switch leg. Figure 2.34 shows key waveforms of phase-a commutation. For convenience of circuit analysis in a switching cycle, the load current i_a is assumed to be positive, and constant at I_a and the middle point of the DC bus is constant in a switching cycle. The load current is commutated from the lower switch's antiparallel diode D_4 to the upper switch S_1 with the help of the auxiliary switch S_{x1}. Figure 2.35 shows the equivalent circuits at different commutation stages.

According to PWM control requirement, the load current is required to transfer from the antiparallel diode D_4 of the lower switch S_4 to the upper switch S_1. To avoid this Type 2 commutation with high loss, the auxiliary branch is used to realize ZVS turn-on of the upper switch S_1. The commutation process is explained as follows.

Before the load commutation ($t < t_1$): Since the load current I_a is positive, the diode D_4 carries the load current I_a as shown in Figure 2.35a:

Stage 1: (t_1, t_2), Pre-charging stage. At t_1, S_{x1} is turned on as shown in Figure 2.35b. Half of the DC bus voltage $v_{dc}/2$ is applied on the resonant inductorL_r. The current i_{Lr} of the resonant inductor linearly increases, while the current of D_4 decreases. Once i_{Lr} is equal to the load current I_a at t_2, this stage ends. As a result, the current of D_4 decreases to zero and D_4 naturally turns off at t_2.

Stage 2: (t_2, t_3), Boost-charging stage. D_4 is turned off at t_2. If S_4 keeps turning on for certain time as shown in Figure 2.35c, then the resonant inductor i_{Lr} will continue to increase by an increment I_{boost}. The increment I_{boost} is designed to boost energy stored in the resonant inductor L_r to realize ZVS turn-on of the upper switch S_1. Once the current i_{Lr} increases to $I_a + I_{boost}$, the lower switch S_4 is turned off, and this stage ends.

Stage 3: (t_3, t_4), Resonant stage. With the lower switch S_4 turning off at time t_3, the circuit configures as shown in Figure 2.35d. Both switches S_1 and S_4 and their antiparalleled diodes are in off-state. It becomes a L-C resonant circuit. The resonant inductor L_r is resonating with capacitor C_{r1} and C_{r4} across the two main switches. Due to the resonance, the voltage on C_{r4} increases, while the voltage on C_{r1} decreases. Once the voltage on C_{r1}, v_{S1}, resonances to zero, the antiparallel diode D_1 of the upper switch S_1 begins to conduct. It creates ZVS turn-on condition for the upper switch S_1 next stage. At time t_4 C_{r1} voltage decreases to zero and D_1 conducts, and the present stage ends.

Stage 4: (t_4, t_5), Discharging stage. When the antiparallel diode D_1 starts to conduct at t_4, a negative $v_{dc}/2$ is applied on the resonant inductor L_r as shown in Figure 2.35e. Thus the current i_{Lr} decreases linearly. Because the voltage on the S_1 is clamped to zero due to the conduction of

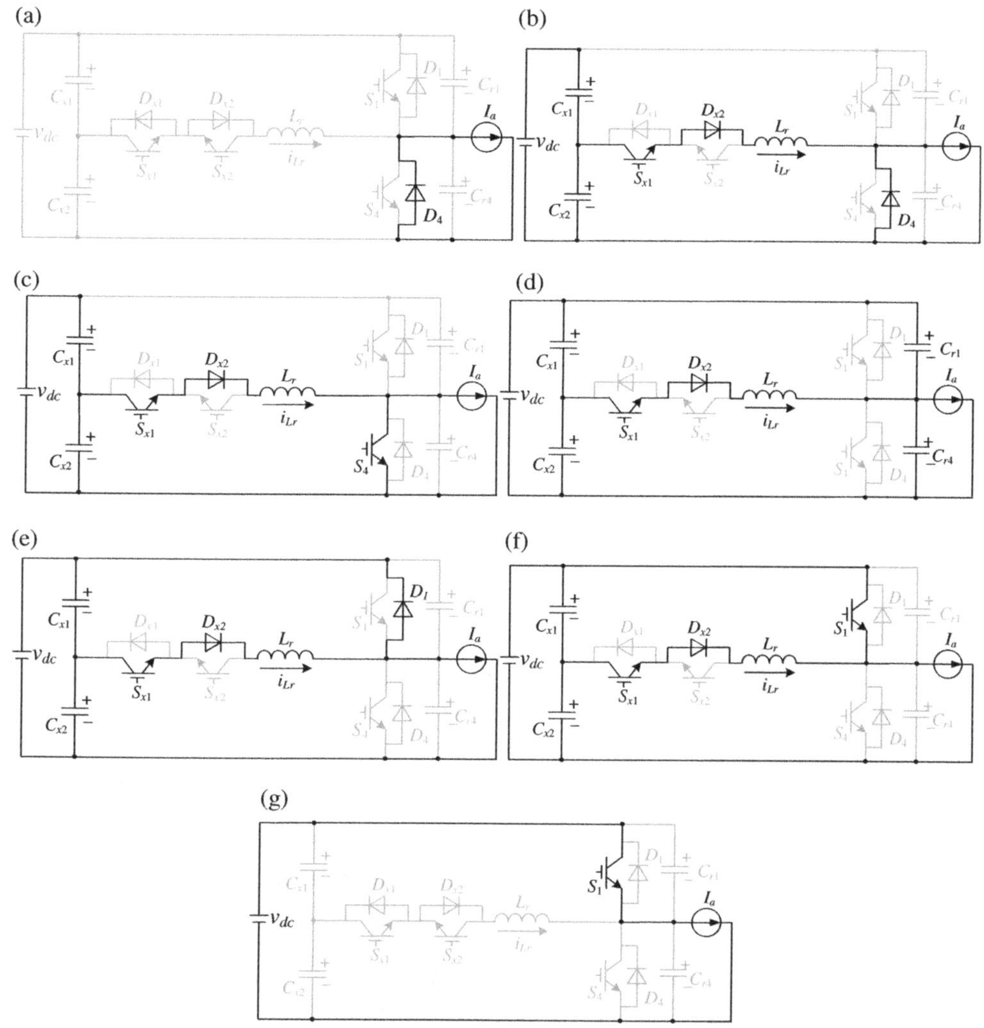

Figure 2.35 Equivalent circuits during commutation stages of ARCP converter. (a) Before the commutation ($t < t_1$). (b) Stage 1: (t_1, t_2), **pre-charging stage.** (c) Stage 2: (t_2, t_3), **boost-charging stage.** (d) Stage 3: (t_3, t_4), **resonant stage.** (e) Stage 4: (t_4, t_5), **discharging stage.** (f) Stage 5: (t_5, t_6), **discharging stage continued.** (g) After the commutation ($t > t_6$).

its antiparallel diode D_1 before t_5, S_1 can be turned on with ZVS without turn-on loss. The current through D_1 decays to zero at t_5 and D_1 naturally turns off when i_{Lr} decreases to the load current I_a.

Stage 5: (t_5, t_6), Discharging stage continued. D_1 turns off at t_5. Then S_1 begins to take over the partial load current with further decreasing of the resonant inductor current i_{Lr} as shown in Figure 2.35f. When i_{Lr} decreases to zero at t_6, the auxiliary switch S_{x1} is turned off naturally under ZCS condition, while the full load current I_a flows through the upper switch S_1. Here the current through S_{x1} cannot reverse the direction due to the blocking of the diode D_{x2}. Now the load current I_a has successfully commutated to the upper switch S_1 as shown in Figure 2.35g. Then the circuit finished the load commutation from the lower switch D_4 to the upper switch S_1. The circuit resumes normal operation of the inverter.

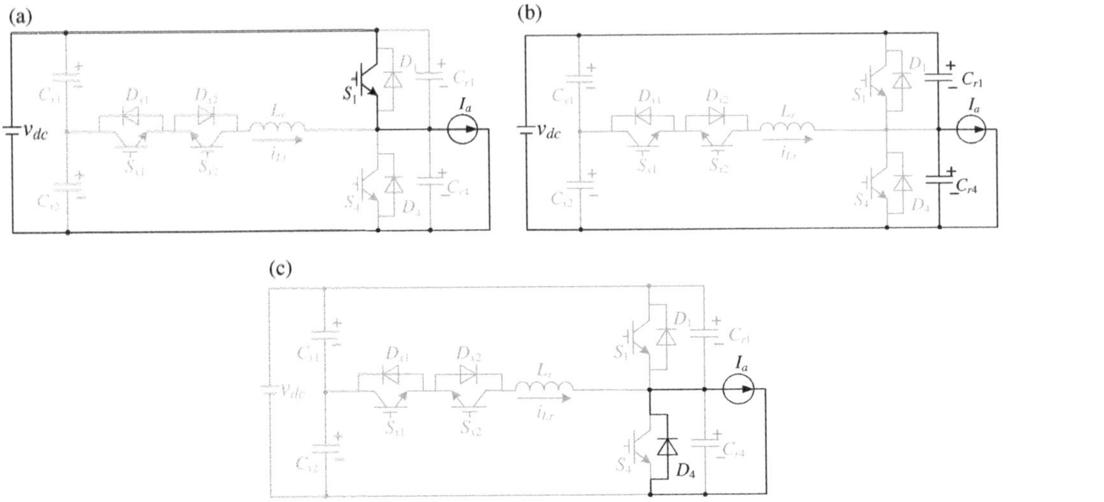

Figure 2.36 Commutation from upper switch S_1 to the lower switch's diode D_4: (a) S_1 conducts load current; (b) charge and discharge of capacitors; (c) D_4 conducts load current.

When the load current needs to commutate from upper switch S_1 to its complimentary switch S_4 antiparallel diode, D_4, in phase-a switching leg, it is not necessary to enable the auxiliary switches, because it is a Type 1 commutation. Before this commutation, the load current I_a flows through the upper switch S_1 shown in Figure 2.36a. When the switch S_1 turns off, the load current I_a will charge the upper capacitor C_{r1} and discharge the lower capacitor C_{r4} as shown in Figure 2.36b. Once the voltage on C_{r4} is reduced to zero, the antiparallel diode D_4 of the lower switch starts to conduct as shown in Figure 2.36c. Up to now the load current is commutated from the upper switch S_1 to the lower switch's diode D_4. It is observed that when the switch S_1 turns off, the capacitors C_{r1} and C_{r4} will slow down the dv/dt on the main switches. Therefore, integral of multiplication of voltage across the switch S_1 and its flowing current over the overlapping time during its turn-off transient is reduced. The turn-off switching loss of the switch S_1 is reduced. Due to the ZVS turn-on capability of the main switches as mentioned before, the energy stored in these capacitors can be recycled.

For Type 1 commutation, it is not needed to enable the auxiliary switches. However, when the load current I_a is so small that C_{r4} is unable to be totally discharged in Figure 2.36b. There is residual voltage on C_{r4} when the lower switch S_4 turns on after a short dead time to the upper switch turn-off. Then S_4 is not turned on with ZVS condition, which results in loss. If we still want to realize ZVS condition for the lower switch S_4, we can leverage the auxiliary circuit. The operation principle is similar to the commutation process from the lower switch's diode D_4 to the upper switch S_1 as mentioned before.

The ARCP converter achieves zero voltage turn-on for main switches and zero-current turn-off for auxiliary switches. Since most of time there is no current flowing through the auxiliary circuits, loss in the auxiliary circuit is reduced. The energy storage of the resonant inductor is able to be adjusted to satisfy soft-switching condition according the load condition. Therefore, the extra loss caused by the auxiliary circuit is optimized.

The disadvantage of the ARCP converter is that it needs more extra components. Besides, the middle point of DC bus voltage is required to be controlled.

2.5.2 Coupled-inductor Zero Voltage-transition (ZVT) Inverter

Figure 2.37 shows the coupled-inductor zero voltage-transition (ZVT) inverter, which was proposed by Fred Lee, Jason Lai et al. [29]. Similar to ARCP inverter, three identical auxiliary circuits are used. Each auxiliary circuit consists of a coupled resonant inductor, two diodes, and two auxiliary switches. Two capacitors are paralleled across the main switches of the switch leg of each phase. The ZVT inverter can achieve zero voltage turn-on for main switches and near-zero-current turn-off for auxiliary switches as the ARCP inverter. In compared with ARCP inverter, issue of control of the middle point of the DC bus is avoided. In addition the auxiliary switches only carry partial of the resonant current because the diodes in the auxiliary circuit take over the other parts. With the coupled inductors with turn ratio of 1:1, the current of auxiliary switches of the ZVT inverter is only half of the ARCP inverters.

For the convenience of analyzing the circuit commutation in a switching cycle, the load current i_a is also assumed to be constant at I_a during one switching cycle.

Similar to the ARCP inverter, the auxiliary switches are in off-state in most time of a switching period, and there is no current flowing through the auxiliary branches. The auxiliary switches S_{x1} and S_{x4} conduct the current only during the load current commutation of the phase-a switch leg. Figure 2.38 shows typical commutation waveforms of phase-a switch leg when the load current I_a is positive. The load current is commutated from the antiparallel diode D_4 of the lower switch to the upper switch S_1 with the help of the auxiliary switch S_{x1}. Figure 2.39 shows the equivalent circuits at different commutation stages.

Before the load commutation ($t < t_1$): Due to the current direction of the load I_a, D_4 carries the load current as shown in Figure 2.39a:

Stage 1: (t_1, t_2), Pre-charging stage. At t_1, S_{x1} is turned on as shown in Figure 2.39b. The DC bus voltage v_{dc} is applied on port L_{r2} of the coupled inductor. The current i_{Lr2} of the resonant inductor L_{r2} linearly increases until it reaches half of the load current, i.e. $I_a/2$. D_4 current is decreased to zero and is turned off naturally at t_2. This stage ends.

Stage 2: (t_2, t_3), Boost-charging stage. D_4 is turned off at t_2. If S_4 keeps turning on certain time as shown in Figure 2.39c, then i_{Lr2} will continue to increase by an increment I_{boost}. The increment I_{boost} can be designed as the ARCP inverter to realize the soft-switching for the switches. Once i_{Lr2} increases to $I_a/2 + I_{boost}$, this stage ends with the lower switch S_4 turned off.

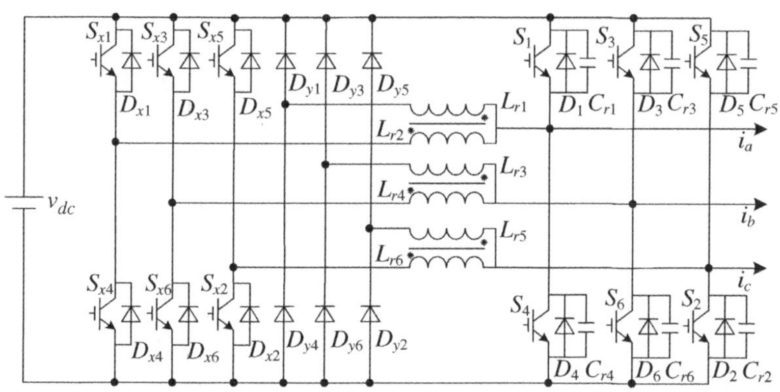

Figure 2.37 Coupled-inductor ZVT inverter. Source: Dong et al. [29]. © 2000 IEEE.

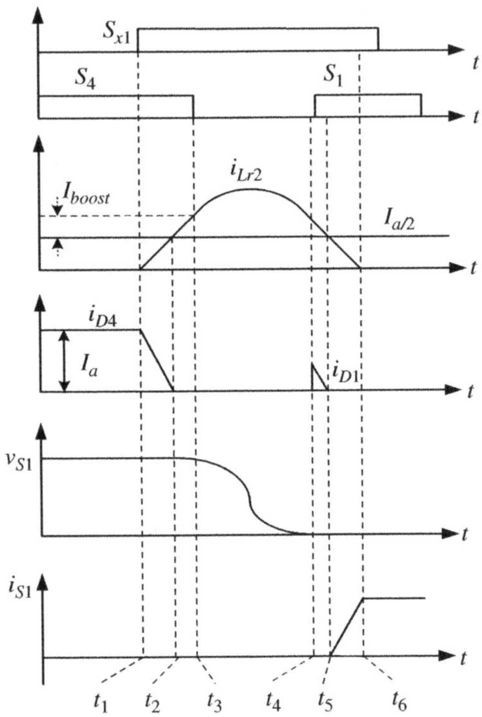

Figure 2.38 Commutation waveforms (from diode to switch) of ZVT converter.

Stage 3: (t_3–t_4), Resonant stage. The lower switch S_4 is turned off at time t_3. Both main switches S_1 and S_4 and their antiparalleled diodes are in off-state in Figure 2.39d. The circuit becomes a *L-C* resonant circuit. Once the voltage of capacitor C_{r1} resonates to zero at time t_4, the antiparallel diode D_1 of the upper switch begins to conduct, and the present stage ends.

Stage 4: (t_4–t_5), Discharging stage. When the antiparallel diode D_1 starts to conduct at t_4, a negative v_{dc} is applied on port L_{r1} of the coupled inductor as shown in Figure 2.39e. Thus the current i_{Lr2} decreases linearly. Before i_{Lr2} decreases to $I_a/2$ at t_5, the main switch S_1 can be turned on under the zero voltage condition.

Stage 5: (t_5–t_6), Discharging stage continued. D_1 naturally turns off at t_5 as shown in Figure 2.39f. S_1 begins to take over the current gradually with further decreasing of the resonant inductor current i_{Lr2}. When the resonant inductor current decreases to zero at t_6, the load current I_a has been successfully commutated to the upper switch S_1. Then the auxiliary switch S_{x1} is turned off automatically under ZCS condition at t_6.

Now the load current I_a has successfully commutated to the upper switch S_1 as shown in Figure 2.39g. Then the circuit finished the load commutation from the antiparallel diode D_4 of the lower switch to the upper switch S_1. The circuit resumes normal operation of the inverter.

As ARCP inverter, the ZVT inverter achieves zero voltage turn-on for main switches and zero-current turn-off for auxiliary switch. Besides the energy storage of the resonant inductor can be adjusted according to the load condition. In addition, the current of auxiliary switches of the ZVT inverter is only half of that of the ARCP inverters.

When D_4 of the lower switch S_4 freewheels load current, it has a voltage drop that will apply on the corresponding the auxiliary circuit. As a result, partial load current will freewheel through D_{x4}

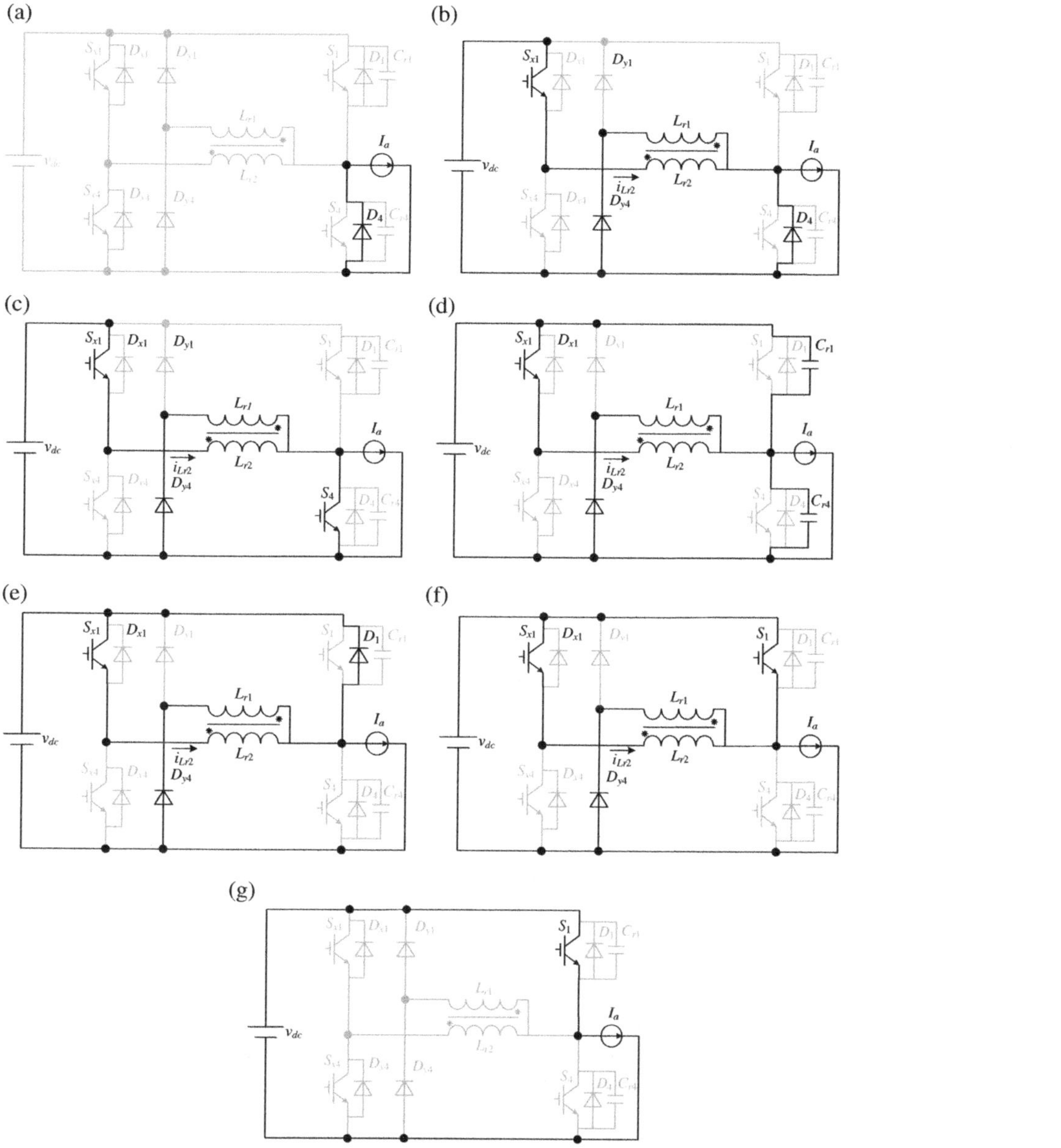

Figure 2.39 Different stages of the commutation in ZVT converter. (a) Before commutation ($t < t_1$). (b) Stage 1: (t_1, t_2), **Pre-charging stage**. (c) Stage 2: (t_2, t_3), **Boost-charging stage**. (d) Stage 3: (t_3–t_4), **Resonant stage.** (e) Stage 4: (t_4–t_5), **Discharging stage.** (f) Stage: (t_5–t_6), **Discharging stage.** (g) After commutation ($t > t_6$).

and D_{y4} as shown in Figure 2.40a. Consequently, during the commutation mentioned before, once the auxiliary switch S_{x1} turns on, D_{x4} will undergo a diode recovery process, which causes extra switching loss on both S_{x1} and D_{x4}. One possible way to avoid this issue is inserting a saturable inductor in series with the coupled inductor to disenable both D_{x4} and D_{y4} to freewheel the load current when D_4 conducts as shown in Figure 2.40b.

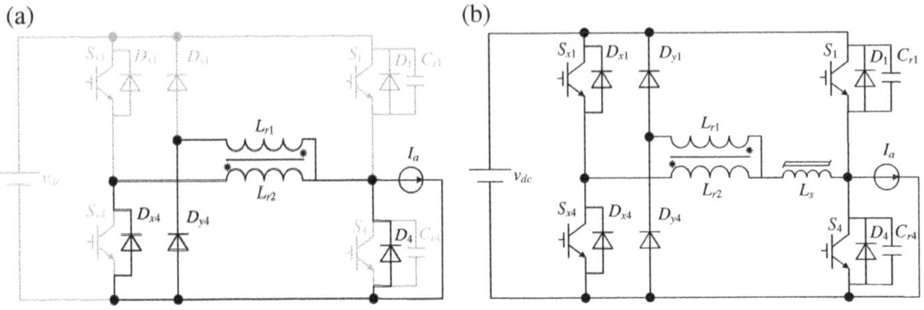

Figure 2.40 Disenable the auxiliary diodes' freewheeling with saturable inductor. (a) Freewheeling through D_{x4} and D_{y4}. (b) Disenable D_{x4} and D_{y4} to freewheel.

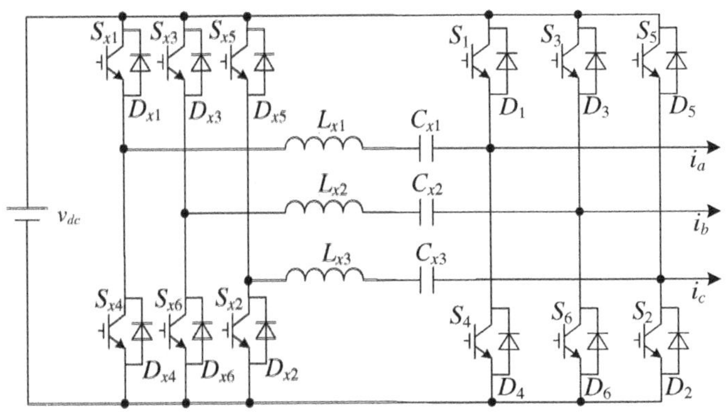

Figure 2.41 Six-switch ZCT inverter. *Source:* Based on Mao et al. [30] and Li et al. [31].

2.5.3 Zero-current Transition (ZCT) Inverter

The zero-current transition (ZCT) inverter, as shown in Figure 2.41, has also three auxiliary circuits [30, 31]. Each one consists of a *L-C* resonant tank and a switch half bridge. It can realize zero-current turn-off for all the switches in both the main circuit and the auxiliary circuits. Thus, the turn-off loss can be almost eliminated, and the diode reverse recovery problem is relieved. One of its drawbacks is that the resonant tank peak current is constant, and is not adjustable according to the load current, which results in lower efficiency in light load. ZCT inverter is more suitable to inverters with bipolar semiconductor devices such as IGBT device, which reduces turn-off loss of the IGBT due to its turn-off tail current.

There are numerous soft-switching inverter topologies with AC side's resonance, which have been investigated by the predecessors who try to simplify the auxiliary circuit with either sharing the auxiliary switch or resonant passive components [32–36]. Due to space limit, this book does not plan the review of all these circuits [37].

2.6 Soft-switching Inverter with TCM Control

Besides the soft-switching converters with DC-side resonance and AC-side resonance mentioned before, there are another type soft-switching converters, which do not need any auxiliary circuit. It is known as the soft-switching converters with triangular current mode (TCM) control. With

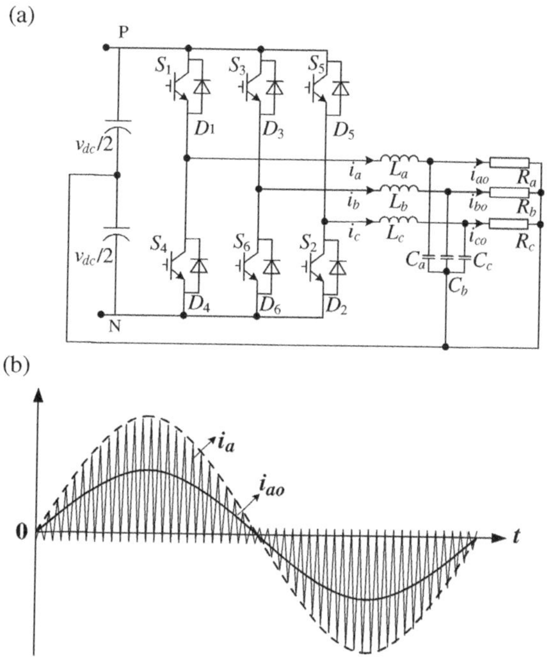

Figure 2.42 Inverter with TCM and its waveforms: (a) three-phase four-wire inverter; (b) phase-a inductor current i_a and load current i_{ao}.

TCM control, AC-side filter inductor currents of three-phase converters appear to be triangle waveform in a switching period. Since energy is stored in the inductor, when one switch in a switch leg is turned off, the voltage of its complementary switch at the same leg will be reset to zero and can be turned on with ZVS. There is no Type 2 commutations after introducing TCM control, i.e. the commutation from antiparallel diode of a switch to its complementary switch in the same leg is avoided. The concept originates from critical conduction mode (CRM) in DC–DC converter to achieve ZVS turn-on. TCM is originally introduced to single-phase totem-pole power factor correction (PFC) converters [38–40] and then extended to three-phase converters [41, 42]. A decoupled control of TCM modulation for the three-phase converter has been proposed by Dr. F. C. Lee et al. [42]. A general ZVS TCM modulation for three-phase AC/DC converter is proposed by J. W. Kolar et al. [43]. In each sector, one phase leg switch state keeps constant, while switches of the other two phases are controlled with TCM.

Figure 2.42 shows an inverter with TCM and its waveforms. For simplicity of analysis, a three-phase four-wire inverter is taken as example, which can be decoupled into three independent half-bridge inverters as shown in Figure 2.43a. In this circuit filter inductor L_a is small, while capacitor C_a is substantial as design of CRM buck converter. The inductor current i_a is a triangle waveform in each switching cycle with its peak profile of lower frequency sinusoidal as shown in Figure 2.42b. Although there is large switching ripple in the inductor current i_a, its moving average is a sinusoidal. Therefore, after filtering of capacitor C_a, phase-a load current i_{ao} becomes sinusoidal.

To further understand TCM control, here takes phase-a half-bridge inverter as an example. Figure 2.43 shows phase-a circuit of the inverter and its waveforms in a switching cycle. Without losing generality, it is assumed that phase-a load current i_{ao} is in the positive half cycle of the lower frequency sinusoidal. Current reference i_{a_ref} is compared with a saw-tooth carrier to generate PWM signal [38]. Actually the current reference decides conducting duration of the upper switch S_1, while the conducting duration of the lower switch S_4 is decided by the inductor current cross

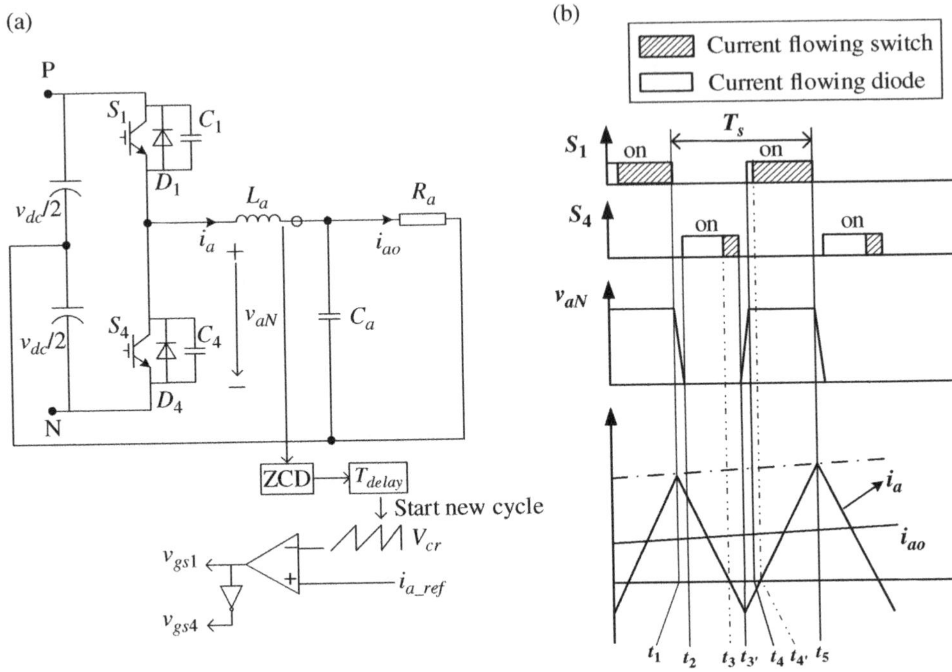

Figure 2.43 Phase-a circuit and key waveforms in a switching cycle: (a) the circuit; (b) key waveforms in a switching cycle.

zero instant, which is determined by zero-current detection (ZCD) circuit. A time delay T_{delay} is added after ZCD instant to store enough energy in the inductor to reset the voltage on the S_1. Once ZCD instant comes, the saw-tooth carrier is reset to zero and restarts a new switching cycle. Therefore, the period of the saw-tooth carrier depends on circuit parameters such as the filter inductance L_a, instant voltage v_{Ca} on capacitor C_a, DC bus voltage, and load. It means switching cycle is variable in the inverter with TCM control.

Switches S_1 and S_4 in the switch leg operate complementally with small dead time as shown in Figure 2.43b. A switching cycle T_s can be divided into six stages. Figure 2.44 shows equivalent circuits of various stages in one switching cycle. The operation of these stages are explained as follows:

Before t_1, upper switch S_1 is conducting, and the midpoint voltage v_{aN} is equal to v_{dc}. The equivalent circuit is shown in Figure 2.44a. The inductor current i_a is linearly increasing. At $t = t_1$, the inductor current i_a reaches the peak value, and S_1 turns off, which starts a switch cycle.

Stage 1: (t_1, t_2), the first resonance stage. Due to turn-off of switch S_1, the circuit becomes LC resonance circuit as shown in Figure 2.44b. The capacitor C_1 of upper switch S_1 is charging from zero, while the capacitor C_4 of the lower switch S_4 is discharging from DC bus voltage v_{dc}. Once the voltage of C_4 is discharged to zero, its antiparallel diode D_4 conducts, and this stage ends. It creates ZVS turn-on condition for S_4 in next stage. Actually it is easy to realize ZVS conduction for S_4 because the inductor current is at peak value at the beginning of this stage.

Stage 2: (t_2–t_3), inductor current decreasing stage. Due to conducting of the antiparallel diode D_4 of the lower switch S_4, the midpoint voltage v_{aN} is equal to 0 as shown in Figure 2.44c. A negative voltage, $v_{La} = -\frac{v_{dc}}{2} - v_{Ca}$, is applied on the inductor L_a because the maximum absolute value of the output capacitor voltage u_{Ca} is always less than $v_{dc}/2$ in this circuit. Thus the inductor

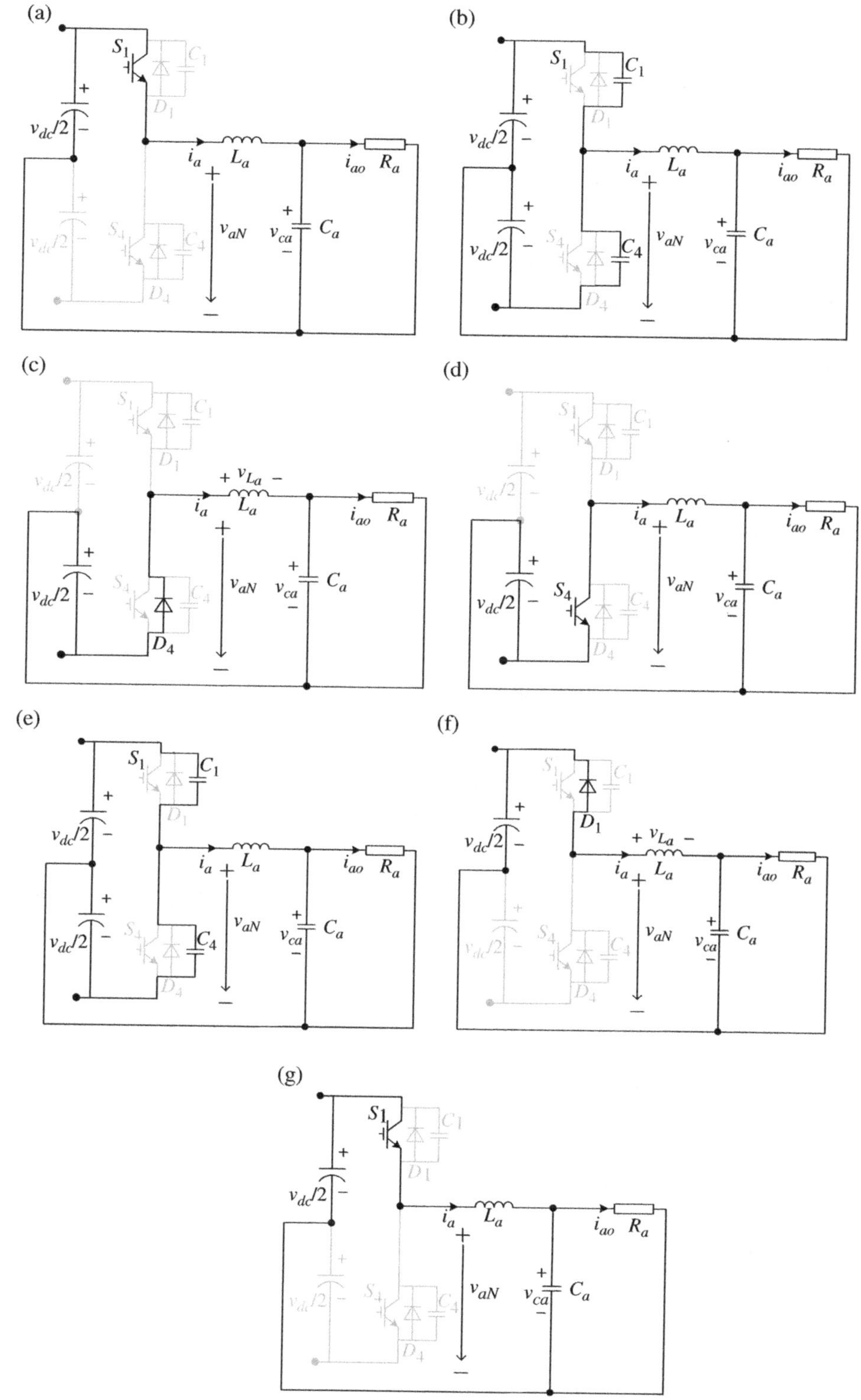

Figure 2.44 Equivalent circuits of various stages in one switching cycle. (a) Initial stage ($t< t_1$). (b) Resonance stage (t_1–t_2). (c) Inductor current decreasing stage (t_2–t_3). (d) Inductor current reversing stage (t_3–t_3). (e) Resonance stage (t_3–t_4). (f) Inductor current increasing stage (t_4–t_4). (g) Inductor current increasing stage continued (t_4–t_5).

current linearly decreases in this stage. As mentioned before, S_4 is required to be turned on during conducting duration of D_4 to realize ZVS turn-on. This stage ends when the inductor current decreases to zero.

Stage 3: (t_3 – t_3'), inductor current reversing stage. At $t = t_3$, the inductor current decreases to zero and reverses its direction as shown in Figure 2.44d. It flows through the lower switch S_4 instead of its antiparallel diode D_4. This stage is used to build up enough energy in the inductor L_a to reset the upper switch S_1 voltage in the next stage. Its duration needs to be minimized to reduce circulation energy in the circuit.

Stage 4: (t_3', t_4), the second resonance stage. The switch S_4 is turned off at $t = t_3'$. The circuit becomes LC resonance circuit as shown in Figure 2.44e. With action of the inductor current i_a, the capacitor C_1 of upper switch S_1 is discharging from v_{dc}, while the capacitor of the lower switch S_4 is charging from zero. Once the voltage on the upper switch S_1 is discharged to zero, its antiparallel diode D_1 turns on automatically, and this stage ends. It creates ZVS turn-on condition for S_1 next stage. It is noted that the negative peak value of the inductor current at $t = t_3'$ needs to be large enough to reset the voltage of S_1 to zero in this stage. Therefore, the design of $i_a(t_3')$ is critical for ZVS operation.

Stage 5: (t_4, t_4'), inductor current increasing stage. Due to conducting of the antiparallel diode D_1, the midpoint voltage v_{aN} is equal to v_{dc}. A positive voltage, $v_{La} = \frac{v_{dc}}{2} - v_{Ca}$, is applied on the inductor as shown in Figure 2.44f. Thus the inductor current linearly increases in this stage. As mentioned before, S_1 is required to be turned on during D_1 conducting state to realize ZVS turn-on. At time t_4' the diode current decays to zero, and this stage ends.

Stage 6: (t_4, t_5), inductor current increasing stage continued. At time t_4' the inductor current reverses direction and flows through the upper switch S_1 instead of its antiparallel diode D_1 as shown in Figure 2.44g. With continuation action of positive voltage on the inductor, inductor current increases linearly toward its peak value. Once it reaches the peak current, this stage ends. Actually the duration of this stage decide the inductor current peak profile, which is related to the load current. At t_5, the switch S_1 is turned off, which starts a new switch cycle.

It is observed that switches S_1 and S_4 in the switch leg operate complementally with small dead time during switch commutations. The dead time is used to reset the voltage on the switches to be turned on with the help of the inductor current. Its antiparallel diode of one switch conducts earlier than itself. Thus there are only Type 1 commutations, i.e. commutations from a switch to its complementary switch's antiparallel diode at the same switch leg. Thus both turn-on loss and reverse recovery loss of the power switch are eliminated.

The soft-switching converters with TCM control have distinct advantages that it does not need extra auxiliary resonant circuit. However, it has drawbacks such as wider range of variable switching frequency, higher turn-off current on the switch due to the triangle current waveform, etc. Due to the triangle current waveforms of the AC side, careful design is needed to satisfy EMC requirement. Combination of CRM, discontinuous conduction mode (DCM), etc. has been studied to compress the switching frequency varying range. Two converters with interleaving switching control is used to cancel higher switching current ripple on the filter inductors due to TCM [39]. The converter with TCM seems more suitable to low or medium power applications.

2.7 Summary

There are two types of commutations in three-phase converters. Among them, the Type 2 commutation results in high switching loss in both power semiconductors switches and diodes.

Soft-switching has been used to reduce the switching loss. The soft-switching three-phase converters can be divided into three classes: soft-switching converters with DC-side auxiliary resonance circuit, soft-switching converters with AC-side auxiliary resonance circuit, and soft-switching converters with TCM control. RDCL converter is a milestone in the evolution of soft-switching three-phase converters, which belongs to the DC-side resonance converter. It only needs one auxiliary resonant circuit regardless of the number of AC phases of the converter. To solve issue of RDCL such as variable switching frequency, high stress on power semiconductor switches, etc., the active-clamping three-phase converter with ZVS-SVM or EA-PWM is developed, which primarily operates as a PWM converter. The ARCP converter is typical representative of the AC-side resonance converter. Distinctive advantage of the AC-side resonance is that the auxiliary circuits are in shunt with the switch bridge, and do not carry the load current. Thus the conduction loss in the auxiliary circuits is smaller. However, more extra components are needed. Concept of soft-switching converters with TCM control is attractive, since it does not need auxiliary circuit. AC-side filter inductor currents in three-phase converters are so controlled that they appear to be triangle waveforms in switching cycles. Enough energy is stored in the inductor so that when one switch of a switch leg is turned off, the voltage of its complementary switch at the same leg is reset to zero. Thus it can be turned on with ZVS. However, it has issue such as wider range of variable switching frequency, higher rms current and turn-off current on the switch, etc. Many interesting works on TCM control have been carried out to try to improve its performance. WBG devices are bringing new opportunities for the soft-switching technology, which will make power electronics equipment having better performance, high efficiency, smaller size, and lower cost.

References

1 N. Mohan, T. M. Undeland, W. P. Robbins, *Power Electronics: Conversion, Applications, and Design*, New York: John Wiley & Son, 2003.

2 Krein, P.T. (2015). *Elements of Power Electronics*. Oxford: Oxford University Press.

3 Bellar, M.D., Wu, T.S., Tchamdjou, A. et al. (1998). A review of soft-switched DC-AC converters. *IEEE Transactions on Industrial Applications* 34 (4): 847–860.

4 Divan, D.M. (1989). The resonant DC link converter – a new concept in static power conversion. *IEEE Transactions on Industry Applications* 25 (2): 317–325.

5 Venkataramanan, G., Divan, D.M., and Jahns, T.M. (1993). Discrete pulse modulation strategies for high frequency inverter systems. *IEEE Transactions on Power Electronics* 8 (3): 279–287.

6 Divan, D.M. and Skibinski, G. (1989). Zero-switching-loss inverters for high-power applications. *IEEE Transactions on Industry Applications* 25 (4): 634–643.

7 He, J. and Mohan, N. (1991). Parallel resonant DC-link circuit – a novel zero switching loss topology with minimum voltage stresses. *IEEE Transactions on Power Electronics* 6 (4): 687–694.

8 J. Cho, H. Kim, and G. Cho, "Novel soft-switching PWM converter using a parallel resonant DC-link," in *Proc. of IEEE Power Electronics Specialist Conference – PESC 1991*, pp. 241–247, 1991.

9 M. Nakaoka, H. Yonemori, and K. Yurugi, "Zero-Voltage soft-switched PDM three-phase AC-DC active power converter operating at unity power factor and sinewave line current", in *Proc. of IEEE Power Electronics Specialist Conference – PESC '93*, pp. 787–794, 1993

10 Agelidis, V., Ziogas, P., and Joos, G. (1994). An optimum modulation strategy for a novel "notch" commutated 3-Φ PWM inverter. *IEEE Transactions on Industry Applications* 30 (1): 52–61.

11 Divan, D.M., Malesani, L., Tenti, P., and Toigo, V. (1993). A synchronized resonant DC link converter for soft-switched PWM. *IEEE Transactions on Industry Applications* 29 (5): 940–948.

12 Milan M. Jovanovic, "A technique for reducing rectifier reverse-recovery-related losses in high-voltage, high-power boost converters," in *Proc. of APEC'97*, pp. 1000–1007. 1997.

13 B. Carsten, "Design techniques for transforment active reset circuits at high frequencies and power levels," in *Proc. of HFPC'90*, pp. 235–246. 1990.

14 Duarte, C.M. and Barbi, I. (1997). A new family of ZVS-PWM active-clamping DC-to-DC boost converters: analysis, design, and experimentation. *IEEE Transactions on Power Electronics* 12 (5): 824–831.

15 S. F. Plano, A. P. Mesquite, "Active clamp for power converter and method of operation thereof," US 6,259,235 B1 U.S. patent.

16 G. Chen, D. Xu, B. Feng and Y. Wang, "A family of compound active-clamping DC-DC converters," in *IEEE Applied Power Electronics Conference and Exposition (APEC)*, pp. 850–856, 2002.

17 Feng, B. and Xu, D. (2005). 1-kW PFC converter with compound active-clamping. *IEEE Transactions on Power Electronics* 20 (2): 324–331.

18 G. Chen, D. Xu, B. Feng, Y. Wang, "Minimum-voltage active-clamping DC-DC converters," in *IEEE 33rd Annual IEEE Power Electronics Specialists Conference*, pp. 403–408, 2002.

19 B. Feng and D. Xu, "Novel ZVS three-phase PFC Converters and zero-voltage-switching space vector modulation (ZVS-SVM) control," in *First International Conference on Power Electronics Systems and Applications (PESA)*, pp. 30–37, 2004.

20 Xu, D., Feng, B., and Li, R. (2007). A Zero Voltage Switching SVM (ZVS–SVM) controlled three-phase boost rectifier. *IEEE Transactions on Power Electronics* 22 (3): 978–986.

21 Li, R., Ma, Z., and Xu, D. (2012). A ZVS grid-connected three-phase inverter. *IEEE Transactions on Power Electronics* 27 (8): 3595–3604.

22 He, N., Chen, M., Wu, J. et al. (2019). 20 kW Zero-Voltage-Switching SiC-MOSFET Grid Inverter with 300 kHz Switching Frequency. *IEEE Transactions on Power Electronics* 34 (6): 5175–5190.

23 M. Mezaroba, D. C. Martins, and I. Barbi, "A ZVS PWM full-bridge inverter with active clamping technique using the reverse recovery energy of the diodes," in *IECON 02*, pp. 621–626, 2002.

24 M. Mezaroba, D. C. Martins, I. Barbi, "A ZVS PWM three-phase inverter with active clamping technique using the reverse recovery energy of the diodes," in *IEEE 35th Annual Power Electronics Specialists Conference*, pp. 4785–4790, 2004.

25 Shi, K., Deng, J., and Xu, D. (2020). A general pulse width modulation method for zero-voltage-switching active-clamping three-phase power converters: Edge Aligned Pulse Width Modulation (EA-PWM). *IEEE Open Journal of Power Electronics* 1: 250–259.

26 He, N., Zhu, Y., Zhao, A., and Xu, D. (2019). Zero-voltage-switching sinusoidal pulse width modulation method for three-phase four-wire inverter. *IEEE Transactions on Power Electronics* 34 (8): 7192–7205.

27 Shi, K., Zhao, A., Deng, J., and Xu, D. (2019). Zero-voltage-switching SiC-MOSFET three-phase four-wire back-to-back converter. *IEEE Journal of Emerging and Selected Topics in Power Electronics* 7 (2): 722–735.

28 R. W. D. Doncker and J. P. Lyons, "The auxiliary commutated resonant pole converter," in *IEEE Industry Applications Society Annual Meeting*, pp. 1228–1235, 1990.

29 W. Dong, D. Peng, H. Yu, F. C. Lee and J. Lai, "A simplified control scheme for zero voltage transition (ZVT) inverter using coupled inductors," in *IEEE Annual Power Electronics Specialists Conference*, pp. 1221–1226, 2000.

30 H. Mao, F.C. Lee, X. Zhou and D. Boroyevich, "Improved zero-current transition converters for high power applications," in *IEEE Industry Applications Conference Thirty-First IAS Annual Meeting*, pp. 1145–1152. 1996.

31 Y. Li, F. C. Lee, J. Lai and D. Boroyevich, "A Novel Three-phase Zero-Current-Transition and Quasi-Zero-Voltage-Transition (ZCT-QZVT) Inverter/Rectifier with Reduced Stresses on Devices and Components," in *IEEE Applied Power Electronics Conference and Exposition (APEC)*, pp. 1030–1036, 2000

32 V. Vlatkovic, D. Borojevic, F. C. Lee, C. Cuadros, and S. Gataric, "A new zero-voltage-transition, three-phase rectifier/inverter," in *Proc. of PESC'93*, pp. 868–873, 1993.

33 Lai, J., Zhang, J., and Yu, H. (2006). Source and load adaptive design for a high-power soft-switching inverter. *IEEE Transactions on Power Electronics* 21 (6): 1667–1675.

34 Yuan, X. and Barbi, I. (2000). Analysis, designing, and experimentation of a transformer-assisted PWM zero-voltage switching pole inverter. *IEEE Transactions on Power Electronics* 15 (1): 72–82.

35 Hoshi, N. and Oguchi, K. (2005). Control schemes for auxiliary switches of three-phase PWM Resonant Snubber inverters. *Electrical Engineering in Japan* 152 (4): 57–67.

36 N. Hoshi, K. Itoga, K. Oguchi, "A novel control strategy for three-phase resonant snubber based soft-switching inverters," in *Proc IPEC-Tokyo*, pp. 536–541, 2000.

37 D. Xu, R. Li, Z. Ma, C. Du, M. Chen, "A family of novel zero-voltage switching three-phase PWM converters topology for distributed generation", in *Proc. of 8th International Conference on Power Electronics – ECCE Asia*, pp. 1–10, 2011

38 Liu, Z., Huang, Z., Lee, F.C., and Li, Q. (2016). Digital-based interleaving control for GaN-based MHz CRM totem-pole PFC. *IEEE Journal of Emerging and Selected Topics in Power Electronics* 4 (3): 808–814.

39 Marxgut, C., Krismer, F., Bortis, D., and Kolar, J.W. (2014). Ultraflat interleaved triangular current mode (TCM) single-phase PFC rectifier. *IEEE Transactions on Power Electronics* 29 (2): 873–882.

40 Knecht, O., Bortis, D., and Kolar, J.W. (2018). ZVS modulation scheme for reduced complexity clamp-switch TCM DC–DC boost converter. *IEEE Transactions on Power Electronics* 33 (5): 4204–4214.

41 Chen, J., Sha, D., Zhang, J., and Liao, X. (Jul. 2019). An SiC MOSFET based three-phase ZVS inverter employing variable switching frequency space vector PWM control. *IEEE Transactions on Power Electronics* 34 (7): 6320–6331.

42 Huang, Z., Liu, Z., Lee, F.C., and Li, Q. (2019). Critical-mode-based soft-switching modulation for high-frequency three-phase bidirectional AC–DC converters. *IEEE Transactions on Power Electronics* 34 (4): 3888–3898.

43 Haider, M., Anderson, J.A., Mirić, S. et al. (2020). Novel ZVS S-TCM modulation of three-phase AC/DC converters. *IEEE Open Journal of Power Electronics* 1: 529–543.

44 M. Lockwood, A. M. Fox, "A novel high power transistor inverter," in *Proceeding of 1983 IPEC*-Tokyo, pp. 637–648, 1983.

3

Soft-switching PWM Control for Active Clamped Three-phase Converters

A general soft-switching control for active clamped three-phase power converters – edge-aligned pulse width modulation (EA-PWM) is introduced. It aligns all the high loss commutations in a switching period at the same instant so that the converter can realize zero-voltage-switching (ZVS) with switching of the resonant auxiliary circuit once each switching period. It is suitable for various modulation methods such as sinusoidal pulse width modulation (SPWM), harmonic injected SPWM, continuous pulse width modulation (CPWM), discontinuous pulse width modulation (DPWM), etc. In this chapter, concept of EA-PWM for ZVS active-clamping three-phase converter is introduced. Then stage analysis of the converter in a switching period is provided. ZVS condition is derived. A control diagram of the soft-switching gird converter with EA-PWM is introduced.

Space vector modulation (SVM) is modified to adapt to soft-switching operation of the active clamped three-phase converters. Vector sequences for each sector are identified with goal to reduce the vector transitions having higher switching loss. Finally, ZVS-SVM scheme having only one high-loss vector transition in a switching cycle is found. This high-loss vector transition is further relieved with the help of the auxiliary circuit. The concept of the ZVS-SVM and characteristics of the active clamped three-phase converters such as voltage stress, DC-side voltage utilization, impact of operating mode, etc. are introduced.

3.1 Introduction

Aiming to reduce the voltage and current overlapping during turn-on or turn-off transients of power semiconductor switches of the three-phase converters, various soft-switching techniques such as soft-switching converters with DC-side resonance, soft-switching converters with AC-side resonance, and soft-switching converters with triangular current mode (TCM) have been developed. Among them, the soft-switching converter with DC-side resonance, especially active-clamping three-phase converters, has distinct advantage as follows:

- Simple auxiliary circuit. One auxiliary circuit is shared among three-phase switching legs of the three-phase converter.
- ZVS for both the main switches and the auxiliary switch is realized. ZVS is suitable to wide-bandgap devices such as Silicon-carbon Metal-oxide-semiconductor field effect transistors (SiC MOSFETs) and Gallium nitride (GaN) devices.
- Voltage stress on power semiconductor switches is reduced.

Soft-Switching Technology for Three-phase Power Electronics Converters, First Edition.
Dehong Xu, Rui Li, Ning He, Jinyi Deng, and Yuying Wu.

Resonant DC link converter is originally controlled with distributed pulse modulation (DPM), and the output tracking accuracy of the converter is constrained by width of the resonant pulse [6–9]. The higher the resonant frequency, the more accurately the output follows a lower frequency sinusoidal reference waveform. Besides, the switch frequency is variable and depends on DC-side instant current of the converter, which is reflected from three-phase AC load currents. The resonant pulses with variable switching frequency complicate the output tracking process. Generally, the DPM is not as efficient as PWM and SVM with respect to following a given low frequency reference signal. It means the converter with DPM is required to operate much higher switching frequency to reach comparable total-harmonic-distortion (THD) specification. Higher frequency switching means more loss during DC-side resonances even though the circuit operates at ZVS condition. For ZVS switching, switch turn-on loss is eliminated but switch turn-off loss still exists, though it is reduced by ZVS turn-off. ZVS turn-off loss depends on types of power semiconductor devices. For SiC MOSFETs, the turn-off loss is reduced substantially by installing parallel capacitors between the drain and source of the device. However, ZVS turn-off is not so effective for the reduction of turn-off loss of the insulated gate bipolar transistor (IGBT) device due to its tail current. In addition, there is high voltage and current stress in the circuit since resonant DC link converter almost operates at resonance state all the time, which is not good for obtaining high power conversion efficiency [19–23].

In this chapter, we explore high efficiency control and design of active-clamping three-phase converter. The idea is to make the converter operate like a PWM converter [10–12]. The converter is expected to have the following characteristics:

- The converter is controlled by PWM with fixed switching frequency, which is similar to conventional SVM or PWM. Thus it results in high power quality in the AC side of the converter.
- ZVS resonance duration to realize soft-switching for all power semiconductor devices is restricted to a small fraction of the switch period, i.e. typical less than 10% of the switching period. It means that the converter operates as PWM converter most of the time.
- The auxiliary semiconductor device only switch once each switching period. Thus extra loss due to the auxiliary circuit operation is minimized.
- Extra voltage stress on power semiconductor switches is small.

To apply PWM to active-clamping converter, we need to review conventional PWM schemes in a three-phase converter and study how to adapt it so that PWM can be used successfully in ZVS active-clamping three-phase converter.

3.2 PWM of Three-phase Converters

Soft-switching three-phase converters with either DC-side resonance or AC-side resonance are shown in Figure 3.1. It is assumed that we use SPWM to control the three-phase converter. Figure 3.2a shows three modulation signals, which can be represented by $v_{ma}(t) = m\sin(\omega t)$, $v_{mb}(t) = m\sin\left(\omega t - \frac{2\pi}{3}\right)$, and $v_{mc}(t) = m\sin\left(\omega t - \frac{4\pi}{3}\right)$, and carrier $v_{car}(t)$. The carrier signal is triangle waveform and its period is T_S, which is known as switching period or switch cycle. By comparing the modulation signals with the carrier, PWM waveforms are generated for each switch leg. Figure 3.2b shows how PWM waveforms are generated. Generally, the repeating frequency of the carrier is much higher than the frequency of the modulation signal. Thus the modulation signals can be approximated as constant in a switching period. The midpoint voltage v_{aN} of phase-a

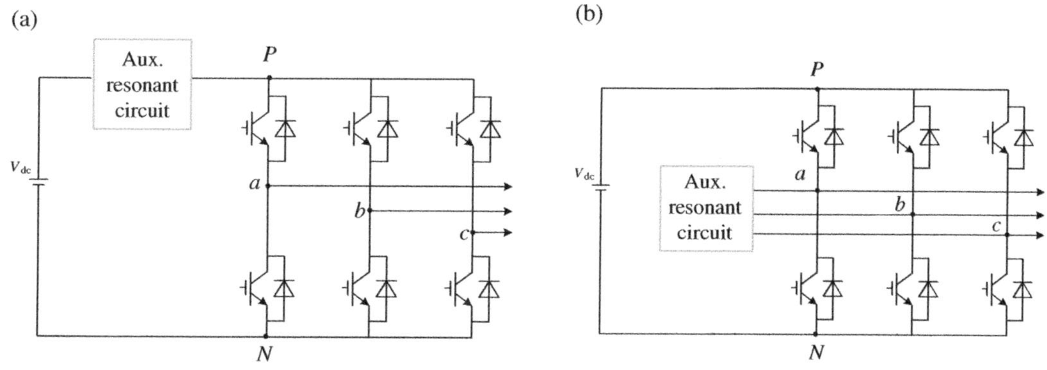

Figure 3.1 Soft-switching converter: (a) with DC-side resonance; (b) with AC-side resonance.

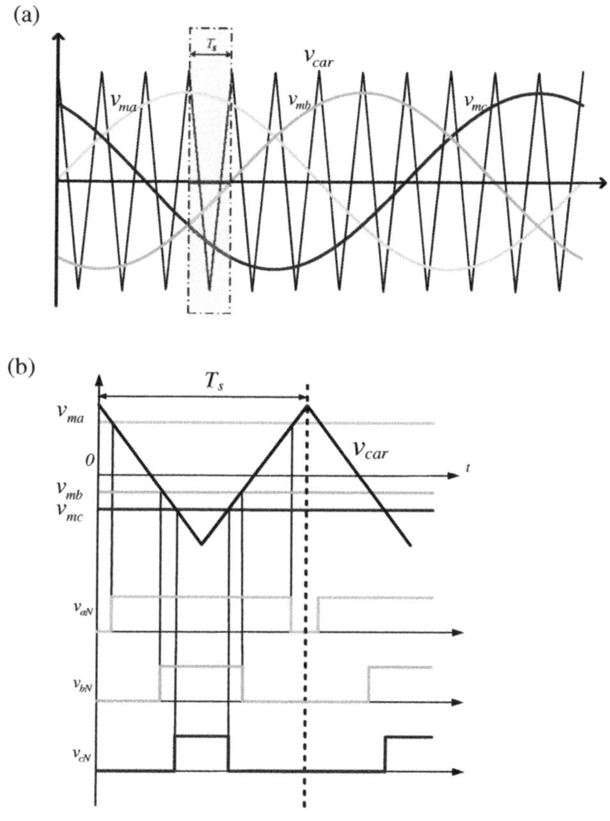

Figure 3.2 SPWM: (a) modulation signals and carrier; (b) PWM waveform in a switch period.

switch leg is high level when $v_{ma}(t) > v_{car}(t)$ while it is equal to zero when $v_{ma}(t) < v_{car}(t)$. Similarly, the midpoint voltages v_{bN} and v_{cN} of phase-*b* and phase-*c* switch legs can be obtained. It observed that each switch leg has two switch commutations in a switching period. Totally there are six commutations in the converter. If either DC-side resonance or AC-side resonance is introduced to realize soft-switching at all these commutations, the auxiliary circuit needs to operate six times each

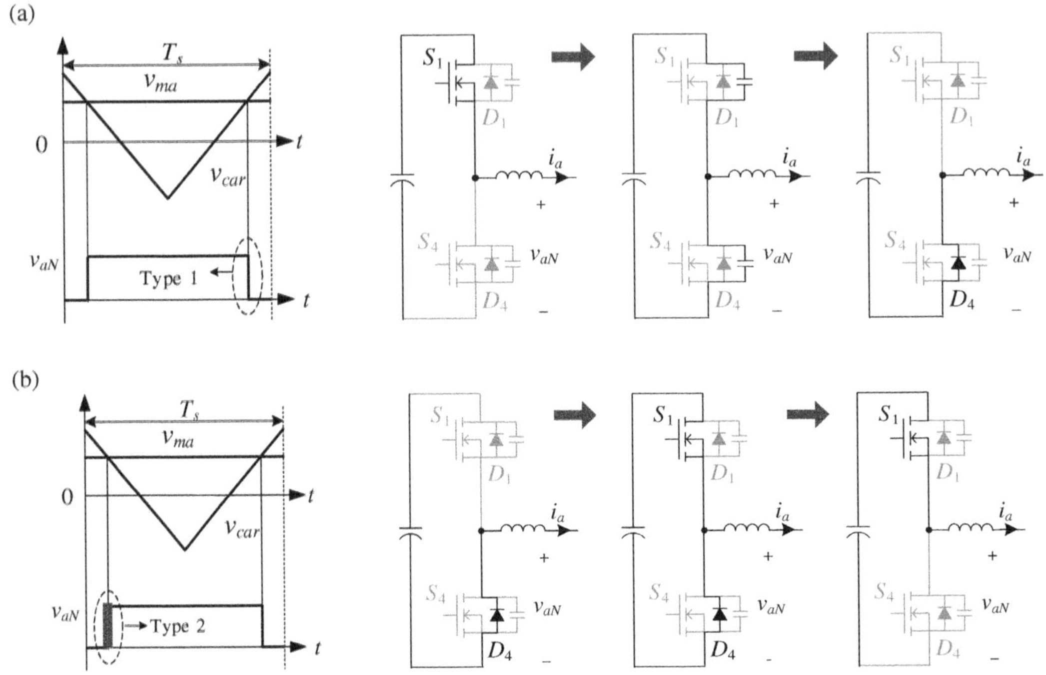

Figure 3.3 Two types of switching commutations of phase-a switch leg: (a) Type 1; (b) Type 2.

switching period. It not only complicates the control but also causes extra circulation loss in the converter due to more frequent operation of the auxiliary circuit.

As mentioned in Chapter 2, there are two types of commutations in a switch leg of the three-phase converter [5]. Figure 3.3 shows two types of switching commutations of phase-a switch leg of the converter. The filter current i_a is assumed as positive. For Type 1 switch commutation, the filter current commutates from switch S_1 to the antiparallel diode D_4 of its complementary switch S_4 as shown in Figure 3.3a. The switch S_1 is ZVS turned off with the help of its paralleled output capacitor. For wide-bandgap device, turn-off loss is much smaller than turn-on loss. Therefore, Type 1 commutation does not need special care. For Type 2 switch commutation, the filter current commutates from the antiparallel diode D_4 to its complementary switch S_1 as shown in Figure 3.3b. S_1 has turn-on loss and D_4 will undergo a reverse recovery process, which results in extra loss in both D_4 and S_1. Besides, it will cause electromagnetic noise (EMI) issue.

Since power loss of Type 2 switch commutation is much larger than that of Type 1 switch commutation, we should try to eliminate Type 2 commutations. How are Type 2 commutations identified? Actually, AC-side filter currents of the converter are thread to identify Type 2 commutations. If the filter current of certain phase is positive, Type 2 commutation occurs when its midpoint voltage steps up. If the filter current of certain phase is negative, Type 2 commutation occurs when its midpoint voltage steps down. Figure 3.4a shows three-phase filter currents ia, ib, ic and modulation signal vma for phase-a. Figure 3.4b shows how PWM waveforms are generated around instant $t1$ of Figure 3.4a. By comparing the modulation waves with the carrier, midpoint voltages v_{aN}, v_{bN}, and v_{cN} are obtained. It is observed that there are six switching commutations in a switching period. For phase-a, the instant current i_a is positive at $t1$ according to Figure 3.4a. Thus step-up instant of v_{aN} is Type 2 commutation as shown with vertical bold bar in Figure 3.4b. For phase-b, the instant current

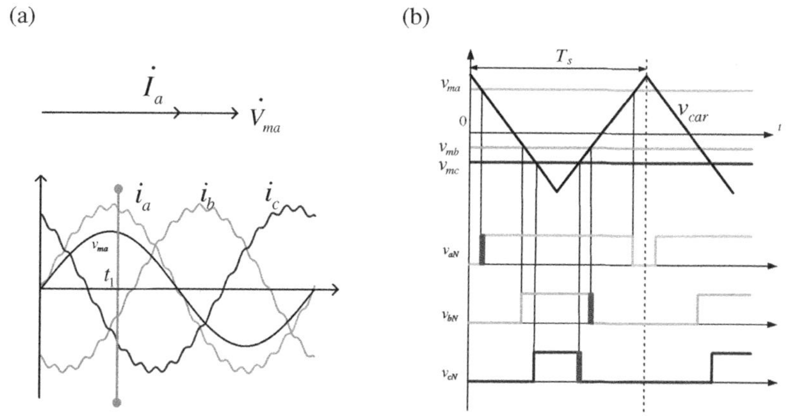

Figure 3.4 Phase current polarity vs. Type 2 commutation edge: (a) filter currents with unit power factor; (b) PWM generation at instant $t1$.

i_b is negative at $t1$ according to Figure 3.4a. Thus step-down instant of v_{bN} is Type 2 commutation. Similarly step-down instant of v_{cN} is also Type 2 commutation because phase-c filter current is also negative at $t1$. Figure 3.5 shows how power factor will affect the Type 2 commutation edges on the PWM waveforms. Around $t1$, although the duty cycle of midpoint voltages v_{aN}, v_{bN}, and v_{cN} keep the same, Type 2 commutation edges change due to changing polarities of the filter currents at $t1$. Step-up instant of v_{aN} and v_{cN} is Type 2 commutation because both phase-a and phase-c filter current are positive while step-down instant of v_{bN} is Type 2 commutation because phase-b filter current is negative.

For each switching period, there are six commutations. Among them, only three commutations are Type 2 commutations and have high loss, which needs special care. If we use this conventional PWM scheme, the auxiliary circuits of either DC-side resonance converters or AC-side resonance

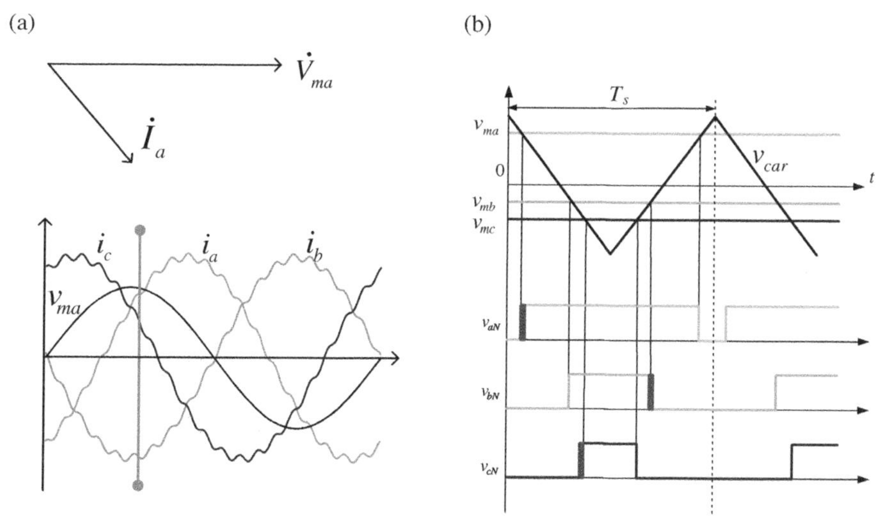

Figure 3.5 Phase current polarity vs. Type 2 commutation edge with non-unit power factor: (a) filter currents with non-unit power factor; (b) PWM generation at instant $t1$.

converters are required to operate at Type 2 commutation instants in each switching period. It means the auxiliary circuits need to operate three times in each switch period. It not only complicates the control but also results in extra circulation loss due to multiple resonant processes on each switching period. Is it possible to align all Type 2 commutations in a switching period in the same instant? If it is so, the auxiliary circuit only needs to operate once. It will significantly simplify the control design and enhance the converter performance.

3.3 Edge-aligned PWM

The conventional PWM scheme is shown in Figure 3.6b. By comparing the modulation voltages v_{ma}, v_{mb}, and v_{mc} with the carrier v_{car}, the switch states of each phase can be determined and midpoint voltages v_{aN}, v_{bN}, and v_{cN} of the converter can be obtained. All the Type 2 commutation instants in a switching period can be easily identified according to the polarities of instant filter currents as shown in Figure 3.6a with vertical bold bar. Actually, whether Type 2 commutation instant will happen in the upper switch or lower switch of a switch leg during each switching cycle is decided by the instant phase current polarity. To achieve soft-switching at all these Type 2 commutation instants, the auxiliary circuit needs to operate three times each switching period, which not only complicates the control but also causes extra circulation loss in the circuit. To overcome this drawback, EA-PWM method is introduced to make all Type 2 commutations in a switching period happen at the same instant [5, 18]. In EA-PWM scheme shown in Figure 3.6c, the original triangular carrier

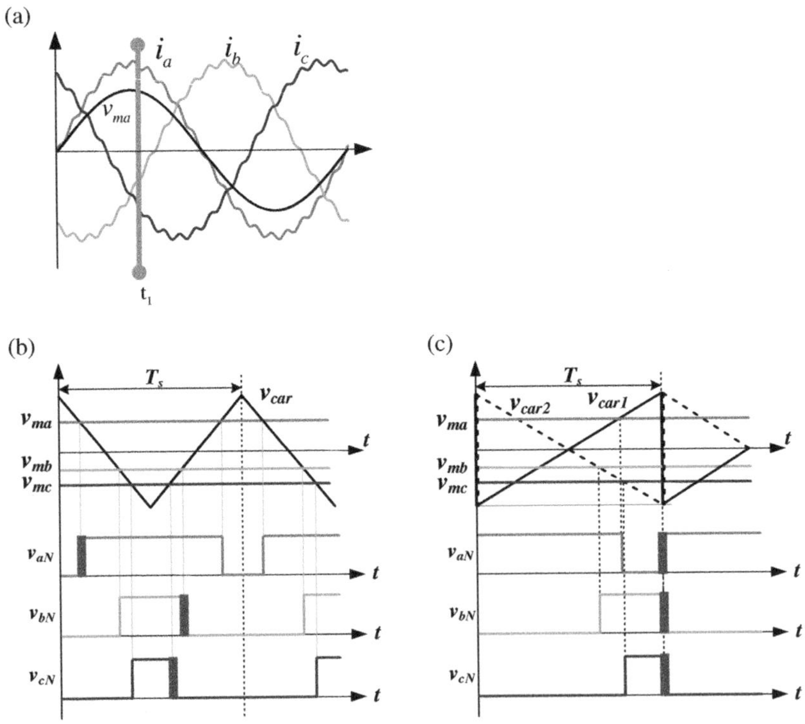

Figure 3.6 PWM scheme vs. EA-PWM scheme: (a) three-phase filter currents; (b) PWM scheme; (c) EA-PWM scheme.

v_{car} in PWM scheme is replaced by two saw-tooth carriers: v_{car1} and v_{car2}. For the positive filter current phase, the ramp-up carrier v_{car1} is selected. For the negative filter current phase, the ramp-down carrier v_{car2} is selected. In Figure 3.6a, at instant t_1, phase-a current i_a is positive. Then the ramp-up carrier v_{car1} is selected to compare with phase-a modulation signal $v_{ma,}$ and midpoint voltage v_{aN} of the converter is obtained as shown in Figure 3.6c. Because phase-a current is positive, step-up instant of v_{aN} is Type 2 commutation, which happens at the terminals of the carrier waveform. In Figure 3.6a, phase-b current i_b at instant $t1$ is negative. Then the ramp-down carrier v_{car2} is selected to compare with phase-b modulation signal v_{mb} and midpoint voltage v_{bN} of the converter can be obtained as shown in Figure 3.6c. Because phase-b current is negative, step-down instant of v_{bN} is Type 2 commutation, which also happens at the terminals of the carrier waveform. Since phase-c current i_c is negative, the ramp-down carrier v_{car2} is selected to compare with phase-c modulation signal v_{mc} and midpoint voltage v_{cN} of the converter is obtained. Step-down instant of v_{cN} is Type 2 commutation, which also happens at the terminals of the carrier waveform. It is observed that all the Type 2 commutations are aligned at the same instant in each switching period.

By introducing EA-PWM, three Type 2 commutations in a switching period are coincident at the same instant. Therefore the auxiliary resonant circuit only needs to operate once during each switching period, and ZVS turn-on can be achieved for all the Type 2 commutation instants. It is noted that the rule of equivalent voltage time-integral principle of PWM in each switching period still holds when we change the triangular carrier to the saw-tooth carrier in the EA-PWM method.

As the conventional PWM, EA-PWM can select modulation signals freely, which encompasses sinusoidal, sinusoidal with third harmonic injection, and other waveforms. In other words, it can be applied to either CPWM or DPWM. EA-PWM is suitable for both three-phase three-wire system and three-phase four-wire system. In EA-PWM, there are two types of carriers. Whether the ramp-up or ramp-down carrier is selected according to the instant current polarity of the phase filter current. In comparison to the conventional PWM method, EA-PWM method needs the information of the instant filter current polarities, which is used to select either the ramp-up carrier or the ramp-down carrier. The big difference between EA-PWM and PWM is that EA-PWM needs filter current polarity information while PWM does not need any filter current information. Similar to traditional PWM method, EA-PWM is suitable to any load condition such as non-unit power factor, load current distortion, and three-phase unbalance.

3.4 ZVS Active-clamping Converter with Edge-aligned PWM

Figure 3.7 shows ZVS active-clamping three-phase converter as introduced in Chapter 2. All high-loss Type 2 switch commutations are eliminated through a resonant process started by the auxiliary resonant circuit to realize ZVS turn-on for all power switches of the converter. There are two kinds of auxiliary circuits: Compound active clamping (CAC) and Minimum voltage active clamping (MVAC) circuit. EA-PWM will be introduced to control the converter so that all Type 2 commutations in a switching period are aligned at the same instant. Thus the auxiliary circuit only needs to operate once each switching period. It not only simplifies the control but also reduces the total circulation loss due to repeating actions of the auxiliary circuit in a switch period.

The feature of the active-clamping converter with EA-PWM is briefly described as follows:

- Before Type 2 commutation happens, bus voltage v_{bus} needs to resonance to zero with the help of the auxiliary circuit.
- With EA-PWM, all the Type 2 commutations in a switching period are co-incident. Thus the auxiliary switch device only switches off once each switching period.

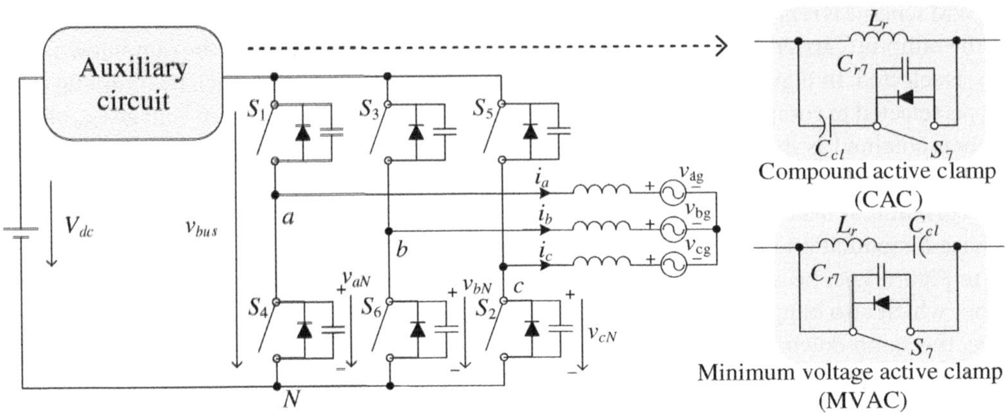

Figure 3.7 Active-clamping ZVS three-phase converter.

- The resonance duration is short, and the auxiliary switch switch-off duty cycle is less than 10% of the switching period.
- EA-PWM is similar to conventional PWM with fixed switching frequency. It is compatible with conventional PWM methods such as SPWM, the third harmonic injection PWM, and CPWM or DPWM.
- There is lower stress on devices because the circuit operates like a PWM converter.

3.4.1 Stage Analysis

This section will make stage analysis of active-clamping three-phase converter with CAC auxiliary circuit in a switching period as shown in Figure 3.8. Three-phase modulation voltages v_{ma}, v_{mb}, and v_{mc} and three-phase filter currents i_a, i_b, and i_c of the converter are depicted in Figure 3.9. An arbitrary time t_s is selected in the figure to illustrate the EA-PWM scheme, when filter current $i_a > 0$, $i_b < 0$, and $i_c < 0$. For simplicity, it is assumed that both modulation signals and the filter currents can been seen as constant inside a switching period.

Key waveforms of the circuit in a switching period is given in Figure 3.10, which happens at instant t_s. According to the principle of the EA-PWM, gate signals u_{g1}–u_{g6} for main power switches

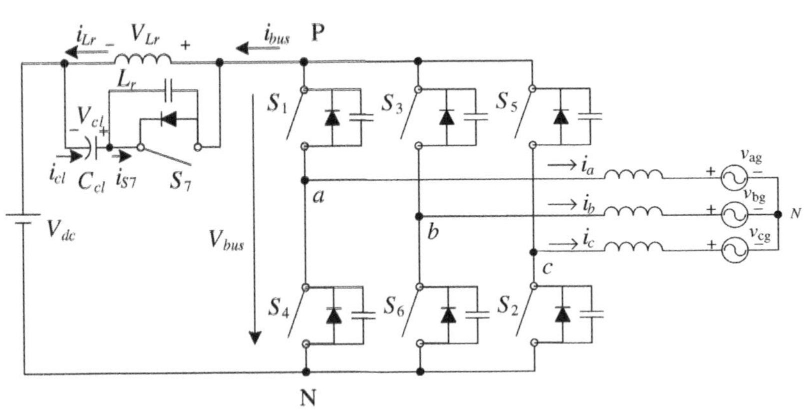

Figure 3.8 Active-clamping three-phase converter with CAC auxiliary circuit.

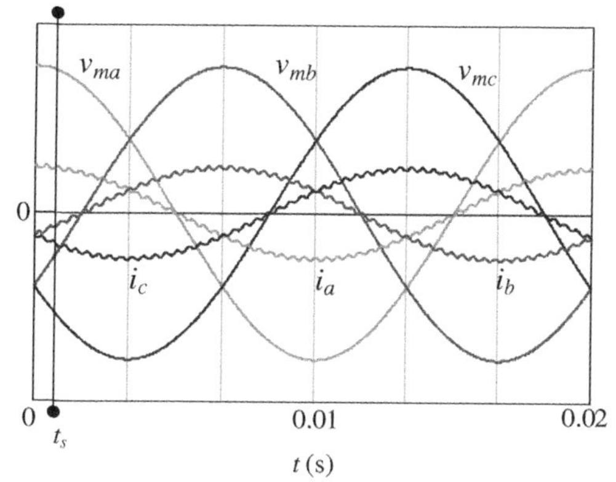

Figure 3.9 Modulation signals and filter current waveforms.

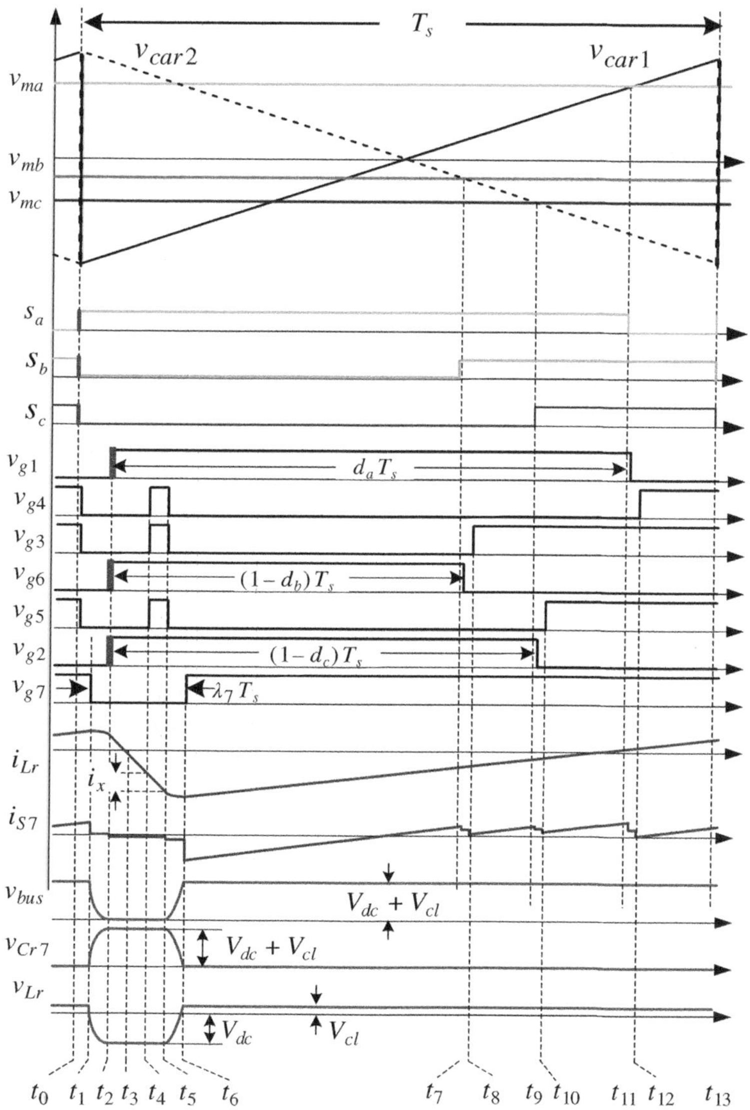

Figure 3.10 Typical waveforms in a switching period.

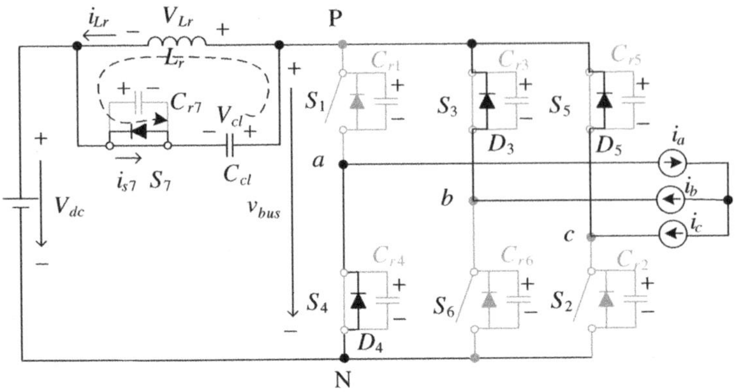

Figure 3.11 Initial state before time t_0.

S1–S6 are basically obtained by comparing three-phase modulation waves v_{ma}, v_{mb}, and v_{mc} with corresponding carrier waves. Later, we will see a small modification is made to tailor soft-switching operation. All Type 2 commutations are aligned at the beginning of the switching period. During t_1–t_6 ($\lambda_7 T_s$), the bus voltage v_{bus} is resonated to zero and power switches, which need to be turned on, are turned on with ZVS condition so that the Type 2 commutations are avoided. The converter operates as the conventional PWM converter in rest of the switching period (t_6–t_{13}).

Without loss of generality, it is assumed that filter current $i_a > 0$, $i_b < 0$, and $i_c < 0$. Besides, they are approximated as constant in a switch period and are represented as current sources. States of the three switch legs are defined by s_a, s_b, and s_c. If $s_x = 1$, $(x = a, b, c)$, the upper switch of leg x is conducting while its lower switch is in off state. If $s_x = 0$, the lower switch of leg x is conducting while its upper switch is in off state. Actually, s_a, s_b, and s_c are generated by EA-PWM as shown Figure 3.10. Their waveforms are similar to the midpoint voltage v_{aN}, v_{bN}, and v_{cN} but with different amplitude. Type 2 commutation instants are marked with bold bar on their waveforms. It is observed that all the Type 2 commutations are aligned. To get rid of high loss of the Type 2 commutations, the auxiliary switch S_7 is turned off before the Type 2 commutation happens. A switch period can be divided into 13 stages as follows:

Initial condition: $t < t_0$

Before time t_0, the states of the three switch legs $s_a = 0$ and $s_b = s_c = 1$. Gate signals for switches S_4, S_3, and S_5 are enabled with high level. According to three filter current directions, i.e. $i_a > 0$, $i_b < 0$, and $i_c < 0$, it is easy to find that for phase-a switch leg the antiparallel diode D_4 of the lower switch conducts, for phase-b switch leg the antiparallel diode of the upper switch D_3 conducts, and for phase-c switch leg the antiparallel diode D_5 of the upper switch conducts as shown in Figure 3.11. Thus all three filter currents flow through the diodes in the switch bridge. At time t_0, the states of the switch legs are expected to change from $s_a = 0$ and $s_b = s_c = 1$ to $s_a = 1$ and $s_b = s_c = 0$. Due to EA-PWM scheme, all three Tpye-2 commutations are aligned at instant t_0.

The auxiliary switch S_7 keeps conducting at this stage. Then the resonant inductor L_r is clamped by V_{cl} and its current i_{Lr} increases linearly with following rate:

$$\frac{di_{Lr}}{dt} = \frac{V_{cl}}{L_r} \tag{3.1}$$

Stage 1 $[t_0, t_1]$: **Dead time**

Type 2 commutations instant is identified at instant t_0. To avoid Type 2 commutation, the switches should not be turned on or turned off with traditional PWM scheme. At t_0, gate signal for

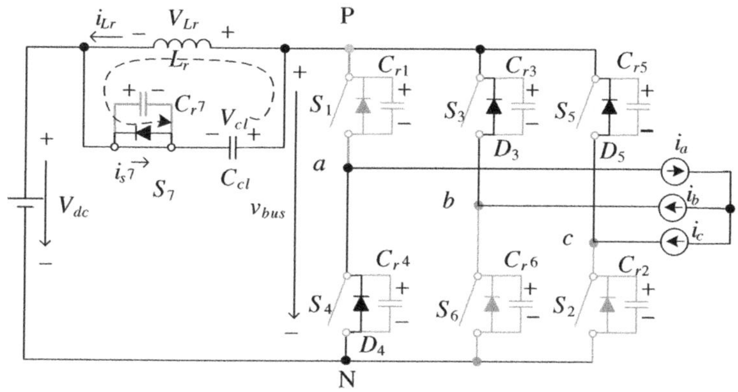

Figure 3.12 Stage 1[t_0, t_1].

conducting switches S_4, S_3, and S_5 are disenabled. However, it will not affect the circuit operation because the filter currents are flowing through their antiparallel diodes. Now all gate signals for the switch bridge are disenabled. The equivalent circuit of the converter at this stage is shown in Figure 3.12. The auxiliary switch S_7 keeps conducting at this stage, and the resonant inductor i_{Lr} increases linearly as before.

This stage ends until the auxiliary switch S_7 turns off, which starts a resonance process.

Stage 2 [t_1, t_2]: First resonance

By observing the states of the switch leg, originally there are Type 2 commutations at instant t_0 if we turn on switch S_1, S_6, and S_2. To avoid this problem, the first resonance process is started by turning off the auxiliary switch S_7. Figure 3.13 shows the equivalent circuit of the converter at this stage. Since the clamping capacitor C_{cl} is large, it can be seen as constant voltage DC voltage source during the switch period. The circuit can be simplified as the second-order system with the resonant inductor L_r being resonant with capacitors paralleled with the blocked switches. During the first resonant stage, the bus voltage v_{bus} is resonance to zero as shown in Figure 3.10.

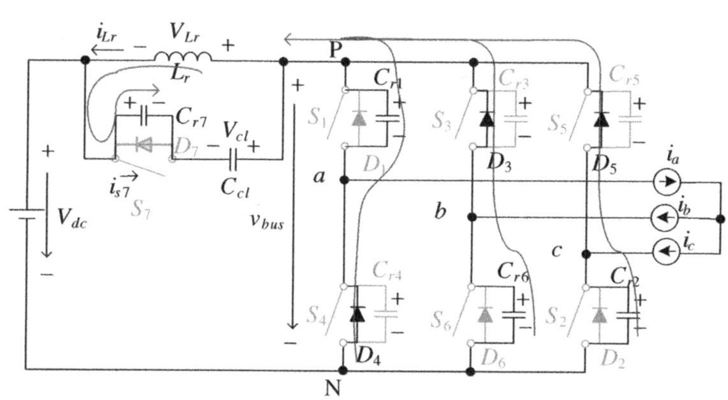

Figure 3.13 Stage 2[t_1, t_2]: first resonance.

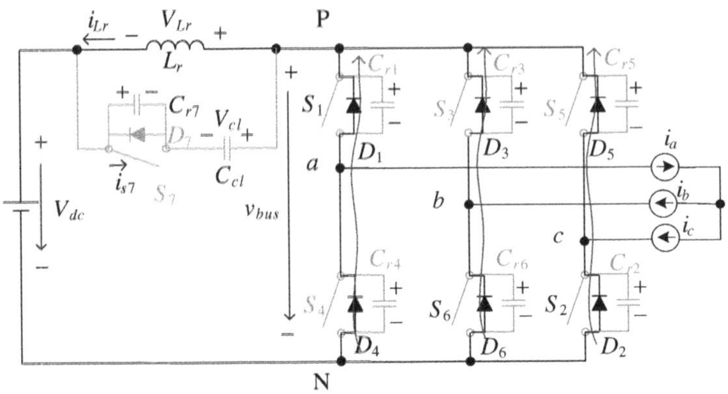

Figure 3.14 Stage 3[t_2, t_3]: resonant inductor discharging duration 1.

The first resonant stage ends when v_{bus}, i.e. voltage on C_{r1}, C_{r2}, and C_{r6}, resonances to zero, which will result in conducting of diodes D_1, D_2, and D_6. It provides ZVS turn-on condition for S_1, S_2, and S_6 in the next stage.

Stage 3 [t_2, t_3]: Resonant inductor discharging duration 1

In this stage $v_{bus} = 0$ and the diodes D_1, D_2, and D_6 are conducting. Figure 3.14 shows the equivalent circuit of the converter at this stage. Since the resonant inductor L_r is clamped by DC source $-V_{dc}$, its current i_{Lr} steeply decreases at the rate

$$\frac{di_{Lr}}{dt} = -\frac{V_{dc}}{L_r} \tag{3.2}$$

In this stage the diodes D_1, D_2, and D_6 are conducting, which provides ZVS turn-on condition for S_1, S_2, and S_6. To realize ZVS turn-on, gate signals for $S1$, $S2$, and $S6$ must be enabled during this stage.

This stage ends when the currents in D_1, D_2, and D_6 decrease to zero.

Stage 4 [t_3, t_4]: Resonant inductor discharging duration 2

Since currents through D_1, D_2, and D_6 decrease to zero, switches antiparalleled with these diodes, S_1, S_2, and S_6, will conduct. Figure 3.15 shows the equivalent circuit of the converter at this stage. Due to resonant inductor L_r, di/dt when D_3, D_4, and D_5 turns off are reduced, which effectively suppresses the reverse recovery process of these diodes. Thus the reverse recovery loss of these diodes is very small and can be neglected.

Since $v_{bus} = 0$, the resonant inductor L_r is still clamped by DC source $-V_{dc}$ and its current continuously decreases linearly as last stage.

This stage ends when currents of diodes D_3, D_4, and D_5 drop to zero. When i_{L_r} decreases to zero, D_3 and D_5 stop conducting. When i_{L_r} decreases to $-ia$, D_4 stops conducting. Generally, their currents do not drop to zero synchronously. As long as there is one diode among them that conducts, the DC bus is still short-circuit and the resonant inductor is still clamped by DC source and its current continuously decreases.

At the end of the stage, the resonant current is represented by

$$i_{Lr}(t_4) = i_{bus}(t_4) = -i_a \tag{3.3}$$

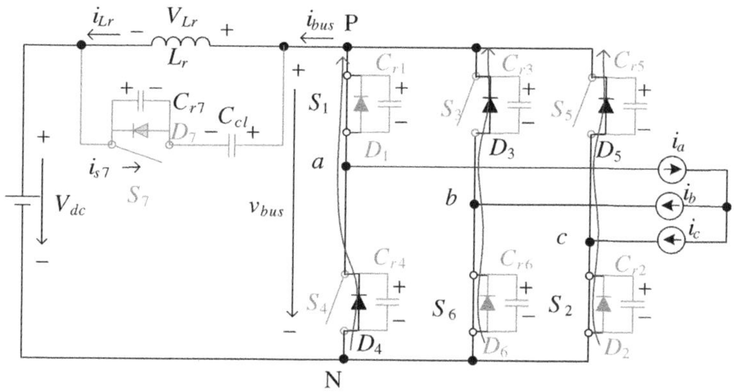

Figure 3.15 Stage 4[t_3, t_4]: resonant inductor discharging duration 2.

Stage 5 [t_4, t_5]: Current boost stage

Sometimes the DC bus voltage v_{bus} is unable to resonance to zero during the first resonance in stage 2. It causes failure of ZVS turn-on in stage 3. Actually, whether v_{bus} can resonance to zero or not depends on resonant inductor current $i_{Lr}(t_1)$ at time $t1$ when the first resonance starts. The larger $i_{Lr}(t_1)$, the easier v_{bus} will resonance to zero. More detailed analysis will be discussed in the next section. To increase the energy storage in the resonant inductor, a current boost stage is added.

By turning on all six switches of the switch bridge for certain duration as shown in Figure 3.16, the resonant inductor current is increased by

$$i_x = \frac{V_{dc} D_{add} T_S}{L_r} \tag{3.4}$$

where $D_{add} = \dfrac{t_5 - t_4}{T_S}$. Figure 3.16b shows the equivalent circuit of the converter at this stage. The duration of the stage depends on the current boost requirement $-i_x$. It will also be discussed later. To realize ZVS turn-on of the switches S_4, S_3, and S_5, they need to be turned on when D_4, D_3, and D_5 are conducting or at least D_4 is conducting in Figure 3.16a. It means that the switches S_4, S_3, and S_5 should be turned on at stage 4.

Actually, it is not necessary to short-circuit all the three switch legs to realize the energy charging of the resonant inductor as shown in Figure 3.16b. We may choose short-circuit one or two switch legs and realize the inductor current boost. Number of short-circuit legs will not affect soft-switching operation. However, selection of short-circuit legs may affect current rating of the switch device and loss distribution.

During the current boost stage, current flows through switches S_3, S_4, and S_5 while the resonant inductor current decreases from $-i_a$ to $-i_a$–i_x. At the end of this stage, S_3, S_4, and S_5 turn off. Then the circuit will start the second resonant stage.

At the end of this stage, the resonant current is represented by

$$i_{Lr}(t_5) = i_{bus}(t_5) = -i_a - i_x \tag{3.5}$$

Stage 6 [t_5, t_6]: Second resonance

When switches S_3, S_4, and S_5 turn off at the end of the inductor current boost stage, the circuit starts in the second resonant stage as shown in Figure 3.17. Similar to the first resonance, it can also be simplified as the second-order system with the resonant inductor L_r being resonant with capacitors

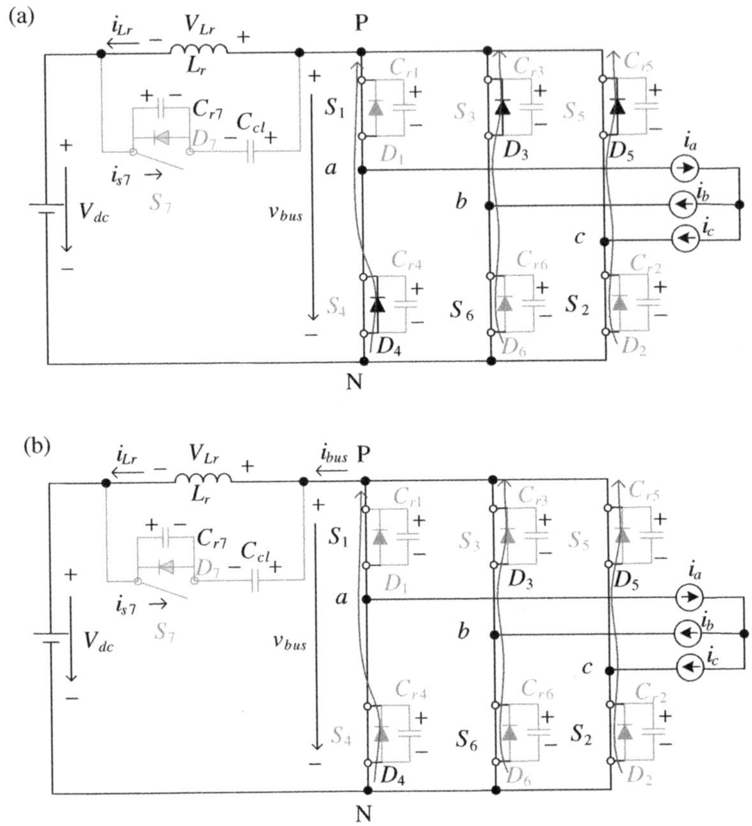

Figure 3.16 Stage 5[t_4, t_5]: current boost stage with three leg short-circuit: (a) at beginning of this stage; (b) current through diodes D_3, D_4, and D_5 reverse direction.

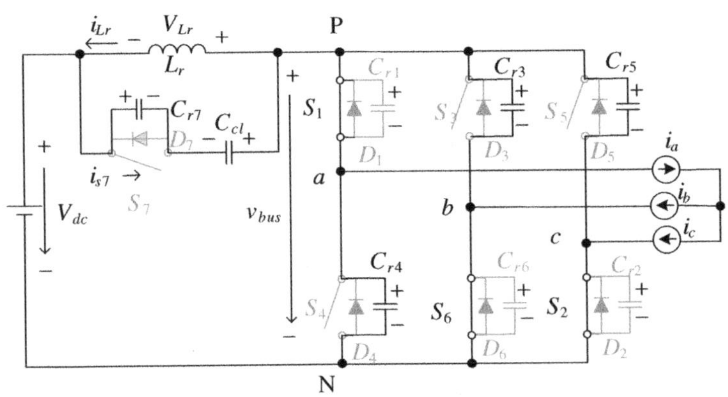

Figure 3.17 Stage 6[t_5, t_6]: second resonance.

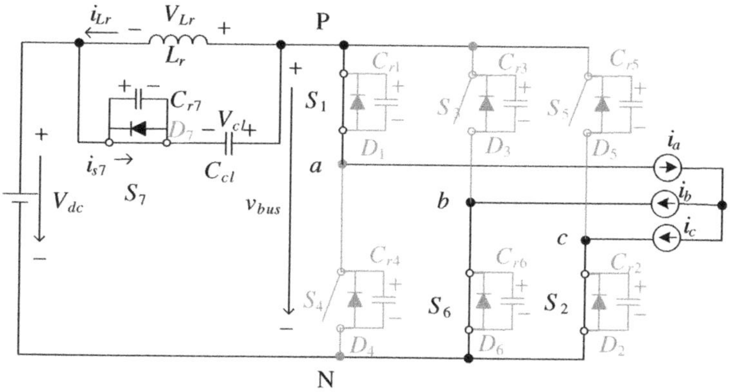

Figure 3.18 Stage 7[t_6, t_7]: PWM duration with state$[s_a \quad s_b \quad s_c] = [1 \quad 0 \quad 0]$.

paralleled with switches. During the second resonant stage, when voltage on capacitor C_{r_7} of the auxiliary switch resonances to zero, diode D_7 will conduct, which creates ZVS turn-on condition for the auxiliary switch S_7 at the next stage.

Once the voltage on capacitor C_{r_7} of the auxiliary switch resonances to zero, this stage ends.

Stage 7 [t_6, t_7]: PWM duration with state $[s_a \quad s_b \quad s_c] = [1 \quad 0 \quad 0]$

At the beginning of this stage, the second resonant stage ends and voltage on capacitor C_{r_7} of the auxiliary switch decreases to zero and diode D_7 conducts as shown in Figure 3.18. Conducting of diode D_7 creates ZVS turn-on conduction for the auxiliary switch S_7. S_7 should be turned on during diode D_7 conducting time. At this stage the bus voltage is a constant and represented as:

$$v_{bus} = V_{dc} + V_{cl} \tag{3.6}$$

At this stage, the resonant inductor L_r is clamped by clamping capacitor voltage V_{cl}, which is approximated as a constant in the switching period. Thus the resonant inductor current increases linearly with rate:

$$\frac{di_{Lr}}{dt} = \frac{V_{cl}}{L_r} \tag{3.7}$$

This stage ends when $S6$ is turned off according to EA-PWM scheme.

Stage 8 [t_7, t_8]: Type 1 commutation

At the beginning of the stage, the lower switch S_6 of phase-b switch leg is turned off in Figure 3.19. A Type 1 commutation happens between S_6 and its complement switch's antiparallel diode D_3 at leg-b. As we mentioned before, S_6 is ZVS turned off and switch loss is reduced. At this stage, the resonant inductor L_r is still clamped by clamping capacitor voltage V_{cl}. Its current increases linearly with the same rate as last stage.

This stage is very short. Once voltage on capacitor C_{r_6} charged to the DC bus voltage $v_{bus} = V_{dc} + V_{cl}$ while the voltage on capacitor C_{r_3} discharged to zero and $D3$ starts to conduct, this stage ends.

Stage 9 [t_8, t_9]: PWM duration with state $[s_a \quad s_b \quad s_c] = [1 \quad 1 \quad 0]$

At the beginning of this stage, the voltage on capacitor C_{r_3} discharged to zero and D_3 starts to conduct in Figure 3.20. Switch bridge state $[s_a \quad s_b \quad s_c] = [1 \quad 1 \quad 0]$. At this stage, S_1, S_2, and D_3 are conducting. The resonant inductor current increases linearly with the same rate as the last stage.

This stage ends when S_2 is turned off according to EA-PWM scheme.

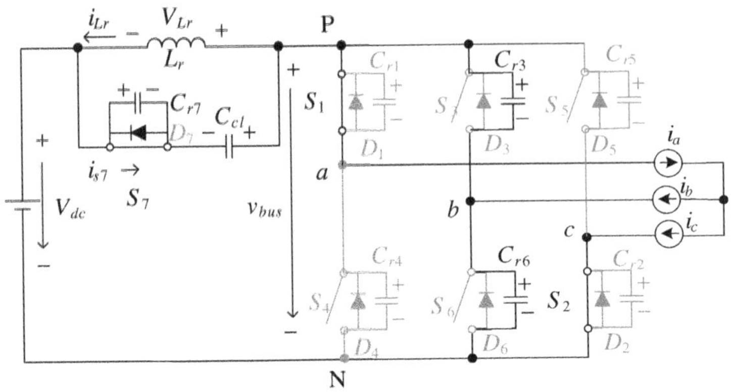

Figure 3.19 Stage 8[t_7, t_8]: Type 1 commutation.

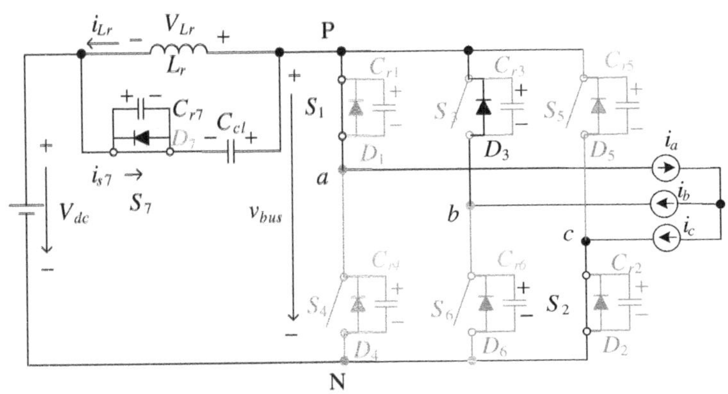

Figure 3.20 Stage 9[t_8, t_9].

Stage 10 [t_9, t_{10}]**: Type 1 commutation**

When S_2 is turned off, a Type 1 commutation happens between S_2 and D_5 at switch leg-c in Figure 3.21. S_2 is ZVS turned off and switch loss is reduced. The resonant inductor current increases linearly as before.

This stage is also very short. Once voltage on capacitor C_{r_2} charged to the DC bus voltage $v_{bus} = V_{dc} + V_{cl}$ while the voltage on capacitor C_{r_5} discharged to zero and diode D_5 starts to conduct, this stage ends.

Stage 11 [t_{10}, t_{11}]**: PWM duration with state** $[s_a \quad s_b \quad s_c] = [1 \quad 1 \quad 1]$

At the beginning of this stage, the voltage on capacitor $Cr5$ discharged to zero and D_5 starts to conduct in Figure 3.22. Switch bridge state $[s_a \quad s_b \quad s_c] = [1 \quad 1 \quad 1]$. At this stage S_1, D_3, and D_5 are conducting. The resonant inductor current increases linearly with the same rate as before. This stage ends when S_1 is turned off according to EA-PWM scheme.

Stage 12 [t_{11}, t_{12}]**: Type 1 commutation**

When S_1 is turned off, a Type 1 commutation happens between S_1 and D_4 at switch leg-a in Figure 3.23. S_1 is ZVS turned off and switch loss is reduced. The resonant inductor current increases linearly as before.

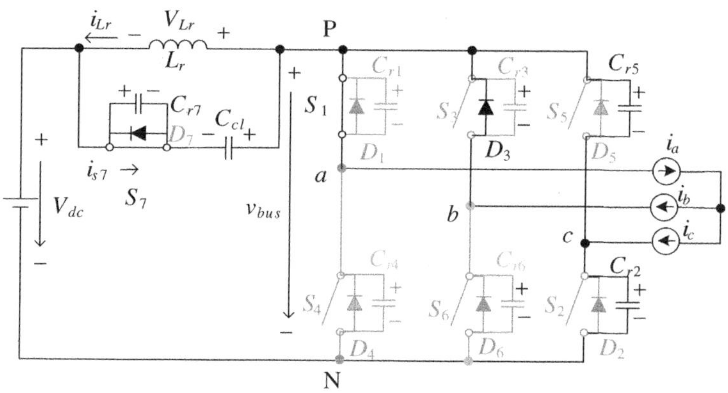

Figure 3.21 Stage 10[t_9, t_{10}]: Type 1 commutation.

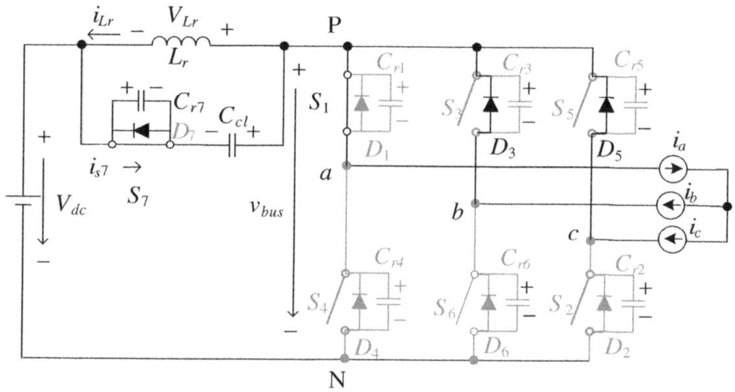

Figure 3.22 Stage 11 [t_{10}, t_{11}]: PWM duration with state[s_a s_b s_c] = [1 1 1].

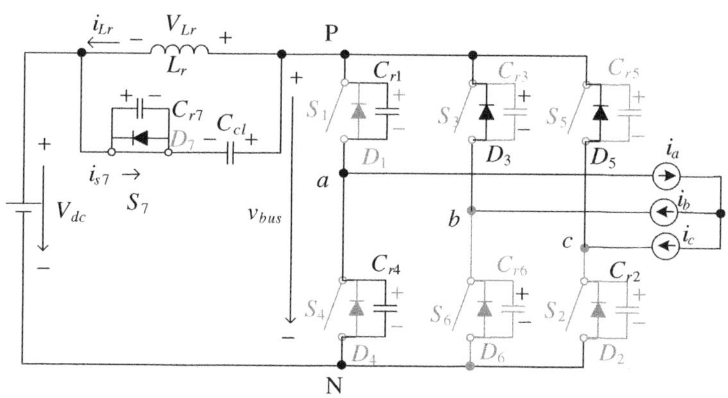

Figure 3.23 Stage 12 [t_{11}, t_{12}]: Type 1 commutation.

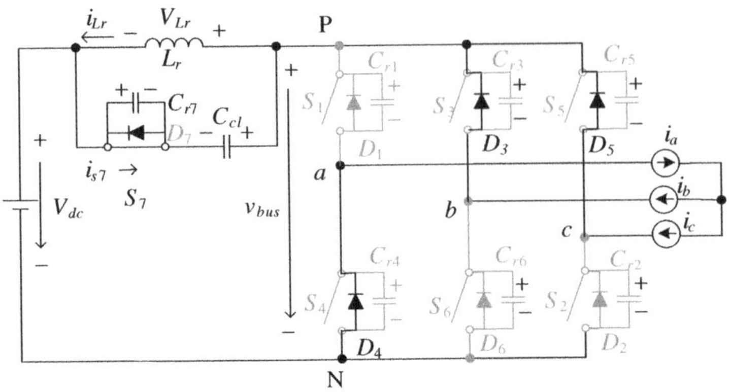

Figure 3.24 Stage 13 $[t_{12}, t_{13}]$: PWM duration with state $[s_a \quad s_b \quad s_c] = [0 \quad 1 \quad 1]$.

This stage is also very short. Once voltage on capacitor C_{r_1} charged to the DC bus voltage $v_{bus} = V_{dc} + V_{cl}$ while the voltage on capacitor C_{r_4} discharged to zero and diode D_4 starts to conduct, this stage ends.

Stage 13 $[t_{12}, t_{13}]$: PWM duration with state $[s_a \quad s_b \quad s_c] = [0 \quad 1 \quad 1]$

At the beginning of this stage, the voltage on capacitor C_{r_4} discharged to zero and *D4* starts to conduct in Figure 3.24. Switch bridge state $[s_a \quad s_b \quad s_c] = [0 \quad 1 \quad 1]$. At this stage D_4, D_3, and D_5 are conducting. The resonant inductor current increases linearly with the same rate as before.

The duration of this stage is decided by EA-PWM scheme. This stage continues until the end of a switch period. From time t_{13}, the converter will start a new switching period.

From above stage analysis, we can see that a switch period can be divided into two time sections. The first time section is relatively short with respect to the switch period and is usually less than 10% of the switch period. It covers stages 1–6 and is used to realize ZVS turn-on for switches in the circuit. Among these six stages, the first resonant stage is used for ZVS of main switches, the second resonant stage is used for ZVS of the auxiliary switch, and the current boost stage is inserted to increase the energy storage in the resonant inductor.

The second time section is main part of EA-PWM in a switch period. It ranges from stages 7–13. It is used to realize PWM of output voltage waveforms, which is similar to conventional PWM.

3.4.2 ZVS Conditions

This section will derive the ZVS condition for active-clamping ZVS three-phase converter with CAC auxiliary circuit [5, 17, 18]. There are two resonant stages in each switching period: the first resonant stage and the second resonant stage. The first resonant stage is used to realize ZVS operation for the main switches, while the second resonant stage is used to realize ZVS operation for the auxiliary switch.

3.4.2.1 The First Resonant Stage

The first resonant stage (t_1–t_2) happens when the auxiliary switch S_7 is turned off. The DC bus voltage v_{bus} will be resonated to zero at the end of the stage. The converter circuit state and its equivalent circuit are given in Figure 3.25. By replacing the conducting diode branches in the three-phase switch bridge with short-circuit, three capacitors C_{r1}, C_{r2}, and C_{r6} are paralleled and can be replaced

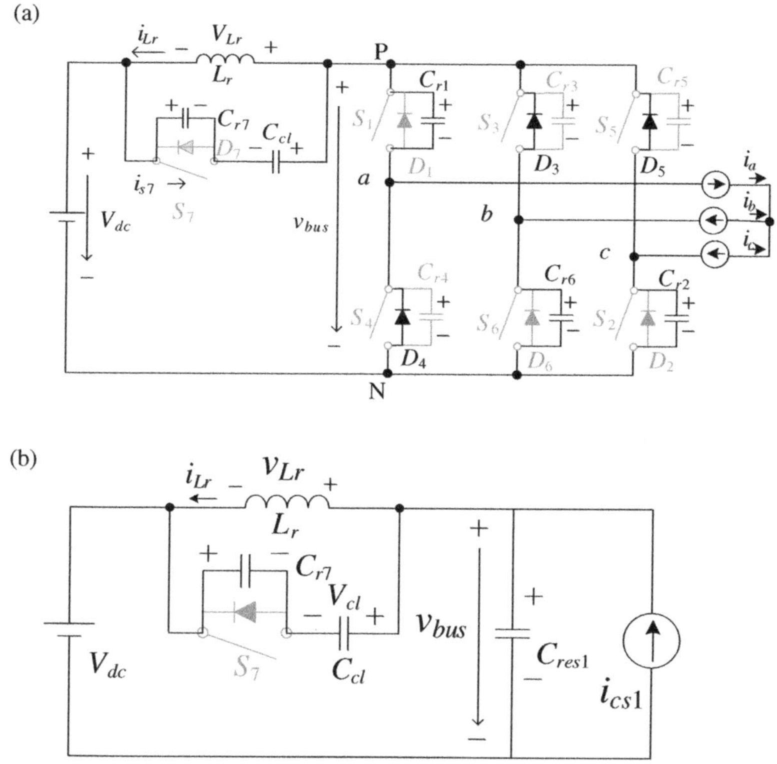

Figure 3.25 Circuit of the first resonant stage: (a) circuit state; (b) equivalent circuit.

by capacitor C_{res1} while the function of three current sources i_a, i_b, and i_c can be replaced by a current source:

$$i_{cs1} = -i_b - i_c \tag{3.8}$$

After aforementioned replacement, Figure 3.25a is simplified into equivalent circuit as shown in Figure 3.25b.

In Figure 3.25a, all three filter currents flow through the diodes. The filter current i_a flows through D_4, i_b flows through D_3, and i_c flows through D_5. Actually, the filter current polarity decides which diode in a switch leg conducts. If phase i's filter current i_i is positive, then the antiparallel diode of the lower switch in switch-leg i will conduct and i_i will not contribute to the current source i_{cs1} in the equivalent circuit in Figure 3.25b. If phase i's filter current i_i is negative, then the anti-parallel diode of the upper switch in switch leg i will conduct and i_i will add to the current source i_{cs1}. Thus all negative filter currents contribute to i_{cs1}. Generally, i_{cs1} is expressed by

$$i_{cs1} = i_a x_a + i_b x_b + i_c x_c \tag{3.9}$$

where $x_i = \begin{cases} -1 & i_i \leq 0 \\ 0 & otherwise \end{cases}$; subscript $i = a,b,c$.

For DPWM there exists one phase whose state is fixed in each 60° duration of a fundamental frequency period. It means the switch-leg state of present discontinuous phase is fixed and its phase filter current keeps flowing through the same switching during the complete 60° duration. If the switch-leg state of the discontinuous phase is "1," i.e. its modulation signal $u_{mi}(t) = 1$, then its filter current will add to current source i_{cs1} in the discontinuous duration. However, if the switch-leg state of the discontinuous phase is "0," i.e. its modulation signal $u_{mi}(t) = -1$, then its filter current does not contribute to i_{cs1} in the discontinuous duration. By considering DPWM situation, Eq. (3.9) needs to be modified as

$$i_{cs1} = i_a x_a + i_b x_b + i_c x_c \tag{3.10}$$

where $x_i = \begin{cases} -1 & [i_i \leq 0; CPWM].or.[u_{mi} = 1; DPWM] \\ 0 & otherwise \end{cases}$; subscript $i = a,b,c$.

Since the clamping capacitor C_{cl} is so large that its voltage can be seen as the constant, it is replaced by a DC voltage source with constant voltage of V_{cl} in Figure 3.25b. According to Kirchhoff's voltage law (KVL) and Kirchhoff's current law (KCL), following state equations are derived from the equivalent circuit of the first resonant stage in Figure 3.25b.

$$\begin{cases} (C_{res1} + C_{r7})\dfrac{dv_{bus}(t)}{dt} = i_{cs1} - i_{Lr}(t) \\ L_r \dfrac{di_{Lr}(t)}{dt} = v_{bus}(t) - V_{dc} \end{cases} \tag{3.11}$$

where $C_{res1} = C_{r1} + C_{r6} + C_{r2} = 3C_r$, $i_{cs1} = -i_b - i_c$. The initial condition of the state variables at t_1 can be written as:

$$\begin{cases} v_{bus}(t_1) = V_{dc} + V_{cl} \\ i_{Lr}(t_1) = I_{Lr,1} \end{cases} \tag{3.12}$$

By solving state equation (3.11) with the initial condition (3.12), the expressions of state variables $v_{bus}(t)$ and $i_{Lr}(t)$ are obtained:

$$\begin{cases} v_{bus}(t) = V_{dc} - \sqrt{V_{cl}^2 + (I_{Lr,1} - i_{cs1})^2 Z_r^2} \sin\left[\omega_r(t - t_1) - \varphi_1\right] \\ i_{Lr}(t) = \sqrt{(I_{Lr,1} - i_{cs1})^2 + \dfrac{V_{cl}^2}{Z_r^2}} \cos\left[\omega_r(t - t_1) - \varphi_1\right] + i_{cs1} \end{cases} \tag{3.13}$$

where ω_r is the resonance angular frequency, Z_r is the characteristic impedance, and φ_1 is the initial angle of the first resonant stage. Their expressions are as follows:

$$\omega_r = 1/\sqrt{L_r(3C_r + C_{r7})} \tag{3.14}$$

$$Z_r = \sqrt{L_r/(3C_r + C_{r7})} \tag{3.15}$$

$$\varphi_1 = \arctan\left[V_{cl}/Z_r(i_{Lr,1} - i_{cs1})\right] \tag{3.16}$$

To realize the ZVS turn-on of the main switches, DC bus voltage $v_{bus}(t)$ is required to resonance to zero. By observing Eq. (3.13), $v_{bus}(t)$ is composed of a constant positive term and a sinusoidal term. To make $v_{bus}(t)$ negative, the amplitude of the sinusoidal term should be larger than the constant term. Thus following inequality needs to be satisfied:

$$V_{dc} < \sqrt{V_{cl}^2 + (I_{Lr,1} - i_{cs1})^2 Z_r^2} \tag{3.17}$$

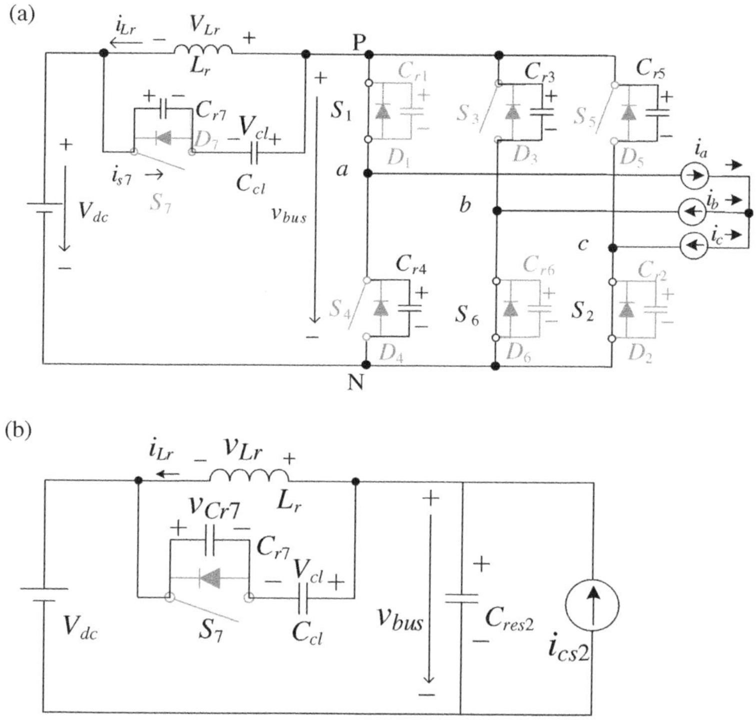

Figure 3.26 The circuit of the second resonant stage: (a) circuit state; (b) equivalent circuit.

The above inequality can be rewritten as

$$I_{Lr,t_1} \geq i_{cs1} + \sqrt{V_{dc}^2 - V_{cl}^2}/Z_r \tag{3.18}$$

The above inequality gives a constraint to the resonant inductor current at time t_1 when the first resonant stage starts. It means that the resonant inductor current at t_1 needs to be large enough to satisfy ZVS condition.

3.4.2.2 The Second Resonant Stage

The second resonant stage happens during (t_5–t_6) when the voltage across the auxiliary switch v_{Cr7} resonates to zero as shown in Figure 3.10. It provides the ZVS turn-on condition for the auxiliary switch S_7. The converter circuit state of the second resonant stage and its equivalent circuit are shown in Figure 3.26.

By replacing the conducting switch branches in the switch bridge with short-circuit, three capacitors, C_{r3}, C_{r4}, and C_{r5}, are paralleled and can be replaced by capacitor C_{res2} while the function of the three-phase currents is represented with following current source:

$$i_{cs2} = -i_a \tag{3.19}$$

After above replacements, Figure 3.26a is simplified into its equivalent circuit shown in Figure 3.26b.

Similar to the first resonant stage analysis, i_{cs2} is generally expressed by

$$i_{cs2} = i_a y_a + i_b y_b + i_c y_c \tag{3.20}$$

$$y_i = \begin{cases} -1 & [i_i \geq 0; CPWM].or.[u_{mi} = 1; DPWM] \\ 0 & otherwise \end{cases}$$; where subscript $i = a,b,c$.

For CPWM scheme, $y_i = -1 - x_i$ according to Eqs. (3.10) and (3.20).

Capacitor C_{res2} is represented by

$$C_{res2} = C_{r4} + C_{r3} + C_{r5} = 3C_r \tag{3.21}$$

Similarly, clamping capacitor C_{cl} is so large that it can be seen as the constant DC voltage source with voltage V_{cl} in Figure 3.26b. According to KVL and KCL, following state equations are derived from the equivalent circuit of the second resonant stage.

$$\begin{cases} (C_{res2} + C_{r7})\dfrac{dv_{Cr7}(t)}{dt} = i_{Lr}(t) - i_{cs2} \\ L_r\dfrac{di_{Lr}(t)}{dt} = V_{cl} - v_{Cr7}(t) \end{cases} \tag{3.22}$$

The initial condition of state equation (3.22) at t_5 is:

$$\begin{cases} i_{Lr}(t_5) = i_{cs2} - i_x \\ v_{Cr7}(t_5) = V_{dc} + V_{cl} \end{cases} \tag{3.23}$$

where i_x is the extra current added to resonant inductor Lr through a short stage (t_4–t_5) in Figure 3.10. By solving state equation (3.22) with initial condition (3.23), the expressions for $v_{Cr7}(t)$ and $i_{Lr}(t)$ during the second resonant stage are obtained:

$$\begin{cases} v_{Cr7}(t) = V_{cl} - \sqrt{V_{dc}^2 + Z_r^2 i_x^2}\sin\left[\omega_r(t - t_5) - \varphi_2\right] \\ i_{Lr}(t) = i_{cs2} - \sqrt{i_x^2 + \dfrac{V_{dc}^2}{Z_r^2}}\cos\left[\omega_r(t - t_5) - \varphi_2\right] \end{cases} \tag{3.24}$$

where φ_2 is expressed by

$$\varphi_2 = \arctan\frac{V_{dc}}{Z_r i_x} \tag{3.25}$$

To realize the ZVS turn-on of the auxiliary switch S_7 during the second resonant stage, $v_{Cr7}(t)$ needs to be resonance to zero. According to Eq. (3.24), the following inequality needs to be satisfied:

$$V_{cl} - \sqrt{V_{dc}^2 + Z_r^2 i_x^2} \leq 0 \tag{3.26}$$

Because V_{cl} is much smaller than V_{dc}, the above inequality is always satisfied. It means the ZVS turn-on for auxiliary switch S_7 is always realized in the second resonant stage.

At time t_6, $v_{Cr7}(t_6)$ is equal to zero. From Eq. (3.24), following equation is obtained

$$v_{Cr7}(t_6) = V_{cl} - \sqrt{V_{dc}^2 + Z_r^2 i_x^2}\sin\left[\omega_r(t_6 - t_5) - \varphi_2\right] = 0 \tag{3.27}$$

Above equation is rewritten as:

$$\sin\left[\omega_r(t_6 - t_5) - \varphi_2\right] = \frac{V_{cl}}{\sqrt{V_{dc}^2 + Z_r^2 i_x^2}} \tag{3.28}$$

The resonant inductor current $i_{Lr}(t_6)$ is calculated with Eq. (3.24):

$$i_{Lr}(t_6) = i_{cs2} - \sqrt{i_x^2 + \frac{V_{dc}^2}{Z_r^2}} \cos\left[\omega_r(t_6 - t_5) - \varphi_2\right] \tag{3.29}$$

The above equation is rewritten as:

$$i_{Lr}(t_6) = i_{cs2} - \sqrt{i_x^2 + \frac{V_{dc}^2}{Z_r^2}} \sqrt{1 - \{\sin\left[\omega_r(t_6 - t_5) - \varphi_2\right]\}^2} \tag{3.30}$$

By substituting Eq. (3.28) into (3.30), the resonant inductor current at t_6, $i_{Lr}(t_6)$, is derived as:

$$i_{Lr}(t_6) = i_{cs2} - \sqrt{\frac{V_{dc}^2 - V_{cl}^2}{Z_r^2} + i_x^2} \tag{3.31}$$

3.4.2.3 Steady Conditions

When the converter has reached steady state, voltage on the resonant inductor L_r satisfies the volt-age-second balance principle in a switching cycle.

$$\int_0^{T_s} v_{Lr}(t)\ dt = 0 \tag{3.32}$$

Except in the first and second resonant stages the resonant inductor voltage v_{Lr} is equal to either V_{cl} or $-V_{dc}$ in a switching period as shown in Figure 3.10. By neglecting these two short durations, Eq. (3.32) is approximated as:

$$\int_0^{T_s} v_{Lr}(t)\ dt \approx V_{cl}(1-\lambda_7)T_s - V_{dc}\lambda_7 T_s = 0 \tag{3.33}$$

where $\lambda_7 = \frac{t_6 - t_1}{T_s}$ is the turn-off duty cycle of the auxiliary switch S_7. Then the clamping capacitor voltage V_{cl} is derived:

$$V_{cl} = \frac{\lambda_7}{1-\lambda_7} V_{dc} \tag{3.34}$$

Voltage stress of all the switches of the converter with CAC can be represented by:

$$V_{s\max} = V_{dc} + V_{cl} = \frac{1}{1-\lambda_7} V_{dc} \tag{3.35}$$

Now let us discuss how the EA-PWM affects the converter output. To simplify following analysis, the dead time stage:[$t0$,$t1$] is neglected. However, the simplification will not affect the conclusion. Due to the effect of λ_7, the duty cycles of the active-clamping converter is slightly different from those in the conventional PWM converter shown in Figure 3.27. For the phase with positive filter current, such as phase-a, the duty cycle of the leg-a midpoint voltage is reduced to $d_a - \lambda_7$. For the phase with negative filter current, such as phase-b and phase-c, the duty cycles of both phase-b and phase-c are the same as before.

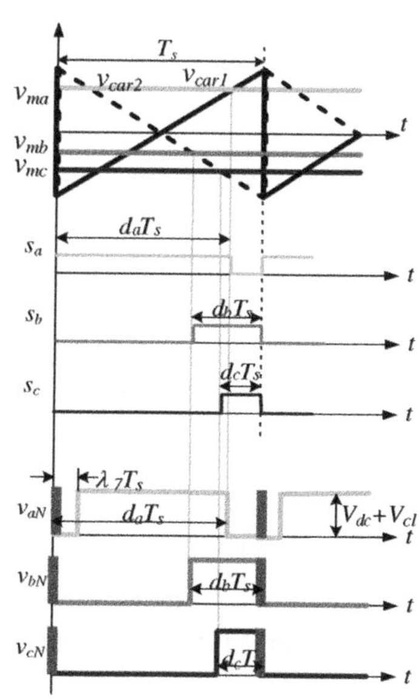

Figure 3.27 Midpoint voltages of three-phase active-clamping converter with CAC.

In addition, amplitude of all midpoint voltages v_{aN}, v_{bN}, and v_{cN} increases from V_{dc} to $V_{dc} + V_{cl}$ due to introduction of the CAC auxiliary circuit.

According to Figure 3.27, the average value of midpoint voltage v_{aN} in a switching period is represented by

$$\langle v_{aN} \rangle_{T_s} = (V_{dc} + V_{cl})(d_a - \lambda_7) \tag{3.36}$$

It is observed that v_{aN} has a duty cycle loss λ_7. The amplitude of v_{aN} has an increment of V_{cl} and partially compensates the duty cycle loss. Average values of midpoint voltages v_{bN} and v_{cN} are represented by

$$\langle v_{bN} \rangle_{T_s} = (V_{dc} + V_{cl}) d_b \tag{3.37}$$

$$\langle v_{cN} \rangle_{T_s} = (V_{dc} + V_{cl}) d_c \tag{3.38}$$

There is no duty cycle loss for both phase-b and phase-c midpoint voltages.

The duty cycle loss results in distortion in low frequency fundamental component of the converter output. Actually, this effect on the output with positive current phase can be compensated by adjusting the modulation signal. The duty cycle of phase-a is so designed that following equation satisfied:

$$\langle v_{aN} \rangle_{T_s} = (V_{dc} + V_{cl})(d_a - \lambda_7) = \frac{V_{dc}(1 + m_a \sin\theta)}{2} \tag{3.39}$$

In above equation $m_a = \frac{V_m}{V_{car}}$ is the amplitude modulation ratio, where V_m is the peak amplitude of the modulation signals, and V_{car} is the amplitude of the carrier signals. The range of m_a depends on PWM scheme and is [0,1] for SPWM. From Eq. (3.39), duty cycle of phase-a is derived as:

$$d_a = \frac{V_{dc}}{V_{dc} + V_{cl}} \frac{(1 + m_a \sin\theta)}{2} + \lambda_7 \tag{3.40}$$

By using Eq. (3.34), the above equation is changed into

$$d_a = \frac{(1 - \lambda_7)(1 + m_a \sin\theta)}{2} + \lambda_7 \tag{3.41}$$

The duty cycle of phase-b is so designed that following equation satisfied:

$$\langle v_{bN} \rangle_{T_s} = (V_{dc} + V_{cl}) d_b = \frac{V_{dc}[1 + m_a \sin(\theta - 2\pi/3)]}{2} \tag{3.42}$$

Then duty cycle of phase-b is derived as follows:

$$d_b = \frac{(1 - \lambda_7)[1 + m_a \sin(\theta - 2\pi/3)]}{2} \tag{3.43}$$

Similarly the duty cycle of phase-c is obtained:

$$d_c = \frac{(1 - \lambda_7)[1 + m_a \sin(\theta - 4\pi/3)]}{2} \tag{3.44}$$

For general PWM scheme, the duty cycles can be represented as:

$$d_a = \frac{(1 - \lambda_7)(1 + v_{ma})}{2} + \lambda_7 \tag{3.45}$$

$$d_b = \frac{(1 - \lambda_7)(1 + v_{mb})}{2} \tag{3.46}$$

$$d_c = \frac{(1-\lambda_7)(1+v_{mc})}{2} \tag{3.47}$$

where v_{ma}, v_{mb}, and v_{mc} are modulation signals, which change according to types of PWM such as SPWM, the third harmonic injection PWM, or either CPWM or DPWM.

When the converter is in steady state, the current through the clamping capacitor C_{cl} obeys the ampere-second balance principle in one switching period.

$$\int_0^{T_s} i_{cl}(t)\ dt = 0 \tag{3.48}$$

During resonance process $[t_1, t_6] = [t_1, t_1 + \lambda_7 T_S]$, i_{Cc} is equal to zero except in the first and second resonant stages. By neglecting these two short resonant stages, integral of clamping capacitor current is simplified as:

$$\int_0^{T_s} i_{cl}(t)\ dt = \int_0^{\lambda_7 T_s} i_{cl}(t)\ dt + \int_{\lambda_7 T_s}^{T_s} i_{cl}(t)\ dt \approx \int_{\lambda_7 T_s}^{T_s} i_{cl}(t)\ dt \tag{3.49}$$

With Eqs. (3.48) and (3.49), following equation is obtained:

$$\int_{\lambda_7 Ts}^{T_s} i_{cl}(t)\ dt = 0 \tag{3.50}$$

From Figure 3.8, according to KCL, i_{cl} is expressed as:

$$i_{cl} = i_{Lr} - i_{bus} \tag{3.51}$$

By introducing integral on both sides of above equation, it changes into

$$\int_{\lambda_7 T_s}^{T_s} i_{cl}(t)\ dt = \int_{\lambda_7 T_s}^{T_s} i_{Lr}(t)\ dt - \int_{\lambda_7 T_s}^{T_s} i_{bus}(t)\ dt \tag{3.52}$$

Because the resonant inductor is clamped with the constant voltage V_{cl} during $[\lambda_7 T_S, T_S]$, its current can be represented by

$$i_{Lr}(t) = i_{Lr}(t_6) + \frac{V_{cl}}{L_r}(t - t_6) \tag{3.53}$$

where $t_6 = \lambda_7 T_s$ is the instant when the antiparallel diode of D_7 of the auxiliary switch S_7 is turned on.

DC side current of the three-phase switch bridge is expressed as follows:

$$i_{bus} = -(s_a i_a + s_b i_b + s_c i_c) \tag{3.54}$$

where s_a, s_b, and s_c are states of the three legs.

By substituting Eqs. (3.53) and (3.54) into Eq. (3.52),

$$\int_{\lambda_7 T_s}^{T_s} i_{cl}(t)\ dt = \int_{\lambda_7 T_s}^{T_s} \left[i_{Lr}(t_6) + \frac{V_{cl}}{L_r}(t - \lambda_7 T_s) \right] dt + \int_{\lambda_7 T_s}^{T_s} (s_a i_a + s_b i_b + s_c i_c)\ dt \tag{3.55}$$

If i_a is approximated as constant in a switch period, then

$$\int_{\lambda_7 T_s}^{T_s} (s_a i_a)\ dt \approx i_a \int_{\lambda_7 T_s}^{T_s} s_a\ dt \tag{3.56}$$

According to EA-PWM shown in Figure 3.27, $s_a = \begin{cases} 1 & (0, d_a T_s) \\ 0 & (d_a T_s, T_s) \end{cases}$, Eq. (3.56) is represented by:

$$\int_{\lambda_7 T_s}^{T_s} (s_a i_a)\ dt \approx i_a \int_{\lambda_7 T_s}^{d_a T_s} 1\ dt = (d_a - \lambda_7) T_s i_a \tag{3.57}$$

Similarly, following integrals are derived as:

$$\int_{\lambda_7 T_s}^{T_s} (s_b i_b)\ dt \approx d_b T_s i_b \tag{3.58}$$

$$\int_{\lambda_7 T_s}^{T_s} (s_c i_c) dt \approx d_c T_s i_c \tag{3.59}$$

Aforementioned three integrals depend on three-phase duty cycles d_a, d_b, and d_c. It is noted that the integral of the phase-a current vs. its switch state product is a little different from other two phases due to the duty cycle loss mentioned before.

By substituting Eqs. (3.57), (3.58), and (3.59) into (3.55), following equation is obtained:

$$\int_{\lambda_7 T_s}^{T_s} i_{cl}\ dt = i_{Lr}(t_6)(1 - \lambda_7) T_s + \frac{V_{cl}}{2L_r}(T_s - \lambda_7 T_s)^2 + [(d_a - \lambda_7) i_a + d_b i_b + d_c i_c] T_s \tag{3.60}$$

By applying duty cycle equations (3.45), (3.46), (3.47) and clamping voltage equation (3.34), above equation changes into

$$\int_{\lambda_7 T_s}^{T_s} i_{cl}\ dt = (1 - \lambda_7) T_s \left\{ i_{Lr}(t_6) + \frac{V_{dc}}{2L_r} \lambda_7 T_s + \left[\frac{(1 + v_{ma})}{2} i_a + \frac{(1 + v_{ma})}{2} i_b + \frac{(1 + v_{ma})}{2} i_c \right] \right\} \tag{3.61}$$

By combining Eq. (3.61) with Eq. (3.50), the turn-off duty cycle λ_7 for the auxiliary switch S_7 is obtained:

$$\lambda_7 = \frac{-\left(\frac{1 + v_{ma}}{2} i_a + \frac{1 + v_{mb}}{2} i_b + \frac{1 + v_{mc}}{2} i_c \right) - i_{Lr}(t_6)}{\frac{V_{dc}}{2L_r} T_s} \tag{3.62}$$

By substituting Eq. (3.31), above equation becomes:

$$\lambda_7 = \frac{-\left(\frac{1 + v_{ma}}{2} i_a + \frac{1 + v_{mb}}{2} i_b + \frac{1 + v_{mc}}{2} i_c \right) - i_{cs2} + \sqrt{\frac{V_{dc}^2 - V_{cl}^2}{Z_r^2} + i_x^2}}{\frac{V_{dc}}{2L_r} T_s} \tag{3.63}$$

By using Eq. (3.19), the turn-off duty cycle is rewritten as:

$$\lambda_7 = \frac{2L_r \left(-\frac{i_a v_{ma}}{2} - \frac{i_b v_{mb}}{2} - \frac{i_c v_{mc}}{2} + \frac{i_a}{2} - \frac{i_b}{2} - \frac{i_c}{2} + \sqrt{\frac{V_{dc}^2 - V_{cl}^2}{Z_r^2} + i_x^2} \right)}{V_{dc} T_s} \tag{3.64}$$

Since the inductor current linearly increases during $[\lambda_7 T_S, T_S]$, the inductor current at the end of the switching period is described as:

$$i_{Lr}(t_{13}) = i_{Lr}(t_1) = i_{Lr}(t_6) + \frac{V_{cl}}{L_r}(1-\lambda_7)T_s \tag{3.65}$$

Above equation is rewritten by substituting Eqs. (3.31) and (3.34) as follows:

$$i_{Lr}(t_1) = \frac{V_{dc}}{L_r}\lambda_7 T_s + i_{cs2} - \sqrt{\frac{V_{dc}{}^2 - V_{cl}{}^2}{Z_r^2} + i_x^2} \tag{3.66}$$

By substituting Eq. (3.63) into the above equation, it is changed to

$$i_{Lr}(t_1) = -(1+v_{ma})i_a - (1+v_{mb})i_b - (1+v_{mc})i_c - i_{cs2} + \sqrt{\frac{V_{dc}{}^2 - V_{cl}{}^2}{Z_r^2} + i_x^2} \tag{3.67}$$

By substituting above equation to inequality (3.18) for the ZVS condition of the first resonant stage, following equality is obtained:

$$-(1+v_{ma})i_a - (1+v_{mb})i_b - (1+v_{mc})i_c - i_{cs2} + \sqrt{\frac{V_{dc}{}^2 - V_{cl}{}^2}{Z_r^2} + i_x^2} > i_{cs1} + \sqrt{V_{dc}{}^2 - V_{cl}{}^2}/Z_r \tag{3.68}$$

By moving i_{cs1} to the left of the equality

$$-(1+v_{ma})i_a - (1+v_{mb})i_b - (1+v_{mc})i_c - (i_{cs1} + i_{cs2}) + \sqrt{\frac{V_{dc}{}^2 - V_{cl}{}^2}{Z_r^2} + i_x^2} > \sqrt{V_{dc}{}^2 - V_{cl}{}^2}/Z_r \tag{3.69}$$

According to Eqs. (3.10) and (3.20), $i_{cs1} = i_a x_a + i_b x_b + i_c x_c$ and $i_{cs2} = i_a y_a + i_b y_b + i_c y_c$
For CPWM, $y_i = -1 - x_i$. Therefore

$$i_{cs1} + i_{cs2} = i_a(x_a + y_a) + i_b(x_b + y_b) + i_c(x_c + y_c) = -i_a - i_b - i_c \tag{3.70}$$

By applying above equation to Eq. (3.69), the ZVS condition of the first resonant stage is described by

$$-v_{ma}i_a - v_{mb}i_b - v_{mc}i_c + \sqrt{\frac{V_{dc}{}^2 - V_{cl}{}^2}{Z_r^2} + i_x^2} > \sqrt{V_{dc}{}^2 - V_{cl}{}^2}/Z_r \tag{3.71}$$

The above equation is rewritten as

$$\sqrt{\frac{V_{dc}{}^2 - V_{cl}{}^2}{Z_r^2} + i_x^2} + 2i_M \geq \frac{\sqrt{V_{dc}{}^2 - V_{cl}{}^2}}{Z_r} \tag{3.72}$$

where i_M is defined as:

$$i_M = -\frac{1}{2}(i_a v_{ma} + i_b v_{mb} + i_c v_{mc}) = -\frac{1}{V_{dc}}\left(i_a v_{ma}\frac{V_{dc}}{2} + i_b v_{mb}\frac{V_{dc}}{2} + i_c v_{mc}\frac{V_{dc}}{2}\right) \tag{3.73}$$

Following variable is defined as:

$$p = i_a v_{ma} \frac{V_{dc}}{2} + i_b v_{mb} \frac{V_{dc}}{2} + i_c v_{mc} \frac{V_{dc}}{2} \tag{3.74}$$

Then Eq. (3.73) is briefly expressed as

$$i_M = -\frac{p}{V_{dc}} \tag{3.75}$$

If SPWM is used, Eq. (3.74) has physical meaning and represents instant power of the converter to the load. Besides, i_M reflects DC side current of the converter. Since it does not have complete switching information, it is not equal to i_{bus}. But it is equal to average value of i_{bus}, i.e. $\langle i_{bus} \rangle_{Ts} = i_M$.

According to inequality (3.72), if $i_M > 0$, ZVS condition is satisfied and no extra current i_x is needed. Otherwise, an extra stage to boost the resonant inductor current is needed to realize ZVS of the first resonant stage. If $i_M < 0$, minimum of the extra current i_x needed can be derived from the equality (3.72). Generally, i_x is expressed as:

$$i_x = \begin{cases} 0 & i_M \geq 0 \\ \sqrt{\left(\frac{\sqrt{V_{dc}^2 - V_{cl}^2}}{Z_r} - 2i_M\right)^2 - \frac{V_{dc}^2 - V_{cl}^2}{Z_r^2}} & i_M < 0 \end{cases} \tag{3.76}$$

Then the duty cycle of an extra stage to boost the resonant inductor current is estimated by

$$D_{add} = \frac{i_x L_r}{V_{dc}} \tag{3.77}$$

The ZVS condition is effective for CPWM scheme. However, when DPWM is used, the current source i_{cs1} and i_{cs2} in Figures 3.25 and 3.26 will be different due to the clamping phase in DPWM scheme.

For DPWM, if phase-a is at clamping duration and $u_{ma} = 1$, then $x_a = y_a = -1$ according to (3.10) and (3.20). For other two phases, $x_b + y_b = -1$ and $x_c + y_c = -1$. Then it is derived:

$$i_{cs1} + i_{cs2} = i_a(x_a + y_a) + i_b(x_b + y_b) + i_c(x_c + y_c) = -2i_a - i_b - i_c \tag{3.78}$$

By substituting $u_{ma} = 1$ and above equation into (3.69)

$$-(1+1)i_a - (1 + v_{mb})i_b - (1 + v_{mc})i_c - (-2i_a - i_b - i_c) + \sqrt{\frac{V_{dc}^2 - V_{cl}^2}{Z_r^2} + i_x^2} > \sqrt{V_{dc}^2 - V_{cl}^2}/Z_r \tag{3.79}$$

Above inequality can be simplified as

$$-v_{mb} i_b - v_{mc} i_c + \sqrt{\frac{V_{dc}^2 - V_{cl}^2}{Z_r^2} + i_x^2} > \sqrt{V_{dc}^2 - V_{cl}^2}/Z_r \tag{3.80}$$

For DPWM, if phase-a is at clamping duration and $u_{ma} = -1$, then $x_a = y_a = 0$ according to (3.10) and (3.20). For other two phases, $x_b + y_b = -1$ and $x_c + y_c = -1$. Then it is derived:

$$i_{cs1} + i_{cs2} = i_a(x_a + y_a) + i_b(x_b + y_b) + i_c(x_c + y_c) = -i_b - i_c \tag{3.81}$$

By substituting $v_{ma} = -1$ and above equation into (3.69)

$$-(1-1)i_a - (1+v_{mb})i_b - (1+v_{mc})i_c - (-i_b - i_c) + \sqrt{\frac{{V_{dc}}^2 - {V_{cl}}^2}{Z_r^2} + i_x^2} > \sqrt{{V_{dc}}^2 - {V_{cl}}^2}/Z_r \tag{3.82}$$

Above inequality can be simplified as

$$-v_{mb}i_b - v_{mc}i_c + \sqrt{\frac{{V_{dc}}^2 - {V_{cl}}^2}{Z_r^2} + i_x^2} > \sqrt{{V_{dc}}^2 - {V_{cl}}^2}/Z_r \tag{3.83}$$

Therefore, for DPWM, if phase-a is clamped to either higher or lower voltage rails, i.e.$|v_{ma}| = 1$, then ZVS condition is changed into

$$-v_{mb}i_b - v_{mc}i_c + \sqrt{\frac{{V_{dc}}^2 - {V_{cl}}^2}{Z_r^2} + i_x^2} > \sqrt{{V_{dc}}^2 - {V_{cl}}^2}/Z_r \tag{3.84}$$

With regard to various clamping phase in DPWM, ZVS condition for DPWM is generalized as:

$$-k_a v_{ma} i_a - k_b v_{mb} i_b - k_c v_{mc} i_c + \sqrt{\frac{{V_{dc}}^2 - {V_{cl}}^2}{Z_r^2} + i_x^2} > \sqrt{{V_{dc}}^2 - {V_{cl}}^2}/Z_r \tag{3.85}$$

where $k_i = \begin{cases} 0 & \text{if } |v_{mi}| = 1 \\ 1 & \text{otherwise} \end{cases} \quad i = a, b, c$

According to Eqs. (3.71) and (3.85), ZVS condition for both CPWM and DPWM is generally expressed:

$$\sqrt{\frac{{V_{dc}}^2 - {V_{cl}}^2}{Z_r^2} + i_x^2} + 2i_M \geq \frac{\sqrt{{V_{dc}}^2 - {V_{cl}}^2}}{Z_r} \tag{3.86}$$

where i_M is generally expressed as

$$i_M = -\frac{1}{2}(k_a i_a v_{ma} + k_b i_b v_{mb} + k_c i_c v_{mc}) = -\frac{1}{V_{dc}}\left(k_a i_a v_{ma}\frac{V_{dc}}{2} + k_b i_b v_{mb}\frac{V_{dc}}{2} + k_c i_c v_{mc}\frac{V_{dc}}{2}\right)$$
$$= -\frac{\sum_{i=a,b,c} k_i i_i v_{mi}\frac{V_{dc}}{2}}{V_{dc}} \tag{3.87}$$

Equation (3.76) to estimate i_x is suitable for both CPWM and DPWM.

It should be noted that for three-phase four-wire system, ZVS condition described by Eqs. (3.86) and (3.76) for i_x is also applicable.

3.4.3 Impact of PWM Scheme and Load on ZVS Condition

As mentioned before, the polarity of i_M is critical to ZVS condition. If i_M is positive, ZVS condition is satisfied and no extra current i_x is needed. Otherwise, an extra stage is needed for boosting the resonant inductor current. According to the expression of i_M in (3.87), i_M depends on load currents, modulation signals, and PWM schemes. In this section we will discuss how these factors will affect the polarity of i_M [18].

The modulation signals are defined as follows:

$$\begin{cases} v_{ma}(t) = m_a \sin(\omega t) + v_z \\ v_{mb}(t) = m_a \sin(\omega t - 2\pi/3) + v_z \\ v_{mc}(t) = m_a \sin(\omega t + 2\pi/3) + v_z \end{cases} \tag{3.88}$$

where v_z is zero-sequence component used to increase voltage utilization in three-phase three-wire systems and is set to zero for three-phase four-wire systems.

For balanced loads, the three-phase loads are defined by

$$\begin{cases} i_a(t) = I_m \sin(\omega t + \theta) \\ i_b(t) = I_m \sin(\omega t + \theta - 2\pi/3) \\ i_c(t) = I_m \sin(\omega t + \theta + 2\pi/3) \end{cases} \tag{3.89}$$

where I_m is the amplitude of the sinusoidal load current and θ is phase angle of phase-a current with respect to the phase-a voltage.

Next, we will discuss ZVS conditions for CPWM or DPWM scheme separately. First, ZVS condition for CPWM is investigated. By substituting Eq. (3.88) into Eq. (3.87) and applying $k_a = k_b = k_c = 1$ for CPWM, following equation is obtained:

$$i_M = -\frac{1}{2}\{i_a[m_a \sin(\omega t)] + i_b[m_a \sin(\omega t - 2\pi/3)] + i_c[m_a \sin(\omega t + 2\pi/3)] + (i_a + i_b + i_c)v_z\} \tag{3.90}$$

By inserting the current expressions of Eq. (3.89) into the above equation, it is simplified as:

$$i_M = -\frac{3}{4} I_m m_a \cos\theta \tag{3.91}$$

We can see that in CPWM scheme, the polarity of i_M depends on the phase angle θ. When the converter works as the rectifier and its θ ranges inside $\theta \in [\pi/2, 3\pi/2]$, i_M is positive and ZVS condition satisfies automatically. However, when the converter works as the inverter and θ ranges inside $[-\pi/2, \pi/2]$, i_M is negative. Thus ZVS condition does not satisfy automatically and current increment i_x is needed for the resonant inductor by introducing current boost stage.

Now let us discuss ZVS condition when DPWM is used. There are different ways to realize DPWM. As an example, following zero-sequence component is used:

$$v_z = [2k(t) - 1] + [k(t) - 1]v_{min} - k(t)v_{max} \tag{3.92}$$

$k(t)$, $v_{\min}$, and $v_{\max}$ are defined as follows:

$$k(t) = \frac{1}{2}(\text{sign}\{\sin[3(\omega t - \pi)]\} + 1) \tag{3.93}$$

$$v_{min} = \min\left(v^{\dagger}_{ma}, v^{\dagger}_{mb}, v^{\dagger}_{mc}\right) \tag{3.94}$$

$$v_{max} = \max\left(v^{\dagger}_{ma}, v^{\dagger}_{mb}, v^{\dagger}_{mc}\right) \tag{3.95}$$

where $v^{\dagger}_{ma} = = m_a \sin(\omega t)$, $v^{\dagger}_{mb} = m_a \sin(\omega t - 2\pi/3)$, and $v^{\dagger}_{mc} = m_a \sin(\omega t + 2\pi/3)$.

Phase-a modulation signals are shown in Figure 3.28. Figure 3.28a shows phase-a modulation signal when amplitude modulation ratio $m_a = 0.9$ and Figure 3.28b shows phase-a modulation signal when $m_a = \frac{2}{\sqrt{3}} = 1.15$. It is observed that amplitude modulation ratio is increased up to 1.15 by introducing zero-sequence component to the sinusoidal modulation signal.

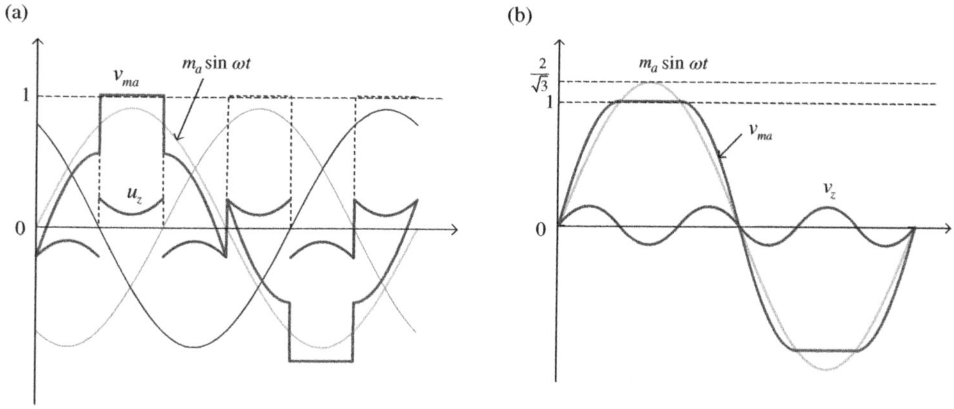

Figure 3.28 Phase-a modulation signal at DPWM: (a) $m_a = 0.9$; (b) $m_a = \frac{2}{\sqrt{3}} = 1.15$.

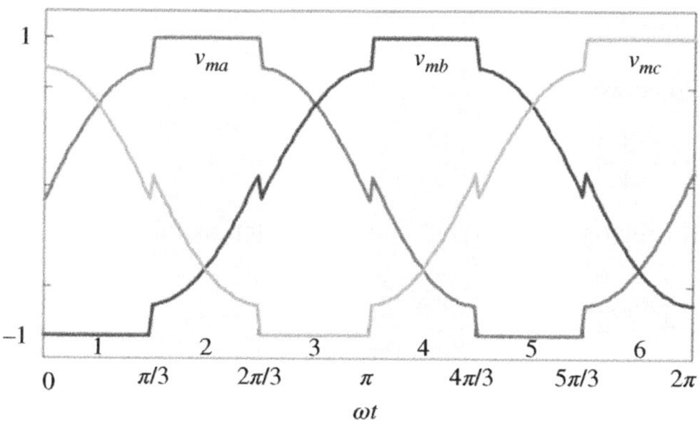

Figure 3.29 Typical three-phase modulation signals of DPWM.

Figure 3.29 shows typical three-phase modulation signals of DPWM in a fundamental period. A fundamental period can be classified into six regions. In each region there is a phase that is clamped to either positive or negative rail of the DC bus.

By substituting Eq. (3.88) into Eq. (3.87),

$$i_M = -\frac{1}{2}\{k_a i_a[m_a \sin(\omega t)] + k_b i_b[m_a \sin(\omega t - 2\pi/3)] + k_c i_c[m_a \sin(\omega t + 2\pi/3)] + (k_a i_a + k_b i_b + k_c i_c)v_z\} \tag{3.96}$$

For clamping region $\left[0, \frac{\pi}{3}\right]$, leg-$b$ output is clamped to the negative rail. Thus $k_b = 0$ and $k_a = k_c = 1$. Equation (3.96) is simplified as

$$i_M = -\frac{1}{2}\{i_a[m_a \sin(\omega t)] + i_c[m_a \sin(\omega t + 2\pi/3)] + (i_a + i_c)v_z\} \tag{3.97}$$

From Eq. (3.93)

$$k(t) = \frac{1}{2}(\text{sign}\{\sin[3(\omega t - \pi)]\} + 1) = 0; \quad \omega t \in \left[0, \frac{\pi}{3}\right] \tag{3.98}$$

From Eq. (3.92)

$$v_z = [2k(t) - 1] + [k(t) - 1]v_{min} - k(t)v_{max} = -1 - v_{min} \tag{3.99}$$

According to Eq. (3.94), for $\omega t \in \left[0, \frac{\pi}{3}\right]$, it is derived

$$v_{min} = \min\left(v_{ma}^{\dagger}, v_{mb}^{\dagger}, v_{mc}^{\dagger}\right) = m_a \sin(\omega t - 2\pi/3) \tag{3.100}$$

By substituting Eq. (3.100) into Eq. (3.99)

$$v_z = -1 - m_a \sin(\omega t - 2\pi/3) \tag{3.101}$$

By substituting Eq. (3.101) into Eq. (3.97)

$$i_M = -\frac{1}{2}\{i_a m_a[\sin(\omega t) - \sin(\omega t - 2\pi/3)] + i_c m_a[\sin(\omega t + 2\pi/3) - \sin(\omega t - 2\pi/3)] - (i_a + i_c)\} \tag{3.102}$$

Then by substituting i_a and i_c expressions in Eq. (3.89), above equation can be simplified as

$$i_M = -\frac{1}{2}I_m \sin(\omega t + \theta - 2\pi/3) - \frac{3}{4}m_a I_m \cos\theta \tag{3.103}$$

The normalized expression of i_M^* is defined as:

$$i_M^* = \frac{i_M}{I_m} = -\frac{1}{2}\sin(\omega t + \theta - 2\pi/3) - \frac{3}{4}m_a \cos\theta \tag{3.104}$$

Similarly, expressions of i_M^* in other five regions can derived and are written as follows:

$$i_M^* = \begin{cases} -\frac{1}{2}\sin(\omega t + \theta - 2\pi/3) - \frac{3}{4}m_a\cos\theta & 0 \le \omega t < \frac{\pi}{3} \\ \frac{1}{2}\sin(\omega t + \theta) - \frac{3}{4}m_a\cos\theta & \frac{\pi}{3} \le \omega t < \frac{2\pi}{3} \\ -\frac{1}{2}\sin(\omega t + \theta + 2\pi/3) - \frac{3}{4}m_a\cos\theta & \frac{2\pi}{3} \le \omega t < \pi \\ \frac{1}{2}\sin(\omega t + \theta - 2\pi/3) - \frac{3}{4}m_a\cos\theta & \pi \le \omega t < \frac{4\pi}{3} \\ -\frac{1}{2}\sin(\omega t + \theta) - \frac{3}{4}m_a\cos\theta & \frac{4\pi}{3} \le \omega t < \frac{5\pi}{3} \\ \frac{1}{2}\sin(\omega t + \theta + 2\pi/3) - \frac{3}{4}m_a\cos\theta & \frac{5\pi}{3} \le \omega t < 2\pi \end{cases} \tag{3.105}$$

It is observed that except ωt, the polarity of i_M^* also depends on initial phase angle θ and the amplitude modulation ratio m_a. In DPWM scheme, when the converter works in inverter mode ($\theta \in [-\pi/2, \pi/2]$), $\cos\theta$ is positive. A larger m_a results in a smaller i_M^*, which means more likely that ZVS condition is not satisfied and an extra current i_x will be needed for the resonant inductor. Figure 3.30 shows i_M^* at θ=0. It is seen that i_M^* is positive as long as m_a is less than $\sqrt{3}/3$.

Figure 3.31 shows i_M^* at $\theta = -\pi/3$ and $\theta = \pi/3$. It is seen that i_M^* alternates between positive and negative values in the output fundamental period. In the region $i_M^* < 0$, an extra current i_x is required to realize the ZVS condition.

When the converter works in the rectifier mode ($\theta \in [\pi/2, 3\pi/2]$), $\cos\theta$ is negative. The larger m_a is, the larger i_M^* will be. Thus it is easier to realize ZVS condition. Figure 3.32 shows i_M^* at $\theta = \pi$. It is observed that if m_a is larger than 2/3, i_M^* is positive in all the fundamental period, and no extra current i_x is needed.

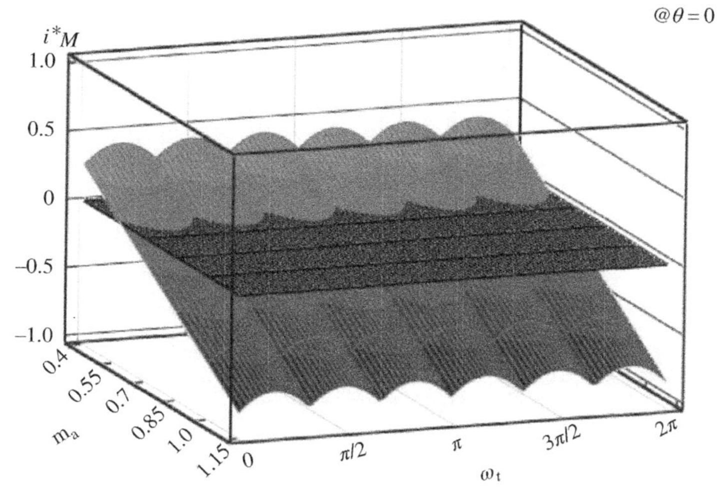

Figure 3.30 i_M^* at θ=0.

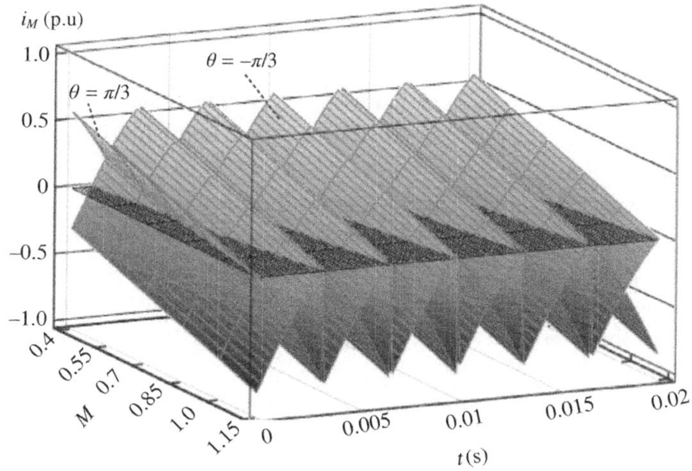

Figure 3.31 i_M^* at $\theta = -\pi/3$ and $\theta = \pi/3$.

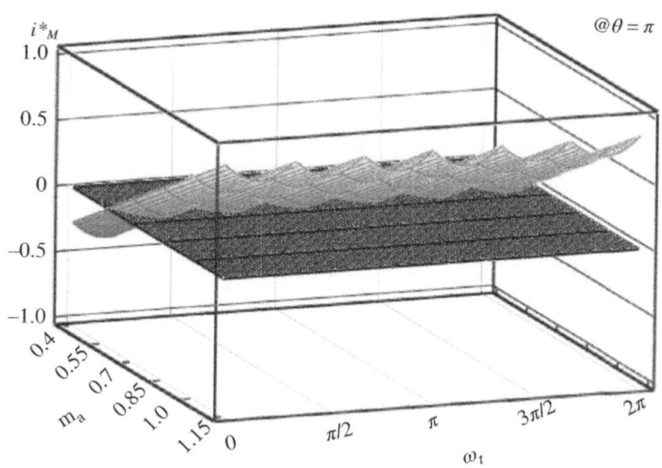

Figure 3.32 i_M^* at $\theta = \pi$.

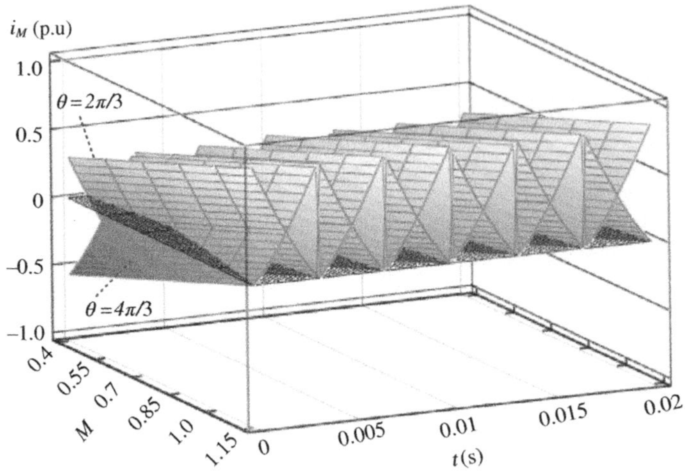

Figure 3.33 i_M^* at $\theta = 2\pi/3$ and $\theta = 4\pi/3$.

Figure 3.33 shows i_M^* at $\theta = 2\pi/3$ and $\theta = 4\pi/3$.It can be seen that when m_a is equal to its maximum value 1.15, i_M^* is always positive. As m_a decreases, the polarity of i_M^* will alternate in the fundamental period. Extra current i_x is needed in the region $i_M^* < 0$.

The ZVS conduction of the converter with CAC auxiliary circuit is summarized in Figure 3.34. If CPWM is adopted, ZVS condition is automatically satisfied when it operates at the rectifier mode, i.e. $\theta\in[\pi/2, 3\pi/2]$ as shown with blank area in Figure 3.34a. However, ZVS condition is not satisfied at the inverter mode, i.e. $\theta\in[-\pi/2, \pi/2]$ as shown with shaded area in Figure 3.34a. It means an extra current i_x is required for adding to the resonant inductor to realize ZVS turn-on at the inverter mode.

If DPWM is adopted, ZVS range becomes narrow as shown in Figure 3.34b. When the converter operates at the rectifier mode, ZVS ranges become narrow in comparison to CPWM. ZVS condition

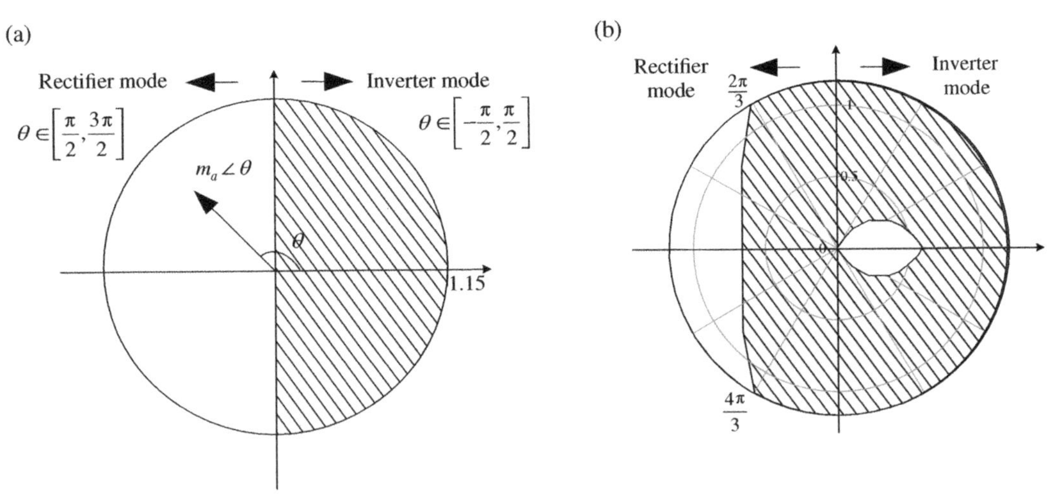

Figure 3.34 ZVS range of CAC converter: (a) with CPWM; (b) with DPWM.

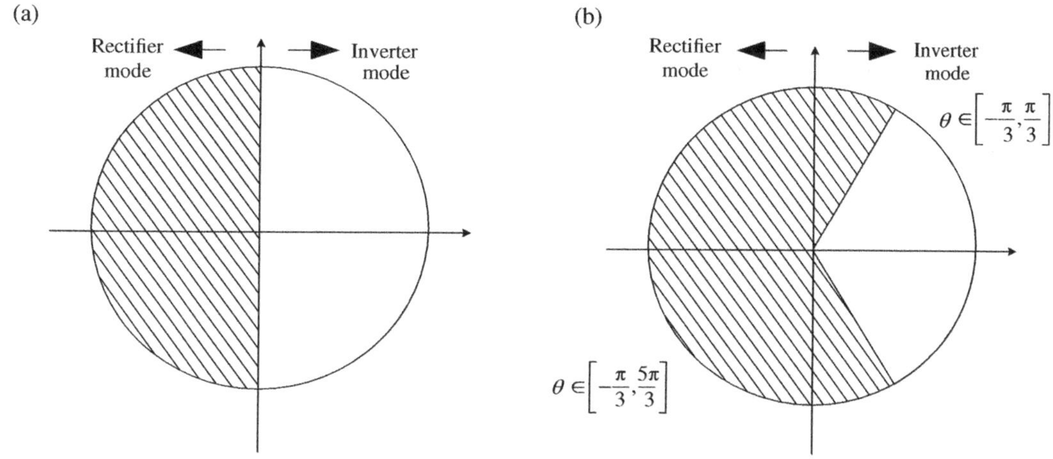

Figure 3.35 ZVS range of MVAC converter: (a) with CPWM; (b) with DPWM.

is satisfied only in certain region where m_a is higher and phase angle ranges inside $\theta \in [2\pi/3, 4\pi/3]$, shown with blank area. When it operates at the inverter mode, ZVS condition is satisfied only in small region as shown in Figure 3.34b, shown with leaf-like blank area. To realize ZVS condition for these shaded areas in the figure, an extra current i_x is required for adding to the resonant inductor.

Similarly, the ZVS range for the active converter with MVAC auxiliary circuit can be derived and shown in Figure 3.35. If CPWM is adopted, ZVS condition is automatically satisfied in the inverter mode as shown in Figure 3.35a, shown as blank area. However, ZVS condition is not satisfied at the rectifier mode. If DPWM is adopted, ZVS range shrinks as shown in Figure 3.35b. When it operates at the rectifier mode as shown as shaded area, ZVS is always not satisfied. It means an extra current i_x is needed to add to the resonant inductor to realize ZVS condition. If it operates at inverter mode, ZVS condition is satisfied when phase angle ranges inside $\theta \in [-\pi/3, \pi/3]$, shown as blank area.

3.5 Control Diagram of the Converter with EA-PWM

Figure 3.36 shows control diagram of the converter with EA-PWM. The *id-iq* current control is used, which outputs *dq* voltage references v_d^{ref} and v_q^{ref}. These *dq* voltage references enter into *dq*/*abc* transformation unit. Then the three-phase modulation signals v_{ma}, v_{mb}, and v_{mc} are obtained. Here, there are different ways to generate these modulation signals according to applications. For three-phase four-wire systems, sinusoidal modulation signals are used. For three-phase three-wire systems, zero-sequence-component-injected sinusoidal modulation signals are used for either CPWM or DPWM. Besides, two key parameters need to be estimated online. The duty cycle D_{add} for i_x is estimated according to Eq. (3.77). The off duty cycle λ_7 of the auxiliary switch S_7 is estimated by Eq. (3.63).

Once we obtain D_{add} and λ_7, EA-PWM can be generated as shown in Figure 3.36. The EA-PWM generator outputs seven gate driving signals v_{g1}–v_{g7} for the switch S_1–S_7. Three-phase modulation signals v_{ma}, v_{mb}, and v_{mc} enter into three EA-PWM modulators with the same structure. The

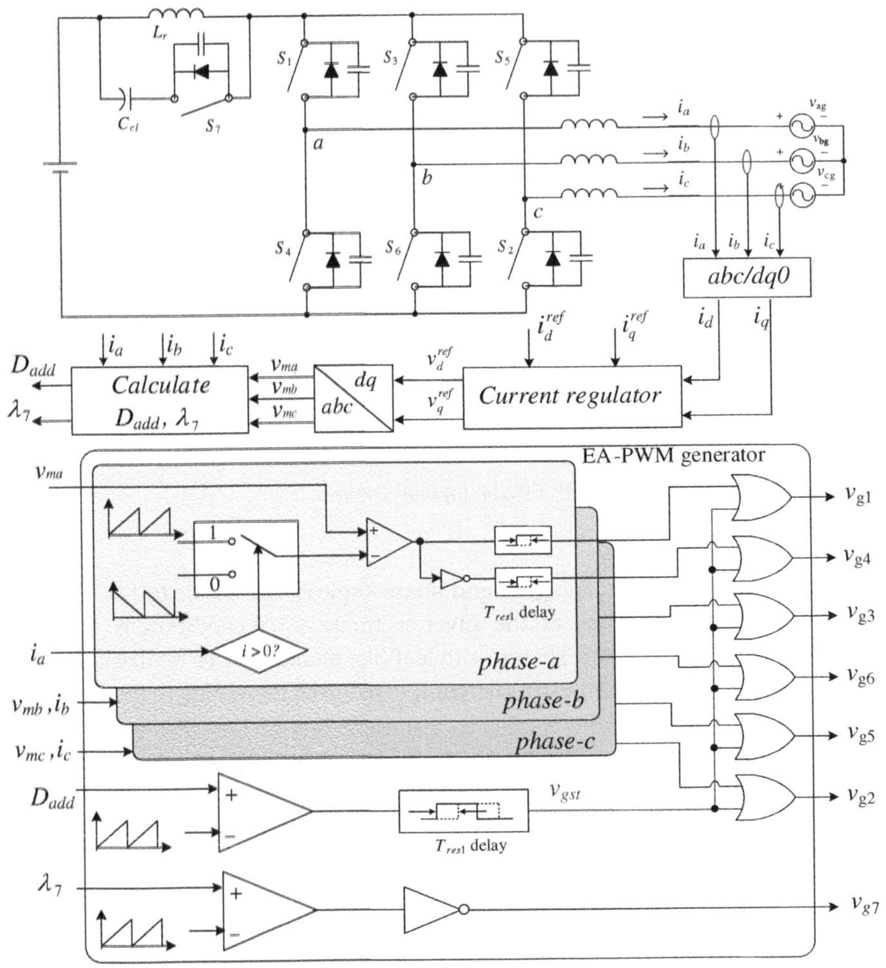

Figure 3.36 Control diagram of three-phase ZVS CAC converter.

polarity information of the sampled current signals i_a, i_b, and i_c are also entering into these modulators for selecting either ramp-up carrier or ramp-down carrier to align all Type 2 commutations in a switching period. Generally, the extra current i_x is needed to realize ZVS condition. Thus a short-circuit pulse generation circuit is added. The short-circuit pulse v_{gst} is generated by comparing D_{add} with the ramp-up saw-tooth carrier and delayed by one-fourth resonant period described as

$$T_{res1} = \frac{1}{4}\frac{2\pi}{\omega_r} = \frac{\pi}{2}\sqrt{L_r(3C_r + C_{r7})} \tag{3.106}$$

v_{gst} is added to gate signals of the six main switches through logical OR operation. Finally, the gate drive signal for the auxiliary circuit is obtained by comparing λ_7 with the ramp-up saw-tooth carrier. All gate drive signals and the short-circuit pulse v_{gst} are shown in Figure 3.37. For other converter topologies, similar EA-PWM scheme implementation approach can be applied.

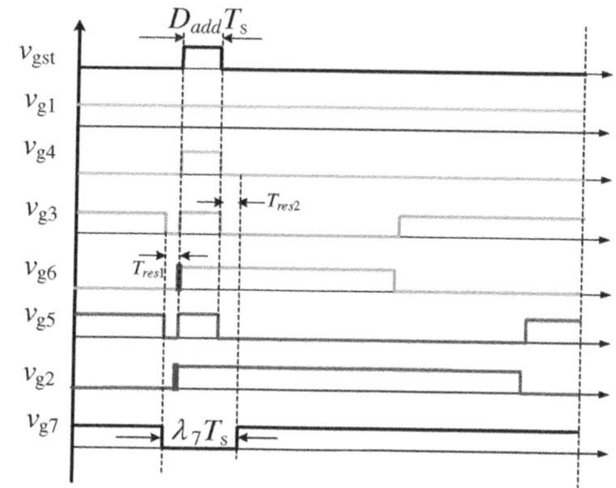

Figure 3.37 Typical waveforms of gate drive signals.

3.6 ZVS-SVM

In this section, we will study how to realize soft-switching of the three-phase converter with SVM [3, 4, 13–16]. The three-phase converter with CAC auxiliary circuit as an example is redrawn in Figure 3.38.

Three-phase grid voltages and three-phase grid currents of a three-phase converter with unit power factor are shown in Figure 3.39. Since the voltage drop on grid filter inductors is quite small, the fundamental components of the output voltages of the three-phase bridges are approximately represented by the three-phase grid voltages. Based on the conventional SVM scheme in Appendix A.1, the line cycle can be divided into six sectors, from sector 1 to sector 6 with 60° phase angle range in each sector [1, 2]. By considering the polarities of filter currents i_a, i_b, and i_c in each phase, each sector is further divided into two subsectors to ensure that all the filter current polarities do not change in the subsectors as shown in Figure 3.39. For example, in sector 1-1, phase-a current i_a is positive and both phase-b current i_b and phase-c current i_c are negative. In sector 1-2, phase-a current i_a and phase-b current i_b are positive while phase-c current i_c is negative. Finally, we get 12 sectors from sector 1-1, sector 1-2, sector 2-1, sector 2-2, ... to sector 6-2 with their three-phase filter current polarities fixed within each sector. It is observed that among three-phase currents, the phase current with the biggest absolute value has the opposite polarity to the other two phase currents in each sector.

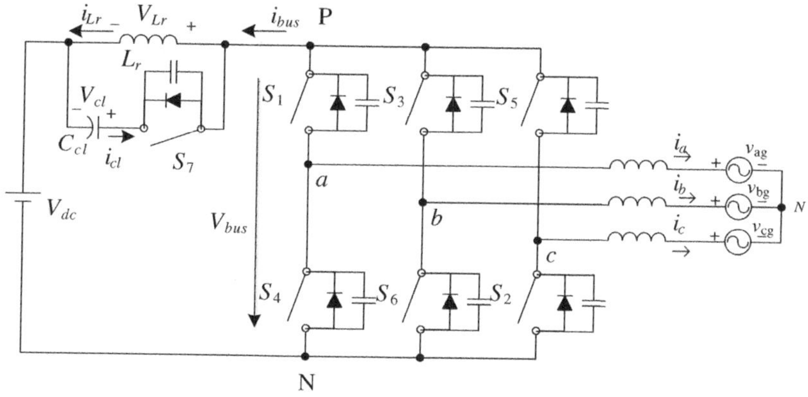

Figure 3.38 Three-phase converter with CAC auxiliary circuit.

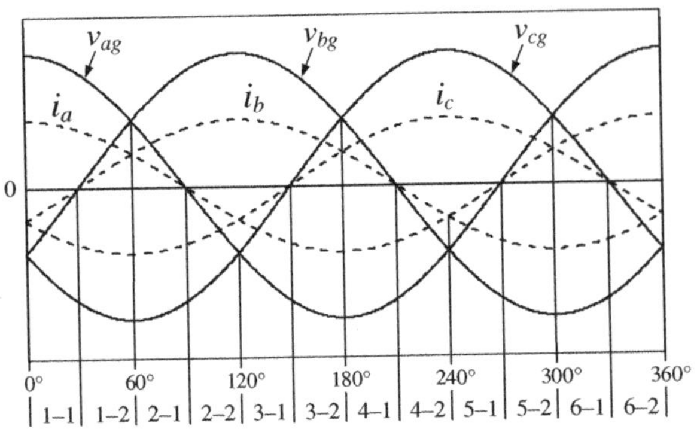

Figure 3.39 Grid voltages and grid currents waveforms of unit power factor hard-switching inverter.

SVM hexagon of the three-phase converter is shown in Figure 3.40. Originally, 60° sectors are further divided into 30° sectors. Twelve sectors in SVM hexagon are corresponding to sectors in the waveforms in Figure 3.39. According to the reference vector composing concept of SVM, two nonzero vectors nearest to the reference vector are used to compose the reference vector. For example, when reference vector $\overrightarrow{V_{ref}}$ locates in sector 1-1 or sector 1-2, nonzero vectors $\overrightarrow{V_1}$ and $\overrightarrow{V_2}$ are nearest to $\overrightarrow{V_{ref}}$ among six nonzero vectors. Thus they are used to compose the reference vector. Besides, the zero vectors are needed to fill remaining time of a switch period so that total executing time of the vectors is equal to the switching period.

By using the nonzero vectors, $\vec{V}_1$ and $\vec{V}_2$, the reference vector $\overrightarrow{V_{ref}}$ in sector 1-1 can be composed as shown in Figure 3.41. The magnitude of $\vec{V}_1$ and $\vec{V}_2$ are the same and equal to $2V_{dc}/3$. φ is the

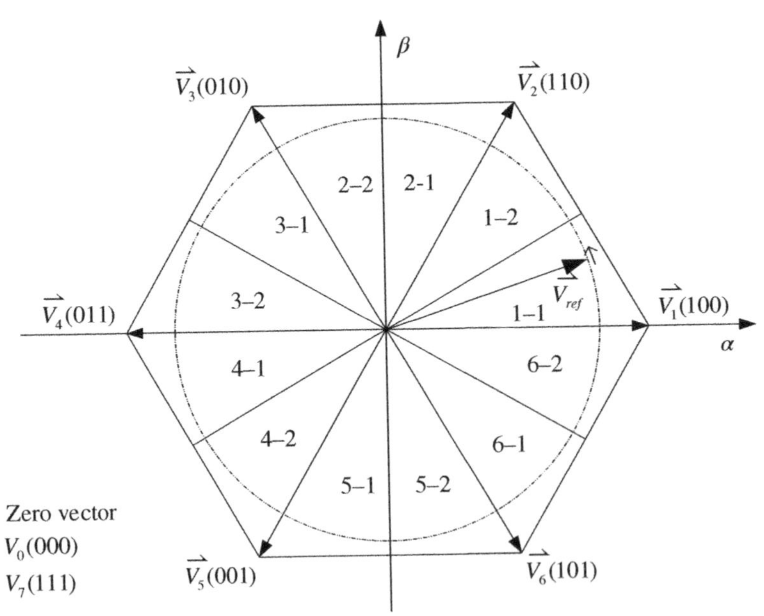

Figure 3.40 SVM hexagon with 12 sectors.

phase angle of the reference vector $\overrightarrow{V_{ref}}$.

Based on Figure 3.41, following equations can be obtained as:

$$\vec{V}_{ref} \cdot T_s = \vec{V}_1 \cdot T_{V1} + \vec{V}_2 \cdot T_{V2} \tag{3.107}$$

where T_{V1} and T_{V2} are the execution duration of voltage vectors $\vec{V}_1$ and $\vec{V}_2$ in a switch cycle, and T_s is switch period. Since $T_{V1} + T_{V2} < T_s$, zero vector is used to fill the remaining time in the switching period. The zero vector executed duration is calculated as:

$$T_{Vz} = T_s - (T_{V1} + T_{V2}) \tag{3.108}$$

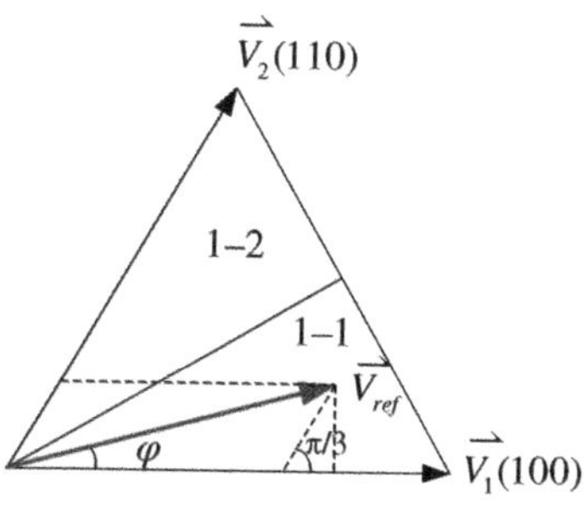

Figure 3.41 Composition of the reference vector in sector 1-1.

Since the amplitude of zero vectors is zero, adding the zero vector will not violate the principle of parallelogram vector synthesis shown in Figure 3.41. By combining Eqs. (3.107) and (3.108), the durations of all the vectors in a switching cycle can be derived as:

$$\begin{cases} T_{V1} = mT_s \sin(\pi/3 - \varphi) \\ T_{V2} = mT_s \sin\varphi \\ T_{Vz} = T_s - T_{V1} - T_{V2} \end{cases} \tag{3.109}$$

where $m = \frac{\sqrt{3}\left|\overrightarrow{V_{ref}}\right|}{V_{dc}}$, and m is called amplitude modulation ratio.

For zero vectors, there are two choices: $\overrightarrow{V_0}$ and $\overrightarrow{V_7}$. How do we select zero vector in a switching period? How do we arrange vector sequence in a switching period? We will discuss these issues in the next section.

3.6.1 Vector Sequence

With regard to reducing switching loss, a vector sequence in a switching period should be selected so that the converter has lower switching loss. As mentioned before, there are two kinds of commutations in the three-phase converters. The Type 1 commutation has lower loss due to ZVS turn-off of the switch. However, the Type 2 commutation has higher loss. Thus the Type 2 commutations in the converter are required to be reduced as less as possible. Actually, both vector sequence and selection of zero vector will affect types of commutations in the converter.

For simplicity, we still take the sector 1-1 as an example. Besides, it is assumed that the converter operates at the unit power factor. In this case $i_a > 0$, $i_b < 0$, and $i_c < 0$ in sector 1-1 as shown in Figure 3.39. There are four vectors available in this sector: two nonzero vectors $\vec{V}_1$ and $\vec{V}_2$ and two zero vectors $\overrightarrow{V_0}$ and $\overrightarrow{V_7}$. Their equivalent circuits of the converter are shown in Figure 3.42, where the AC side filter currents are replaced by current sources for simplicity. A transition from one vector to another vector reflects that the converter state changes. As we known, converter state changing accompanies switching commutations, which are either Type 1 or Type 2 commutations.

Let us investigate vector transitions. For the vector transition from $\overrightarrow{V_1}$ to $\overrightarrow{V_2}$, switch leg-a and switch leg-c of the converter are not changed as shown in Figure 3.42a and b. Thus there are no commutations in switch leg-a and switch leg-c during the vector transition. Before the vector transition, the circuit state corresponds to $\overrightarrow{V_1}$, shown in Figure 3.42a. Current source i_b flows into the

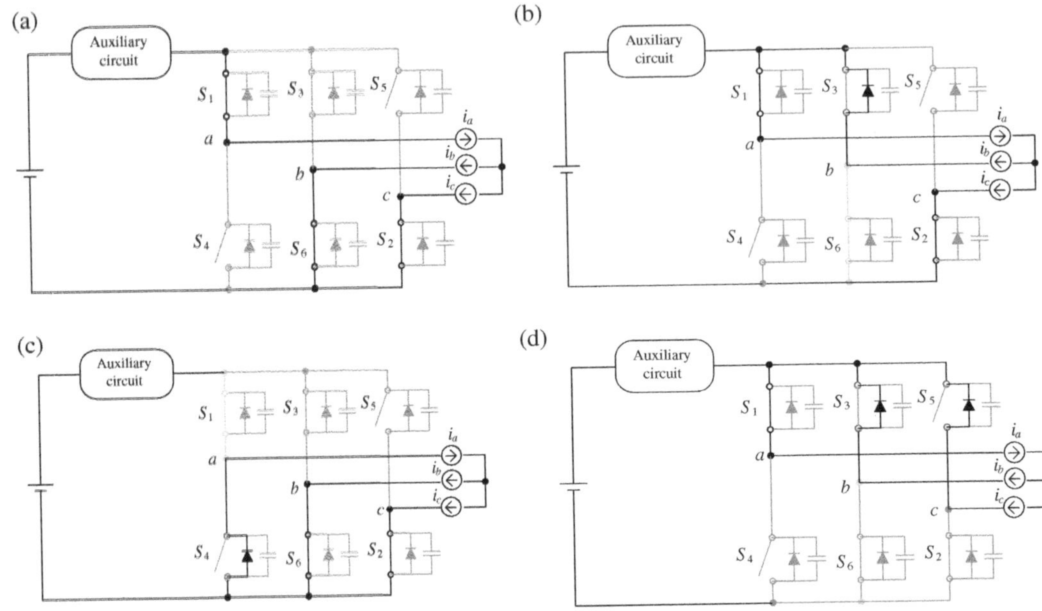

Figure 3.42 Vectors available in sector 1-1. (a) Nonzero vector $\vec{V}_1$:state 100. (b) Nonzero vector $\vec{V}_2$:state 110. (c) Zero vector $\overrightarrow{V_0}$: state 000. (d) Zero vector $\overrightarrow{V_7}$: state 111.

leg-b lower switch S_6. After this vector transition, the circuit state changes to Figure 3.42b, which corresponds to $\overrightarrow{V_2}$. Current source i_b flows into the antiparallel diode of the switch leg-b upper switch due to the polarity of i_b. It is observed that in switch leg-b, lower switch S_6 is turned off and phase-b current i_b is commutated to the antiparallel diode of switch S_3 at the same leg. This is a commutation from a switch to its complement diode in the same leg, known as Type 1 commutation. S_6 is turned off with ZVS condition, which has lower loss. Thus the vector transition from $\overrightarrow{V_1}$ to $\overrightarrow{V_2}$ has lower loss and is a welcomed vector transition.

Now we reverse above vector transition sequence. For the vector transition from $\overrightarrow{V_2}$ to $\overrightarrow{V_1}$, there are also no commutations in switch leg-a and switch leg-c. However, in switch leg-b, phase-b current i_b is commutated from the antiparallel diode of upper switch S_3 to lower switch S_6 at the same leg. This is a commutation from a diode to its complement switch in the same leg, known as Type 2 commutation. It results in higher loss. Therefore the vector transition from $\overrightarrow{V_2}$ to $\overrightarrow{V_1}$ causes higher loss and needs to be avoided if possible.

For the vector transition from $\overrightarrow{V_7}$ to $\overrightarrow{V_1}$, there are no commutations in switch leg-a as shown in Figure 3.42a and d. However, in switch leg-b, phase-b current i_b is commutated from the antiparallel diode of upper switch S_3 to lower switch S_6 at the same leg. This is a commutation from a diode to its complement switch in the same leg, known as Type 2 commutation. In switch leg-c, phase-c current i_c is commutated from the antiparallel diode of upper switch S_5 to lower switch S_2 at the same leg. This is also a Type 2 commutation. Type 2 commutations happen at both leg-b and leg-c, which results in higher switching loss. Measure should be taken to suppress these losses.

Similarly, we can analyze switch-loss performance of all possible vectors transitions among the four vectors in sector 1-1. Figure 3.43 shows all the Type 2 commutations between vectors in sector 1-1. When a vector sequence is selected for reference vector composition, vectors transitions shown in this figure should be avoided if possible.

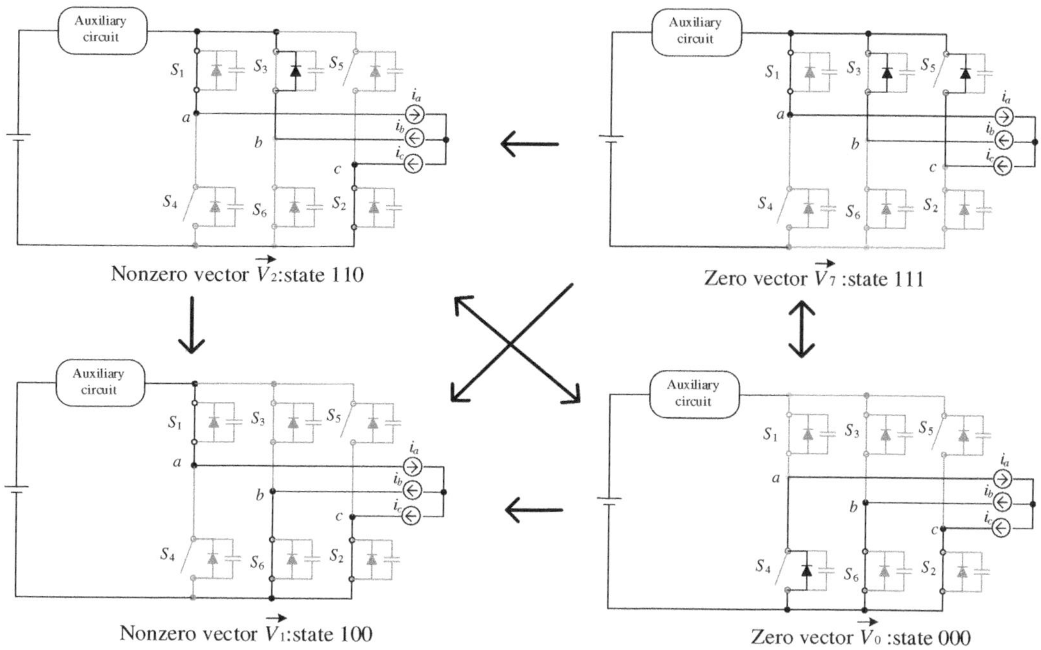

Figure 3.43 Type 2 commutations between vectors in sector 1-1.

To compose the reference vector in a switch cycle in sector 1-1, both $\overrightarrow{V_1}$ and $\overrightarrow{V_2}$ need to be used. Besides, at least one zero vector is used to fill the switch period. After careful study, it is found that it is impossible to find a vector sequence that only includes the Type 1 commutation. Finally, an optimal vector sequence: $\overrightarrow{V_1} - \overrightarrow{V_2} - \overrightarrow{V_7}$ is selected for the reference vector composition in sector 1-1 [3]. To understand commutation feature of the vector sequence, the roadmap of converter state changing is shown in Figure 3.44. The vector transition from $\overrightarrow{V_1}$ to $\overrightarrow{V_2}$ is Type 1 commutation, i.e. ZVS turn-off. The vector transition from $\overrightarrow{V_2}$ to $\overrightarrow{V_7}$ is also Type 1 commutation. However, the vector transition from $\overrightarrow{V_7}$ to $\overrightarrow{V_1}$ has Type 2 commutations, which happens in both switch leg-b and leg-c. Fortunately, these two Type 2 commutations occur at the same instant. They can be relieved by only one action of the auxiliary circuit. Both switch leg-b and switch leg-c can realize ZVS turn-on and the diode reverse recovery process is also suppressed. In the case the auxiliary circuit only needs to switch once per switching cycle. It not only simplifies the control but also reduces circulation loss in the circuit.

3.6.2 ZVS-SVM Scheme

For preferred vector sequence $\overrightarrow{V_1} - \overrightarrow{V_2} - \overrightarrow{V_7}$ in sector 1-1, according to Eq. (3.109), execution duration of vector $\vec{V}_1$, $\vec{V}_2$, and $\overrightarrow{V_7}$ are estimated asT_{V1}, T_{V2}, and T_{Vz}. Figure 3.45 shows waveforms of three-phase switch bridge state, auxiliary switch's state, and v_{bus} of the active clamped three-phase converter in a switch period. In the figure, although zero vector comes first, the vector sequence is the same as $\overrightarrow{V_1} - \overrightarrow{V_2} - \overrightarrow{V_7}$ because the vector sequence is periodical. In the figure $s_a = 1$ means that switch leg-a keeps constant state in sector 1-1 and has no switching loss. In each switching cycle both leg-b and leg-c have two switch commutations, one of which with bold bar is

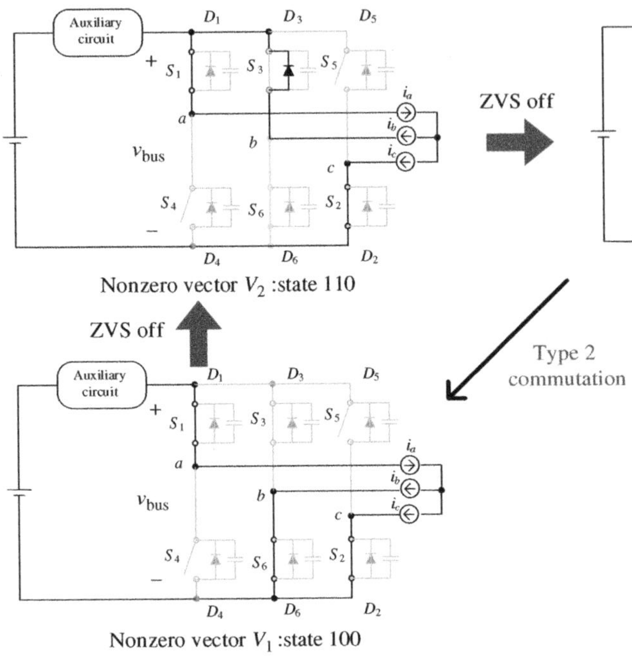

Figure 3.44 Converter state changing roadmap for vector sequence: $\overrightarrow{V_1} - \overrightarrow{V_2} - \overrightarrow{V_7}$.

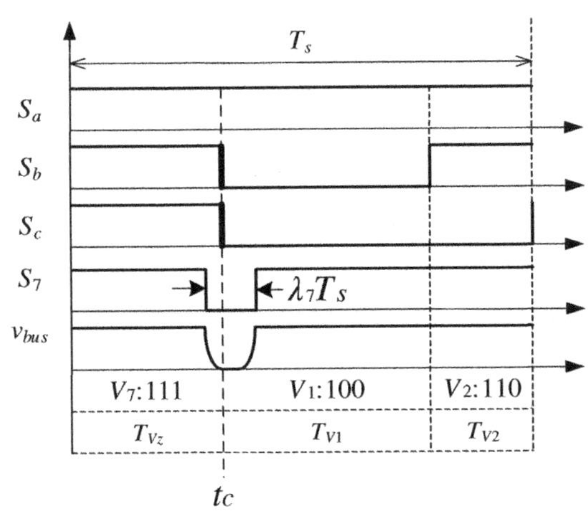

Figure 3.45 Waveforms of three-phase switch bridge state, auxiliary switch state, and v_{bus}.

the Type 2 commutations. The Type 2 commutations of leg-*b* and leg-*c* are due to transition from $\overrightarrow{V_7}$ to $\vec{V}_1$ and coincident at instant t_c. As shown in Figure 3.44, the currents of upper diodes D_3 and D_5 in both leg-b and leg-c are expected to commutate to their complement switches in lower side S_6 and S_2 during the transition from $\overrightarrow{V_7}$ to $\vec{V}_1$. To suppress the loss caused by these Type 2 commutations, the auxiliary switch is turned off before time t_c to start a resonance process in the circuit. The bus voltage v_{bus} eventually will resonance to zero. Once v_{bus} reaches zero, it makes ZVS condition for both lower switch S_6 and S_2 in leg-*b* and leg-*c*. The main switch can change state with ZVS

Table 3.1 Vector sequences in each sector.

	Zero vector	Nonzero vector		Zero vector
Sector	V_z →	V_k →	V_{k+1} →	V_z
Sector 1–1	111	100	110	111
Sector 1–2	000	110	100	000
Sector 2–1	000	110	010	000
Sector 2–2	111	010	110	111
Sector 3–1	111	010	011	111
Sector 3–2	000	011	010	000
Sector 4–1	000	011	001	000
Sector 4–2	111	001	011	111
Sector 5–1	111	001	101	111
Sector 5–2	000	101	001	000
Sector 6–1	000	101	100	000
Sector 6–2	111	100	101	111

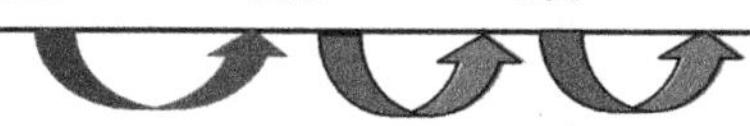

condition according to SVM requirement. In addition, the reverse recovery process of the diodes D_3 and D_5 to be blocked are suppressed. Since two Type 2 commutations are coincident, the auxiliary switch S_7 only needs to switch off once each switching cycle.

Turn-off duty ratio of the auxiliary switch S_7 is defined as λ_7. It is usually designed less than 10% of the switching period. Therefore the converter basically behaves as SVM controlled converter.

Above analysis on sector 1-1 can be extended to all the 12 sectors. Vector sequence in various sectors are listed in Table 3.1. As sector 1-1, vector sequence in each sector has three vectors: two nonzero vectors and one zero vector. In the table, the first column of the vector is zero vector, V_z; second and third columns are nonzero vectors, V_k and V_{k+1}; and the last column is zero vector that actually belongs to a new switch cycle. The vector transition from the zero vector V_z to nonzero vector V_k is Type 2 commutation while the vector transitions from the nonzero vector V_k to nonzero vector V_{k+1} and from the nonzero vector V_{k+1} to zero vector V_z are Type 1 commutations, which are ZVS turn-off of a switch and have lower loss.

As mentioned before, to avoid Type 2 commutations from the zero vector V_z to the nonzero vector V_k, the auxiliary circuit is introduced and ZVS turn-on of the main switches is realized. In addition, the reverse recovery process of the diodes is relieved. By modification, the conventional SVM can be tailored to the active clamped three-phase converters. The modified SVM is called as zero-voltage-switching space vector modulation (ZVS-SVM).

3.6.3 Characteristics of the Converter with ZVS-SVM

Similar to EA-PWM, there are two kinds of active clamped three-phase converters with ZVS-SVM according to the auxiliary circuits: converter with CAC and converter with MVAC, as shown in Figure 3.46. Both converters use almost the same ZVS-SVM scheme. During the vector transition

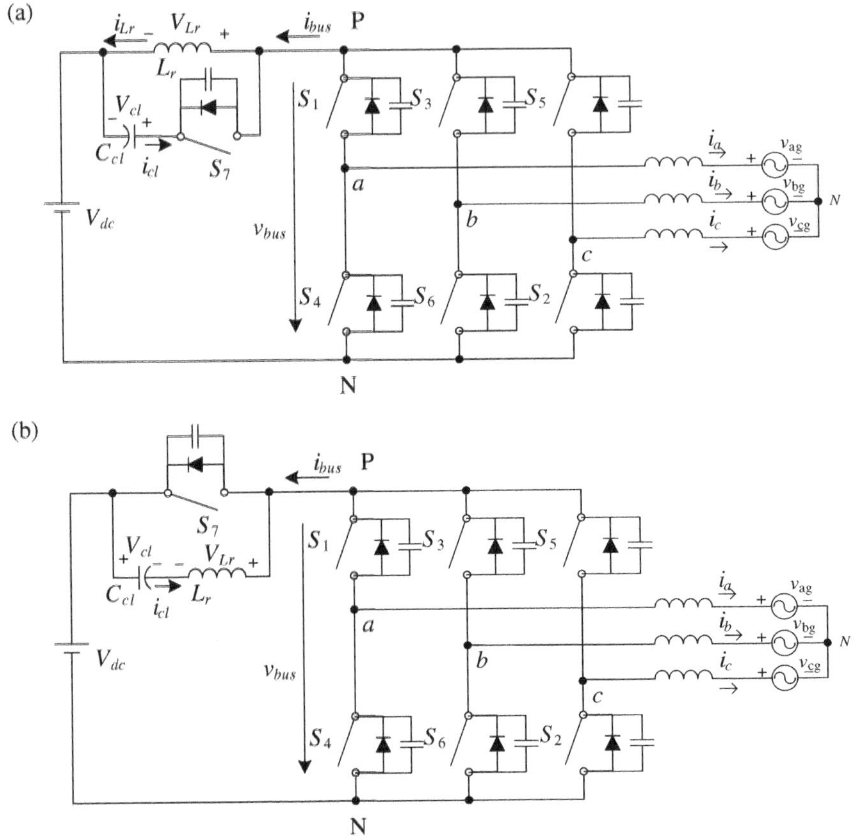

Figure 3.46 Active clamped three-phase converters: (a) with CAC; (b) with MVAC.

from the zero vector V_z to the nonzero vector V_k in each switching cycle, the auxiliary switch S_7 turns off to make v_{bus} resonance to zero voltage. It creates ZVS turn-on condition for the main switches in the converter.

Similar to the converter with EA-PWM, a switch-period stage analysis will be made in later chapters of this book. The voltage on the clamping capacitor is described by

$$V_{cl} = \frac{\lambda_7}{1-\lambda_7} V_{dc} \tag{3.110}$$

According to the above equation, if λ_7 is 0.1, V_{cl} is about 10% of DC voltage source V_{dc}.

For the CAC converter, the voltage stress on all the devices is equal to $V_{dc} + V_{cl}$, which is $1.1V_{dc}$ if λ_7 is 0.1. Although the switch bridge DC side voltage v_{bus} is set to zero within duration of $\lambda_7 T_s$ due to the resonance, it increases to $V_{dc} + V_{cl}$ in duration of $(1 - \lambda_7)T_s$. Thus the switching period average value of v_{bus} is still equal to V_{dc}. It means that the utilization of DC side voltage of the CAC converter is as good as the original three-phase converter.

For MVAC converter, the voltage stress on all the devices is equal to V_{dc}, which is the same as the original three-phase converter. The switch bridge DC side voltage v_{bus} is set to zero within duration of $\lambda_7 T_s$ due to the resonance. However, it is equal to V_{dc} in the remaining switching period. Thus the switching period average value of v_{bus} is $(1 - \lambda_7)V_{dc}$, which reduces DC side voltage utilization by

Table 3.2 Characteristics of the converters vs. auxiliary circuit and operating mode.

Auxiliary circuit /mode	Characteristics:
	Voltage stress, ZVS condition, DC voltage utilization
CAC/ inverter mode	• Voltage stress of all switches is $V_{dc}+V_{cl}$ • A boost charging stage is needed • Well DC side voltage utilization
CAC/ rectifier mode	• Voltage stress of all switches is $V_{dc}+V_{cl}$ • ZVS condition is automatically satisfied • Well DC side voltage utilization
MVAC/inverter mode	• Voltage stress of all switches is V_{dc} • ZVS condition is automatically satisfied • Decrease of DC side voltage utilization by$\lambda_7 \cdot 100\%$
MVAC/ rectifier mode	• Voltage stress of all switches is V_{dc} • A boost charging stage is needed • Decrease of DC side voltage utilization by$\lambda_7 \cdot 100\%$

$\lambda_7 \cdot 100\%$. If λ_7 is designed less than 0.1, decrease of DC side voltage utilization of the MVAC converter is 10%.

Another issue is soft-switching condition of the active clamped converters. Actually, it depends on the operation condition. One of the key parameters is power factor. Here we will provide conclusions, and more detailed analysis will be given in later chapters of the book. For CAC converter, if it operates in rectifier mode, ZVS condition is automatically satisfied. If it operates in inverter mode, a boost charging stage for the resonant inductor is required to realize ZVS condition.

For MVAC converter, if it operates in inverter mode, ZVS condition is automatically satisfied. If it operates in rectifier mode, a boost charging stage for the resonant inductor is required. Table 3.2 gives a brief summary of the characteristics of the active clamped converters with ZVS-SVM. Both CAC and MVAC converters have the following characteristics:

- All the switches realize ZVS turn-on, which include both the main switches and the auxiliary switch.
- Reverse recovery of all the diodes is suppressed.
- Both main switches and auxiliary switch operate in the same and fixed switching frequency.
- The auxiliary switch only turns off once per switch cycle.
- The resonance duration is 10% less than the switching period. The soft-switching converter operates like PWM converter.
- Lower voltage stress on all the devices and lower current stress on main switches.

3.7 Summary

A generic soft-switching PWM, EA-PWM, for the active clamped three-phase converters is introduced. It is applicable not only to three-phase three-wire active clamped converters but also to three-phase four-wire active clamped converters and other combination converters. It is compatible to almost all modulation methods: SPWM and PWM with the third harmonic injection. Either it is CPWM or DPWM. In addition it is not limited by load conditions such as power factor, three-phase balanced or unbalanced, and linear or nonlinear load. However, either the load conditions or the PWM schemes will affect the automatic ZVS ranges. Generally speaking, the active clamped converters are able to realize ZVS operation by adding an extra current in the resonant inductor during the current boost stage.

ZVS-SVM is developed from the conventional SVM to tailor to ZVS operation of the active-clamping three-phase converters. Vector sequences for each sector are identified with goal to reduce the vector transitions having high switching loss (Type 2 commutation). The ZVS-SVM scheme has only one high-loss vector transition in a switching cycle. This transition can be relieved with the help of the auxiliary circuit, which is very similar to the soft-switching mechanism of EA-PWM.

Essentially, ZVS-SVM is equivalent to the EA-PWM with the third harmonic injected modulation waves. For each ZVS-SVM, you may find an equivalent EA-PWM scheme. The active clamped three-phase converters have the same characteristics whether you use EA-PWM or ZVS-SVM as long as they are equivalent. On theory, EA-PWM is more generic than ZVS-SVM. Besides, it is easy to use. In EA-PWM, the duty cycles are calculated as the conventional PWM except the AC filter current polarity is used to select ramp-up or ramp-down saw-tooth carrier in a switching cycle. However, for ZVS-SVM, if the power factor is not equal to 1 or −1, sector division becomes complex, which depends on both output voltage and current information.

References

1 N. Mohan, T. M. Undeland, W. P. Robbins, *Power Electronics: Conversion, Applications, and Design*, John Wiley &Son, 2003.

2 P. T. Krein, *Elements of Power Electronics*, Oxford University Press, 2015

3 B. Feng and D. Xu, "Novel ZVS three-phase PFC converters and zero-voltage-switching space vector modulation (ZVS-SVM) control," First International Conference on Power Electronics Systems and Applications (PESA), pp. 30–37, 2004.

4 D. Xu, B. Feng, R. Li "A zero voltage switching SVM (ZVS–SVM) controlled three-phase boost rectifier" *IEEE Transactions on Power Electronics*, vol. **22**, no. 3, pp. 978–986, 2007.

5 K. Shi, J. Deng, D. Xu,"A general pulse width modulation method for zero-voltage-switching active-clamping three-phase power converters: edge aligned pulse width modulation (EA-PWM)", *IEEE Open Journal of Power Electronics*, vol. **1**, 2020.

6 D. M. Divan, "The resonant DC link converter-a new concept in static power conversion," *IEEE Transactions on Industry Applications*, vol. **25**, no. 2, pp. 317–325, 1989.

7 G. Venkataramanan, D. M. Divan, and T. M. Jahns, "Discrete pulse modulation strategies for high frequency inverter systems," *IEEE Transactions on Power Electronics*, vol. **8**, no. 3, pp. 279–287, 1993.

8 D. M. Divan and G. Skibinski, "Zero-switching-loss inverters for high-power applications," *IEEE Transactions on Industry Applications*, vol. **25**, no. 4, pp. 634–643, 1989.

9 D. M. Divan, L. Malesani, P. Tenti and V. Toigo, "A synchronized resonant DC link converter for soft-switched PWM," *IEEE Transactions on Industry Applications*, vol. **29**, no. 5, pp. 940–948, 1993.

10 M. Mezaroba, D. C. Martins, I. Barbi. "A ZVS PWM three-phase inverter with active clamping technique using the reverse recovery energy of the diodes, " IEEE 35th Annual Power Electronics Specialists Conference, pp. 4785–4790, 2004.

11 N. Hoshi and K. Oguchi, "Control schemes for auxiliary switches of three-phase pwm resonant snubber inverters," *Electrical Engineering in Japan*, vol. **152**, no. 4, pp. 57–67, 2005.

12 N. Hoshi, K. Itoga, K. Oguchi, "A novel control strategy for three-phase resonant snubber based soft-switching inverters, " Proceedings of IPEC-Tokyo 2000, pp. 536–541.

13 R. Li and D. Xu, "A zero-voltage switching three-phase inverter," *IEEE Transactions on Power Electronics*, vol. **29**, no. 3, pp. 1200–1210, March 2014.

14 Z. Ma, D. Xu, R. Li, C. Du and X. Zhang, "A novel DC-side zero-voltage switching (ZVS) three-phase boost PWM rectifier controlled by an improved SVM method," *IEEE Transactions on Power Electronics*, vol. **27**, no. 11, pp. 4391–4408, 2012.

15 N. He, Y. Zhu, A. Zhao and D. Xu, "Zero-voltage-switching sinusoidal pulse width modulation method for three-phase four-wire inverter," *IEEE Transactions on Power Electronics*, vol. **34**, no. 8, pp. 7192–7205, 2019.

16 N. He, M. Chen, J. Wu, N. Zhu, and D. Xu, "20 kW zero-voltage-switching SiC-MOSFET grid inverter with 300 kHz switching frequency", *IEEE Transactions on Power Electronics*, vol. **34**, no. 6, 2019.

17 K. Shi, A. Zhao, J. Deng, and D. Xu, "Zero-voltage-switching SiC-MOSFET three-phase four-wire back-to-back converter," *IEEE Journal of Emerging and Selected Topics in Power Electronics*, vol. **7**, no. 2, pp. 722–735, June 2019.

18 K. Shi, J. Deng, and D. Xu, "Edge aligned PWM for soft-switching three-phase power converters", 2020 IEEE 9th International Power Electronics and Motion Control Conference (IPEMC2020-ECCE Asia), Nanjing, 2020.

19 M. Nakaoka, H. Yonemori, and K. Yurugi, "Zero-voltage soft-switched PDM three-phase AC-DC active power converter operating at unit power factor and sinewave line current", Proceedings of IEEE Power Electronics Specialist Conference – PESC '93, pp. 787–794, 1993.

20 V. Agelidis, P. Ziogas, and G. Joos, "An optimum modulation strategy for a novel "notch" commutated 3-Φ PWM inverter," *IEEE Transactions on Industry Applications*, vol. **30**, no. 1, pp 52–61, 1994.

21 J. He and N. Mohan, "Parallel resonant DC-link circuit – a novel zero switching loss topology with minimum voltage stresses," *IEEE Transactions on Power Electronics*, vol. **6**, no.4, pp. 687–694, 1991.

22 J. Cho, H. Kim, and G. Cho, "Novel soft-switching PWM converter using a parallel resonant DC-link," Proceedings of IEEE Power Electronics Specialist Conference – PESC1991, pp. 241–247, 1991.

23 M. D. Bellar, T. S. Wu, A. Tchamdjou, J. Mahdavi, and M. Ehsani, "A review of soft-switched DC-AC converters", *IEEE Transactions on Industrial Applications*, vol. **34**, no. 4, pp. 847–860, 1998.

Part 2

ZVS-SVM Applied to Three-phase Rectifiers

4

Three-phase Rectifier with Compound Active-clamping Circuit

Soft-switching technology is an effective way to achieve high efficiency, high power density, and low electromagnetic noise. In this chapter, a zero-voltage-switching (ZVS) three-phase rectifier with compound active-clamping (CAC) circuit is introduced, including the operation principle, the circuit analysis, and the design guidelines of the CAC rectifier [1]. First, the operation principle of the proposed space vector modulation (SVM) scheme is introduced. Then, soft-switching conditions will be discussed to help readers understand this circuit. Finally, the design guidelines of the three-phase rectifier with CAC circuit will be introduced and a design example based on a 40 kW prototype will be provided [1, 2].

4.1 Introduction

Circuit topology of the CAC rectifier is shown in Figure 4.1. It consists of a standard three-phase rectifier and a CAC circuit. The three-phase rectifier is composed of grid side input inductors L_a, L_b, and L_c; six switches S_1, S_2, S_3, S_4, S_5, and S_6; active switches' antiparallel diodes D_1, D_2, D_3, D_4, D_5, and D_6. The clamping circuit has an active switch S_7, a resonant inductor L_r, and a clamping capacitor C_{cl}. Capacitors C_1, C_2, C_3, C_4, C_5, C_6, and C_7 are paralleled with the corresponding switches.

With the ZVS-SVM scheme introduced in Chapter 3, all the switches, including bridge switches and auxiliary switch can be turned on and turned off under zero voltage condition. The reverse recovery of all the diodes in the CAC rectifier is well suppressed. The voltage stress of all the switching devices is restricted around 1–1.1 times of the DC bus voltage. In this chapter, the operation principle of the ZVS-SVM scheme will be introduced. Soft-switching condition will be derived. Design guidelines will be introduced, and a design example of a 40 kW CAC rectifier prototype will be discussed.

Soft-Switching Technology for Three-phase Power Electronics Converters, First Edition.
Dehong Xu, Rui Li, Ning He, Jinyi Deng, and Yuying Wu.

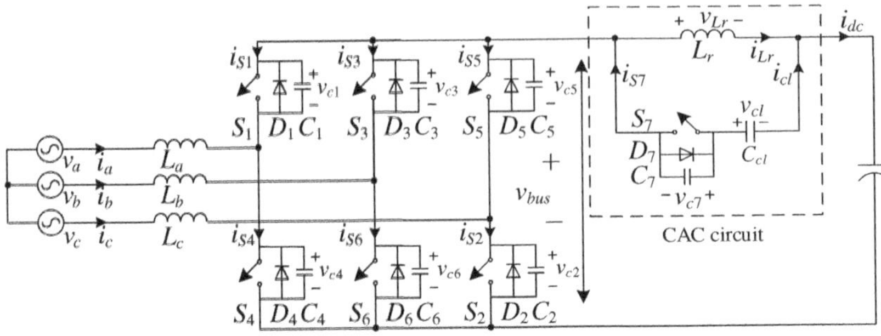

Figure 4.1 Circuit topology of the CAC rectifier.

4.2 Operation Principle of CAC Rectifier

4.2.1 Space Vector of Three-phase Grid Voltage

The grid side three-phase voltage v_a, v_b, and v_c of three-phase converter are expressed by (4.1),

$$\begin{cases} v_a = V_m \cos(\omega_l t) \\ v_b = V_m \cos\left(\omega_l t - \dfrac{2\pi}{3}\right) \\ v_c = V_m \cos\left(\omega_l t + \dfrac{2\pi}{3}\right) \end{cases} \tag{4.1}$$

where V_m and ω_l are the amplitude and the angular frequency of the grid side voltage, respectively.

For a symmetrical three-phase system, the three-phase voltages v_a, v_b, and v_c satisfy the relationship $v_a + v_b + v_c = 0$. In this case, all the three-dimensional space vectors satisfying this condition are located on a plane in the three-dimensional space. Therefore, three-phase voltage can be simply expressed as a complex vector $\vec{v}$ in the complex plane with the following transformation.

$$\vec{v} = \frac{2}{3}\begin{bmatrix} 1 & \alpha & \alpha^2 \end{bmatrix}\begin{bmatrix} v_a \\ v_b \\ v_c \end{bmatrix} = \frac{2}{3}(v_a + \alpha v_b + \alpha^2 v_c) \tag{4.2}$$

where $\alpha = e^{j2\pi/3}$ and $\alpha^2 = e^{j4\pi/3}$. Constant 2/3 of expression (4.2) is chosen to scale the complex vector to have the same amplitude as that of the three-phase voltage. Complex vector $\vec{v}$ is composed of real part v_α and imaginary part v_β

$$\vec{v} = v_\alpha + jv_\beta \tag{4.3}$$

According to Eqs. (4.2) and (4.3), the relationship between v_a, v_b, v_c and v_α, v_β are derived as:

$$\begin{cases} v_\alpha = \dfrac{2}{3}\left(v_a + v_b \cos\dfrac{2}{3}\pi + v_c \cos\dfrac{4}{3}\pi\right) \\ v_\beta = \dfrac{2}{3}\left(v_b \sin\dfrac{2}{3}\pi + v_c \sin\dfrac{4}{3}\pi\right) \end{cases} \tag{4.4}$$

Equation (4.4) can be rewritten in the matrix form as:

$$\begin{bmatrix} v_\alpha \\ v_\beta \end{bmatrix} = \frac{2}{3} \begin{bmatrix} 1 & -\frac{1}{2} & -\frac{1}{2} \\ 0 & \frac{\sqrt{3}}{2} & -\frac{\sqrt{3}}{2} \end{bmatrix} \cdot \begin{bmatrix} v_a \\ v_b \\ v_c \end{bmatrix} \tag{4.5}$$

The above transformation is called Clarke transformation. The following matrix is called Clarke transformation matrix.

$$T_C = \frac{2}{3} \begin{bmatrix} 1 & -\frac{1}{2} & -\frac{1}{2} \\ 0 & \frac{\sqrt{3}}{2} & -\frac{\sqrt{3}}{2} \end{bmatrix} \tag{4.6}$$

The inverse Clarke transformation can be derived as:

$$\begin{bmatrix} v_a \\ v_b \\ v_c \end{bmatrix} = \frac{3}{2} \begin{bmatrix} \frac{2}{3} & 0 \\ -\frac{1}{3} & \frac{\sqrt{3}}{3} \\ -\frac{1}{3} & \frac{\sqrt{3}}{3} \end{bmatrix} \cdot \begin{bmatrix} v_\alpha \\ v_\beta \end{bmatrix} \tag{4.7}$$

According to Eqs. (4.1), (4.3), and (4.4), the complex vector $\vec{v}$ of the symmetrical three-phase voltages can be derived as:

$$\vec{v} = V_m e^{j\omega_l t} \tag{4.8}$$

The complex vector $\vec{v}$ rotates counterclockwise at a constant angular frequency ω_l in complex plane as shown in Figure 4.2. The trajectory of the complex vector $\vec{v}$ is a perfect circle; v_α and v_β are the projection of the complex vector $\vec{v}$ on the α-axis and β-axis in the complex plane, respectively.

Figure 4.2 Vector representation in complex plane with $\alpha\beta$ coordinate.

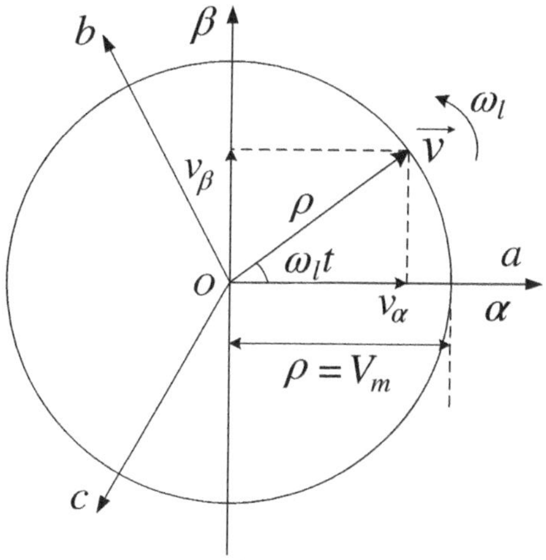

4.2.2 Space Vector Modulation of Three-phase Converter

According to the invertibility principle of complex space vector transformation, if the output voltage complex space vector of the three-phase converter is a trajectory close to a circle, the three-phase output voltage of the converter would be close to symmetrical sinusoidal wave. Topology of the three-phase converter is shown in Figure 4.3. This topology consists of six active switches S_1, S_2, S_3, S_4, S_5, and S_6 and DC-link capacitor C_{dc}. Ignore the current ripple of the grid side filter inductor, the grid side current of the converter can be represented by current sources i_a, i_b, and i_c.

In order to ensure proper operation of the circuit, only one switch in each leg is in conduction state, while the other one is in off state. According to those principles, there are eight switch states in the converter. Further, each switch state can be represented by complex voltage space vector, as shown in Table 4.1.

As shown in Table 4.1, if the voltage space vector is $\vec{V}_1(100)$, S_1, S_2, and S_6 are conducting and S_3, S_4, and S_5 are in off state. Three phases' bridge voltage $v_{an} = v_{dc}$, $v_{bn} = 0$, and $v_{cn} = 0$. According to Eq. (4.2), complex vector $\vec{V}_1(100)$ is derived as follows:

$$\vec{V}_1(100) = \frac{2}{3}\left(v_{dc} + 0 \cdot \alpha + 0 \cdot \alpha^2\right) = 2v_{dc}/3 \tag{4.9}$$

Then the direction of $\vec{V}_1(100)$ is the same as the α-axis and the module value of $\vec{V}_1(100)$ is $2v_{dc}/3$. The complex vectors of other voltage space vectors can be calculated in the same way.

$$\vec{V}_2(110) = \frac{2}{3}\left(v_{dc} + v_{dc} \cdot \alpha + 0 \cdot \alpha^2\right) = 2v_{dc}e^{j\pi/3}/3 \tag{4.10}$$

$$\vec{V}_3(010) = \frac{2}{3}\left(0 + v_{dc} \cdot \alpha + 0 \cdot \alpha^2\right) = 2v_{dc}e^{j2\pi/3}/3 \tag{4.11}$$

$$\vec{V}_4(011) = \frac{2}{3}\left(0 + v_{dc} \cdot \alpha + v_{dc} \cdot \alpha^2\right) = 2v_{dc}e^{j\pi}/3 \tag{4.12}$$

$$\vec{V}_5(001) = \frac{2}{3}\left(0 + 0 \cdot \alpha + v_{dc} \cdot \alpha^2\right) = 2v_{dc}e^{j4\pi/3}/3 \tag{4.13}$$

$$\vec{V}_6(101) = \frac{2}{3}\left(v_{dc} + 0 \cdot \alpha + v_{dc} \cdot \alpha^2\right) = 2v_{dc}e^{j5\pi/3}/3 \tag{4.14}$$

$$\vec{V}_7(111) = \frac{2}{3}\left(v_{dc} + v_{dc} \cdot \alpha + v_{dc} \cdot \alpha^2\right) = 0 \tag{4.15}$$

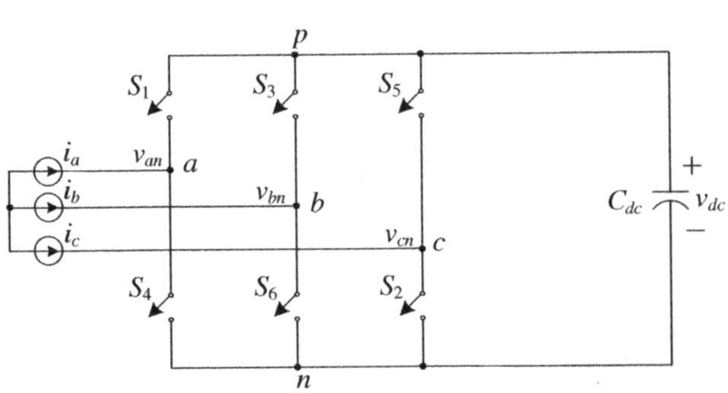

Figure 4.3 Topology of the three-phase converter.

Table 4.1 Eight switch states.

Complex vector	Switches in on-state	v_{an}	v_{bn}	v_{cn}	Complex vector value
$\vec{V}_1(100)$	S_1, S_2, S_6	v_{dc}	0	0	$2v_{dc}/3$
$\vec{V}_2(110)$	S_1, S_2, S_3	v_{dc}	v_{dc}	0	$2v_{dc}e^{j\pi/3}/3$
$\vec{V}_3(010)$	S_2, S_3, S_4	0	v_{dc}	0	$2v_{dc}e^{j2\pi/3}/3$
$\vec{V}_4(011)$	S_3, S_4, S_5	0	v_{dc}	v_{dc}	$2v_{dc}e^{j\pi}/3$
$\vec{V}_5(001)$	S_4, S_5, S_6	0	0	v_{dc}	$2v_{dc}e^{j4\pi/3}/3$
$\vec{V}_6(101)$	S_1, S_5, S_6	v_{dc}	0	v_{dc}	$2v_{dc}e^{j5\pi/3}/3$
$\vec{V}_7(111)$	S_1, S_3, S_5	v_{dc}	v_{dc}	v_{dc}	0
$\vec{V}_0(000)$	S_2, S_4, S_6	0	0	0	0

$$\vec{V}_0(000) = \frac{2}{3}\left(0 + 0 \cdot \alpha + 0 \cdot \alpha^2\right) = 0 \tag{4.16}$$

According to Table 4.1, there are six nonzero voltage vectors $\vec{V}_1(100)$, $\vec{V}_2(110)$, $\vec{V}_3(010)$, $\vec{V}_4(011)$, $\vec{V}_5(001)$, and $\vec{V}_6(101)$ and two zero voltage vectors $\vec{V}_0(000)$ and $\vec{V}_7(111)$. The six nonzero voltage vectors divide the complex plane into six sectors, as shown in Figure 4.4.

A reference voltage vector $\vec{v}_{ref}$ can be generated by integrating two adjacent nonzero voltage vectors in a sector and the zero voltage vectors by the parallelogram rule. For instance, if the reference vector is located in sector 1, the corresponding nonzero voltage vectors $\vec{V}_1(100)$, $\vec{V}_2(110)$ and the zero voltage vectors $\vec{V}_0(000)$ or $\vec{V}_7(111)$ are used. The reference voltage vector $\vec{v}_{ref}$ is expressed as:

$$\vec{v}_{ref} = \left|\vec{V}_{ref}\right| e^{j\theta} \tag{4.17}$$

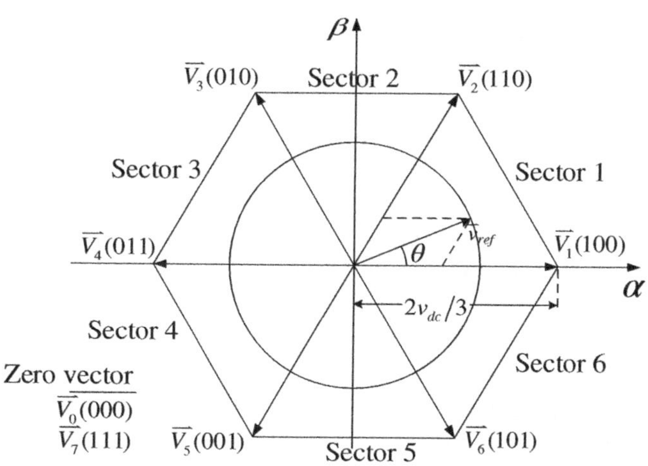

Figure 4.4 Sectors and voltage vectors in space complex plane.

where $\left|\vec{V}_{ref}\right|$ is the module value of $\vec{v}_{ref}$ and θ is the phase angle of $\vec{v}_{ref}$. The reference voltage vector $\vec{V}_{ref}$ can be realized by the voltage vectors $\vec{V}_1(100)$, $\vec{V}_2(110)$, and $\vec{V}_0(000)$ as follows:

$$d_{V1}T_s\vec{V}_1 + d_{V2}T_s\vec{V}_2 + d_{V0}T_s\vec{V}_0 = T_s\vec{v}_{ref} \tag{4.18}$$

where d_{V1}, d_{V2}, and d_{V0} are the duty ratio of voltage vectors $\vec{V}_1(100)$, $\vec{V}_2(110)$, and $\vec{V}_0(000)$, respectively. These duty ratios are constrained as follows:

$$d_{V1} + d_{V2} + d_{V0} = 1 \tag{4.19}$$

By substituting Eqs. (4.9), (4.10), (4.16), (4.17), and (4.19) into Eq. (4.18), the duty ratios of $\vec{V}_1$, $\vec{V}_2$, and $\vec{V}_0$ can be derived as:

$$\begin{cases} d_{V1} = \dfrac{\sqrt{3}}{2} m \sin(\pi/3 - \theta) \\ d_{V2} = \dfrac{\sqrt{3}}{2} m \sin\theta \\ d_{V0} = 1 - \dfrac{\sqrt{3}}{2} m \sin\left(\dfrac{\pi}{3} + \theta\right) \end{cases} \tag{4.20}$$

where m is the voltage modulation ratio. It is expressed as follows:

$$m = 2\left|\vec{V}_{ref}\right| / v_{dc} \tag{4.21}$$

The reference voltage vector $\vec{v}_{ref}$ is composed of two adjacent nonzero voltage vectors in the same sector in each switching cycle. Normally, the sum of the duty ratio of these two adjacent voltage vectors is less than 1, then the zero voltage vector fills the rest of the switching cycle. In a switching cycle, the voltage space vector acts in a time-sharing manner to form a space vector sequence in time. There are many ways to organize space vector sequences. According to the symmetry of the space vector sequence, it can be divided into symmetrical and asymmetrical space vector sequences. According to the number of commutation phases in a switching cycle, it can be divided into two-phase switching commutation and three-phase switching commutation. Two typical organization methods of space vector sequence are introduced next.

We take sector 1 as an example. Two kinds of space vector sequences are shown in Figure 4.5. Figure 4.5a shows a symmetrical voltage vector sequence. Here only zero vector $\vec{V}_7(111)$ is used. There are four switching commutations in a switching cycle. The output pulse width modulation (PWM) waveforms are symmetrical. Figure 4.5b shows an asymmetrical voltage vector sequence. It also has four switching commutations in a switching cycle. The output PWM waveforms are not symmetrical in a switch period. Both vector sequences in Figure 4.5 belong to two-phase switching commutation, which is usually called discontinuous PWM (DPWM).

4.2.3 Switching Scheme of CAC Rectifier

In a switching cycle of three-phase rectifier, different switching commutation processes bring different switching loss [3, 4]. Here leg-a is taken as an example to investigate the switching commutation process. The inductor current i_a is positive as shown in Figure 4.6. The key waveforms are demonstrated in Figure 4.7.

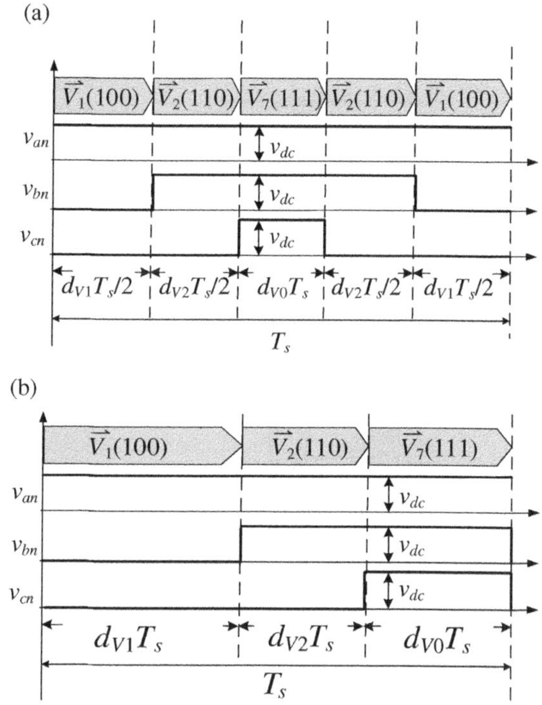

Figure 4.5 Two kinds of space vector sequences. (a) Symmetrical voltage vector sequence. (b) Asymmetrical voltage vector sequence.

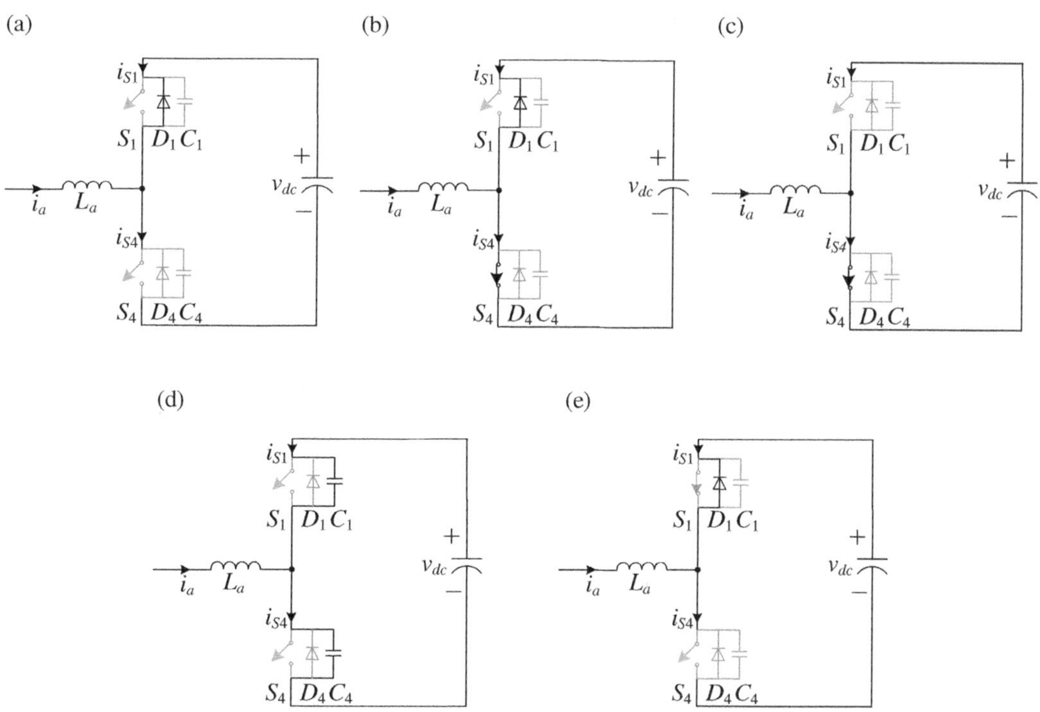

Figure 4.6 Switching commutation process of a leg in three-phase rectifier. (a) Stage 1 (t_0–t_1); (b) stage 2 (t_1–t_2); (c) stage 3 (t_2–t_3); (d) stage 4 (t_3–t_4); (e) stage 5 (t_4–t_5).

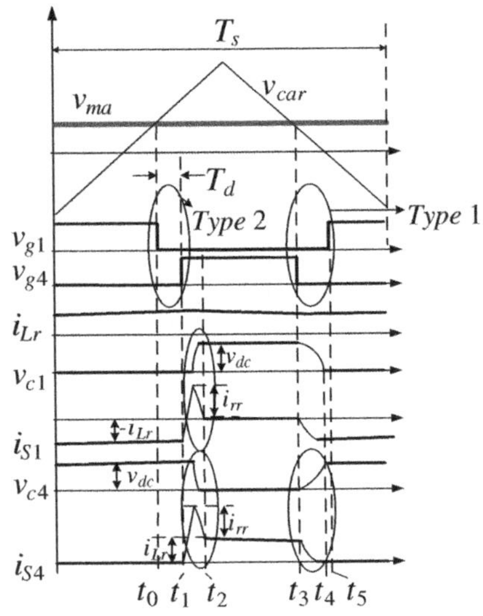

Figure 4.7 Driving logic and key waveforms of a leg in three-phase converter.

There exists one hard-switching turn-on (commutation from D_1 to S_4) for lower switch S_4 as shown in Figure 4.6b. As soon as S_4 is turned on at t_1, the diode D_1 experiences the reverse recovery process. The reverse recovery current of the diode D_1 flows not only through D_1 but also through the switch S_4, which brings D_1's reverse recovery process loss E_{rr} and S_4's turn-on loss E_{on}. At t_3, S_4 is turned off. The input current i_a charges C_4 and discharges C_1. With the buffer of the parasitic capacitors C_1 and C_4, S_4 is turned off under ZVS condition. The larger the capacitance of C_1 and C_4, the lower the turn-off loss E_{off}. According to the above analysis, due to the presence of switch's parasitic capacitors C_1 and C_4, the turn-off loss of S_4 in the commutation from the switch to the diode is relatively small. This switching process is defined as switch commutation Type 1. Turn-on loss of the switch S_4 and reverse recovery loss of the diode D_1 only occur in the process of commutation from the diode to the switch. This switching process is defined as switch commutation Type 2.

The grid side current of the three-phase rectifier is expressed as:

$$\begin{cases} i_a = I_m \cos(\omega_l t + \varphi_0) \\ i_b = I_m \cos\left(\omega_l t - \dfrac{2\pi}{3} + \varphi_0\right) \\ i_c = I_m \cos\left(\omega_l t + \dfrac{2\pi}{3} + \varphi_0\right) \end{cases} \tag{4.22}$$

where φ_0 is the grid current power factor angle, and I_m is the amplitude of the grid current.

Considering the rectifier operates under unit power factor, the power factor angle φ_0 is zero. The voltage and current waveforms in a grid line cycle are demonstrated in Figure 4.8. Based on the basic SVM knowledge, a grid line cycle is first equally divided into six sectors, from sector 1 to sector 6, as shown in Figure 4.4. Considering the current direction of different phases, each sector is further divided into two subsectors. Therefore, a line cycle is finally equally divided into 12 sectors as shown in Figure 4.9.

We take sector 1-1 as an example to demonstrate the principle of the proposed ZVS-SVM. As shown in Figure 4.8, in this sector, three-phase current meets $i_a > 0$, $i_b < 0$, and $i_c < 0$. The corresponding vectors in this sector are nonzero voltage vectors $\vec{V}_1(100)$ and $\vec{V}_2(110)$ and zero vector $\vec{V}_0(000)$, and $\vec{V}_7(111)$. Since different switching commutation processes bring different switching

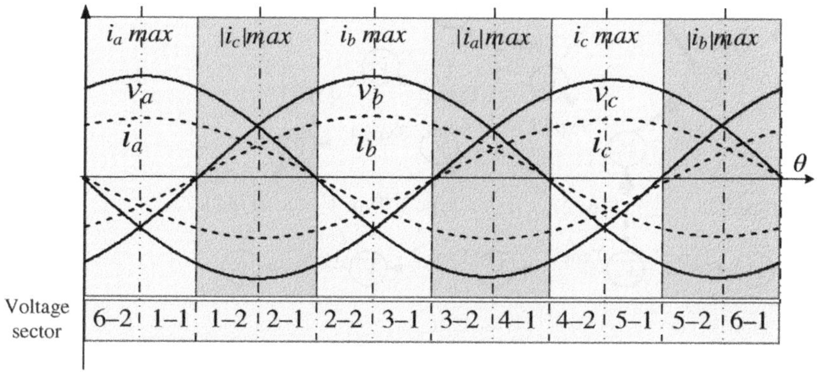

Figure 4.8 The voltage and current waveforms in a utility cycle.

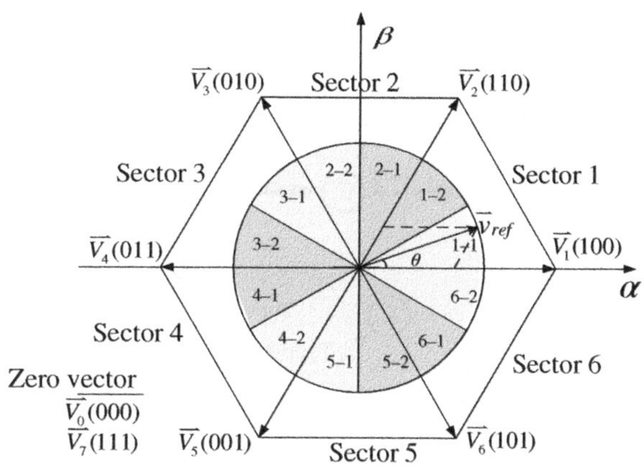

Figure 4.9 Sectors in space vector diagram.

loss, to find the voltage vector sequence with lowest switching loss, the dendrogram analysis of switching commutation sequences in sector 1-1 is made as shown in Figure 4.10. Because the switching sequence of a switching cycle is periodical, the beginning voltage vector and the ending voltage vector should be the same in the dendrogram.

For the three-phase rectifier, there exists two types of switching commutations with different switching loss. The commutation from the switch to the antiparallel diode of its complement switch has relatively small loss, which is defined as switch commutation Type 1. The switch's turn-on loss and the corresponding diode's reverse recovery loss occur in the process of commutation from the diode to the complement switch, which is defined as switch commutation Type 2. Once a Type 2 current commutation happens in the CAC ZVS three-phase rectifier, the auxiliary switch S_7 is required to turn off to cause the resonance between the resonant inductor L_r and the resonant capacitors C_1 and C_4. As a result, the DC side voltage of the three-phase bridges v_{bus} will be resonated to zero, it brings the opportunity of ZVS turn-on for main switches. Auxiliary switch is not required to turn off when a Type 1 commutation happens. With the ZVS-SVM scheme, all the switches, including bridge switches and auxiliary switch, can be turned on and turned off under zero voltage condition. The reverse recovery process of all the diodes in the CAC ZVS rectifier is also suppressed. In addition, CAC rectifier has lower voltage stress on the devices. The voltage stress of all the switching devices is about 1.1 times of DC source voltage.

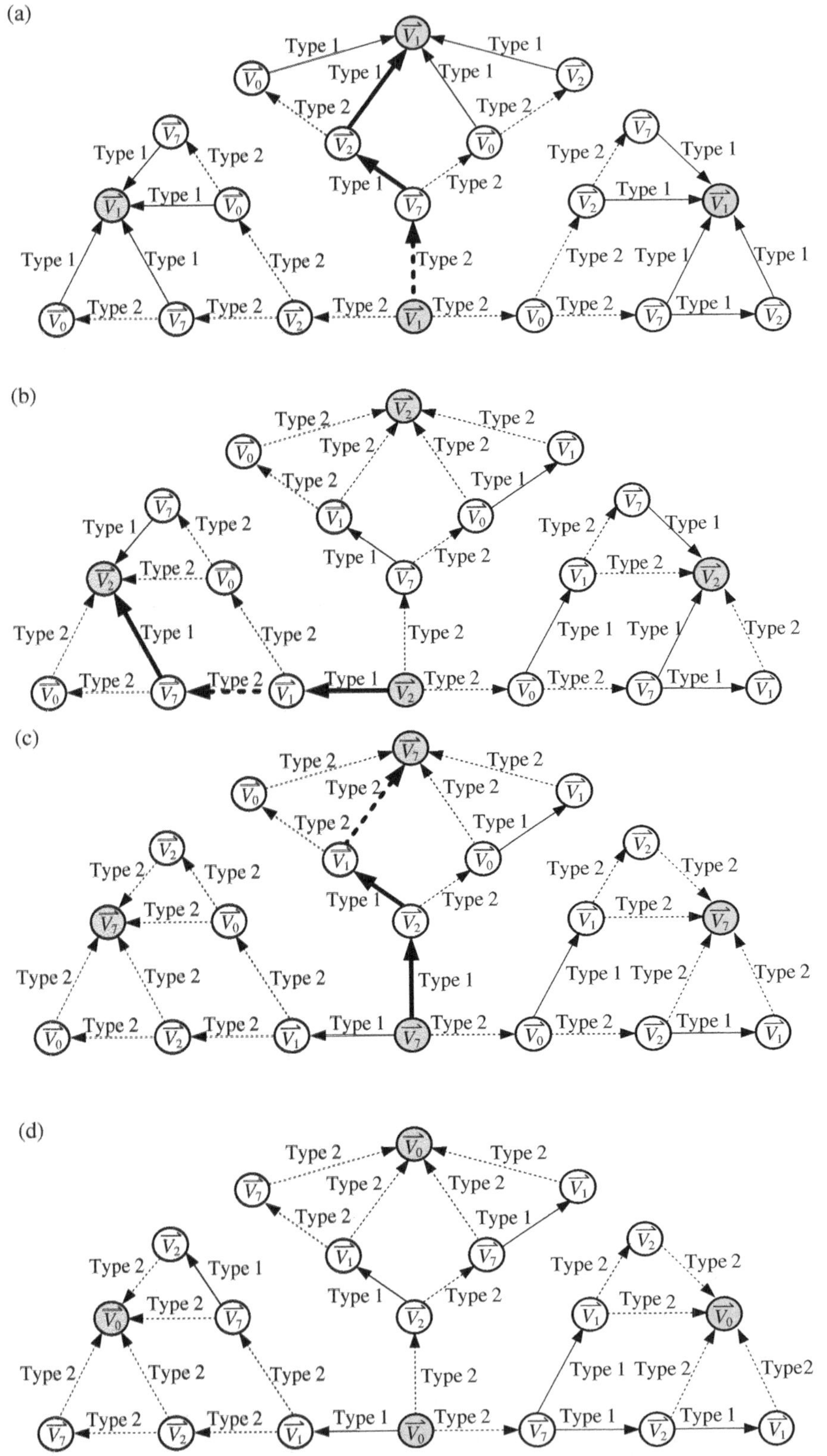

Figure 4.10 The dendrogram analysis of switching commutation sequences between different voltage vectors in sector 1-1. (a) Beginning with nonzero vector $\vec{V}_1(100)$. (b) Beginning with nonzero vector $\vec{V}_2(110)$. (c) Beginning with zero vector $\vec{V}_7(111)$. (d) Beginning with zero vector $\vec{V}_0(000)$.

Now let's investigate dendrogram of the vector sequence. It is assumed that vectors in sector 1-1 can only be used once in each switch cycle. In Figure 4.10a, $\vec{V}_1(100)$ is the beginning voltage vector and the ending voltage vector. After $\vec{V}_1(100)$ acts for some time, the adjacent voltage vector can be $\vec{V}_2(110)$, $\vec{V}_7(111)$, or $\vec{V}_0(000)$. Supposing $\vec{V}_7(111)$ is selected as the adjacent voltage vector. After $\vec{V}_7(111)$ acts for some time, its adjacent voltage vector can be $\vec{V}_2(110)$ or $\vec{V}_0(000)$. If $\vec{V}_2(110)$ is selected as the adjacent voltage vector and acts for some time, the adjacent voltage vector can be $\vec{V}_1(100)$ or $\vec{V}_0(000)$. Suppose that $\vec{V}_0(000)$ is the adjacent voltage vector. After $\vec{V}_0(000)$ acts for some time, the adjacent voltage vector can only be $\vec{V}_1(100)$, which is the end of this switching cycle.

In Figure 4.10a, all the vector sequences begin from $\vec{V}_1(100)$. Among all vector sequences, there is only one commutation sequence with the lowest switching loss, which is $\vec{V}_1(100)$-$\vec{V}_7(111)$-$\vec{V}_2(110)$-$\vec{V}_1(100)$. It has only one Type 2 switching commutation in a switching cycle.

In Figure 4.10b, the first vector is $\vec{V}_2(110)$. The best vector sequence is $\vec{V}_2(110)$-$\vec{V}_1(100)$-$\vec{V}_7(111)$-$\vec{V}_2(110)$, which has only one Type 2 switching commutation. In Figure 4.10c, the first vector is $\vec{V}_7(111)$. The best vector sequence is $\vec{V}_7(111)$-$\vec{V}_2(110)$-$\vec{V}_1(100)$-$\vec{V}_7(111)$. In Figure 4.10d, we cannot find vector sequence with one Type 2 switching commutation.

Figure 4.11 shows the three vector sequences with one Type 2 commutation process. By comparing these vector sequences, we find that they are actually the same. The only difference is which vector is selected as the beginning voltage vector in a switching cycle.

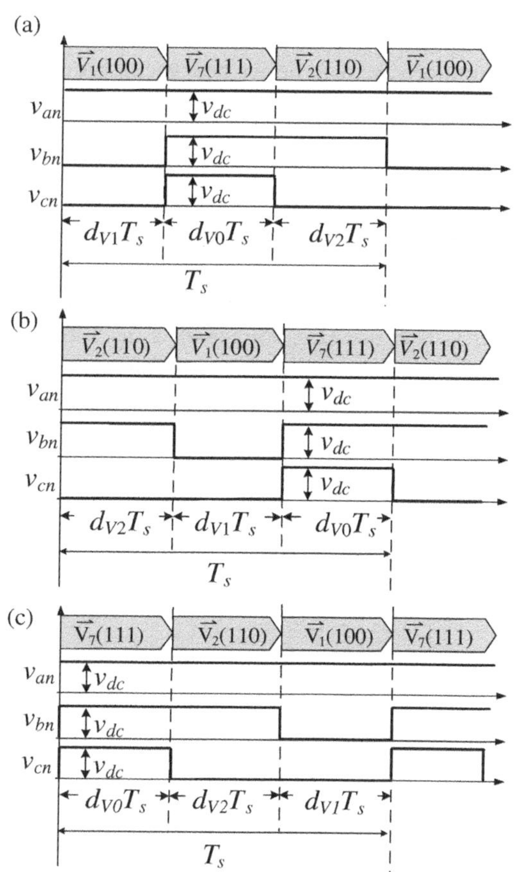

Figure 4.11 Three kinds of space vector sequences with one Type 2 commutation process. (a) $\vec{V}_1(100)$-$\vec{V}_7(111)$-$\vec{V}_2(110)$-$\vec{V}_1(100)$. (b) $\vec{V}_2(110)$-$\vec{V}_1(100)$-$\vec{V}_7(111)$-$\vec{V}_2(110)$. (c) $\vec{V}_7(111)$-$\vec{V}_2(110)$-$\vec{V}_1(100)$-$\vec{V}_7(111)$.

Table 4.2 Switching sequence in each sector.

Sector	First vector	Zero vector	Second vector	First vector
Sector 1-1	$\vec{V}_1(100)$	$\vec{V}_7(111)$	$\vec{V}_2(110)$	$\vec{V}_1(100)$
Sector 1-2	$\vec{V}_2(110)$	$\vec{V}_0(000)$	$\vec{V}_1(100)$	$\vec{V}_2(110)$
Sector 2-1	$\vec{V}_2(110)$	$\vec{V}_0(000)$	$\vec{V}_3(010)$	$\vec{V}_2(110)$
Sector 2-2	$\vec{V}_3(010)$	$\vec{V}_7(111)$	$\vec{V}_2(110)$	$\vec{V}_3(010)$
Sector 3-1	$\vec{V}_3(010)$	$\vec{V}_7(111)$	$\vec{V}_4(011)$	$\vec{V}_3(010)$
Sector 3-2	$\vec{V}_4(011)$	$\vec{V}_0(000)$	$\vec{V}_3(010)$	$\vec{V}_4(011)$
Sector 4-1	$\vec{V}_4(011)$	$\vec{V}_0(000)$	$\vec{V}_5(001)$	$\vec{V}_4(011)$
Sector 4-2	$\vec{V}_5(001)$	$\vec{V}_7(111)$	$\vec{V}_4(011)$	$\vec{V}_5(001)$
Sector 5-1	$\vec{V}_5(001)$	$\vec{V}_7(111)$	$\vec{V}_6(101)$	$\vec{V}_5(001)$
Sector 5-2	$\vec{V}_6(101)$	$\vec{V}_0(000)$	$\vec{V}_5(001)$	$\vec{V}_6(101)$
Sector 6-1	$\vec{V}_6(101)$	$\vec{V}_0(000)$	$\vec{V}_1(100)$	$\vec{V}_6(101)$
Sector 6-2	$\vec{V}_1(100)$	$\vec{V}_7(111)$	$\vec{V}_6(101)$	$\vec{V}_1(100)$

Similarly, we can derive best vector sequences for other sectors, which are listed in Table 4.2. In each sector, the Type 2 switching commutation happens between the first vector and the zero vector.

Compared with the standard six-switch three-phase rectifier, the CAC rectifier has an additional CAC auxiliary circuit, which is shown in Figure 4.1. The clamping circuit is composed of an active switch S_7, a resonant inductor L_r, and a clamping capacitor C_{cl}. By ignoring the short switching commutation process. The auxiliary circuit has two switching states, on-state of active switch S_7 and off-state of active switch S_7, which is shown in Figure 4.12a and b, respectively.

If S_7 is in on-state, voltage on S_7 $v_{c7} = 0$, and the bus voltage $v_{bus} = v_{dc} + v_{cl}$. If S_7 is in off-state, voltage on S_7 $v_{c7} = v_{dc} + v_{cl}$, and the bus voltage $v_{bus} = 0$. If Type 2 switching commutation occurs in off-state of S_7, ZVS turn-on condition for the bridge switches can be realized.

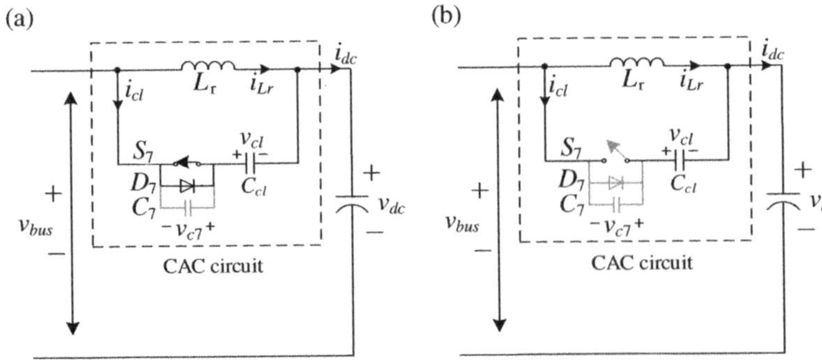

Figure 4.12 Switching state of the CAC circuit: (a) S_7's on-state; (b) S_7's off-state.

Figures 4.13 and 4.14 show the switch commutation processes and key waveforms of the CAC rectifier in sector 1-1. Before the Type 2 switching commutation occurs from $\vec{V}_1(100)$ to $\vec{V}_7(111)$, S_7 is turned off to create the ZVS turn-on condition for the bridge switches. In the Type 1 switching commutation process from $\vec{V}_7(111)$ to $\vec{V}_2(110)$ and from $\vec{V}_2(110)$ to $\vec{V}_1(100)$, the active switch S_7 keeps in conduction state.

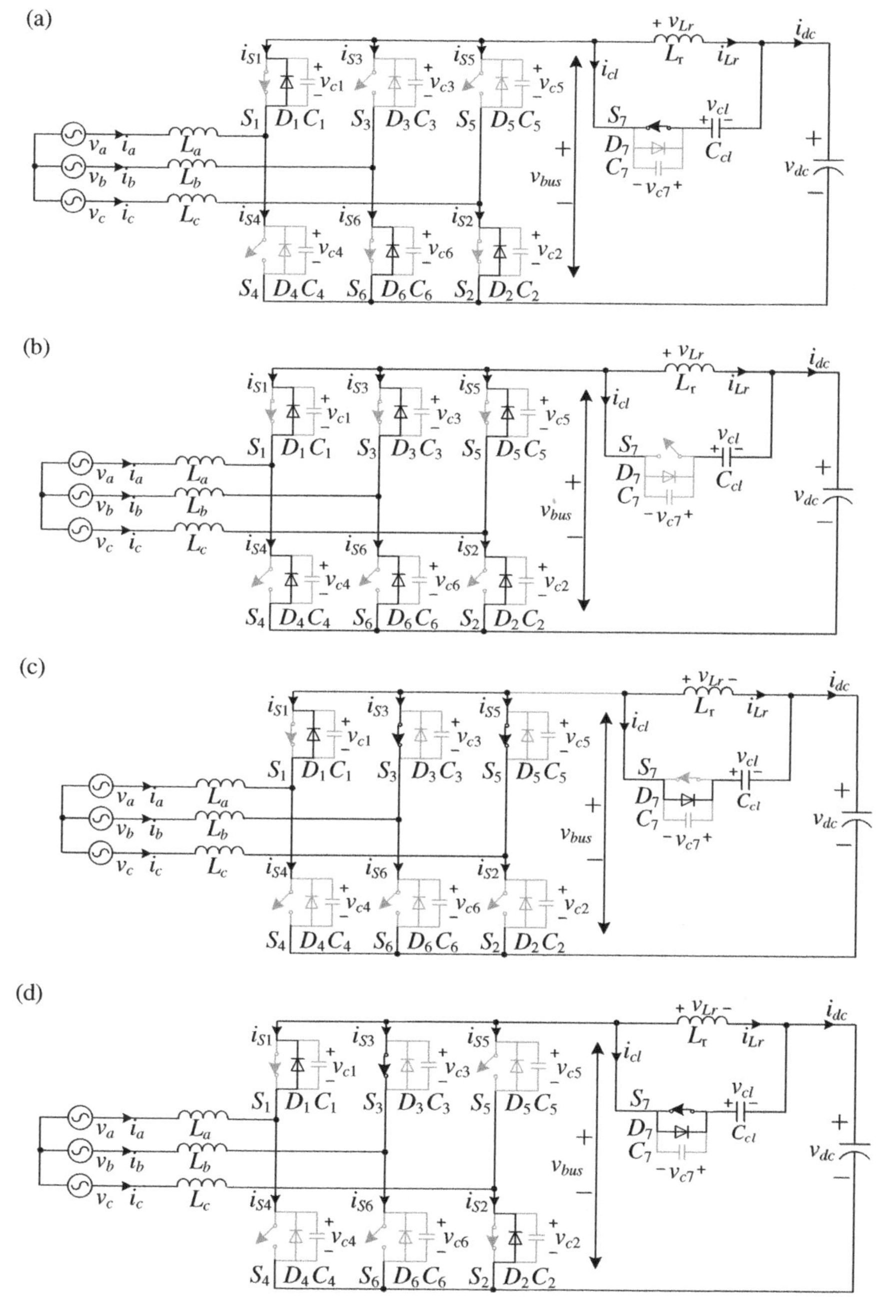

Figure 4.13 Switch commutation processes of CAC rectifier in sector 1-1. (a) Equivalent circuit for $\vec{V}_1(100)$. (b) Equivalent circuit for S_7 is off. (c) Equivalent circuit for $\vec{V}_7(111)$. (d) Equivalent circuit for $\vec{V}_2(110)$.

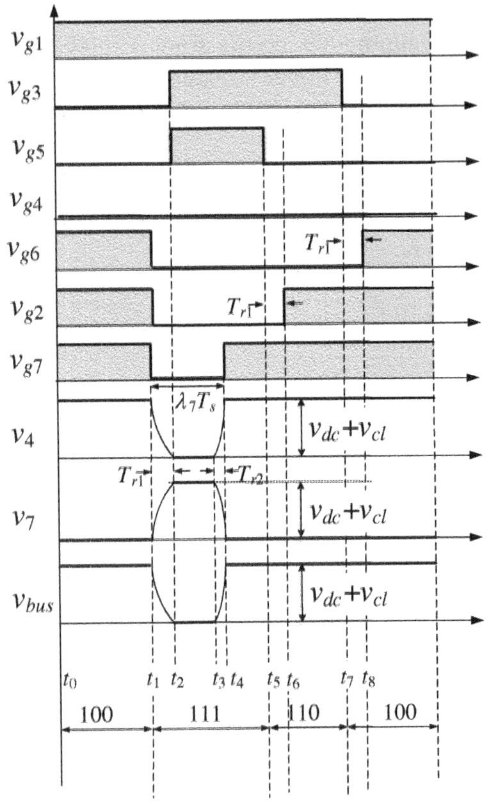

Figure 4.14 Key waveforms in sector 1-1.

With the modulation scheme as shown in Figure 4.14, the auxiliary switch S_7 needs to be turned on and off once in a switching cycle to realize the ZVS condition for the bridge switches. Then the auxiliary switch operates in the same frequency with bridge switches. The detailed circuit analysis will be given in the next section.

4.3 Circuit Analysis

4.3.1 Operation Stage Analysis

In order to simplify the analysis, following assumptions are made:

1) All of the switching devices are treated as ideal switches.
2) Filter inductors at the grid side are so large that the filter inductor currents can be seen as constant in a switching cycle T_s.
3) Parasitic capacitance of a switch is included in its paralleled capacitor.
4) All paralleled capacitors on the switches are the same, i.e. $C_1 = C_2 = C_3 = C_4 = C_5 = C_6 = C_7 = C_r$.
5) The clamping capacitor C_{cl} is so large that the clamping capacitor voltage v_{cl} can be treated as a constant value V_{cl} in a switching cycle.

Sector 1-1 is taken as an example to investigate the operation stages of the CAC rectifier. The operation of the rectifier includes eight stages. Figure 4.15 shows the waveforms of the key components in the rectifier in one switching period. The operation stages are explained as follows.

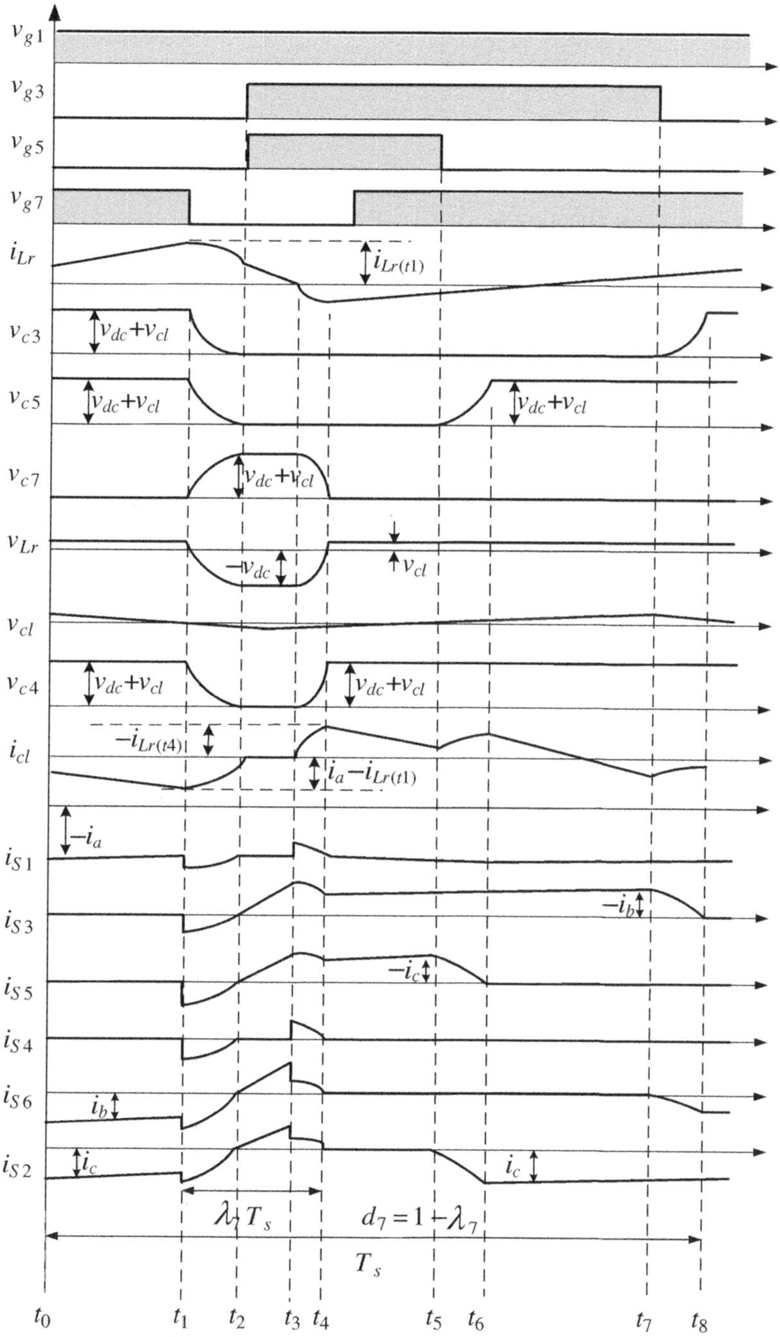

Figure 4.15 Driving logic and key waveforms in sector 1-1.

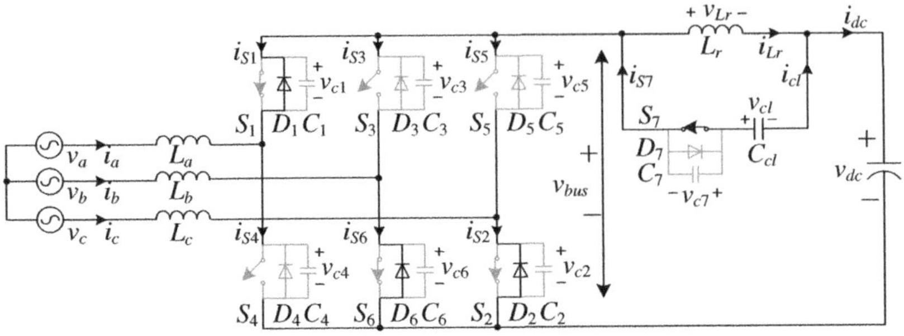

Figure 4.16 Equivalent circuit of stage 1: the initial stage.

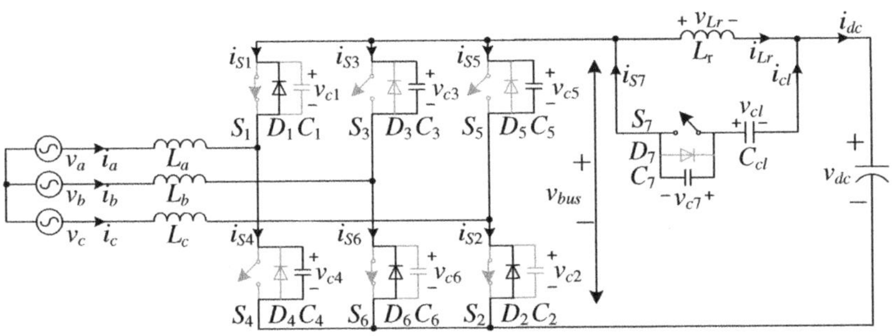

Figure 4.17 Equivalent circuit of stage 2: the first resonance stage.

Stage 1 (t_0–t_1): Initial stage. In this stage, the operation state is in vector $\vec{V}_1(100)$. The equivalent circuit is shown in Figure 4.16. Considering the current direction of AC side filters, D_1, D_2, and D_6 are conducting. Auxiliary switch S_7 conducts in this stage, and the resonant inductor L_r is charged by the clamping capacitor C_{cl}. Therefore, the current through L_r increases linearly at the rate of

$$\frac{di_{Lr}}{dt} = \frac{V_{cl}}{L_r} \tag{4.23}$$

Stage 2 (t_1–t_2): The first resonance stage. The equivalent circuit is shown in Figure 4.17. At t_1, S_7 is turned off. Resonance between the resonant inductor L_r and the resonant capacitors C_3, C_4, C_5, and C_7 is provoked. During this stage, the bus voltage v_{bus} will resonance to zero. Once the bus voltage resonances to zero, it creates ZVS condition for switches S_3 and S_5. This stage ends at t_2, when v_{bus} decreases to zero and v_{c7} increases to be $V_{dc} + V_{cl}$.

Stage 3 (t_2–t_3): The freewheeling stage. The equivalent circuit is shown in Figure 4.18. At t_2, v_{bus} oscillates to zero and D_3, D_5, and D_4 begin to conduct. S_3 and S_5 can be turned on under ZVS condition. Currents in phase b and phase c begin to transfer from D_6 and D_2 to S_3 and S_5, respectively. During this period, the current changing rates of i_{D6} and i_{D2} are restrained to a small value by L_r, so that the reverse recovery losses on D_6 and D_2 can be suppressed. i_{D6} and i_{D2} are reduced to zero at time t_3. In stage 3, the resonant inductor current i_{Lr} decreases with a rate of

$$\frac{di_{Lr}}{dt} = -\frac{V_{dc}}{L_r} \tag{4.24}$$

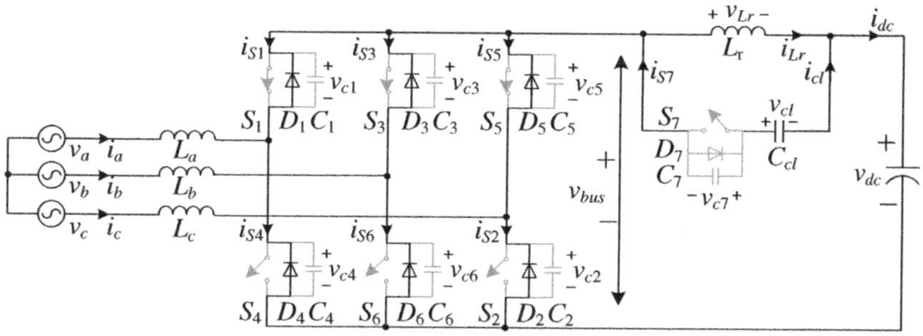

Figure 4.18 Equivalent circuit of stage 3: the freewheeling stage.

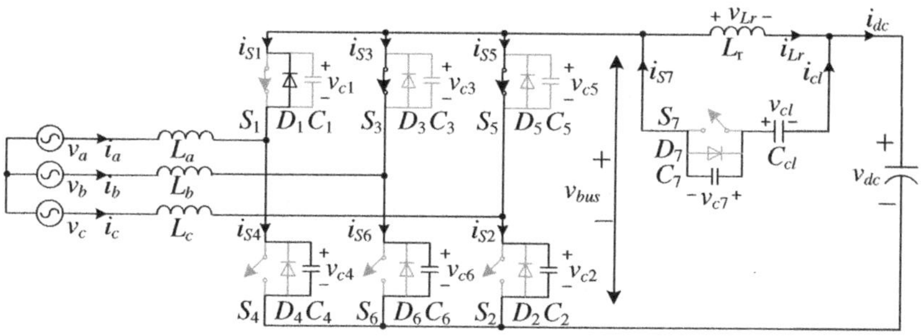

Figure 4.19 Equivalent circuit of stage 4: the second resonance stage.

Stage 4 (t_3–t_4): The second resonance stage. The equivalent circuit is shown in Figure 4.19. At t_3, i_{D6} and i_{D2} are reduced to zero and D_6 and D_2 are off. Another resonance occurs between L_r and C_2, C_4, C_6, and C_7. During this resonance, v_{bus} increases and v_{c7} decreases. At time t_4, v_{bus} oscillates to $V_{dc} + V_{cl}$ while v_{c7} reduces to zero and the antiparallel diode of S_7 starts to conduct. Once D_7 conducts at t_4, this stage ends.

Stage 5 (t_4–t_5): The second steady stage. The equivalent circuit is shown in Figure 4.20. At t_4, D_7 conducts. Then S_7 can be turned on with ZVS condition. The rectifier operates in vector $\vec{V}_7(111)$ in stage 5. S_1, S_3, S_5, and S_7 are conducting. The voltage across L_r is clamped by V_{cl} in this stage, and the resonant inductor current i_{Lr} increases with a rate of

$$\frac{di_{Lr}}{dt} = \frac{V_{cl}}{L_r} \tag{4.25}$$

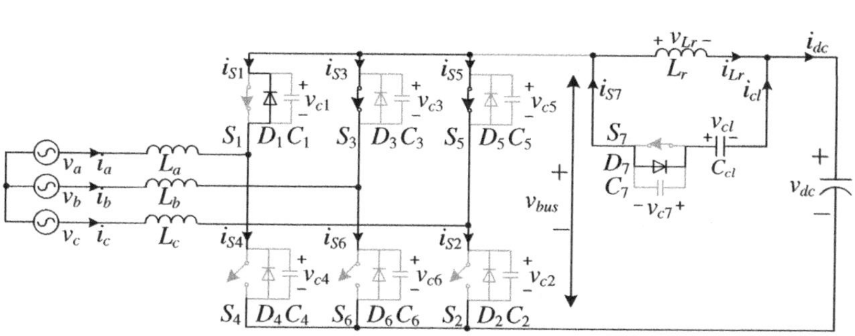

Figure 4.20 Equivalent circuit of stage 5: the second steady stage.

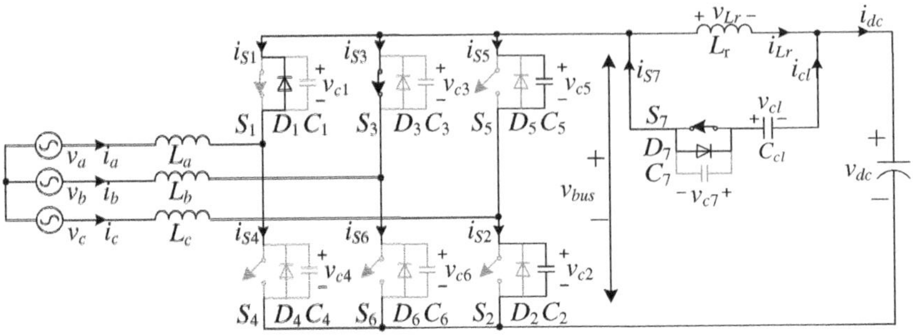

Figure 4.21 Equivalent circuit of stage 6: the Type 1 commutation stage.

Stage 6 (t_5–t_6): Type 1 commutation. The equivalent circuit is shown in Figure 4.21. At t_5, S_5 turns off. The input current i_c charges C_5 and discharges C_2. With the buffer of the resonant capacitors C_2 and C_5, S_5 is turned off under ZVS condition. This stage ends at t_6 when v_{c2} reduces to zero and D_2 conducts.

Stage 7 (t_6–t_7): The third steady stage. The equivalent circuit is shown in Figure 4.22. In this stage, rectifier is operated in voltage vector $\vec{V}_2(110)$. The resonant inductor is still clamped by v_{cl}. The resonant inductor current i_{Lr} increases with a rate of

$$\frac{di_{Lr}}{dt} = \frac{V_{cl}}{L_r} \tag{4.26}$$

Stage 8 (t_7–t_8): Type 1 commutation. The equivalent circuit is shown in Figure 4.23. At t_7, S_3 is turned off. The input current i_b charges C_3 and discharges C_6. With the buffer of the resonant capacitors C_3 and C_6, S_3 is turned off under ZVS conditions. This stage ends at t_8 when v_{c6} reduces to zero and D_6 conducts.

After stage 8, the rectifier returns to stage 1 of a new switching cycle.

4.3.2 Resonant Stages Analysis

There are two resonant stages in a switching cycle. The first one is used for ZVS turn-on for the bridge switches, and the second one realizes the ZVS turn-on for the auxiliary switch. Besides,

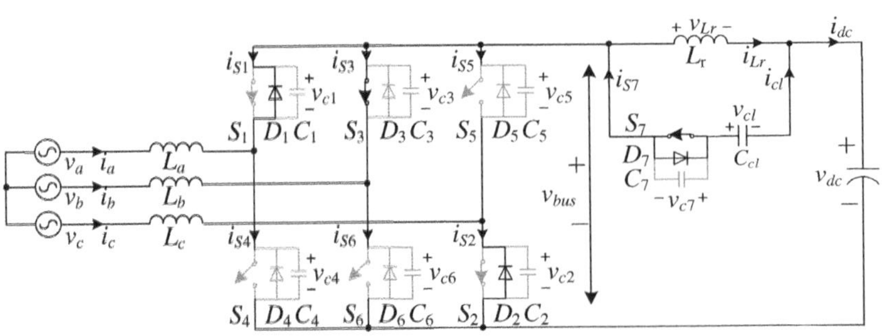

Figure 4.22 Equivalent circuit of stage 7: the third steady stage.

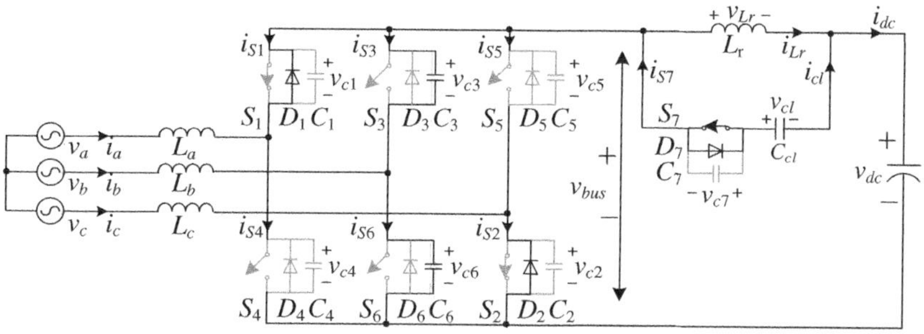

Figure 4.23 Equivalent circuit of stage 8: the Type 1 commutation stage.

the reverse recovery of the freewheeling diodes is suppressed by restraining the current changing rate with the resonant inductor. In addition, the turn-off actions for the switches are under ZVS conditions. The first resonance stage happens at stage 2. The circuit of this stage is illustrated in Figure 4.24a. It can be simplified into an equivalent circuit illustrated in Figure 4.24b. Here the AC side current is assumed as constant.

The first resonance stage begins at t_1. The initial values of the capacitor voltage $v_{c3}(t_1) = v_{c4}(t_1) = v_{c5}(t_1) = V_{dc} + V_{cl}$; $v_{c7}(t_1) = 0$. The initial values of the resonant inductor current and bus voltage at time t_1 are expressed by

$$\begin{cases} i_{Lr}(t_1) = i_{Lr(t1)} \\ v_{bus}(t_1) = V_{dc} + V_{cl} \end{cases} \tag{4.27}$$

Considering $v_{c7} + v_{bus} = V_{dc} + V_{cl}$, v_{bus} keeps reducing while v_{c7} increases during the first resonance stage. This stage ends at t_2, when v_{bus} decreases to zero and v_{c7} increases to $V_{dc} + V_{cl}$. According to the equivalent circuit in Figure 4.24b, following differential equation is derived:

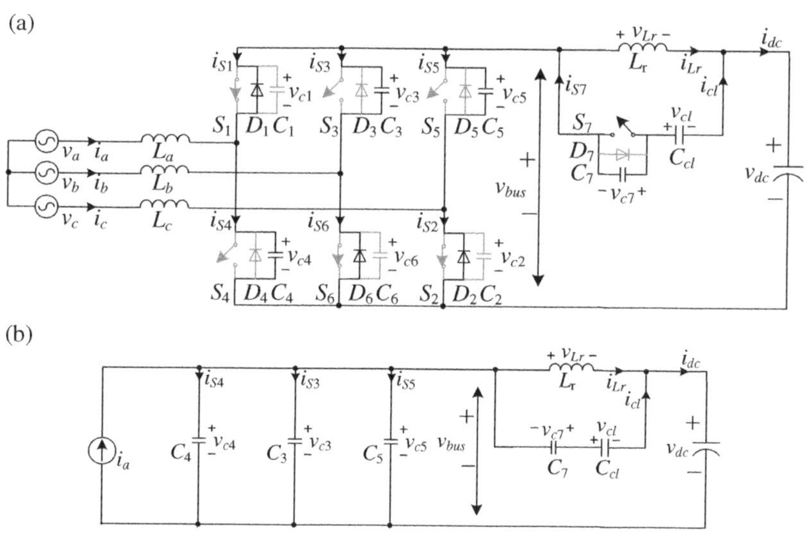

Figure 4.24 Resonance circuit and its equivalent circuit of stage 2 (t_1–t_2): (a) circuit state of first resonance stage; (b) equivalent circuit.

$$\begin{cases} L_r \dfrac{di_{Lr}}{dt} = v_{bus} - V_{dc} \\ 4C_r \dfrac{dv_{bus}}{dt} = -i_{Lr} + i_a \end{cases} \tag{4.28}$$

Solving the differential equation (4.28) with the initial value in (4.27), i_{Lr} and v_{bus} are derived as

$$\begin{cases} i_{Lr} = i_a + \dfrac{V_{cl}}{Z_r} \sin \omega_r t - \left(i_a - i_{Lr(t1)}\right) \cos \omega_r t \\ v_{bus} = V_{dc} + V_{cl} \cos \omega_r t + \left(i_a - i_{Lr(t1)}\right) \cdot Z_r \sin \omega_r t \end{cases} \tag{4.29}$$

where ω_r is the resonant angular frequency and is equal to

$$\omega_r = 1/\sqrt{4L_r C_r} \tag{4.30}$$

The duration of the first resonant stage T_{r1} can be estimated as:

$$T_{r1} = \frac{1}{4f_r} = \frac{\pi}{2\omega_r} = \pi\sqrt{C_r L_r} \tag{4.31}$$

where f_r is the resonant frequency and equal to $\omega_r/(2\pi)$.

Z_r is the resonant impedance and is defined as:

$$Z_r = \sqrt{L_r/4C_r} \tag{4.32}$$

Equation (4.29) can be rewritten as:

$$\begin{cases} i_{Lr}(t) = i_a - \sqrt{V_{cl}^2/Z_r^2 + \left(i_a - i_{Lr(t1)}\right)^2} \cos\left(\omega_r t + \tau\right) \\ v_{bus}(t) = V_{dc} + \sqrt{V_{cl}^2 + \left(i_a - i_{Lr(t1)}\right)^2 Z_r^2} \sin\left(\omega_r t + \tau\right) \end{cases} \tag{4.33}$$

where

$$\tau = \arctan \frac{V_{cl}}{\left|i_a - i_{Lr(t1)}\right| Z_r} \tag{4.34}$$

According to resonant stage analysis, to realize ZVS turn-on for the main switches, it is necessary that the bus voltage v_{bus} will resonance to zero. v_{bus} in Eq. (4.33) consists of two terms: DC component and a sinusoidal component. If v_{bus} needs to become negative, the amplitude of the sinusoidal term must be larger than the DC component. Thus we get

$$\sqrt{V_{cl}^2 + \left(i_a - i_{Lr(t1)}\right)^2 Z_r^2} > V_{dc} \tag{4.35}$$

By further simplifying Eq. (4.35), the ZVS turn-on condition of the bridge switches can be rewritten as:

$$i_{Lr(t1)} > i_a + \frac{\sqrt{V_{dc}^2 - V_{cl}^2}}{Z_r} \tag{4.36}$$

The second resonance stage occurs at stage 4. The circuit state of this stage is illustrated in Figure 4.25a. Since D_1, S_3, and S_5 are in conduction, the grid side current does not influence the circuit resonance. The equivalent circuit of this stage is illustrated in Figure 4.25b. During this resonance, resonance occurs among L_r, C_2, C_4, C_6, and C_7, and v_{bus} increases and v_{c7} decreases.

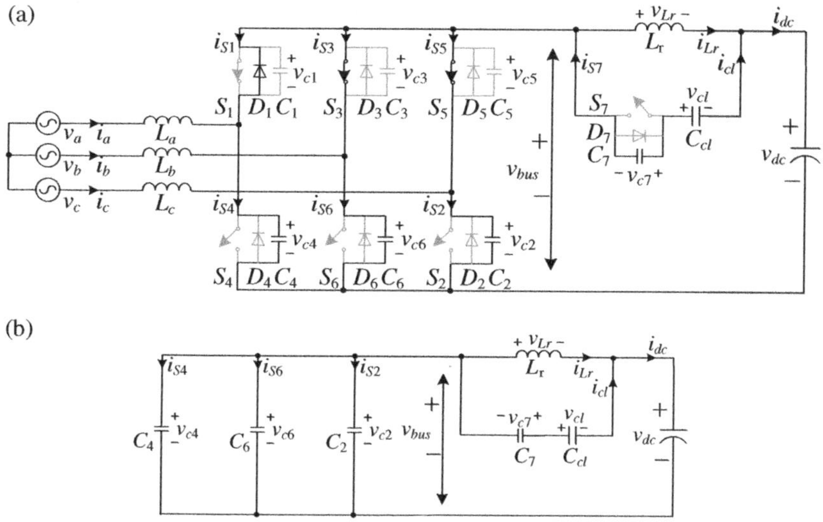

Figure 4.25 Equivalent circuit of stage 4 (t_3–t_4): (a) circuit state of the second resonance; (b) equivalent circuit.

The second resonance stage begins at t_3. The initial values of the capacitor voltage $v_{c2}(t_3) = v_{c4}(t_3) = v_{c6}(t_3) = 0$; $v_{c7}(t_3) = V_{dc} + V_{cl}$. The initial values of the resonant inductor current and bus voltage at time t_3 are expressed by

$$\begin{cases} i_{Lr}(t_3) = 0 \\ v_{bus}(t_3) = 0 \end{cases} \tag{4.37}$$

According to the equivalent circuit in Figure 4.25b, following differential equation is derived.

$$\begin{cases} L_r \dfrac{di_{Lr}}{dt} = v_{bus} - V_{dc} \\ 4C_r \dfrac{dv_{bus}}{dt} = -i_{Lr} \end{cases} \tag{4.38}$$

By solving the differential equation (4.38) with the initial value of (4.37), following state variables are deducted:

$$\begin{cases} i_{Lr}(t) = -\dfrac{V_{dc}}{Z_r} \sin \omega_r t \\ v_{bus}(t) = V_{dc} - V_{dc} \cos \omega_r t \end{cases} \tag{4.39}$$

Considering $v_{bus} + v_{c7} = V_{dc} + V_{cl}$, combining Eq. (4.39), v_{c7} can be derived as:

$$v_{c7}(t) = V_{cl} + V_{dc} \cos \omega_r t \tag{4.40}$$

To realize ZVS turn-on of the auxiliary switch S_7, v_{c7} should resonance to zero.

Since the clamping capacitor voltage V_{cl} is much lower than the DC side voltage V_{dc}, v_{c7} can always reach to zero according to Eq. (4.40). The resonant time of the second resonant stage T_{r2} is extremely small in comparison to the resonant time of first resonant stage T_{r1}, it can be ignored in calculation.

The second resonance stage stops at t_4, combining Eq. (4.39), the resonant inductor current i_{Lr} and capacitor voltage v_{bus} at time t_4 can be derived as:

$$i_{Lr}(t_4) = -\frac{V_{dc}}{Z_r} \cdot \sqrt{1 - \frac{V_{cl}^2}{V_{dc}^2}} \tag{4.41}$$

$$v_{bus}(t_4) = V_{dc} + V_{cl} \tag{4.42}$$

Actually, i_{Lr} reaches its minimum value i_{Lr_min} at t_4. Then (4.41) can be rewritten as:

$$i_{Lr}(t_4) = i_{Lr\_\text{min}} = -\frac{V_{dc}}{Z_r} \cdot \sqrt{1 - \frac{V_{cl}^2}{V_{dc}^2}} \tag{4.43}$$

4.3.3 Steady State Analysis

When the rectifier is operated in steady state, the mean voltage of the resonant inductor in a switching cycle is zero. That is,

$$\frac{1}{T_s}\int_{t_0}^{t_0 + T_s} v_{Lr}dt = 0 \tag{4.44}$$

The voltage waveform of the resonant inductor v_{Lr} is shown in Figure 4.15. Since the intervals (t_1–t_2) and (t_3–t_4) during the resonance process are very short in a switch cycle, the integrations of the v_{Lr} during (t_1–t_2) and (t_3–t_4) can be ignored. Thus, Eq. (4.44) can be written as:

$$\frac{1}{T_s}\int_0^{T_s} v_{Lr}dt = \frac{1}{T_s}\left(\int_0^{t_1} v_{Lr}dt + \int_{t_2}^{t_3} v_{Lr}dt + \int_{t_4}^{t_8} v_{Lr}dt\right) \tag{4.45}$$

According to Figure 4.15, during (t_0–t_1) and (t_4–t_8), the resonant inductor v_{Lr} is expressed as:

$$v_{Lr} = V_{cl} \tag{4.46}$$

During (t_2–t_3), the resonant inductor v_{Lr} is expressed as:

$$v_{Lr} = -V_{dc} \tag{4.47}$$

By substituting Eqs. (4.46) and (4.47) into Eq. (4.45), the following equation can be obtained:

$$\frac{1}{T_s}\int_0^{T_s} v_{Lr}dt = V_{cl}d_7T_s - V_{dc}(1 - d_7)T_s \tag{4.48}$$

where d_7 is duty ratio of the auxiliary switch

$$d_7 = \frac{T_s - (t_4 - t_1)}{T_s} \tag{4.49}$$

Following equation is obtained by combining Eqs. (4.44) and (4.48):

$$V_{cl}d_7T_s - V_{dc}(1 - d_7)T_s = 0 \tag{4.50}$$

Define λ_7 to be the auxiliary switch turn-off duty cycle:

$$\lambda_7 = 1 - d_7 \tag{4.51}$$

By substituting Eqs. (4.50) and (4.51) into Eq. (4.49), the expression of the clamping capacitor voltage is

$$V_{cl} = \frac{(1-d_7)V_{dc}}{d_7} = \frac{\lambda_7 V_{dc}}{1-\lambda_7} \tag{4.52}$$

Because the auxiliary switch turn-off duty cycle λ_7 is 5–10%, the clamping capacitor voltage is less than 0.1 times of the DC-link voltage.

When the rectifier is operated in steady state, the mean current through clamping capacitor is zero. That is,

$$\frac{1}{T_s}\int_{t_0}^{t_0+T_s} i_{cl}dt = 0 \tag{4.53}$$

The current waveform of the clamping capacitor i_{cl} is shown in Figure 4.15. Since the intervals during $(t_1–t_2)$, $(t_3–t_4)$, $(t_5–t_6)$, and $(t_7–t_8)$ are the resonance process, these intervals are very short in a switch cycle. To simplify the expression, the integrations of the i_{cl} during $(t_1–t_2)$, $(t_3–t_4)$, $(t_5–t_6)$, and $(t_7–t_8)$ are ignored in the analysis. The Eq. (4.53) can be expressed as:

$$\frac{1}{T_s}\int_{t_0}^{t_0+T_s} i_{cl}dt = \frac{1}{T_s}\left(\int_{t_0}^{t_1} i_{cl\_stage\ 1}(t)dt + \int_{t_2}^{t_3} i_{cl\_stage\ 3}(t)dt + \int_{t_4}^{t_5} i_{cl\_stage\ 5}(t)dt + \int_{t_6}^{t_7} i_{cl\_stage\ 7}(t)dt\right) = 0 \tag{4.54}$$

According to resonant stage analysis, in stage 3 $(t_2–t_3)$, S_7 is off, then the current through clamping capacitor is zero from t_2 to t_3, $i_{cl\_stage3} = 0$.

According to resonant stage analysis, combining Eq. (4.43), i_{Lr} reaches its minimum value i_{Lr_min} at t_4. In stage 5 $(t_4–t_5)$, the rectifier is operated in voltage vector $\vec{V}_7(111)$, $i_{dc} = 0$, $i_{cl\_stage\ 5} = -i_{Lr\_stage\ 5}$. Combining Eqs. (4.25) and (4.43), the current through clamping capacitor can be expressed as:

$$i_{cl\_stage\ 5}(t) = -i_{Lr\_stage\ 5} = -\left(i_{Lr\_min} + \frac{V_{cl}}{L_r}(t-t_4)\right) = \frac{V_{dc}}{Z_r}\cdot\sqrt{1-\frac{V_{cl}^2}{V_{dc}^2}} - \frac{V_{cl}}{L_r}(t-t_4) \tag{4.55}$$

In stage 7 $(t_6–t_7)$, the rectifier is operated in voltage vector $\vec{V}_2(110)$, $i_{dc} = |i_c|$, $i_{cl} = -i_{Lr} + |i_c|$. Combining Eqs. (4.25) and (4.43), the current through clamping capacitor can be expressed as:

$$i_{cl\_stage\ 7}(t) = -i_{Lr\_stage\ 7} + |i_c| = \frac{V_{dc}}{Z_r}\cdot\sqrt{1-\frac{V_{cl}^2}{V_{dc}^2}} - \frac{V_{cl}}{L_r}(t-t_4) + |i_c| \tag{4.56}$$

In stage 1 $(t_0–t_1)$, the rectifier is operated in voltage vector $\vec{V}_1(100)$, $i_{dc} = i_a$, $i_{cl\_stage\ 1} = -i_{Lr\_stage\ 1} + i_a$. Combining Eqs. (4.25) and (4.43), the current through clamping capacitor can be expressed as:

$$i_{cl\_stage\ 1}(t) = -i_{Lr\_stage\ 1} + i_a = \frac{V_{dc}}{Z_r}\cdot\sqrt{1-\frac{V_{cl}^2}{V_{dc}^2}} - \frac{V_{cl}}{L_r}(t-t_0+T_s-t_4) + i_a \tag{4.57}$$

Since the intervals during *stage 1* $(t_0–t_1)$, *stage 5* $(t_4–t_5)$, and *stage 7* $(t_6–t_7)$ are $d_{V1}T_s$, $[d_{V0}-(1-d_7)]T_s$, and $d_{V2}T_s$, respectively, combining Eqs. (4.55), (4.56), and (4.57), the Eq. (4.54) is derived as:

$$\frac{V_{dc}}{Z_r}\cdot\sqrt{1-\frac{V_{cl}^2}{V_{dc}^2}}\cdot d_7 T_s - \frac{V_{dc}(1-d_7)}{2L_r}d_7 T_s^2 + |i_c|d_{V2}T_s + i_a d_{V1}T_s = 0 \tag{4.58}$$

By substituting Eq. (4.52) into Eq. (4.58), the following equation can be obtained:

$$\frac{V_{dc}}{Z_r}\cdot\sqrt{2d_7-1}\cdot T_s-\frac{V_{dc}(1-d_7)}{2L_r}d_7T_s^2+|i_c|d_{V2}T_s+i_a d_{V1}T_s=0 \tag{4.59}$$

Considering the auxiliary switch S_7's duty ratio d_7 is very close to one, then

$$\sqrt{2d_7-1}\approx 1 \tag{4.60}$$

By substituting preceding equation into Eq. (4.59), the equation can be derived as:

$$d_7=1-\frac{2L_r\left(i_a d_{V1}+|i_c|d_{V2}+\dfrac{V_{dc}}{Z_r}\right)}{V_{dc}T_s} \tag{4.61}$$

Considering the rectifier operates under unit power factor and the grid current power factor angle φ_0 is zero, combining Eq. (4.20), the expression of grid side current and the duty ratio of nonzero voltage vector are rewritten as follows:

$$\begin{cases} i_a=I_m\cos\theta \\ i_c=I_m\cos\left(\theta+\dfrac{2\pi}{3}\right)\end{cases} \tag{4.62}$$

$$\begin{cases} d_{V1}=\dfrac{\sqrt{3}}{2}m\sin(\pi/3-\theta) \\ d_{V2}=\dfrac{\sqrt{3}}{2}m\sin\theta\end{cases} \tag{4.63}$$

where θ is the phase angle of the reference voltage vector, and m is the voltage modulation ratio.

By substituting Eqs. (4.62) and (4.63) into Eq. (4.61), following equation is derived:

$$d_7=1-\frac{2L_r\left(\dfrac{3}{4}mI_m+\dfrac{V_{dc}}{Z_r}\right)}{T_sV_{dc}} \tag{4.64}$$

By substituting the preceding equation into Eq. (4.52), the clamping capacitor voltage equation can be derived as:

$$V_{cl}=\frac{2L_rV_{dc}\left(\dfrac{3}{4}mI_m+\dfrac{V_{dc}}{Z_r}\right)}{T_sV_{dc}-2L_r\left(\dfrac{3}{4}mI_m+\dfrac{V_{dc}}{Z_r}\right)} \tag{4.65}$$

4.3.4 Soft-switching Condition

According to the earlier resonance stage analysis, from t_4 to t_1, the resonance inductor L_r is clamped by the clamping capacitor C_{cl}, then the inductor current at t_1 can be expressed as:

$$i_{Lr(t1)}=i_{Lr(t4)}+\frac{V_{cl}}{L_r}(T_s-t_4+t_1) \tag{4.66}$$

By substituting the preceding equation (4.49) into Eq. (4.66), the equation can be derived as:

$$i_{Lr(t1)}=i_{Lr(t4)}+\frac{V_{cl}}{L_r}d_7T_s \tag{4.67}$$

By substituting Eqs. (4.43) and (4.61) into Eq. (4.67), the equation can be derived as:

$$i_{Lr(t1)} = -\frac{V_{dc}}{Z_r} \cdot \sqrt{1 - \frac{V_{cl}^2}{V_{dc}^2}} + \frac{3}{2} m I_m + \frac{2V_{dc}}{Z_r} \tag{4.68}$$

Actually, i_{Lr} reaches its maximum value i_{Lr_max} at t_1. By substituting the preceding equation to the main switches' ZVS turn-on condition (4.36), the ZVS turn-on condition of bridge switches can be rewritten as:

$$\frac{3}{2} m I_m - i_a + \frac{2\left(V_{dc} - \sqrt{V_{dc}^2 - V_{cl}^2}\right)}{Z_r} \geq 0 \tag{4.69}$$

By considering $V_{dc} - \sqrt{V_{dc}^2 - V_{cl}^2}$ is always larger than zero, to satisfy the inequality (4.69), following condition should be satisfied as:

$$\frac{3}{2} m I_m - i_a \geq 0 \tag{4.70}$$

where I_m is the amplitude of the grid side current i_a, if the rectifier modulation ratio m is larger than 2/3, inequality (4.70) can be satisfied. Combining m's expression in Eq. (4.21), when the rectifier AC line to line voltage is 380 V RMS, if $V_{dc} < 933$ V, $m > 2/3$, then the soft-switching condition for the bridge switches in stage 2 can be satisfied. Normally, when the AC line to line voltage of the three-phase rectifier is 380 V RMS, the DC voltage is generally lower than 800 V. Therefore, the soft-switching condition of CAC rectifier's bridge switches can easily be satisfied.

4.3.5 Control Technique of Compound Active-clamping Three-phase Rectifier

Once the reference sector is obtained by the control loop, the location and the magnitude of the reference sector can also be determined. With the modulation sequences of the ZVS-SVM in Table 4.1 and Eq. (4.20), the used vectors and their durations, d_{V0}, d_{V1}, and d_{V2}, can be easily obtained. The duty ratio of the auxiliary switch d_7 can be calculated with the Eq. (4.61). Take sector 1-1 as an example to introduce the procedures of gate signal derivation, the related waveforms are shown in Figure 4.26.

In Figure 4.26, v_{ca} is the saw-tooth carrier wave, and one count up timer period is equal to T_s. According to the procedures of driving sequences derivation of sector 1-1, four modulation indexes for four PWM modules based on d_{V0}, d_{V1}, d_{V2}, and d_7 can be deduced as:

$$\begin{cases} m_a = 0 \\ m_b = d_{V1} \\ m_c = 1 - d_{V0} \\ m_{d7} = d_7 \end{cases} \tag{4.71}$$

With the specific action scheme applying to the count up timer and counter-compare values, the switching functions can be obtained. With inverting and turn-on delay, each switching function will be expanded into two complementary primary driving sequences. For example, with inverting and turn-on delay T_{r1}, modulation index m_b is expanded into v_{g3} and v_{g6}. All the time delay for the primary driving sequences of three-phase bridges are set as T_{r1}.

The control block diagram of the three-phase CAC rectifier based on the synchronous rotating coordinate system is shown in Figure 4.27 [5–8]. The output of the DC voltage control loop is the reference of active current i_d; the reference of reactive current i_q is zero. Through grid voltage feed

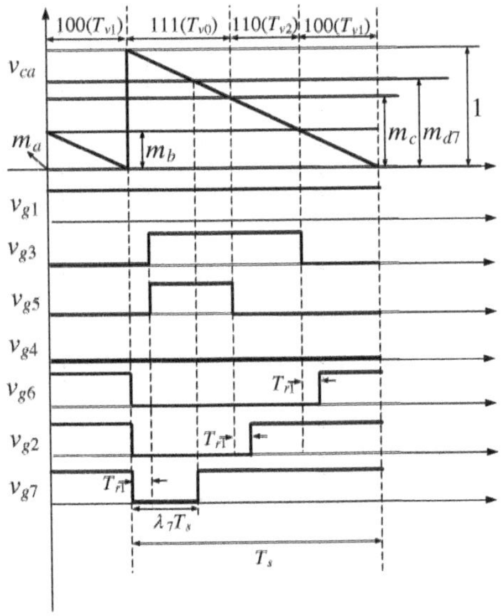

Figure 4.26 Gate signal derivation in sector 1-1.

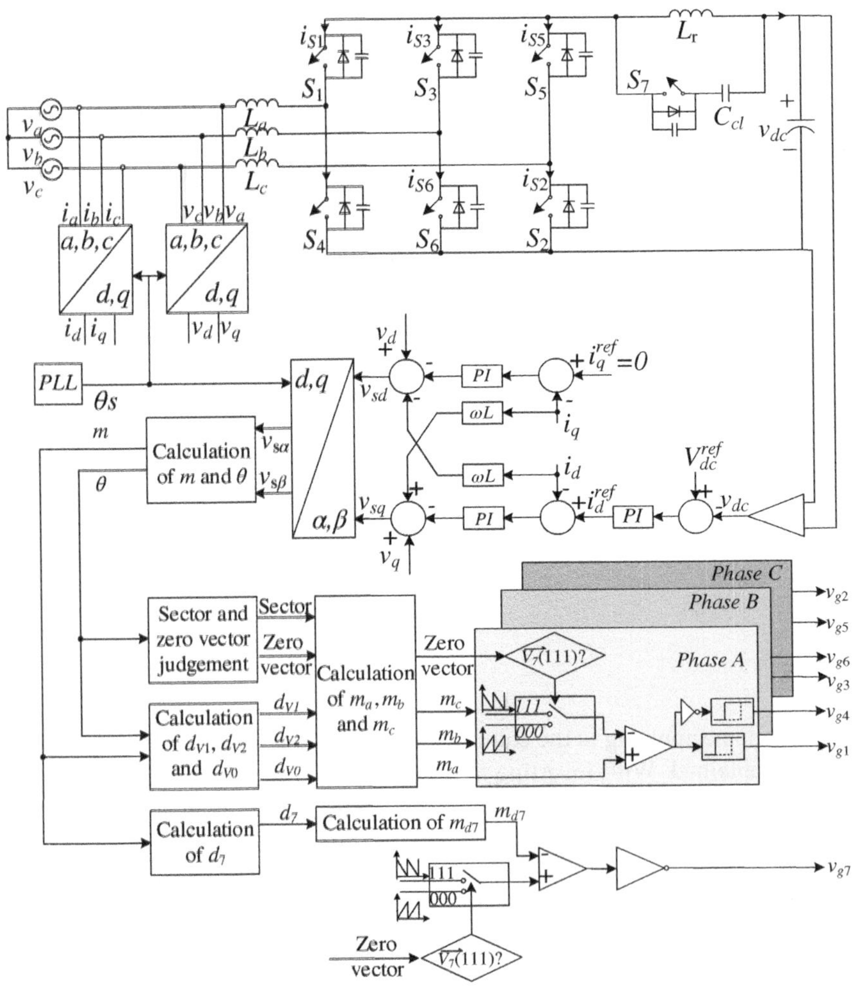

Figure 4.27 Control block diagram of the rectifier.

forward and inductor current decoupling control, the modulation ratio m and the rotation vector angle θ can be obtained.

With the Eq. (4.20), the duty ratio of the nonzero voltage vector and zero voltage vector, d_{V0}, d_{V1}, and d_{V2}, can be calculated through the modulation ratio m and the rotation vector angle θ. The duty ratio of the auxiliary switch d_7 can be calculated with the Eq. (4.61). With Eq. (4.20), four modulation indexes m_a, m_b, m_c, and m_{d7} based on d_{V0}, d_{V1}, d_{V2}, and d_7 can be obtained.

With the modulation sequences of the ZVS-SVM in Table 4.2, voltage sector and zero voltage vector can be selected according to the rotation vector angle θ. According to sector selection, zero voltage vector judgment, the saw-tooth carrier wave v_c can also be chosen. Through comparing of the modulation indexes m_a, m_b, m_c, and m_{d7} with the carrier wave, all the switches' driving signals can finally be gained.

4.4 Prototype Design

The previous sections introduce the operation principles of the proposed ZVS rectifier and the modulation scheme. The theoretical derivations show the essence and relationships among different parameters. In this section, considerations and design guidelines will be provided through a design example.

4.4.1 Specifications of a 40 kW Rectifier

In order to illuminate the design considerations, a 40 kW ZVS grid-connected rectifier prototype was introduced. The specifications of the rectifier shown in Figure 4.28 are listed in Table 4.3. Insulated Gate Bipolar Transistor (IGBT) modules with capacitance of 1200 V/200 A are selected for switching devices S_1–S_7. It is assumed that $C_1 = C_2 = C_3 = C_4 = C_5 = C_6 = C_7 = C_r$.

4.4.2 Parameter Design

CAC rectifier parameter design needs to consider many factors, not only soft-switching condition but also the resonant impedance, resonant period, and the duty ratio of the auxiliary switch.

The resonant parameters of the CAC rectifier are designed as follows:

1) Set the minimum value of the resonant inductance to ensure the ability to suppress the reverse recovery current of the diode. The resonant inductor suppresses the reverse recovery energy of

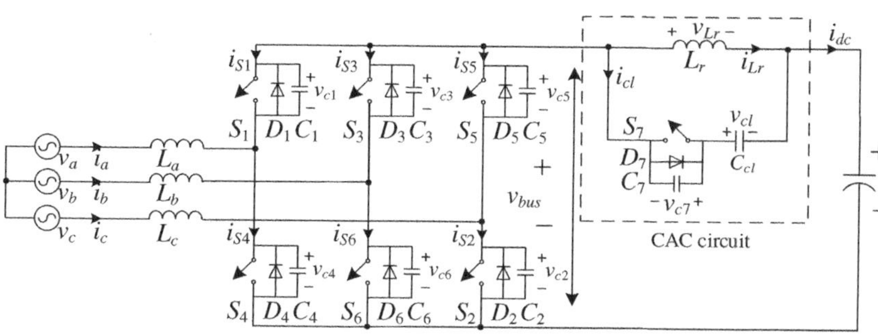

Figure 4.28 Topology of the CAC rectifier.

Table 4.3 Specifications of the ZVS rectifier.

Parameters	Descriptions	Value
P_o	Output power	40 kW
V_a, V_b, V_c	Phase voltage RMS value	220 Vrms
f_g	Frequency of the grid	50 Hz
f_s	Switching frequency	16 kHz
L_a, L_b, L_c	Filter inductance	0.3 mH
V_{dc}	DC source voltage	700 V

the freewheeling diodes in stage 3. In order to suppress the reverse recovery losses of the diodes, the di_{Lr}/dt during the turn-off process of the diodes is selected to be lower than 100 A/μs. In stage 3, it is expressed as:

$$\frac{di_{Lr}}{dt} = \frac{V_{dc}}{L_r} \leq 100\ \text{A}/\mu\text{s} \tag{4.72}$$

Therefore, in this case, the inductance should satisfy

$$L_r \geq 7\ \mu\text{H} \tag{4.73}$$

2) To avoid loss of duty ratio, the auxiliary switch turn-off duty cycle λ_7 should be less than d_{V0}. Combining Eq. (4.20), the expression of the zero voltage vector duty ratio d_{V0} is rewritten as follows:

$$d_{V0} = 1 - \frac{\sqrt{3}}{2} m \sin\left(\frac{\pi}{3} + \theta\right) \tag{4.74}$$

where θ is the phase angle of the reference voltage vector, and m is the voltage modulation ratio. Due to the maximum value of $\sin(\pi/3 + \theta)$ is one, the minimum value of d_{V0} can be derived as:

$$d_{V0\,min} = 1 - \frac{\sqrt{3}}{2} m \tag{4.75}$$

By substituting Eq. (4.21) into Eq. (4.75), the following equation is derived:

$$d_{V0\,min} = 1 - \sqrt{3}\frac{\left|\vec{V}_{ref}\right|}{V_{dc}} \tag{4.76}$$

Ignore the voltage drop of the grid side filter inductor; rectifier's reference voltage vector module $\left|\vec{V}_{ref}\right|$ is equal to the amplitude value of the grid side voltage V_m. In this case, by substituting $V_{dc} = 700$ V and $\left|\vec{V}_{ref}\right| = V_m = 220 \cdot \sqrt{2} = 311$ V into Eq. (4.76), the minimum value of d_{V0} can be calculated, $d_{V0min} = 0.23$. Considering λ_7 should be less than d_{V0}, the following inequality can be derived:

$$\lambda_7 < 0.23 \tag{4.77}$$

By substituting Eqs. (4.21) and (4.64) into Eq. (4.51), λ_7 is derived as follows:

$$\lambda_7 = \frac{L_r\left(3\frac{\left|\vec{V}_{ref}\right|}{V_{dc}}I_m + 2\frac{V_{dc}}{Z_r}\right)}{T_s V_{dc}} \tag{4.78}$$

Equation (4.78) can be rewritten as:

$$L_r = \frac{\lambda_7 T_s V_{dc}}{\left(3\frac{\left|\vec{V}_{ref}\right|}{V_{dc}}I_m + 2\frac{V_{dc}}{Z_r}\right)} \tag{4.79}$$

In this case, the rated power of the CAC rectifier is $P_o = 40$ kW, ignore the current ripple of th- e grid side filter inductor, and the amplitude value of the grid side current $I_m = \sqrt{2}P_o/3 \cdot 220 = 85.7\text{A}$. The switching cycle T_s=1/f_s=62.5 μs. By substituting $V_{dc} = 700$ V, I_m = 85.7 A, T_s = 62.5 μs, and $\left|\vec{V}_{ref}\right| = 311$ V into Eq. (4.79), the relationship between resonance impedance Z_r and resonance inductance L_r under different λ_7 is shown in Figure 4.29. The upper boundary of the available λ_7 is mainly determined by the value of L_r. When Z_r is between 40 and 80 Ω, the upper limits of L_r is between 60 and 68 μH.

3) Combining operation stage analysis in Section 4.3.1, i_{Lr} reaches its maximum value i_{Lr_max} at t_1. Then (4.68) can be rewritten as:

$$i_{Lr(t1)} = i_{Lr\_max} = -\frac{V_{dc}}{Z_r}\cdot\sqrt{1-\frac{V_{cl}^2}{V_{dc}^2}} + \frac{3}{2}mI_m + \frac{2V_{dc}}{Z_r} \tag{4.80}$$

Because the clamping capacitor voltage V_{cl} is normally less than 0.1 times of the DC-link voltage V_{dc}, by further simplifying Eq. (4.80), V_{cl} is ignored in calculation, the maximum value of i_{Lr} can be rewritten as:

$$i_{Lr(t1)} = i_{Lr\_max} \approx \frac{3}{2}mI_m + \frac{V_{dc}}{Z_r} \tag{4.81}$$

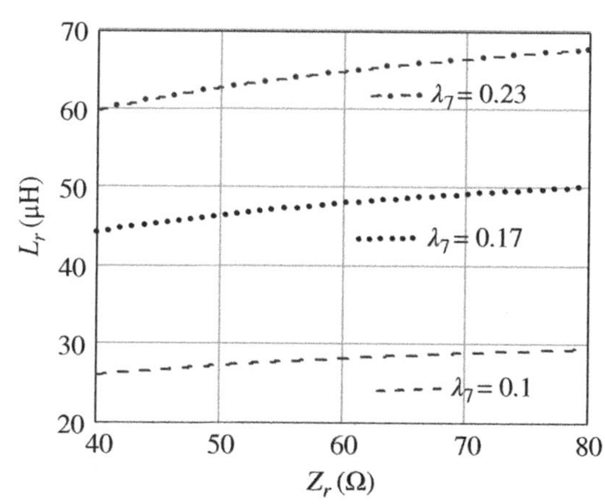

Figure 4.29 λ_7 versus Z_r.

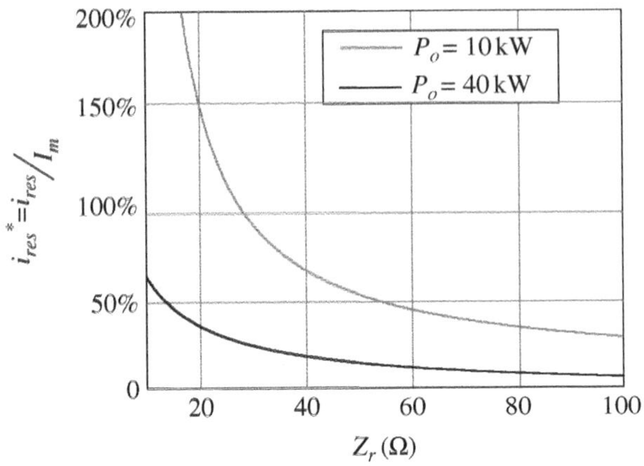

Figure 4.30 i_{res}/I_m versus resonant impedance Z_r.

In stage 1 (t_0–t_1), the current of the auxiliary switch S_7 is $i_{S7} = i_{Lr} - i_a$. By substituting equation $i_{S7} = i_{Lr} - i_a$ into Eq. (4.81), i_{S7} at time t_1 is expressed by

$$i_{S7(t1)} = -i_{cl(t1)} = -i_a + \frac{3}{2}mI_m + \frac{V_{dc}}{Z_r} \tag{4.82}$$

Actually, i_{S7} reaches its maximum value i_{S7_max} at t_1. When the rectifier is operated in steady state, the mean current through auxiliary switch S_7 and clamping capacitor C_{cl} in a switching cycle is zero. Therefore, the larger the $i_{S7(t1)}$, the greater the conduction loss of S_7 is. The Eq. (4.82) consists of two parts, $-i_a + 3mI_m/2$ is independent of resonant parameters, only V_{dc}/Z_r is affected by resonant parameters. Defining V_{dc}/Z_r is the additional current i_{res}, which is expressed by

$$i_{res} = \frac{V_{dc}}{Z_r} \tag{4.83}$$

The relationship between the i_{res}/I_m and resonance impedance Z_r under different power condition is shown in Figure 4.30. The lighter the load is, the larger i_{res}/I_m is. In this case, set i_{res}/I_m under rated power less than 20%, combining Eq. (4.83), i_{res}/I_m should meet the following inequality as:

$$\frac{i_{res}}{I_m} = \frac{V_{dc}}{I_m \cdot Z_r} \leq 20\% \tag{4.84}$$

By substituting $V_{dc} = 700$ V and $I_m = 85.7$ A to inequality (4.84), the minimum value of Z_r is limited to 41Ω. According to the resonant impedance Z_r in Eq. (4.32), the relationship between resonant impedance Z_r versus resonant parameters is shown in Figure 4.31.

4) In the ZVS rectifiers, since all of the switching devices can be turned on under ZVS conditions, turn-on losses of the switches can be eliminated. Besides, all the switches are turned off under ZVS condition with the buffering by resonant capacitors. The turn-off losses reduction depends on the capacitance of the resonant capacitors. The turn-off losses under ZVS condition of IGBT module CM200DU-24NFH is demonstrated in Figure 4.32. This figure implies that for a given

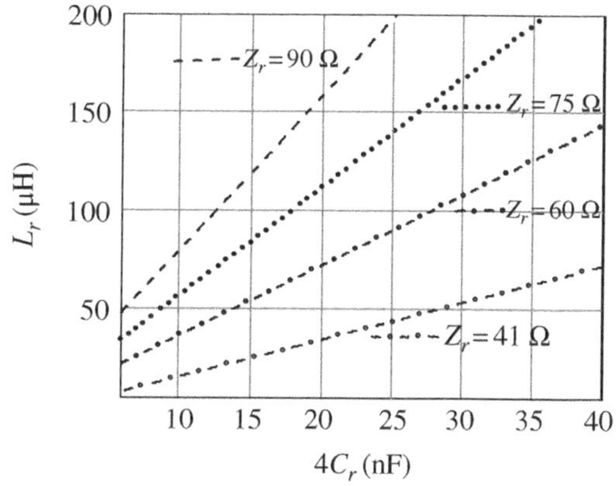

Figure 4.31 Resonant impedance Z_r versus resonant parameters.

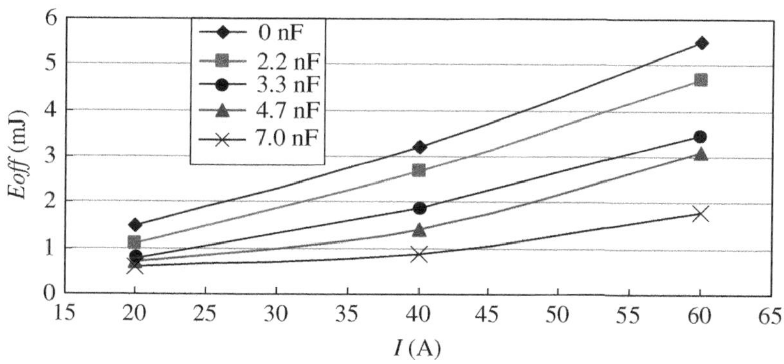

Figure 4.32 CM200DU-24NFH IGBT turn-off loss under ZVS conditions.

turn-off current of 60 A, when C_r is 2.2 nF, the turn-off loss is reduced by 8%; when C_r is 3.3 nF, the turn-off loss is reduced by 38%; when C_r is 4.7 nF, the turn-off loss is reduced by 45%; and when C_r is 7 nF, the turn-off loss is reduced by 61%. In order to obtain an obvious turn-off loss reduction, the capacitance of the resonant capacitor $C_r > 3.3$ nF is preferred.

5) In order to maintain the power quality, the resonant time T_{r1} in (4.31) is designed to be smaller than 2.5 % of the switching period:

$$T_{r1} = \frac{\pi}{2\omega_r} = \pi\sqrt{L_r C_r} \leq 0.025 T_s = 1.5\ \mu\text{s} \tag{4.85}$$

With the aforementioned five requirements, the appropriate region for selecting resonant parameters can be obtained as shown in Figure 4.33. The resonant parameters in shadow area meet all the requirements discussed before. The dark region shown in Figure 4.33 can be used to select the appropriate resonant inductance and resonant capacitance.

4.4.3 Experiment Platform and Testing Results

As shown in Figure 4.34, a 40 kW CAC rectifier prototype was built to verify the designed parameters and the proposed modulation scheme of the circuit.

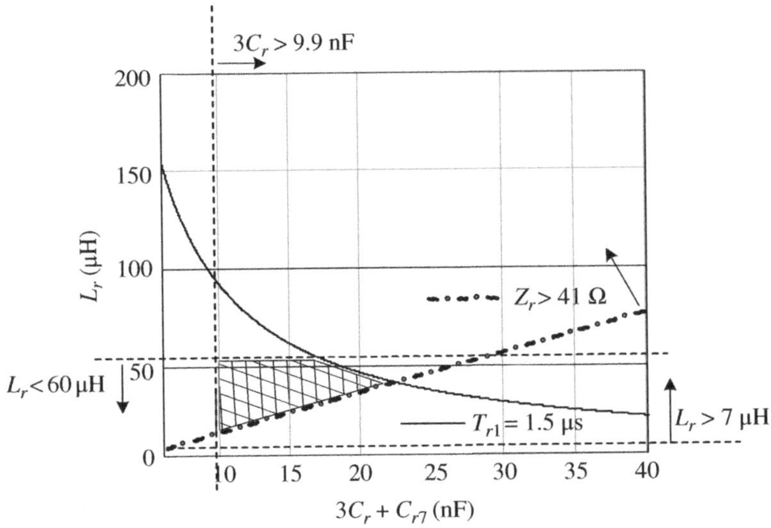

Figure 4.33 Recommended resonant parameters area.

Figure 4.34 The 40 kW ZVS rectifier prototype.

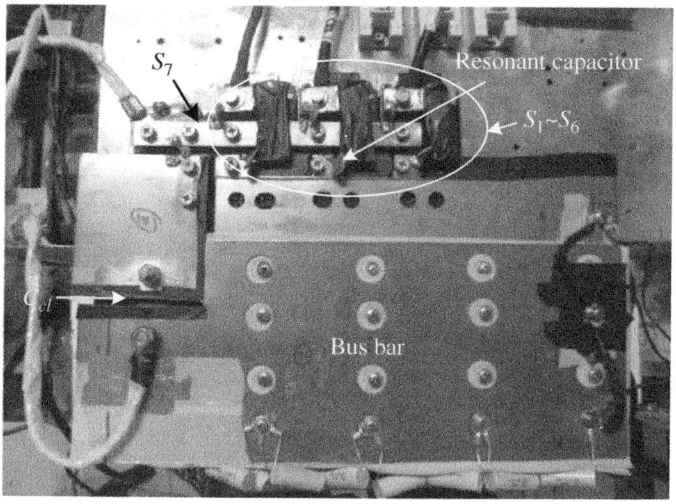

Figure 4.35 Bus bar and structure 40 kW ZVS rectifier prototype.

Table 4.4 Resonant parameters selected.

Parameters	Descriptions	Value
L_r	Inductance of the resonant inductor	26 μH
C_r	Capacitance of the resonant capacitors	4.7 nF
C_{cl}	Capacitance of the clamping capacitor	100 μF

The main structure of the bus bar and switching bridges is shown in Figure 4.35. IGBT modules CM200DU-24NFH are selected for bridge switches and auxiliary switch. In the circuit, ceramic capacitors are utilized as the resonant capacitors C_1–C_6. In order to meet the requirements of current, voltage, and capacitance, film capacitors are connected in parallel to operate as the clamping capacitor C_{cl}.

The resonant parameters are shown in Table 4.4.

The grid-connected voltage and current are demonstrated in Figure 4.36. The measured current total harmonic distortion (THD) is 2.9% under 40 kW condition.

The voltage and current of the bridge switch S_4 are shown in Figure 4.37. In this figure, current begins to increase from the negative value after the voltage drops to zero. Therefore, the bridge switches turn on in ZVS condition as required.

The voltage and current of the bridge switch's antiparallel diode D_1 are shown in Figure 4.38. It can be seen from this figure that the diode reverse recovery current has been suppressed well.

The voltage and current of the auxiliary switch S_7 are shown in Figure 4.39. When S_7 turns on, the current through the module goes negative, which means the freewheeling diode begins to conduct. Therefore, the ZVS condition for the auxiliary switch is realized.

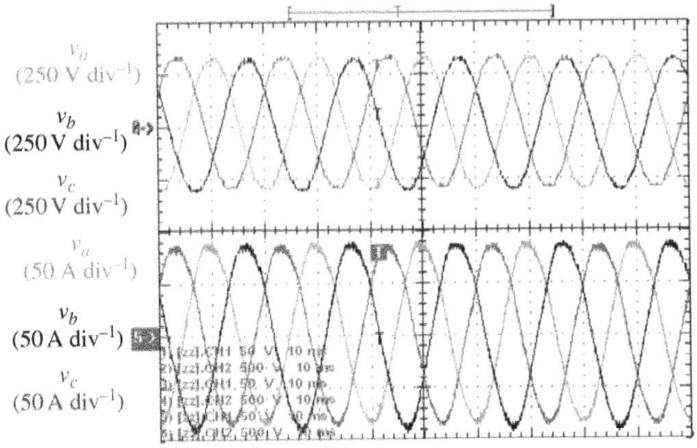

Figure 4.36 Output current and grid voltage

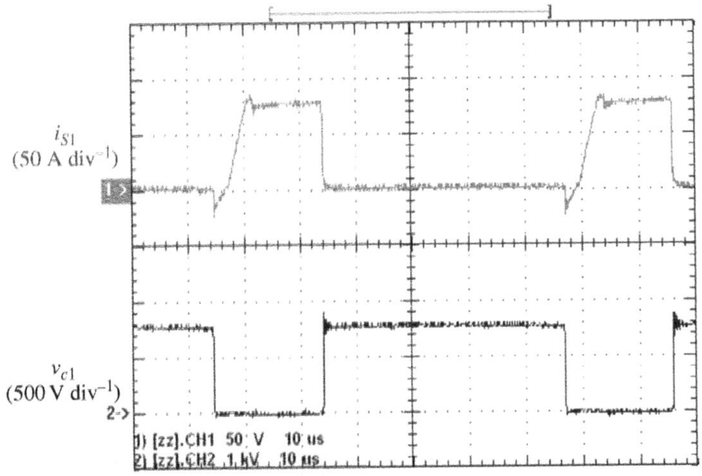

Figure 4.37 Voltage and current waveforms of the bridge switch.

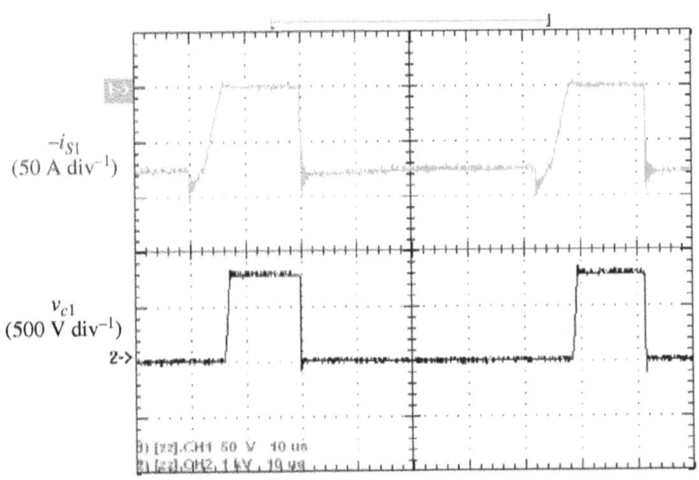

Figure 4.38 Voltage and current waveforms of the bridge switch's antiparallel diode.

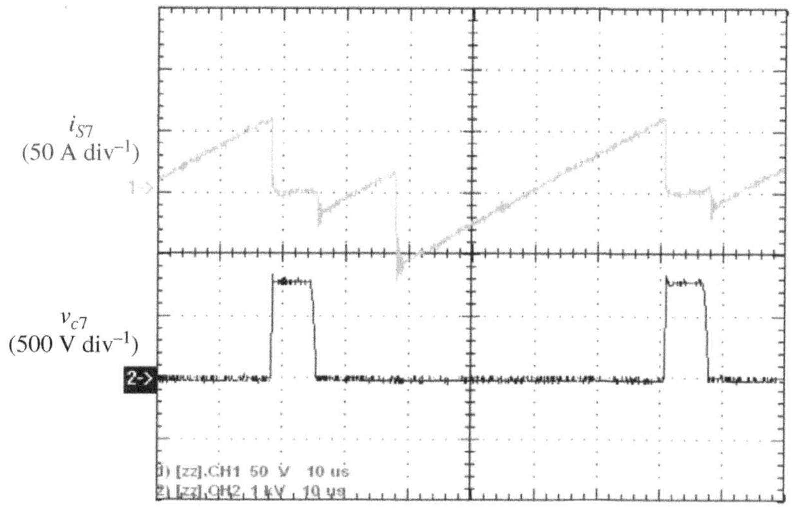

Figure 4.39 Voltage and current waveforms of the auxiliary switch.

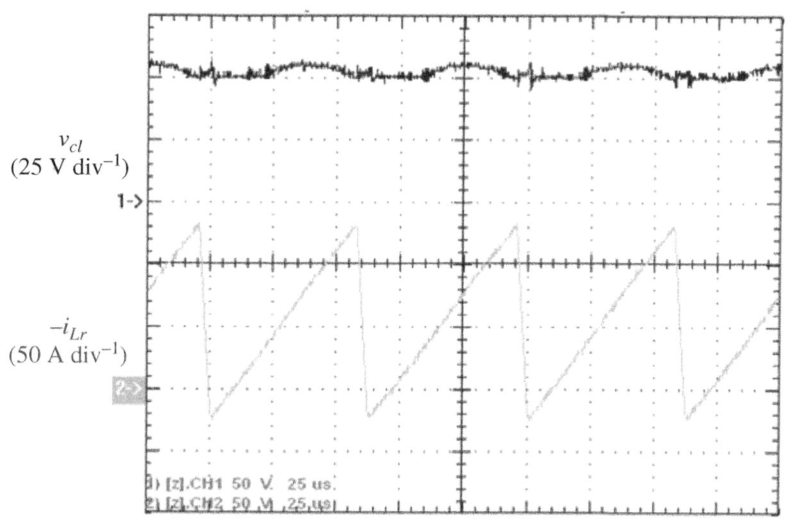

Figure 4.40 Voltage across C_{cl} and current through L_r.

The voltage across the clamping capacitor C_{cl} and the current through resonant inductor L_r are shown in Figure 4.40. The average voltage across clamping capacitor C_{cl} is 50 V. The voltage ripple across C_{cl} is 5 V, which is less than 10% of the mean voltage. The peak value of i_{Lr} is 130 A.

The voltage and current of the bridge switch in a line cycle are shown in Figure 4.41. It can be seen from this figure that the diode reverse recovery current has been suppressed well.

The conversion efficiency of the ZVS rectifier is measured by power analyzer WT1800. And the results are shown in Figure 4.42. The maximum efficiency is 97.9%.

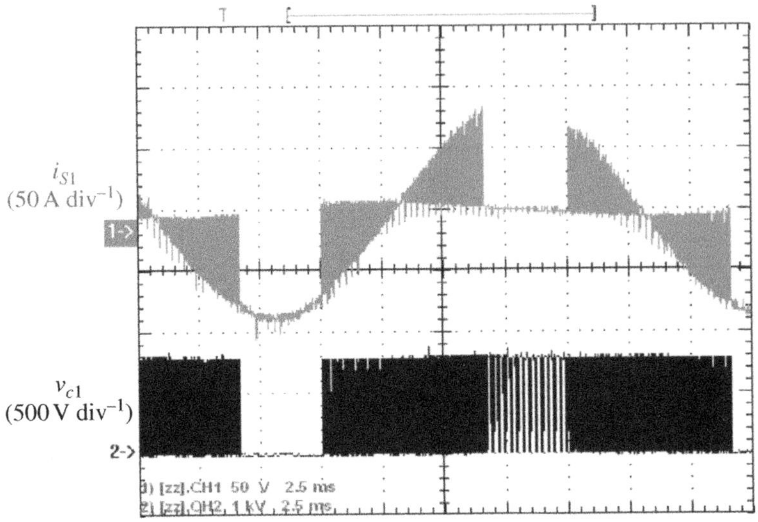

Figure 4.41 Voltage and current waveforms of the bridge switch S_1 in a line cycle.

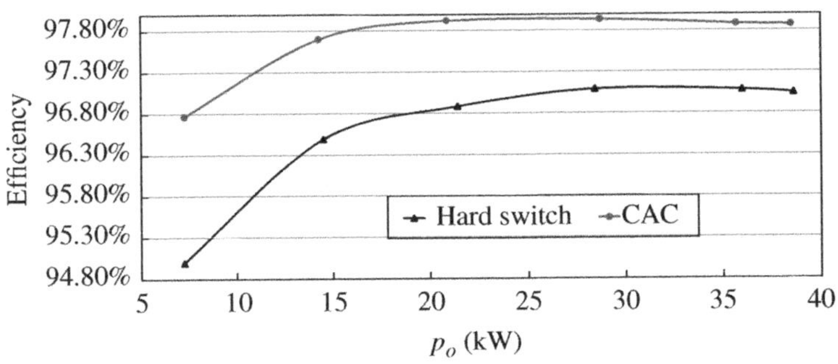

Figure 4.42 Measured efficiency.

4.5 Summary

CAC rectifier is introduced in this chapter. CAC rectifier can realize ZVS operation for all switching devices; the antiparallel diode reverse recovery current of all switch devices is suppressed well. To realize the ZVS condition, a ZVS-SVM scheme is presented, and the detailed design considerations are shown in this chapter. All the switch devices can be operated at the fixed switching frequency. And the switching voltage stress of all the power switch devices is only a little higher than the DC link voltage. The rectifier can achieve ZVS in all the switches. A 40-kW prototype has been built and realizes desirable ZVS soft-switching features.

References

1 D. Xu, B. Feng, and R. Li, "A zero voltage switching SVM (ZVS-SVM) controlled three-phase boost rectifier," *IEEE Transactions on Power Electronics*, vol. 22, no. 3, pp. 978–986, May 2007.

2 Z. Ma, D. Xu, R. Li, C. Du and X. Zhang, "A novel DC-side zero-voltage switching (ZVS) three-phase boost PWM rectifier controlled by an improved SVM method," *IEEE Transactions on Power Electronics*, vol. 27, no. 11, pp. 4391–4408, Nov. 2012.

3 R. Li, Z. Ma and D. Xu, "A ZVS grid-connected three-phase inverter," *IEEE Transactions on Power Electronics*, vol. 27, no. 8, pp. 3595–3604, Aug. 2012.

4 R. Li and D. Xu, "A zero-voltage switching three-phase inverter," *IEEE Transactions on Power Electronics*, vol. 29, no. 3, pp. 1200–1210, March 2014.

5 C. Du, D. Xu, N. He and N. Zhu, "Modeling and optimization of a zero-voltage switching inverter for high efficiency and miniaturization," *IEEE Transactions on Power Electronics*, vol. 32, no. 1, pp. 150–163, Jan. 2017.

6 M. Chen, J. Xi, Y. Chen and D. Xu, "Safe startup method for zero-voltage-switching MOSFET grid inverter," *IEEE Transactions on Power Electronics*, vol. 34, no. 9, pp. 8749–8761, September 2019.

7 K. Shi, A. Zhao, J. Deng and D. Xu, "Zero-voltage-switching SiC-MOSFET three-phase four-wire back-to-back converter," *IEEE Journal of Emerging and Selected Topics in Power Electronics*, vol. 7, no. 2, pp. 722–735, June 2019.

8 N. He, Y. Zhu, A. Zhao and D. Xu, "Zero-voltage-switching sinusoidal pulsewidth modulation method for three-phase four-wire inverter," *IEEE Transactions on Power Electronics*, vol. 34, no. 8, pp. 7192–7205, August 2019.

5

Three-phase Rectifier with Minimum Voltage Active-clamping Circuit

In this chapter, a zero-voltage-switching (ZVS) three-phase rectifier with minimum voltage active clamping (MVAC) circuit is introduced, including the operation principle, the circuit analysis, and the design guidelines of the MVAC rectifier [1, 2]. First, the operation principle of the proposed space-vector-modulation (SVM) scheme is introduced. Then, soft-switching condition will be derived. The design guidelines of three-phase rectifier with MVAC will be introduced, and a design example based on a 30 kW MVAC rectifier prototype will be provided finally [2–4].

5.1 Introduction

Circuit topology of the MVAC rectifier is shown in Figure 5.1. It consists of a standard three-phase rectifier and a MVAC circuit. The three-phase rectifier is composed of grid side input inductors L_a, L_b, and L_c; six switches S_1, S_2, S_3, S_4, S_5, and S_6; and active switches' antiparallel diodes D_1, D_2, D_3, D_4, D_5, and D_6. The clamping circuit has an active switch S_7, a resonant inductor L_r, and a clamping capacitor C_{cl}. Capacitors C_1, C_2, C_3, C_4, C_5, C_6, and C_7 are paralleled with the corresponding switches.

With the ZVS-SVM scheme introduced in Chapter 3, all the switches, including bridge switches and auxiliary switch can be turned on and turned off under zero voltage condition. The reverse recovery of all the diodes in the MVAC rectifier is well suppressed. The voltage stress of all the switching devices is equal to the DC bus voltage. In this chapter, the operation principle of the ZVS-SVM scheme will be introduced. Soft-switching condition will be derived. Design guidelines will be introduced, and a design example of a 30 kW MVAC rectifier prototype will be discussed.

5.2 Operation Principle of MVAC Rectifier

5.2.1 Space Vector Modulation of Three-phase Converter

The grid side three-phase voltages v_a, v_b, and v_c of three-phase converter are rewritten as,

$$\begin{cases} v_a = V_m \cos(\omega_l t) \\ v_b = V_m \cos\left(\omega_l t - \dfrac{2\pi}{3}\right) \\ v_c = V_m \cos\left(\omega_l t + \dfrac{2\pi}{3}\right) \end{cases} \tag{5.1}$$

Soft-Switching Technology for Three-phase Power Electronics Converters, First Edition.
Dehong Xu, Rui Li, Ning He, Jinyi Deng, and Yuying Wu.

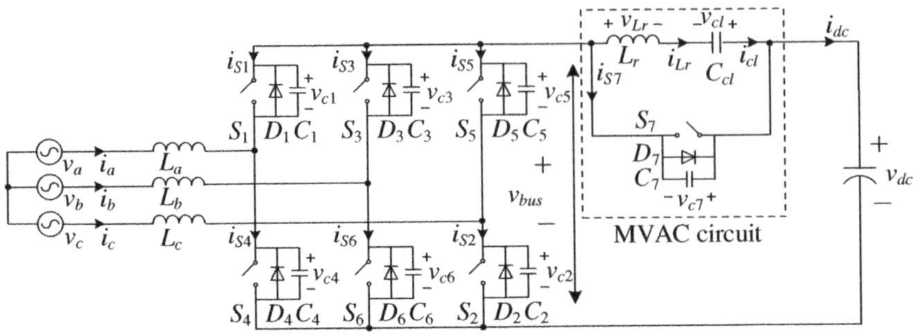

Figure 5.1 Circuit topology of the MVAC rectifier.

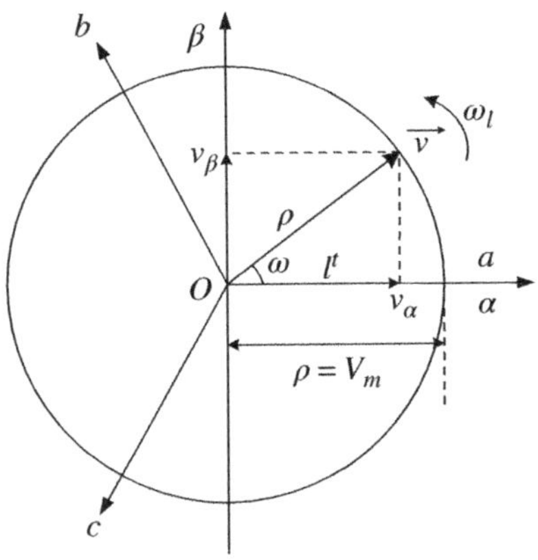

Figure 5.2 Vector representation in complex plane with $\alpha\beta$ coordinate.

where V_m and ω_l are the amplitude and the angular frequency of the grid side voltage, respectively.

According to Section 4.2.1, the three-phase voltage described by equation (5.1) can be represented by the complex vector $\vec{v}$ shown in Figure 5.2. The complex vector $\vec{v}$ rotates counterclockwise at a constant angular frequency ω_l in complex plane. The trajectory of the complex vector $\vec{v}$ is a circle, v_α and v_β are the projections of the complex vector $\vec{v}$ on the α-axis and β-axis in the complex plane, respectively.

According to the invertibility principle of complex space vector transformation, if the output voltage complex space vector of the three-phase converter is a trajectory close to a circle, the three-phase output voltage of the converter would be close to symmetrical sinusoidal waveforms. As shown in Figure 5.3, the three-phase converter consists of six active switches S_1, S_2, S_3, S_4, S_5, S_6, and the DC-link capacitor. By ignoring the current ripple of the grid side filter inductors, the grid side current of the converter can be represented by current sources i_a, i_b, and i_c.

As the analysis in Section 4.2.1, there are eight switch states in the rectifier. Each switch state can be represented by a complex voltage space vector, as shown in Table 5.1.

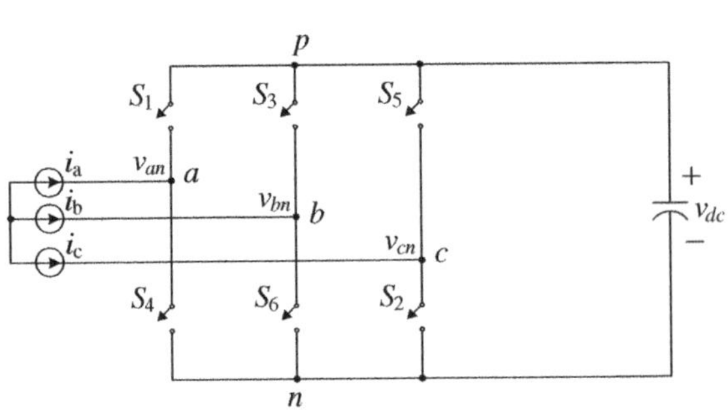

Figure 5.3 Topology of the three-phase converter.

Table 5.1 Eight switch states.

Complex vector	Switches in on-state	v_{an}	v_{bn}	v_{cn}	Complex vector value
$\vec{V}_1(100)$	S_1, S_2, S_6	v_{dc}	0	0	$2v_{dc}/3$
$\vec{V}_2(110)$	S_1, S_2, S_3	v_{dc}	v_{dc}	0	$2v_{dc}e^{j\pi/3}/3$
$\vec{V}_3(010)$	S_2, S_3, S_4	0	v_{dc}	0	$2v_{dc}e^{j2\pi/3}/3$
$\vec{V}_4(011)$	S_3, S_4, S_5	0	v_{dc}	v_{dc}	$2v_{dc}e^{j\pi}/3$
$\vec{V}_5(001)$	S_4, S_5, S_6	0	0	v_{dc}	$2v_{dc}e^{j4\pi/3}/3$
$\vec{V}_6(101)$	S_1, S_5, S_6	v_{dc}	0	v_{dc}	$2v_{dc}e^{j5\pi/3}/3$
$\vec{V}_7(111)$	S_1, S_3, S_5	v_{dc}	v_{dc}	v_{dc}	0
$\vec{V}_0(000)$	S_2, S_4, S_6	0	0	0	0

According to Table 5.1, there are six nonzero voltage vectors $\vec{V}_1(100)$, $\vec{V}_2(110)$, $\vec{V}_3(010)$, $\vec{V}_4(011)$, $\vec{V}_5(001)$, and $\vec{V}_6(101)$ and two zero voltage vectors $\vec{V}_0(000)$ and $\vec{V}_7(111)$. The six nonzero voltage vectors divide the complex plane into six sectors, as shown in Figure 5.4.

A reference voltage vector $\vec{v}_{ref}$ can be generated by integrating two adjacent nonzero voltage vectors in a sector by the parallelogram rule. For instance, if the reference vector is located in sector 1, the corresponding nonzero voltage vectors are $\vec{V}_1(100)$ and $\vec{V}_2(110)$. The reference voltage vector $\vec{v}_{ref}$ is expressed as,

$$\vec{v}_{ref} = \left|\vec{V}_{ref}\right| e^{j\theta} \tag{5.2}$$

where $\left|\vec{V}_{ref}\right|$ is the module value of $\vec{v}_{ref}$ and θ is the phase angle of $\vec{v}_{ref}$. The reference voltage vector $\vec{v}_{ref}$ may be realized by the voltage vector $\vec{V}_1(100)$, $\vec{V}_2(110)$, and $\vec{V}_0(000)$ as follows:

$$d_{V1}T_s\vec{V}_1 + d_{V2}T_s\vec{V}_2 + d_{V0}T_s\vec{V}_0 = T_s\vec{v}_{ref} \tag{5.3}$$

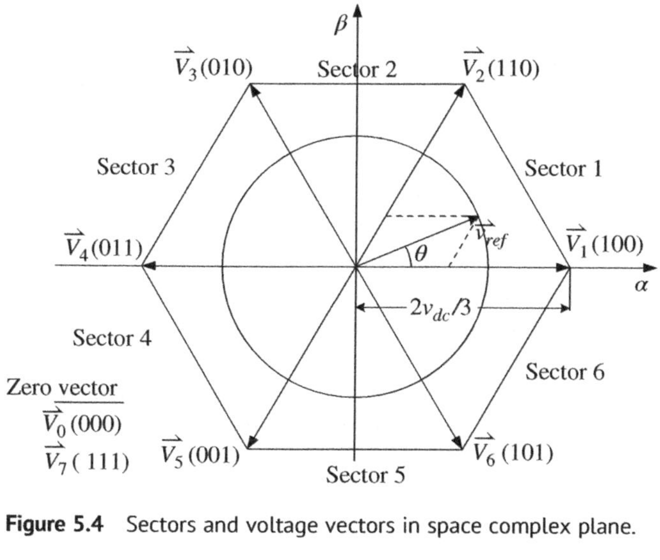

Figure 5.4 Sectors and voltage vectors in space complex plane.

where d_{V1}, d_{V2}, and d_{V0} are the duty ratio of voltage vector $\vec{V}_1(100)$, $\vec{V}_2(110)$, and $\vec{V}_0(000)$, respectively. The duty ratios are constrained as follows:

$$d_{V1} + d_{V2} + d_{V0} = 1 \tag{5.4}$$

The duty ratio of $\vec{V}_1$, $\vec{V}_2$, and $\vec{V}_0$ can be derived as,

$$\begin{cases} d_{V1} = \dfrac{\sqrt{3}}{2} m \sin(\pi/3 - \theta) \\ d_{V2} = \dfrac{\sqrt{3}}{2} m \sin\theta \\ d_{V0} = 1 - \dfrac{\sqrt{3}}{2} m \sin\left(\dfrac{\pi}{3} + \theta\right) \end{cases} \tag{5.5}$$

where m is the voltage modulation ratio. It is defined as follows:

$$m = 2\left|\vec{V}_{ref}\right| / v_{dc} \tag{5.6}$$

The reference voltage vector $\vec{v}_{ref}$ is composed of two adjacent nonzero voltage vectors in the same sector in each switching cycle. Since the sum of the duty ratio of these two adjacent voltage vectors is less than 1, then the zero-voltage vector is used to fill the rest of the switching cycle. In a switching cycle, the voltage space vector acts in a time-sharing manner to form a space vector sequence in time. There are many ways to organize the space vector sequences, as introduced in Chapter 4.

5.2.2 Switching Scheme of MVAC Rectifier

The grid side current of the three-phase rectifier is rewritten as follows:

$$\begin{cases} i_a = I_m \cos(\omega_l t + \varphi_0) \\ i_b = I_m \cos\left(\omega_l t - \dfrac{2\pi}{3} + \varphi_0\right) \\ i_c = I_m \cos\left(\omega_l t + \dfrac{2\pi}{3} + \varphi_0\right) \end{cases} \tag{5.7}$$

where φ_0 is the grid current power factor angle, and I_m is the amplitude of the grid current.

Considering the rectifier operates under unit power factor, the power factor angle φ_0 is zero. The voltage and current waveforms in a utility cycle are demonstrated in Figure 5.5. Under the symmetrical condition of the three-phase grid voltage, in any sector, there is always one phase grid voltage with opposite polarity to the other two phases' grid voltages. Considering the current direction of different phases, each voltage sector in Figure 5.4 can be divided into two smaller sectors. Consequently, a utility cycle is divided into 12 sectors, which is shown in Figures 5.5 and 5.6.

With the SVM scheme analysis introduced in Section 4.2.2, for each leg in the three-phase rectifier, there exist two types of switching commutations. The turn-off loss of the commutation from the switch to the antiparallel diode of its complement switch has relatively small loss, which is known as switch commutation Type 1. The switch's turn-on loss and the corresponding diode's reverse recovery loss occur in the commutation from the diode to its complement switch, which is known as switch commutation Type 2. Once there is Type 2 current commutation in the MVAC ZVS three-phase rectifier as shown in Figure 5.1, the auxiliary switch S_7 is required to turn off to start a resonance between the resonant inductor L_r and the resonant capacitors in the circuit. As a

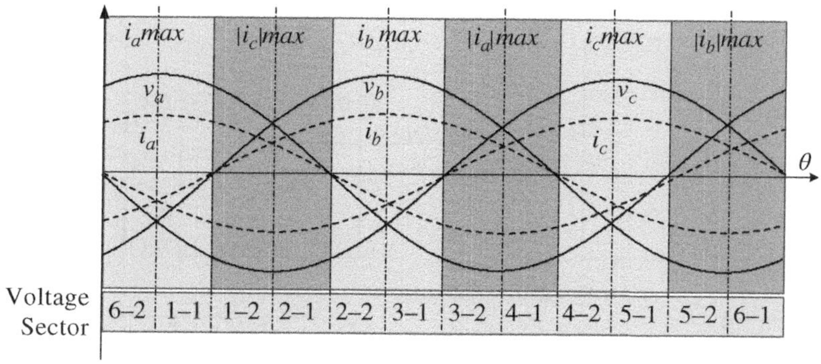

Figure 5.5 The voltage and current waveforms in a utility cycle.

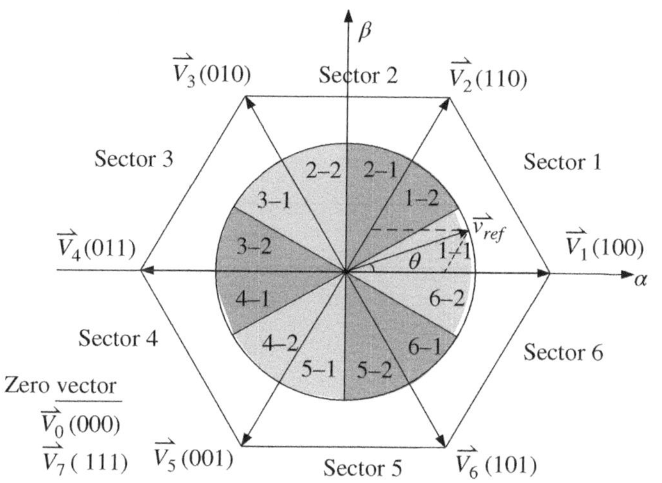

Figure 5.6 Sectors in space vector diagram.

result, the DC-side voltage of the three-phase bridges v_{bus} will be resonated to zero, and it brings the opportunity of ZVS turn-on for main switches. Auxiliary switch is not required to turn off when Type 1 switch commutation comes due to its smaller loss. With the ZVS-SVM scheme, all the switches, including bridge switches and auxiliary switch, can be turned on and turned off under zero voltage condition. The reverse recovery process of all the diodes in the MVAC rectifier is also suppressed. In addition, MVAC rectifier has lower voltage stress on the devices. The voltage stress of all the switching devices is equal to the output DC voltage.

Now let us investigate dendrogram of the vector sequence. It is assumed that vectors in sector 1-1 can only be used once in each switch cycle. In Figure 5.5, three-phase current $i_a>0$, $i_b<0$, and $i_c<0$. The corresponding vectors in this sector are nonzero voltage vectors $\vec{V}_1(100)$ and $\vec{V}_2(110)$ and zero vectors $\vec{V}_0(000)$ and $\vec{V}_7(111)$. The equivalent circuits of the four vectors in sector 1-1 are shown in Figure 5.7.

Different switching commutation processes bring different switching loss. In order to obtain the switching modulation sequence with less switching loss, the switching commutation sequences between different vectors in Sector 1-1 are analyzed, which is listed in Table 5.2.

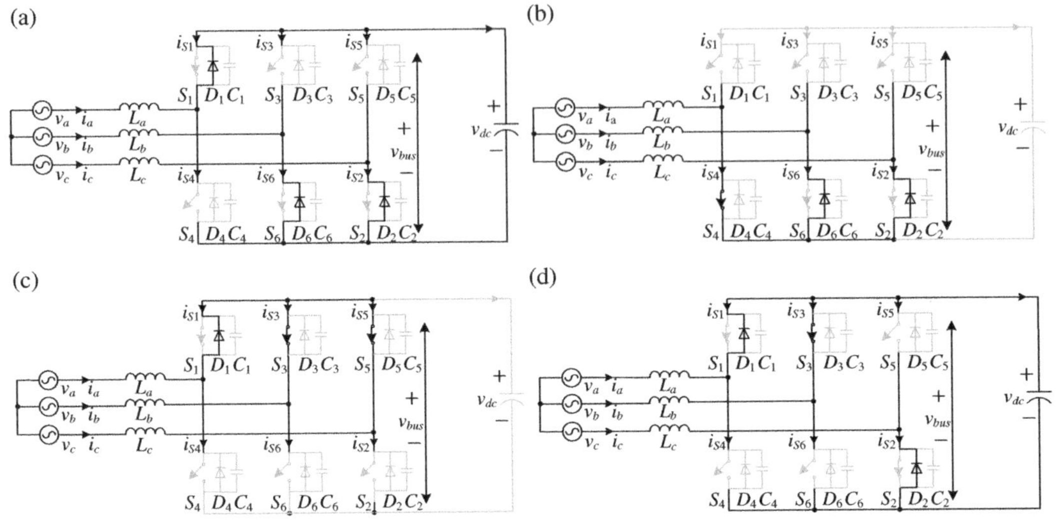

Figure 5.7 Three-phase main bridges' equivalent circuits of four vectors in sector 1-1. (a) Vector 100. (b) Vector 000. (c) Vector 111. (d) Vector 110.

Table 5.2 Modulation sequences and current commutation type.

Modulation sequences	Type 1 switch commutation	Type 2 switch commutation
$\vec{V}_7(111) \rightarrow \vec{V}_0(000)$	Exist	Exist
$\vec{V}_7(111) \rightarrow \vec{V}_1(100)$	Exist	Non-exist
$\vec{V}_7(111) \rightarrow \vec{V}_2(110)$	Exist	Non-exist
$\vec{V}_0(000) \rightarrow \vec{V}_7(111)$	Exist	Exist
$\vec{V}_0(000) \rightarrow \vec{V}_1(100)$	Exist	Non-exist
$\vec{V}_0(000) \rightarrow \vec{V}_2(110)$	Exist	Exist
$\vec{V}_1(100) \rightarrow \vec{V}_7(111)$	Non-exist	Exist
$\vec{V}_1(100) \rightarrow \vec{V}_0(000)$	Non-exist	Exist
$\vec{V}_1(100) \rightarrow \vec{V}_2(110)$	Exist	Exist
$\vec{V}_2(110) \rightarrow \vec{V}_7(111)$	Exist	Exist
$\vec{V}_2(110) \rightarrow \vec{V}_0(000)$	Exist	Exist
$\vec{V}_2(110) \rightarrow \vec{V}_1(100)$	Exist	Non-exist

In order to find the voltage vector sequence with lowest switching loss, the dendrogram analysis of switching commutation sequences between different voltage vectors in sector 1-1 is shown in Figure 5.8. Since the switching sequence in a current switching cycle should be periodical, the beginning voltage vector and the ending voltage vector should be the same in the dendrogram.

In Figure 5.8a, $\vec{V}_1(100)$ is the beginning voltage vector and the ending voltage vector. From $\vec{V}_1(100)$, its adjacent voltage vector can be $\vec{V}_2(110)$, $\vec{V}_7(111)$, or $\vec{V}_0(000)$. If $\vec{V}_7(111)$ is selected as the adjacent voltage vector, $\vec{V}_7(111)$'s adjacent voltage vector can be $\vec{V}_2(110)$ or $\vec{V}_0(000)$.

And if $\vec{V}_2(110)$ is selected as the adjacent voltage vector, its adjacent voltage vector can be $\vec{V}_1(100)$ or $\vec{V}_0(000)$. Suppose that $\vec{V}_0(000)$ is selected as the adjacent voltage vector, after $\vec{V}_0(000)$ acts for some time, the adjacent voltage vector of $\vec{V}_0(000)$ can only be $\vec{V}_1(100)$, which is the end of this switching cycle. Here it is assumed the four vectors in a sector only use once in a switch cycle.

In Figure 5.8a, all the vector sequences begin from $\vec{V}_1(100)$. Among all vector sequences, there is only one commutation sequence with the lowest switching loss, which is $\vec{V}_1(100)$, $\vec{V}_7(111)$, $\vec{V}_2(110)$, $\vec{V}_1(100)$. It has only one Type 2 switching commutation in a switching cycle.

In Figure 5.8b, the first vector is $\vec{V}_2(110)$. The best vector sequence is $\vec{V}_2(110)$, $\vec{V}_1(100)$, $\vec{V}_7(111)$, $\vec{V}_2(110)$, which has only one Type 2 switching commutation. In Figure 5.8c, the first vector is $\vec{V}_7(111)$. The best vector sequence is $\vec{V}_7(111)$, $\vec{V}_2(110)$, $\vec{V}_1(100)$, $\vec{V}_7(111)$. However, Figure 5.8d has no commutation sequence with only one Type 2 switching commutation.

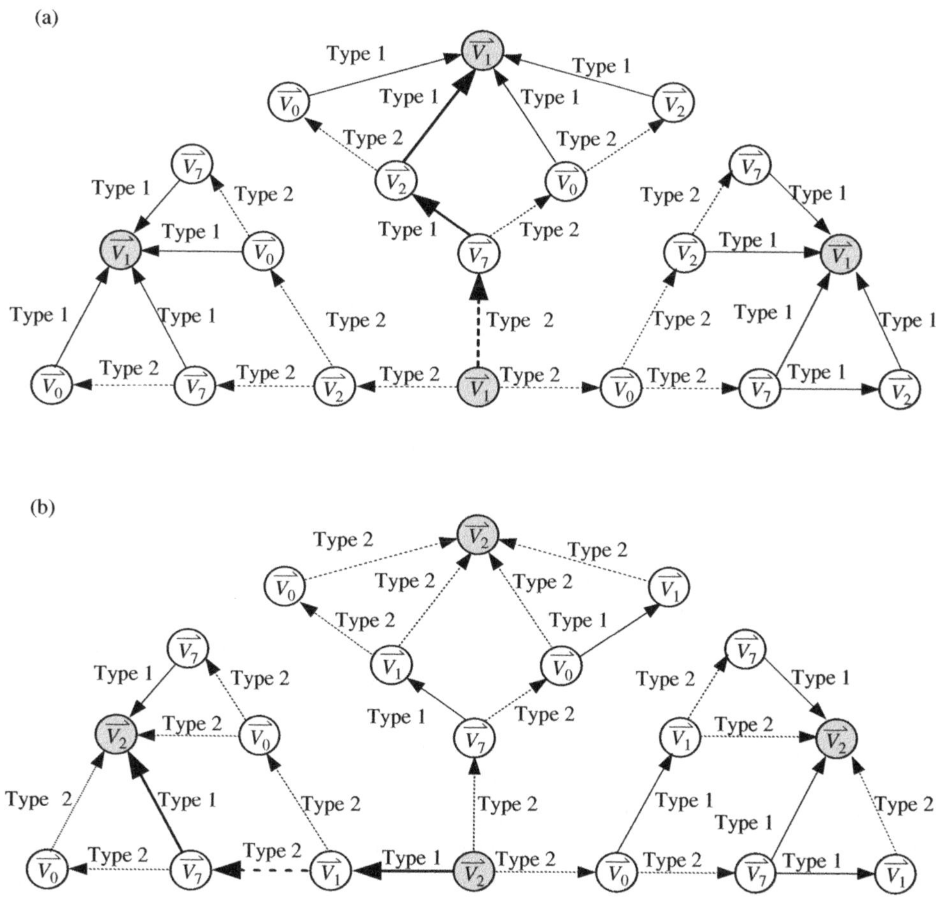

Figure 5.8 The dendrogram analysis of switching commutation sequences between different voltage vectors in Sector 1-1. (a) Beginning with nonzero vector $\vec{V}_1(100)$. (b) Beginning with nonzero vector $\vec{V}_2(110)$. (c) Beginning with zero vector $\vec{V}_7(111)$. (d) Beginning with zero vector $\vec{V}_0(000)$.

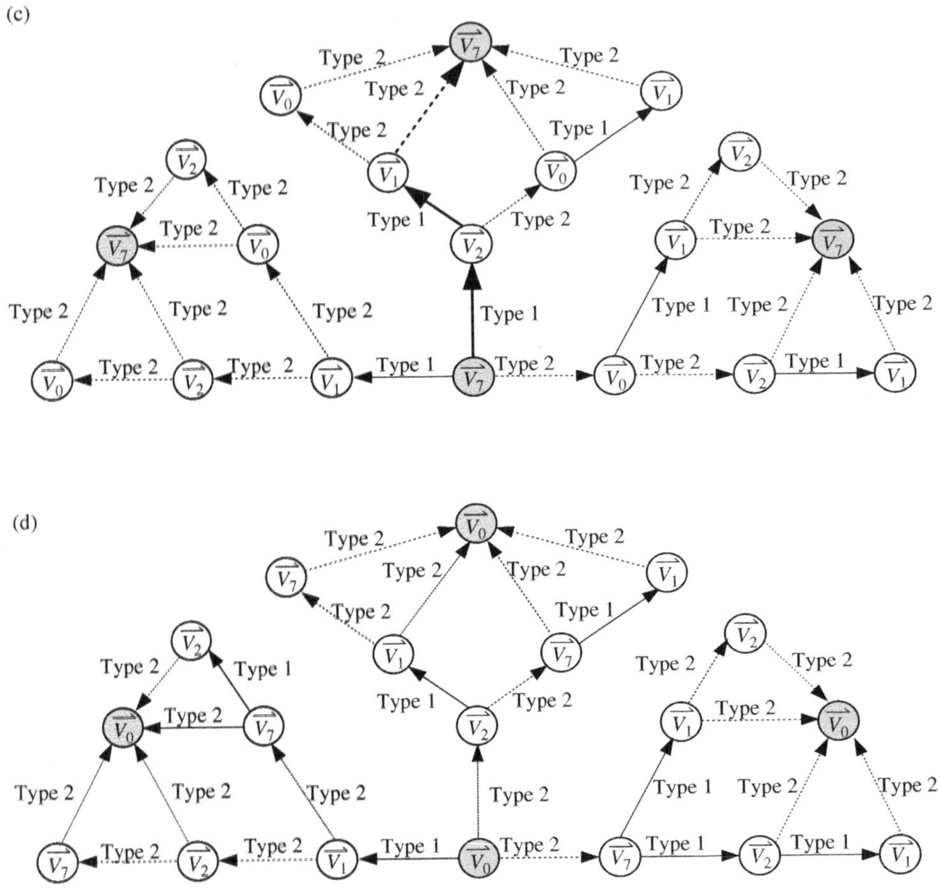

Figure 5.8 (Continued)

By comparing these three space vector sequences in Figure 5.9, they are actually the same. The difference is due to selection of the beginning voltage vector in a switching cycle.

Similarly, we can derive best vector sequences for other sectors, which are listed in Table 5.3. In each sector, the Type 2 switching commutation only happens between the first vector and the zero vector. For commutation from the zero vector to the second vector or from the second vector to the first vector, there is only the Type 1 commutation, which has less switch loss.

Gate drive signals of the six main switches and the auxiliary switch in Sector 1-1 are shown in Figure 5.10. Before the Type 2 switching commutation process occurs from $\vec{V}_1(100)$ to $\vec{V}_7(111)$, S_7 is turned off to create the ZVS turn-on condition for the circuit. The auxiliary switch S_7 is turned off with a resonant time (T_{r1}) ahead of the turn-on of the main switches, S_3 and S_5. The turn-off of auxiliary switch S_7 causes the resonance between the resonant inductor L_r and some of the resonant capacitors. As a result, the voltage across the three-phase main bridges v_{bus} will be resonated to zero, which brings the opportunity of ZVS turn-on for main switches. Besides, there will be another extra added short-circuit pulse for all main switches when the v_{bus} is clamped at zero, which is used to store adequate energy for the resonant inductor. In the Type 1 switching commutation process from $\vec{V}_7(111)$ to $\vec{V}_2(110)$ and from $\vec{V}_2(110)$ to $\vec{V}_1(100)$, the active switch S_7 is kept in conduction state. The detailed operation stages of the ZVS-SVM will be investigated in the following section.

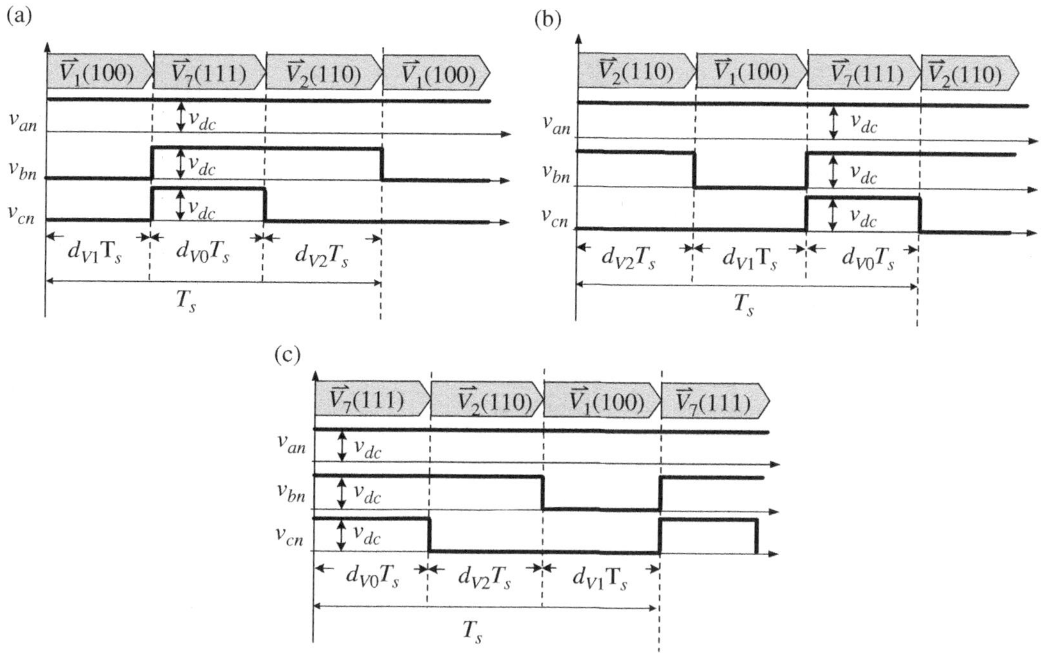

Figure 5.9 The space vector sequence with one Type 2 commutation process. (a) $\vec{V}_1(100)$, $\vec{V}_7(111)$, $\vec{V}_2(110)$, $\vec{V}_1(100)$. (b) $\vec{V}_2(110)$, $\vec{V}_1(100)$, $\vec{V}_7(111)$, $\vec{V}_2(110)$. (c) $\vec{V}_7(111)$, $\vec{V}_2(110)$, $\vec{V}_1(100)$, $\vec{V}_7(111)$.

Table 5.3 Switching sequence in each sector.

Sector	First vector	Zero vector	Second vector	First vector
Sector 1-1	$\vec{V}_1(100)$	$\vec{V}_7(111)$	$\vec{V}_2(110)$	$\vec{V}_1(100)$
Sector 1-2	$\vec{V}_2(110)$	$\vec{V}_0(000)$	$\vec{V}_1(100)$	$\vec{V}_2(110)$
Sector 2-1	$\vec{V}_2(110)$	$\vec{V}_0(000)$	$\vec{V}_3(010)$	$\vec{V}_2(110)$
Sector 2-2	$\vec{V}_3(010)$	$\vec{V}_7(111)$	$\vec{V}_2(110)$	$\vec{V}_3(010)$
Sector 3-1	$\vec{V}_3(010)$	$\vec{V}_7(111)$	$\vec{V}_4(011)$	$\vec{V}_3(010)$
Sector 3-2	$\vec{V}_4(011)$	$\vec{V}_0(000)$	$\vec{V}_3(010)$	$\vec{V}_4(011)$
Sector 4-1	$\vec{V}_4(011)$	$\vec{V}_0(000)$	$\vec{V}_5(001)$	$\vec{V}_4(011)$
Sector 4-2	$\vec{V}_5(001)$	$\vec{V}_7(111)$	$\vec{V}_4(011)$	$\vec{V}_5(001)$
Sector 5-1	$\vec{V}_5(001)$	$\vec{V}_7(111)$	$\vec{V}_6(101)$	$\vec{V}_5(001)$
Sector 5-2	$\vec{V}_6(101)$	$\vec{V}_0(000)$	$\vec{V}_5(001)$	$\vec{V}_6(101)$
Sector 6-1	$\vec{V}_6(101)$	$\vec{V}_0(000)$	$\vec{V}_1(100)$	$\vec{V}_6(101)$
Sector 6-2	$\vec{V}_1(100)$	$\vec{V}_7(111)$	$\vec{V}_6(101)$	$\vec{V}_1(100)$

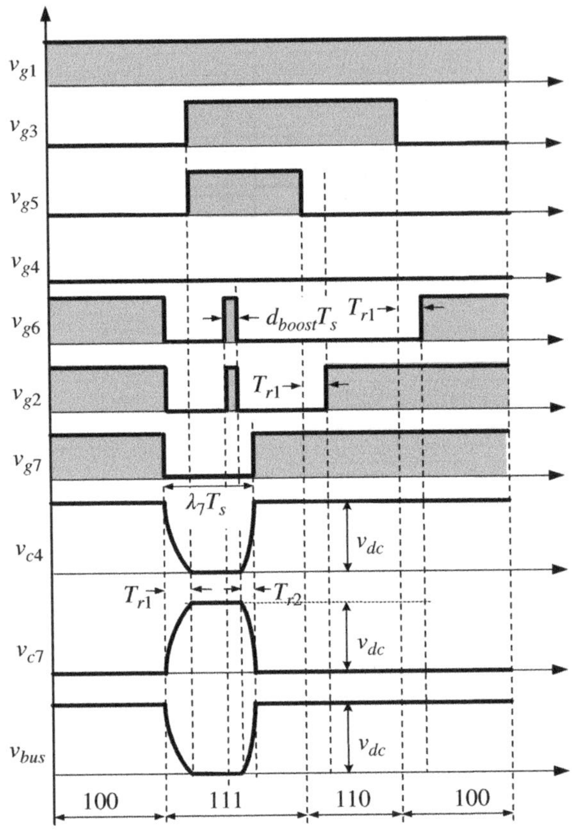

Figure 5.10 Switching state of the MVAC circuit.

5.3 Circuit Analysis of MVAC Rectifier

5.3.1 Operation Stage Analysis

In order to simplify the analysis, following assumptions are made:

1) All of the switching devices are ideal switches.
2) Filter inductors at the grid side are so large that the filter inductor currents can be seen as constant in a switching cycle.
3) Parasitic capacitance of all the switches is included in the capacitor paralleled with the active switches.
4) All paralleled capacitors on the switches are the same, i.e. $C_1 = C_2 = C_3 = C_4 = C_5 = C_6 = C_7 = C_r$.
5) The clamping capacitor C_{cl} is so large that the v_{cl} can be treated as a constant value V_{cl} in a switching cycle.

Sector 1-1 is taken as an example to investigate the operation stages of the MVAC rectifier. The operation of the rectifier includes nine stages. Figure 5.11 shows the waveforms of the key components in the rectifier in one switching period [5–8]. The operation stages are explained as follows.

Stage 1 (t_0–t_1): Initial stage. In this stage, the operation state is in vector $\vec{V}_1(100)$. Equivalent circuit is shown in Figure 5.12. Considering the current direction of AC-side filters, D_1, D_2, and D_6 are

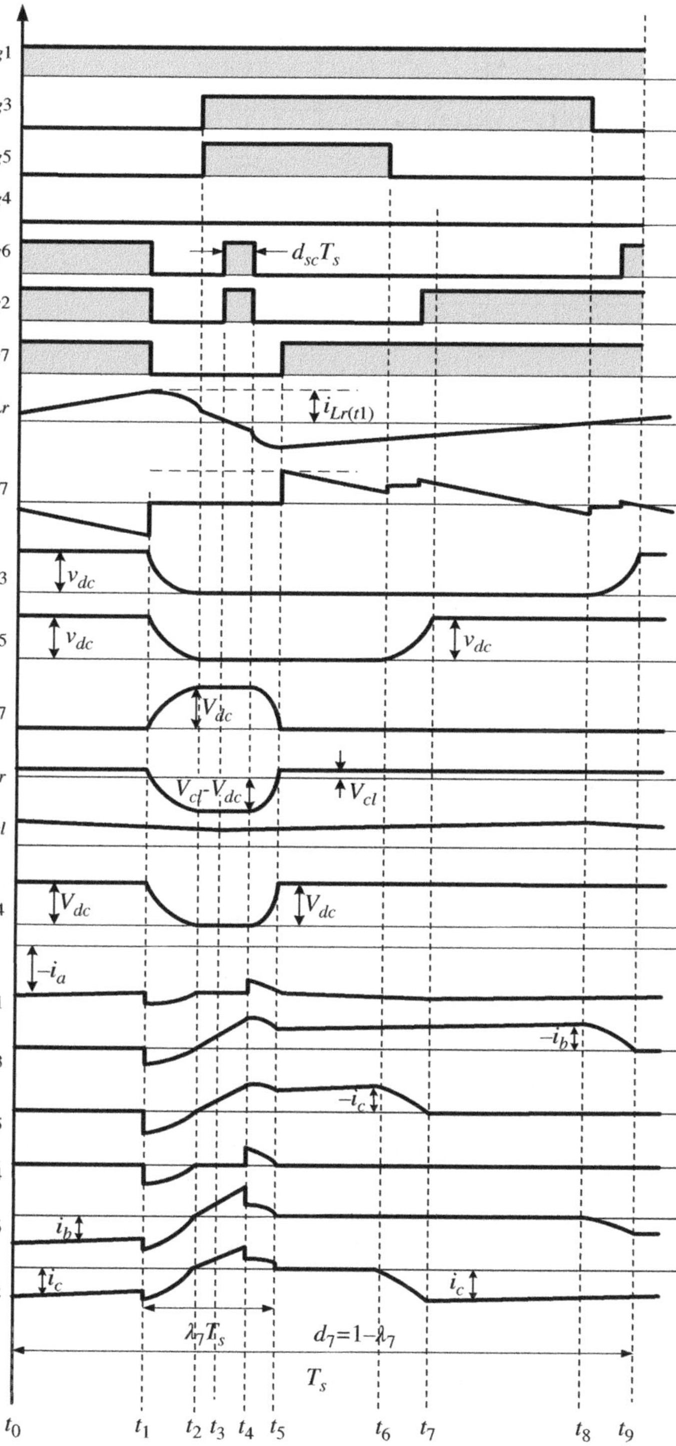

Figure 5.11 Driving logic and key waveforms in Sector 1-1.

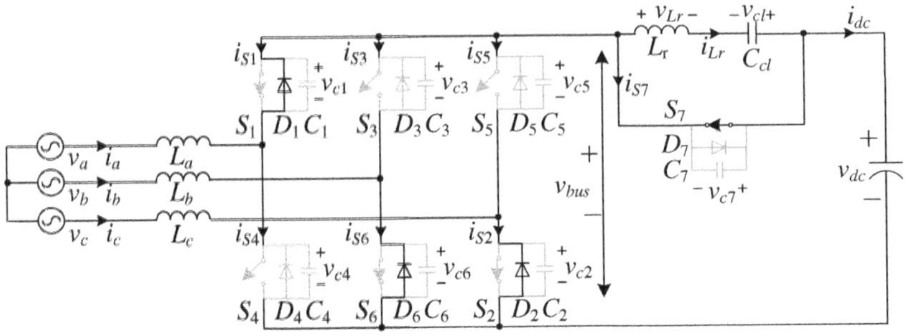

Figure 5.12 Equivalent circuit of stage 1: the initial stage.

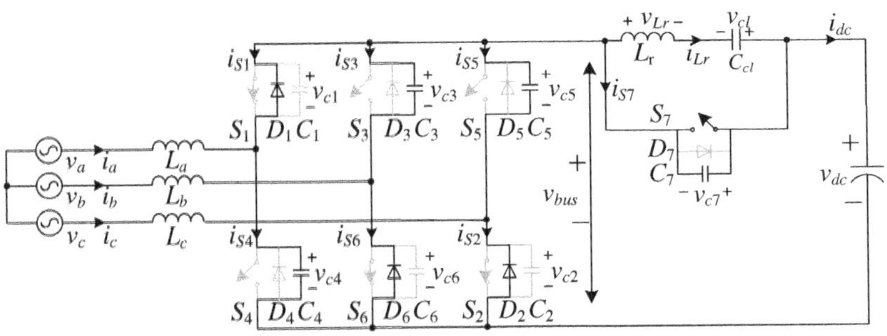

Figure 5.13 Equivalent circuit of stage 2: the first resonance stage.

conducting. Auxiliary switch S_7 conducts in this stage, and the resonant inductor L_r is charged by the clamping capacitor C_{cl}. Therefore, the current through L_r increases linearly at following rate:

$$\frac{di_{Lr}}{dt} = \frac{V_{cl}}{L_r} \tag{5.8}$$

Stage 2 (t_1–t_2): The first resonance stage. Equivalent circuit is shown in Figure 5.13. At t_1, S_7 is turned off, which starts a resonance between the resonant inductor L_r and the resonant capacitors C_3, C_4, C_5, and C_7. During this stage, the bus voltage v_{bus} will resonate to zero. Once the bus voltage resonates to zero, it creates ZVS condition for switches S_3 and S_5. This stage ends at t_2, when v_{bus} decreases to zero.

Stage 3 (t_2–t_3): The freewheeling stage. Equivalent circuit is shown in Figure 5.14. At t_2, v_{bus} oscillates to zero, D_3, D_5 and D_4 begins to conduct. S_3 and S_5 can be turned on under ZVS condition. Currents in phase-b and phase-c begin to transfer from D_6 and D_2 to S_3 and S_5, respectively. During this period, the current changing rates of i_{D6} and i_{D2} are restrained to a small value by L_r so that the reverse recovery losses on D_6 and D_2 can be suppressed. i_{D6} and i_{D2} are reduced to zero at time t_3. At t_3, $i_{Lr(t3)} = 0$. In stage 3, the resonant inductor current i_{Lr} decreases with a rate of,

$$\frac{di_{Lr}}{dt} = -\frac{V_{dc} - V_{cl}}{L_r} \tag{5.9}$$

Stage 4 (t_3–t_4): The short-circuit stage. Equivalent circuit is shown in Figure 5.15. At t_3, i_{D6}, i_{D2}, with i_{Lr} are reduced to zero. To store more energy in L_r, S_6 and S_2 are turned on at t_3 under ZVS

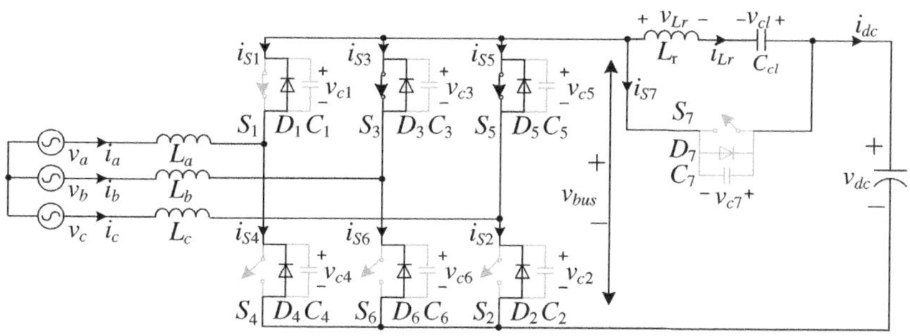

Figure 5.14 Equivalent circuit of stage 3: the freewheeling stage.

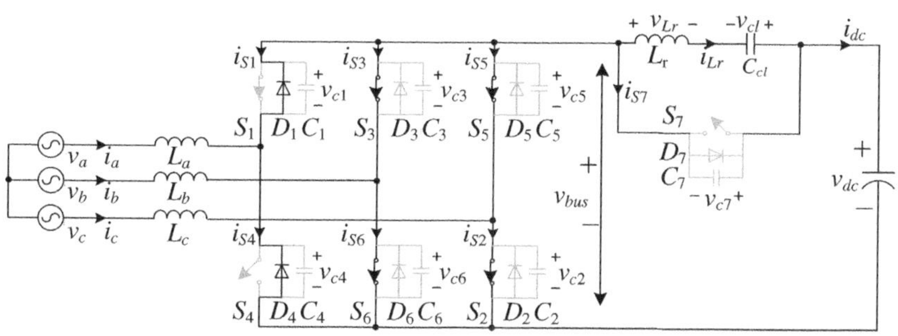

Figure 5.15 Equivalent circuit of stage 4: the short-circuit stage.

condition. During this stage, v_{bus} keeps in zero and the resonant inductor decreases by i_x. In stage 3, the resonant inductor current i_{Lr} decreases with a rate of,

$$\frac{di_{Lr}}{dt} = -\frac{V_{dc} - V_{cl}}{L_r} \tag{5.10}$$

The resonant inductor current at t_4 is expressed by,

$$i_{Lr}(t_4) = -i_x \tag{5.11}$$

Stage 5 (t_4–t_5): The second resonance stage. Equivalent circuit is shown in Figure 5.16. At t_4, S_6 and S_2 are turned off. The second resonance occurs between L_r and C_2, C_4, C_6, and C_7. During this resonance, v_{bus} increases while v_{c7} decreases. At t_5, v_{bus} increases to V_{dc} while v_{c7} reduces to zero, and the antiparallel diode of S_7 starts to conduct. It creates ZVS turn-on condition for S_7. At t_5, i_{Lr} reaches its minimum value i_{Lr_min}. Once the antiparalleled diodes of S_7 conducts at t_5, this stage ends.

Stage 6 (t_5–t_6): The second steady stage. Equivalent circuit is shown in Figure 5.17. At this stage, S_7 can be turned on with ZVS condition since v_{c7} equals zero. The rectifier operates in vector $\vec{V}_7(111)$ in this stage. The voltage across L_r is clamped by V_{cl} in this stage, the resonant inductor current i_{Lr} increases with a rate of,

$$\frac{di_{Lr}}{dt} = \frac{V_{cl}}{L_r} \tag{5.12}$$

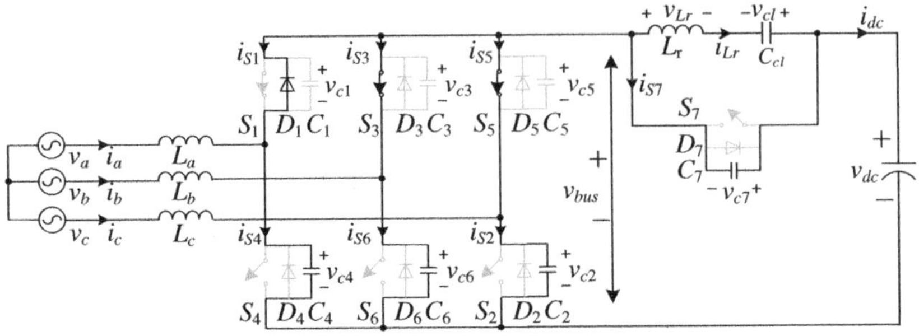

Figure 5.16 Equivalent circuit of stage 5: the second resonance stage.

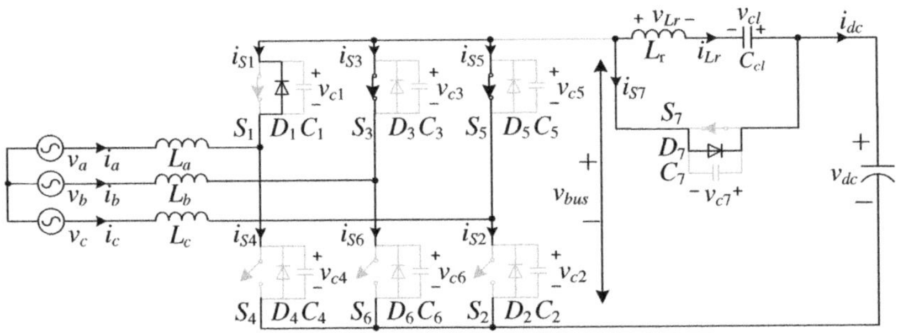

Figure 5.17 Equivalent circuit of stage 6: the second steady stage.

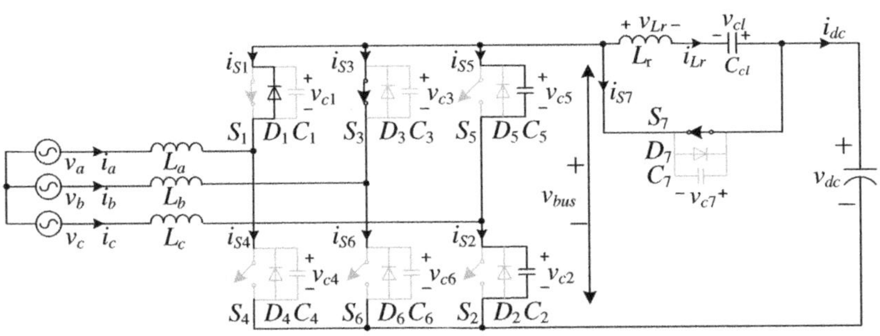

Figure 5.18 Equivalent circuit of stage 7: the Type 1 commutation stage.

Stage 7 (t_6–t_7): The first Type 1 commutation stage. Equivalent circuit is shown in Figure 5.18. At t_6, S_5 turns off. The input current i_c charges C_5 and discharges C_2. Once v_{c2} changes to zero, diode D_2 becomes conductivity. With the buffer of C_5 and C_2, S_5 is turned off under ZVS condition. This stage ends at t_7 when v_{c2} reduces to zero and D_2 conducts.

Stage 8 (t_7–t_8): The third steady stage. Equivalent circuit is shown in Figure 5.19. In this stage, rectifier is operated in voltage vector $\overrightarrow{V_2}(110)$. The resonant inductor is still clamped by V_{cl}. The current through the resonant inductor can be expressed as,

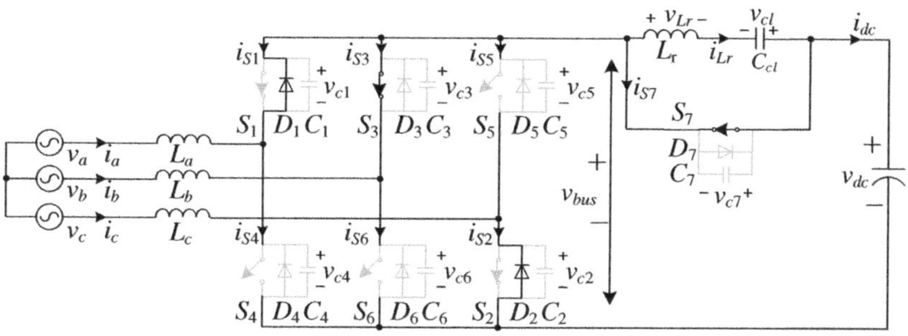

Figure 5.19 Equivalent circuit of stage 8: the third steady stage.

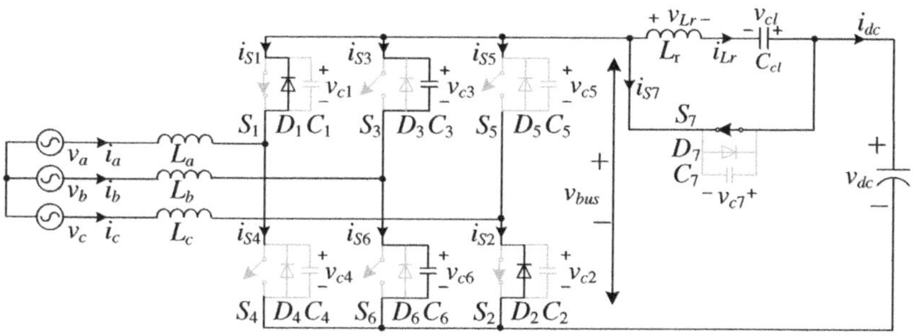

Figure 5.20 Equivalent circuit of stage 9: the Type 1 commutation stage.

$$\frac{di_{Lr}}{dt} = \frac{V_{cl}}{L_r} \tag{5.13}$$

Stage 9 (t_8–t_9): Type 1 commutation stage. Equivalent circuit is shown in Figure 5.20. At t_8, S_3 is turned off. The input current i_b charges C_3 and discharges C_6. With the buffer of C_3 and C_6, S_3 is turned off under ZVS conditions. This stage ends at t_9 when v_{c6} reduces to zero and D_6 conducts.

After stage 9, the rectifier begins a stage 1 of a new switching cycle.

In brief, there are two resonant stages in a switching cycle. The first one realizes ZVS turn-on for the bridge switches and the second one realizes the ZVS turn-on for the auxiliary switch. Besides, the reverse recovery of the freewheeling diodes is suppressed by restraining the current changing rate with the resonant inductor. In addition, the turn-off actions for the switches are under ZVS conditions.

5.3.2 Resonant Stages Analysis

As shown in Figure 5.11, there are two resonant stages in a switching cycle. The first resonance happens at stage 2. The circuit of this stage is illustrated in Figure 5.21a. It can be simplified into an equivalent circuit illustrated in Figure 5.21b. Here the AC-side current are assumed as constant.

According to the equivalent circuit in Figure 5.21b, following state equation is derived:

$$\begin{cases} L_r \dfrac{di_{Lr}}{dt} = v_{bus} - V_{dc} + V_{cl} \\ 4C_r \dfrac{dv_{bus}}{dt} = -i_{Lr} + i_a \end{cases} \tag{5.14}$$

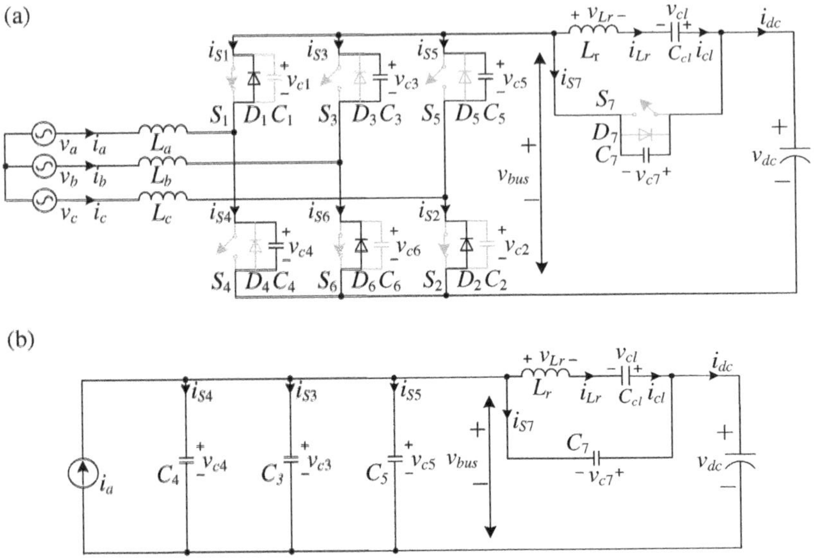

Figure 5.21 Resonance circuit and its equivalent circuit of stage 2 (t_1–t_2). (a) Circuit state of first resonance stage. (b) Equivalent circuit.

The initial values of the resonant inductor current and bus voltage at time t_1 are expressed by,

$$\begin{cases} i_{Lr}(t_1) = I_{Lr(t1)} \\ v_{bus}(t_1) = V_{dc} \end{cases} \tag{5.15}$$

By solving the state equation (5.14) with the initial value, i_{Lr} and v_{bus} are derived,

$$\begin{cases} i_{Lr} = i_a + \dfrac{V_{cl}}{Z_r}\sin\omega_r t - \left(i_a - i_{L(t1)}\right)\cos\omega_r t \\ v_{bus} = V_{dc} - V_{cl} + V_{cl}\cos\omega_r t + \left(i_a - i_{L(t1)}\right)\cdot Z_r \sin\omega_r t \end{cases} \tag{5.16}$$

where ω_r is the resonant angular frequency and is equal to

$$\omega_r = 1/\sqrt{4L_r C_r} \tag{5.17}$$

Z_r is the characteristic impedance and is defined as

$$Z_r = \sqrt{L_r/4C_r} \tag{5.18}$$

Equation (5.16) can be rewritten as

$$\begin{cases} i_{Lr} = i_a - \sqrt{V_{cl}^2/{Z_r}^2 + \left(i_a - i_{Lr(t1)}\right)^2}\cos(\omega_r t + \tau) \\ v_{bus} = V_{dc} - V_{cl} + \sqrt{V_{cl}^2 + \left(i_a - i_{Lr(t1)}\right)^2 {Z_r}^2}\sin(\omega_r t + \tau) \end{cases} \tag{5.19}$$

where

$$\tau = \arctan\frac{V_{cl}}{\left|i_a - i_{Lr(t1)}\right| Z_r} \tag{5.20}$$

To realize ZVS turn-on for the main switches, v_{bus} needs to resonance to zero. According to (5.19), v_{bus} is required to satisfy following inequality.

$$\sqrt{V_{cl}^2 + \left(i_a - i_{Lr(t1)}\right)^2 {Z_r}^2} > V_{dc} - V_{cl} \tag{5.21}$$

By further simplifying the above equation (5.21), the ZVS turn-on condition of the bridge switches is rewritten as,

$$i_{Lr(t1)} > i_a + \frac{\sqrt{V_{dc}^2 - 2V_{dc}V_{cl}}}{Z_r} \tag{5.22}$$

The second resonance happens at stage 4. The circuit state of this stage is illustrated in Figure 5.22a. Since the main switch D_1, S_3, and S_5 are in conduction, the grid side current does not influence the circuit resonance. The equivalent circuit of this stage is simplified into Figure 5.22b. During the second resonance, resonance occurs among L_r, C_2, C_4, C_6, and C_7, v_{bus} increases and v_{c7} decreases.

The second resonance stage begins at t_4. The initial values of the resonant inductor current and bus voltage at time t_4 are expressed by,

$$\begin{cases} i_{Lr}(t_4) = -i_x \\ v_{bus}(t_4) = 0 \end{cases} \tag{5.23}$$

where the boosted current i_x in short-circuit stage is equal to

$$i_x = \frac{V_{dc} - V_{cl}}{L_r}(t_4 - t_3) \tag{5.24}$$

According to the equivalent circuit in Figure 5.22b, following state equation is derived.

(a)

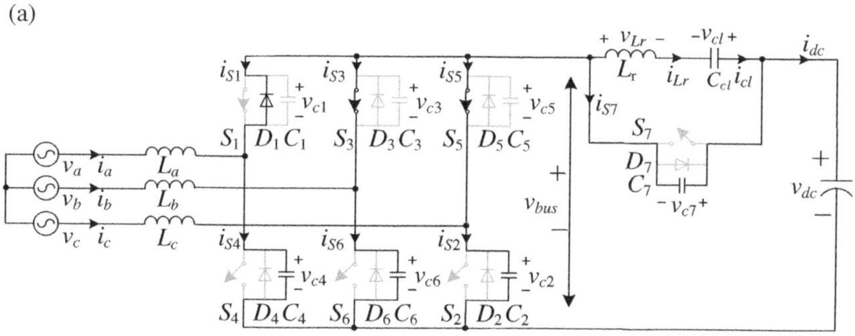

(b)

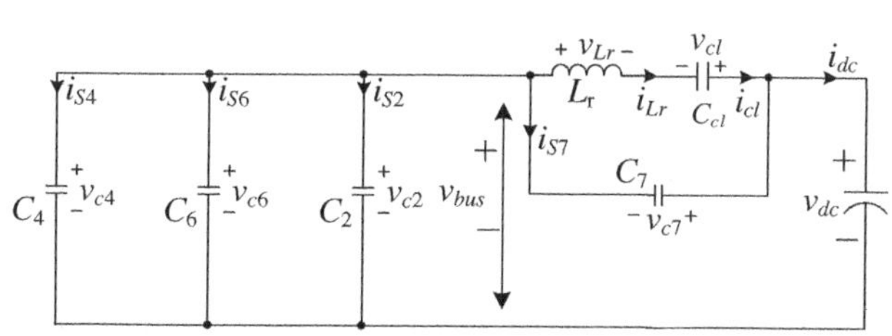

Figure 5.22 Equivalent circuit of stage 4 (t_4–t_5). (a) Circuit state of the second resonance. (b) Equivalent circuit.

$$\begin{cases} L_r \dfrac{di_{Lr}}{dt} = v_{bus} - V_{dc} + V_{cl} \\ 4C_r \dfrac{dv_{bus}}{dt} = -i_{Lr} \end{cases} \tag{5.25}$$

By solving above state equation (5.25) with the initial value of (5.23), following state variables are deducted,

$$\begin{cases} i_{Lr}(t) = \dfrac{V_{cl} - V_{dc}}{Z_r} \sin \omega_r t + i_x \cos \omega_r t \\ v_{bus}(t) = (V_{cl} - V_{dc}) \cos \omega_r t - i_x Z_r \sin \omega_r t - (V_{cl} - V_{dc}) \end{cases} \tag{5.26}$$

Equation (5.26) can be rewritten as

$$\begin{cases} i_{Lr} = \sqrt{(V_{cl} - V_{dc})^2 / Z_r{}^2 + i_x^2} \sin(\omega_r t + \sigma) \\ v_{bus} = (V_{dc} - V_{cl}) + \sqrt{(V_{cl} - V_{dc})^2 + i_x^2 Z_r{}^2} \cos(\omega_r t + \sigma) \end{cases} \tag{5.27}$$

where

$$\sigma = \arctan\left(\frac{i_x Z_r}{(V_{dc} - V_{cl})}\right) \tag{5.28}$$

Considering $v_{bus} + v_{c7} = V_{dc}$, v_{c7} can be derived as

$$v_{C7}(t) = V_{cl} - \sqrt{(V_{cl} - V_{dc})^2 + i_x^2 Z_r{}^2} \cos(\omega_r t + \sigma) \tag{5.29}$$

To realize ZVS turn-on of the auxiliary switch S_7, $v_{c7}(t)$ should resonance to zero.

According to second resonant stage analysis, the auxiliary switch S_7 can realize ZVS turn-on condition when the voltage v_{c7} decreases to zero. According to equation (5.29), the ZVS turn-on condition of the auxiliary switch can be expressed as,

$$\sqrt{(V_{cl} - V_{dc})^2 + i_x^2 Z_r{}^2} > V_{cl} \tag{5.30}$$

Since the clamping capacitor voltage V_{cl} is much smaller than the DC-side voltage V_{dc}, above inequality is always satisfied.

The second resonance stage stops at t_5 when v_{bus} is equal to V_{dc}; values of the capacitor voltage v_{bus} at time t_5 can be derived as

$$v_{bus}(t_5) = (V_{dc} - V_{cl}) + \sqrt{(V_{cl} - V_{dc})^2 + i_x^2 Z_r{}^2} \cos(\omega_r t_5 + \sigma) = V_{dc} \tag{5.31}$$

Actually, i_{Lr} reaches its minimum value i_{Lr_min} at t_5. By combining equation (5.27) with (5.31), the value of i_{Lr_min} can be rewritten as,

$$i_{Lr}(t_5) = i_{Lr\_\min} = -\frac{\sqrt{V_{dc}^2 - 2V_{dc}V_{cl} + i_x^2 Z_r{}^2}}{Z_r} \tag{5.32}$$

During (t_5–t_1), the resonant inductor L_r is clamped by V_{cl}, then

$$i_{Lr}(t_1) = i_{Lr(t5)} + \frac{V_{cl}}{L_r} d_7 T_s \tag{5.33}$$

where d_7 is duty ratio of the auxiliary switch,

$$d_7 = \frac{T_s - (t_5 - t_1)}{T_s} \tag{5.34}$$

Define λ_7 to be the auxiliary switch turn-off duty cycle,

$$\lambda_7 = 1 - d_7 \tag{5.35}$$

By substituting equation (5.35) into equation (5.33), the expression of the resonant inductor current at t_1 can be rewritten as

$$i_{Lr}(t_1) = i_{Lr(t5)} + \frac{V_{cl}}{L_r}(1 - \lambda_7)T_s \tag{5.36}$$

5.3.3 Steady State Analysis

When the rectifier is operated in steady state, the mean voltage of the resonant inductor in a switching cycle is zero. That is,

$$\frac{1}{T_s}\int_0^{T_s} v_{Lr}dt = 0 \tag{5.37}$$

The voltage waveform of the resonant inductor v_{Lr} is shown in Figure 5.11. Since the resonance intervals (t_1–t_2) and (t_4–t_5) during the resonance process are very short in a switch cycle, the integrations of the v_{Lr} during (t_1–t_2) and (t_4–t_5) can be ignored. Thus, equation (5.37) can be written as:

$$\frac{1}{T_s}\int_0^{T_s} v_{Lr}dt = \frac{1}{T_s}\left(\int_0^{t_1} v_{Lr}dt + \int_{t_2}^{t_4} v_{Lr}dt + \int_{t_5}^{t_9} v_{Lr}dt\right) \tag{5.38}$$

According to Figure 5.11, during (t_0–t_1) and (t_5–t_9), the resonant inductor v_{Lr} is expressed as

$$v_{Lr} = V_{cl} \tag{5.39}$$

During t_2–t_4, the resonant inductor v_{Lr} is expressed as

$$v_{Lr} = V_{cl} - V_{dc} \tag{5.40}$$

By substituting above equations (5.39) and (5.40) into equation (5.38), the following equation is obtained:

$$\frac{1}{T_s}\int_0^{T_s} v_{Lr}dt = V_{cl}(1 - \lambda_7)T_s + (V_{cl} - V_{dc})\lambda_7 T_s \tag{5.41}$$

Following equation is obtained by combining equations (5.37) and (5.41):

$$V_{cl}(1 - \lambda_7)T_s + (V_{cl} - V_{dc})\lambda_7 T_s = 0 \tag{5.42}$$

Thus, the clamping capacitor voltage is derived,

$$V_{cl} = \lambda_7 V_{dc} \tag{5.43}$$

Because the auxiliary switch turn-off duty cycle λ_7 is 5–10%, the clamping capacitor voltage is less than 5–10% of the DC-link voltage.

When the rectifier is operated in steady state, the mean current through clamping capacitor is zero. That is,

$$\frac{1}{T_s}\int_{t_0}^{t_0+T_s} i_{cl}dt = 0 \tag{5.44}$$

The current waveform of the clamping capacitor i_{cl} is shown in Figure 5.11. Since two resonance intervals during $(t_1 - t_2)$, $(t_4 - t_5)$, and two Type 1 commutation intervals $(t_6 - t_7)$ and $(t_8 - t_9)$ are very short in a switch cycle. To simplify the expression, their integrations are ignored in the analysis as follows:

$$\frac{1}{T_s}\int_{t_0}^{t_0+T_s} i_{cl}dt = \frac{1}{T_s}\left(\int_{t_0}^{t_1} i_{cl\_\text{stage }1}(t)dt + \int_{t_2}^{t_4} i_{cl\_\text{stage }3,4}(t)dt + \int_{t_5}^{t_6} i_{cl\_\text{stage }6}(t)dt + \int_{t_7}^{t_8} i_{cl\_\text{stage }8}(t)dt\right) = 0 \tag{5.45}$$

According to resonant stage analysis, combining equation (5.32), i_{Lr} reaches its minimum value i_{Lr_min} at t_5. In stage 6 (t_5–t_6), the resonant inductor L_r is clamped by the clamping capacitor C_{cl}, $i_{cl\_stage\ 6}=i_{Lr\_stage\ 6}$. Combining equations (5.12) and (5.32), the current through clamping capacitor can be expressed as,

$$i_{cl\_\text{stage }6}(t) = i_{Lr\_\text{stage }6} = i_{Lr(t5)} + \frac{V_{cl}}{L_r}(t - t_5) \tag{5.46}$$

In stage 8 (t_7–t_8), the resonant inductor L_r is still clamped by the clamping capacitor C_{cl}, $i_{cl\_stage\ 8}=i_{Lr\_stage\ 8}$. Combining equations (5.13) and (5.32), the current through clamping capacitor can be expressed as,

$$i_{cl\_stage\ 8}(t) = i_{Lr\_stage\ 8} = i_{Lr(t5)} + \frac{V_{cl}}{L_r}(t - t_5) \tag{5.47}$$

In stage 1 (t_0–t_1), the resonant inductor L_r is still clamped by the clamping capacitor C_{cl}, $i_{cl\_stage\ 1}=i_{Lr\_stage\ 1}$. Combining equations (5.8) and (5.32), the current through clamping capacitor can be expressed as,

$$i_{cl\_stage\ 1}(t) = i_{Lr\_stage\ 1} = i_{Lr(t5)} + \frac{V_{cl}}{L_r}(t - t_0 + T_s - t_5) \tag{5.48}$$

From stage 3 (t_2–t_3) to stage 4 (t_3–t_4), the resonant inductor L_r is clamped by $V_{cl}-V_{dc}$. By ignoring the short resonance interval (t_1–t_2), the current through clamping capacitor can be expressed as

$$i_{cl\_\text{stage }3,4}(t) = i_{Lr\_\text{stage }3,4}(t) = i_{Lr(t1)} + \frac{V_{cl} - V_{dc}}{L_r}(t - t_1) \tag{5.49}$$

Since the intervals during stage 1 $(t_0 - t_1)$; stage 3, 4 (t_2–t_4); stage 6 (t_5–t_6); and stage 8 (t_7–t_8) are $d_{V1}T_s$; $\lambda_7 T_s$; $[d_{V0}$-$\lambda_7]T_s$; and $d_{V2}T_s$, respectively, combining equations (5.36), (5.46), (5.47), (5.48), and (5.49), the equation (5.45) is derived as,

$$I_{Lr(t5)}T_s + \frac{V_{cl}}{L_r}\lambda_7 d_7 T_s^2 - \frac{V_{dc} - V_{cl}}{2L_r}\lambda_7^2 T_s^2 + \frac{V_{cl}}{2L_r}d_7^2 T_s^2 = 0 \tag{5.50}$$

By combining (5.32) and (5.50), the solution can be expressed as,

$$\lambda_7 \approx \frac{2L_r\sqrt{V_{dc}^2 + i_x^2 Z_r^2}}{Z_r V_{dc} T_s} \tag{5.51}$$

By combining equations (5.32), (5.33), and (5.51), the expression of the resonant inductor current at t_1 is derived as,

$$i_{Lr(t1)} = -\frac{\sqrt{V_{dc}^2 - 2V_{dc}V_{cl} + i_x^2 Z_r^2}}{Z_r} + \frac{V_{cl}}{L_r} d_7 T_s = \frac{\sqrt{V_{dc}^2 - 2V_{dc}V_{cl} + i_x^2 Z_r^2}}{Z_r} \tag{5.52}$$

5.3.4 Soft-switching Condition

According to the steady state analysis, the auxiliary switch S_7's ZVS turn-on condition can always be realized. ZVS turn-on condition for the bridge switches is expressed by inequality (5.22). By substituting equation (5.52) into inequality (5.22), following inequality is obtained,

$$\frac{\sqrt{V_{dc}^2 - 2V_{dc}V_{cl} + i_x^2 Z_r^2}}{Z_r} > i_a + \frac{\sqrt{V_{dc}^2 - 2V_{dc}V_{cl}}}{Z_r} \tag{5.53}$$

The above inequality can rewritten as

$$i_x \geq \sqrt{\frac{2i_a\sqrt{V_{dc}^2 - 2V_{dc}V_{cl}}}{Z_r} + i_a^2} \tag{5.54}$$

Inequality (5.54) describes the ZVS condition for ZVS turn-on of main switches. The extra added current i_x should be controlled to meet (5.54) so that there is enough stored energy in resonant inductor to realize ZVS condition in the first resonant stage.

According to (5.7), the grid side current i_a is written as,

$$i_a = I_m \cos(\omega_l t + \varphi_0) \tag{5.55}$$

As before, sector 1-1 is taken as an example. In sector 1-1, it is assumed the rectifier operates with unit power factor, i.e. $\varphi_0 = 0$. Then the range of phase current i_a in sector 1-1 is as follows:

$$\frac{\sqrt{3}}{2} I_m \leq i_a \leq I_m \tag{5.56}$$

To satisfy the ZVS turn-on condition in sector 1-1, the inequality (5.54) needs to satisfy with maximum of i_a in the sector. By substituting $i_a = I_m$ to the inequality (5.54), the extra added current i_x should meet the following inequality:

$$i_x \geq \sqrt{\frac{2I_m\sqrt{V_{dc}^2 - 2V_{dc}V_{cl}}}{Z_r} + I_m^2} \tag{5.57}$$

By substituting inequality (5.57) into the equation (5.24), the short-circuit duration is estimated as

$$(t_4 - t_3) \geq \frac{L_r}{V_{dc} - V_{cl}} \sqrt{\frac{2I_m\sqrt{V_{dc}^2 - 2V_{dc}V_{cl}}}{Z_r} + I_m^2} \tag{5.58}$$

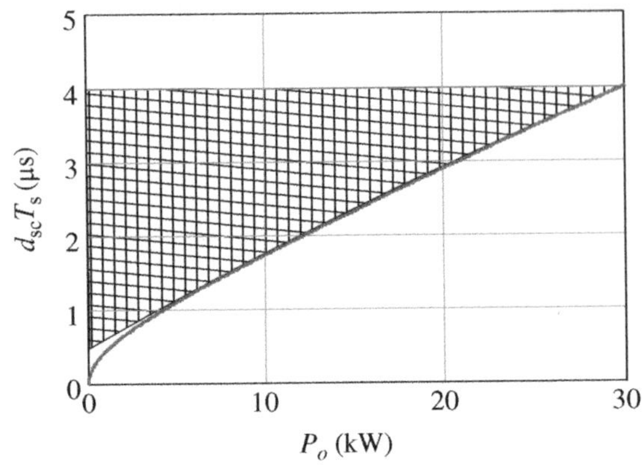

Figure 5.23 Short-circuit duration ($t_4 - t_3$) vs. P_O.

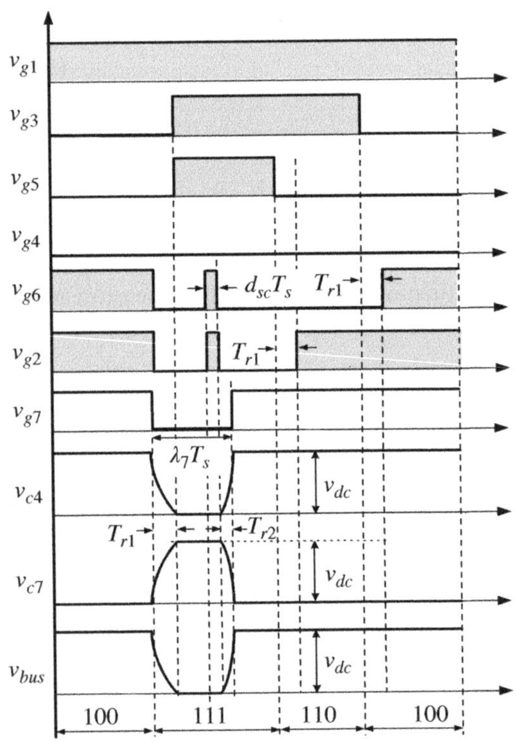

Figure 5.24 Driving logic waveform in sector 1-1.

From (5.58), in order to realize the ZVS turn-on condition of the main switches, the duration time of stage 4 must be long enough to store enough energy in the resonant inductor. By defining the short-circuit duty ratio, $d_{sc} = (t_4 - t_3)/T_s$, the inequality (5.58) is rewritten as,

$$d_{sc} \geq \frac{L_r}{(V_{dc} - V_{cl})T_s}\sqrt{\frac{2I_m\sqrt{V_{dc}^2 - 2V_{dc}V_{cl}}}{Z_r} + I_m{}^2} \tag{5.59}$$

To satisfy the bridge switches' ZVS turn-on condition, the duration time of stage 4 should be designed so long to store enough energy in the resonant inductor. When $f_s = 16$ kHz, $V_{dc} = 700$ V DC, the relationship between P_0 and short-circuit duration time ($t_4 - t_3$) is shown in Figure 5.23. The short-circuit duration ($t_4 - t_3$) increases with the increasing power of the converter.

Figure 5.24 shows the driving logic waveform of MVAC rectifier in sector 1-1. Before the Type 2 switching commutation occurs from $\vec{V}_1(100)$ to $\vec{V}_7(111)$, S_7 is turned off first to create the ZVS turn-on condition for the bridge switches and restrain the freewheeling diode's reverse recovery current. After 1/4 of the resonant period $T_r/4$, the corresponding main switch can be turn-on under ZVS condition. A short-circuit switch state is added in this commutation process. The duration time of the short-circuit time and the auxiliary switch's turn-off time are $d_{sc}T_s$ and $\lambda_7 T_s$, respectively. In the Type 1 switching commutation process from $\vec{V}_7(111)$ to $\vec{V}_2(110)$ and from $\vec{V}_2(110)$ to $\vec{V}_1(100)$, the active switch S_7 is kept in conduction state.

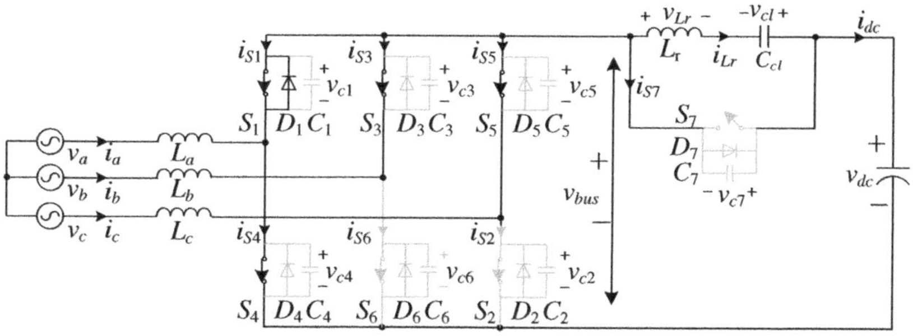

Figure 5.25 The short-circuit mode 1.

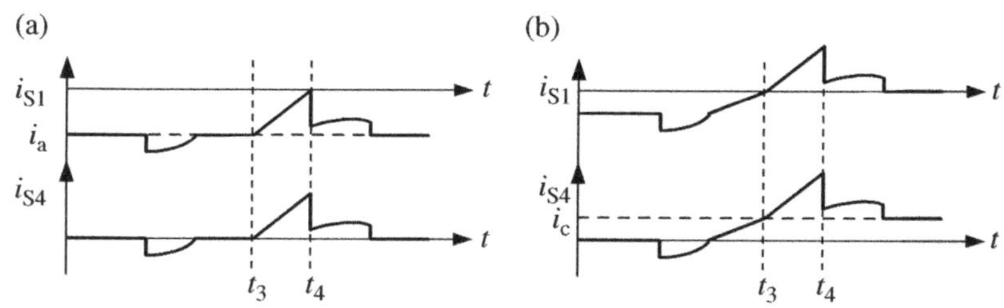

Figure 5.26 The short-circuit switches' current. (a) Sector 1-1. (b) Sector 1-2.

For MVAC converter, a short-circuit period is necessary to satisfy the ZVS condition for the main switches. It can be realized by only one bridge switch or by three-phase bridge switches. Different short-circuit modes bring different losses.

For example, in Figure 5.25, the short-circuit switches are always fixed in phase-a bridge, and the corresponding current waveform of phase-a bridge switches is shown in Figure 5.26. The short-circuit modulation scheme with fixed phase bridge switches is named as short-circuit mode 1.

In Figure 5.27, the short-circuit switches are always fixed in three-phase switches, and the corresponding current waveform of phase-a bridge switches is shown in Figure 5.28. The short-circuit modulation scheme with fixed three-phase switches is named as short-circuit mode 2.

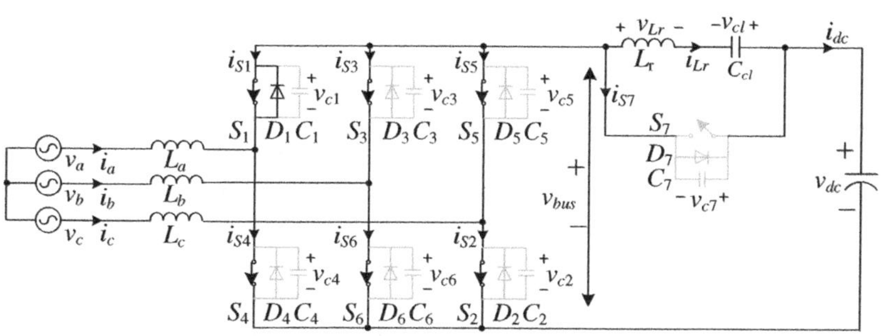

Figure 5.27 The short-circuit mode 2.

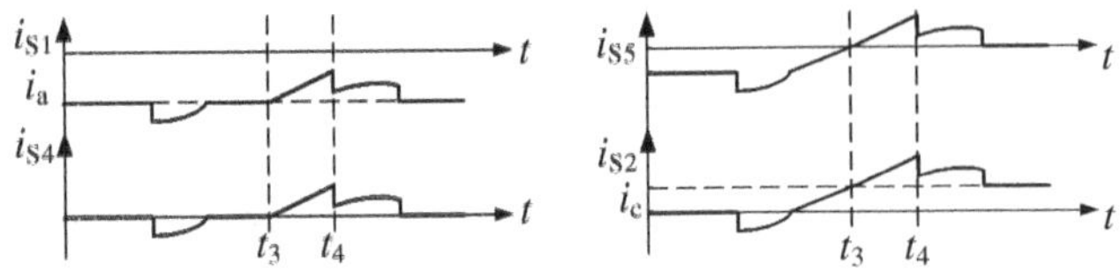

Figure 5.28 The short-circuit switches' current. (a) Sector 1-1. (b) Sector 1-2.

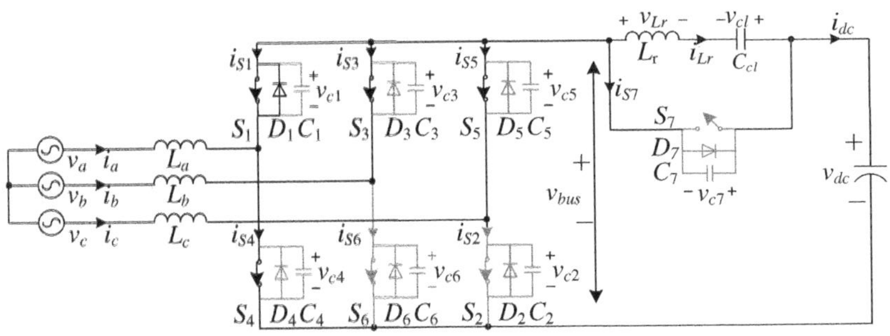

Figure 5.29 The short-circuit mode 3.

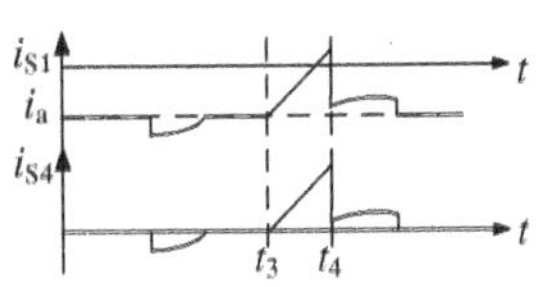

Figure 5.30 The short-circuit switches' current.

In Figure 5.29, the short-circuit switches are always fixed into the bridge switches with the largest phase current, and the corresponding current waveform of phase-a bridge switches is shown in Figure 5.30. The short-circuit modulation scheme fixed into the bridge switches with the largest phase current is named as short-circuit mode 3.

Evaluate the above three short-circuit modulation schemes; in terms of current stress, short-circuit mode 2 has the lowest current stress, short-circuit mode 1 has the largest current stress, and the switch thermal stress distribution of short-circuit mode 1 is not uniform. Though short-circuit mode 3 also has the largest current stress, its thermal stress distribution is uniform and its switch conduction loss is almost same as short-circuit mode 2. In terms of turn-off loss, compared with short-circuit mode 1 and short-circuit mode 3, short-circuit mode 2 produces three times the turn-off loss. Based on the above analysis, the short-circuit mode 3 is the most suitable scheme, it produces the smallest turn-off loss, and also the thermal stress is evenly distributed.

5.3.5 Control Technique of Minimum Voltage Active-clamping Three-phase Rectifier

The control block diagram of the three-phase MVAC rectifier based on the synchronous rotating coordinate system is shown in Figure 5.31. The output of the DC voltage control loop is the reference of active current i_d, and the reference of reactive current i_q is zero. Through grid voltage feed forward and inductor current decoupling control, the modulation ratio m and the reference vector angle θ can be obtained. With the equation (5.5), the duty ratio of the nonzero voltage vector and zero voltage vector, d_{V0}, d_{V1}, and d_{V2}, can be calculated through the modulation ratio m and the rotation vector angle θ. The duty ratio of the auxiliary switch d_7 can be calculated with the equation (5.51). With equation (5.5), four modulation indexes m_a, m_b, m_c, and m_{d7} based on d_{V0}, d_{V1}, d_{V2}, and d_7, can be obtained. The short-circuit time can be calculated through equation (5.54). With the modulation sequences of the ZVS-SVM in Table 5.3, voltage sector and zero voltage vector can be

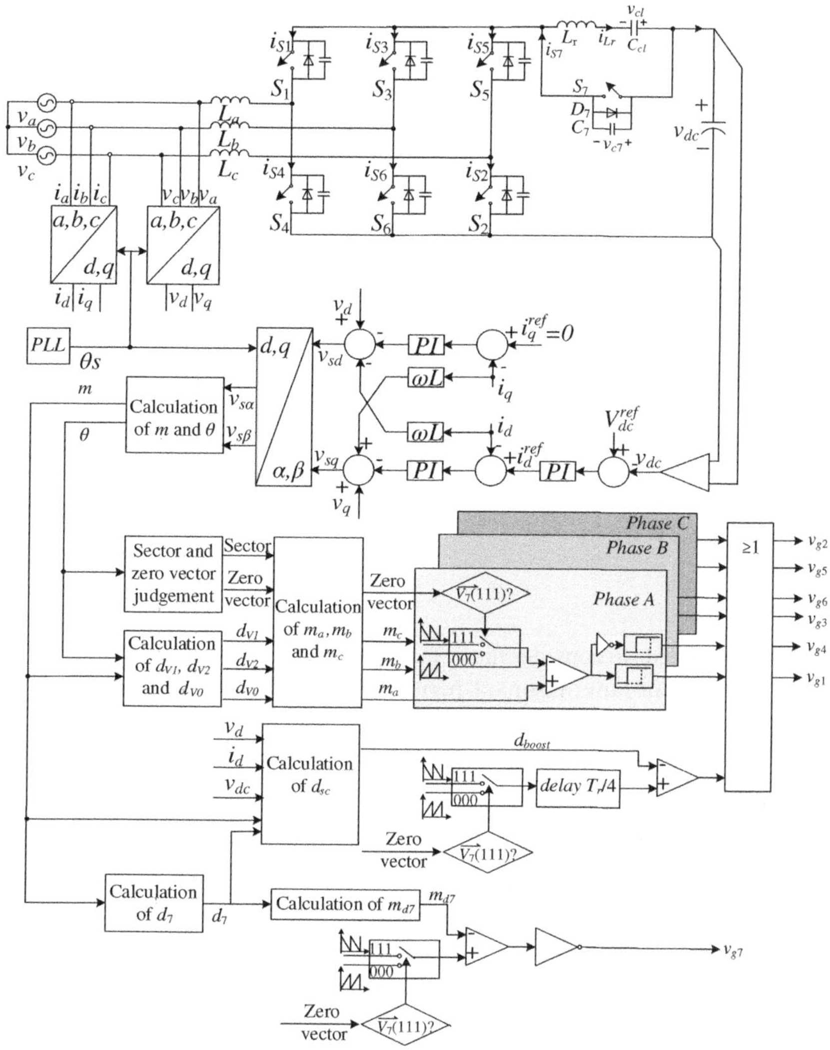

Figure 5.31 Control block diagram of the rectifier.

Table 5.3 Specifications of the ZVS rectifier.

Parameters	Descriptions	Value
P_0	Output power	30 kW
V_a, V_b, V_c	Phase voltage RMS value	220 Vrms
f_g	Frequency of the grid	50 Hz
f_s	Switching frequency	16 kHz
L_a, L_b, L_c	Filter inductance	0.3 mH
V_{dc}	DC source voltage	700 V

selected according to the rotation vector angle θ. According to sector selection, zero voltage vector judgment, the saw-tooth carrier wave v_c can also be chosen. Through comparing of the modulation indexes m_a, m_b, m_c, and m_{d7} with the carrier wave, all the switches' driving signals can finally be obtained.

5.4 Prototype Design

The previous sections introduced the operation principles of the proposed ZVS rectifier and the modulation scheme. In this section, design guidelines will be provided through a design example.

5.4.1 Specifications of a 30 kW Rectifier

In order to illuminate the design, a 30 kW ZVS grid-connected rectifier prototype was introduced. The specifications of the rectifier shown in Figure 5.32 are listed in Table 5.3. IGBT modules with capacitance of 1200 V/200 A are selected for switching devices S_1–S_7. It is assumed that $C_1 = C_2 = C_3 = C_4 = C_5 = C_6 = C_7 = C_r$.

5.4.2 Parameter Design

MVAC rectifier parameter design needs to consider many factors, not only soft-switching condition but also the resistance of the resonant tank, resonant period, and the duty cycle of the auxiliary switch.

The resonant parameters of the MVAC rectifier are designed as follows:

1) Set the minimum value of the resonant inductance to ensure the ability to suppress the reverse recovery current of the diode. The resonant inductor suppresses the reverse recovery energy of the freewheeling diodes in stage 3. In order to suppress the reverse recovery losses of the diodes, the di_{Lr}/dt during the turn-off process of the diodes is selected to be lower than 100 A/μs. In stage 3, it is expressed as,

$$\frac{di_{Lr}}{dt} = \frac{V_{dc}}{L_r} \leq 100\text{A}/\mu\text{s} \tag{5.60}$$

Therefore, in this case, the inductance should satisfy,

$$L_r \geq 7\mu\text{H} \tag{5.61}$$

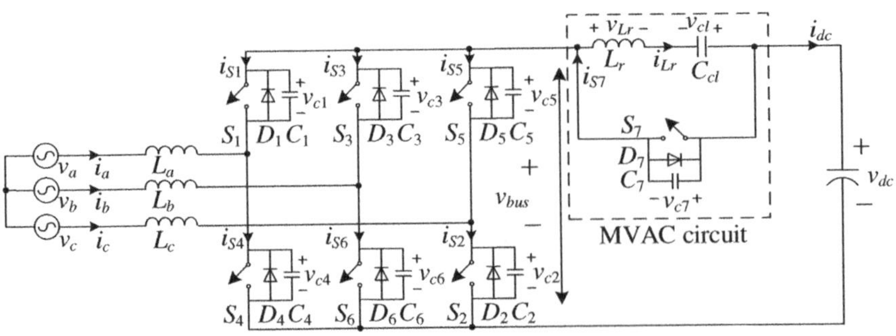

Figure 5.32 Topology of the MVAC rectifier.

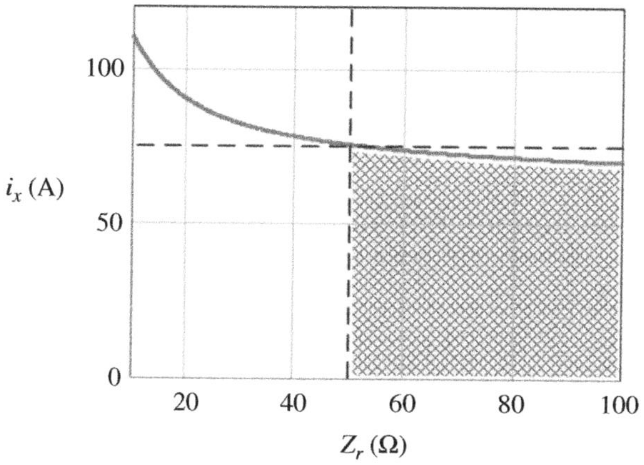

Figure 5.33 The short-circuit current i_x vs. Zr.

2) The short-circuit duration of stage 4 is used to store enough energy in the resonant inductor so that ZVS condition is satisfied for the main switches. However, longer short-circuit duration means higher extra added current in the resonant inductor i_x, which will cause higher turn-off loss of the main switches. Thus the minimum short-circuit duration is selected. In this case, the rated power of the MVAC rectifier is $P_0 = 30kW$, $V_{dc} = 700$ V, according to the equation (5.57); the relationship between Z_r and short-circuit current i_x is shown in Figure 5.33. The characteristic impedance Z_r is chosen as 50 Ω. If Z_r further increases after 50 Ω, i_x value does not decrease significantly. Ignore the current ripple of the grid side filter inductor; the amplitude value of the grid side current $I_m = \sqrt{2}P_0/(3{\cdot}220) = 64.3$A. By substituting $V_{dc} = 700$ V, $I_m = 64.3$ A, and $Z_r = 50\ \Omega$ into the equation (5.57), $i_x = 76.4$ A.
3) To avoid loss of duty ratio, the auxiliary switch turn-off duty cycle λ_7 should be less than d_{V0}. The expression of the zero-voltage vector duty ratio d_{V0} in equation (5.5) is rewritten here:

$$d_{V0} = 1 - \frac{\sqrt{3}}{2} m \sin\left(\frac{\pi}{3} + \theta\right) \tag{5.62}$$

where θ is the phase angle of the reference vector, and m is the voltage modulation ratio. Due to the maximum value of sin(π/3+θ) is one, the minimum value of d_{V0} is

$$d_{V0\,\min} = 1 - \frac{\sqrt{3}}{2} m \tag{5.63}$$

By substituting equations (5.6) into (5.63), following equation is derived:

$$d_{V0\,\min} = 1 - \sqrt{3}\frac{\left|\vec{V}_{ref}\right|}{V_{dc}} \tag{5.64}$$

By ignoring the voltage drop of the grid side filter inductor, the amplitude of the reference voltage vector $\left|\vec{V}_{ref}\right|$ is approximately equal to the amplitude of the grid source voltage V_m. In this case, by substituting $V_{dc} = 700$ V and $\left|\vec{V}_{ref}\right| = V_m = 220{\cdot}\sqrt{2} = 311V$ to equation (5.64), the minimum value of d_{V0} is calculated, $d_{V0\min}$=0.23. If the auxiliary switch turn-off duty cycle λ_7 should be less than d_{V0}, then the following inequality needs to satisfy:

$$\lambda_7 < 0.23 \tag{5.65}$$

According to the stages analysis in Section 5.3.2, λ_7 is approximately expressed by,

$$\lambda_7 = \frac{t_5 - t_2 + T_r/4}{T_s} \tag{5.66}$$

where T_r is the resonant period frequency and is equal to

$$T_r = 4\pi\sqrt{L_r C_r} \tag{5.67}$$

In order to maintain the power quality, the resonant period T_r is designed to be smaller than10% of the switching period T_s, T_r<6.25 μs.

The λ_7 expression of equation (5.51) is rewritten here.

$$\lambda_7 \approx \frac{2L_r\sqrt{V_{dc}^2 + i_x^2 Z_r{}^2}}{Z_r V_{dc} T_s} \tag{5.68}$$

In this case, the switching cycle $T_s = 1/f_s = 62.5$ μs. By substituting $V_{dc} = 700$ V, $I_m = 64.3$ A, $T_s = 62.5$ μs and $i_x = 76.4$ A into the above equation, the relationship between resonance impedance Z_r and resonance inductance L_r under different λ_7 is shown in Fig. 5.34. The upper boundary of the available λ_7 is mainly determined by the value of L_r. When Z_r is between 50 Ω and 80 Ω, the upper limits of L_r is between 53 and 54 μH.

According to the stages analysis in Section 5.3.2, λ_7 is composed of the first resonance, second resonance, flywheel, and short-circuit process. The resonant period $T_r = 4\pi\sqrt{L_r C_r}$. In order to limit λ_7, the resonant period T_r is designed to be smaller than10% of the switching period T_s, $T_r < 6.25$ μs.

4) In the ZVS rectifiers, turn-on losses of the switches are eliminated. However, turn-off losses of all of the switches still exist. They can be reduced by the paralleled capacitors. The turn-off losses reduction depends on the capacitance of the paralleled capacitors. The turn-off losses under ZVS condition of IGBT module CM200DU-24NFH is demonstrated in Figure 5.35. For a given turn-off current of 60 A, when C_r is 2.2 nF, the turn-off loss is reduced by 8%; when C_r is 3.3 nF, the turn-off loss is reduced by 38%; when C_r is 4.7 nF, the turn-off loss is reduced by 45%; and when C_r is 7 nF, the turn-off loss is reduced by 61%. In order to obtain an obvious turn-off loss reduction, the capacitance of the resonant capacitor C_r>3.3 nF is preferred.

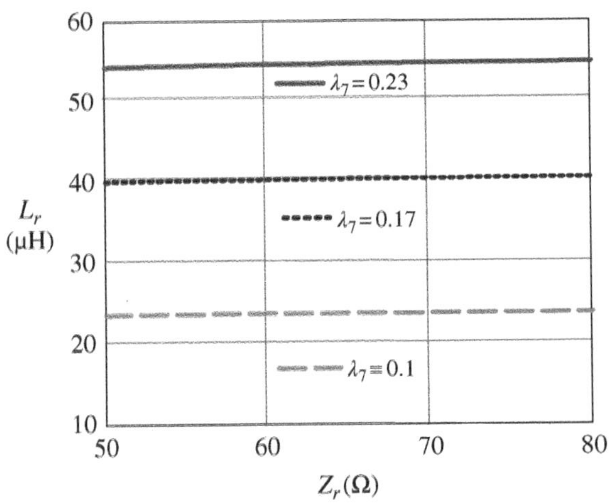

Figure 5.34 *L_r versus Z_r.*

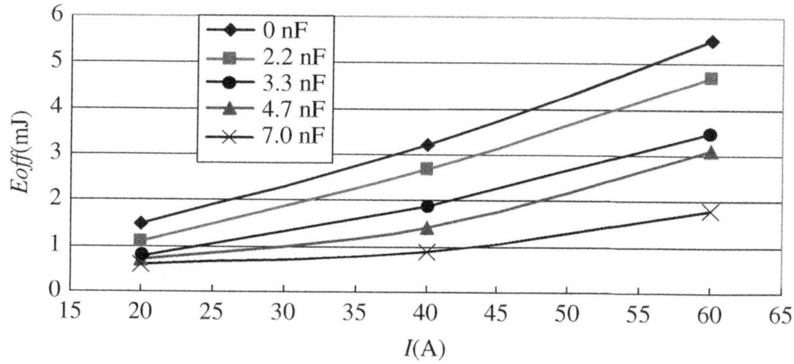

Figure 5.35 CM200DU-24NFH IGBT turn-off loss under ZVS conditions.

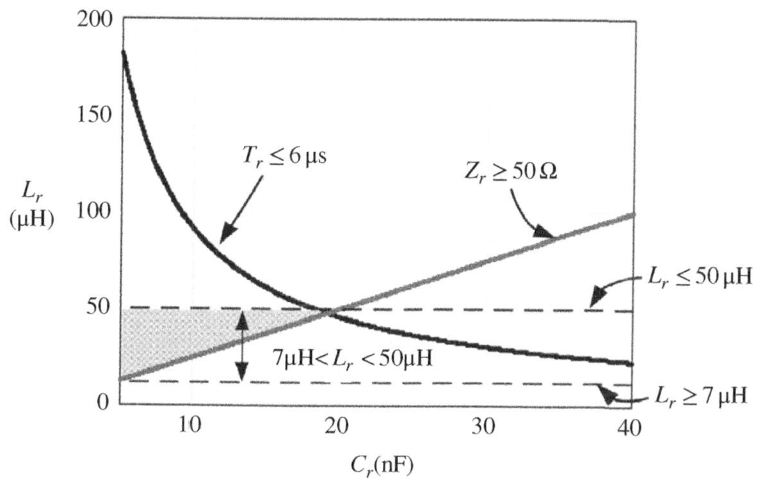

Figure 5.36 Recommended resonant parameters area.

Based on the analysis above, the feasible solution of the resonant parameters will be confined to a small region shown as Figure 5.36. The resonant parameters in shadow area meet all the requirements discussed before. The dark region shown in Figure 5.36 can be used to select the appropriate resonant inductance and resonant capacitance.

5.4.3 Experiment Platform and Testing Results

As shown in Figure 5.37, a 30 kW MVAC rectifier prototype with switch frequency 16 kHz was designed.

The main structure of the bus bar and switching bridges is shown in Figure 5.38. IGBT modules CM200DU-24NFH are selected for bridge switches and auxiliary switch. In the circuit, ceramic capacitors are utilized as the resonant capacitors C_1–C_6. Film capacitors are connected in parallel to operate as the clamping capacitor C_{cl}.

The resonant parameters are shown in Table 5.4.

The grid connected voltage and current are demonstrated in Figure 5.39. The measured current Total Harmonic Distortion (THD) is 2.9% under 30 kW condition.

The driving signal of main switches S_1 and S_4 is shown in Figure 5.40b; the driving signals of S_1 and S_4 are both high to make the main switch leg of phase-a in short circuit in stage 5.

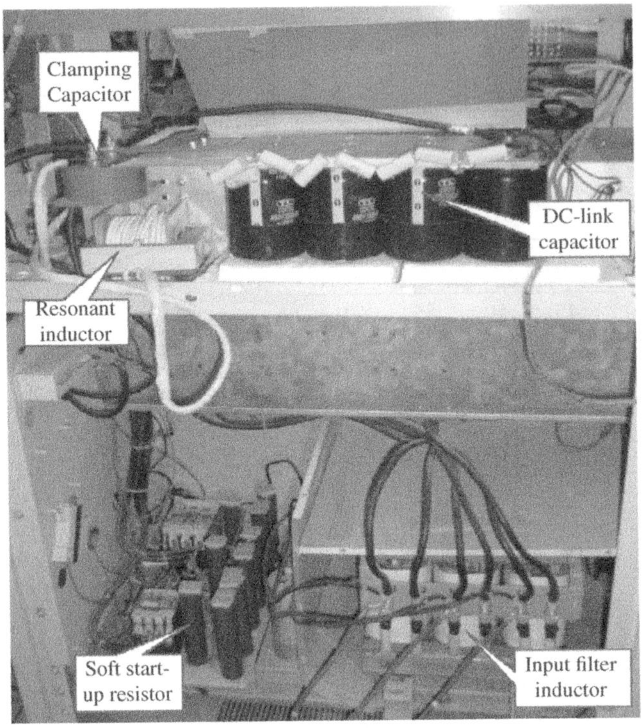

Figure 5.37 30 kW ZVS rectifier prototype.

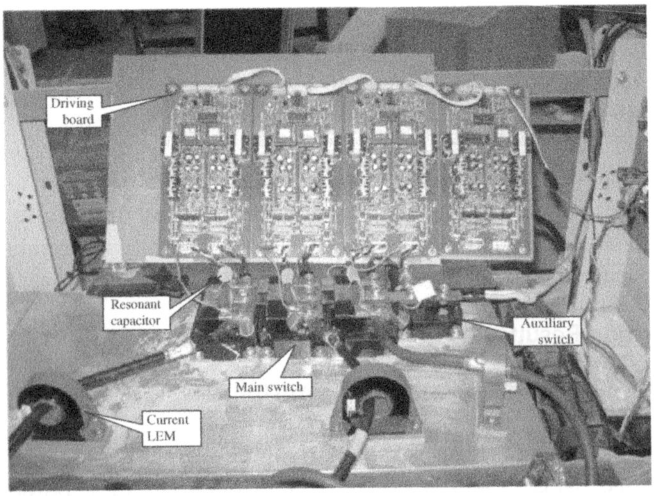

Figure 5.38 Bus bar and structure 30 kW ZVS rectifier prototype.

Table 5.4 Resonant parameters selected.

Parameters	Descriptions	Value
L_r	Inductance of the resonant inductor	45 μH
C_r	Capacitance of the resonant capacitors	4.7 nF
C_{cl}	Capacitance of the clamping capacitor	100 μF

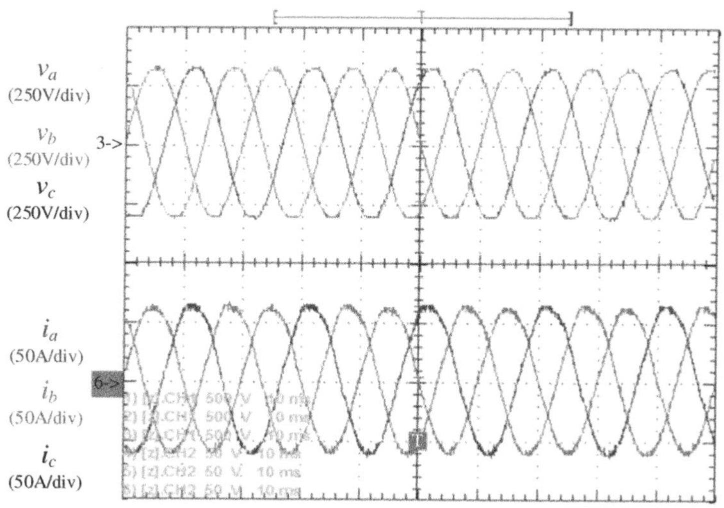

Figure 5.39 Output current and grid voltage.

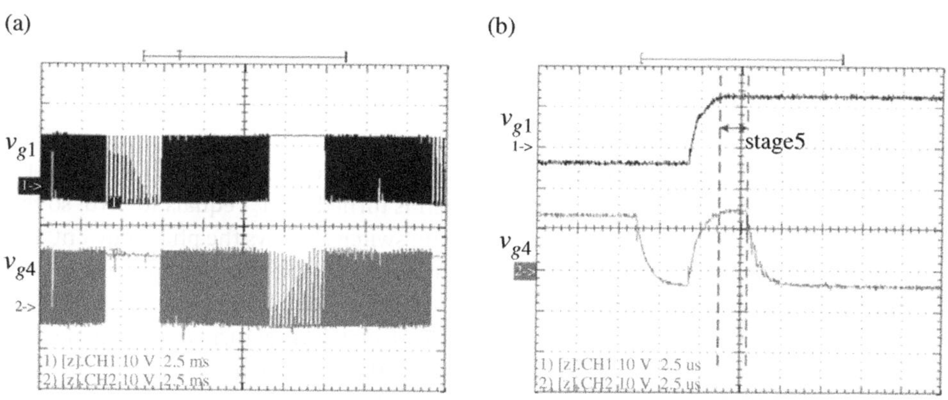

Figure 5.40 The driving signal of main switches S_1 and S_4. (a) Time: 2.5 ms/div. (b) Time: 2.5 μs/div.

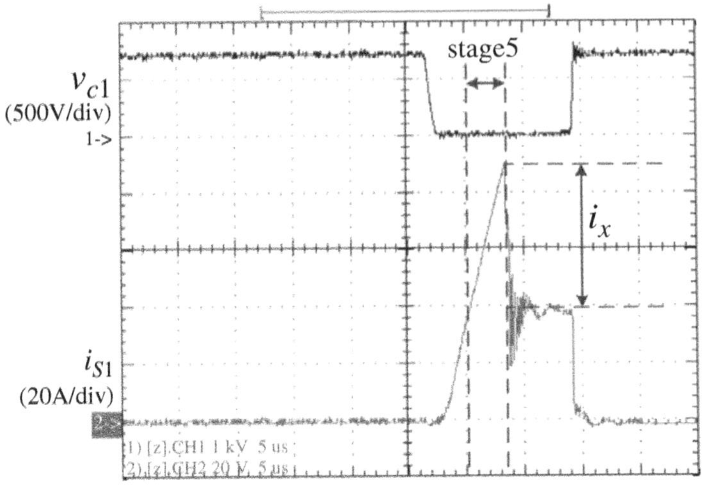

Figure 5.41 v_{c1} and i_{S1} (time 5 μs/div).

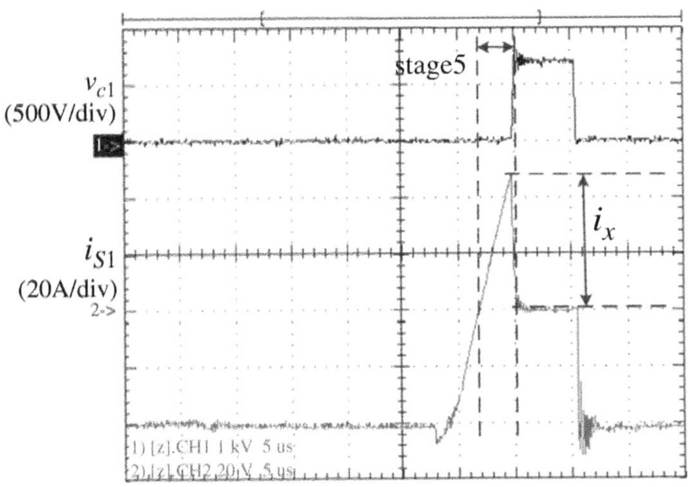

Figure 5.42 v_{c1} and i_{S1} (time 5 μs/div).

When the output power is 20 kW, the waveform of v_{c1} and i_{S1} in sector 3-1 is shown in Figure 5.41. It can be seen from Figure 5.41 that before the main switch is turned on, v_{c1} equals to zero, so the main switch is turn-on under ZVS condition. When the main switch S_1 flows the phase current, the additional energy storage stage 5 is inserted and switch S_1 will carry the short-circuit current i_x.

When the output power is 20 kW, the waveform of v_{c1} and i_{S1} in sector 1-2 is shown in Figure 5.42. When the current of the diode decreases to zero, the IGBT of switch S_1 is turned on to make the switch leg of phase-a in short circuit and switch S_1 will carry the short-circuit current i_x.

The waveform of the auxiliary switch current and voltage is shown in Figure 5.43. It can be seen that the antiparallel diode is in conduction before the auxiliary switch is turned on. So the auxiliary switch is turned on under ZVS condition.

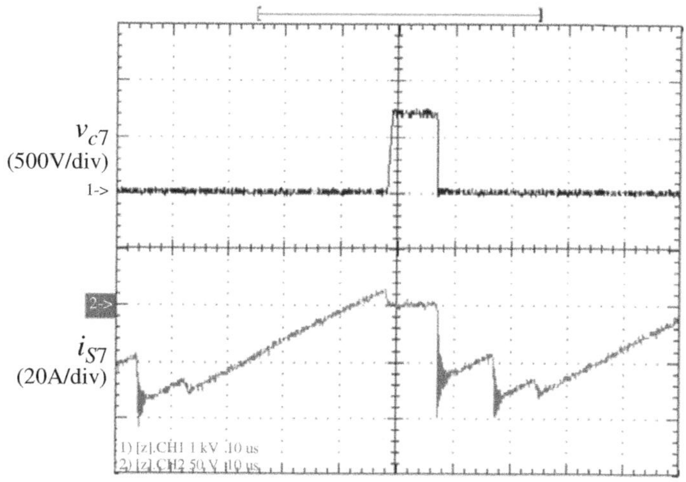

Figure 5.43 v_{c7} and i_{S7} (time 10 μs/div).

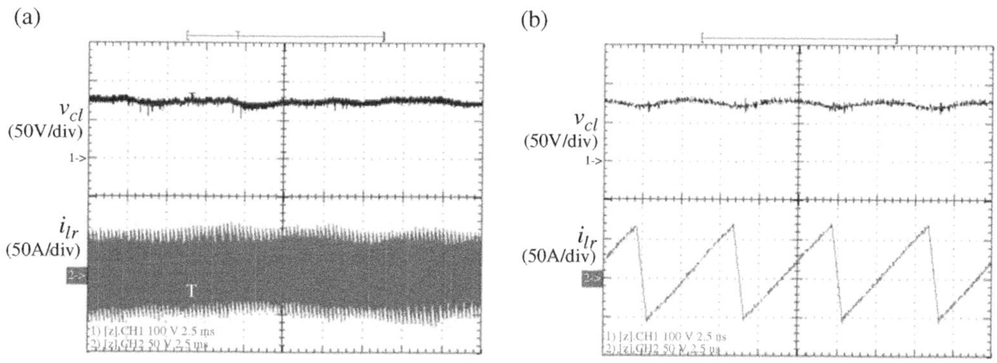

Figure 5.44 Current flowing L_r (i_{Lr}) and voltage across C_{cl} (V_{cl}). (a) Time: 2.5 ms/div. (b) Time: 25 μs/div.

The waveforms of voltage across the clamping capacitor and current flowing the resonant inductor are shown in Figure 5.44. It can be seen that V_{cl} is stable and less than 80 V.

The conversion efficiency of the ZVS rectifier is measured by power analyzer WT1800 and is shown in Figure 5.45.

5.5 Summary

MVAC rectifier is introduced in this chapter. MVAC rectifier can realize ZVS operation for all switching devices, and the antiparallel diode reverse recovery current of all switch devices is suppressed well. To realize the ZVS condition, a ZVS-SVM scheme is presented. All the switch devices can be operated at the fixed switching frequency. The switching voltage stress of all the power switch devices is equal to the DC-link voltage. A 30-kW prototype has been built and realizes desirable ZVS soft-switching features.

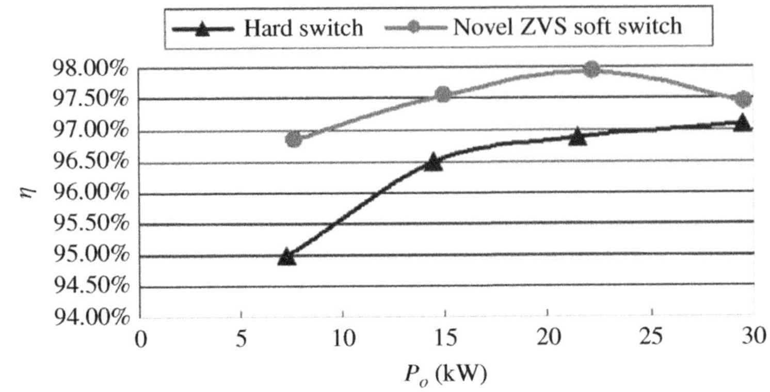

Figure 5.45 Measured efficiency.

MVAC rectifier has advantage of low voltage stress on devices. However, it has disadvantage of lower DC voltage utilization. Its DC voltage utilization is about 10% lower than the conventional pulse-width-modulation (PWM) rectifier if the auxiliary switch turn-off duty cycle λ_7 is 10%. To reduce this side effect, it is preferred to select lower λ_7.

References

1 D. Xu, B. Feng, and R. Li, "A zero voltage switching SVM (ZVS-SVM) controlled three-phase boost rectifier," *IEEE Transactions on Power Electronics*, vol. 22, no. 3, pp. 978–986, 2007.

2 Z. Ma, D. Xu, R. Li, C. Du and X. Zhang, "A novel DC-Side Zero-Voltage Switching (ZVS) three-phase boost PWM rectifier controlled by an improved SVM method," *IEEE Transactions on Power Electronics*, vol. 27, no. 11, pp. 4391–4408, 2012.

3 R. Li, Z. Ma and D. Xu, "A ZVS grid-connected three-phase inverter," *IEEE Transactions on Power Electronics*, vol. 27, no. 8, pp. 3595–3604, 2012.

4 R. Li and D. Xu, "A zero-voltage switching three-phase inverter," *IEEE Transactions on Power Electronics*, vol. 29, no. 3, pp. 1200–1210, 2014.

5 C. Du, D. Xu, N. He and N. Zhu, "Modeling and optimization of a zero-voltage switching inverter for high efficiency and miniaturization," *IEEE Transactions on Power Electronics*, vol. 32, no. 1, pp. 150–163, 2017.

6 M. Chen, J. Xi, Y. Chen and D. Xu, "Safe startup method for zero-voltage-switching MOSFET grid inverter," *IEEE Transactions on Power Electronics*, vol. 34, no. 9, pp. 8749–8761, 2019.

7 K. Shi, A. Zhao, J. Deng and D. Xu, "Zero-voltage-switching SiC-MOSFET three-phase four-wire back-to-back converter," *IEEE Journal of Emerging and Selected Topics in Power Electronics*, vol. 7, no. 2, pp. 722–735, 2019.

8 N. He, Y. Zhu, A. Zhao and D. Xu, "Zero-voltage-switching sinusoidal pulsewidth modulation method for three-phase four-wire inverter," *IEEE Transactions on Power Electronics*, vol. 34, no. 8, pp. 7192–7205, 2019.

Part 3

ZVS-SVM Applied to Three-phase Grid Inverters

6

Three-phase Grid Inverter with Minimum Voltage Active-clamping Circuit

In this chapter, a three-phase inverter with minimum voltage active-clamping (MVAC) circuit is introduced, including the operation principle, the circuit analysis, and the design guidelines of the MVAC inverter. First, the modulation scheme and stage analysis are introduced. Then, the soft-switching condition will be derived to help readers to understand this circuit. The design guidelines will be introduced, and a design example based on a 30-kW prototype will be shown [1, 2].

6.1 Introduction

The circuit topology of the MVAC ZVS three-phase inverter is shown in Figure 6.1 [3, 4]. It consists of a standard three-phase inverter and a minimum voltage active-clamping (MVAC) auxiliary circuit. The three-phase inverter consists of grid side filter inductors L_a, L_b, and L_c; six active switches S_1, S_2, S_3, S_4, S_5, and S_6; and switches' antiparallel diodes D_1, D_2, D_3, D_4, D_5, and D_6. The MVAC auxiliary circuit is composed of an active switch S_7, a resonant inductor L_r, and a clamping capacitor C_{cl}. Capacitors C_1, C_2, C_3, C_4, C_5, C_6, and C_7 are paralleled with the active switches.

In this chapter, the operation principle of the Zero Voltage Switching (ZVS) Space Vector Modulation (SVM) will be introduced. Then, the stage analysis of the circuit will be provided. Soft-switching condition will be derived. In addition, a steady-state analysis is carried out to get stress of the switches. After the theoretical analysis, the design guidelines will be introduced, and a design example based on a 30-kW prototype will be given.

6.2 Operation Principle of MVAC Inverter

6.2.1 Space Vector of Three-phase Grid Voltage

The grid side three-phase voltages v_a, v_b, and v_c of three-phase inverter are expressed by

$$\begin{cases} v_a = V_m \cos(\omega_l t) \\ v_b = V_m \cos\left(\omega_l t - \dfrac{2\pi}{3}\right) \\ v_c = V_m \cos\left(\omega_l t + \dfrac{2\pi}{3}\right) \end{cases} \tag{6.1}$$

where V_m and ω_l are the amplitude and the angular frequency of the grid side voltage, respectively.

Soft-Switching Technology for Three-phase Power Electronics Converters, First Edition.
Dehong Xu, Rui Li, Ning He, Jinyi Deng, and Yuying Wu.

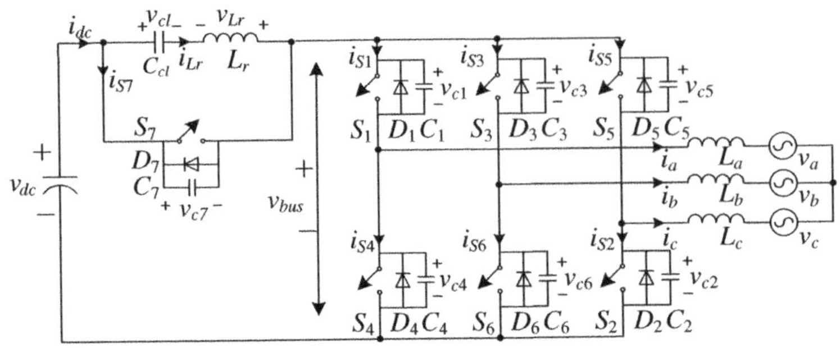

Figure 6.1 Topology of MVAC ZVS three-phase inverter.

For a symmetrical three-phase system, the three-phase voltages v_a, v_b, and v_c satisfy the relationship $v_a + v_b + v_c = 0$. In this case, all the three-dimensional space vectors satisfying this condition are located on a plane in the three-dimensional space. Therefore, the three-phase voltage can be expressed as a complex vector:

$$\vec{v} = \frac{2}{3}\left(v_a + \alpha v_b + \alpha^2 v_c\right) = \frac{2}{3}\begin{bmatrix}1 & \alpha & \alpha^2\end{bmatrix}\begin{bmatrix}v_a \\ v_b \\ v_c\end{bmatrix} \tag{6.2}$$

where $\alpha = e^{j2\pi/3}$ and $\alpha^2 = e^{j4\pi/3}$. Constant 2/3 of expression (6.2) is chosen to scale the complex vector to have the same amplitude as that of the three-phase voltage. Complex vector $\vec{v}$ is composed of real-part v_α and imaginary part v_β:

$$\vec{v} = v_\alpha + jv_\beta \tag{6.3}$$

According to Equations (6.2) and (6.3), the relationship between v_a, v_b, v_c, and v_α, v_β is derived as

$$\begin{cases} v_\alpha = \dfrac{2}{3}\left(v_a + v_b \cos\dfrac{2}{3}\pi + v_c \cos\dfrac{4}{3}\pi\right) \\ x_\beta = \dfrac{2}{3}\left(v_b \sin\dfrac{2}{3}\pi + v_c \sin\dfrac{4}{3}\pi\right) \end{cases} \tag{6.4}$$

Equation (6.4) can be rewritten in the matrix form

$$\begin{bmatrix}v_\alpha \\ v_\beta\end{bmatrix} = \frac{2}{3}\begin{bmatrix}1 & -\dfrac{1}{2} & -\dfrac{1}{2} \\ 0 & \dfrac{\sqrt{3}}{2} & -\dfrac{\sqrt{3}}{2}\end{bmatrix} \cdot \begin{bmatrix}v_a \\ v_b \\ v_c\end{bmatrix} \tag{6.5}$$

The previously mentioned transformation is called Clarke transformation. The following matrix is called Clarke transformation matrix:

$$T_C = \frac{2}{3}\begin{bmatrix}1 & -\dfrac{1}{2} & -\dfrac{1}{2} \\ 0 & \dfrac{\sqrt{3}}{2} & -\dfrac{\sqrt{3}}{2}\end{bmatrix} \tag{6.6}$$

The inverse Clarke transformation can be derived as

$$\begin{bmatrix} v_a \\ v_b \\ v_c \end{bmatrix} = \frac{3}{2} \begin{bmatrix} \frac{2}{3} & 0 \\ -\frac{1}{3} & \frac{\sqrt{3}}{3} \\ -\frac{1}{3} & \frac{\sqrt{3}}{3} \end{bmatrix} \cdot \begin{bmatrix} v_\alpha \\ v_\beta \end{bmatrix} \tag{6.7}$$

According to Equations (6.1), (6.3), and (6.4), the complex vector $\vec{v}$ is derived as

$$\vec{v} = V_m e^{j\omega_l t} \tag{6.8}$$

The complex vector $\vec{v}$ in α–β plane is shown in Figure 6.2. The complex vector $\vec{v}$ rotates counterclockwise at a constant angular frequency ω_l in complex plane $\alpha\beta$. The trajectory of the complex vector $\vec{v}$ on the complex plane is a perfect circle; v_α and v_β are the projection of the complex vector $\vec{v}$ on the α-axis and β-axis in the complex plane, respectively.

6.2.2 Space Vector Modulation of Three-phase Inverter

According to the invertibility principle of complex space vector transformation, if the output voltage complex space vector of the three-phase inverter is a trajectory close to a circle, the three-phase output voltage of the inverter would be close to symmetrical sinusoidal waveforms. The topology of the three-phase inverter is shown in Figure 6.3. This topology consists of six active switches S_1, S_2, S_3, S_4, S_5, and S_6, DC-link capacitor. The grid side current of the inverter is represented by current sources i_a, i_b, and i_c.

In order to ensure proper operation of the circuit, only one switch in each leg is in conduction state, while the other one in the same leg is in off-state. Thus, there are eight switch states for the inverter. Further, each switch state can be represented by a complex voltage space vector, as shown in Table 6.1.

In Table 6.1, for vector $\vec{V_1}(100)$, S_1, S_2, and S_6 are conducting, while S_3, S_4, and S_5 are in off-state. Midpoint voltages of the three legs $v_{an} = v_{dc}$, $v_{bn} = 0$, $v_{cn} = 0$. According to Equation (6.2), vector $\vec{V_1}(100)$ is derived as

Figure 6.2 Vector representation in complex plane with $\alpha\beta$ coordinate.

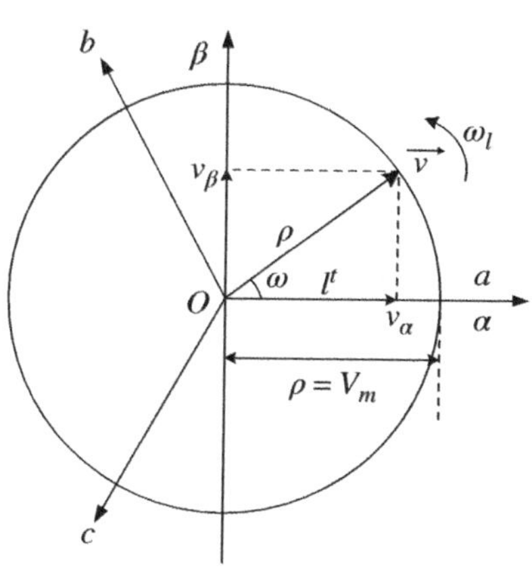

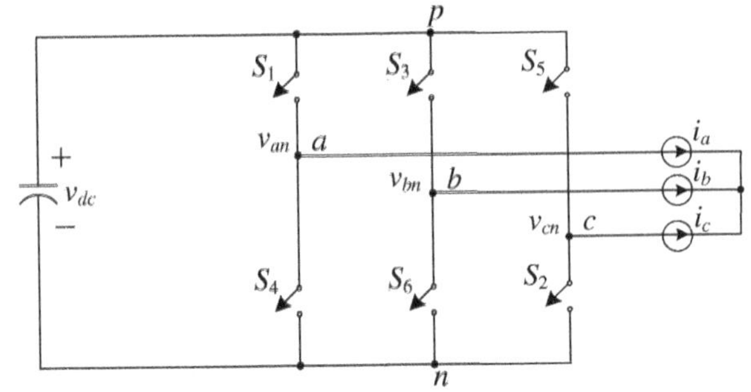

Figure 6.3 Topology of the three-phase inverter.

Table 6.1 Eight switch states.

Complex vector	Switches in on-state	v_{an}	v_{bn}	v_{cn}	Complex vector value
$\vec{V}_1(100)$	S_1, S_2, S_6	v_{dc}	0	0	$2v_{dc}/3$
$\vec{V}_2(110)$	S_1, S_2, S_3	v_{dc}	v_{dc}	0	$2v_{dc}e^{j\pi/3}/3$
$\vec{V}_3(010)$	S_2, S_3, S_4	0	v_{dc}	0	$2v_{dc}e^{j2\pi/3}/3$
$\vec{V}_4(011)$	S_3, S_4, S_5	0	v_{dc}	v_{dc}	$2v_{dc}e^{j\pi}/3$
$\vec{V}_5(001)$	S_4, S_5, S_6	0	0	v_{dc}	$2v_{dc}e^{j4\pi/3}/3$
$\vec{V}_6(101)$	S_1, S_5, S_6	v_{dc}	0	v_{dc}	$2v_{dc}e^{j5\pi/3}/3$
$\vec{V}_7(111)$	S_1, S_3, S_5	v_{dc}	v_{dc}	v_{dc}	0
$\vec{V}_0(000)$	S_2, S_4, S_6	0	0	0	0

$$\vec{V}_1(100) = \frac{2}{3}\left(v_{dc} + 0{\cdot}\alpha + 0{\cdot}\alpha^2\right) = 2v_{dc}/3 \tag{6.9}$$

Then, the direction of $\vec{V}_1(100)$ is the same as the α-axis, and the module value of $\vec{V}_1(100)$ is $2v_{dc}/3$. Similarly, other vectors can be calculated as follows:

$$\vec{V}_2(110) = \frac{2}{3}\left(v_{dc} + v_{dc}{\cdot}\alpha + 0{\cdot}\alpha^2\right) = 2v_{dc}e^{j\pi/3}/3 \tag{6.10}$$

$$\vec{V}_3(010) = \frac{2}{3}\left(0 + v_{dc}{\cdot}\alpha + 0{\cdot}\alpha^2\right) = 2v_{dc}e^{\,j2\pi/3}/3 \tag{6.11}$$

$$\vec{V}_4(100) = \frac{2}{3}\left(0 + v_{dc}{\cdot}\alpha + v_{dc}{\cdot}\alpha^2\right) = 2v_{dc}e^{j\pi}/3 \tag{6.12}$$

$$\vec{V}_5(001) = \frac{2}{3}\left(0 + 0{\cdot}\alpha + v_{dc}{\cdot}\alpha^2\right) = 2v_{dc}e^{\,j4\pi/3}/3 \tag{6.13}$$

$$\vec{V}_6(101) = \frac{2}{3}\left(v_{dc} + 0{\cdot}\alpha + v_{dc}{\cdot}\alpha^2\right) = 2v_{dc}e^{\,j5\pi/3}/3 \tag{6.14}$$

$$\vec{V}_7(111) = \frac{2}{3}\left(v_{dc} + v_{dc}{\cdot}\alpha + v_{dc}{\cdot}\alpha^2\right) = 0 \tag{6.15}$$

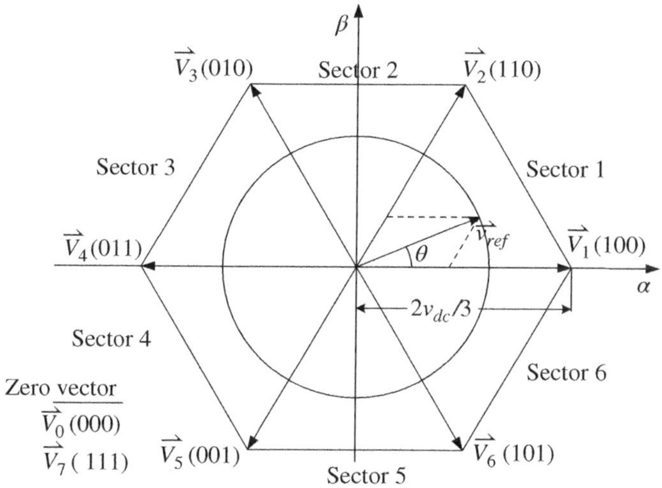

Figure 6.4 Sectors and voltage vectors in space complex plane.

$$\vec{V}_0(000) = \frac{2}{3}\left(0 + 0{\cdot}\alpha + 0{\cdot}\alpha^2\right) = 0 \tag{6.16}$$

In Table 6.1, there are totally six nonzero vectors $\vec{V}_1(100)$, $\vec{V}_2(110)$, $\vec{V}_3(010)$, $\vec{V}_4(011)$, $\vec{V}_5(001)$, and $\vec{V}_6(101)$ and two zero vectors $\vec{V}_0(000)$ and $\vec{V}_7(111)$. These six nonzero voltage vectors divide the complex plane into six sectors, as shown in Figure 6.4.

A reference voltage vector $\vec{v}_{ref}$ can be generated by integrating two adjacent nonzero voltage vectors in a sector and the zero voltage vectors by the parallelogram rule. For instance, if the reference vector is located in Sector 1, the corresponding nonzero voltage vectors $\vec{V}_1(100)$, $\vec{V}_2(110)$ and the zero voltage vectors $\vec{V}_0(000)$ or $\vec{V}_7(111)$ are used. The reference voltage vector $\vec{v}_{ref}$ is expressed as

$$\vec{v}_{ref} = \left|\vec{V}_{ref}\right| e^{j\theta} \tag{6.17}$$

where $\left|\vec{V}_{ref}\right|$ is the module value of $\vec{v}_{ref}$ and θ is the phase angle of $\vec{v}_{ref}$. The reference voltage vector $\vec{v}_{ref}$ can be realized by the voltage vector $\vec{V}_1(100)$, $\vec{V}_2(110)$ and zero vector as follows:

$$d_{V1}T_s\vec{V}_1 + d_{V2}T_s\vec{V}_2 + d_{V0}T_s\vec{V}_0 = T_s\vec{v}_{ref} \tag{6.18}$$

where d_{V1}, d_{V2}, and d_{V0} are the duty ratio of voltage vector $\vec{V}_1(100)$, $\vec{V}_2(110)$, and $\vec{V}_0(000)$, respectively. Here, zero vector $\vec{V}_0(000)$ is used. These duty ratios need to meet the following condition:

$$d_{V1} + d_{V2} + d_{V0} = 1 \tag{6.19}$$

By substituting Equations (6.9), (6.10), (6.16), (6.17), and (6.19) to Equation (6.18), the duty ratio of $\vec{V}_1$, $\vec{V}_2$, and $\vec{V}_0$ is derived:

$$\begin{cases} d_{V1} = \dfrac{\sqrt{3}}{2} m \sin(\pi/3 - \theta) \\ d_{V2} = \dfrac{\sqrt{3}}{2} m \sin\theta \\ d_{V0} = 1 - \dfrac{\sqrt{3}}{2} m \sin\left(\dfrac{\pi}{3} + \theta\right) \end{cases} \tag{6.20}$$

where m is the voltage modulation ratio. It is expressed by

$$m = 2\left|\vec{V}_{ref}\right|/v_{dc} \tag{6.21}$$

The reference voltage vector $\vec{v}_{ref}$ is composed of two adjacent nonzero voltage vectors in the same sector in each switching cycle. Since the sum of the duty ratios of two adjacent voltage vectors is less than 1, the zero voltage vector is needed to fill the rest of the switching cycle. In a switching cycle, the three vectors act in a time-sharing manner to form a space vector sequence.

6.2.3 Switching Scheme of MVAC Inverter Under Unit Power Factor

Considering the inverter operates under unit power factor, the grid side current of the three-phase inverter is expressed as

$$\begin{cases} i_a = I_m \cos(\omega_l t) \\ i_b = I_m \cos\left(\omega_l t - \dfrac{2\pi}{3}\right) \\ i_c = I_m \cos\left(\omega_l t + \dfrac{2\pi}{3}\right) \end{cases} \tag{6.22}$$

The voltage and current waveforms in a utility cycle are demonstrated in Figure 6.5. The phase current has the same phase angle with the same phase grid voltage. If three-phase grid voltage is symmetrical, there is always one-phase grid voltage with opposite polarity to the other two-phase grid voltages in a sector. By further considering the current polarity of different phases, each voltage sector in Figure 6.4 can be divided into two smaller sectors. Consequently, a utility cycle is divided into 12 sectors, as shown in Figure 6.6. It is observed that both the phase voltage and current polarity are fixed in each sector.

As mentioned in Chapter 2, there are two types of switch commutations in the three-phase inverter. The switch commutation from a switch to the antiparallel diode of its complement switch is defined as the Type 1 current commutation, which has less loss. The current commutation from a diode to its complement switch in the same leg is defined as the Type 2 commutation, which has high loss. Once there is a Type 2 commutation in the MVAC three-phase inverter as shown in Figure 6.1, the auxiliary switch S_7 is required to turn off to cause the resonance between the resonant inductor L_r and the resonant capacitors. As a result, DC-side voltage of the inverter v_{bus} will be resonated to zero, which brings ZVS turn-on for main switches. Auxiliary switch is not required

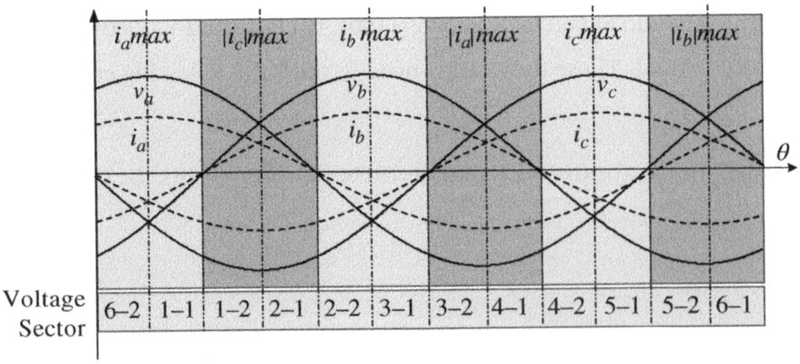

Figure 6.5 The voltage and current waveforms in a utility cycle.

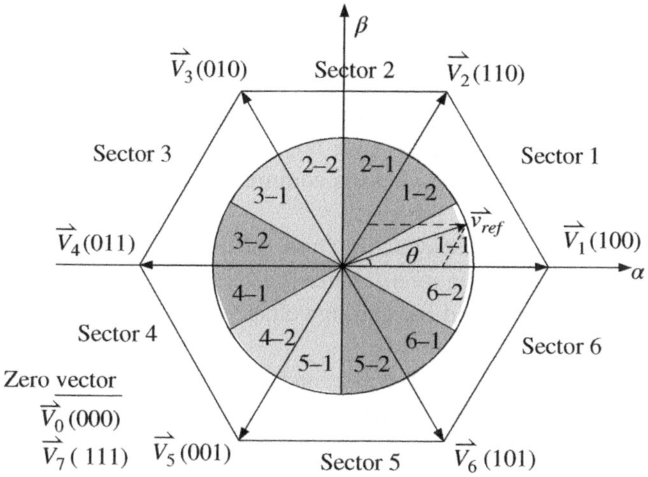

Figure 6.6 Sectors in space vector diagram.

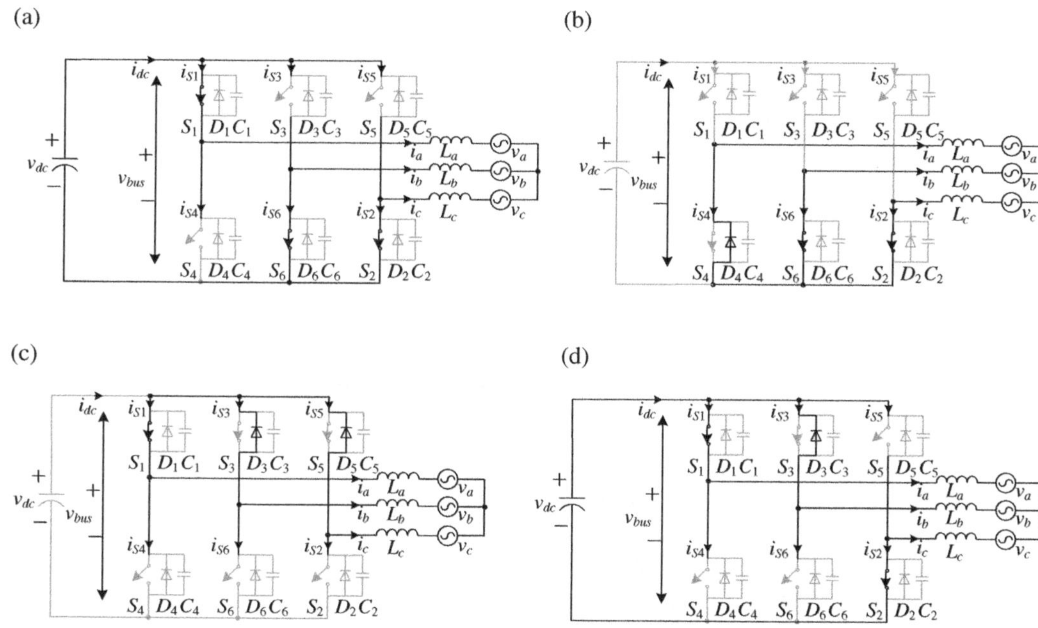

Figure 6.7 Three-phase main bridges' equivalent circuits of four vectors in Sector 1-1. (a) Vector 100. (b) Vector 000. (c) Vector 111. (d) Vector 110.

to turn off when Type 1 current commutation comes. With the ZVS space vector modulation scheme, all the switches, including bridge switches and auxiliary switch, can be turned on and turned off under zero voltage condition. The reverse recovery process of all diodes in the MVAC ZVS grid inverter is also suppressed. The voltage stress of all the switching devices is equal to DC source voltage.

We take Sector 1-1 as an example to demonstrate the principles of the ZVS SVM, where the three-phase current $i_a > 0$, $i_b < 0$, and $i_c < 0$, as shown in Figure 6.5. The corresponding voltage vectors in this sector are nonzero voltage vector $\vec{V}_1(100)$ and $\vec{V}_2(110)$ and zero vector $\vec{V}_0(000)$ and $\vec{V}_7(111)$. The equivalent circuits of the four vectors in Sector 1-1 are shown in Figure 6.7.

Since switching commutation processes affect switching loss, we hope to obtain the switching modulation sequence with minimum switching loss. The switching commutation between different vectors in Sector 1-1 is analyzed, which is shown in Table 6.2

In order to find the voltage vector sequence with lowest switching loss, the dendrogram analysis of vector sequences in Sector 1-1 deduced as shown in Figure 6.7. Since the vector sequence is periodical in a switch cycle, the beginning voltage vector and the ending voltage vector should be the same in the dendrogram.

Now, let us investigate dendrogram of the vector sequence. It is assumed that vectors in Section 1-1 can only be used once in each switch cycle. In Figure 6.8a, $\vec{V}_1(100)$ is the beginning voltage vector and the ending voltage vector. After $\vec{V}_1(100)$ acts for some time, the adjacent voltage vector can be $\vec{V}_2(110)$, $\vec{V}_7(111)$, or $\vec{V}_0(000)$. Supposing $\vec{V}_7(111)$ is selected as the adjacent voltage vector, after $\vec{V}_7(111)$ acts for some time, the adjacent voltage vector can be $\vec{V}_2(110)$ or $\vec{V}_0(000)$. And if

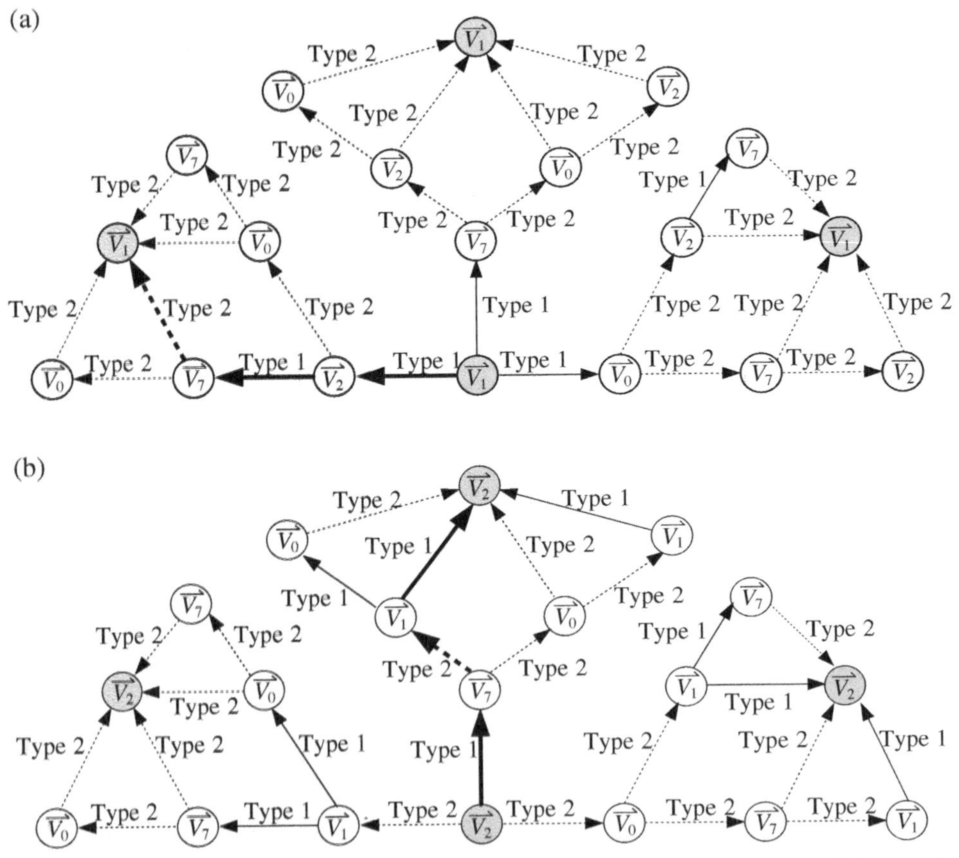

Figure 6.8 The dendrogram analysis of switching commutation sequences between different voltage vectors in Sector 1-1. (a) Beginning with nonzero vector $\vec{V}_1(100)$. (b) Beginning with nonzero vector $\vec{V}_2(110)$. (c) Beginning with zero vector $\vec{V}_7(111)$. (d). Beginning with zero vector $\vec{V}_0(000)$.

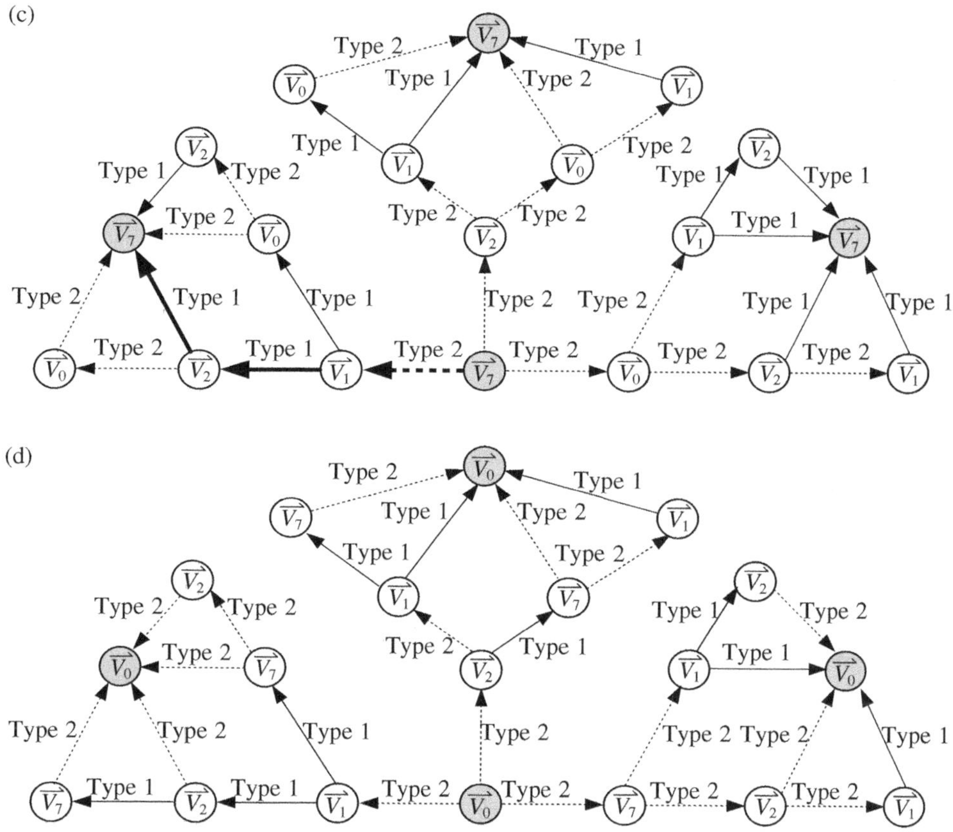

Figure 6.8 (Continued)

Table 6.2 Modulation sequences and current commutation type.

Modulation sequences	Type 1 current commutation	Type 2 current commutation
$\vec{V}_7(111) \rightarrow \vec{V}_0(000)$	Exist	Exist
$\vec{V}_7(111) \rightarrow \vec{V}_1(100)$	Non-exist	Exist
$\vec{V}_7(111) \rightarrow \vec{V}_2(110)$	Non-exist	Exist
$\vec{V}_0(000) \rightarrow \vec{V}_7(111)$	Exist	Exist
$\vec{V}_0(000) \rightarrow \vec{V}_1(100)$	Non-exist	Exist
$\vec{V}_0(000) \rightarrow \vec{V}_2(110)$	Exist	Exist
$\vec{V}_1(100) \rightarrow \vec{V}_7(111)$	Exist	Non-exist
$\vec{V}_1(100) \rightarrow \vec{V}_0(000)$	Exist	Non-exist
$\vec{V}_1(100) \rightarrow \vec{V}_2(110)$	Exist	Non-exist
$\vec{V}_2(110) \rightarrow \vec{V}_7(111)$	Exist	Non-exist
$\vec{V}_2(110) \rightarrow \vec{V}_0(000)$	Exist	Exist
$\vec{V}_2(110) \rightarrow \vec{V}_1(100)$	Non-exist	Exist

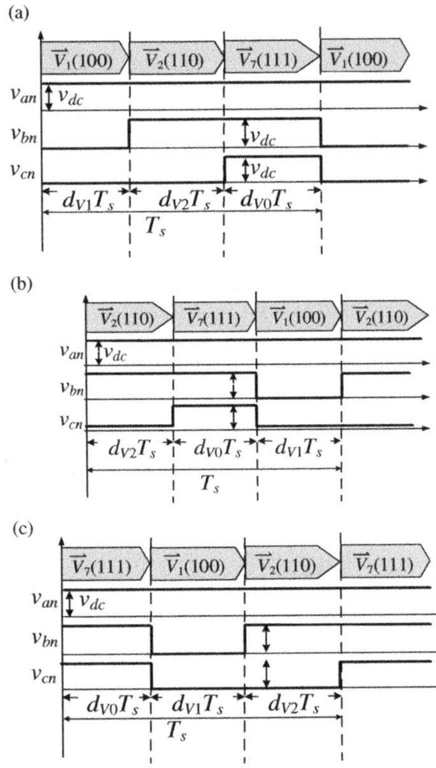

Figure 6.9 Three kinds of space vector sequences with one Type 2 commutation process. (a) $\vec{V}_1(100)$, $\vec{V}_2(110)$, $\vec{V}_7(111)$, $\vec{V}_1(100)$. (b) $\vec{V}_2(110)$, $\vec{V}_7(111)$, $\vec{V}_1(100)$, $\vec{V}_2(110)$. (c) $\vec{V}_7(111)$, $\vec{V}_1(100)$, $\vec{V}_2(110)$, $\vec{V}_7(111)$.

$\vec{V}_2(110)$ is selected as the adjacent voltage vector, after $\vec{V}_2(110)$ acts for some time, the adjacent voltage vector can be $\vec{V}_1(100)$ or $\vec{V}_0(000)$. Suppose that $\vec{V}_0(000)$ is selected as the adjacent voltage vector, after $\vec{V}_0(000)$ acts for some time, the adjacent voltage vector can only be $\vec{V}_1(100)$, which is the end of this switching cycle. In Figure 6.9a, all the possible switching commutation sequences beginning with $\vec{V}_1(100)$. Among all the vector sequences, there is only one commutation sequence with the lowest switching loss, which is $\vec{V}_1(100)$, $\vec{V}_7(111)$, $\vec{V}_2(110)$, $\vec{V}_1(100)$. It has only one Type 2 switching commutation in a switching cycle.

In Figure 6.8b, there is only one commutation sequence with one Type 2 switching commutation, $\vec{V}_2(110)$, $\vec{V}_1(100)$, $\vec{V}_7(111)$, $\vec{V}_2(110)$. In Figure 6.8c, the only one commutation sequence with one Type 2 switching commutation is $\vec{V}_7(111)$, $\vec{V}_2(110)$, $\vec{V}_1(100)$, $\vec{V}_7(111)$. However, Figure 6.8d has no commutation sequence with only one Type 2 switching commutation.

Figure 6.9 shows the three vector sequences with one Type 2 commutation process. By comparing these vector sequences, we find that they are actually the same. The only difference is which vector is the beginning voltage vector in a switching cycle.

Similarly, we can derive best vector sequences for other sectors, which are listed in Table 6.3. In each sector, the Type 2 switching commutation happens between the zero vector and the first vector.

Once the modulation of three-phase main bridges is derived, the driving sequences of the six main switches and the auxiliary switch in Sector 1-1 can be easily obtained as shown in Figure 6.10. Before the Type 2 switching commutation process occurs from $\vec{V}_7(111)$ to $\vec{V}_1(100)$, S_7 is turned off to create the ZVS turn-on condition for the bridge switches and restrain the free-wheeling diode's reverse recovery current. The auxiliary switch S_7 is turned off resonant time (T_{r1})

Figure 6.10 Driving logic waveform in Sector 1-1.

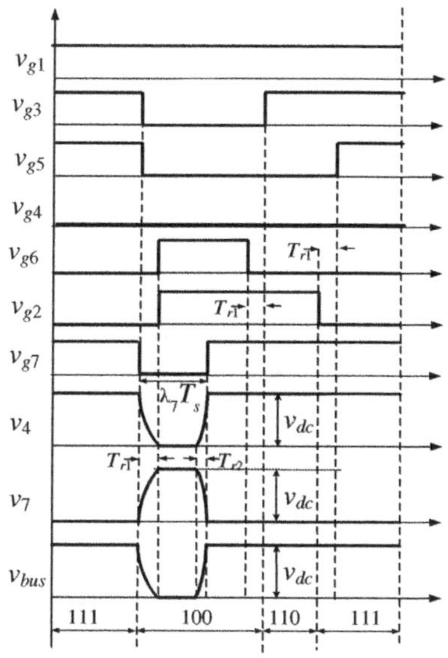

Table 6.3 Switching sequence in each sector.

Sector	Zero vector	First vector	Second vector	Zero vector
Sector 1-1	$\vec{V}_7(111)$	$\vec{V}_1(100)$	$\vec{V}_2(110)$	$\vec{V}_7(111)$
Sector 1-2	$\vec{V}_0(000)$	$\vec{V}_2(110)$	$\vec{V}_1(100)$	$\vec{V}_0(000)$
Sector 2-1	$\vec{V}_0(000)$	$\vec{V}_2(110)$	$\vec{V}_3(010)$	$\vec{V}_0(000)$
Sector 2-2	$\vec{V}_7(111)$	$\vec{V}_3(010)$	$\vec{V}_2(110)$	$\vec{V}_7(111)$
Sector 3-1	$\vec{V}_7(111)$	$\vec{V}_3(010)$	$\vec{V}_4(011)$	$\vec{V}_7(111)$
Sector 3-2	$\vec{V}_0(000)$	$\vec{V}_4(011)$	$\vec{V}_3(010)$	$\vec{V}_0(000)$
Sector 4-1	$\vec{V}_0(000)$	$\vec{V}_4(011)$	$\vec{V}_5(001)$	$\vec{V}_0(000)$
Sector 4-2	$\vec{V}_7(111)$	$\vec{V}_5(001)$	$\vec{V}_4(011)$	$\vec{V}_7(111)$
Sector 5-1	$\vec{V}_7(111)$	$\vec{V}_5(001)$	$\vec{V}_6(101)$	$\vec{V}_7(111)$
Sector 5-2	$\vec{V}_0(000)$	$\vec{V}_6(101)$	$\vec{V}_5(001)$	$\vec{V}_0(000)$
Sector 6-1	$\vec{V}_0(000)$	$\vec{V}_6(101)$	$\vec{V}_1(100)$	$\vec{V}_0(000)$
Sector 6-2	$\vec{V}_7(111)$	$\vec{V}_1(100)$	$\vec{V}_6(101)$	$\vec{V}_7(111)$

ahead of the turn-on of the main switches, S_2 and S_6. The turn-off of auxiliary switch S_7 causes the resonance between the resonant inductor L_r and some of the resonant capacitors. As a result, the voltage across the three-phase main bridges v_{bus} will be resonated to zero, which brings the opportunity of ZVS turn-on for main switches if the main switches are turned on during the small interval

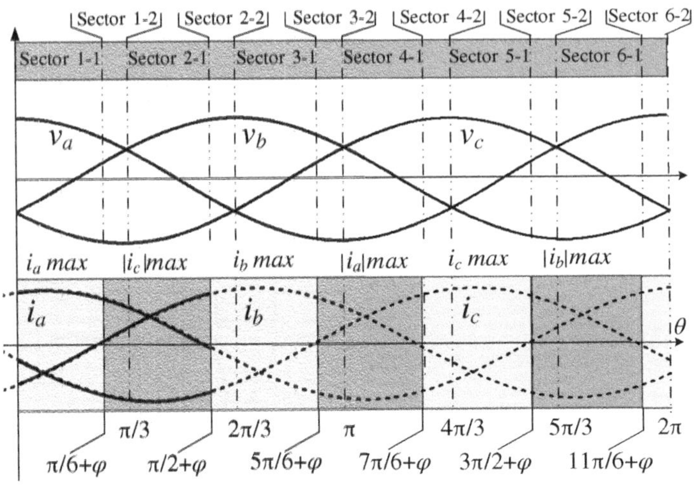

Figure 6.11 Voltage and current waveforms, current polarity regions, and sector definitions when $0 < \varphi < \pi/6$.

when v_{bus} is clamped at zero. In the Type 1 switching commutation process from $\vec{V}_1(100)$ to $\vec{V}_1(100)$ and from $\vec{V}_2(110)$ to $\vec{V}_7(111)$, the active switch S_7 is kept in conduction state. The detailed operation stages of the ZVS SVM will be investigated in the following paragraph.

6.2.4 Generalized Space Vector Modulation Method of MVAC Inverter with Arbitrary Output

With regard to the inverter operating at non-unity power factor, the grid side current of the inverter are expressed as

$$\begin{cases} i_a = I_m \cos(\omega_l t - \varphi) \\ i_b = I_m \cos\left(\omega_l t - \dfrac{2\pi}{3} - \varphi\right) \\ i_c = I_m \cos\left(\omega_l t + \dfrac{2\pi}{3} - \varphi\right) \end{cases} \tag{6.23}$$

where φ is the grid current power factor angle. When $0 < \varphi < \pi/6$, the voltage and current waveforms in a utility cycle are demonstrated in Figure 6.11 [5]. For the symmetrical condition of three-phase grid voltage, there is always one-phase grid voltage with opposite polarity to the other two-phase grid voltages in a sector. By further consideration of the polarity of the three-phase grid currents, a utility cycle can be divided into 12 sectors as shown in Figure 6.11. Two adjacent active vectors plus one zero vector are selected to synthesize the reference vector. For Sector 1-1 the modulation sequence $\vec{V}_7(111)$, $\vec{V}_1(100)$, $\vec{V}_2(110)$, $\vec{V}_7(111)$ is used. Type 2 commutation is synchronized at the same instant.

However, this method is only applicable to care $0 < \varphi < \pi/6$. When $\pi/6 < \varphi < \pi/2$, the voltage and current waveforms in a utility cycle are demonstrated in Figure 6.12. It is found that no matter how to arrange the vector sequence, the auxiliary switch S_7 needs to operate at least twice per cycle. Multiple operations of S_7 will cause additional turn-off loss and circulation current loss in the circuit. Therefore, the auxiliary switch S_7 is preferred to operate once per switching cycle.

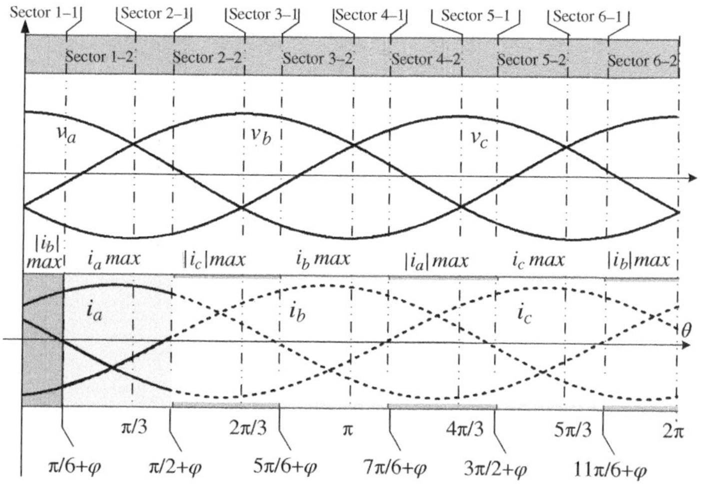

Figure 6.12 Voltage and current waveforms, current polarity regions, and sector definitions when $\pi/6 < \varphi < \pi/2$.

Comparing Figures 6.11 and 6.12, it can be noticed that the only difference is that new current polarity combination appears in the same sector. For example, in Sector1 when $0<\varphi<\pi/6$, only i_a and $-i_c$ can achieve the maximum value. By contrast, when $\pi/6<\varphi< \pi/2$, a new maximum value of $-i_b$ appears in Sector 1. Therefore, the problem remained to be solved is how to select qualified vectors and arrange the vector sequence when new polarity appears. Still taking Sector 1 for an example. In Sector1-1 of Figure 6.12, unlike traditional method, three active vectors are selected to synthesize the reference vector v_{ref} to realize the goal mentioned before. As shown in Figure 6.13, three nonzero vectors $\vec{V}_1(100)$, $\vec{V}_2(110)$, and $\vec{V}_3(010)$ are used to synthesize the reference vector v_{ref} when it is located in Sector 1. The preferred vector sequence is $\vec{V}_3(010)$, $\vec{V}_1(100)$, $\vec{V}_2(110)$, $\vec{V}_3(010)$. The corresponding state transitions are drawn in Figure 6.14 with the preferred vector sequence.

State transitions can be classified into hard transition and soft transition. Transitions including Type 2 commutations in three-phase legs are hard transitions, while transitions only including

Figure 6.13 Synthesis of v_{ref} in Sector 1-1 when $\pi/6 < \varphi < \pi/2$.

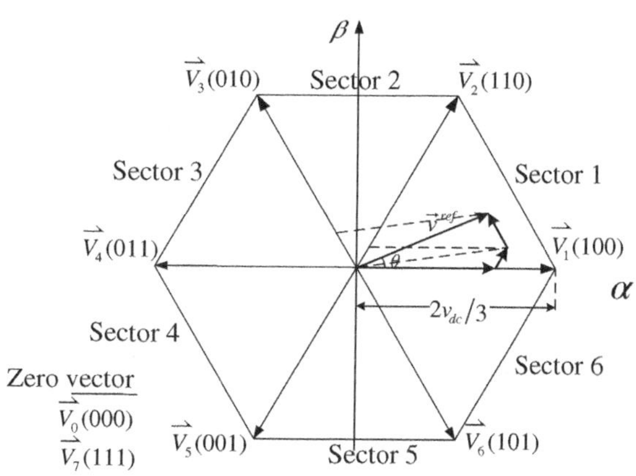

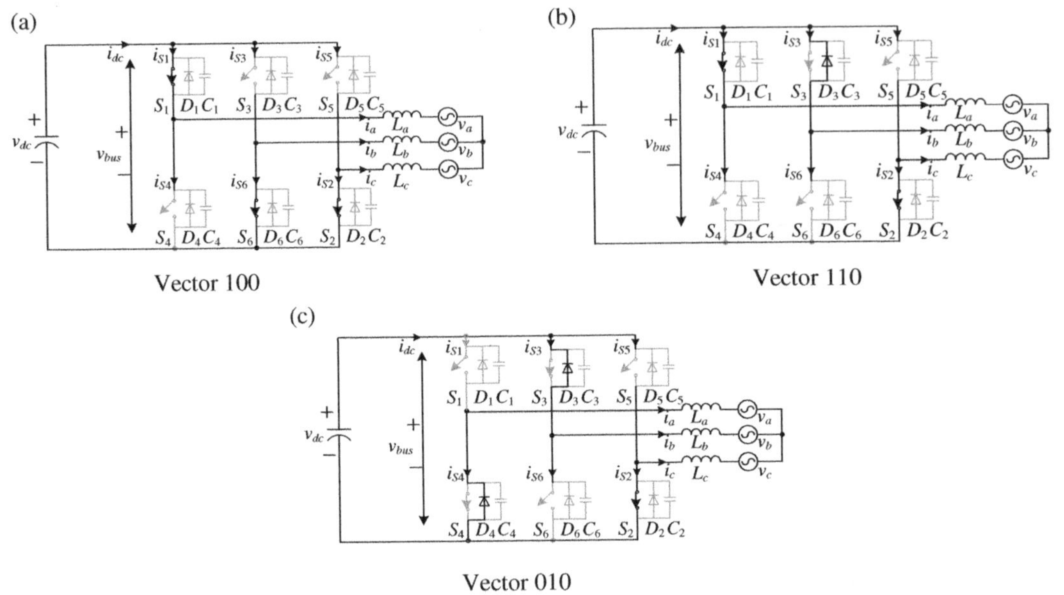

Figure 6.14 State transitions with preferred vector sequence in Sector 1-1 when $\pi/6 < \varphi < \pi/2$. (a) Vector 100. (b) Vector 110. (c) Vector 010.

Type 1 commutations in three-phase legs are soft transitions. As shown in Figure 6.14, there are three transitions in a switching cycle. The leg state of phase c keeps constant in a switching cycle, while the leg states of phase a and b change according to vector sequence. For a phase leg with positive current polarity, Type 2 commutation occurs from 0 →1 (i.e. lower diode to upper switch), while for a phase leg with negative current polarity, Type 2 commutation occurs from 1→0 (i.e. upper diode to lower switch).

From $\vec{V}_1(100)$ to $\vec{V}_2(110)$ transition, only the leg state of phase b changes from 0 to 1 during this transition, which is a Type 1 commutation since the current of phase b is negative. It has low loss due to ZVS turn-off. $\vec{V}_2(110)$, $\vec{V}_3(010)$ is also soft transition. Only the leg state of phase a changes from 1 to 0 during this transition, which is a Type 1 commutation since the current of phase a is positive. It also has low loss due to ZVS turn-off.

However, $\vec{V}_3(010)$, $\vec{V}_1(100)$ is hard transition. The leg state of phase a changes from 0 to 1 with positive current polarity, and the leg state of phase b changes from 1 to 0 with negative current polarity, which are both Type 2 commutations. Because two Type 2 commutations in phase a and phase a are synchronized, the auxiliary resonant circuit only needs to act once in each switching cycle to resonate the DC bus voltage to zero and creates the ZVS condition for switches in phase a and phase b. Therefore, the auxiliary switch can work at the same frequency as the main switches.

The reference voltage vector $\vec{V}_{ref}$ is integrated by vector $\vec{V}_1$, $\vec{V}_2$, and $\vec{V}_3$ as follows:

$$d_{V1}T_s\vec{V}_1 + d_{V2}T_s\vec{V}_2 + d_{V3}T_s\vec{V}_3 = T_s\vec{v}_{ref} \tag{6.24}$$

where d_{V1}, d_{V2}, and d_{V3} are the duty ratio of voltage vectors $\vec{V}_1(100)$, $\vec{V}_2(110)$, and $\vec{V}_3(010)$, respectively. In addition, these duty ratios need to meet the following equation:

$$d_{V1} + d_{V2} + d_{V3} = 1 \tag{6.25}$$

By substituting Equations (6.9), (6.10), (6.16), (6.17), and (6.25) to Equation (6.24), the duty ratio of nonzero voltage vector and zero voltage vector can be derived:

$$\begin{cases} d_{V1} = 1 - m\sin\theta \\ d_{V2} = 1 - m\sin\left(\dfrac{\pi}{3} + \theta\right) \\ d_{V3} = \sqrt{3}m\sin\left(\dfrac{\pi}{6} + \theta\right) - 1 \end{cases} \tag{6.26}$$

where m is the voltage modulation ratio. It is be expressed by

$$m = \sqrt{3}\left|\vec{V}_{ref}\right|/v_{dc} \tag{6.27}$$

In certain cases, d_{V2} will be lower than zero. This is irrational in the practical system. In this case, vector $\vec{V}_2(110)$ should be replaced by $\vec{V}_0(000)$. The preferred sequence is changed to $\vec{V}_3(010)$, $\vec{V}_1(100)$, $\vec{V}_0(000)$, $\vec{V}_3(010)$, and the corresponding duration time for each sector is recalculated as

$$\begin{cases} d_{V1} = m\sin\left(\dfrac{\pi}{3} + \theta\right) \\ d_{V3} = m\sin\theta \\ d_{V0} = 1 - \sqrt{3}m\sin\left(\dfrac{\pi}{6} + \theta\right) \end{cases} \tag{6.28}$$

It is shown that with the vector sequence of $\vec{V}_3(010)$, $\vec{V}_1(100)$, $\vec{V}_0(000)$, $\vec{V}_3(010)$, there is also only one Type 2 commutation in a switching cycle, and does not affect the operation of S_7. The previously mentioned analyses can be extended to other sectors, and the preferred vector sequences in each sector when $\pi/6 < \varphi < \pi/2$ are summarized in Table 6.4. In Table 6.4, only transition from the first vector to the second vector is hard transition; the remaining two transitions are soft transitions. The third vector is either nonzero vector or zero vector.

Table 6.4 Switching sequence in each sector.

Sector	First vector	Second vector	Third vector	Fourth vector
Sector 1-1	$\vec{V}_3(010)$	$\vec{V}_1(100)$	$\vec{V}_2(110)$ *or* $\vec{V}_0(000)$	$\vec{V}_3(010)$
Sector 1-2	$\vec{V}_3(010)$	$\vec{V}_1(100)$		$\vec{V}_3(010)$
Sector 2-1	$\vec{V}_4(011)$	$\vec{V}_2(110)$	$\vec{V}_3(010)$ *or* $\vec{V}_7(111)$	$\vec{V}_4(011)$
Sector 2-2	$\vec{V}_4(011)$	$\vec{V}_2(110)$		$\vec{V}_4(011)$
Sector 3-1	$\vec{V}_5(001)$	$\vec{V}_3(010)$	$\vec{V}_4(011)$ *or* $\vec{V}_0(000)$	$\vec{V}_5(001)$
Sector 3-2	$\vec{V}_5(001)$	$\vec{V}_3(010)$		$\vec{V}_5(001)$
Sector 4-1	$\vec{V}_6(101)$	$\vec{V}_4(011)$	$\vec{V}_5(001)$ *or* $\vec{V}_7(111)$	$\vec{V}_6(101)$
Sector 4-2	$\vec{V}_6(101)$	$\vec{V}_4(011)$		$\vec{V}_6(101)$
Sector 5-1	$\vec{V}_1(100)$	$\vec{V}_5(001)$	$\vec{V}_6(101)$ *or* $\vec{V}_0(000)$	$\vec{V}_1(100)$
Sector 5-2	$\vec{V}_1(100)$	$\vec{V}_5(001)$		$\vec{V}_1(100)$
Sector 6-1	$\vec{V}_2(110)$	$\vec{V}_6(101)$	$\vec{V}_1(100)$ *or* $\vec{V}_7(111)$	$\vec{V}_2(110)$
Sector 6-2	$\vec{V}_2(110)$	$\vec{V}_6(101)$		$\vec{V}_2(110)$

6.3 Circuit Analysis

6.3.1 Operation Stage Analysis

In order to simplify the analysis, the following assumptions are made:

1) All of the switching devices are ideal switches.
2) Filter inductor at the grid side is so large that the filter inductor current can be seen as constant in a switching cycle.
3) Parasitic capacitance of a switch is included in its paralleled capacitor of the active switch.
4) All capacitors paralleled to the switches are the same, i.e. $C_1 = C_2 = C_3 = C_4 = C_5 = C_6 = C_r$.
5) The clamping capacitor C_{cl} is so large that the v_{cl} can be treated as a constant value V_{cl} in a switching cycle.
6) AC-side current $i_a>0$, $i_b<0$, and $i_c<0$.

Sector 1-1 is taken as an example to investigate the operation stages of the MVAC inverter in a switch cycle. Figure 6.15 shows the waveforms of the key components in the inverter in one switching period. The operation stages for ZVS are analyzed as follows.

Stage 1 (t_0–t_1): $\overrightarrow{V_7}$ stage. In this stage, the operation state is in vector $\overrightarrow{V_7}$ (111). The equivalent circuit of this stage is shown in Figure 6.16. The directions of the AC-side filter current indicate that S_1, D_3, and D_5 are conducting. The auxiliary switch S_7 is conducting in this stage, and the resonant inductor L_r is charged by the clamping capacitor C_{cl}. v_{bus} is clamped by V_{dc} and i_{S7} is equal to $-i_{Lr}$. The current through L_r decreases linearly at the rate of

$$\frac{di_{Lr}}{dt} = -\frac{V_{cl}}{L_r} \tag{6.29}$$

Stage 2 (t_1–t_2): The first resonance stage. The equivalent circuit of this stage is shown in Figure 6.17. At t_1, S_7 is turned off. Resonance between the resonant inductor L_r and the resonant capacitors C_2, C_4, C_6, and C_7 is provoked. During this stage, v_{bus} is decreasing, while v_{c7} is increasing. Once v_{bus} resonates to zero, it creates ZVS turn-on condition for main switches to be turned on. This stage ends at t_2, when v_{bus} is reduced to zero and v_{c7} reaches V_{dc}.

Stage 3 (t_2–t_3): The diode freewheeling stage. The equivalent circuit of this stage is shown in Figure 6.18. At t_2, v_{bus} resonances to zero. Since L_r still has residual energy to keep current i_{Lr} negative, D_4, D_6, and D_2 start to conduct in freewheeling mode. Thus, S_6 and S_2 are turned on under ZVS condition. i_{Lr} increases with a rate of

$$\frac{di_{Lr}}{dt} = \frac{V_{dc} - V_{cl}}{L_r} \tag{6.30}$$

Stage 4 (t_3–t_4): Diode reverse recovery stage. The equivalent circuit of this stage is shown in Figure 6.19. At t_3, i_{Lr} reduces to zero. Currents in phase b and phase c begin to transfer from D_3 and D_5 to S_6 and S_2, respectively. During this period, the changing rates of i_{s3} and i_{s5} are restrained to a small value by L_r so that the reverse recovery process of D_3 and D_5 can be suppressed. i_{s3} and i_{s5} are reduced to zero by time t_4.

Stage 5 (t_4–t_5): The second resonance stage. The equivalent circuit of the second resonance stage is shown in Figure 6.20. Another resonance occurs among L_r, C_3, C_4, C_5, and C_7. During this resonance, v_{bus} increases and v_{c7} decreases. At time t_5, v_{bus} resonances to V_{dc}, while v_{c7} reduces to zero. At t_5, i_{Lr} reaches its maximum value i_{Lr_max}, and the current through S_7 is

$$i_{s7} = -i_{Lr} - i_a \tag{6.31}$$

Figure 6.15 Key waveforms in Sector 1-1.

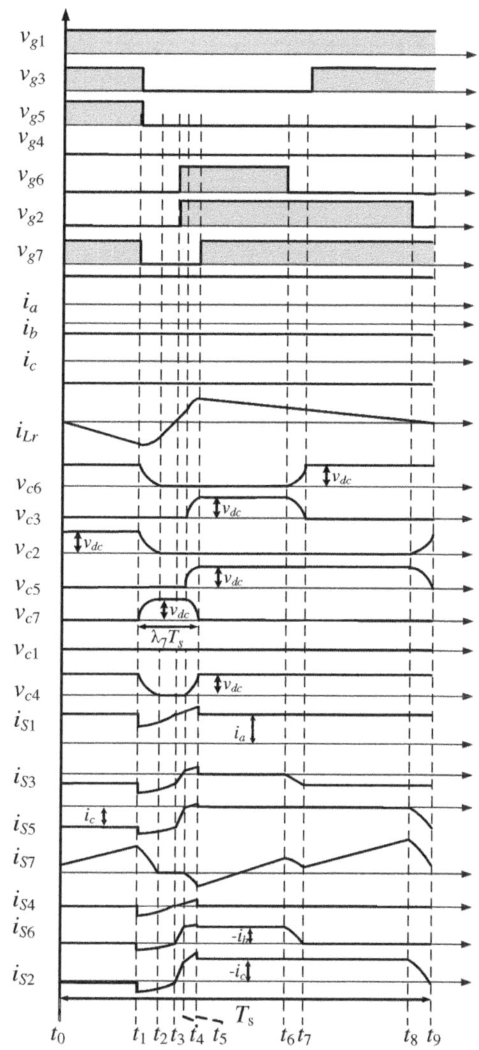

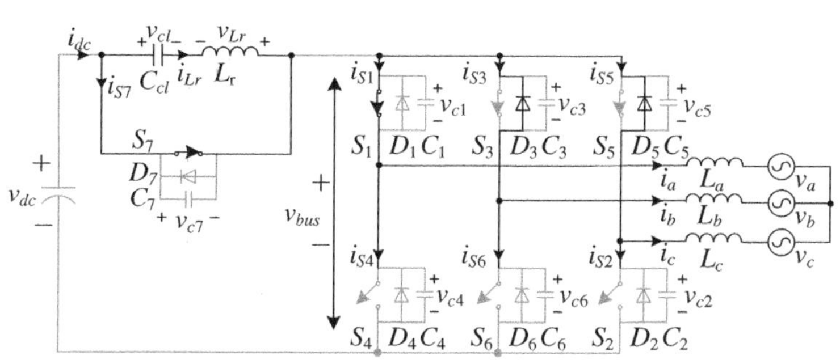

Figure 6.16 Equivalent circuit of stage 1: $\overrightarrow{V_7}$ stage.

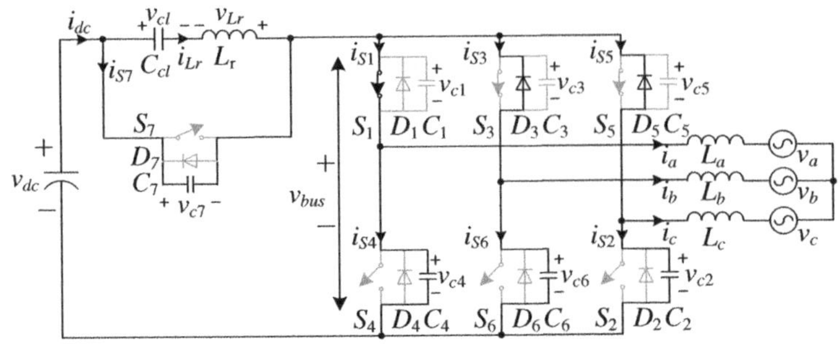

Figure 6.17 Equivalent circuit of stage 2: the first resonance stage.

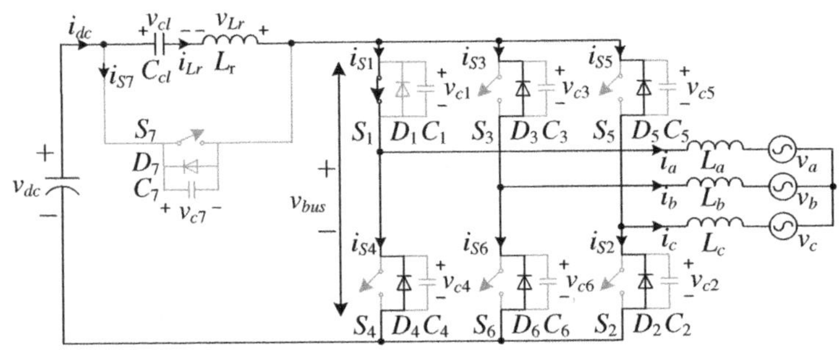

Figure 6.18 Equivalent circuit of stage 3: the diode freewheeling stage.

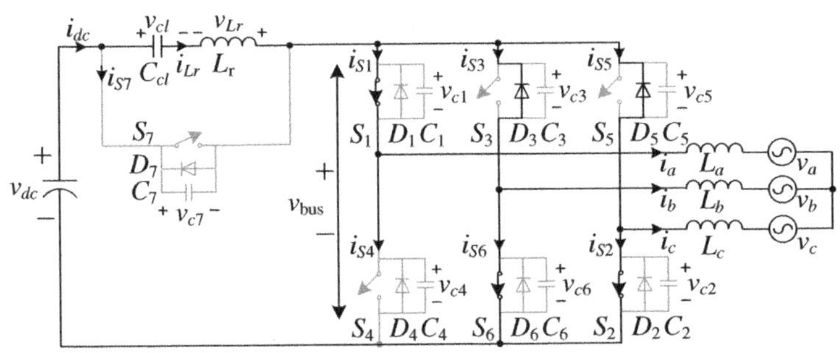

Figure 6.19 Equivalent circuit of stage 4: the diode reverse recovery stage.

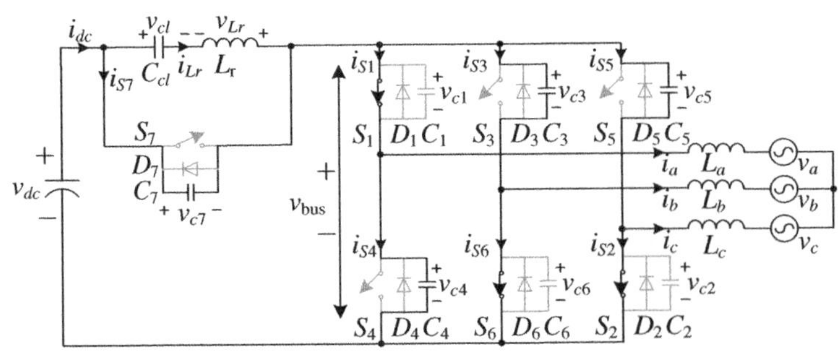

Figure 6.20 Equivalent circuit of stage 5: the second resonance stage.

Stage 6 (t_5–t_6): $\overrightarrow{V_1}$ stage. The equivalent circuit of this stage is shown in Figure 6.21. At t_5, S_7 can be turned on under ZVS condition since v_{c7} is equal to zero. The inverter outputs vector $\overrightarrow{V_1}$ (100) in this stage. The voltage across L_r is clamped by $-V_{cl}$ in this stage.

Stage 7 (t_6–t_7): The first Type 1 commutation stage. The equivalent circuit of this stage is shown in Figure 6.22. At t_6, S_6 is turned off. The output current i_b charges C_6 and discharges C_3. With the buffer of the resonant capacitor, S_6 is turned off under ZVS condition. This stage ends at t_7 when v_{c3} reduces to zero and D_3 conducts.

Stage 8 (t_7–t_8): $\overrightarrow{V_2}$ stage. The equivalent circuit of this stage is shown in Figure 6.23. In this stage, the inverter is operated in vector $\overrightarrow{V_2}$(110). The resonant inductor is clamped by V_{cl}. i_{Lr} increases with a rate of

$$\frac{di_{Lr}}{dt} = \frac{V_{dc} - V_{cl}}{L_r} \tag{6.32}$$

Stage 9 (t_8–t_9): The second Type 1 commutation stage. The equivalent circuit of this stage is shown in Figure 6.24. At t_8, S_2 is turned off. The output current i_c charges C_2 and discharges C_5 simultaneously. With these capacitors, S_2 is turned off under ZVS conditions. This stage ends at t_9 when v_{c5} reduces to zero and D_5 conducts.

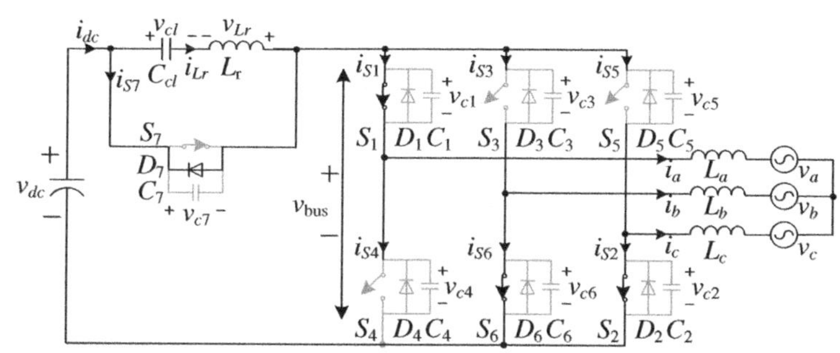

Figure 6.21 Equivalent circuit of stage 6: $\overrightarrow{V_1}$ stage.

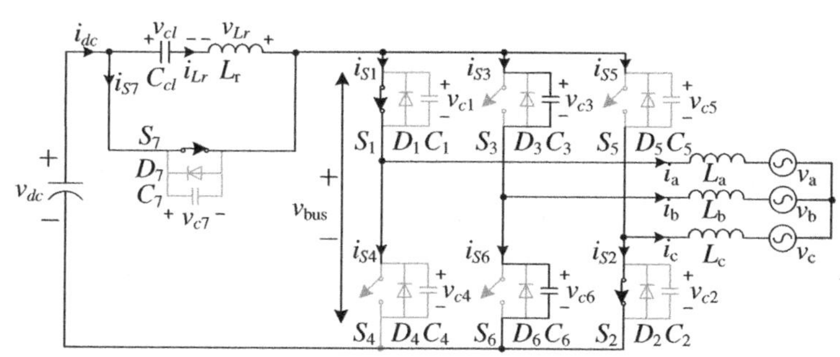

Figure 6.22 Equivalent circuit of stage 7: the first Type 1 commutation stage.

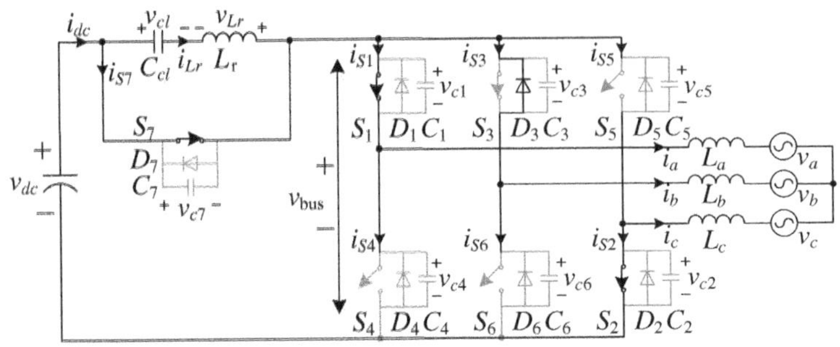

Figure 6.23 Equivalent circuit of stage 8: $\overrightarrow{V_2}$ stage.

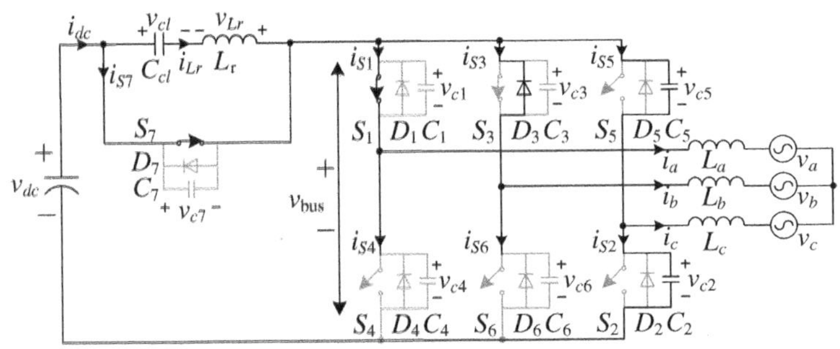

Figure 6.24 Equivalent circuit of stage 9: the second Type 1 commutation stage.

After this, the inverter returns to stage 1 of a new switching cycle.

In brief, there are two resonant stages in a switching cycle. The first resonance stage is used for ZVS turn-on for the main bridge switches, while the second resonant stage is used to for ZVS turn-on for the auxiliary switch. Besides, the reverse recovery of the freewheeling diodes is suppressed by reduction of the current changing rate of the diodes to be turned off. It means there are almost no turn-on loss and diode reverse recovery loss in MVAC inverter. In addition, the turn-off actions for the switches are under ZVS conditions.

6.3.2 Resonant Stages Analysis

During the first resonance stage, resonance occurs among L_r, C_2, C_4, C_6, and C_7. The circuit of the first resonance stage is illustrated in Figure 6.25a. Since the main switches S_1, S_3, and S_5 are in conduction, the grid side current does not influence the circuit resonance. Its equivalent circuit is obtained as shown in Figure 6.25b. Once S_7 is turned off at t_1, a resonance between the resonant inductor L_r and the resonant capacitors C_2, C_4, C_6, and C_7 is provoked. The initial values of the capacitor voltage $v_{c2}(t_1) = v_{c4}(t_1) = v_{c6}(t_1) = V_{dc}$, $v_{c7}(t_1) = 0$. The initial values of the resonant inductor current and bus voltage at time t_1 are expressed by

$$\begin{cases} i_{Lr}(t_1) = I_{Lr(t_1)} \\ v_{bus}(t_1) = V_{dc} \end{cases} \tag{6.33}$$

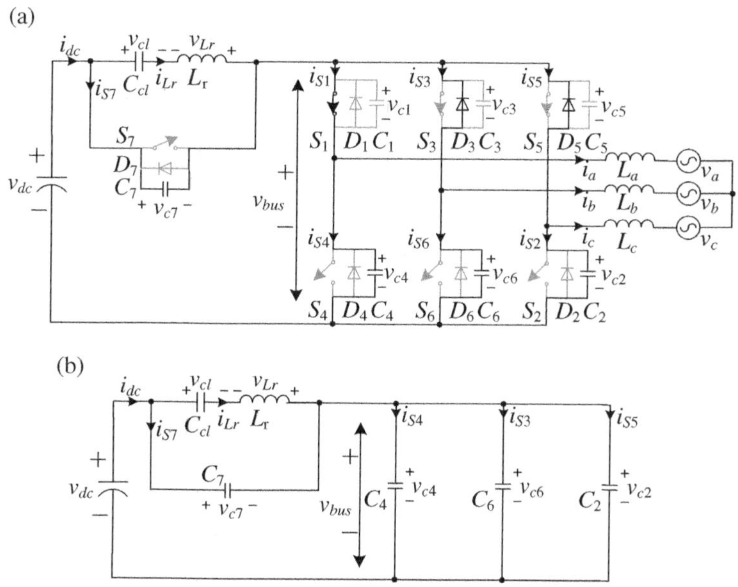

Figure 6.25 Equivalent circuit of stage 2 (t_1–t_2): the first resonance stage. (a) Circuit state of the first resonance stage. (b) Equivalent circuit.

According to Figure 6.25b, the following differential equation is derived:

$$\begin{cases} L_r \dfrac{di_{Lr}}{dt} = V_{dc} - V_{cl} - v_{bus} \\ (3C_r + C_7)\dfrac{dv_{bus}}{dt} = i_{Lr} \end{cases} \tag{6.34}$$

By solving the differential Equation (6.34) with the initial value in (6.33), i_{Lr} and v_{bus} are derived:

$$\begin{cases} i_{Lr} = -\dfrac{\left|i_{Lr(t1)}Z_r\right|}{Z_r}\cos\omega_r t - \dfrac{V_{cl}}{Z_r}\sin\omega_r t \\ v_{bus} = V_{dc} - V_{cl} - \left|i_{Lr(t1)}Z_r\right|\sin\omega_r t + V_{cl}\cos\omega_r t \end{cases} \tag{6.35}$$

where ω_r is the resonant angular frequency and is equal to

$$\omega_r = 1/\sqrt{L_r(3C_r + C_7)} \tag{6.36}$$

Z_r is the characteristic impedance and is defined as

$$Z_r = \sqrt{L_r/(3C_r + C_7)} \tag{6.37}$$

Equation (6.35) can be rewritten as

$$\begin{cases} i_{Lr} = -\dfrac{\sqrt{V_{cl}^2 + i_{Lr(t1)}^2 Z_r^2}}{Z_r}\cos(\omega_r t - \sigma) \\ v_{bus} = V_{dc} - V_{cl} - \sqrt{V_{cl}^2 + i_{Lr(t1)}^2 Z_r^2}\sin(\omega_r t - \sigma) \end{cases} \tag{6.38}$$

where

$$\sigma = \arctan\left(\frac{V_{cl}}{\left|i_{Lr(t1)}Z_r\right|}\right) \tag{6.39}$$

According to resonant stage analysis, to realize ZVS turn-on for the main switches, it is necessary that the bus voltage v_{bus} will resonance to zero. v_{bus} in Equation (6.38) consists of two terms: DC component and sinusoidal component. If we want v_{bus} to be negative, then the amplitude of the sinusoidal term must be larger than the DC component, i.e.

$$\sqrt{V_{cl}^2 + i_{Lr(t1)}^2 Z_r^2} > V_{dc} - V_{cl} \tag{6.40}$$

By further simplifying the previously mentioned equation, the ZVS turn-on condition for the bridge switches can be rewritten as

$$i_{Lr(t1)} \leq -\frac{\sqrt{(V_{dc} - V_{Cl})^2 - V_{cl}^2}}{Z_r} \tag{6.41}$$

The second resonance happens at stage 5. The second resonance circuit is illustrated in Figure 6.26a. The equivalent circuit of this stage is illustrated in Figure 6.26b. Ignoring the grid side filter inductor current ripple, the inductor current can be replaced by a current source i_a.

In stage 5, S_4 is turned off at t_4. A resonance occurs among L_r, C_3, C_4, C_5, and C_7. v_{bus} increases and v_{c7} decreases. At time t_5, v_{bus} resonances to V_{dc}, while v_{c7} reduces to zero. The initial values of the capacitor voltage $v_{c3}(t_4) = v_{c4}(t_4) = v_{c5}(t_4) = 0$, $v_{c7}(t_4) = V_{dc}$. The initial values of the resonant inductor current and capacitor voltage at time t_4 are expressed by

$$\begin{cases} v_{bus}(t_4) = 0 \\ i_{Lr}(t_4) = i_a \end{cases} \tag{6.42}$$

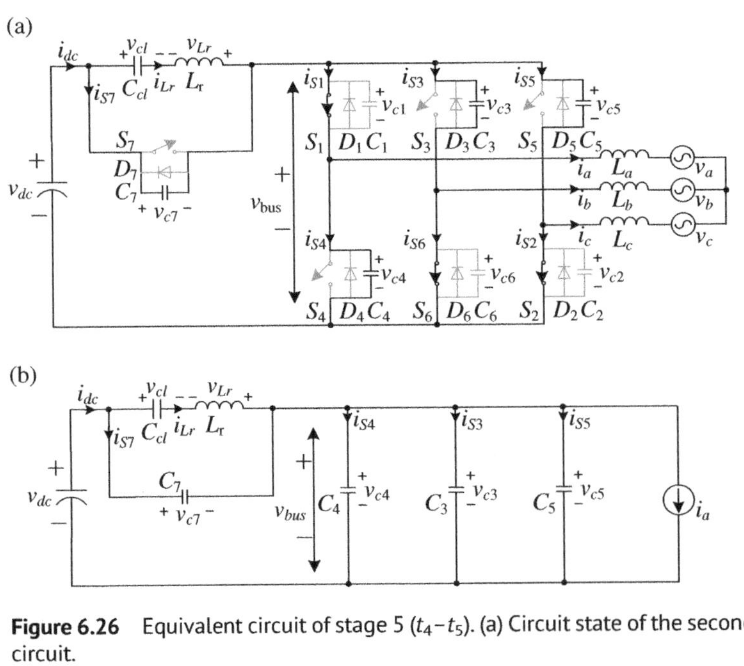

Figure 6.26 Equivalent circuit of stage 5 (t_4–t_5). (a) Circuit state of the second resonance stage. (b) Equivalent circuit.

According to the equivalent circuit in Figure 6.26b, the following differential equation is derived:

$$\begin{cases} L_r \dfrac{di_{Lr}}{dt} = V_{dc} - V_{cl} - v_{bus} \\ (3C_r + C_7)\dfrac{dv_{bus}}{dt} = i_{Lr} - I_a \end{cases} \tag{6.43}$$

By solving differential Equation (6.43) with the initial value of (6.42), the following state variables are deducted:

$$\begin{cases} i_{Lr}(t) = I_a + \dfrac{(V_{dc} - V_{cl})}{Z_r} \sin \omega_r t \\ v_{bus}(t) = (V_{dc} - V_{cl}) - (V_{dc} - V_{cl}) \cos \omega_r t \end{cases} \tag{6.44}$$

Considering $v_{bus} + v_{c7} = V_{dc}$, v_{c7} can be derived as

$$v_{C7}(t) = V_{cl} + (V_{dc} - V_{cl}) \cos \omega_r t \tag{6.45}$$

To realize ZVS turn-on of the auxiliary switch S_7, $v_{c7}(t)$ should resonance to zero.

Since the clamping capacitor voltage V_{cl} is much lower than the DC-side voltage V_{dc}, v_{c7} can always reach to zero according to Equation (6.45). Besides, the resonant time of the second resonant stage T_{r2} is extremely small in comparison to the resonant time of the first resonant stage T_{r1}.

The second resonance stage stops at t_5, values of the resonant capacitor voltage v_{bus} at time t_5 can be derived as

$$v_{bus}(t_5) = V_{dc} \tag{6.46}$$

Combining Equation (6.46) and Equation (6.44), values of the resonant inductor current i_{Lr} at time t_5 can be derived as

$$i_{Lr}(t_5) = I_a + \frac{(V_{dc} - V_{cl})}{Z_r} \sqrt{1 - \left(\frac{V_{cl}}{V_{dc} - V_{cl}}\right)^2} \tag{6.47}$$

6.3.3 Steady-state Analysis

When the inverter is operated in steady state, the average values of the resonant inductor voltage in a switch cycle are zero. That is,

$$\frac{1}{T_s}\int_0^{T_s} v_{Lr} dt = 0 \tag{6.48}$$

Since the intervals during (t_1–t_2) and (t_4–t_5) are very short in a switch cycle, they are ignored in the analysis. Thus, Equation (6.48) can be written as

$$\frac{1}{T_s}\int_0^{T_s} v_{Lr} dt \approx \int_0^{t_1} -V_{cl} dt + \int_{t_2}^{t_4} (V_{dc} - V_{cl}) dt + \int_{t_5}^{t_9} -V_{cl} dt = V_{dc}(1 - d_7)T_s - V_{cl} T_s = 0 \tag{6.49}$$

where d_7 is duty ratio of the auxiliary switch. Define λ_7 to be the auxiliary switch turn-off duty cycle, i.e. $\lambda_7 = 1\text{-}d_7$. From Equation (6.49), the expression of the clamping capacitor voltage is derived as

$$V_{cl} = \lambda_7 V_{dc} \tag{6.50}$$

In the steady state, the average current through the clamping capacitor is zero. That is,

$$\frac{1}{T_s}\int_0^{T_s} i_{cl} dt = 0 \tag{6.51}$$

Neglecting short intervals (t_1–t_2), (t_4–t_5), (t_6–t_7), and (t_8–t_9), Equation (6.51) is rewritten as

$$\frac{1}{T_s}\int_0^{T_s} i_{cl}dt \approx \frac{1}{T_S}\int_{t_0}^{t_1} i_{cl}dt + \frac{1}{T_S}\int_{t_2}^{t_4} i_{cl}dt + \frac{1}{T_S}\int_{t_5}^{t_8} i_{cl}dt = 0 \tag{6.52}$$

Because the auxiliary switch turn-off duty cycle λ_7 is 5–10%, the clamping capacitor voltage is less than 0.1 times of the DC-link voltage. Combining Equation (6.47), the current through clamping capacitor at t_5 can be simplified as

$$i_{cl}(t_5) = i_a + (V_{dc} - V_{cl})\sqrt{\frac{3C_r + C_{r7}}{L_r}} \tag{6.53}$$

From t_5 to t_9, L_r is clamped by $-V_{cl}$, hence the current through clamping capacitor C_{cl} during (t_5–t_9) is

$$i_{cl} = i_a + (V_{dc} - V_{cl})\sqrt{\frac{3C_r + C_{r7}}{L_r}} - \frac{V_{cl}}{L_r}(t - t_5) \tag{6.54}$$

From t_0 to t_1, L_r is clamped by $-V_{cl}$, hence the current through clamping capacitor C_{cl} during (t_0–t_1) is

$$i_{cl} = i_a + (V_{dc} - V_{cl})\sqrt{\frac{3C_r + C_{r7}}{L_r}} - \frac{V_{cl}}{L_r}(t - t_0 + T_S - t_5) \tag{6.55}$$

Combining Equation (6.55), the current through clamping capacitor at t_1 can be calculated as

$$i_{cl}(t_1) = i_a + (V_{dc} - V_{cl})\sqrt{\frac{3C_r + C_{r7}}{L_r}} - \frac{V_{cl}}{L_r}d_7T_S \tag{6.56}$$

Neglecting several short intervals (t_1–t_2), (t_4–t_5), from t_2 to t_4, L_r is clamped by V_{dc}–V_{cl}, the current during (t_1–t_5) can be re-expressed as

$$\begin{aligned} i_{cl}(t) &= i_{cl}(t_1) + \frac{V_{dc} - V_{cl}}{L_r}(t - t_1) \\ &= i_a + (V_{dc} - V_{cl})\sqrt{\frac{3C_r + C_{r7}}{L_r}} - \frac{V_{cl}}{L_r}d_7T_s + \frac{V_{dc} - V_{cl}}{L_r}(t - t_1) \end{aligned} \tag{6.57}$$

Combining Equations (6.54), (6.55), and (6.57), (6.52) is derived as

$$i_a + V_{dc}d_7\sqrt{\frac{3C_r + C_{r7}}{L_r}} + \frac{V_{dc}}{2L_r}T_s(1 - d_7)^2 - \frac{V_{dc}(1 - d_7)}{2L_r}T_s = 0 \tag{6.58}$$

Then, the duty ratio of the auxiliary switch is obtained:

$$\begin{cases} d_7 \approx 1 - \dfrac{(i_a + V_{dc}/Z_r)2L_r}{T_SV_{dc} + 2L_rV_{dc}/Z_r} \\ \lambda_7 \approx \dfrac{(i_a + V_{dc}/Z_r)2L_r}{T_SV_{dc} + 2L_rV_{dc}/Z_r} \end{cases} \tag{6.59}$$

$$V_{cl} = V_{dc}(1 - d_7) = \frac{(i_aZ_r + V_{dc})2L_r}{T_sZ_r + 2L_r} \tag{6.60}$$

6.3.4 Soft-switching Condition

ZVS condition for main switch in Equation (6.40) is rewritten as

$$\sqrt{V_{cl}^2 + i_{Lr(t1)}^2 Z_r^2} > V_{dc} - V_{cl} \tag{6.61}$$

From Equation (6.56), the resonant inductor current $i_{Lr}(t_1)$ is

$$i_{Lr}(t_1) = i_a + (V_{dc} - V_{cl})\sqrt{\frac{3C_r + C_{r7}}{L_r}} - \frac{V_{cl}}{L_r} d_7 T_S \tag{6.62}$$

By substituting Equation (6.59), (6.62) to inequality (6.61), ZVS condition can be expressed as

$$V_{cl}^2 + i_a^2 Z_r^2 + 2 i_a Z_r (V_{dc} - V_{cl}) > 0 \tag{6.63}$$

By considering that i_a is always larger than zero in Sector 1-1, Equation (6.63) can always be satisfied. Therefore, the soft-switching condition of MVAC inverter's bridge switches can always be satisfied.

6.3.5 Control Technique of MVAC Inverter

Once the reference sector $\vec{V}_{ref}$ is obtained by the control loop, the location and the magnitude of the reference sector can also be determined. With the modulation sequences of the ZVS SVM in Table 6.2 and Equation (6.26), the used vectors and their durations, d_{V0}, d_{V1}, and d_{V2}, can be easily obtained. The duty ratio of the auxiliary switch d_7 can be calculated with Equation (6.59).

Take Sector 1-1 as an example to introduce the procedures of gate signal derivation; the related waveforms are shown in Figure 6.27. In Figure 6.27, v_c is the saw-tooth carrier wave; one count-up timer period is equal to T_S. According to the procedures of driving sequences derivation of Sector 1-1, four modulation indexes for four Pulse Width Modulation (PWM) modules based on d_{V0}, d_{V1}, d_{V2}, and d_7 can be deduced as

$$\begin{cases} m_a = 0 \\ m_b = d_{V1} \\ m_c = 1 - d_{V0} \\ m_{d7} = \lambda_7 \end{cases} \tag{6.64}$$

With the specific action scheme applying to the count-up timer and counter-compare values, the switching functions can be obtained. With inverting and turn-on delay, each switching function will be expanded into two complementary primary driving sequences. For example, with inverting

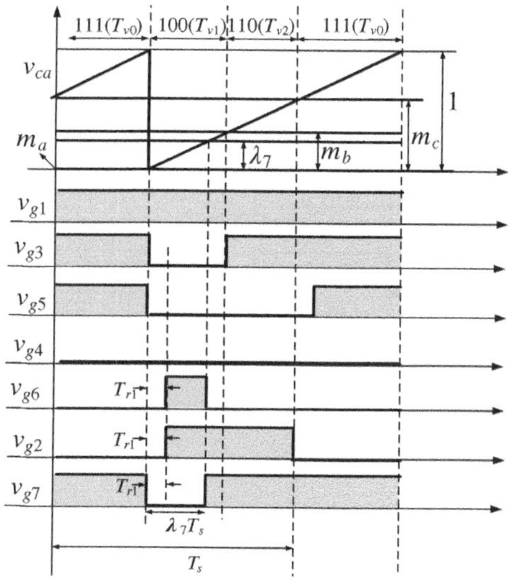

Figure 6.27 Gate signal derivation in Sector 1-1.

and turn-on delay T_{r1}, modulation index m_b is expanded into v_{g3} and v_{g6}. All the time delays for the primary driving sequences of three-phase bridges are set as T_{r1}.

The control block diagram of the three-phase MVAC inverter based on the synchronous rotating coordinate system is shown in Figure 6.28 [6, 7]. The references of active current and reactive current are i_d^{ref} and i_q^{ref}, respectively. Through grid voltage feed forward and inductor current decoupling control, the modulation ratio m and the rotation vector angle θ can be obtained.

With Equation (6.26), the duty ratio of the nonzero voltage vector and zero voltage vector, d_{V0}, d_{V1}, and d_{V2} can be calculated through the modulation ratio m and the rotation vector angle θ. The duty ratio of the auxiliary switch d_7 can be calculated with Equation (6.59). With Equation (6.26),

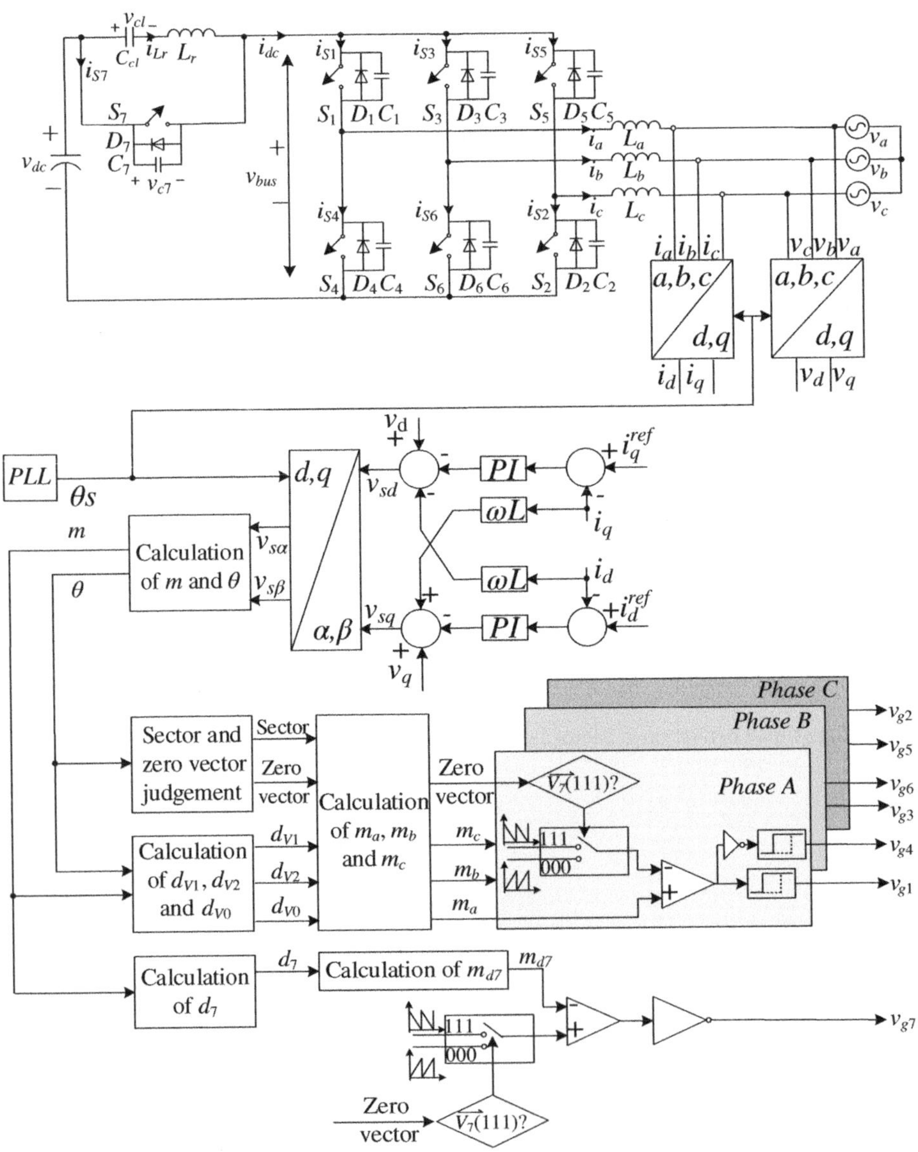

Figure 6.28 Control block diagram of the MVAC inverter.

four modulation indexes m_a, m_b, m_c, and m_{d7} based on d_{V0}, d_{V1}, d_{V2}, and d_7 can be obtained. With the modulation sequences of the ZVS SVM in Table 6.2, voltage sector and zero voltage vector can be selected according to the rotation vector angle θ. According to sector selection, zero voltage vector judgment, the saw-tooth carrier wave v_c can also be chosen. Through comparison of the modulation indexes m_a, m_b, m_c, and m_{d7} with the carrier wave, all the switches' driving signals can finally be gained.

6.4 Design Prototype

The previous sections introduce the operation principles of the MVAC ZVS inverter. In this section, a design will be provided through an example.

6.4.1 Specifications of a 30-kW Inverter

A 30-kW ZVS grid-connected inverter prototype was introduced. The specifications of the inverter shown in Figure 6.29 are listed in Table 6.5. IGBT modules with 1200 V/200 A rating are selected for switching devices S_1–S_7.

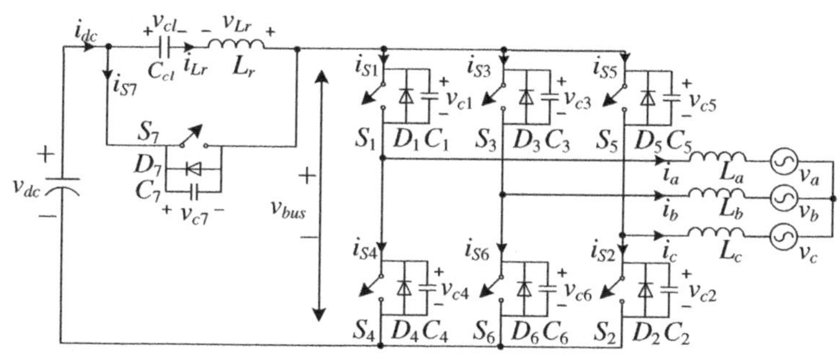

Figure 6.29 Topology of the ZVS inverter.

Table 6.5 Specifications of the ZVS inverter.

Parameters	Descriptions	Value
P_o	Output power	30 kW
V_a, V_b, V_c	Phase voltage of the grid	220 Vrms
f_g	Frequency of the grid	50 Hz
L_a, L_b, L_c	Filter inductance	0.3 mH
V_{dc}	DC source voltage	620 V
fs	Switching frequency	15kHz

6.4.2 Parameter Design

MVAC inverter parameter design needs to consider many factors, not only soft-switching condition but also the resonant impedance, resonant period, and the duty ratio of the auxiliary switch.

The resonant parameters of the MVAC inverter are designed as follows:

1) Set the minimum value of the resonant inductance to ensure the ability to suppress the reverse recovery current of the diode. The resonant inductor suppresses the reverse recovery energy of the freewheeling diodes in stage 4. In order to suppress the reverse recovery losses of the diodes, the di_{Lr}/dt during the turn-off process of the diodes is selected to be lower than 100 A/μs. In stage 4, it is expressed as

$$\frac{di_{Lr}}{dt} = \frac{V_{dc}}{L_r} \leq 100\text{A}/\mu\text{s} \tag{6.65}$$

Therefore, in this case, the inductance should satisfy

$$L_r \geq 6.2\mu\text{H} \tag{6.66}$$

2) By substituting equation (6.50) to equation (6.35), the bus voltage in stage 2 can be rewritten as,

$$v_{bus} = V_{dc}(1 - \lambda_7 + \lambda_7 \cos \omega_r t) - \left|i_{Lr(t1)} Z_r\right| \sin \omega_r t \tag{6.67}$$

Since the off duty cycle of auxiliary switch λ_7 is usually designed smaller than 0.1, $(1-\lambda_7+\lambda_7\cos\omega_r t)$ is very close to 1. As a result, the resonant time of the first resonant stage T_{r1} can be derived as:

$$T_{r1} \approx \frac{\arcsin\left(V_{dc}/\left|i_{Lr(t_1)}\right|Z_r\right)}{\omega_r} = \sqrt{L_r(3C_r + C_{r7})}\arcsin\left(\frac{V_{dc}}{\left|i_{Lr(t_1)}\right|Z_r}\right) \tag{6.68}$$

Combining soft switching condition analysis in 6.3.4, the soft switching condition of MVAC inverter's bridge switches can always be satisfied, so $0 \leq \arcsin(V_{dc}/|i_{Lr}(t_1)|Z_r) \leq \pi/2$, $0 \leq T_{r1} \leq \pi/2\omega_r$. In order to maintain the power quality, the maximum duration of the first resonant stage T_{r1} is designed to be smaller than 2.5 % of the switching period:

$$T_{r1\_max} = \frac{\pi}{2\omega_r} = \frac{\pi\sqrt{L_r(3C_r + C_7)}}{2} \leq 0.025T_s = 1.5\mu\text{s} \tag{6.69}$$

The maximum duration of the first resonant stage T_{r1} versus resonant inductor and resonant capacitors are illustrated in Figure 6.30.

3) To avoid loss of duty ratio, the auxiliary switch turn-off duty cycle λ_7 should be less than d_{V0}. From Equation (6.20), the expression of the zero voltage vector duty ratio d_{V0} is rewritten as follows:

$$d_{V0} = 1 - \frac{\sqrt{3}}{2} m \sin\left(\frac{\pi}{3} + \theta\right) \tag{6.70}$$

where θ is the phase angle of the reference voltage vector and m is the voltage modulation ratio. Due to the maximum value of $\sin(\pi/3+\theta)$ is one, the minimum value of d_{V0} is

$$d_{V0\min} = 1 - \frac{\sqrt{3}}{2} m \tag{6.71}$$

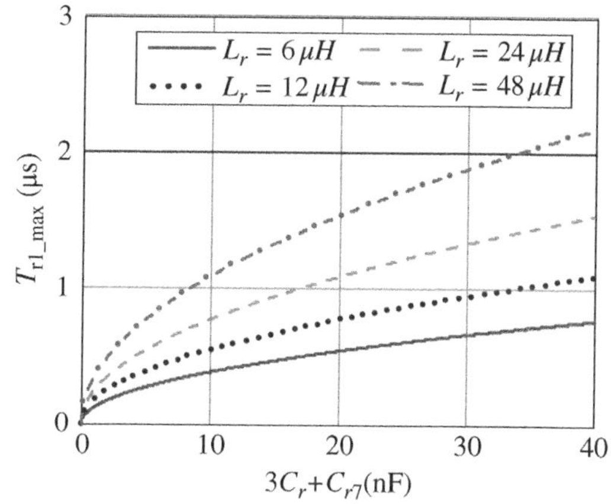

Figure 6.30 T_{r1} versus resonant parameters.

By substituting Equation (6.21) to Equation (6.71), the following equation is derived:

$$d_{V0\,\min} = 1 - \sqrt{3}\frac{\left|\vec{V}_{ref}\right|}{V_{dc}} \tag{6.72}$$

By ignoring the voltage drop of the grid side filter inductor, the reference voltage vector module $\left|\vec{V}_{ref}\right|$ is approximately equal to the amplitude of the grid source voltage V_m. By substituting $V_{dc} = 620$ V and $\left|\vec{V}_{ref}\right| = V_m = 220\cdot\sqrt{2} = 311V$ to Equation (6.72), the minimum value of d_{V0} can be calculated, $d_{V0\min} = 0.13$. Considering λ_7 should be less than d_{V0}, the following inequality is derived:

$$\lambda_7 < 0.13 \tag{6.73}$$

The turn-off duty cycle of the auxiliary switch λ_7 of Equation (6.59) is rewritten as

$$\lambda_7 = \frac{(i_a + V_{dc}/Z_r)2L_r}{T_s V_{dc} + 2L_r V_{dc}/Z_r} \tag{6.74}$$

If the rated power of the MVAC inverter is $P_0 = 30$ kW and the current ripple of the grid side filter inductor is ignored, the amplitude of the grid current $I_m = \sqrt{2}P_0/3\cdot 220 = 64.3$A. The switching cycle $T_s = 1/f_s = 62.5$μs. By substituting $V_{dc} = 620$ V, $I_m = 64.3$ A, $T_s = 62.5$ μs, and $\left|\vec{V}_{ref}\right| = 311V$ to Equation (6.74), the relationship between resonance impedance Z_r and resonance inductance L_r under different λ_7 is shown in Figure 6.31. The upper boundary of the available λ_7 is mainly determined by the value of L_r. When $\lambda_7 = 0.13$, Z_r is between 60 and 100 Ω, the upper limits of L_r is between 34 and 36 μH.

4) As stage analysis in Section 6.3.1, i_{Lr} reaches its maximum value i_{Lr_max} at t_1. Then, Equation (6.62) is rewritten as

$$\left|i_{Lr(t1)}\right| = i_{Lr\_\max} = i_a + (V_{dc} - V_{cl})\sqrt{\frac{3C_r + C_{r7}}{L_r}} - \frac{V_{cl}}{L_r}d_7 T_S \tag{6.75}$$

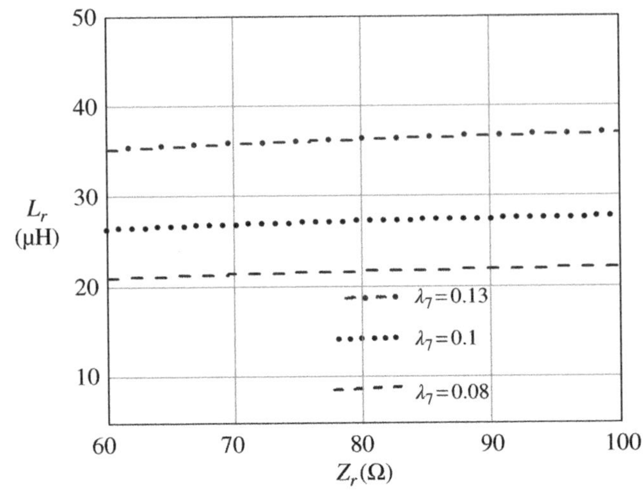

Figure 6.31 λ_7 versus Z_r.

Because the clamping capacitor voltage V_{cl} is much smaller than the DC source voltage V_{dc}, the previously mentioned equation is simplified by neglecting V_{cl} as follows:

$$\left|i_{Lr(t1)}\right| = i_{Lr\_\max} = i_a + \frac{V_{dc}}{Z_r} \tag{6.76}$$

In stage 1 ($t_0 - t_1$), the current of the auxiliary switch S_7 $i_{S7} = -i_{Lr}$. According to (6.76), i_{S7} at time t_1 is expressed by

$$i_{S7(t1)} = -i_{Lr(t1)} = i_{Lr\_\max} = i_a + \frac{V_{dc}}{Z_r} \tag{6.77}$$

Actually, i_{S7} reaches its maximum value i_{S7_max} at t_1. When the inverter is operated in steady state, the mean current through auxiliary switch S_7 is equal to the mean current of i_{dc} in a switching cycle. Therefore, the larger the $i_{S7(t1)}$, the greater the conduction loss of S_7 is. Equation (6.77) consists of two parts. The first term i_a is independent of resonant parameters. The second term V_{dc}/Z_r depends on resonant impedance, which is defined as

$$i_{res} = \frac{V_{dc}}{Z_r} \tag{6.78}$$

The relationship between the i_{res}/I_m and the resonance impedance Z_r under different power condition is shown in Figure 6.32. The lighter the load is, the larger i_{res}/I_m is. In this case, set i_{res}/I_m at rated power less than 20%, combining Equation (6.78), i_{res}/I_m should meet the following inequality:

$$\frac{i_{res}}{I_m} = \frac{V_{dc}}{I_m \cdot Z_r} \leq 20\% \tag{6.79}$$

By substituting $V_{dc} = 620$ V and $I_m = 64.3$ A to inequality (6.79), the minimum value of Z_r is 45 Ω. According to the resonant impedance Z_r in Equation (6.37), the relationship between resonant impedance Z_r versus resonant parameters is shown in Figure 6.33.

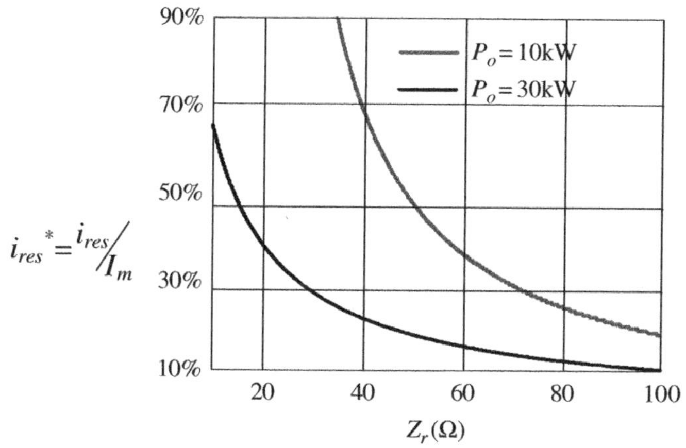

Figure 6.32 i_{res}/I_m versus resonant impedance Z_r.

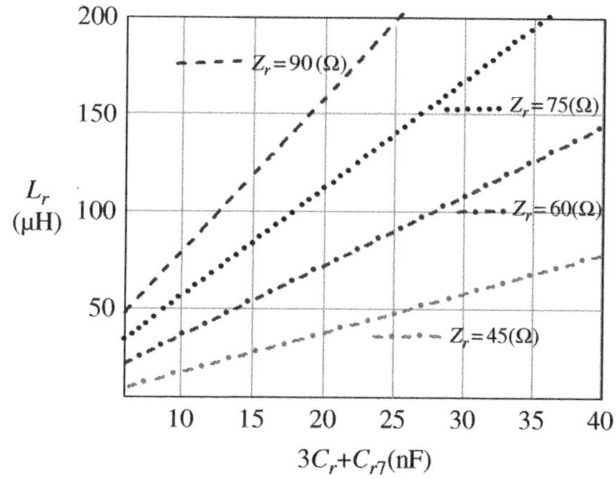

Figure 6.33 Resonant impedance Z_r versus resonant parameters.

5) In the ZVS inverter, turn-on losses of the switches are eliminated. However, turn-off loss of all of the switches still exists. They can be reduced by the paralleled capacitors. The turn-off loss reduction depends on the capacitance of the paralleled capacitors. The turn-off losses under ZVS condition of IGBT module CM200DU-24NFH is demonstrated in Figure 6.34. This figure implies that for a given turn-off current of 60 A, when C_r is 2.2 nF, the turn-off loss is reduced by 8%; when C_r is 3.3 nF, the turn-off loss is reduced by 38%; when C_r is 4.7 nF, the turn-off loss is reduced by 45%; when C_r is 7 nF, the turn-off loss is reduced by 61%. In order to obtain an obvious turn-off loss reduction, the capacitance of the resonant capacitor $C_r > 3.3$ nF is preferred.

With the previously mentioned five requirements, the appropriate region for selecting resonant parameters can be obtained as shown in Figure 6.35. The resonant parameters in shadow area meet all the requirements discussed before. The dark region in the figure can be used to select the appropriate resonant inductance and resonant capacitance. So the optimized resonant parameters can be selected as, $L_r = 30$ μH, $C_r = 3.3$nF.

6.4.3 Experiment Results

In this case, the designed resonant parameters are listed in Table 6.6.

The experimental circuit is shown in Figure 6.36.

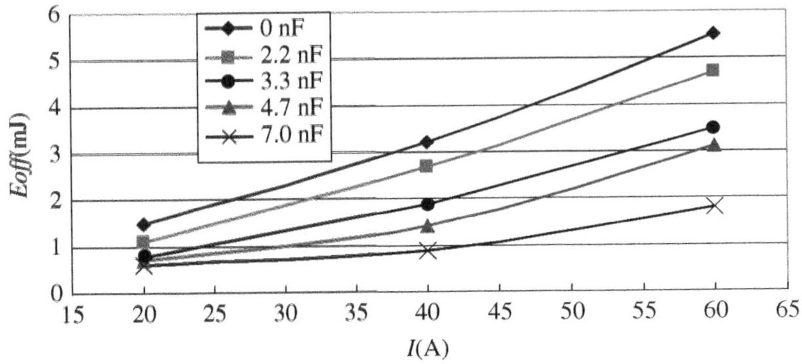

Figure 6.34 CM200DU-24NFH IGBT turn-off loss under ZVS conditions.

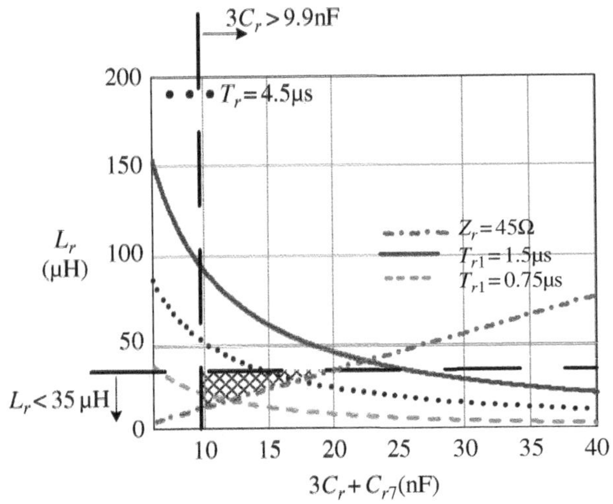

Figure 6.35 Recommended resonant parameters area.

Table 6.6 Resonant parameters selected.

Parameters	Descriptions	Value
L_r	Inductance of the resonant inductor	30 μH
C_r	Capacitance of the resonant capacitors	3.3 nF
C_7	Capacitance of S_7	0 nF
C_{cl}	Capacitance of the clamping capacitor	110 μF
S_1–S_7	Main switches and auxiliary switch	CM200DU-24NFH

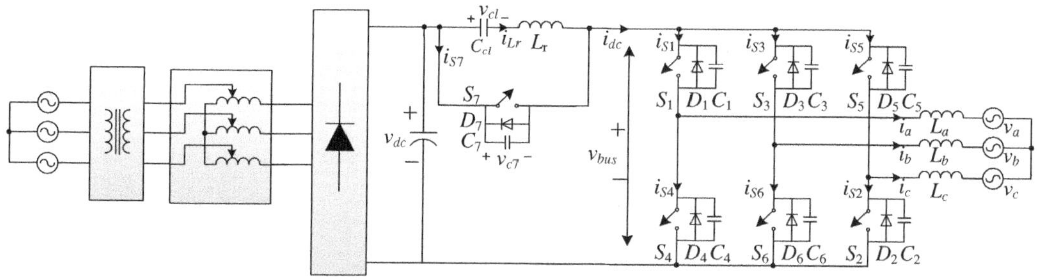

Figure 6.36 Experimental circuit.

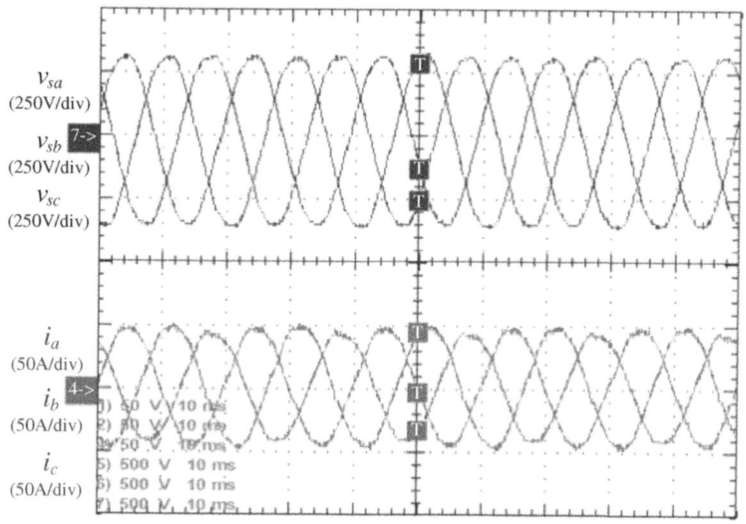

Figure 6.37 Output current and grid voltage.

The grid-connected voltage and current are demonstrated in Figure 6.37. The measured current THD is 2.9% under 30-kW condition.

The voltage and current of the bridge switch S_1 are shown in Figure 6.38. In this figure, current begins to increase from the negative value after the voltage drops to zero. The antiparallel diode conducts firstly, then the load current transfers to its main switch. Therefore, the bridge switches turn on in ZVS condition as required.

The voltage and current of the auxiliary switch S_7 are shown in Figure 6.39. When S_7 turns on, the current through the module goes negative, which means the freewheeling diode begins to conduct firstly, then the load current transfers to its main switch. Therefore, the ZVS condition for the auxiliary switch is realized.

The resonant branch current and auxiliary switch voltage are shown in Figure 6.40. The resonant inductor current i_{Lr} is symmetrical current. The mean value of the current in S_7 is equal to the mean value of the DC-side current i_{dc}. It means that the larger the inverter's output power, the more the conduction losses of auxiliary switch S_7.

The voltage across the clamping capacitor C_{cl} and the current through resonant inductor L_r are shown in Figure 6.41. The average voltage across C_{cl} is 30 V. The peak value of i_{Lr} is 22 A.

The conversion efficiency of the ZVS inverter is measured by power analyzer WT1800. And the results are shown in Figure 6.42. The maximum efficiency is 97.5%. It is observed that MVAC inverter has high efficiency in light load condition.

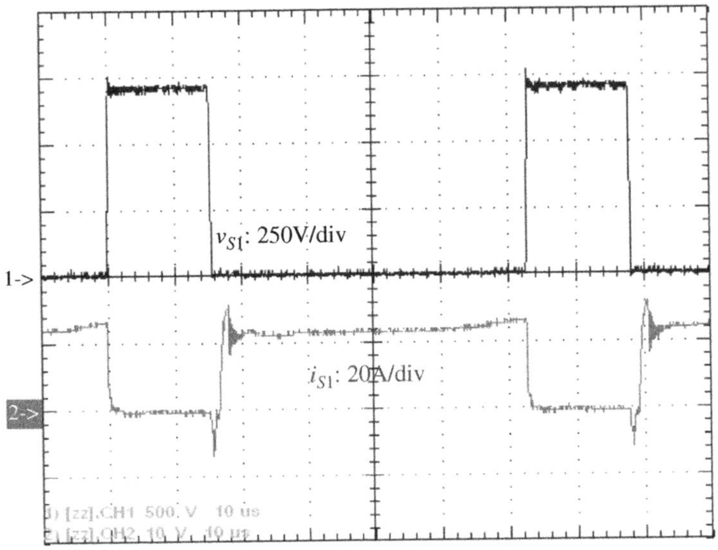

Figure 6.38 Voltage and current waveforms of the bridge switch.

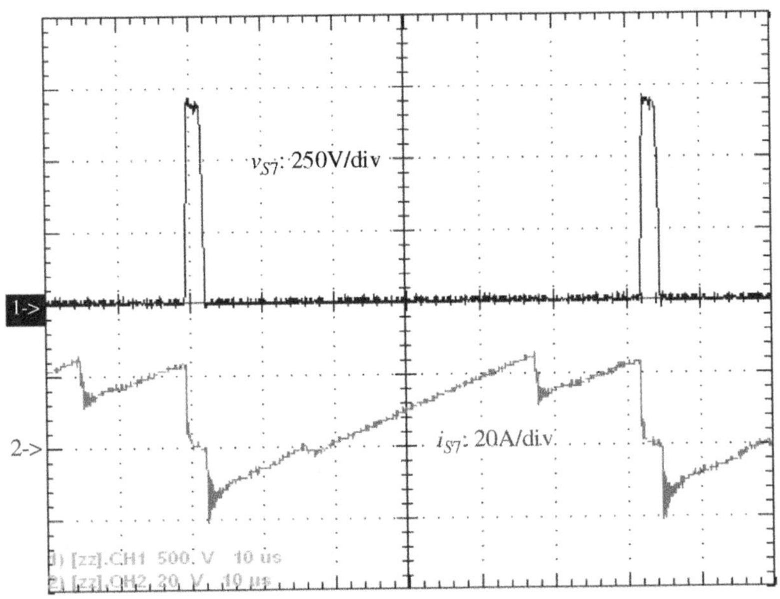

Figure 6.39 Voltage and current waveforms of the auxiliary switch.

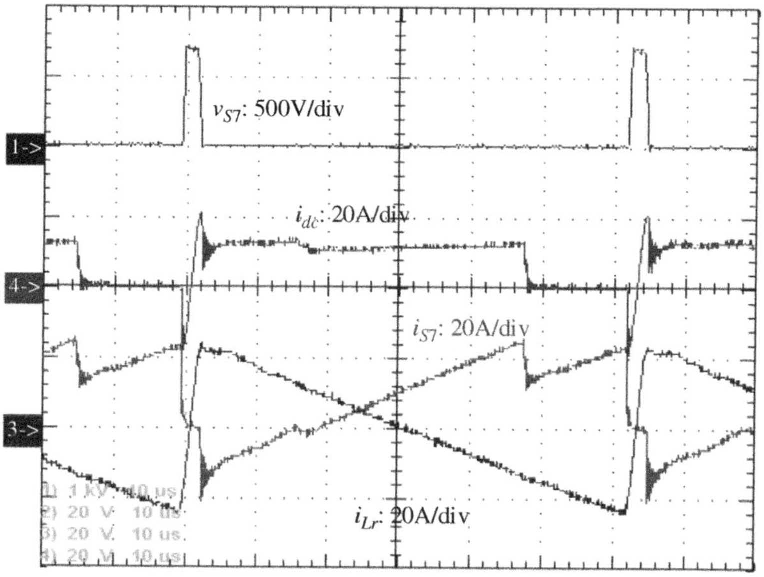

Figure 6.40 The resonant branch current and auxiliary switch voltage.

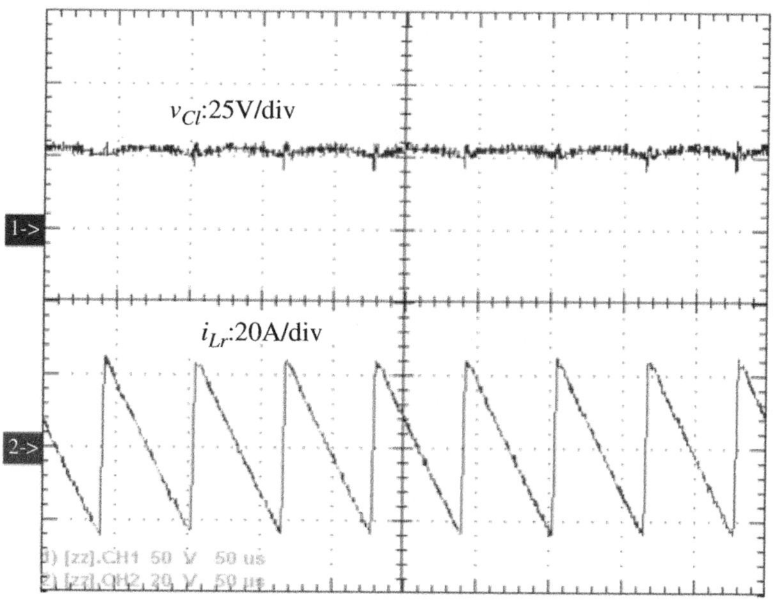

Figure 6.41 Voltage across C_{Cl} and current through L_r.

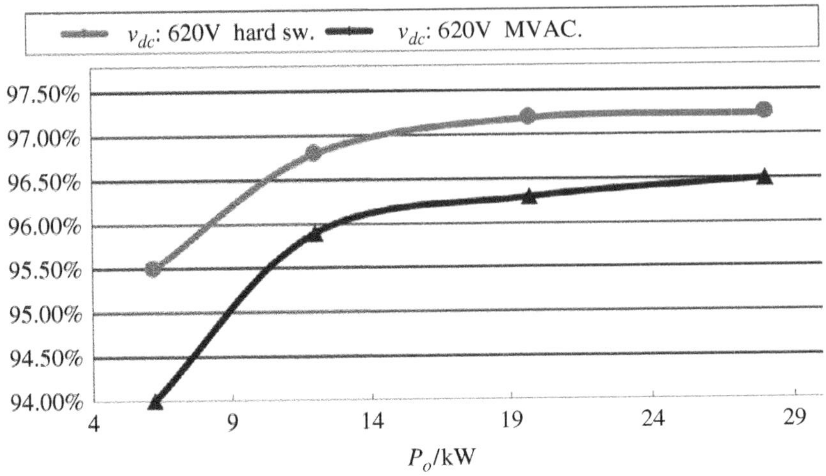

Figure 6.42 Measured efficiency.

6.5 Summary

MVAC inverter is introduced in this chapter. MVAC inverter can realize ZVS operation for all switching devices; the antiparallel diode reverse recovery current of all switch devices is suppressed well. The control scheme is required to modify the traditional SVM algorithm. All the switch devices can be operated at the fixed switching frequency. The switching voltage stress of all the power switch devices is equal to the DC source voltage. The inverter can achieve ZVS in all the switches under the grid current power factor angle from $-\pi/2$ to $\pi/2$. A 30-kW prototype has been built and realizes desirable ZVS soft-switching features.

References

1 R. Li, Z. Ma and D. Xu, "A ZVS grid-connected three-phase inverter," *IEEE Transactions on Power Electronics*, vol. 27, no. 8, pp. 3595–3604, 2012.
2 R. Li and D. Xu, "A zero-voltage switching three-phase inverter," *IEEE Transactions on Power Electronics*, vol. 29, no. 3, pp. 1200–1210, 2014.
3 D. M. Divan and G. Skibinski, "Zero-switching-loss inverters for high-power applications," *IEEE Transactions on Industry Applications*, vol. 25, no. 4, pp. 634–643, 1989.
4 M. Mezaroba, D. C. Martins, and I. Barbi, "A ZVS PWM three-phase inverter with active clamping technique using the reverse recovery energy of the diodes," in *Proc. IEEE Conf. Power Electron. Spec.*, pp. 689–694, 2004.
5 Y. Wu, N. He, and D. Xu, "Generalized space-vector-modulation method for soft-switching three-phase inverters," *IEEE Transactions on Power Electronics*, vol. 36, no. 5, pp. 6030–6045, 2021.
6 C. Du, D. Xu, N. He and N. Zhu, Modeling and optimization of a zero-voltage switching inverter for high efficiency and miniaturization. *IEEE Transactions on Power Electronics* 2017;32(1):150–163.
7 M. Chen, J. Xi, Y. Chen and D. Xu, Safe startup method for zero-voltage-switching MOSFET grid inverter. *IEEE Transactions on Power Electronics* 2019;34(9):8749–8761.

7

Three-phase Inverter with Compound Active-clamping Circuit

In this chapter, a zero voltage switching (ZVS) three-phase inverter with compound active-clamping (CAC) circuit is introduced, including the operation principle, the circuit analysis, and the design guidelines. First, the operation principle of the proposed modulation scheme is introduced. Next, based on circuit stage analysis, soft-switching conditions will be derived. Besides, the implementation of ZVS-SVM in digital controller is also provided. Finally, the design guidelines will be summarized, and a design example based on a 30 kW prototype will be provided.

7.1 Introduction

The topology of the compound active-clamping (CAC) zero voltage switching (ZVS) three-phase inverter is shown in Figure 7.1. A CAC auxiliary circuit is installed in the DC side of the inverter [1–6]. The auxiliary circuit is composed of a resonant inductor L_r, a clamping capacitor C_{cl}, and an auxiliary switch S_7. The clamping capacitor and the auxiliary switch are connected in series and then in parallel with the resonant inductor. Resonant capacitors, C_1, C_2, C_3, C_4, C_5, C_6, and C_7, are paralleled to the main switches and the auxiliary switch S_7.

As mentioned before, there are two types of switch commutations in the three-phase main bridges. The commutation from a switch to the antiparalleled diode of its complementary switch in a leg is defined as the Type 1 switch commutation, which has a small loss. The switch commutation from a diode to its complementary switch in a leg is defined as the Type 2 switch commutation, which has higher loss. Once there comes up a Type 2 switch commutation in the CAC ZVS three-phase inverter as shown in Figure 7.1, the auxiliary switch S_7 is required to turn off to start resonance between the resonant inductor L_r and some of the resonant capacitors. As a result, the voltage across the three-phase main bridges v_{bus} will be resonated to zero, which brings the opportunity of ZVS turn-on for main switches. The main switches are turned on during the small interval when v_{bus} is clamped at zero. Auxiliary switch is not required to turn off when a Type 1 switch commutation comes. According to the ZVS space vector modulation (SVM) scheme introduced in Chapter 3, all switches, including main switches and auxiliary switch, can be turned on under zero voltage condition. The reverse recovery process of freewheeling diodes in the CAC ZVS grid inverter is also suppressed.

Soft-Switching Technology for Three-phase Power Electronics Converters, First Edition.
Dehong Xu, Rui Li, Ning He, Jinyi Deng, and Yuying Wu.

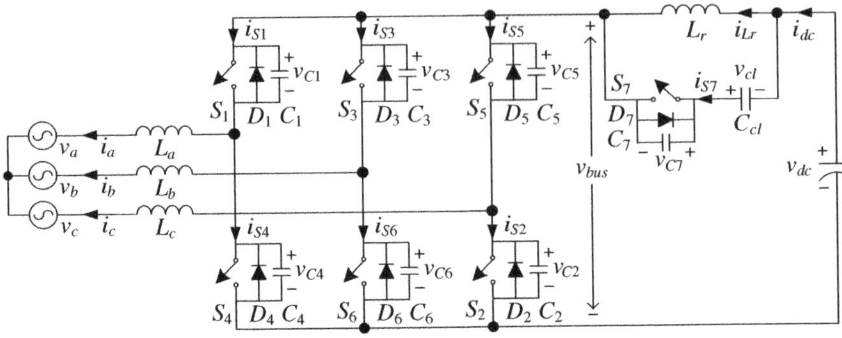

Figure 7.1 Topology of the CAC ZVS three-phase inverter.

In this chapter, the operation principle of the ZVS-SVM will be introduced. Then, the circuit analysis will be provided in order to derive the soft-switching conditions. In the end, the design guidelines will be introduced, and a design example based on a 30 kW prototype will be given.

7.2 Scheme of ZVS-SVM

7.2.1 Switch Commutations in Main Bridges of Three-phase Inverter

It is assumed that the CAC ZVS three-phase inverter shown in Figure 7.1 is operating under unity power factor. Its alternating current (AC) side current and voltage waveforms in a line cycle are demonstrated in Figure 7.2a. Based on the basic SVM knowledge in Appendix A.1, a line cycle is first evenly divided into six sectors, from sector 1 to sector 6. Considering the current direction of different phases, each sector is further divided into two subsectors. Therefore, a line cycle is finally divided into 12 sectors as shown in Figure 7.2.

We take sector 1-1 as an example to demonstrate the principles of the ZVS-SVM. Here are some requirements for the modulation method. First of all, only the closest basic vectors in each sector are utilized. In sector 1-1, vectors $\vec{V}_1(100)$, $\vec{V}_2(110)$, $\vec{V}_0(000)$, and $\vec{V}_7(111)$ are selected to compose the reference vector $\vec{V}_{ref}$. Furthermore, the auxiliary switch is preferred to be turned off only once in each switching period to simplify the control and avoid extra loss. Last but not least, the number of total switching actions in a switch cycle should be as few as possible to reduce the switching loss.

With regard to the current directions, $i_a > 0$, $i_b < 0$, and $i_c < 0$, in Figure 7.2a, the equivalent circuits of the four vectors in sector 1-1 are drawn in Figure 7.3. The switch commutations of three-phase main bridges between two vectors are summarized as follows:

From vector $\vec{V}_7(111)$ to vector $\vec{V}_1(100)$, there exist two Type 2 switch commutations.

From vector $\vec{V}_7(111)$ to vector $\vec{V}_2(110)$, there exists one Type 2 switch commutation.

From vector $\vec{V}_7(111)$ to vector $\vec{V}_0(000)$, there exist two Type 2 switch commutations and one Type 1 switch commutation.

From vector $\vec{V}_0(000)$ to vector $\vec{V}_1(100)$, there exists one Type 2 switch commutation.

From vector $\vec{V}_0(000)$ to vector $\vec{V}_2(110)$, there exist one Type 2 switch commutation and one Type 1 switch commutation.

From vector $\vec{V}_0(000)$ to vector $\vec{V}_7(111)$, there exist one Type 2 switch commutation and two Type 1 switch commutations.

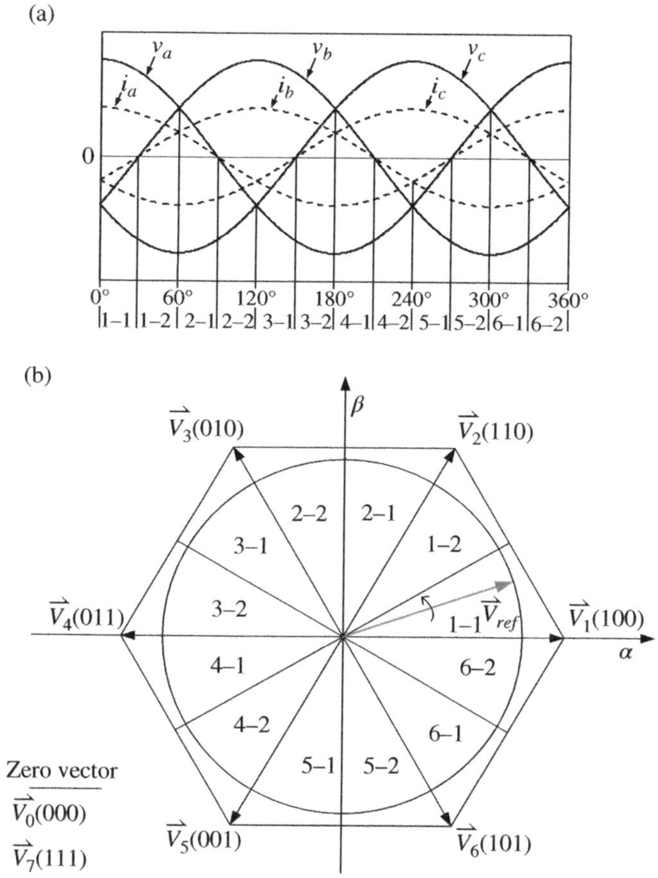

Figure 7.2 Sectors of space vector diagram.

From vector $\vec{V}_1(100)$ to vector $\vec{V}_7(111)$, there exist two Type 1 switch commutations.
From vector $\vec{V}_1(100)$ to vector $\vec{V}_2(110)$, there exists one Type 1 switch commutation.
From vector $\vec{V}_1(100)$ to vector $\vec{V}_0(000)$, there exists one Type 1 switch commutation.
From vector $\vec{V}_2(110)$ to vector $\vec{V}_1(100)$, there exists one Type 2 switch commutation.
From vector $\vec{V}_2(110)$ to vector $\vec{V}_7(111)$, there exists one Type 1 switch commutation.
From vector $\vec{V}_2(110)$ to vector $\vec{V}_0(000)$, there exist one Type 2 switch commutation and one Type 1 switch commutation.

7.2.2 Derivation of ZVS-SVM

After the commutations of three-phase main bridges between two vectors have been identified, the numbers of Type 2 (T2) and Type 1 (T1) commutations can be counted of each vector sequence composed of the corresponding four vectors. Since two nonzero vectors $\vec{V}_1(100)$,$\vec{V}_2(110)$ and at least one zero vector are needed, there will be three combinations of the four vectors. The first combination set uses all four vectors, and their all possible vector sequences are shown in Figure 7.4. The

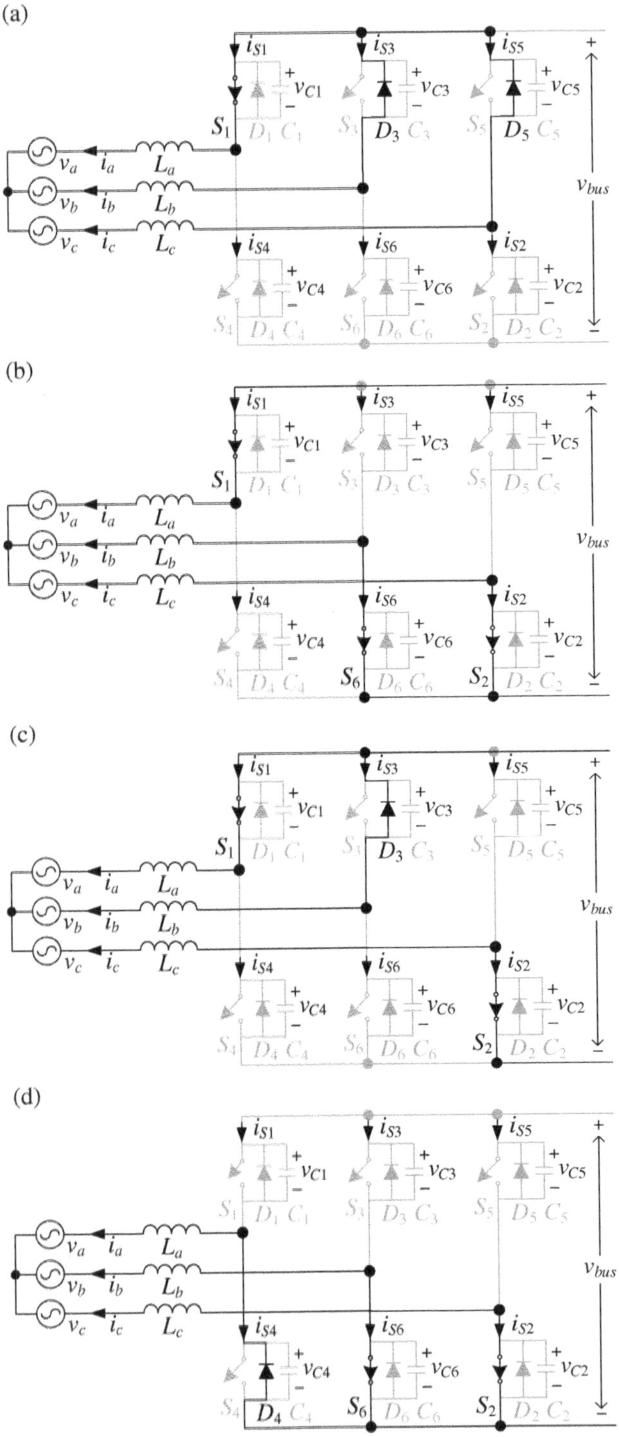

Figure 7.3 Equivalent circuits of four vectors in sector 1-1: (a) vector 111, (b) vector 100, (c) vector 110, and (d) vector 000.

second combination set uses two nonzero vectors and zero vector $\vec{V}_0(000)$, and their all possible vector sequences are shown in Figure 7.5. The last combination set uses two nonzero vectors and zero vector $\vec{V}_7(111)$, and their all possible vector sequences are shown in Figure 7.6.

By considering the three requirements for the ZVS-SVM, it is observed that there are always at least two Type 2 commutations during one switching period in the possible vector sequences shown in Figures 7.4 and 7.5. As shown in Figure 7.6a, there is only one vector sequence that can meet the requirement with only one Type 2 commutation during one switching period, which is $\vec{V}_7(111)$, $\vec{V}_1(100)$, $\vec{V}_2(110)$, $\vec{V}_7(111)$. As shown in Figure 7.6b, there is also only one vector sequence that can meet the requirement with only one Type 2 commutation during one switching period, which is $\vec{V}_1(100)$, $\vec{V}_2(110)$, $\vec{V}_7(111)$, $\vec{V}_1(100)$. As shown in Figure 7.6c, only one vector sequence can meet the requirement with only one Type 2 commutation during one switching

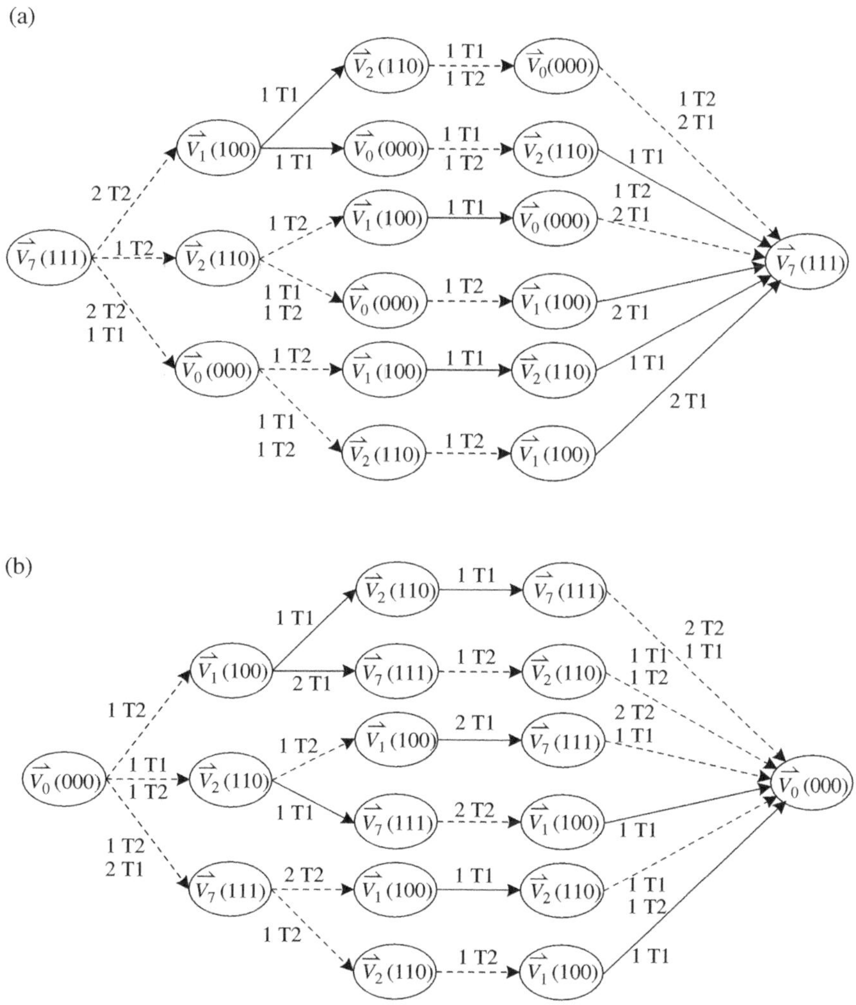

Figure 7.4 Vector sequences with four vectors: starting with (a) 111, (b) 000, (c) 100, and (d) 110.

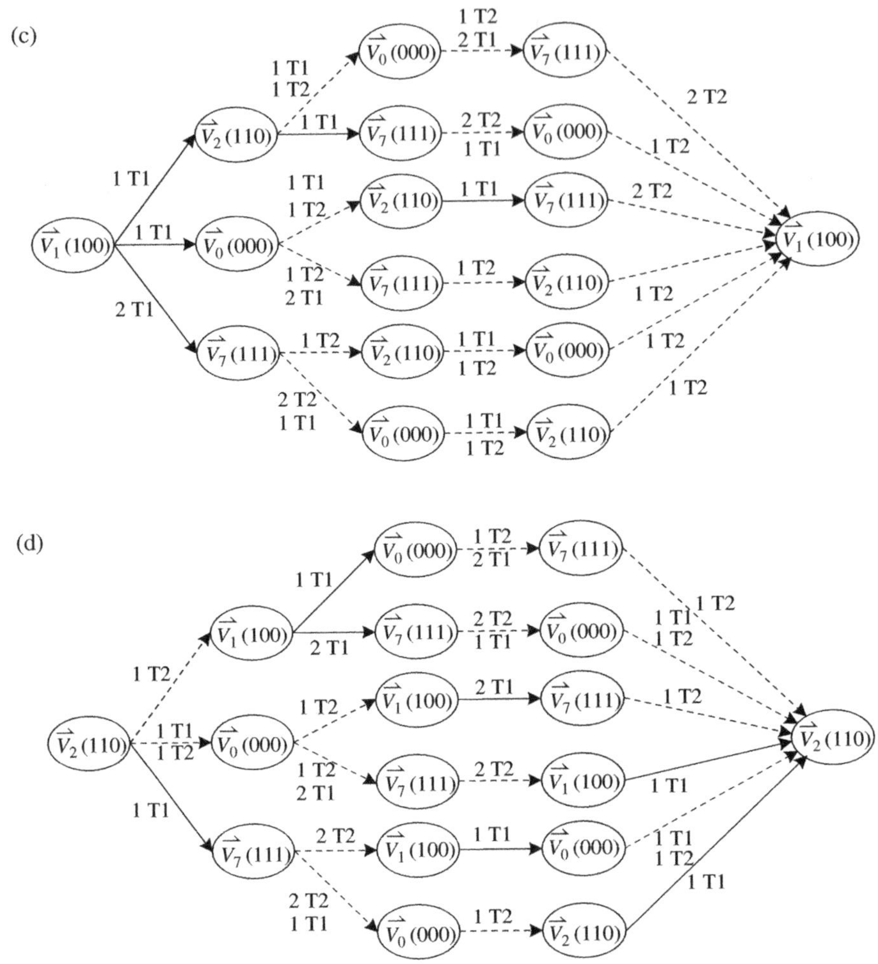

Figure 7.4 (Continued)

period, which is $\vec{V}_2(110)$, $\vec{V}_7(111)$, $\vec{V}_1(100)$, $\vec{V}_2(110)$. Obviously, the three feasible vector sequences in Figure 7.6 are the same.

With the similar derivations, the vector sequences of the ZVS-SVM in entire 12 sectors are obtained as shown in Table 7.1.

Once the vector sequences of three-phase main bridges are derived, the driving sequences of the six main switches and the auxiliary switch in sector 1-1 can be easily obtained as shown in Figure 7.7. The auxiliary switch S_7 is turned off with resonant time (T_{r1}) ahead of the turn-on of the main switches, S_2 and S_6. The turn-off of auxiliary switch S_7 starts the resonance between the resonant inductor L_r and some of the resonant capacitors. As a result, the voltage across the three-phase main bridges v_{bus} will be resonated to zero, which brings the opportunity of ZVS turn-on for main switches. The main switches are turned on during the small interval when v_{bus} is clamped at zero. Besides, there will be another extra added short circuit pulse for all main switches when v_{bus} is clamped at zero, which is used to store adequate energy for the resonant inductor. The detailed operation stages of the ZVS-SVM will be investigated in the following paragraph.

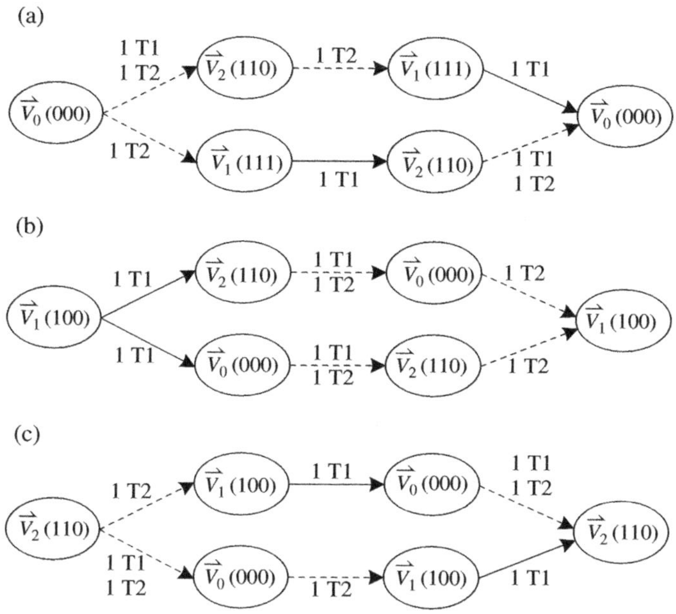

Figure 7.5 Vector sequences with two nonzero vectors and zero vector 000: starting with (a) 000, (b) 100, and (c) 110.

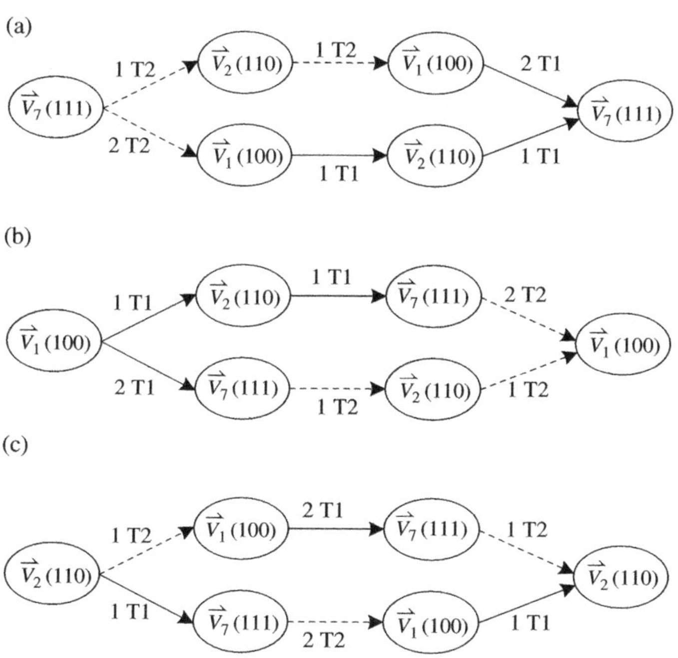

Figure 7.6 Vector sequences with two nonzero vectors and zero vector 111: starting with (a) 111, (b) 100, and (c) 110.

Table 7.1 Vector sequences of the ZVS-SVM in entire 12 sectors.

Sector	Modulation sequences
1-1	$\vec{V}_7(111) \to \vec{V}_1(100) \to \vec{V}_2(110) \to \vec{V}_7(111)$
1-2	$\vec{V}_0(000) \to \vec{V}_2(110) \to \vec{V}_1(100) \to \vec{V}_0(000)$
2-1	$\vec{V}_0(000) \to \vec{V}_2(110) \to \vec{V}_3(010) \to \vec{V}_0(000)$
2-2	$\vec{V}_7(111) \to \vec{V}_3(010) \to \vec{V}_2(110) \to \vec{V}_7(111)$
3-1	$\vec{V}_7(111) \to \vec{V}_3(010) \to \vec{V}_4(011) \to \vec{V}_7(111)$
3-2	$\vec{V}_0(000) \to \vec{V}_4(011) \to \vec{V}_3(010) \to \vec{V}_0(000)$
4-1	$\vec{V}_0(000) \to \vec{V}_4(011) \to \vec{V}_5(001) \to \vec{V}_0(000)$
4-2	$\vec{V}_7(111) \to \vec{V}_5(001) \to \vec{V}_4(011) \to \vec{V}_7(111)$
5-1	$\vec{V}_7(111) \to \vec{V}_5(001) \to \vec{V}_6(101) \to \vec{V}_7(111)$
5-2	$\vec{V}_0(000) \to \vec{V}_6(101) \to \vec{V}_5(001) \to \vec{V}_0(000)$
6-1	$\vec{V}_0(000) \to \vec{V}_6(101) \to \vec{V}_1(100) \to \vec{V}_0(000)$
6-2	$\vec{V}_7(111) \to \vec{V}_1(100) \to \vec{V}_6(101) \to \vec{V}_7(111)$

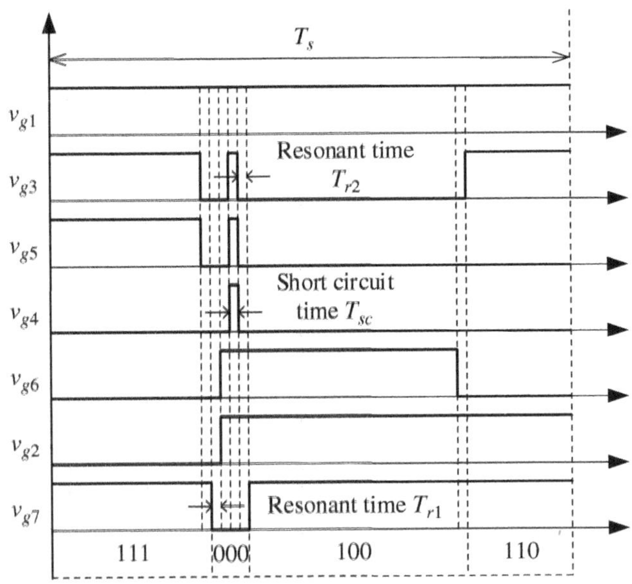

Figure 7.7 Driving sequence of ZVS-SVM in sector 1-1 within one switching period.

7.3 Circuit Analysis

7.3.1 Operation Stage Analysis

In order to simplify the analysis, the following assumptions are made:

1) All switching devices are treated as ideal switches.
2) Grid filter inductor is so large that the filter inductor current can be seen as constant in a switching period.

3) Parasitic capacitance of each switch is included in the capacitor paralleled to it.
4) The value of clamping capacitor C_{cl} is so large that the v_{cl} can be treated as constant in a switching period.

Sector 1-1 is taken as an example to describe the operation stages in one switching period. Figure 7.8 shows the waveforms of the key components in the soft-switching three-phase inverter in one switching period. The operation stages are analyzed as follows:

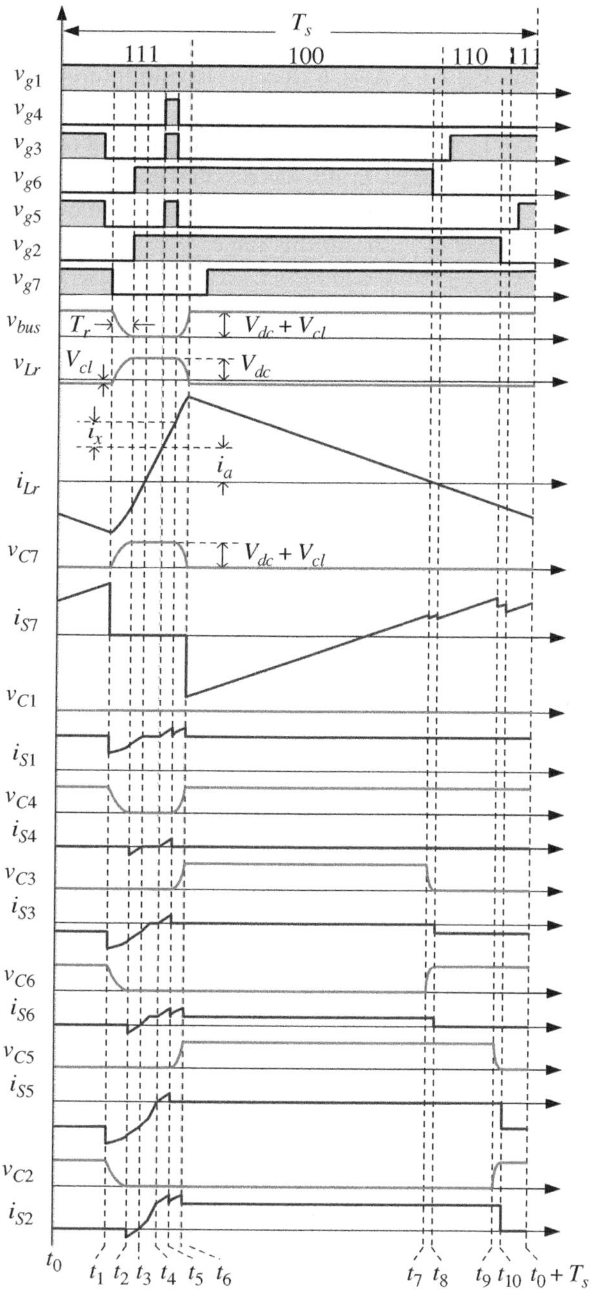

Figure 7.8 Key waveforms in sector 1-1.

Stage 1 (t_0–t_1) is the first steady stage. In this stage, the inverter outputs vector $\overrightarrow{V_7}$ (111). The equivalent circuit of this stage is shown in Figure 7.9. The current direction of each phase indicates that S_1, D_3, and D_5 are in conducting state. Auxiliary switch S_7 is in conducting state in this stage, and the resonant inductor L_r is discharged by the clamping capacitor C_{cl}. Therefore, the current through L_r decreases with a rate of V_{cl}/L_r. In this stage, v_{bus} is clamped to be $V_{dc}+V_{cl}$ and i_{S7} is equal to $-i_{Lr}$.

Stage 2 (t_1–t_2) is the first resonant stage. At t_1, S_7 is turned off. A resonance between the resonant inductor L_r and the resonant capacitors C_2, C_4, C_6, and C_7 is provoked. The equivalent circuit of this stage is shown in Figure 7.10. During this stage, v_{bus} keeps reducing, while v_{C7} increases. This stage ends at t_2, when v_{bus} resonates to zero and v_{C7} reaches $V_{dc}+V_{cl}$. At t_2, i_{Lr} reaches its minimum value i_{Lr_min}.

Stage 3 (t_2–t_3) is the diode freewheeling stage. After t_2, v_{bus} is equal to zero. The equivalent circuit of this stage is shown in Figure 7.11. After last stage's resonance, L_r still has residual energy to keep current i_{Lr} negative. D_4, D_6, and D_2 conduct in freewheeling mode. S_6 and S_2 are turned on during this stage under ZVS condition. i_{Lr} increases with a rate of V_{dc}/L_r in this stage.

Stage 4 (t_3–t_4) is the Type 2 commutation stage. At t_3, i_{Lr} reduces to zero. Currents in phase-b and phase-c begin to transfer from D_3 and D_5 to S_6 and S_2, respectively. The equivalent circuit of this stage is shown in Figure 7.12. During this period, the changing rate of i_{S3} and i_{S5} is restrained to a small value by L_r so that the reverse recovery losses on D_3 and D_5 can be eliminated. i_{S3} and i_{S5} reduce to zero by time t_4. Meanwhile, the resonant inductor current at t_4 is equal to i_a.

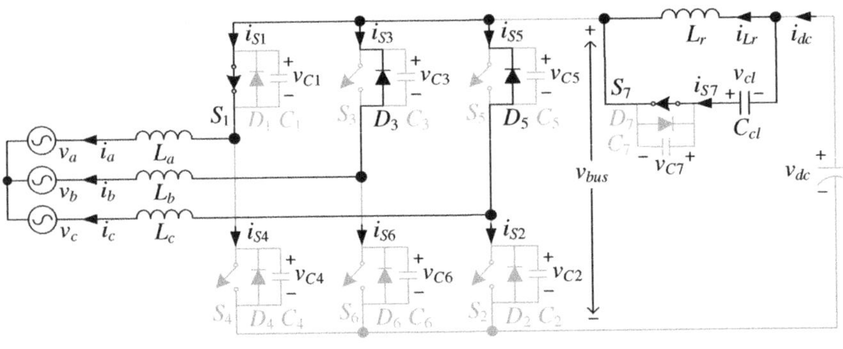

Figure 7.9 Equivalent circuit of the first steady stage.

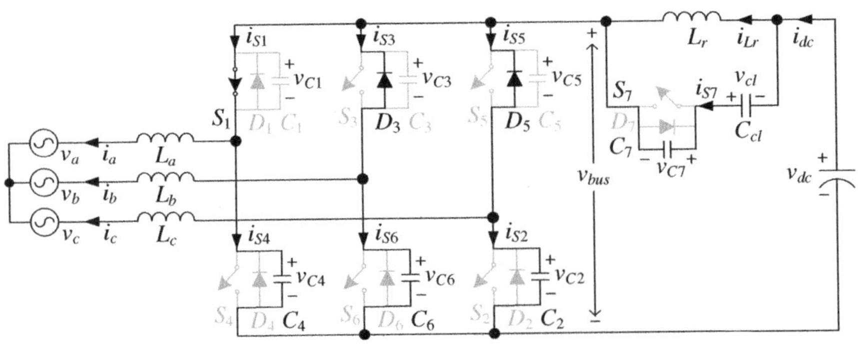

Figure 7.10 Equivalent circuit of the first resonant stage.

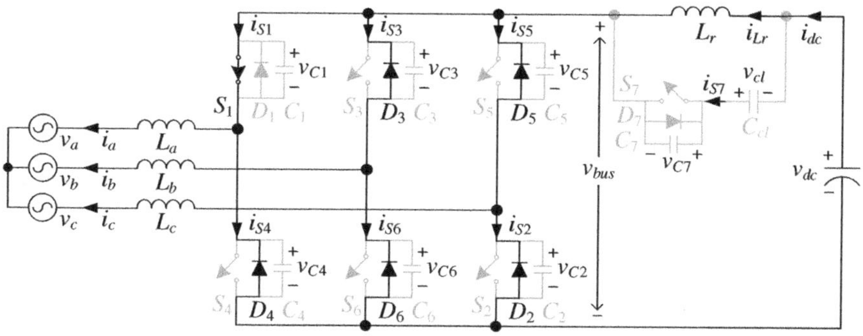

Figure 7.11 Equivalent circuit of diode freewheeling stage.

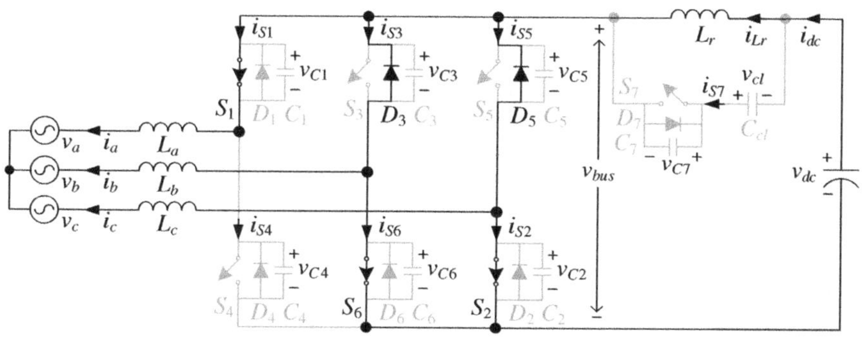

Figure 7.12 Equivalent circuit of Type 2 commutation stage.

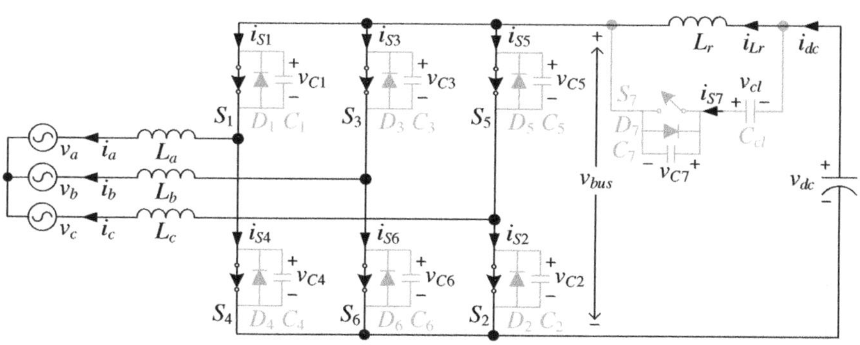

Figure 7.13 Equivalent circuit of short-circuit stage.

Stage 5 (t_4–t_5) is the short-circuit stage. At t_4, S_4, S_3, and S_5 are ZVS turned on to construct short-circuit paths. The equivalent circuit of this stage is shown in Figure 7.13. L_r is charged during this stage to store enough energy to fulfill the ZVS condition for the next switching period. The extra added current in this stage for i_{Lr} is defined as i_x.

Stage 6 (t_5–t_6) is the second resonant stage. After the short circuit stage, S_4, S_3, and S_5 are turned off at t_5. Resonance occurs among L_r, C_3, C_4, C_5, and C_7. The equivalent circuit of this stage is shown in Figure 7.14. During this resonance, v_{bus} increases and v_{C7} decreases. At time t_6, v_{bus} reaches to $V_{dc} + V_{cl}$, while v_{C7} reduces to zero. At t_6, i_{Lr} reaches its maximum value i_{Lr_max}.

Stage 7 (t_6–t_7) is the second steady stage. After the second resonance, S_7 can be turned on under ZVS condition since v_{C7} is equal to zero. The inverter outputs vector $\overrightarrow{V_1}$ (100) in this stage. The equivalent circuit of this stage is shown in Figure 7.15. The resonant inductor current i_{Lr} decreases with a rate of V_{cl}/L_r in this stage.

Stage 8 (t_7–t_8) is the first Type 1 commutation stage. At t_7, S_6 turns off. The output current i_b charges C_6 and discharges C_3. The equivalent circuit of this stage is shown in Figure 7.16. This stage ends at t_8 when v_{C3} reduces to zero and D_3 conducts in freewheeling mode.

Stage 9 (t_8–t_9) is the third steady stage. In this stage, inverter outputs vector $\overrightarrow{V_2}$ (110). The equivalent circuit of this stage is shown in Figure 7.17. The resonant inductor current i_{Lr} decreases with a rate of V_{cl}/L_r in this stage.

Stage 10 (t_9–t_{10}) is the second Type 1 commutation stage. At t_9, S_2 is turned off. The output current i_c charges C_2 and discharges C_5 simultaneously. The equivalent circuit of this stage is shown in Figure 7.18. This stage ends at t_{10} when v_{C5} reduces to zero and D_5 conducts in freewheeling mode.

After Stage 10, the inverter begins with stage 1 of the next switching period.

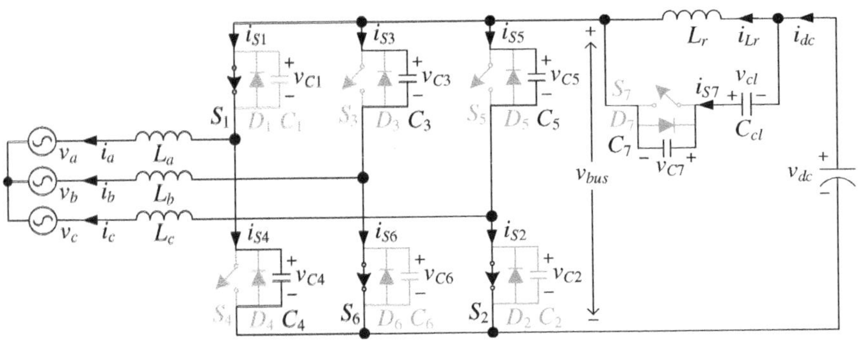

Figure 7.14 Equivalent circuit of the second resonant stage.

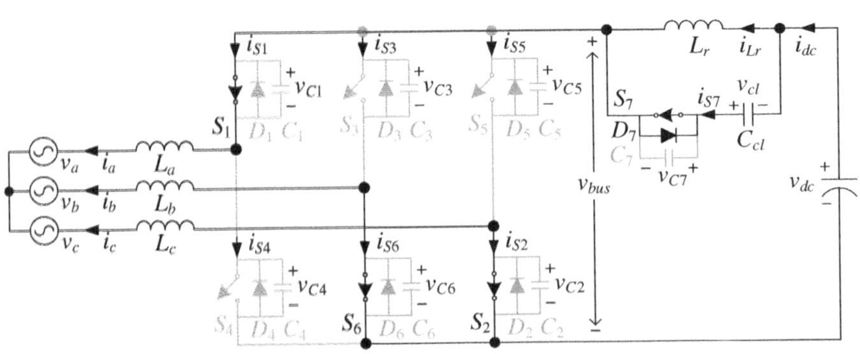

Figure 7.15 Equivalent circuit of the second steady stage.

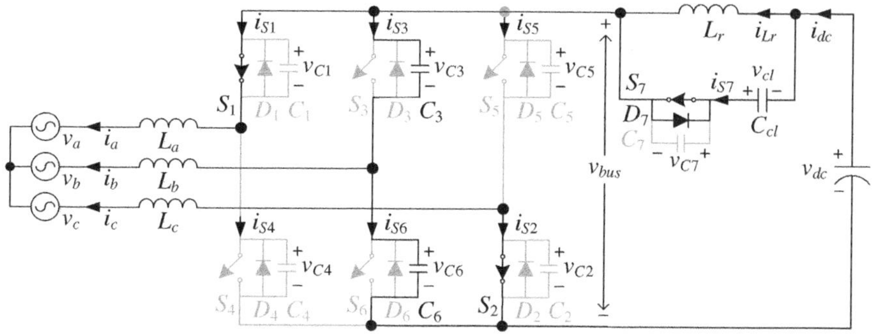

Figure 7.16 Equivalent circuit of the first Type 1 commutation stage.

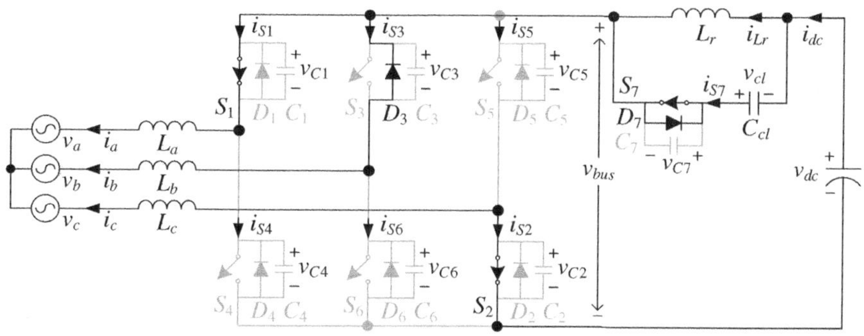

Figure 7.17 Equivalent circuit of the third steady stage.

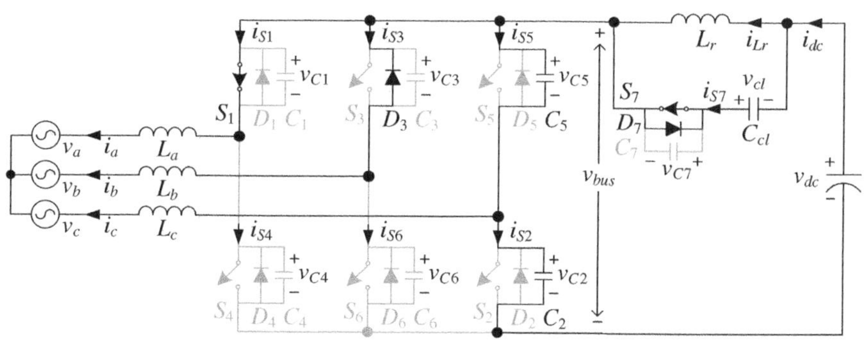

Figure 7.18 Equivalent circuit of the second Type 1 commutation stage.

7.3.2 Resonant Stages Analysis

There are two resonant stages in each switching period. The first resonant stage, stage 2, is used to realize ZVS turn-on conditions for the main bridge switches, while the second resonant stage, stage 6, is used to realize the ZVS turn-on condition for the auxiliary switch. Besides, the reverse recovery of the freewheeling diodes is relieved by reduction of the di/dt with the resonant inductor. It means

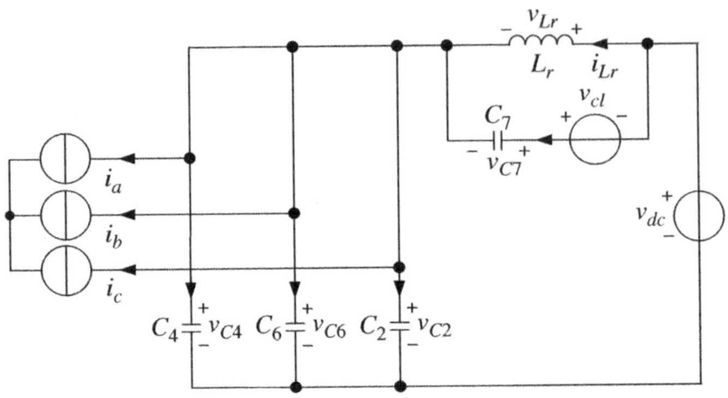

Figure 7.19 Simplified equivalent circuit of the first resonant stage.

there are no turn-on loss and diode reverse recovery loss. The losses of the switches that need to be considered are only conducting loss and turn-off loss.

It is assumed that the paralleled capacitors of main switches are equal as follows:

$$C_1 = C_4 = C_3 = C_6 = C_5 = C_2 = C_r \tag{7.1}$$

The equivalent circuit in Figure 7.10 can be simplified as shown in Figure 7.19.

The following equations can be obtained from Figure 7.19:

$$\begin{cases} i_{Lr} + C_7\dfrac{dv_{C7}}{dt} = C_r\dfrac{dv_{C4}}{dt} + i_a + C_r\dfrac{dv_{C6}}{dt} + i_b + C_r\dfrac{dv_{C2}}{dt} + i_c \\ v_{Lr} = L_r\dfrac{di_{Lr}}{dt} \end{cases} \tag{7.2}$$

where

$$v_{C4} = v_{C6} = v_{C2} = V_{dc} + V_{cl} - v_{C7} \tag{7.3}$$

$$V_{dc} = v_{Lr} + v_{C4} \tag{7.4}$$

$$i_a + i_b + i_c = 0 \tag{7.5}$$

By substituting (7.3), (7.4), and (7.5) to (7.2), the equations of state variables, resonant inductor current i_{Lr}, and resonant capacitor voltage v_{C4} can be simplified as

$$\begin{cases} (3C_r + C_7)\dfrac{dv_{C4}}{dt} = i_{Lr} \\ L_r\dfrac{di_{Lr}}{dt} = -v_{C4} + V_{dc} \end{cases} \tag{7.6}$$

The initial condition can be derived as

$$\begin{cases} v_{C4}(t_1) = V_{cl} + V_{dc} \\ i_{Lr}(t_1) = i_{Lr}(t_1) \end{cases} \tag{7.7}$$

By solving the differential Equation (7.6) with the initial condition (7.7), the expressions of resonant capacitor voltage v_{C4} and resonant inductor current i_{Lr} can be obtained as

$$\begin{cases} v_{C4}(t) = V_{cl}\cos(\omega_r t) + i_{Lr}(t_1)Z_r\sin(\omega_r t) + V_{dc} \\ i_{Lr}(t) = -\dfrac{V_{cl}}{Z_r}\sin(\omega_r t) + i_{Lr}(t_1)\cos(\omega_r t) \end{cases} \tag{7.8}$$

where the characteristic impedance Z_r and the resonant angular frequency ω_r are defined as

$$Z_r = \sqrt{\frac{L_r}{3C_r + C_7}} \tag{7.9}$$

$$\omega_r = \frac{1}{\sqrt{L_r(3C_r + C_7)}} \tag{7.10}$$

The resonant capacitor voltage v_{C4} in Equation (7.8) can be rewritten as

$$v_{C4}(t) = V_{dc} + \sqrt{V_{cl}^2 + i_{Lr}(t_1)^2 Z_r^2} \cos(\omega_r t + \sigma) \tag{7.11}$$

where

$$\sigma = \arctan\left(\frac{|i_{Lr}(t_1)|Z_r}{V_{cl}}\right) \tag{7.12}$$

According to (7.11), the ZVS turn-on can be realized unless the minimum value of resonant capacitor v_{C4} is smaller than zero. Thus following inequality should be satisfied as

$$\sqrt{V_{cl}^2 + i_{Lr}(t_1)^2 Z_r^2} \geq V_{dc} \tag{7.13}$$

The ZVS condition for main switches can be further simplified as

$$i_{Lr}(t_1) \leq -\sqrt{\frac{V_{dc}^2 - V_{cl}^2}{Z_r^2}} \tag{7.14}$$

Stage 6 is the second resonant stage for the ZVS turn-on of auxiliary switch. The equivalent circuit in Figure 7.14 can be simplified as shown in Figure 7.20.

The following equations can be obtained from Figure 7.20:

$$\begin{cases} i_{Lr} + C_7 \dfrac{dv_{C7}}{dt} = C_r \dfrac{dv_{C4}}{dt} + i_a + C_r \dfrac{dv_{C6}}{dt} + C_r \dfrac{dv_{C2}}{dt} \\ v_{Lr} = L_r \dfrac{di_{Lr}}{dt} \end{cases} \tag{7.15}$$

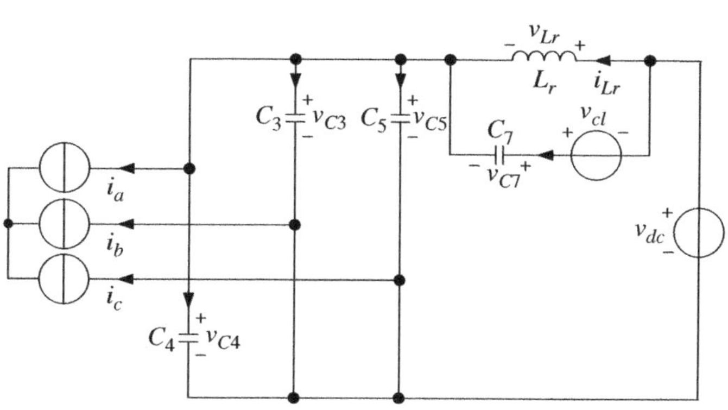

Figure 7.20 Simplified equivalent circuit of the second resonant stage.

where

$$v_{C4} = v_{C3} = v_{C5} = V_{dc} + V_{cl} - v_{C7} \tag{7.16}$$

$$V_{cl} + v_{C4} = v_{Lr} \tag{7.17}$$

By substituting (7.16) and (7.17) to (7.15), the equations of state variables, resonant inductor current i_{Lr}, and resonant capacitor voltage v_{C7} can be simplified as

$$\begin{cases} (3C_r + C_7)\dfrac{dv_{C7}}{dt} = -i_{Lr} + i_a \\ L_r\dfrac{di_{Lr}}{dt} = v_{C7} - V_{cl} \end{cases} \tag{7.18}$$

The initial condition can be derived as

$$\begin{cases} i_{Lr}(t_5) = i_a + i_x \\ v_{C7}(t_5) = V_{dc} + V_{cl} \end{cases} \tag{7.19}$$

By solving the differential Equation (7.18) with the initial condition (7.19), the expressions of resonant capacitor voltage v_{C7} and resonant inductor current i_{Lr} are obtained as

$$\begin{cases} v_{C7}(t) = V_{cl} + V_{dc}\cos(\omega_r t) - i_x Z_r \sin(\omega_r t) \\ i_{Lr}(t) = \dfrac{V_{dc}}{Z_r}\sin(\omega_r t) + i_x\cos(\omega_r t) + i_a \end{cases} \tag{7.20}$$

The resonant capacitor voltage v_{C7} can be rewritten as

$$v_{C7}(t) = V_{cl} + \sqrt{V_{dc}^2 + i_x^2 Z_r^2}\cos(\omega_r t + \tau) \tag{7.21}$$

where

$$\tau = \arctan\left(\frac{i_x Z_r}{V_{dc}}\right) \tag{7.22}$$

According to (7.21), since the clamping capacitor voltage V_{cl} is much smaller than V_{dc}, v_{C7} can reach negative. Therefore, the ZVS condition for auxiliary switch is easy to satisfy.

At the end of stage 6 at t_6, the resonant capacitor voltage v_{C7} and resonant inductor current i_{Lr} meet:

$$v_{C7}(t_6) = V_{cl} + V_{dc}\cos(\omega_r t_6) - i_x Z_r \sin(\omega_r t_6) = 0 \tag{7.23}$$

$$i_{Lr}(t_6) = \frac{V_{dc}}{Z_r}\sin(\omega_r t_6) + i_x\cos(\omega_r t_6) + i_a \tag{7.24}$$

Equations (7.23) and (7.24) can be changed as

$$[V_{dc}\cos(\omega_r t_6) - i_x Z_r \sin(\omega_r t_6)]^2 = V_{cl}^2 \tag{7.25}$$

$$[V_{dc}\sin(\omega_r t_6) + i_x Z_r \cos(\omega_r t_6)]^2 = Z_r^2[i_{Lr}(t_6) - i_a]^2 \tag{7.26}$$

By adding (7.25) with (7.26), the following equation is obtained:

$$V_{dc}^2 + i_x^2 Z_r^2 = V_{cl}^2 + Z_r^2[i_{Lr}(t_6) - i_a]^2 \tag{7.27}$$

Therefore, the resonant inductor current i_{Lr} at t_6 is derived as

$$i_{Lr}(t_6) = \sqrt{\frac{V_{dc}^2 - V_{cl}^2}{Z_r^2} + i_x^2} + i_a \tag{7.28}$$

7.3.3 Steady-state Analysis

According to the voltage second balance principle of the resonant inductor, the mean value of voltage across the resonant inductor in a switching period is zero. That is,

$$\frac{1}{T_s}\int_{t_0}^{t_0+T_s} v_{Lr}dt = 0 \tag{7.29}$$

The voltage waveform of the resonant inductor v_{Lr} is shown in Figure 7.8. Since the resonant stages during intervals (t_1–t_2) and (t_5–t_6) are relatively short in a switching period, the integrations of the v_{Lr} during intervals (t_1–t_2) and (t_5–t_6) are ignored to simplify the calculation as

$$\frac{1}{T_s}\int_{t_0}^{t_0+T_s} v_{Lr}\ dt = \frac{1}{T_s}\left(\int_{t_0}^{t_1} v_{Lr}\ dt + \int_{t_2}^{t_5} v_{Lr}\ dt + \int_{t_6}^{t_0+T_s} v_{Lr}\ dt\right) = 0 \tag{7.30}$$

By substituting the voltage of resonant inductor at different intervals as shown in Figure 7.8 to (7.30), Equation (7.30) is rewritten as

$$\frac{1}{T_s}\int_{0}^{T_s} v_{Lr}\ dt = V_{cl}(1-\lambda_7)T_s - V_{dc}\lambda_7 T_s = 0 \tag{7.31}$$

where λ_7 is the turn-off duty cycle of the auxiliary switch S_7. $\lambda_7 T_s$ can be defined as the time interval from t_2 to t_5 in Figure 7.8. The voltage of clamping capacitor V_{cl} can be obtained by simplifying (7.31) as

$$V_{cl} = \frac{\lambda_7}{1-\lambda_7}V_{dc} \tag{7.32}$$

According to the current second balance principle of clamping capacitor, the mean value of current through clamping capacitor is zero during a switching period. That is,

$$\frac{1}{T_s}\int_{t_0}^{t_0+T_s} i_{cl}\ dt = 0 \tag{7.33}$$

By ignoring two short resonant stages and two Type 1 commutation stages, (7.33) can be expanded as

$$\begin{aligned}&\int_{t_0}^{t_1} i_{cl\_stage\ 1}(t)\ dt + \int_{t_2}^{t_5} i_{cl\_stage\ 3,4,5}(t)\ dt + \int_{t_6}^{t_7} i_{cl\_stage\ 7}(t)\ dt \\ &\quad + \int_{t_8}^{t_9} i_{cl\_stage\ 9}(t)\ dt + \int_{t_{10}}^{t_0+T_s} i_{cl\_stage\ 1}(t)\ dt \approx 0\end{aligned} \tag{7.34}$$

The current through clamping capacitor i_{cl} is equal to $-i_{S7}$ as shown in Figure 7.1. Therefore, i_{cl} in stage 3, stage 4, and stage 5 is equal to zero:

$$i_{cl\_stage\ 3,4,5}(t) = 0 \tag{7.35}$$

In other stages, i_{cl} is obtained with the DC bus current i_{dc} and the resonant inductor current i_{Lr} as

$$i_{cl} = i_{Lr} - i_{dc} \tag{7.36}$$

With the initial value at t_6 in (7.28) and the deceasing rate of V_{cl}/L_r, the resonant inductor current i_{Lr} from t_6 to t_1 of the next switching period shown in Figure 7.8 can be expressed as

$$\begin{aligned} i_{Lr}(t) &= i_{Lr}(t_6) - \frac{V_{cl}}{L_r}(t - t_6) \\ &= \sqrt{\frac{V_{dc}^2 - V_{cl}^2}{Z_r^2} + i_x^2} + i_a - \frac{V_{cl}}{L_r}(t - t_6) \end{aligned} \tag{7.37}$$

Since the outputs vectors of inverter in stage 7 (t_6–t_7), stage 9 (t_8–t_9), and stage 1 (t_0–t_1, t_{10}–$t_0 + T_s$) are $\overrightarrow{V_0}$ (000), $\overrightarrow{V_1}$ (100), and $\overrightarrow{V_2}$ (110), the DC bus current i_{dc} in these stages is obtained as

$$\begin{cases} i_{dc\_stage\ 7}(t) = i_a \\ i_{dc\_stage\ 9}(t) = i_a + i_b \\ i_{dc\_stage\ 1}(t) = 0 \end{cases} \tag{7.38}$$

By substituting Equations (7.37) and (7.38) to (7.36), i_{cl} in stage 7, stage 9, and stage 1 can be derived as

$$\begin{cases} i_{cl\_stage\ 7}(t) = \sqrt{\dfrac{V_{dc}^2 - V_{cl}^2}{Z_r^2} + i_x^2} - \dfrac{V_{cl}}{L_r}(t - t_6) \\ i_{cl\_stage\ 9}(t) = \sqrt{\dfrac{V_{dc}^2 - V_{cl}^2}{Z_r^2} + i_x^2} - \dfrac{V_{cl}}{L_r}(t - t_6) - i_b \\ i_{cl\_stage\ 1}(t) = \sqrt{\dfrac{V_{dc}^2 - V_{cl}^2}{Z_r^2} + i_x^2} - \dfrac{V_{cl}}{L_r}(t - t_6) + i_a \end{cases} \tag{7.39}$$

As the equivalent vector within the time interval from t_2 to t_5 ($\lambda_7 T_s$) is zero vector $\overrightarrow{V_0}$ and the DC voltage at other stages is $V_{dc} + V_{cl}$, the soft-switching inverter can be viewed as a hard-switching inverter with DC voltage at $V_{dc} + V_{cl}$ when calculating the duration of stage 1, stage 7, and stage 9. The basic vectors in stage 1, stage 7, and stage 9 are $\overrightarrow{V_0}$, $\overrightarrow{V_1}$, and $\overrightarrow{V_2}$, respectively. Their durations, T_{V0}, T_{V1}, and T_{V2}, can be obtained as

$$\begin{cases} T_{V1} = MT_s \sin\left(\dfrac{\pi}{3} - \varphi\right) \\ T_{V2} = MT_s \sin(\varphi) \\ T_{V0} = T_s - \lambda_7 T_s - T_{V1} - T_{V2} \\ \varphi = \omega_l t = 2\pi f_l t \end{cases} \tag{7.40}$$

where M is the equivalent ZVS-SVM modulation ratio of soft-switching inverter, φ is the phase angle of reference vector, which can also be viewed as the phase angle of the vector of three-phase grid voltages if the soft-switching inverter is operating at unity power factor, ω_l is the angle frequency of grid voltages, and f_l is the line frequency of grid voltages. M can be expressed as

$$M = \frac{\sqrt{3}V_m}{V_{dc} + V_{cl}} = \frac{\sqrt{3}V_m}{V_{dc}}(1 - \lambda_7) = m(1 - \lambda_7) \tag{7.41}$$

where m is the SVM modulation ratio of hard-switching inverter and V_m is the amplitude of grid voltage. If the soft-switching inverter is operating at unity power factor, the grid currents meet:

$$\begin{cases} i_a = I_m \cos(\varphi) \\ i_b = I_m \cos(\varphi - 2\pi/3) \\ i_c = I_m \cos(\varphi + 2\pi/3) \end{cases} \tag{7.42}$$

Substituting Equations (7.35), (7.39), and (7.40) to Equation (7.34),

$$\begin{aligned}
0 &\approx \int_{t_0}^{t_1} i_{cl_stage\ 1}(t)\ dt + \int_{t_6}^{t_7} i_{cl_stage\ 7}(t)dt + \int_{t_8}^{t_9} i_{cl_stage\ 9}(t)\ dt + \int_{t_{10}}^{t_0+T_s} i_{cl_stage\ 1}(t)\ dt \\
&= \int_{t_6}^{t_6+T_{V1}} \left[\sqrt{\frac{V_{dc}^2 - V_{cl}^2}{Z_r^2} + i_x^2} - \frac{V_{cl}}{L_r}(t-t_6)\right] dt \\
&+ \int_{t_8}^{t_8+T_{V2}} \left[\sqrt{\frac{V_{dc}^2 - V_{cl}^2}{Z_r^2} + i_x^2} - \frac{V_{cl}}{L_r}(t-t_6) - i_b\right] dt \\
&+ \int_{t_{10}}^{t_{10}+T_{V0}} \left[\sqrt{\frac{V_{dc}^2 - V_{cl}^2}{Z_r^2} + i_x^2} - \frac{V_{cl}}{L_r}(t-t_6) + i_a\right] dt \\
&= \int_{t_6}^{t_6+T_s-\lambda_7 T_s} \left[\sqrt{\frac{V_{dc}^2 - V_{cl}^2}{Z_r^2} + i_x^2} - \frac{V_{cl}}{L_r}(t-t_6)\right] dt - \int_{t_8}^{t_8+T_{V2}} i_b dt + \int_{t_{10}}^{t_{10}+T_{V0}} i_a dt \\
&= \sqrt{\frac{V_{dc}^2 - V_{cl}^2}{Z_r^2} + i_x^2}(1-\lambda_7)T_s - \frac{V_{cl}}{2L_r}(1-\lambda_7)^2 T_s^2 - i_b T_{V2} + i_a(T_s - \lambda_7 T_s - T_{V1} - T_{V2})
\end{aligned} \tag{7.43}$$

With Equations (7.32), (7.40), (7.41), and (7.42), Equation (7.43) can be further simplified as

$$\begin{aligned}
0 &\approx \sqrt{\frac{V_{dc}^2 - V_{cl}^2}{Z_r^2} + i_x^2}(1-\lambda_7)T_s - \frac{V_{cl}}{2L_r}(1-\lambda_7)^2 T_s^2 - i_b T_{V2} + i_a(T_s - \lambda_7 T_s - T_{V1} - T_{V2}) \\
&= \sqrt{\frac{V_{dc}^2 - V_{cl}^2}{Z_r^2} + i_x^2}(1-\lambda_7)T_s - \frac{\lambda_7 V_{dc}}{2L_r}(1-\lambda_7)T_s^2 + i_a(1-\lambda_7)T_s \\
&- I_m M T_s\left[\cos\left(\varphi - \frac{2\pi}{3}\right)\sin(\varphi) + \cos(\varphi)\sin(\varphi) + \cos(\varphi)\sin\left(\frac{\pi}{3} - \varphi\right)\right] \\
&= \sqrt{\frac{V_{dc}^2 - V_{cl}^2}{Z_r^2} + i_x^2}(1-\lambda_7)T_s - \frac{\lambda_7 V_{dc}}{2L_r}(1-\lambda_7)T_s^2 + i_a(1-\lambda_7)T_s - \frac{\sqrt{3}}{2} m I_m (1-\lambda_7)T_s \\
&= (1-\lambda_7)T_s\left(\sqrt{\frac{V_{dc}^2 - V_{cl}^2}{Z_r^2} + i_x^2} + \frac{V_{dc}T_s}{2L_r}\lambda_7 + i_a - \frac{\sqrt{3}}{2} m I_m\right)
\end{aligned} \tag{7.44}$$

The turn-off duty cycle of the auxiliary switch λ_7 can be derived from (7.44) as

$$\lambda_7 \approx \frac{\sqrt{\dfrac{V_{dc}^2 - V_{cl}^2}{Z_r^2} + i_x^2} + i_a - \dfrac{\sqrt{3}}{2} m I_m}{\dfrac{V_{dc}T_s}{2L_r}} \tag{7.45}$$

7.3.4 Soft-switching Condition

With Equations (7.32), (7.37), and (7.45) and the durations of stage 1, stage 7, and stage 9, the resonant inductor i_{Lr} at t_1 can be calculated as

$$\begin{aligned} i_{Lr}(t_1) &= i_{Lr}(t_6) - \frac{V_{cl}}{L_r}(1-\lambda_7)T_s \\ &= \sqrt{\frac{V_{dc}^2 - V_{cl}^2}{Z_r^2} + i_x^2} + i_a - \frac{V_{dc}\lambda_7 T_s}{L_r} \\ &= \sqrt{\frac{V_{dc}^2 - V_{cl}^2}{Z_r^2} + i_x^2} + i_a - 2\left(\sqrt{\frac{V_{dc}^2 - V_{cl}^2}{Z_r^2} + i_x^2} + i_a - \frac{\sqrt{3}}{2} m I_m\right) \\ &= -\sqrt{\frac{V_{dc}^2 - V_{cl}^2}{Z_r^2} + i_x^2} - i_a + \sqrt{3} m I_m \end{aligned} \tag{7.46}$$

By substituting (7.46) to (7.14), extra added current i_x meets

$$-\sqrt{\frac{V_{dc}^2 - V_{cl}^2}{Z_r^2} + i_x^2} - i_a + \sqrt{3} m I_m \le -\sqrt{\frac{V_{dc}^2 - V_{cl}^2}{Z_r^2}} \tag{7.47}$$

The extra added current i_x can be obtained as

$$i_x \ge \sqrt{\left(\sqrt{\frac{V_{dc}^2 - V_{cl}^2}{Z_r^2}} - i_a + \sqrt{3} m I_m\right)^2 - \frac{V_{dc}^2 - V_{cl}^2}{Z_r^2}} \tag{7.48}$$

Equation (7.48) describes the ZVS condition for ZVS turn-on of main switches. The duration of stage 5 should be controlled to meet (7.48) and store enough energy in the resonant inductor. By substituting the minimum value of extra added current i_x in (7.48) to (7.14) and (7.28), the minimum resonant current at t_1 and the maximum resonant current at t_6 can be simplified as

$$\begin{cases} i_{Lr}(t_1) = -\sqrt{\dfrac{V_{dc}^2 - V_{cl}^2}{Z_r^2}} \approx -\dfrac{V_{dc}}{Z_r} \\ i_{Lr}(t_6) = \sqrt{\dfrac{V_{dc}^2 - V_{cl}^2}{Z_r^2}} + \sqrt{3} m I_m \approx \dfrac{V_{dc}}{Z_r} + \sqrt{3} m I_m \end{cases} \tag{7.49}$$

The turn-off duty cycle of the auxiliary switch λ_7 in (7.45) is further simplified as

$$\lambda_7 \approx \frac{2L_r}{V_{dc}T_s}\left(\frac{V_{dc}}{Z_r} + \frac{\sqrt{3}}{2} m I_m\right) \tag{7.50}$$

In the following paragraph, the minimum value of the extra added current i_x is used.

7.3.5 Resonant Time Comparison

The expressions of voltage across main switch S_4 during the first resonant stage in (7.11) and the voltage across auxiliary switch S_7 during the second resonant stage in (7.21) can be rewritten as

$$v_{C4}(t) = V_{dc} + v_{r1}(t) \tag{7.51}$$

$$v_{C7}(t) = V_{cl} + v_{r2}(t) \tag{7.52}$$

where the two sinusoidal terms v_{r1} and v_{r2} are expressed by

$$v_{r1}(t) = \sqrt{V_{cl}^2 + i_{Lr}(t_1)^2 Z_r^2} \cos(\omega_r t + \sigma) \approx V_{dc} \cos(\omega_r t + \sigma) \tag{7.53}$$

$$v_{r2}(t) = \sqrt{V_{dc}^2 + i_x^2 Z_r^2} \cos(\omega_r t + \tau) \tag{7.54}$$

It is observed that the two sinusoidal terms have the same resonant frequency. The amplitude of v_{r2} is larger than that of v_{r1} if the extra added current i_x is nonzero. Since the initial values of both v_{C4} and v_{C7} are $V_{dc} + V_{cl}$ while their final value after resonant stage are zero, the initial values and final values of the two resonant terms v_{r1} and v_{r2} can be obtained as

$$\begin{cases} v_{r1}(0) = V_{cl} \\ v_{r1}(T_{r1}) = -V_{dc} \end{cases} \tag{7.55}$$

$$\begin{cases} v_{r2}(0) = V_{dc} \\ v_{r2}(T_{r2}) = -V_{cl} \end{cases} \tag{7.56}$$

where T_{r1} and T_{r2} are the resonant time of the two resonant stages. If the extra add current i_x is assumed to be zero, the amplitudes of the two resonant terms v_{r1} and v_{r2} are the same and equal to V_{dc}. Certainly, the two resonant times T_{r1} and T_{r2} are identical due to the symmetry of the cosine function. Since the extra add current i_x always exists for the requirement of ZVS, the amplitude of v_{r2} is always larger than that of v_{r1}. As a result, the actual resonant time T_{r2} will be smaller than its value with zero extra added current. The resonant time of the first resonant stage T_{r1} will be always larger than the resonant time of the second resonant stage T_{r2}.

With (7.51), the resonant time of the first resonant stage T_{r1} can be solved by the following equation:

$$v_{C4}(T_{r1}) = V_{dc} + V_{dc} \cos(\omega_r T_{r1} + \sigma) = 0 \tag{7.57}$$

T_{r1} can be derived as

$$T_{r1} = \frac{1}{\omega_r}[\arccos(-1) - \sigma] = \frac{1}{\omega_r}(\pi - \sigma) \tag{7.58}$$

where the phase angle σ can be simplified as

$$\sigma = \arctan\left(\frac{|i_{Lr}(t_1)|Z_r}{V_{cl}}\right) = \arctan\left(\frac{V_{dc}}{V_{cl}}\right) = \arctan\left(\frac{1-\lambda_7}{\lambda_7}\right) \tag{7.59}$$

Since the turn-off duty cycle of auxiliary switch λ_7 is usually designed smaller than 0.1, the phase angle σ is very close to $\pi/2$. As a result, the resonant time of the first resonant stage T_{r1} in (7.58) can be derived as

$$T_{r1} \approx \frac{\pi}{2\omega_r} \tag{7.60}$$

With (7.52), the resonant time of the second resonant stage T_{r2} can be solved by the following equation:

$$v_{C7}(T_{r2}) = V_{cl} + \sqrt{V_{dc}^2 + i_x^2 Z_r^2} \cos(\omega_r T_{r2} + \tau) = 0 \tag{7.61}$$

T_{r2} is expressed as

$$T_{r2} = \frac{1}{\omega_r}\left[\arccos\left(\frac{-V_{cl}}{\sqrt{V_{dc}^2 + i_x^2 Z_r^2}}\right) - \tau\right] \tag{7.62}$$

7.4 Implementation of ZVS-SVM

7.4.1 Regulation of Short Circuit Stage

As shown in Figure 7.8, the width of the short circuit stage T_{sc} can be calculated as

$$T_{sc} = t_5 - t_4 = \frac{L_r}{V_{dc}} i_x \tag{7.63}$$

As described in Section 7.3.1, stage 3 (t_2–t_3) is the diode freewheeling stage. The resonant inductor i_{Lr} may still have residual energy to keep current i_{Lr} negative, if the calculation of T_{sc} is not that precise due to the existence of the tolerance and the nonlinear characteristics of the parameters. The duration of this stage T_{stage3} can be calculated as

$$T_{stage3} = t_3 - t_2 = \frac{L_r}{V_{dc}} |i_{Lr}(t_2)| \tag{7.64}$$

Stage 4 (t_3–t_4) is the Type 2 commutation stage. As the waveforms shown in Figure 7.8, the resonant inductor currents at t_3 and t_4 are equal to zero and i_a, respectively. The duration of this stage T_{stage4} can be calculated as

$$T_{stage4} = t_4 - t_3 = \frac{L_r}{V_{dc}} i_a = \frac{L_r}{V_{dc}} I_m \cos(\omega_l t) \tag{7.65}$$

As shown in (7.65), T_{stage4} varies with the phase angle $\omega_l t$ in different switching periods. Obviously, the stage durations T_{stage3} and T_{stage4} are both difficult to identify in each switching period. As a result, the short circuit pulse shown in Figure 7.8 is hard to be added to the main switches precisely. Besides, the bus voltage across the main bridges is already clamped at zero from t_2 after the first resonant stage. By considering those two reasons, the instant of adding short circuit pulse to the main switches can be moved up from t_4 to t_2 and aligned with the turn-on instant of the main switch. The enlarged width of short circuit pulse will not impact the operations of diode freewheeling and current commutation. With the new timing of the short circuit pulse, the driving sequences in Figure 7.8 can be updated as shown in Figure 7.21. Therefore, the width of the short circuit pulse T_{sc} can be modified as

$$T_{sc} = t_5 - t_2 \approx T_{off7} - T_{r1} - T_{r2} = \lambda_7 T_s - T_{r1} - T_{r2} \tag{7.66}$$

7.4.2 Implementation in Digital Controller

In the following paragraph, we will introduce the implementation of ZVS-SVM in practical application with digital controller such as the Texas Instruments (TI) digital signal processors (DSPs). Once the reference sector $\vec{V}_{ref}$ is calculated by the control loop, the location and the magnitude of

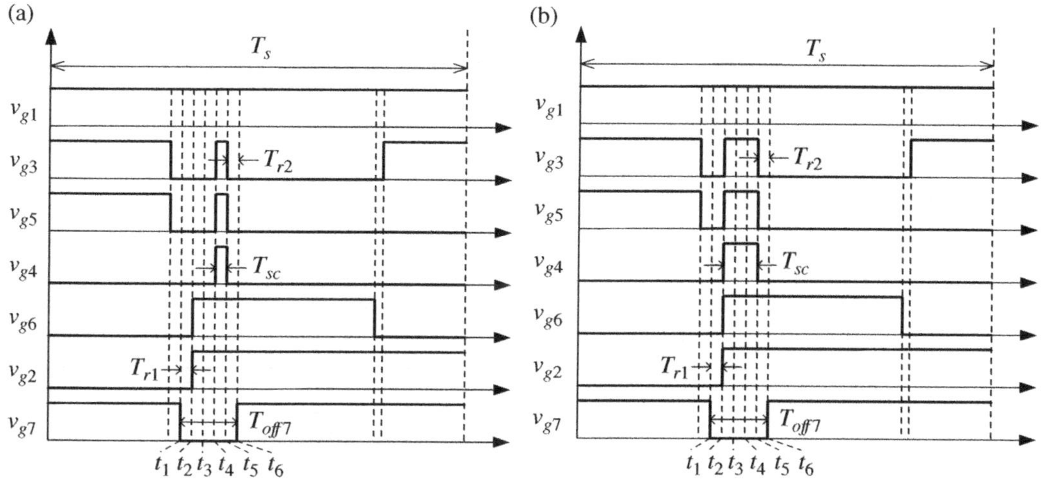

Figure 7.21 Modification of short circuit pulse: (a) previous driving sequences and (b) updated driving sequences.

the reference sector can also be determined. With the modulation sequences of the ZVS-SVM in Table 7.1 and Equation (7.40), the used vectors and their durations, T_{V0}, T_{V1}, and T_{V2}, can be easily obtained.

Take sector 1-1 as an example to introduce the procedures of driving sequences derivation, and the related waveforms are shown in Figure 7.22a. In this example, one count up timer with period at T_s, four PWM modules (PWM1, PWM2, PWM3, PWM4) each with two counter-compare channels (channel A and channel B) and two counter-compare values (COMPxA, COMPxB, x = 1,2,3,4) are used. Channel A's counter-compare value COMPxA is not allowed to be larger than channel B's counter-compare value COMPxB.

First, the eight counter-compare values for four PWM modules based on T_{V0}, T_{V1}, and T_{V2} should be calculated.

Next, with the specific action scheme applying to the count up timer and eight counter-compare values, the switching functions can be obtained. The switching functions of three-phase bridges are defined as s_1, s_2, and s_3, while the switching function for short circuit pulse and auxiliary switch is defined as s_4. In sector 1-1, when the timer is equal to channel A's counter-compare value COMPxA, the switching functions of three-phase bridges will be cleared to low level, and when the timer is equal to channel B's counter-compare value COMPxB, the switching functions of three-phase bridges will be set to high level. However, when the timer is equal to channel A's counter-compare value COMPxA, the switching function of the short circuit pulse and auxiliary switch s_4 will be set to high level, and when the timer is equal to channel B's counter-compare value COMPxB, the switching function of the short circuit pulse and auxiliary switch will be cleared to low level.

Later, with inverting and turn-on delay, each switching function will be expanded into two complementary primary driving sequences. For example, with inverting and turn-on delay T_{d1}, switching function s_1 is expanded into s_{1a} and s_{1b}. All the time delays for the primary driving sequences of three-phase bridges are set as T_{d1}. However, the time delay for the primary short circuit pulse s_{4a} is T_{r1}, while it is T_{r2} for the primary driving sequence of auxiliary switch s_{4b}.

Finally, through the OR logic operation, the primary short circuit pulse s_{4a} is added to the six primary driving sequences for three-phase bridges, s_{1a}, s_{1b}, s_{2a}, s_{2b}, s_{3a}, and s_{3b}. Then, the final driving sequences for three-phase bridges and the auxiliary switch, v_{g1}, v_{g4}, v_{g3}, v_{g6}, v_{g5}, v_{g2}, and v_{g7}, are derived.

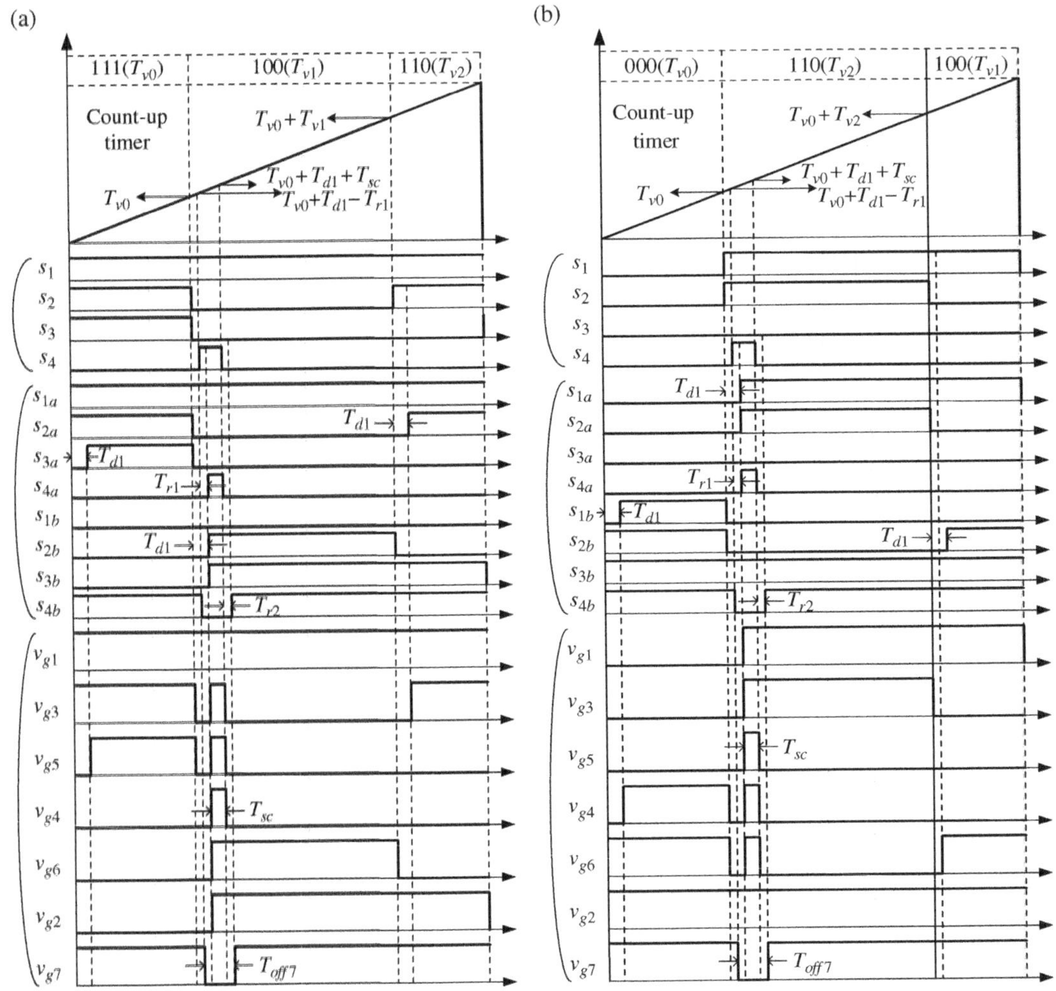

Figure 7.22 Driving sequences derivation: (a) sector 1-1 and (b) sector 1-2.

With the similar method, the driving sequences in sector 1-2 can be derived as shown in Figure 7.22b. However, the action scheme in sector 1-2 is different from that in sector 1-1. In sector 1-2, when the timer is equal to channel A's counter-compare value COMPxA, the switching functions of the three-phase bridges will be set to high level, and when the timer is equal to channel B's counter-compare value COMPxB, the switching function of the three-phase bridges will be cleared to low level. The differences in the action scheme are caused by the different zero vector utilization in sector 1-1 and sector 1-2. In sectors with zero vector $\vec{V}_7(111)$, their action scheme will be the same as that in sector 1-1, while in sectors with zero vector $\vec{V}_0(000)$, their action scheme will be the same as that in sector 1-2. The action scheme for the switching function of the short circuit pulse and auxiliary switch in sector 1-2 is the same to that in sector 1-1.

Table 7.2 Counter-compare values for three-phase bridges.

Sector	COMP1A	COMP1B	COMP2A	COMP2B	COMP3A	COMP3B
1-1	T_s	T_s	T_{V0}	$T_{V0}+T_{V1}$	T_{V0}	T_s
1-2	T_{V0}	T_s	T_{V0}	$T_{V0}+T_{V2}$	T_s	T_s
2-1	T_{V0}	$T_{V0}+T_{V1}$	T_{V0}	T_s	T_s	T_s
2-2	T_{V0}	$T_{V0}+T_{V2}$	T_s	T_s	T_{V0}	T_s
3-1	T_{V0}	T_s	T_s	T_s	T_{V0}	$T_{V0}+T_{V1}$
3-2	T_s	T_s	T_{V0}	T_s	T_{V0}	$T_{V0}+T_{V2}$
4-1	T_s	T_s	T_{V0}	$T_{V0}+T_{V1}$	T_{V0}	T_s
4-2	T_{V0}	T_s	T_{V0}	$T_{V0}+T_{V2}$	T_s	T_s
5-1	T_{V0}	$T_{V0}+T_{V1}$	T_{V0}	T_s	T_s	T_s
5-2	T_{V0}	$T_{V0}+T_{V2}$	T_s	T_s	T_{V0}	T_s
6-1	T_{V0}	T_s	T_s	T_s	T_{V0}	$T_{V0}+T_{V1}$
6-2	T_s	T_s	T_{V0}	T_s	T_{V0}	$T_{V0}+T_{V2}$

According to the procedures of driving sequences derivation of sector 1-1 as shown in Figure 7.22a, the eight counter-compare values, COMP1A, COMP1B, COMP2A, COMP2B, COMP3A, COMP3B, COMP4A, and COMP4B, can be deduced as

$$\begin{aligned}
&\text{COMP1A} = T_s\\
&\text{COMP1B} = T_s\\
&\text{COMP2A} = T_{V0}\\
&\text{COMP2B} = T_{V0} + T_{V1}\\
&\text{COMP3A} = T_{V0}\\
&\text{COMP3B} = T_s\\
&\text{COMP4A} = T_{V0} + T_{d1} - T_{r1}\\
&\text{COMP4B} = T_{V0} + T_{d1} + T_{sc}
\end{aligned} \tag{7.67}$$

Obviously, if the starting vector in each switching period is the zero vector, the counter-compare values in PWM4, COMP4A, and COMP4B will be the same in entire 12 sectors as shown in (7.67). With the similar derivation procedures, the counter-compare values for three-phase bridges in entire 12 sectors can be summarized as shown in Table 7.2.

7.4.3 Control Block Diagram with ZVS-SVM

The control block diagram of the CAC ZVS three-phase inverter is shown in Figure 7.23. The control loop is based on the synchronous rotating coordinate system. The output of the DC voltage control loop is the reference of active current i_d; the reference of reactive current i_q is zero. Through grid voltage feed forward and inductor current decoupling control, the reference vector $\vec{V}_{ref}$ can be obtained. The next steps include the vectors selection, vectors duration calculation, the compare values calculation, the counter-compare and action, the inverting and turn-on delay, and the OR logic operation, which have been explained in the last section.

7.5 Prototype Design

7.5.1 Specifications of a 30-kW Inverter

A 30 kW CAC ZVS three-phase grid inverter prototype was designed and implemented. The specifications of the inverter are listed in Table 7.3.

7.5.2 Parameter Design

The resonant parameters in soft-switching SiC inverter include resonant inductor L_r, resonant capacitor C_r for each three-phase main switch, and resonant capacitor C_7 for auxiliary switch. They are designed to meet the following requirements.

7.5.2.1 Requirement of Diode Reverse Recovery Suppression

The resonant inductance is required to suppress the reverse recovery process of the diode in stage 4. Therefore, the current slop in resonant inductor di_{Lr}/dt during the turn-off process of the diodes should meet the following inequality:

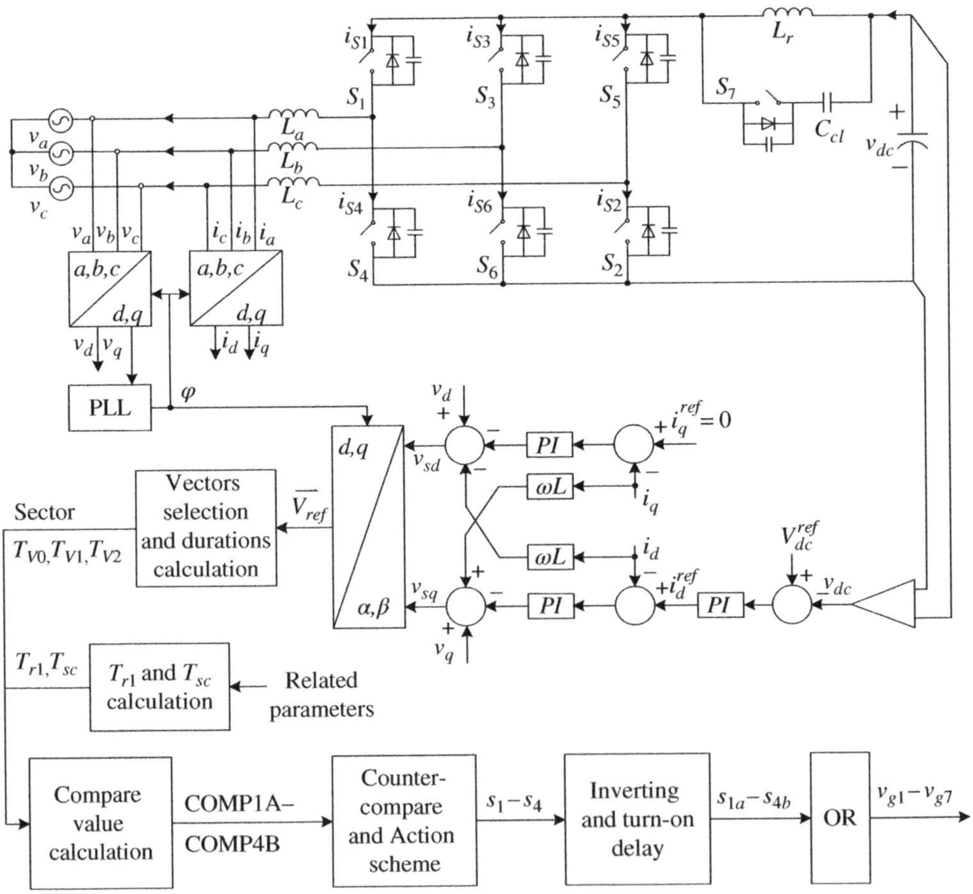

Figure 7.23 Control block diagram of CAC ZVS three-phase inverter.

Table 7.3 Specifications of the ZVS inverter.

Parameters	Descriptions	Value
P	Output power	30 kW
v_a, v_b, v_c	Phase voltage of the grid	220 Vrms
f_l	Frequency of the grid voltage	50 Hz
f_s	Switching frequency	16 kHz
L_a, L_b, L_c	Filter inductance	0.77 mH
V_{dc}	DC source voltage	600 V

$$\frac{di_{Lr}}{dt} = \frac{V_{dc}}{L_r} \leq 100 \text{ A } \mu\text{s}^{-1} \tag{7.68}$$

With the parameter in Table 7.3, the resonant inductance should satisfy

$$L_r \geq 6 \ \mu\text{H} \tag{7.69}$$

7.5.2.2 Requirement of Voltage Stress

Conventionally, the off duty of auxiliary switch λ_7 is designed smaller than 0.1 to avoid high voltage stress on all switches. With Equation (7.50), the resonant parameters are requirement to meet

$$\lambda_7 \approx \frac{2L_r}{V_{dc}T_s}\left(\frac{V_{dc}}{Z_r} + \frac{\sqrt{3}}{2}mI_m\right) \leq 0.1 \tag{7.70}$$

7.5.2.3 Requirement of reducing turn-off loss in auxiliary switch

Since the turn-on loss of main switches has been eliminated in the CAC ZVS three-phase inverter, the turn-off loss becomes the main source of the switching loss. The reduction of turn-off loss in main switches will be introduced in the following section. Here, we focus on the reduction of turn-off loss in auxiliary switch.

As shown in Figure 7.9, the turn-off current of auxiliary switch at t_1 is equal to the resonant inductor current. With equation (7.49), the turn-off current i_{S7} at t_1 can be derived as

$$i_{S7}(t_1) = -i_{Lr}(t_1) = \frac{V_{dc}}{Z_r} \tag{7.71}$$

It is observed that the turn-off current depends on the resonant impedance Z_r.

According to the input and output parameters in Table 7.3, Insulated gate bipolar transistor (IGBT) module SKM200GB12T4 at 1200 V/ 200 A rating is selected for all seven switching devices including the six three-phase main switches and the auxiliary switch. In order to reduce the turn-off loss in auxiliary switch, the turn-off current in auxiliary switch is limited to be smaller than $0.25I_m$ as follows:

$$\frac{V_{dc}}{Z_r} \leq 0.25I_m \tag{7.72}$$

7.5.2.4 Requirement of Minimum Resonant Capacitance

Since all seven switching devices can be turned on under ZVS conditions, the turn-on losses of the switches can be eliminated. Besides, all switches are turned off with the buffering of the resonant capacitors. The turn-off losses reduction depends on the capacitance of the resonant capacitors. The turn-off losses of IGBT module SKM200GB12T4 with paralleled buffer capacitor C_b are demonstrated in Figure 7.24. This figure implies that when C_b is 1.1 nF, the turn-off loss is reduced by 8%; when C_b is 2.4 nF, the turn-off loss is reduced by 20%; when C_b is 4.7 nF, the turn-off loss is reduced by 30%; when C_b is 10.2 nF, the turn-off loss is reduced by 50%; and when C_b is 21.9 nF, the turn-off loss is reduced by 60%. In order to obtain a significant turn-off loss reduction, the capacitance of the resonant capacitor for each main switch beyond 2.4 nF is preferred.

7.5.2.5 Requirement of Resonant Time

As described in Section 7.3.5, the first resonant time T_{r1} is larger than the second resonant time T_{r2}. In order to maintain the power quality, the resonant time T_{r1} in (7.60) is designed to be smaller than 2.5% of the switching period:

$$T_{r1} \approx \frac{\pi}{2\omega_r} = \frac{\pi\sqrt{L_r(3C_r + C_7)}}{2} \leq 0.025T_s \tag{7.73}$$

With the parameters in Table 7.3, for satisfying the five requirements mentioned previously, the suitable region for selecting resonant parameters is obtained as shown in Figure 7.25. The dark region in Figure 7.25 is used to select the appropriate resonant inductance and resonant capacitance.

The parameters of the black point shown in Figure 7.25 are chosen as the resonant parameters. The resonant inductance L_r is 30 μH; the total resonant capacitance is 9.9 nF; the resonant capacitance for each main switch is 3.3 nF; no external paralleled capacitor is added to auxiliary switch. The resonant impedance Z_r can be calculated as 55 Ω.

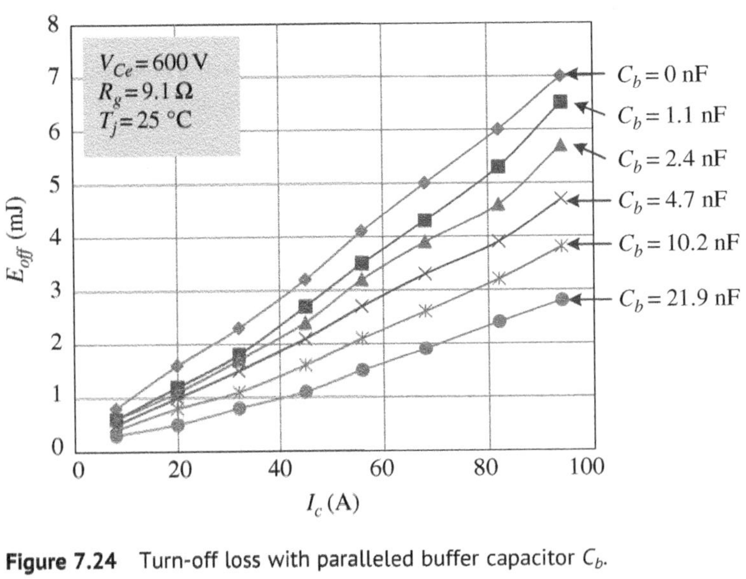

Figure 7.24 Turn-off loss with paralleled buffer capacitor C_b.

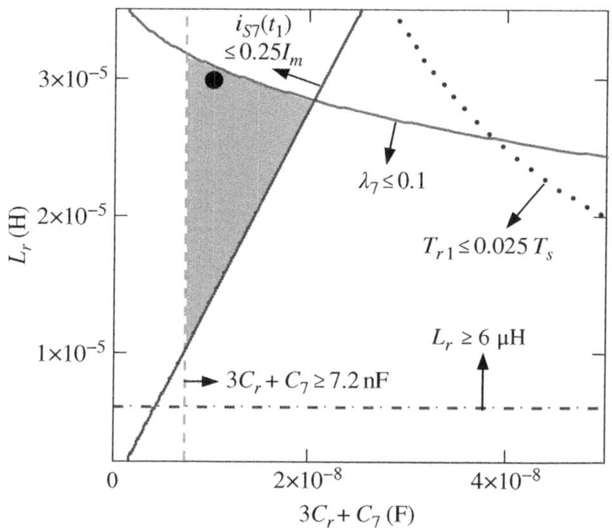

Figure 7.25 Region for selecting resonant parameters.

7.5.3 Experiment Platform and Testing Results

As shown in Figure 7.26, a 30 kW ZVS inverter prototype was built to verify the designed parameters and the proposed modulation scheme.

The structure of the bus bar and switching bridges is shown in Figure 7.27. IGBT modules SKM200GB12T4 are selected for bridge switches and auxiliary switch. In the circuit, capacitors with 1200 V rating are utilized as the resonant capacitors for six main switches. Five pieces of 22 μF EPCOS film capacitors are connected in parallel to operate as the clamping capacitor C_{cl}.

The prototype of the resonant inductor is shown in Figure 7.28. Two sets of magnetic powder cores with relative permeability of 26μ are utilized as the core of the resonant inductor. The winding is composed of 2000 strands of litz wire with diameter of 0.1 mm.

A three-phase filter inductor is implemented in the prototype to meet the grid-connected power quality, which is shown in Figure 7.29. The magnetic core of this inductor is made of silicon steel laminations.

The experimental circuit is shown in Figure 7.30. The DC source voltage is generated and regulated by rectifier bridge.

The grid-connected voltage and current are demonstrated in Figure 7.31. The measured current THD is 2.9% under 30 kW condition.

The voltage and current of the bridge switch S_4 are shown in Figure 7.32. In this figure, current begins to increase after the voltage drops to zero. Therefore, the bridge switches turn on if ZVS condition is realized.

The voltage and current of the auxiliary switch S_7 are shown in Figure 7.33. When S_7 turns on, the current through the module goes negative, which means the freewheeling diode begins to conduct earlier than the action switch. Therefore, the ZVS condition for the auxiliary switch is realized.

The voltage across the clamping capacitor C_{cl} and the current through resonant inductor L_r are shown in Figure 7.34. The average voltage across C_{cl} is 70 V. The voltage ripple across clamping capacitor is 5 V, which is less than 10% of the mean voltage. The peak value of i_{Lr} is 125 A.

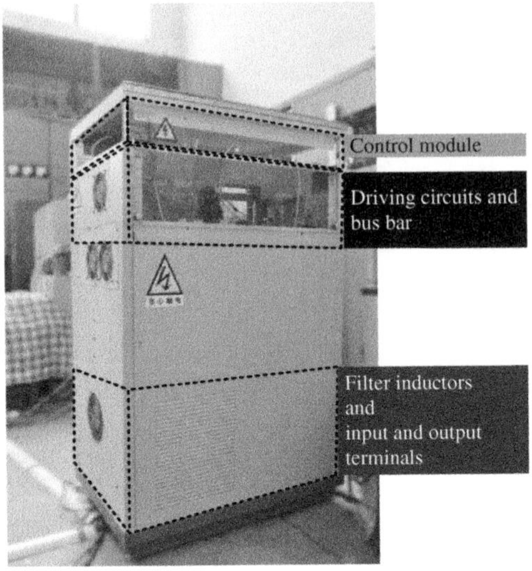

Figure 7.26 30 kW ZVS inverter prototype.

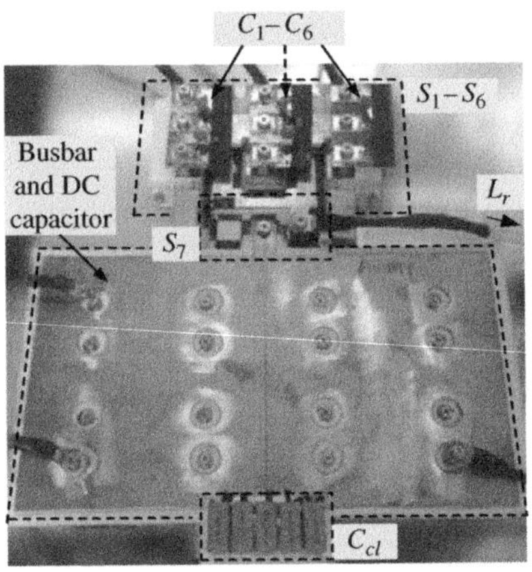

Figure 7.27 Bus bar and structure 30 kW ZVS inverter prototype.

Figure 7.28 Prototype of resonant inductor.

Figure 7.29 Filter inductor.

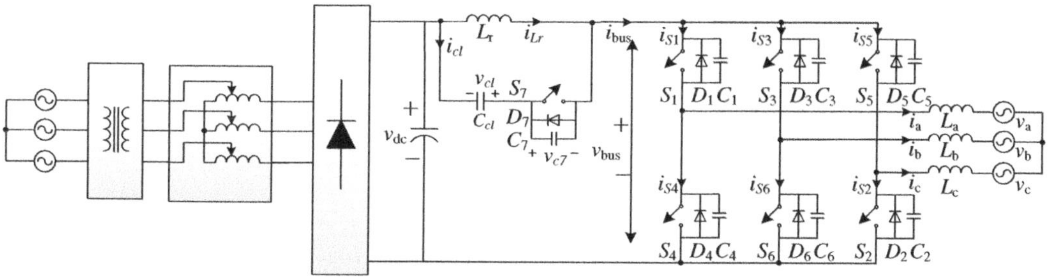

Figure 7.30 Experimental circuit.

Figure 7.31 Output current and grid voltage.

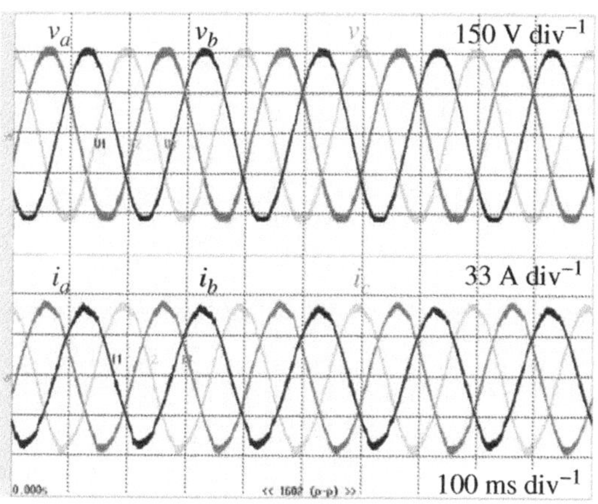

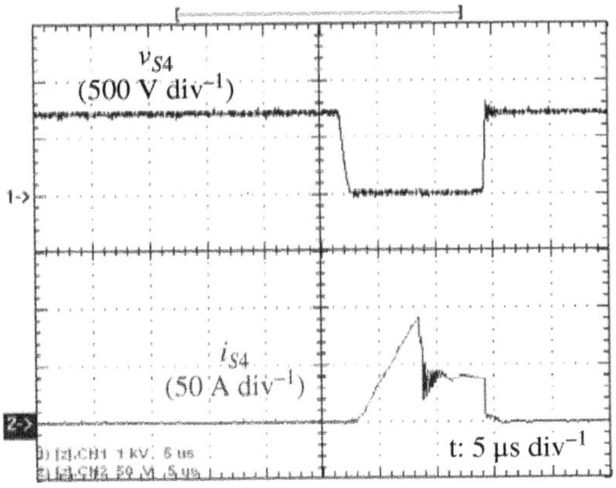

Figure 7.32 Voltage and current waveforms of the bridge switch.

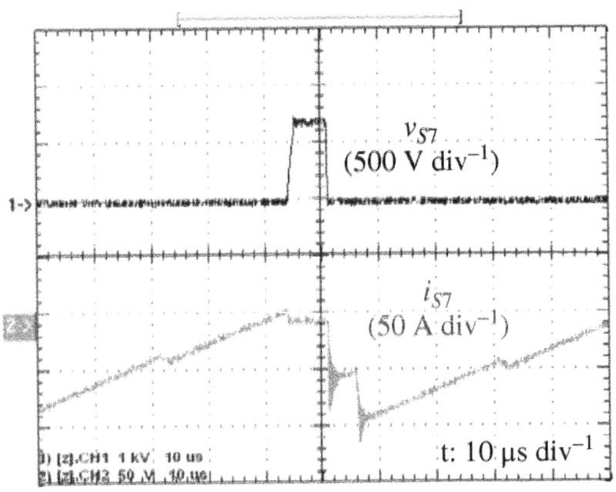

Figure 7.33 Voltage and current waveforms of the auxiliary switch.

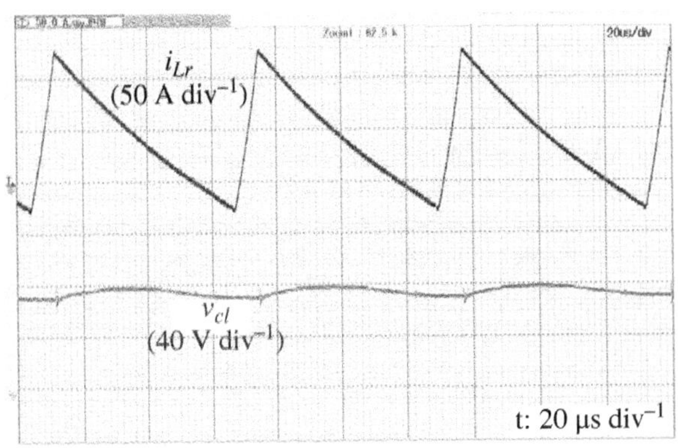

Figure 7.34 Voltage across C_{cl} and current through L_r.

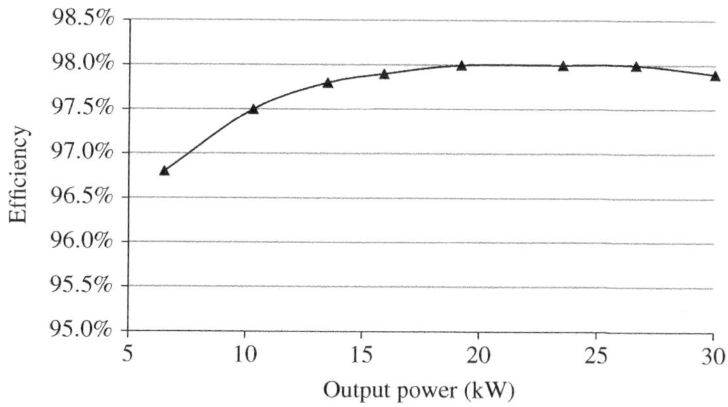

Figure 7.35 Measured efficiency vs. output power.

The conversion efficiency of the ZVS inverter is measured by power analyzer WT1800. And the results are shown in Figure 7.35. The maximum efficiency is 98%.

7.6 Summary

CAC ZVS three-phase inverter is introduced in this chapter. CAC inverter can realize ZVS operation for all switching devices and suppress the antiparallel diode reverse recovery of all switching devices. The ZVS-SVM is derived based on the different arrangement of the vectors from traditional SVM. To realize the ZVS condition, another short circuit stage is added to charge enough energy for the resonant inductor. The switching voltage stress of all switching devices is less than 1.1 times of the DC-link voltage. The implementation of ZVS-SVM with digital controller is also provided. A 30-kW prototype has been built to verify the circuit and ZVS-SVM control.

References

1 D. M. Divan and G. Skibinski, "Zero-switching-loss inverters for high-power applications," *IEEE Trans. Ind. Appl.*, vol. 25, no. 4, pp. 634–643, Jul./Aug. 1989.

2 M. Mezaroba, D. C. Martins, and I. Barbi, "A ZVS PWM three-phase inverter with active clamping technique using the reverse recovery energy of the diodes," in *Proc. IEEE Conf. Power Electron. Spec.*, pp. 689–694, 2004.

3 R. Li, Z. Ma, and D. Xu, "A ZVS grid-connected three-phase inverter," *IEEE Trans. Power Electron.*, vol. 27, no. 8, pp. 3595–3604, Aug. 2012.

4 R. Li and D. Xu, "A zero-voltage switching three-phase inverter," *IEEE Trans. Power Electron.*, vol. 29, no. 3, pp. 1200–1210, 2014.

5 D. Xu, B. Feng, and R. Li, "A zero voltage switching SVM (ZVS-SVM) controlled three-phase boost rectifier," *IEEE Trans. Power Electron.*, vol. 22, no. 3, pp. 978–986, May 2007.

6 Z. Ma, D. Xu, R. Li, C. Du, and X. Zhang, "A novel DC-side zero-voltage switching (ZVS) three-phase boost PWM rectifier controlled by an improved SVM method," *IEEE Trans. Power Electron.*, vol. 27, no. 11, pp. 4391–4408, Nov. 2012.

8

Loss Analysis and Optimization of a Zero-voltage-switching Inverter

In this chapter, the loss model of compound active-clamping (CAC) zero-voltage-switching (ZVS) three-phase grid inverter is thoroughly derived as well as the dimension model of the critical components such as the resonant inductor, the filter inductors, the clamping capacitors, and the heat sink. Based on the loss and dimension models, an optimized design methodology using mathematical optimization technique for a CAC ZVS three-phase grid inverter with objectives of both high efficiency and miniaturization is introduced.

8.1 Introduction

The soft-switching technique requires additional components such as capacitors, inductors, and auxiliary switches. In order to evaluate the effectiveness of ZVS space vector modulation (ZVS-SVM) technique, it is necessary to investigate the loss model of key components in the CAC ZVS three-phase grid inverter. In addition to the conversion efficiency, power density is another metric to evaluate the performance of soft-switching technique. Therefore, the dimension models of the critical components such as the resonant inductor, the filter inductors, the clamping capacitors, and the heat sink are required.

With the loss models and dimension models, it is still difficult to design an optimal soft-switching three-phase inverter with both high efficiency and high power density because there are so many parameters such as semiconductor power devices, resonant parameters, switching frequency, etc. Mathematical optimization technique provides an effective way to allow all design variables and constrained conditions to be considered simultaneously. With the aid of modern computers' high processing speed, the optimum design results can be computed in short time, and particularly, without subjective influence of the designer.

In this chapter, first the basic operating principles of the CAC ZVS three-phase grid inverter are reviewed to obtain the key waveforms of the switching devices and resonant components. Second, the loss and dimension models of the components in the CAC ZVS three-phase grid inverter are derived. With these models, the design issue is transformed into searching the minimum solution of a constrained nonlinear multivariable function. Finally, corresponding experimental results on a 30 kW CAC ZVS inverter prototype are provided.

Soft-Switching Technology for Three-phase Power Electronics Converters, First Edition.
Dehong Xu, Rui Li, Ning He, Jinyi Deng, and Yuying Wu.

8.2 Basic Operation Principle of the CAC ZVS Inverter

In order to build the loss models of the ZVS-SVM three-phase grid inverter as shown in Figure 8.1, the basic operation principles will be briefly introduced in this section [1].

It is assumed that the inverter operates under unity power factor; the current and voltage waveforms in a line cycle are demonstrated in Figure 8.2a. According to SVM, a line cycle is divided into six sectors, from sector 1 to sector 6. Considering the current direction of different phases, each sector is divided into two subsectors. Therefore, a line cycle is divided into 12 sectors as shown in Figure 8.2b.

For instance, in sector 1-1, grid currents $i_a > 0$, $i_b < 0$, and $i_c < 0$. The nonzero vectors in this sector are $\overrightarrow{V_1}(100)$, $\overrightarrow{V_2}(110)$ with selected zero vector $\overrightarrow{V_7}(111)$. For convenience, the antiparalleled diodes of switches $S_1 - S_7$ are named as $D_1 - D_7$, respectively. Since the switching frequency is much higher than the line frequency and the ripple current through the filter inductors is relatively small, the grid currents in a switching period can be treated as constant. The switching period in sector 1-1 of the ZVS inverter includes several stages. The corresponding key waveforms are shown in Figure 8.3.

8.2.1 Operation Stage Analysis

Stage 1 ($t_0 - t_1$) is the first steady stage. In this stage, the inverter outputs vector $\overrightarrow{V_7}(111)$. The equivalent circuit of this stage is shown in Figure 8.4. The current direction of each phase indicates that S_1, D_3, and D_5 are in conducting state. Auxiliary switch S_7 is in conducting state in this stage, and the resonant inductor L_r is charged by the clamping capacitor C_{cl}. The capacitance of C_{cl} is large enough that V_{cl}, the voltage across C_{cl}, can be treated as constant during the switching period. Therefore, the current through L_r changes linearly at the rate of $-V_{cl}/L_r$. In this stage, v_{bus} is clamped to be $V_{dc} + V_{cl}$ and $i_{S7} = -i_{Lr}$.

Stage 2 ($t_1 - t_2$) is the first resonant stage. At t_1, S_7 is turned off. Oscillation between the resonant inductor L_r and the resonant capacitors C_2, C_4, C_6, and C_7 is provoked. The equivalent circuit of this stage is shown in Figure 8.5. During this stage, v_{bus} decreases while v_{C7} increases. With the help of the resonant capacitors, S_7 is turned off under ZVS condition. This stage ends at t_2, when v_{bus} resonates to zero and v_{C7} reaches $V_{dc} + V_{cl}$. At around t_2, i_{Lr} reaches its minimum value i_{Lr_min}.

Stage 3 ($t_2 - t_3$) is the diode freewheeling stage. After t_2, v_{bus} equals zero. The equivalent circuit of this stage is shown in Figure 8.6. After the first resonance, L_r still has residual energy to keep current

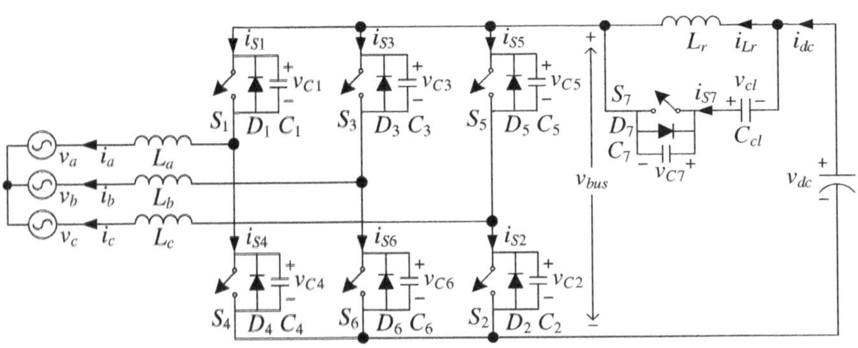

Figure 8.1 Topology of the CAC ZVS three-phase grid inverter. Source: Modified from Li and Xu [1].

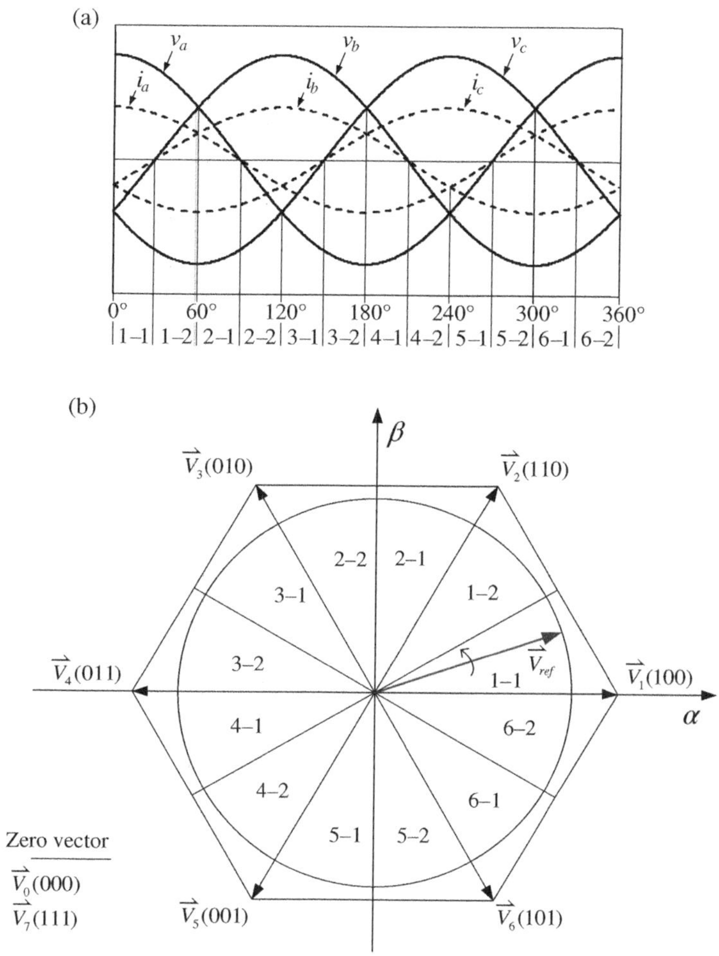

Figure 8.2 Sectors of space vector diagram.

i_{Lr} negative. D_4, D_6, and D_2 conduct in freewheeling mode. Thus, S_6 and S_2 can be turned on during this stage under ZVS condition. i_{Lr} increases with a rate of V_{dc}/L_r in this stage.

Stage 4 ($t_3 - t_4$) is the Type 2 commutation stage. At t_3, i_{Lr} increases to zero. Currents in phase-b and phase-c begin to transfer from D_3 and D_5 to S_6 and S_2, respectively. The equivalent circuit of this stage is shown in Figure 8.7. During this stage, the changing rate of i_{S3} and i_{S5} is restrained to a small value by L_r so that the reverse recovery losses on D_3 and D_5 can be eliminated. i_{S3} and i_{S5} reduce to zero by time t_4.

Stage 5 ($t_4 - t_5$) is the short-circuit stage. At t_4, S_4, S_3, and S_5 are turned on to construct short-circuit paths. The equivalent circuit of this stage is shown in Figure 8.8. L_r is charged during this stage to store enough energy to fulfill the ZVS condition for the next switching period. During this stage, because v_{bus} equals zero, S_4, S_3, and S_5 are turned on under ZVS condition.

Stage 6 ($t_5 - t_6$) is the second resonant stage. After enough energy has been charged to resonant inductor L_r, S_4, S_3, and S_5 are turned off at t_5. With the buffer of the resonant capacitors, S_4, S_3, and S_5 are turned off under ZVS conditions. Another resonance occurs among L_r, C_3, C_4, C_5, and C_7. The equivalent circuit of this stage is shown in Figure 8.9. During this resonance, v_{bus} increases and v_{C7}

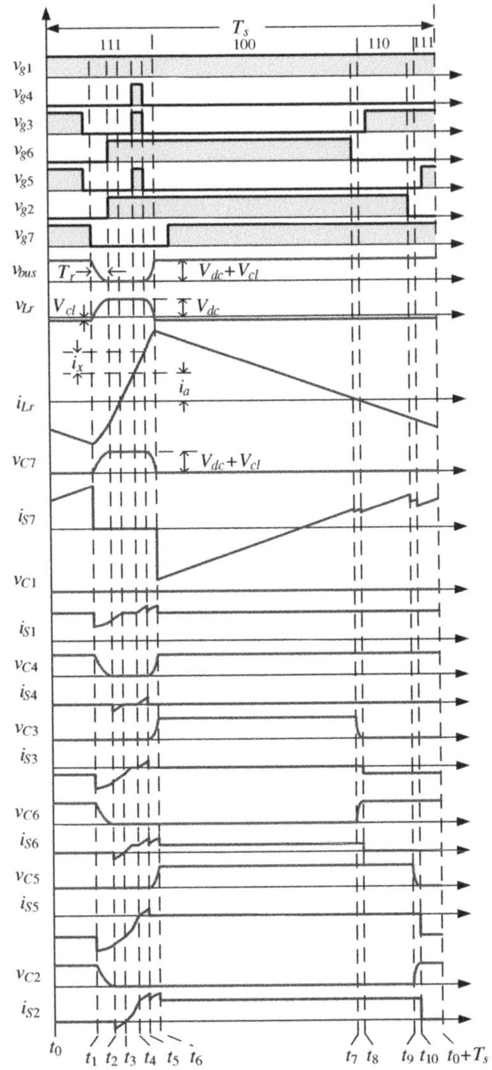

Figure 8.3 Key waveforms in sector 1-1.

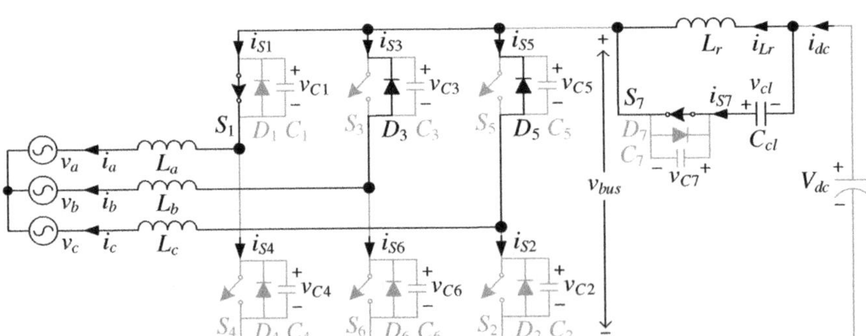

Figure 8.4 Equivalent circuit of the first steady stage.

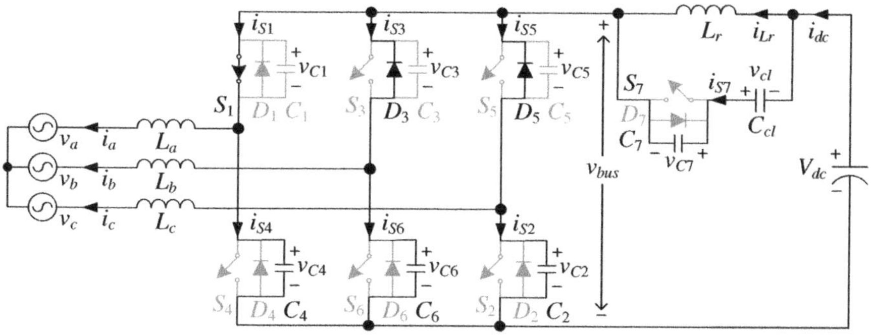

Figure 8.5 Equivalent circuit of the first resonant stage.

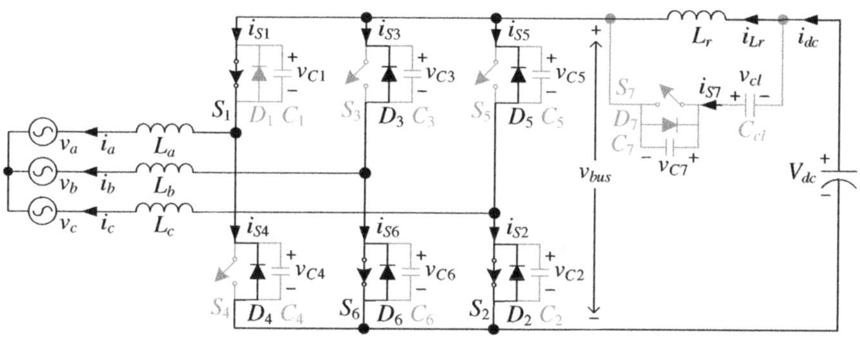

Figure 8.6 Equivalent circuit of the diode freewheeling stage.

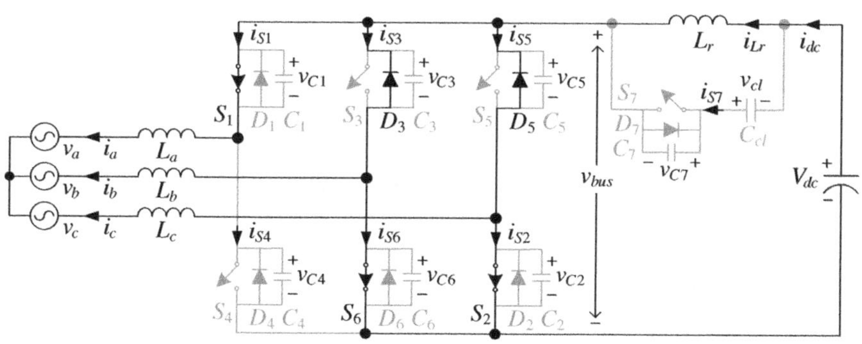

Figure 8.7 Equivalent circuit of the Type 2 commutation stage.

decreases. At time t_6, v_{bus} resonances to $V_{dc} + V_{cl}$, while v_{C7} reduces to zero. At around t_6, i_{Lr} reaches its maximum value i_{Lr_max}.

Stage 7 ($t_6 - t_7$) is the second steady stage. The inverter outputs vector $\overrightarrow{V_1}$ (100) in this stage. After the second resonance, S_7 can be turned on under ZVS condition since v_{C7} equals zero. The equivalent circuit of this stage is shown in Figure 8.10.

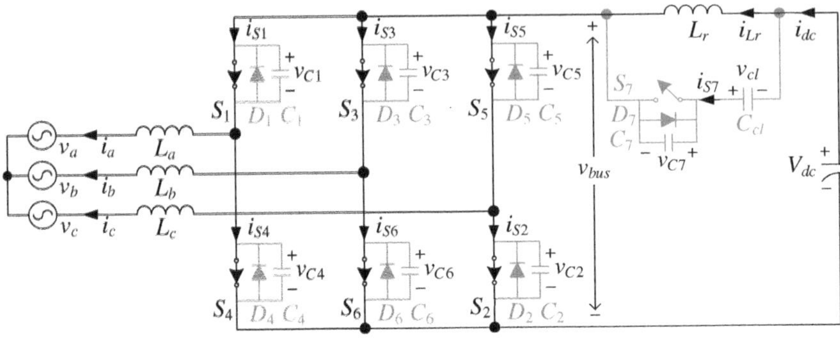

Figure 8.8 Equivalent circuit of the short-circuit stage.

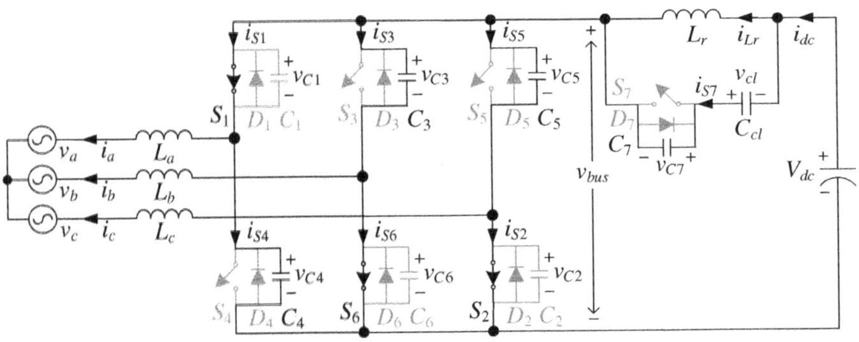

Figure 8.9 Equivalent circuit of the second resonant stage.

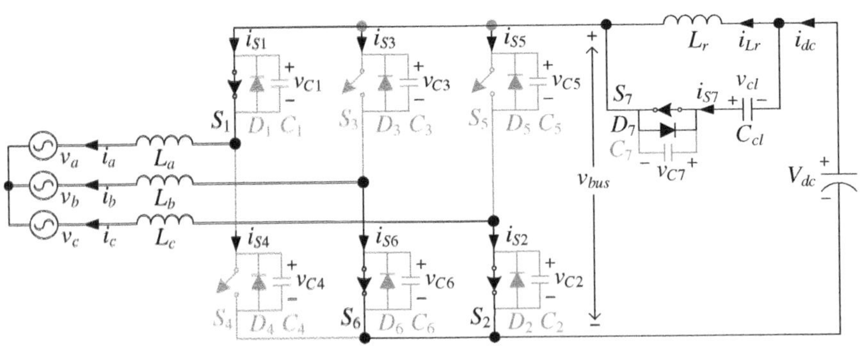

Figure 8.10 Equivalent circuit of the second steady stage.

Stage 8 ($t_7 - t_8$) is the first Type 1 commutation stage. At t_7, S_6 turns off. The output current i_b charges C_6 and discharges C_3. The equivalent circuit of this stage is shown in Figure 8.11. With the buffer of the resonant capacitor, S_6 achieves a ZVS-off. This stage ends at t_8 when v_{C3} reduces to zero and D_3 conducts in freewheeling mode.

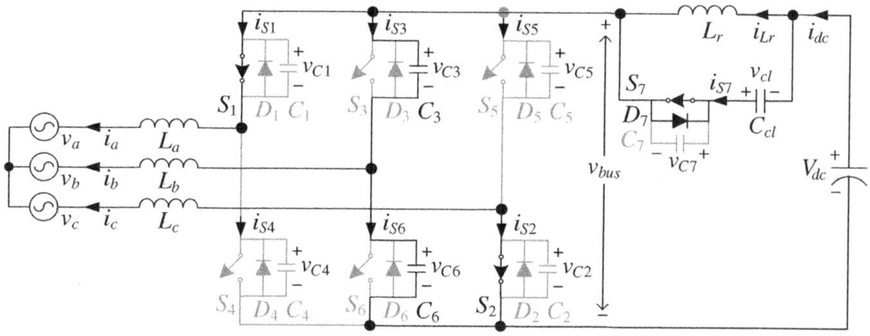

Figure 8.11 Equivalent circuit of the first Type 1 commutation stage.

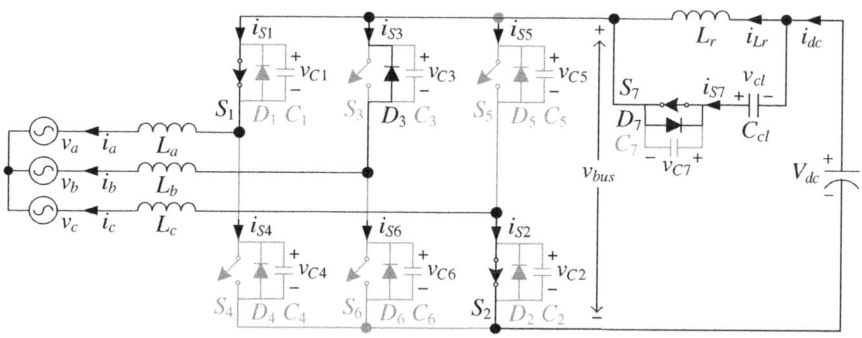

Figure 8.12 Equivalent circuit of the third steady stage.

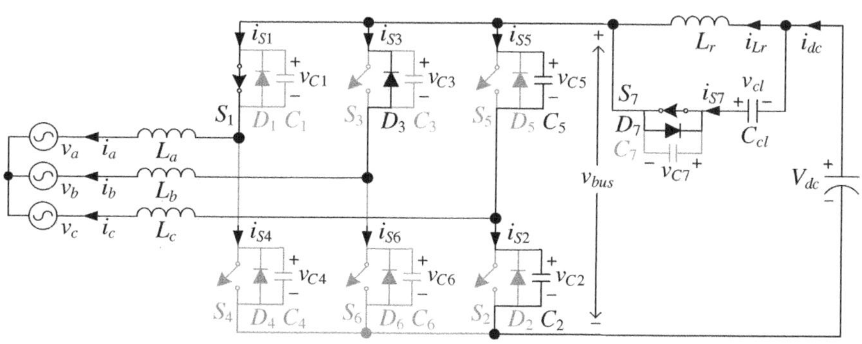

Figure 8.13 Equivalent circuit of second Type 1 commutation stage.

Stage 9 ($t_8 - t_9$) is the third steady stage. In this stage, inverter outputs vector $\overrightarrow{V_2}$ (110) and $i_{S7} = -i_c - i_{Lr}$. The equivalent circuit of this stage is shown in Figure 8.12.

Stage 10 ($t_9 - t_{10}$) is the second Type 1 commutation stage. At t_9, S_2 is turned off. The output current i_c charges C_2 and discharges C_5 simultaneously. The equivalent circuit of this stage is shown in Figure 8.13. With these capacitors, S_2 is turned off under ZVS condition. This stage ends at t_{10} when

v_{C5} reduces to zero and D_5 conducts. After stage 10, the inverter returns to stage 1 of a new switching period.

There are two resonant stages in each switching period. The first resonant stage, stage 2, is used to realize ZVS-on conditions for the bridge switches while the second resonant stage, stage 6, is used to realize the ZVS-on condition for the auxiliary switch. Besides, the reverse recovery of the freewheeling diodes is relieved by reduction of the current changing rate with the resonant inductor. It means there are no turn-on loss and diode reverse recovery loss. The losses of the switches that need to be considered are only conducting loss and turn-off loss.

8.2.2 ZVS Condition Derivation

It is assumed that the resonant capacitors of main switches are:

$$C_1 = C_4 = C_3 = C_6 = C_5 = C_2 = C_r \tag{8.1}$$

According to the equivalent circuit of the first resonance in Figure 8.5, following equations are derived as:

$$\begin{cases} i_{Lr} + C_7 \dfrac{dv_{C7}}{dt} = C_r \dfrac{dv_{C4}}{dt} + i_a + C_r \dfrac{dv_{C6}}{dt} + i_b + C_r \dfrac{dv_{C2}}{dt} + i_c \\ v_{Lr} = L_r \dfrac{di_{Lr}}{dt} \end{cases} \tag{8.2}$$

where

$$\begin{cases} v_{C4} = v_{C6} = v_{C2} = V_{dc} + V_{cl} - v_{C7} \\ V_{dc} = v_{Lr} + v_{C4} \\ i_a + i_b + i_c = 0 \end{cases} \tag{8.3}$$

By substituting (8.3) into (8.2), the equation of state variables, resonant inductor current i_{Lr}, and resonant capacitor voltage v_{C4} can be simplified as:

$$\begin{cases} (3C_r + C_7)\dfrac{dv_{C4}}{dt} = i_{Lr} \\ L_r \dfrac{di_{Lr}}{dt} = -v_{C4} + V_{dc} \end{cases} \tag{8.4}$$

with the initial condition:

$$\begin{cases} v_{C4}(t_1) = V_{cl} + V_{dc} \\ i_{Lr}(t_1) = i_{Lr}(t_1) \end{cases} \tag{8.5}$$

By solving the state equation (8.4) with its initial condition, the expressions of resonant capacitor voltage v_{C4} and resonant inductor current i_{Lr} are obtained:

$$\begin{cases} v_{C4}(t) = V_{cl}\cos(\omega_r t) + i_{Lr}(t_1) Z_r \sin(\omega_r t) + V_{dc} \\ i_{Lr}(t) = -\dfrac{V_{cl}}{Z_r}\sin(\omega_r t) + i_{Lr}(t_1)\cos(\omega_r t) \end{cases} \tag{8.6}$$

where the characteristic impedance Z_r and the resonant angular frequency ω_r are:

$$\begin{cases} Z_r = \sqrt{\dfrac{L_r}{3C_r + C_7}} \\ \omega_r = \dfrac{1}{\sqrt{L_r(3C_r + C_7)}} \end{cases} \tag{8.7}$$

where f_h is the frequency of harmonic component in filter inductor current, μ_0 is the vacuum permeability, and μ_r is the relative permeability.

The winding loss is calculated by:

$$P_{L\_winding} = \sum \frac{I_h^2}{2} R_h(N_f, d_f, a_{wf}, h_f) \tag{8.66}$$

where I_h is the amplitude of the different harmonic frequency components in the filter inductor current.

The volume of the core Vol_{L_core} can be estimated as:

$$Vol_{L\_core} = \pi\left[(r_{if} + d_f)^2 - r_{if}^2\right]h_f \tag{8.67}$$

Since the core loss in filter inductor is mainly caused by the switching frequency component, it can be calculated by the Steinmetz equation as [5]:

$$P_{L\_core} = 0.301 \times f_s^{1.5} B_{ac}^{1.8} Vol_{L\_core} \tag{8.68}$$

where B_{ac} is the amplitude of flux density caused by the switching frequency component of the filter inductor with the unit in T, f_s is the switching frequency with the unit in kHz, and Vol_{L_core} is the volume of the core with the unit of m^3.

If the window area of the filter inductor is assumed fully utilized, the volume of the filter inductor can be calculated by:

$$Vol_L = \pi\left(r_{if} + d_{rf} + 2\sqrt{\frac{a_{wf}}{\pi}}\right)^2\left(h_f + 4\sqrt{\frac{a_{wf}}{\pi}}\right) \tag{8.69}$$

8.3.4 Dimension Model of Other Components

8.3.4.1 Clamping Capacitor

Clamping capacitor C_{cl} cascades with auxiliary switch S_7, therefore the current through C_{cl} equals i_{S7}. The electrical charge in C_{cl} during a cycle may be taken as constant. Then the charge fluctuation in a cycle can be expressed as:

$$\Delta Q = \frac{\int_0^{T_s} |i_{S7}| dt}{2} \tag{8.70}$$

The peak-peak value of the voltage ripple across C_{cl} is restricted to 25% of the average value, and the minimum capacitance required is:

$$C_{cl} \geq \frac{\Delta Q}{25\% \times V_{cl}} \tag{8.71}$$

The clamping capacitor is assembled by a certain amount of film capacitors. A serial chain of EPCOS B32524 capacitors with 250 V withstand voltage are selected as the design reference. As Figure 8.21 shows, the average value of the volume over capacitance r_{vc} is calculated to be the reference for estimating the total volume of the clamping capacitor by knowing the value of capacitance.

The volume of the clamping capacitor can be estimated by:

$$Vol_{cl} = C_{cl} r_{vc} \tag{8.72}$$

The resonant capacitor voltage v_{C4} can be further modified as:

$$v_{C4}(t) = V_{dc} + \sqrt{V_{cl}^2 + i_{Lr}(t_1)^2 Z_r^2} \sin(\omega_r t + \sigma) \tag{8.8}$$

where

$$\sigma = \arctan\left(\frac{V_{cl}}{i_{Lr}(t_1) Z_r}\right) \tag{8.9}$$

According to (8.8), the ZVS turn-on can be realized unless the minimum value of resonant capacitor v_{C4} is smaller than zero. Thus the following inequality should be satisfied as:

$$\sqrt{V_{cl}^2 + i_{Lr}(t_1)^2 Z_r^2} \geq V_{dc} \tag{8.10}$$

In order to reduce the voltage stress of switches, which equals $V_{dc} + V_{cl}$, the clamping capacitor voltage V_{cl} is designed much smaller than V_{dc}. The ZVS condition (8.10) for main switches can be further simplified as:

$$i_{Lr}(t_1) \leq -\sqrt{\frac{V_{dc}^2 - V_{cl}^2}{Z_r^2}} \tag{8.11}$$

Stage 6 is the second resonant stage for the ZVS turn-on of auxiliary switch. The mathematical model of equivalent circuit in Fig. 8.9 can be derived as:

$$\begin{cases} i_{Lr} + C_7 \dfrac{dv_{C7}}{dt} = C_r \dfrac{dv_{C4}}{dt} + i_a + C_r \dfrac{dv_{C6}}{dt} + C_r \dfrac{dv_{C2}}{dt} \\ v_{Lr} = L_r \dfrac{di_{Lr}}{dt} \end{cases} \tag{8.12}$$

where

$$\begin{cases} v_{C4} = v_{C3} = v_{C5} = V_{dc} + V_{cl} - v_{C7} \\ V_{cl} + v_{C4} = v_{Lr} \end{cases} \tag{8.13}$$

By substituting (8.13) into (8.12), the equation of state variables, resonant inductor current i_{Lr}, and resonant capacitor voltage v_{C7} can be simplified as:

$$\begin{cases} (3C_r + C_7) \dfrac{dv_{C7}}{dt} = -i_{Lr} + i_a \\ L_r \dfrac{di_{Lr}}{dt} = v_{C7} - V_{cl} \end{cases} \tag{8.14}$$

with the initial condition:

$$\begin{cases} i_{Lr}(t_5) = i_a + i_x \\ v_{C7}(t_5) = V_{dc} + V_{cl} \end{cases} \tag{8.15}$$

By solving the state equation (8.14) with its initial condition, the expressions of resonant capacitor voltage v_{C7} and resonant inductor current i_{Lr} are obtained:

$$\begin{cases} v_{C7}(t) = V_{cl} + V_{dc} \cos(\omega_r t) - i_x Z_r \sin(\omega_r t) \\ i_{Lr}(t) = \dfrac{V_{dc}}{Z_r} \sin(\omega_r t) + i_x \cos(\omega_r t) + i_a \end{cases} \tag{8.16}$$

The resonant capacitor voltage v_{C7} can be further modified as:

$$v_{C7}(t) = V_{cl} + \sqrt{V_{dc}^2 + i_x^2 Z_r^2} \cos(\omega_r t + \tau) \tag{8.17}$$

where

$$\tau = \arctan\left(\frac{i_x Z_r}{V_{dc}}\right) \tag{8.18}$$

According to (8.17), since the clamping capacitor voltage V_{cl} is much smaller than V_{dc}, the minimum value of v_{C7} is always smaller than zero. Therefore, the ZVS condition for auxiliary switch is easy to be satisfied. At the end of stage 6 at t_6, the resonant capacitor voltage v_{C7} and resonant inductor current i_{Lr} meet:

$$\begin{cases} v_{C7}(t_6) = V_{cl} + V_{dc}\cos(\omega_r t_6) - i_x Z_r \sin(\omega_r t_6) = 0 \\ i_{Lr}(t_6) = \dfrac{V_{dc}}{Z_r}\sin(\omega_r t_6) + i_x \cos(\omega_r t_6) + i_a \end{cases} \tag{8.19}$$

Equation (8.19) can be modified as:

$$[V_{dc}\cos(\omega_r t_6) - i_x Z_r \sin(\omega_r t_6)]^2 = V_{cl}^2 \tag{8.20}$$

$$[V_{dc}\sin(\omega_r t_6) + i_x Z_r \cos(\omega_r t_6)]^2 = Z_r^2[i_{Lr}(t_6) - i_a]^2 \tag{8.21}$$

By adding (8.20) and (8.21), following equation is obtained:

$$V_{dc}^2 + i_x^2 Z_r^2 = V_{cl}^2 + Z_r^2[i_{Lr}(t_6) - i_a]^2 \tag{8.22}$$

Thus the resonant inductor current i_{Lr} at t_6 is derived as:

$$i_{Lr}(t_6) = \sqrt{\frac{V_{dc}^2 - V_{cl}^2}{Z_r^2} + i_x^2} + i_a \tag{8.23}$$

According to the waveform of the voltage across resonant inductor v_{Lr} and the voltage-second balance principle of the resonant inductor, the voltage across clamping capacitor V_{cl} can be obtained as:

$$V_{cl} = \frac{\lambda_7}{1 - \lambda_7} V_{dc} \tag{8.24}$$

where λ_7 is the turn-off duty cycle of the auxiliary switch S_7. $\lambda_7 T_s$ can be defined as the time interval from t_2 to t_5 as shown in Figure 8.3 if the two short resonant stages are ignored. By ignoring the resonant stages and two Type 1 commutation stages, the current in the clamping capacitor C_{cl} in stage 7 ($t_6 - t_7$), stage 9 ($t_8 - t_9$), and stage 1 ($t_{10} - t_1$) can be derived as:

$$\begin{cases} i_{cl\_stage\ 7}(t) = i_{Lr}(t) - i_a = \sqrt{\dfrac{V_{dc}^2 - V_{cl}^2}{Z_r^2} + i_x^2} - \dfrac{V_{cl}}{L_r}(t - t_6) \\ i_{cl\_stage\ 9}(t) = i_{Lr}(t) - i_a - i_b = \sqrt{\dfrac{V_{dc}^2 - V_{cl}^2}{Z_r^2} + i_x^2} - \dfrac{V_{cl}}{L_r}(t - t_6) - i_b \\ i_{cl\_stage\ 1}(t) = i_{Lr}(t) = \sqrt{\dfrac{V_{dc}^2 - V_{cl}^2}{Z_r^2} + i_x^2} - \dfrac{V_{cl}}{L_r}(t - t_6) + i_a \end{cases} \tag{8.25}$$

According to the current-second balance principle of the clamping capacitor, the average current through the clamping capacitor C_{cl} in one switching period should satisfy:

$$\int_{t_6}^{t_7} i_{cl\_stage\ 7}(t)dt + \int_{t_8}^{t_9} i_{cl\_stage\ 9}(t)dt + \int_{t_{10}}^{t_1} i_{cl\_stage\ 1}(t)dt \approx 0 \tag{8.26}$$

Since the equivalent vector within the time interval from t_2 to t_5 ($\lambda_7 T_s$) is zero vector $\overrightarrow{V_7}$ and the DC voltage at other stages are $V_{dc} + V_{cl}$, the soft-switching inverter can be viewed as an hard-switching inverter with DC voltage at $V_{dc} + V_{cl}$ when calculating the duration of stage 7 and stage 9. The basic vectors in stage 1, stage 7, and stage 9 are $\overrightarrow{V_7}$, $\overrightarrow{V_1}$, and $\overrightarrow{V_2}$, respectively. Their durations, T_{V0}, T_{V1}, and T_{V2}, can be obtained as:

$$\begin{cases} T_{V1} = MT_s \sin\left(\frac{\pi}{3} - \varphi\right) \\ T_{V2} = MT_s \sin(\varphi) \\ T_{V0} = T_s - \lambda_7 T_s - T_{V1} - T_{V2} \\ \varphi = \omega_l t = 2\pi f_l t \end{cases} \tag{8.27}$$

where M is the equivalent SVM modulation ratio of soft-switching inverter, φ is the phase angle of reference vector, ω_l is the angle frequency of grid voltages, and f_l is the line frequency of grid voltages. M can be expressed as:

$$M = \frac{\sqrt{3}V_m}{V_{dc} + V_{cl}} = \frac{\sqrt{3}V_m}{V_{dc}}(1 - \lambda_7) = m(1 - \lambda_7) \tag{8.28}$$

where m is the SVM modulation ratio of hard-switching inverter and V_m is the amplitude of grid voltage. If the soft-switching inverter is at unity power factor, then the grid currents meet:

$$\begin{cases} i_a = I_m \cos(\varphi) \\ i_b = I_m \cos(\varphi - 2\pi/3) \\ i_c = I_m \cos(\varphi + 2\pi/3) \end{cases} \tag{8.29}$$

By combining (8.25), (8.26), (8.27), (8.28), and (8.29), the turn-off duty cycle of the auxiliary switch λ_7 can be derived as:

$$\lambda_7 \approx \frac{\sqrt{\frac{V_{dc}^2 - V_{cl}^2}{Z_r^2} + i_x^2} + i_a - \frac{\sqrt{3}}{2} m I_m}{\frac{V_{dc} T_s}{2L_r}} \tag{8.30}$$

With (8.23), (8.24), and (8.30), the resonant inductor i_{Lr} at t_1 can be calculated as:

$$i_{Lr}(t_1) = i_{Lr}(t_6) - \frac{V_{cl}}{L_r}(1 - \lambda_7)T_s = -\sqrt{\frac{V_{dc}^2 - V_{cl}^2}{Z_r^2} + i_x^2} - i_a + \sqrt{3} m I_m \tag{8.31}$$

By substituting (8.31) into (8.11), extra added current i_x meets:

$$-\sqrt{\frac{V_{dc}^2 - V_{cl}^2}{Z_r^2} + i_x^2} - i_a + \sqrt{3} m I_m \le -\sqrt{\frac{V_{dc}^2 - V_{cl}^2}{Z_r^2}} \tag{8.32}$$

The minimum i_x can be obtained as:

$$i_x \ge \sqrt{\left(\sqrt{\frac{V_{dc}^2 - V_{cl}^2}{Z_r^2}} - i_a + \sqrt{3} m I_m\right)^2 - \sqrt{\frac{V_{dc}^2 - V_{cl}^2}{Z_r^2}}} \tag{8.33}$$

Equation (8.33) is the critical condition for ZVS turn-on of main switches. The duration of stage 5 should be controlled to meet (8.33) and store enough energy in resonant inductor for the next resonant stage. By substituting the minimum value of extra added current i_x as shown in (8.33) into (8.23) and (8.31), the minimum resonant current at t_1 and the maximum resonant current at t_6 can be derived as:

$$\begin{cases} i_{Lr}(t_1) = -\sqrt{\dfrac{V_{dc}{}^2 - V_{cl}{}^2}{Z_r{}^2}} \approx -\dfrac{V_{dc}}{Z_r} \\ i_{Lr}(t_6) = \sqrt{\dfrac{V_{dc}{}^2 - V_{cl}{}^2}{Z_r{}^2}} + \sqrt{3}mI_m \approx \dfrac{V_{dc}}{Z_r} + \sqrt{3}mI_m \end{cases} \tag{8.34}$$

The turn-off duty cycle of the auxiliary switch λ_7 can be further simplified as:

$$\lambda_7 \approx \frac{2L_r}{V_{dc}T_s}\left(\frac{V_{dc}}{Z_r} + \frac{\sqrt{3}}{2}mI_m\right) \tag{8.35}$$

8.3 Loss and Dimension Models

By neglecting the power dissipations in capacitors, the total loss of the inverter include loss in semiconductor devices, loss in resonant inductor, and loss in filter inductors. The loss in semiconductor devices contains conduction loss and turn-off loss. The loss in resonant inductor and filter inductors contain winding loss and core loss. In order to attain the optimum parameters, the models of these losses will be described in this section. Besides, the volume of the passive components including clamping capacitors, resonant inductors, and filter inductors are also modeled for the dimension considerations. Since the calculation is the same in each SVM sector, sector 1-1 is taken as example.

8.3.1 Loss Model of IGBT Devices

8.3.1.1 Conduction Loss of IGBT Devices

According to the loss model of ZVS-SVM three-phase grid inverter in Appendix A.4, the conduction loss P_{con} can be divided into conduction loss in main switches P_{con3} and conduction loss in auxiliary switch P_{conS7} as:

$$P_{con} = P_{con3} + P_{conS7} \tag{8.36}$$

Since the turn-off duration of auxiliary circuit is very short, typically below 10% of switching period, its impact on conduction loss of main switches P_{con3} can be ignored. Therefore, the conduction loss of main switches in ZVS-SVM three-phase grid inverter can also be calculated by using the loss model of hard-switching three-phase grid inverter.

The conduction loss of main switches P_{con3} can be further divided into the conduction loss of three-phase main switches' channel P_{conQ3} and the conduction loss of three-phase main switches' antiparalleled diodes P_{conD3} as:

$$P_{con3} = P_{conQ3} + P_{conD3} \tag{8.37}$$

By considering the symmetricity of three phases, P_{conQ3} and P_{conD3} can be calculated by:

$$P_{conQ3} = 3 \cdot f_l \cdot 4 \sum_{n=1}^{3p} \left[V_{ce}(i_a(n)) \cdot i_a(n) \cdot T_{up}(n)\right] \tag{8.38}$$

$$P_{conD3} = 3 \cdot f_l \cdot 4 \sum_{n=1}^{3p} \left[V_f(i_a(n)) \cdot i_a(n) \cdot \left[T_s - T_{up}(n)\right]\right] \tag{8.39}$$

where f_l is the grid line frequency, p is the number of switching periods in each sector, $V_{ce}(i_a(n))$ is the conducting voltage at grid current $i_a(n)$, $V_f(i_a(n))$ is the forward voltage of antiparalleled diode at grid current $i_a(n)$, and $T_{up}(n)$ is the conducting time of upper switch after considering dead time. If the grid current ripple is ignored and the grid current is assumed to be constant in each switching period, the grid current $i_a(n)$ in the nth switching period can be expressed as:

$$i_a(n) = I_m \cos\left(\frac{2\pi f_l n}{f_s}\right) \tag{8.40}$$

According to the loss model in Appendix A.4, the conducting time of upper switch $T_{up}(n)$ within the first 1/4 grid line cycle can be deduced as:

$$T_{up}(n) = \begin{cases} T_s & \text{Sector 1-1} \\ T_{V1}(n) + T_{V2}(n) - T_{dead} & \text{Sector 1-2} \\ T_{V1}(n) - T_{dead} & \text{Sector 2-1} \end{cases} \tag{8.41}$$

where $T_{V1}(n)$ and $T_{V2}(n)$ are the theoretical durations of two effective basic vectors in the nth switching period. Since the three-phase output voltage vector is 000 when auxiliary switch is off, the operation of auxiliary circuit will not impact the output voltage. Besides, due to the existence of clamping capacitor the equivalent DC-side voltage has been increased to $V_{cl} + V_{dc}$ when auxiliary switch is on. Therefore, the operation time of three basic vectors should be modified as:

$$\begin{cases} T_{V1}(n) = MT_s \sin\left[\frac{\pi}{3} - 2\pi f_l\left(\frac{n - k \cdot 2p}{f_s}\right)\right] \\ T_{V2}(n) = MT_s \sin\left[2\pi f_l\left(\frac{n - k \cdot 2p}{f_s}\right)\right] \\ T_{V0}(n) = T_s - T_{V1}(n) - T_{V2}(n) \\ \mathrm{M} = \frac{\sqrt{3}V_m}{V_{dc} + V_{cl}} \end{cases} \tag{8.42}$$

As shown in (8.38) and (8.39), it is necessary to measure the conducting voltage V_{ce} and the forward voltage of antiparalleled diode V_f of Insulated Gate Bipolar Transistors (IGBT) device at different conducting current under the designed gate voltage. Five different IGBT devices with nominal rating at 1200 V/200 A were measured. The driving voltage is 15 V and the chip junction temperature is 25 °C. The conducting voltage V_{ce} and the forward voltage of antiparalleled diode V_f can be fitted as functions of the current through the switch I_c as:

$$V_{ce}(I_c) = q_0 + q_1 \lg(I_c + q_2) \tag{8.43}$$

$$V_f(I_f) = k_0 + k_1 \lg(I_f + k_2) \tag{8.44}$$

where q_0, q_1, q_2, k_0, k_1, and k_2 are the coefficients of the fitted functions. The coefficient values for each device are listed in Table 8.1. Here, the normalized root-mean-square (RMS) error of each device is less than 3%.

Table 8.1 Fitting coefficients for conduction voltage of IGBT device and forward voltage of its antiparalleled diode.

Coefficients	Device 1	Device 2	Device 3	Device 4	Device 5
q_0	−2.258	−8.965	−4.151	−3.533	−2.533
q_1	7.435×10^{-1}	2.512	1.439	9.567×10^{-1}	7.792×10^{-1}
q_2	4.998×10	5.126×10	3.051×10	8.502×10	6.043×10
Normalized root-mean-square error	2.01%	1.97%	2.33%	2.46%	2.24%
k_0	−1.589	−1.751	-9.744×10^{-1}	-8.623×10^{-1}	−1.119
k_1	7.386×10^{-1}	7.661×10^{-1}	5.910×10^{-1}	5.677×10^{-1}	5.273×10^{-1}
k_2	2.085×10	2.480×10	1.609×10	1.504×10	2.871×10
Normalized root-mean-square error	2.35%	2.14%	1.83%	2.03%	2.64%

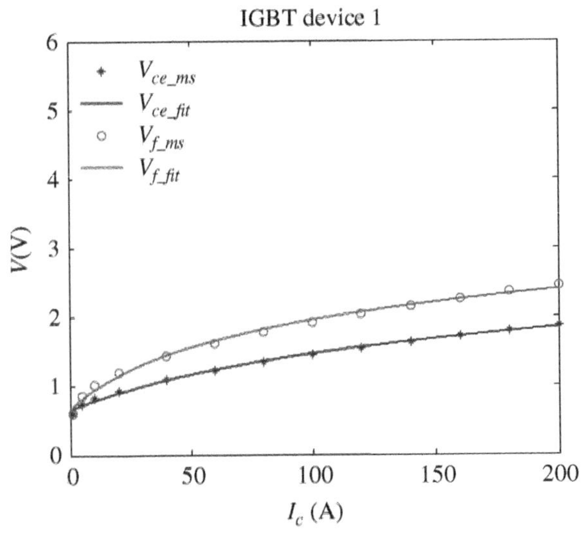

Figure 8.14 Demonstration of conduction loss models.

Conduction voltage and forward voltage functions of device 1 are drawn in Figure 8.14. The operation current ranges from 0 to 200 A. V_{ce_ms} and V_{f_ms} are the measured points while V_{ce_fit} and V_{f_fit} are the curves of fitted functions.

8.3.1.2 Switching Loss of the IGBT Devices

All IGBT devices are operated under ZVS turn-on conditions. However, turn-off loss still exists. According to the loss model of ZVS-SVM three-phase grid inverter in Appendix A.4, the switching loss P_{off} can be divided into turn-off loss in main switches P_{off3} and turn-off loss in auxiliary switch P_{offS7} as:

$$P_{off} = P_{off3} + P_{offS7} \tag{8.45}$$

By considering the symmetricity of three phases, the turn-off loss in main switches P_{off3} can be further expressed as:

$$P_{off3} = 3\left(P_{offA} + P_{offA\_sc}\right) \tag{8.46}$$

where P_{offA} is the turn-off loss of phase A's main switches during Type 1 commutation and P_{offA_sc} is the turn-off loss of phase A's main switches after short-circuit stage. Their expressions are:

$$P_{offA} = f_l \left[\frac{V_{dc} + V_{cl}}{V_{st}} \cdot 4 \cdot \sum_{n=p+1}^{3p} E_{off}\left(|i_a(n)|, 2C_r\right)\right] \tag{8.47}$$

$$P_{offA\_sc} = f_l \left[\frac{V_{dc} + V_{cl}}{V_{st}} \cdot 12 \cdot \sum_{n=1}^{p} E_{off}\left(\frac{i_x(n)}{3}, \frac{(3C_r + C7)}{3}\right)\right] \tag{8.48}$$

where $V_{cl} + V_{dc}$ is the equivalent DC-side voltage for main switches, E_{off} is the turn-off energy of main switch measured with testing DC voltage V_{st}, and $i_x(n)$ is the extra added current on resonant inductor. The turn-off loss in auxiliary switch P_{offS7} is expressed as:

$$P_{offS7} = f_l \cdot \frac{V_{dc} + V_{cl}}{V_{st}} \cdot \sum_{n=1}^{12p} E_{off}\left(i_{S7}(t_1), (3C_r + C7)\right) \tag{8.49}$$

where $i_{S7}(t_1)$ is the turn-off current in auxiliary switch and is equal to $-i_{Lr}(t_1)$. According to (8.34), $i_{S7}(t_1)$ is constant in every switching period.

In order to guarantee the accuracy of the loss model, turn-off energy E_{off} under different current and different buffer capacitance C_b conditions are measured. The measurement circuit and key waveforms are shown in Figure 8.15. A single pulse v_g is used to turn on the switch. Once the current i_c through the switch reaches the given value, the switch is turned off [2]. The time integration of product of collector-emitter voltage v_{ce} and collector current i_c during turn-off process are obtained as the turn-off energy E_{off}.

Five different IGBT devices were measured. The nominal rating of all the devices is 1200 V/ 200 A.

A three-dimensional (3D) turn-off loss model can be fitted with the measured data. All the loss data are tested at V_{dc} = 600V and chip junction temperature T_j = 25°C. The turn-off energy of each switching action can be fitted as a function of the buffer capacitor C_b and current through the switch I_c:

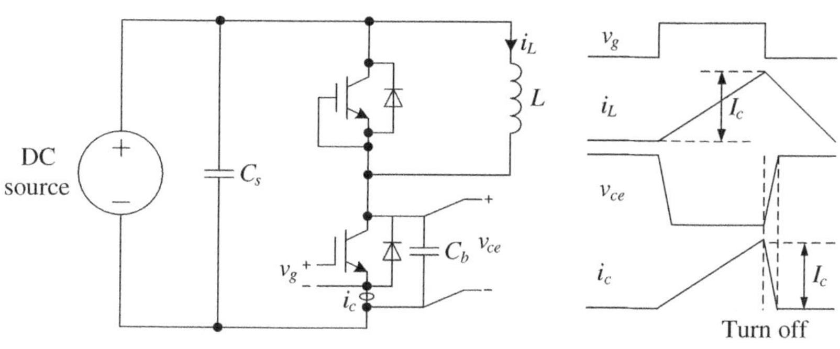

Figure 8.15 Turn-off loss measurement circuit and key waveforms.

$$E_{off}(I_c, C_b) = p_0 + p_1 \cdot C_b + p_2 \cdot C_b^2 + p_3 \cdot I_c + p_4 \cdot I_c^2 + p_5 \cdot C_b I_c + p_6 \cdot C_b^2 I_c \quad (8.50)$$

where $p_0 - p_6$ are the coefficients of the fitted function.

The coefficient values for each device are listed in Table 8.2. Here, the normalized RMS error of each device is less than 2%.

Turn-off loss function of device 1 is drawn in Figure 8.16. The operation current ranges from 0 to 100 A and the capacitance ranges from 0 to 20 nF.

Table 8.2 Fitting coefficients for turn-off energy.

Coefficients	Device 1	Device 2	Device 3	Device 4	Device 5
p_0	1.340	-3.442×10^{-2}	1.039×10^{-1}	2.772×10^{-1}	1.658
p_1	-5.728×10^{-2}	4.659×10^{-2}	-7.29×10^{-3}	-1.215×10^{-2}	-9.947×10^{-3}
p_2	1.156×10^{-3}	-1.510×10^{-3}	1.140×10^{-3}	3.456×10^{-4}	1.016×10^{-3}
p_3	1.432×10^{-1}	3.664×10^{-2}	6.887×10^{-2}	5.623×10^{-2}	1.213×10^{-1}
p_4	-3.57×10^{-4}	1.783×10^{-4}	1.063×10^{-4}	1.671×10^{-4}	-2.881×10^{-5}
p_5	-8.551×10^{-4}	-4.511×10^{-3}	-4.026×10^{-3}	-3.159×10^{-3}	-2.421×10^{-3}
p_6	2.537×10^{-5}	1.417×10^{-4}	8.815×10^{-5}	8.586×10^{-5}	6.976×10^{-5}
Normalized root-mean-square error	1.15%	1.93%	1.70%	1.36%	1.27%

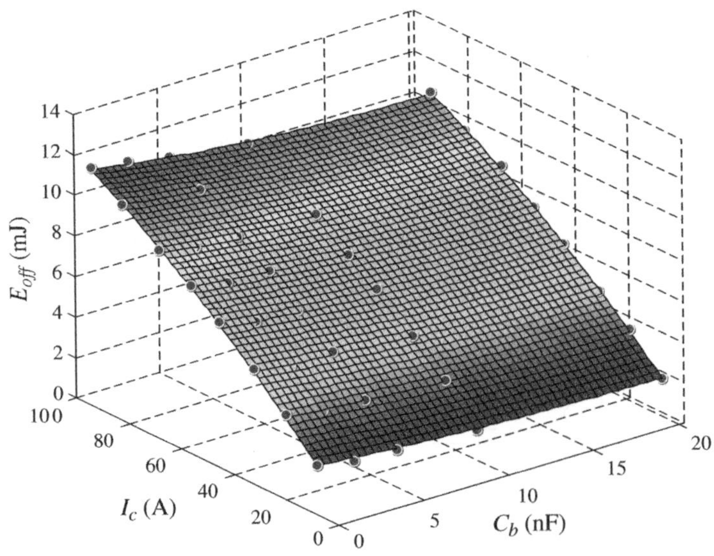

Figure 8.16 Fitted turn-off loss function of device 1.

8.3.2 Loss and Dimension Models of Resonant Inductor

Resonant inductor loss includes the winding loss and the core loss. "EE" shaped ferrite core is utilized in this section for its low core loss at high exciting frequency. According to the typical EE cores, their dimensions can be modeled as shown in Figure 8.17 with unit in centimeter. The thickness, the window width, and the central leg width of the cores is defined as d_r. The width and length of the side leg are defined as $d_r/2$ and $3d_r$, respectively. The cross area of the winding is defined as a_{wr}. The number of winding turns is defined as N_r. The air gap is defined as g_r. The practical core dimensions will be selected according to the optimization results.

The inductance can be deduced as:

$$L_r(N_r, g_r, d_r) = \frac{0.4\pi N_r^2 A_c \times 10^{-8}}{g_r + \frac{MPL}{\mu_m}} \approx \frac{0.4\pi \times 10^{-8} N_r{}^2 d_r{}^2}{g_r} \quad \text{(H)} \tag{8.51}$$

where A_c is the effective cross area of core, MPL is the magnetic path length, and μ_m is the permeability of core material. Since the typical μ_m of ferrite material is 2500, MPL/μ_m can be neglected.

The flux density B is obtained by:

$$B(i_{Lr}, N_r, g_r) = \frac{0.4\pi N_r i_{Lr} \times 10^{-4}}{g_r + \frac{MPL}{\mu_m}} \approx \frac{0.4\pi \times 10^{-4} N_r i_{Lr}}{g_r} \quad \text{(T)} \tag{8.52}$$

The volume of the inductor core for core loss calculation can be expressed as:

$$Vol_{Lr\_core} = (4d_r)^2 d_r - 2\cdot d_r \cdot 3d_r \cdot d_r \tag{8.53}$$

According to the analysis in the second section, the current i_{Lr} can be treated as a triangular waveform with a rise time of $\lambda_7 T_s$ in a switching period, shown in Figure 8.3. The core losses were measured under triangular current stimulation. B-H loop measurement method was utilized [3]. The setup of magnetic core loss measurement is demonstrated in Figure 8.18. The principles are as

Figure 8.17 Definitions of the unified inductor model.

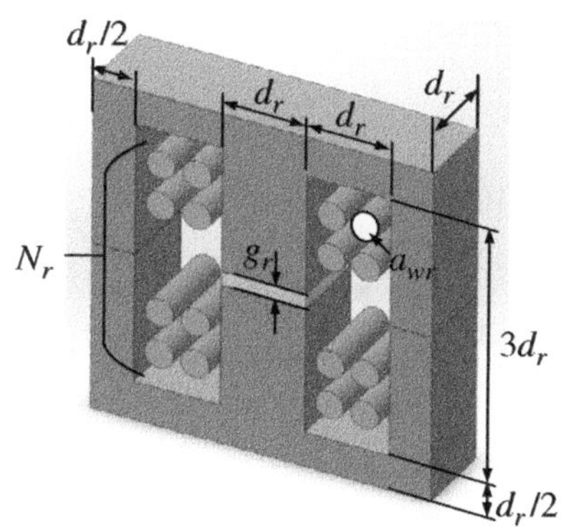

Figure 8.18 Measurement setup of magnetic core loss.

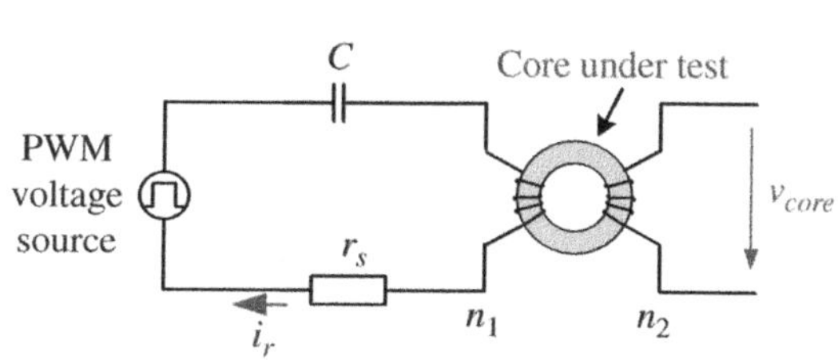

follows: a pulse-width-modulation (PWM) voltage source is to generate a PWM waveform with arbitrary duty cycle. The core under test is placed with two windings. The voltage across secondary winding v_{core} is to measure the flux density B. The current in the primary winding i_r is proportional to the magnetic field strength H. The turns of both primary and secondary windings n_1 and n_2 equal 50 in this case. The power loss per volume of the core P_v under test can be expressed as:

$$P_v = \frac{1}{T_s}\int HdB \tag{8.54}$$

where

$$H = \frac{n_1 i_r}{MPL} \tag{8.55}$$

$$v_{core} = n_2\frac{d\phi}{dt} = n_2 A_c\frac{dB}{dt} \tag{8.56}$$

With (8.55) and (8.56), the power loss per volume of the core P_v can be derived as:

$$P_v = \frac{1}{MPLA_cT_s}\int i_r v_{core}dt \tag{8.57}$$

Core losses per volume at $D = 0.5$ were measured under different frequency conditions and shown in Figure 8.19.

The measured results in Figure 8.19 can be fitted as:

$$P_v = K_c f_s^{\ \alpha}\Delta B^{\beta} \tag{8.58}$$

where $K_c = 3.84\times10^{-4}$, $\alpha = 1.564$, and $\beta = 2.084$ are coefficients of the loss calculation. Based on (8.58), core losses with different duty cycle stimulation are evaluated by [4]:

$$P_{Lr\_core} = K_c\left(\frac{\lambda_7}{(2\lambda_7/f_s)^{\alpha}} + \frac{1-\lambda_7}{(2(1-\lambda_7)/f_s)^{\alpha}}\right)\left(\frac{B(i_{Lr\_max}) - B(i_{Lr\_min})}{2}\right)^{\beta}\cdot Vol_{Lr\_core} \tag{8.59}$$

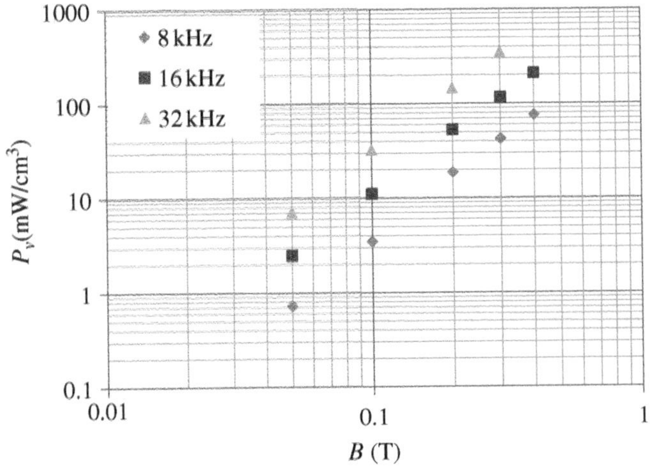

Figure 8.19 Core losses per volume under different frequencies.

Winding resistance is calculated by:

$$r_{Lr\_winding}(N_r, a_{wr}, d_r) = MLT_r \cdot N_r \cdot \rho_{20}(1 + \alpha_{20}(T - 20^{\circ}\text{C}))/a_{wr} \tag{8.60}$$

where MLT_r is the mean length of a turn and defined as $K_p \times 4d_r$, K_p is a coefficient and selected as 2 here, and ρ_{20} presents copper resistivity at 20°C. α_{20} represents the temperature coefficient of the resistivity at 20°C. T represents the temperature of the winding, which is presumed to be 80°C. With the resistance, winding loss is obtained by:

$$P_{Lr\_winding} = {i_{Lr\_rms}}^2 r_{winding}(N_r, a_{wr}, d_r) \tag{8.61}$$

where i_{Lr_rms} is the RMS value of resonant inductor current.

If the winding thickness is estimated as d_r, the volume of the resonant inductor is evaluated by:

$$Vol_{Lr} = (4d_r)^2 \cdot 3d_r \tag{8.62}$$

8.3.3 Loss and Dimension Models of the Filter Inductor

In this application, silicon iron powder core of toroidal shape is utilized in filter inductors and demonstrated in Figure 8.20. The shape of the cross-sectional area of the core is assumed as square, of which the side length is defined as d_f and the thickness is h_f. The internal radius is represented as r_{if}. The conduction area of the winding is defined as a_{wf}, and the number of winding turns is represented by N_f. The relative permeability is fixed to 60.

The AC resistance of the winding by only considering the skin effect can be expressed as:

$$R_{ac}(N_f, d_f, a_{wf}, h_f) = \frac{N_f \cdot MLT_f \cdot \rho_{Cu}}{a_{wf} - \pi\left(\sqrt{a_{wf}/\pi} - \delta\right)^2} \tag{8.63}$$

where MLT_f is the mean length of a turn and can be estimated as:

$$MLT_f = 1.5 \times 2(d_f + h_f) \tag{8.64}$$

and δ is the skin depth and its expression is:

$$\delta = \sqrt{\frac{\rho_{Cu}}{\pi f_h \mu_0 \mu_r}} \tag{8.65}$$

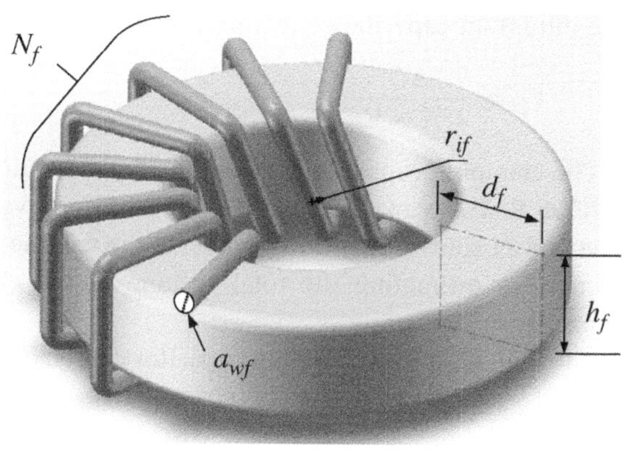

Figure 8.20 Demonstration of the filter inductor.

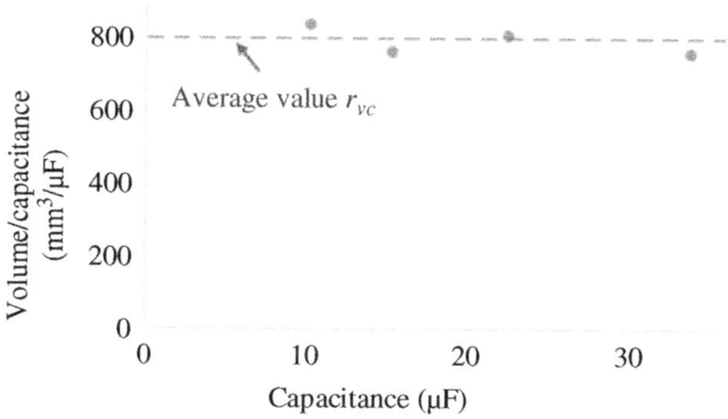

Figure 8.21 Average value of the volume over capacitance.

8.3.4.2 Heat Sink

Since forced air cooling structure is utilized in the system, only two types of heat transfer, heat conduction and heat convection, are considered [6]. The selected shape of the heat sink is shown in Figure 8.22. Fin duty cycle k, groove hydraulic diameter d_h, and average Reynolds number of air flowing in the groove Re_m are defined as:

$$k = \frac{s}{b/n} \tag{8.73}$$

$$d_h = \frac{2sc}{s+c} \tag{8.74}$$

$$Re_m = \frac{2V}{n(s+c)\varepsilon} \tag{8.75}$$

where V represents the air volume and ε is the air viscosity.

With (8.74), the convection heat transfer coefficient h_{HS} can be expressed as:

$$h_{HS} = \frac{Nu_m \lambda_{air}}{d_h} \tag{8.76}$$

Figure 8.22 Shape and dimensions of selected heat sink.

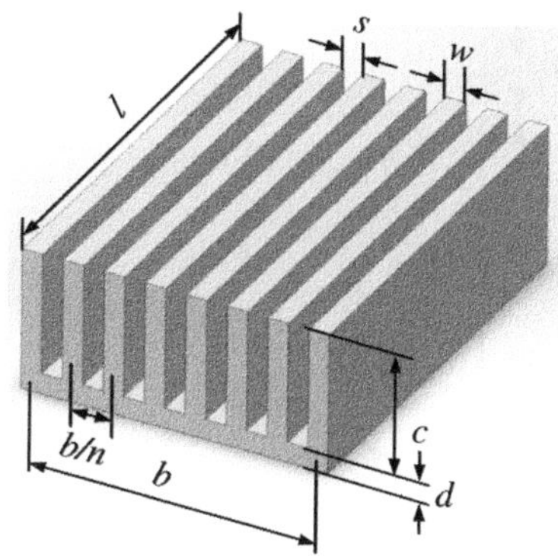

where λ_{air} is the thermal conductivity of air, and Nu_m is the average Nusselt number. Nu_m meets:

$$Nu_m = \begin{cases} \dfrac{3.657\left[\tanh\left(2.264X^{1/3} + 1.7X^{2/3}\right)\right]^{-1} + \dfrac{0.0499}{X}\tanh(X)}{\tanh\left(2.432Pr^{1/6}X^{1/6}\right)} & Re_m < 2300 \\ \dfrac{\left[8(0.79\ln(Re_m) - 1.64)^2\right]^{-1}(Re_m - 1000)Pr}{1 + 12.7\sqrt{\left[8(0.79\ln(Re_m) - 1.64)^2\right]^{-1}}\left(Pr^{2/3} - 1\right)}\left[1 + \left(\dfrac{d_h}{L}\right)^{2/3}\right] & Re_m \geq 2300 \end{cases} \tag{8.77}$$

where Pr is the Prandtl number of air. The variable X meets:

$$X = \frac{L}{d_h Re_m Pr} \tag{8.78}$$

It is clear that once the air volume V is derived, the parameters in above equations can be easily calculated based on the physical dimensions of the selected heat sink.

The thermal model of the heat sink can be derived as shown in Figure 8.23. $R_{th,d}$ and $R_{th,Fin}$ are the thermal resistances of base plate and fin caused by heat conduction, respectively. $R_{th,A}$ is the thermal resistance from fin to air caused by heat convection. They can be calculated by:

$$R_{th,d} = \frac{d}{\frac{A_{HS}}{n}\lambda_{HS}} \tag{8.79}$$

$$R_{th,Fin} = \frac{c}{wl\lambda_{HS}} \tag{8.80}$$

$$R_{th,A} = \frac{1}{clh_{HS}} \tag{8.81}$$

where A_{HS} is the area of the cooling plate and λ_{HS} is the thermal conductivity of heat sink. Another source of thermal resistance is the average temperature rise of the air in the channel due to the heat transported away via convection, and its expression is:

$$R_{th,c} = \frac{1}{2\rho_{air}c_{P,air}V} \tag{8.82}$$

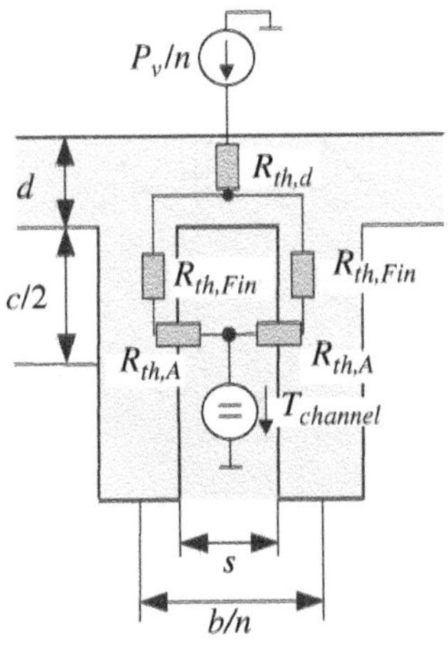

Figure 8.23 Thermal model of the heat sink.

where ρ_{air} is the density of the air, and $c_{P,\,air}$ is the specific heat capacity of the air.

The total thermal resistance of the heat sink can be derived as:

$$R_{th,S-a}^{HS} = \frac{1}{n}\left[\frac{1}{2}(R_{th,Fin} + R_{th,A}) + R_{th,d}\right] + R_{th,c} \tag{8.83}$$

By substituting (8.79), (8.80), (8.81), and (8.82) into (8.83), $R_{th,S-a}^{HS}$ can be rewritten as:

$$R_{th,S-a}^{HS} = \frac{1}{2n}\left(\frac{c}{wl\lambda_{HS}} + \frac{1}{clh_{HS}}\right) + \frac{d}{A_{HS}\lambda_{HS}} + \frac{1}{2\rho_{air}c_{P,air}V} \tag{8.84}$$

Similarly, once the air volume V is derived, the total thermal resistance of the selected heat sink $R_{th,S-a}^{HS}$ can be calculated based on the physical dimensions.

San Ace 60 type fans are selected for the forced air cooling system. The characteristics of the fan is fitted by:

$$\begin{aligned}\Delta p_{fan}(V) &= 125 - 16053V + 9\times 10^{6}V^{2} - 4\times 10^{9}V^{3}\\ &\quad + 5\times 10^{11}V^{4} - 3\times 10^{13}V^{5} + 8\times 10^{14}V^{6}\end{aligned} \tag{8.85}$$

For a given heat sink geometry, the air flow through channels creates a pressure drop along the heat sink length l under laminar flow condition:

$$\Delta p_{lam}(V) = \begin{cases} 1.5\dfrac{32\rho_{air}\varepsilon l}{n(sc)d_h^2}V & Re_m < 2300 \\[2ex] \dfrac{\frac{1}{2}\left(\frac{V}{n(sc)}\right)^{2} l\frac{s+c}{2sc}\rho_{air}}{\left[0.79\ln\left(\frac{2V}{n(s+c)\varepsilon}\right) - 1.64\right]^{2}} & Re_m \geq 2300 \end{cases} \tag{8.86}$$

According to the balance of the pressure, there is:

$$\Delta p_{lam}(V) = k\Delta p_{fan}(V) \tag{8.87}$$

By solving (8.87) with the physical dimensions of the selected heat sink, the air volume V could be obtained. Then the total thermal resistance could also be calculated.

In this application, the temperature on the heat sink plate is restricted to 80°C and the temperature rise is restricted to 40°C. The thermal resistance of the heat sink should satisfy the conditions:

$$T_{HS\_rise} = R_{th,S-a}^{HS}(P_{con} + P_{off}) \leq 40^{\circ}\mathrm{C} \tag{8.88}$$

3D model of the heat sink with IGBT devices is shown in Figure 8.24, and the total volume of the heat sink could be expressed as a function of the geometry as:

$$Vol_{HS} = (b + w)(c + d)l \tag{8.89}$$

The heat sink with minimum volume can be investigated by using the following solution:

$$\begin{aligned}&\min\ Vol_{HS}(b, w, l, c, d, n)\\ &s.t.\begin{cases} T_{HS\_rise} \leq 40^{\circ}C \\ n \in Z^{+} \\ 0 < k < 1 \\ l \geq l_{case} + b_{case} \\ b + w \geq 3b_{case} \end{cases}\end{aligned} \tag{8.90}$$

Figure 8.24 3D model of the heat sink with IGBT devices.

where l_{case} is length of IGBT module, and b_{case} is width of the IGBT module. By changing the dimensions of the heat sink, the length l, the thickness of base plate d, the length of fins c, the number of fins n, and the fin duty cycle k, each combination of these dimensions leads to one volume of the corresponding heat sink. After solving above optimized problem in (8.90), the heat sink with minimum volume is obtained.

8.4 Parameters Optimization and Design Methodology

The optimization design procedures are outlined in Figure 8.25. In order to utilize the optimization algorithm to solve the design issue, the total target, which is conversion efficiency in this application, should be described as a function of all the design variables. Besides, all of the constrained conditions are also needed to be expressed as a function of the design variables. The optimization algorithm is utilized to get the optimum results with each alternative IGBT devices. Based on the lowest IGBT device losses under different constrained conditions, the volumes of the heat sink are calculated. By combining the volume of the resonant inductor, clamping capacitor, and filter inductor, the volume of the inverter is also considered simultaneously. Finally, select the optimum design parameters according to the optimization Pareto front.

8.4.1 Objective Function

The design object is the total loss of the inverter. The design object is expressed as the function of design variables. The goal is to obtain the minimum value of the design object, which is expressed by:

$$\begin{aligned}\min .P_{total} = & P_{con}(f_s, C_r, N_r, g_r, d_r) \\ & + P_{off}(f_s, C_r, N_r, g_r, d_r) \\ & + P_{Lr\_core}(f_s, C_r, N_r, g_r, d_r) \\ & + P_{Lr\_winding}(f_s, C_r, N_r, a_{wr}, g_r, d_r) \\ & + P_{L\_core}(f_s, r_{if}, N_f, d_f, h_f) \\ & + P_{L\_winding}(f_s, N_f, d_f, a_{wf}, h_f)\end{aligned} \tag{8.91}$$

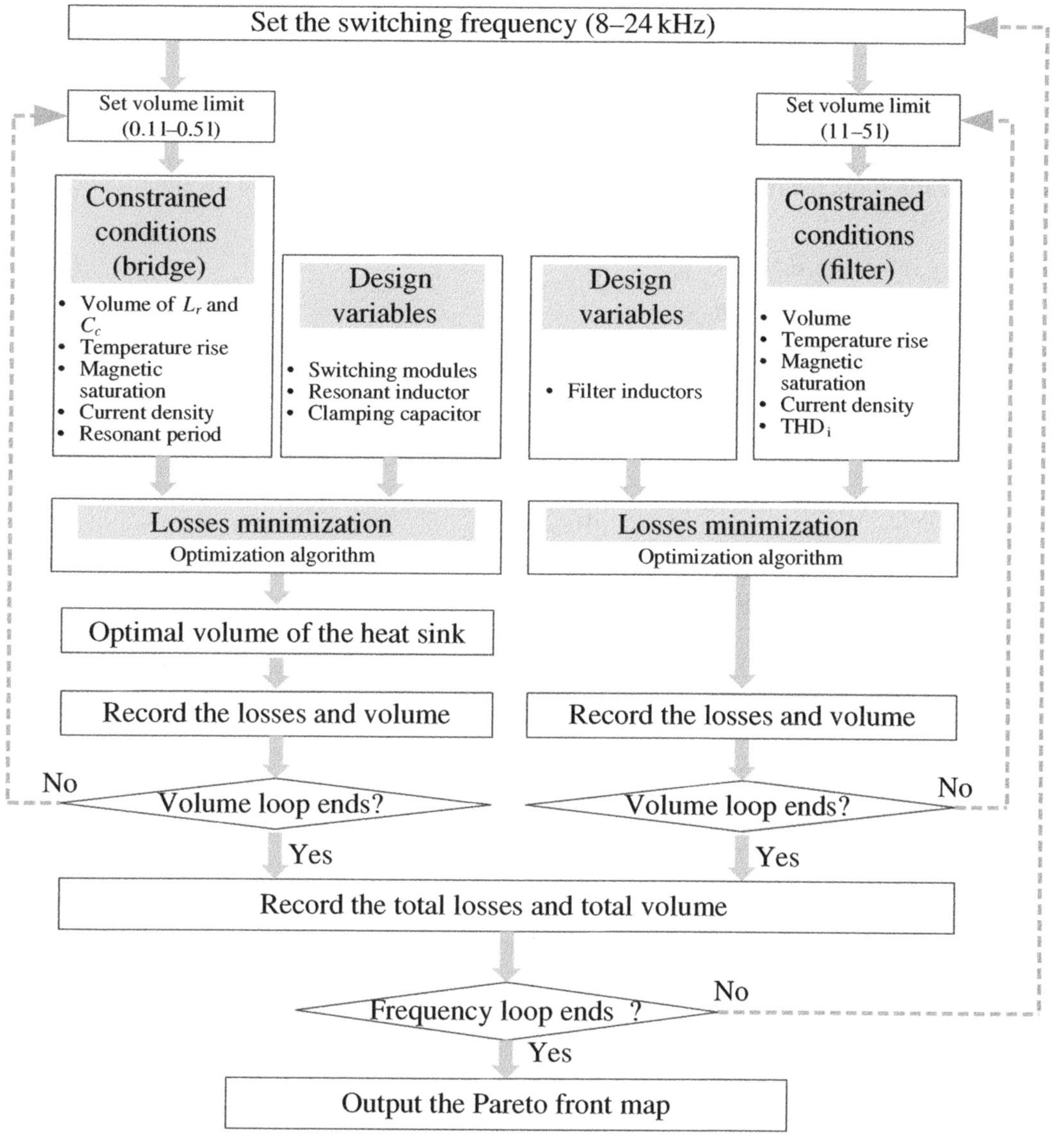

Figure 8.25 Procedures of optimization design.

The total volume of the resonant inductor, clamping capacitors, heat sink, and filter inductor is defined as the limit in the design procedure, which can also be expressed by the design variables:

$$Vol_{total} = Vol_{Lr}(d_r) + Vol_L(r_{if}, d_{rf}, a_{wf}, h_f) + Vol_{cl}(C_{cl}) + Vol_{HS}(b, c, d, l, w) \tag{8.92}$$

8.4.2 Constrained Conditions

In order to guarantee the functionality of the design results and meet the specifications, several constrained conditions are added to set boundaries for the design process.

The output current should meet the power quality specifications. The current total harmonic distortion (THD) is restricted by:

$$THD_{i(2nd\sim 33rd)} \leq 4.5\% \tag{8.93}$$

for the 2nd to 33rd harmonics.

To avoid saturation, the magnetic flux density of the resonant inductor and the filter inductor is limited by:

$$B_{Lr}(f_s, C_r, N_r, g_r, d_r) \leq B_{Lr\_max} \tag{8.94}$$

$$B_L(f_s, N_f, d_f, r_{if}, h_f) \leq B_{L\_max} \tag{8.95}$$

The temperature rise of the resonant inductor and filter inductor is estimated by [7]:

$$T_{Lr\_rise}(f_s, C_r, N_r, a_{wr}, g_r, d_r) = \frac{P_{Lr\_core} + P_{Lr\_winding}}{At_{Lr\_core} \cdot h_c} \leq T_{Lr\_rise\_max} \tag{8.96}$$

$$T_{L\_rise}(f_s, N_f, a_{wf}, r_{if}, h_f, d_r) = \frac{P_{L\_core} + P_{L\_winding}}{At_{L\_core} \cdot h_c} \leq T_{L\_rise\_max} \tag{8.97}$$

where At_{Lr_core} and At_{L_core} are the surface areas of the cores of resonant inductor and filter inductor, respectively, and h_c is the coefficient of heat transfer and is typically 10 W/(mK) in this application.

The current density in both resonant inductor and filter inductor windings is limited to 5 A/mm^2.

Based on experience, the window utilization ratio of the inductor cores should be limited to guarantee the implementation:

$$u_{ratio\_Lr} = \frac{N_r a_{wr}}{3d_r^2} \leq u_{ratio\_Lr\_max} \tag{8.98}$$

$$u_{ratio\_L} = \frac{N_f a_{wf}}{\pi r_{if}^2} \leq u_{ratio\_L\_max} \tag{8.99}$$

In order to eliminate the reverse recovery loss, the current changing rate in the diodes should be restrained:

$$\frac{di}{dt} = \frac{V_{dc}}{L_r(N_r, r_i, d_r)} \leq \left.\frac{di}{dt}\right|_{max} \tag{8.100}$$

During the operation of the inverter, long resonant time is not expected, therefore the limitation of the resonant period is limited as:

$$T_{resonant}(C_r, C_7, N_r, r_i, d_r) = 2\pi\sqrt{L_r(3C_r + C_7)} \leq T_{resonant\_max} \tag{8.101}$$

In order to achieve optimum efficiency with relatively small volume of the inverter, the volume of main passive components including clamping capacitors, resonant inductor, and filter inductors is set as a constrained condition for the design. With the results of the optimum efficiency under different volume limitations, the final design parameters could be selected.

The detailed value of each constrained condition is listed in Table 8.3.

8.4.3 Optimization Design

Aiming to design a 30 kW three-phase ZVS inverter, the source voltage V_{dc} is set to be 600 V. Capacitor C_7 is omitted. In this section, the function "fmincon" in Matlab with sequential quadratic

Table 8.3 Constrained conditions.

Parameters	Descriptions	Value
B_{Lr_max}	Maximum magnetic density of the resonant inductor core	0.35 T
B_{L_max}	Maximum magnetic density of the filter inductor core	1.5 T
$T_{Lr\_rise\_max}$	Maximum temperature rise of the resonant inductor	70 °C
$T_{L\_rise\_max}$	Maximum temperature rise of the filter inductor	70 °C
J_{max}	Maximum current density in the windings	5 A/mm^2
$u_{ratio\_Lr\_max}$	Maximum window utilization ratio of resonant inductor	0.2
$u_{ratio\_L\_max}$	Maximum window utilization ratio of filter inductor	0.35
di/dt_{max}	Maximum current changing rate	100 A/μs
$T_{resonant\_max}$	Maximum resonant period	5 μs

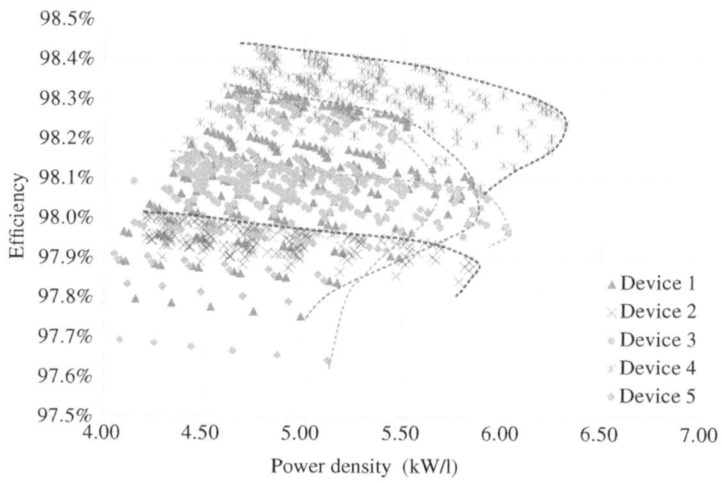

Figure 8.26 Optimization results of different switching modules.

programming algorithm is utilized to attain the optimization results. The optimized results of different switching modules are shown in Figure 8.26.

In Figure 8.26, the horizon axis is the power density of the ZVS inverter, including heat sink, resonant inductor, clamping capacitor, and filter inductor; the vertical axis is the optimum efficiencies under different power density limits and different frequency conditions. The dot lines in this figure show the optimization Pareto front of different IGBT devices. The results reveal that the optimum efficiency is restricted by the power density. Since the case with IGBT device 4 shows highest efficiency and the highest power density limit, this IGBT device is selected for further test. The design goal is to achiéve the highest power density with reasonable efficiency [8, 9].

The theoretical design variables and the corresponding variables utilized in the experiment are listed in Table 8.4.

Table 8.4 Optimization design results.

Variables or conditions	Descriptions	Theoretical values
f_s	Switching frequency	16 kHz
C_r	Resonant capacitor	4.8 nF
N_r	Number of the winding turns (L_r)	7
a_{wr}	Cross-sectional area of the winding (L_r)	16 mm^2
g_r	Air gap (L_r)	2.9 mm
d_r^2	Cross-sectional area of the core (L_r)	467 mm^2
N_f	Number of the winding turns (L)	77
a_{wf}	Cross-sectional area of the winding (L)	14.7 mm^2
r_{if}	Internal radius of the core (L)	32.1 mm
d_f	Width of the core (L)	33.8 mm
h_f	Height of the legs (L)	43.1 mm
L	Inductance of the filter inductor	1.08 mH
Vol_{Cr}	Volume of the clamping capacitors	0.09 l
Vol_{Lr}	Volume of the resonant inductor	0.12 l
Vol_L	Volume of the filter inductor	2.3 l
Vol_{HS}	Volume of the heat sink	2.14 l

8.5 Prototype and Experimental Results

A 30 kW inverter prototype was built according to the optimization design parameters. Ferrite EE55A core of KeDa Magnetoelectricity Company is used in the resonant inductor. Two paralleled Litz wires are selected to implement the windings. Each Litz wire has 1000 strands with individual strand diameters of 0.1 mm. A former prototype was designed by traditional method in [1].

The comparison between the former prototype and improved resonant inductor is shown in Figure 8.27. The volume of the improved resonant inductor is 61% smaller than the former one.

As Figure 8.28 shows, the filter inductor in the prototype is composed of silicon iron powder magnetic cores and flat copper wires made by Tamura Corporation. The volume of the improved filter inductor is only 27% of that of the former prototype.

In Figure 8.29, the improved clamping capacitor is presented. The volume of the clamping capacitor has been reduced by 76% (from 0.26 to 0.09 l) due to the optimization.

The variables of the prototype are listed in Table 8.5. It could be observed that the parameters of the prototype follow the theoretical optimized results.

The experimental circuit is shown in Figure 8.30. The parameters are listed in Table 8.6.

The measured current THD is 3.0%, which meets the design requirement. The efficiency was measured by Yokogawa WT1800 power analyzer, and the measured efficiencies at different output power levels are shown in Figure 8.31. The results reveal that the highest efficiency is 98.3%. Compared to the prototype without optimization, the efficiencies are 0.5% higher at the low output power condition and nearly 0.2% at high output power condition.

Calculated by the volume of resonant inductor, clamping capacitor, filter inductors, and heat sink, the power density of the prototype is demonstrated in Figure 8.32. The measured point is close to the design point. The power density of the inverter prototype after optimization is 6.27 kW/l, which is 2.2 times of the former one.

Figure 8.27 Prototype of the improved resonant inductor.

Figure 8.28 Prototype of the improved filter inductor.

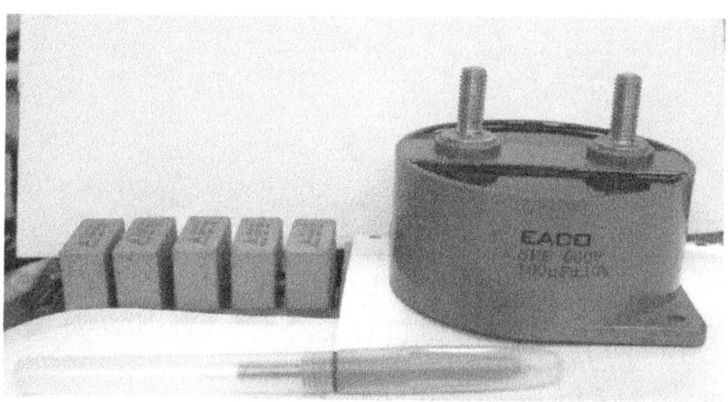

Figure 8.29 Prototype of the improved clamping capacitor.

Table 8.5 Parameters of the inverter prototype.

Variables	Descriptions	Design value	Values in prototype
f_s	Switching frequency	16 kHz	16 kHz
C_r	Resonant capacitor	4.8 nF	4.8 nF
N_r	Number of the winding turns (L_r)	7	6
a_{wr}	Cross-sectional area of the winding (L_r)	16 mm^2	15.7 mm^2
g_r	Air gap (L_r)	2.9 mm	3.0 mm
d_r^2	Cross-sectional area of the core (L_r)	467 mm^2	424 mm^2
L_r	Inductance of the resonant inductor	9.5 μH	10 μH
N_f	Number of the winding turns (L)	77	70
a_{wf}	Cross-sectional area of the winding (L)	14.7 mm^2	12 mm^2
r_{if}	Internal radius of the core (L)	32.1 mm	28.6 mm
d_f	Width of the core (L)	33.8 mm	22.2 mm
h_f	Height of the legs (L)	43.1 mm	14.9 mm × 3
L	Inductance of the filter inductor	1.08 mH	1.1 mH
Vol_{Cr}	Volume of the clamping capacitors	0.09 l	0.09 l
Vol_{Lr}	Volume of the resonant inductor	0.12 l	0.14 l
Vol_L	Volume of the filter inductor	2.3 l	2.1 l
Vol_{HS}	Volume of the heat sink	2.14 l	2.45 l

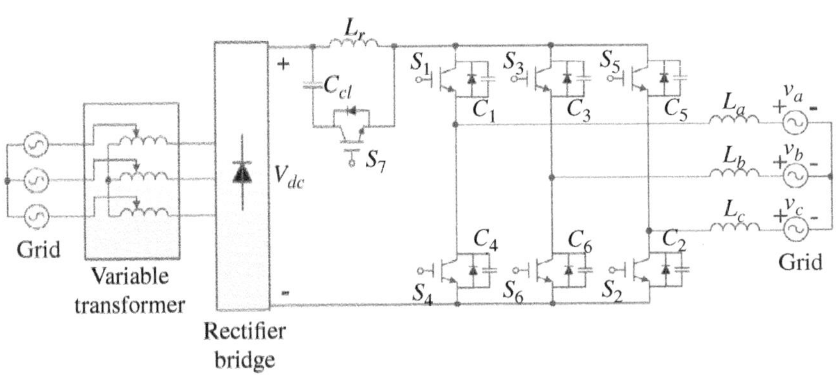

Figure 8.30 Experimental circuit.

Table 8.6 Experiment parameters.

Parameters	Descriptions	Value
P	Output power	30 kW
v_a, v_b, v_c	Phase voltage of the grid	220 Vrms
f_l	Frequency of the grid	50 Hz
L_a, L_b, L_c	Filter inductance	1.1 mH
V_{dc}	Source voltage	600 V
C_{cl}	Capacitance of the clamping capacitor	110 μF

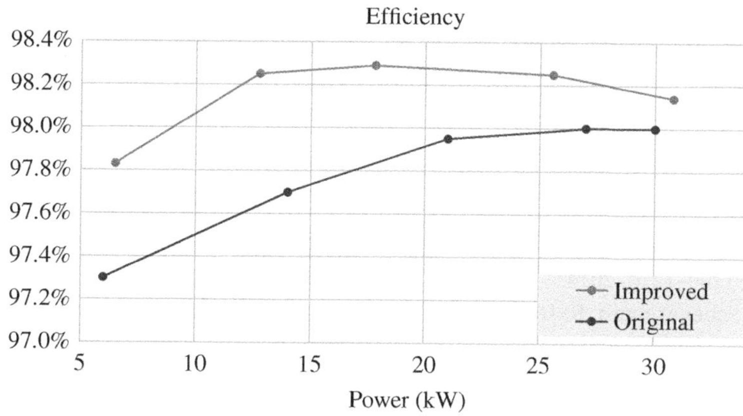

Figure 8.31 Measured conversion efficiency.

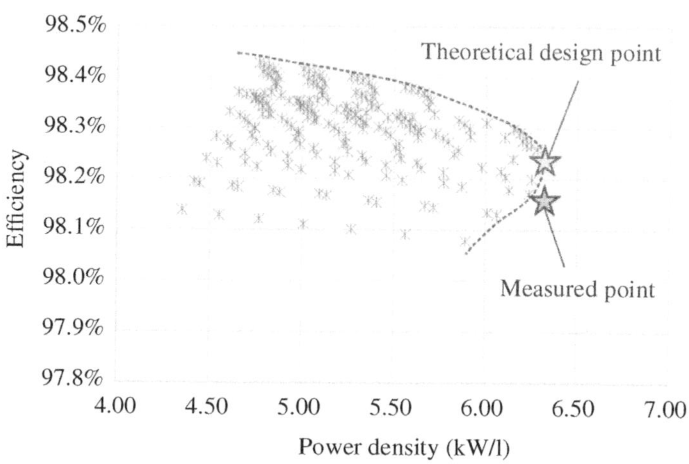

Figure 8.32 Measured efficiency and power density.

8.6 Summary

Conventional design methodologies for a ZVS inverter involve a subset of design variables and tuning several of the remaining design variables and observing the results. This step is repeated, which requires extensive efforts to obtain acceptable results. This book utilizes an optimization algorithm to optimize the efficiency and dimensions of the ZVS inverter. In order to evaluate the turn-off loss of the IGBT devices, measured data of turn-off loss under different conducting currents and buffering capacitances are utilized to build the loss models. The dimensions of the main passive components in the circuits such as the resonant inductors, clamping capacitors, and filter inductors are all modeled and considered in the design procedure. The optimum efficiencies under different switching frequencies and volumes of the passive components for five different IGBT devices are obtained with the proposed design method. With the optimum selection of the switching module, switching frequency, a 30 kW prototype was built and tested. According to the optimized design, the efficiency of the prototype has been increased by 0.2 – 0.5% compared to the prototype without optimization. Besides, the volume of the resonant inductor, clamping capacitor, and the filter inductor has been reduced by 61, 76, and 73%, respectively, compared to the former prototype.

References

1 R. Li and D. Xu, "A zero-voltage switching three-phase inverter," *IEEE Transactions on Power Electronics*, vol. **29**, no. 3, pp. 1200–1210, 2014.

2 J. Lutz, H. Schlangenotto, U. Scheuermann, R. De Doncker, *Semiconductor Power Devices*, Germany: Springer, 2011.

3 V. J. Thottuvelil, T. G. Wilson, H. A. Owen Jr., "High-frequency measurement techniques for magnetic cores", *IEEE Transactions on Power Electronics*, vol. **5**, no. 1, pp. 41–53, 1990.

4 J. Liu, T. G. Wilson, R. C. Wong, R. Wunderlich, F. C. Lee, "A method for inductor core loss estimation in power factor correction applications," in *Proc. 17th Annu. IEEE Appl. Power Electron. Conf. Expo. (APEC)*, pp. 439–445, 2002.

5 C. P. Steinmetz, "On the law of hysteresis", *AIEE Transactions*, vol. **9**, pp. 3–64, 1892. Reprinted under the title "A Steinmemtz contribution to the ac power revolution", introduction by J. E. Brittain, in Proc. IEEE, vol. 72, no. 2, pp. 196–221, 1984.

6 U. Drofenik, G. Laimer, J. W. Kolar, "Theoretical converter power density limits for forced convection cooling," in *Proc. Int. PCIM Europe*, pp. 608–619, 2005.

7 W. G. Hurley and W. H. Wolfle, *Transformers and Inductors for Power Electronics. Theory, Design and Applications*, United Kingdom: Wiley, 2013.

8 C. Du, D. Xu, N. He and N. Zhu, "Modeling and optimization of a zero-voltage switching inverter for high efficiency and miniaturization," *IEEE Transactions on Power Electronics*, 2017, vol. **1**, no. 32, pp. 150–163.

9 C. Du, W. G. Hurley, D. Xu, "Design methodology of resonant inductor in a ZVS inverter," *IEEE Journal of Emerging and Selected Topics in Power Electronics*, vol. **3**, no. 4, pp. 1142–1150, 2015.

9

Design of the Resonant Inductor

The resonant inductor flows through a high-frequency triangle current. Design of the resonant inductor is critical for both loss and size reduction. In this chapter, the design of the resonant inductor is introduced. Different winding structures of the inductor are investigated with respect to skin effect, proximity effect, and fringing effect. The optimization procedure of the resonant inductor is also discussed.

9.1 Introduction

The resonant inductor is an indispensable part of the zero-voltage-switching (ZVS) converters, which will not only affect the soft-switching operation but also increase volume and loss for the system. The resonant inductor is not a standard component like switching devices. High-frequency triangle current flows through the resonant inductor, which is different from the traditional AC filter inductor.

The typical current and voltage waveforms of the resonant inductor L_r in the ZVS converter are shown in Figure 9.1. During the turn-on duration of the auxiliary switch, a constant clamping capacitor voltage V_{Cl} is applied on the inductor and its current i_{Lr} decreases linearly. During the turn-off duration of the auxiliary switch, a high voltage V_{dc} is applied on the inductor and its current i_{Lr} increases very steeply. The resonant inductor current i_{Lr} can be simplified as a triangular wave approximately with the peak value I_1 and valley value I_2.

In this chapter, the design of the resonant inductor is introduced. For the calculation of skin effect, proximity effect, and eddy current effect, finite element analysis (FEA) is used. Different winding structures are compared, such as winding position and winding thickness. Finally, a design of the resonant inductor applied in a 30 kW ZVS inverter is introduced.

9.2 Fundamental of Inductor

In the inductor design, the ferrite core is selected, since its core loss is smaller in the high-frequency operation compared with other core materials. However, the ferrite has a low saturation flux density and is sensitive to working temperature [1–4].

The simple structure of an inductor with air gap is shown in Figure 9.2. The magnetic core is made up of magnetic material, and the winding is composed of N turns conductor. It is assumed

Soft-Switching Technology for Three-phase Power Electronics Converters, First Edition.
Dehong Xu, Rui Li, Ning He, Jinyi Deng, and Yuying Wu.

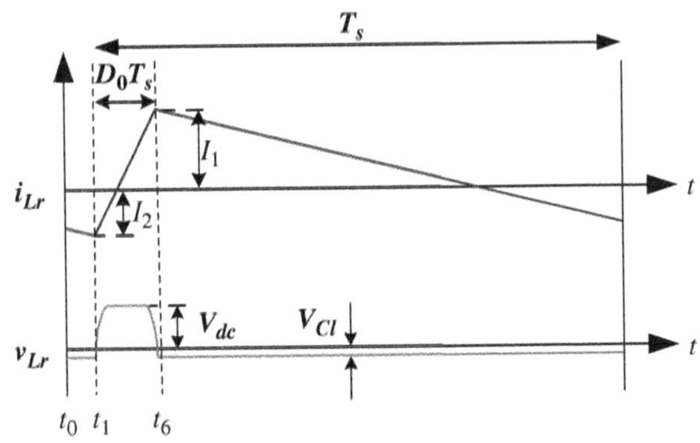

Figure 9.1 Current and voltage waveforms of the resonant inductor.

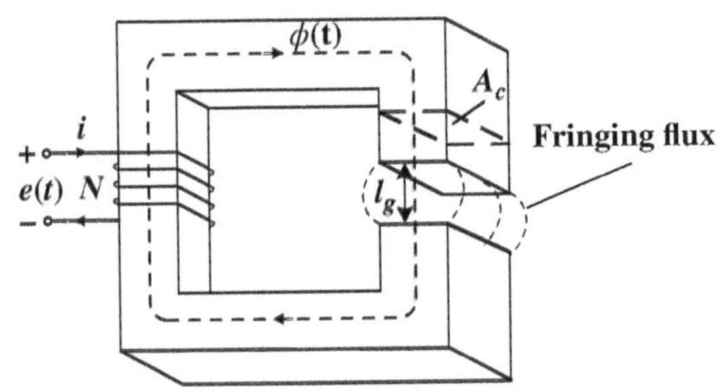

Figure 9.2 Inductor with air gap.

that the instantaneous current through the winding is i. By applying the Ampere-turns equilibrium law to the closed magnetic loop around a magnetic core with air gap, the following equation can be established.

$$H_c l_c + H_g l_g = Ni \tag{9.1}$$

where H_c is the magnetic field intensity in the magnetic core, H_g is the magnetic field intensity in the air gap, l_c is the loop length of the core, and l_g is the length of the air gap.

The H_c and H_g also can be represented by magnetic flux density.

$$H_c = B_c/(\mu_r \mu_0) \tag{9.2}$$

$$H_g = B_g/\mu_0 \tag{9.3}$$

where $\mu_0 = 4\pi \times 10^{-7}$H/m, μ_r is the relative permeability, B_c is the magnetic flux density in the magnetic core, and B_g is the magnetic flux density in the air gap.

The magnetic flux ϕ is mainly concentrated in the magnetic core and air gap, and the magnetic flux density is the magnetic flux in each unit cross section. Therefore,

$$B_c = \phi/A_c \tag{9.4}$$

$$B_g = \phi/A_g \tag{9.5}$$

where A_c and A_g are the cross-section areas of the core and air gap, respectively.

Usually, the A_g is slightly larger than A_c because of the fringing effect in the air gap. The magnitude of the fringing flux is related to the length of the air gap. It can assume $A_g = A_c$ to neglect the influence of fringing flux and simplify the calculation.

Combining Eqs. (9.1)–(9.5), the following equation is obtained.

$$\left(\frac{l_c}{\mu_r \mu_0 A_c} + \frac{l_g}{\mu_0 A_g}\right)\phi = Ni \tag{9.6}$$

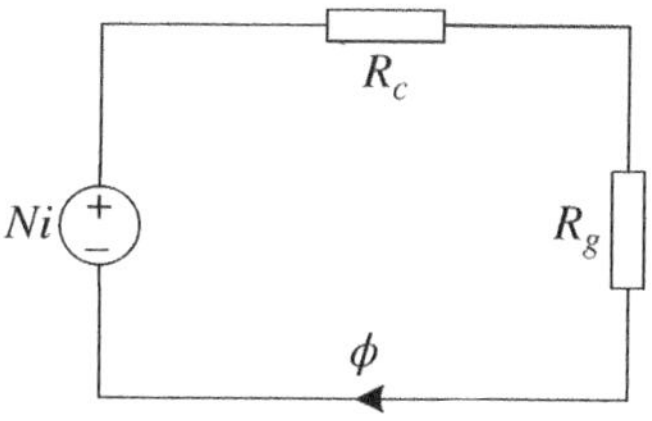

Figure 9.3 Equivalent magnetic circuit model.

By introducing reluctance of the core R_c and reluctance of the air gap R_g, Eq. (9.6) is rewritten as:

$$(R_c + R_g)\phi = Ni \tag{9.7}$$

where

$$R_c = \frac{l_c}{\mu_r \mu_0 A_c} \tag{9.8}$$

$$R_g = \frac{l_g}{\mu_0 A_g} \tag{9.9}$$

In terms of Eq. (9.7), an equivalent magnetic circuit model is drawn in Figure 9.3. Ni represents magnetic potential, and each magnetic resistance has a corresponding magnetic voltage drop. It is observed that the magnetic circuit also satisfies the Kirchhoff's voltage law.

Since the magnetic flux ϕ stimulates all N turns winding, the flux linkage ψ is calculated by $N\phi$. The flux linkage ψ can also be calculated by Li. Therefore, the following relationship is established.

$$\psi = N\phi = Li \tag{9.10}$$

By combining Eqs. (9.7) and (9.10), the inductance L can be represented as:

$$L = \frac{\psi}{i} = \frac{N\phi}{i} = \frac{N^2}{R_c + R_g} \tag{9.11}$$

So far, important instantaneous magnetic relationship equations are established (i.e. magnetic field intensity H, magnetic density B, magnetic flux ϕ, magnetic linkage ψ). Faraday's law reveals the relationship between the voltage of the winding and time-varying magnetic linkage ψ, which is written as:

$$e(t) = \frac{d\psi(t)}{dt} = N\frac{d\phi(t)}{dt} \tag{9.12}$$

where $e(t)$ is the induced voltage, and the direction relationship between the induced voltage $e(t)$ and magnetic flux $\phi(t)$ is shown in Figure 9.2.

9.3 Design Methodology

Core selection is the first step for a designer and usually can be achieved by estimating the A_p value, which is the product of the window area A_e and cross-section area A_c of the core [5, 6]. The areas of A_e and A_c are shown in Figure 9.4.

9.3.1 Cross-section Area of the Core A_c

Since the magnetic flux ϕ equals BA_c, Eq. (9.10) is rewritten as:

$$Li = N\phi = NBA_c \tag{9.13}$$

By rearranging Eq. (9.13), the cross-section area A_c is derived as:

$$A_c = \frac{Li}{NB} \tag{9.14}$$

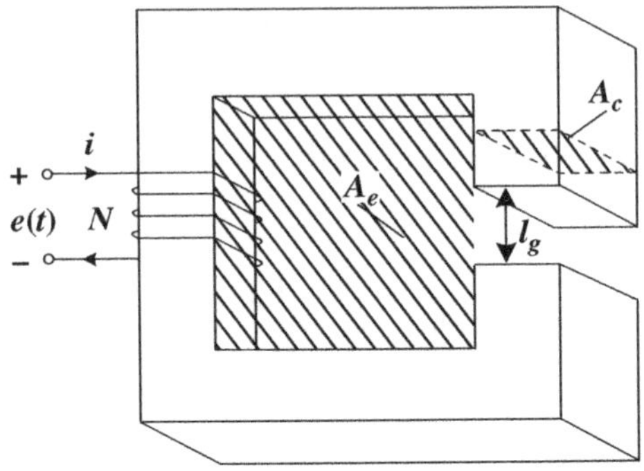

Figure 9.4 Window area A_e and cross-section area A_c.

Magnetic material has a limit on the maximum magnetic density B_{max}, which is given in the magnetic material datasheet. It is necessary to ensure that work magnetic density does not exceed B_{max}. Otherwise, it will cause higher core loss. It is specified that the maximum current I_{max} corresponds to B_{max}. Thus, the A_c is rewritten as:

$$A_c = \frac{LI_{\max}}{NB_{\max}} \tag{9.15}$$

9.3.2 Window Area A_e

In the window area, the winding is filled. The window utilization factor K_w is introduced to describe the proportion coefficient between the cross-section area of winding A_w and the window area A_e. The K_w varies between 0.2 and 0.8 depending on the type of winding.

$$A_w = K_w A_e \tag{9.16}$$

The maximum current density of the winding J_0 is calculated by the maximum root mean square (RMS) current I_{rms} and the A_w. It is assumed that the RMS current of the N conductors is the same and uniformly distributed. Then, the J_0 is expressed as:

$$J_0 = \frac{NI_{rms}}{A_w} \tag{9.17}$$

Combining the Eq. (9.16) and (9.17), the A_e is solved as:

$$A_e = \frac{NI_{rms}}{K_w J_0} \tag{9.18}$$

9.3.3 Area-product A_p

The area-product A_p is obtained by multiplying A_c with A_e, which is derived as:

$$A_p = A_c A_e = \frac{LI_{\max}}{NB_{\max}} \times \frac{NI_{rms}}{K_w J_0} = \frac{LI_{\max} I_{rms}}{B_{\max} K_w J_0} \tag{9.19}$$

The above equation shows that the magnetic core size is independent of N. An increase in L leads to an increase in core size, and a larger B_{max} is helpful to decrease the core size. However, this selection is far away from the best design because the skin effect, proximity effect, fringing effect are ignored.

9.3.4 Turns of Winding N

According to the calculation of A_p, a core with an approximate A_p can be selected from the core datasheet. Once a core is determined, the next step is to calculate the turns of winding N as follows.

$$N = round\left(\frac{LI_{\max}}{A_c B_{\max}}\right) \tag{9.20}$$

9.3.5 Length of the Air Gap l_g

The air gap is introduced to avoid the change of inductance L with the relative permeability μ_r at different work currents, especially in the nonlinear region of μ_r. In the ferrite core, the μ_r is much higher than 1. Therefore, compared with air gap reluctance R_g, the core reluctance R_c is so small that it can be ignored. Then, Eq. (9.11) is simplified as:

$$L \approx \frac{N^2}{R_g} \approx \frac{N^2 \mu_0 A_g}{l_g} \tag{9.21}$$

Then, the l_g can be solved by making A_g equal to A_c approximately.

$$l_g \approx \frac{N^2 \mu_0 A_c}{L} \tag{9.22}$$

9.3.6 Winding Loss P_{dc}

According to the triangular inductor current shown in Figure 9.5, the average value I_{dc} and peak-to-peak value ΔI of the inductor current are derived as:

$$I_{dc} = \frac{I_1 + I_2}{2} \tag{9.23}$$

$$\Delta I = I_1 - I_2 \tag{9.24}$$

Therefore, the instantaneous current of the resonant inductor $i_{Lr}(t)$ in a switching period can be represented as following.

$$i_{Lr}(t) = \begin{cases} I_{dc} - \dfrac{\Delta I}{2} + \dfrac{\Delta I}{T_0} t & 0 \le t \le T_0 \\ I_{dc} + \dfrac{\Delta I}{2} - \dfrac{\Delta I}{T_s - T_0}(t - T_0) & T_0 \le t \le T_s \end{cases} \tag{9.25}$$

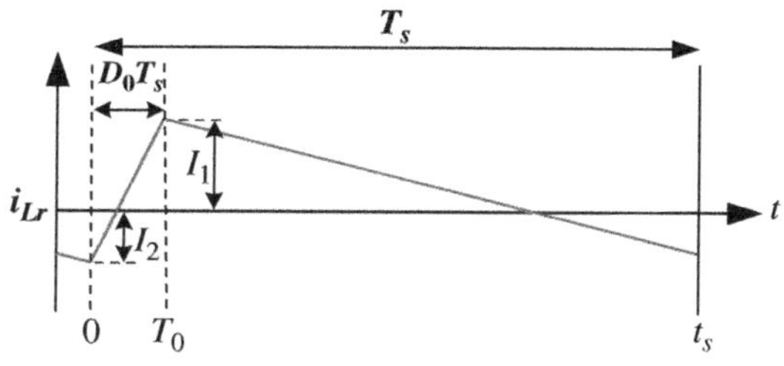

Figure 9.5 Triangular inductor current.

The RMS value of the resonant inductor current i_{Lr} is calculated as:

$$I_{rms} = \sqrt{\frac{1}{T_s}\int_0^{T_s} i_{Lr}^2(t)dt} \tag{9.26}$$

By substituting Eq. (9.25) into (9.26), the RMS value of i_{Lr} is derived as:

$$\begin{aligned} I_{rms} &= \sqrt{\frac{1}{T_s}\left\{\int_0^{T_0}\left(I_{dc} - \frac{\Delta I}{2} + \frac{\Delta I}{T_0}t\right)^2 dt + \int_{T_0}^{T_s}\left[I_{dc} + \frac{\Delta I}{2} - \frac{\Delta I}{T_s - T_0}(t - T_0)\right]^2 dt\right\}} \\ &= \sqrt{I_{dc}^2 + \frac{\Delta I^2}{12}} \end{aligned} \tag{9.27}$$

The resistance of the winding R_{dc} is expressed as:

$$R_{dc} = \rho_w \frac{N \times \text{MLT}}{A_w} \tag{9.28}$$

where ρ_w is the electrical conductivity of the winding material, and MLT is the average length of each turn.

Then the winding loss P_{dc} is obtained by combining Eqs. (9.27) and (9.28).

$$P_{dc} = R_{dc}I_{\text{rms}}^2 = \rho_w \frac{\text{NMLT}\left(I_{\text{dc}}^2 + \frac{\Delta I^2}{12}\right)}{A_w} \tag{9.29}$$

This winding loss calculation method is not accurate since the loss caused by skin effect, proximity effect, and fringing effect in high frequency is ignored [7, 8].

9.3.7 Core Loss P_{core}

The core loss is estimated by the Steinmetz's equation [9]. In our application, core loss per unit volume is calculated as:

$$P_{fe} = K_c f^{\alpha}\left(\frac{\Delta B}{2}\right)^{\beta} \tag{9.30}$$

Therefore, core loss P_{core} is obtained by multiplying P_{fe} with core volume V_c.

$$P_{core} = P_{fe}V_c = K_c f^{\alpha}\left(\frac{\Delta B}{2}\right)^{\beta} V_c \tag{9.31}$$

where K_c, α, β are the core coefficient, which are provided by the core manufacturer.

ΔB is working magnetic density swing. In our application, it can be calculated by combining Eqs. (9.1)–(9.5), (9.24).

$$\Delta B = \frac{\mu_0 N \Delta I}{l_g} \tag{9.32}$$

Combining Eqs. (9.31) and (9.32), P_{core} is derived as:

$$P_{core} = K_c f^{\alpha}\left(\frac{\mu_0 N \Delta I}{2l_g}\right)^{\beta} V_c \tag{9.33}$$

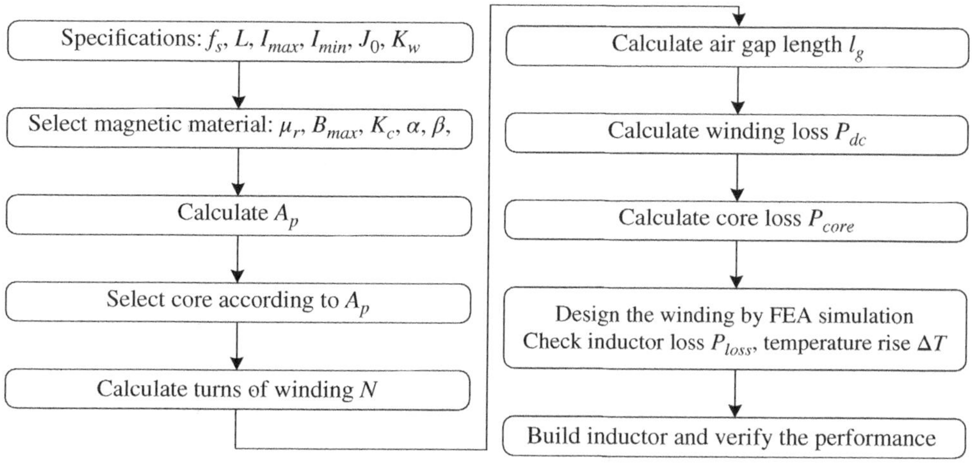

Figure 9.6 Design flowchart.

9.3.8 Design Procedure

Combining the theories and equations introduced above, a design flowchart is shown in Figure 9.6 and the design procedure can be summarized as follows [10, 17]:

Step 1: specify the parameters of the designed inductor.
Step 2: select one magnetic material which ferrites are preferred.
Step 3: calculate A_p to choose a proper core.
Step 4: according to the A_p value, a core is chosen based on the magnetic core manufacturer' datasheet.
Step 5: calculate turns of winding N.
Step 6: calculate air gap length l_g.
Step 7: calculate winding loss P_{dc}.
Step 8: calculate core loss P_{core}.
So far, one predesigned inductor has been finished.
Step 9: To estimate the inductor loss P_{loss} and temperature rise ΔT more accurately, FEA simulation is used.
Step 10: build an inductor and verify its performance in a ZVS converter prototype.

9.4 Design Example

In this section, a resonant inductor applied in a 30 kW ZVS three-phase grid inverter is introduced. The inductance $L = 10\,\mu$H. The maximum inductor current in a switching period is shown in Figure 9.1, where the peak value $I_1 = 138$ A, the valley value $I_2 = -38$ A, the RMS value $I_{rms} = 71.3$ A, the turn-off duty of auxiliary switch $D_0 = 0.027$, and the switching frequency $f_s = 16$ kHz. It is assumed that the designed current density $J_0 = 5$ A/mm^2 and window utilization factor $K_w = 0.3$. Therefore, the cross-section area of winding $A_w = I_{rms}/J_0 \approx 14.2$ mm^2.

The next step is to choose a magnetic material. In this design, the Mn-Zn ferrite DMR95 core produced by DMEGC, Inc., is selected. The DMR95 has maximum magnetic density $B_{max} = 0.33$ T and the relative permeability $\mu_r = 3300$ [11].

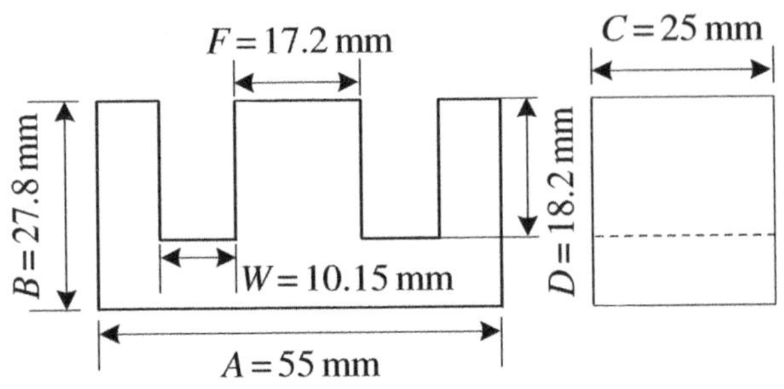

Figure 9.7 EE55A shape size.

According to the design parameters, a preferred area-product A_p is calculated as:

$$A_p = \frac{LI_{\max} I_{rms}}{B_{\max} K_w J_0} = \frac{10 \times 10^{-6} \times 138 \times 71.3}{0.33 \times 0.3 \times 5 \times 10^6} \times 10^{12} = 198730\,\text{mm}^4 \tag{9.34}$$

From DMEGC's core type catalog, a pair of E-type core is finally chosen, which is EE55A and its A_p is about 158868 mm^4. The core shape and size are shown in Figure 9.7.

Based on the EE55A core, the turns of winding N is calculated according to Eq. (9.20).

$$N = round\left(\frac{LI_{\max}}{A_c B_{\max}}\right) = round\left(\frac{10 \times 10^{-6} \times 138}{424 \times 10^{-6} \times 0.33}\right) = 10 \tag{9.35}$$

Then the air gap length l_g is also obtained according to Eq. (9.22).

$$l_g \approx \frac{N^2 \mu_0 A_c}{L} = \frac{10^2 \times 1.257 \times 10^{-6} \times 424 \times 10^{-9}}{10 \times 10^{-6}} \approx 5.3\text{mm} \tag{9.36}$$

Next, the winding loss P_{dc} can be preliminarily evaluated according to Eq. (9.29).

$$P_{\text{dc}} = \rho_w \frac{\text{NMLT}I_{\text{rms}}^2}{A_w} = 1.75 \times 10^{-8} \frac{10 \times 0.201 \times 71.3^2}{14.257 \times 10^{-6}} = 12.54\,\text{W} \tag{9.37}$$

Here, copper foil is used as the winding. To avoid the skin effect in the winding, it is recommended that the thickness of the copper foil should be at least less than twice the skin depth. The skin depth ΔCu is calculated as:

$$\Delta Cu = \frac{1}{\sqrt{\pi f_s \mu_0 \sigma_{Cu}}} \tag{9.38}$$

where $\sigma_{Cu} = 58 \times 10^6$ is the copper conductivity. In this application, $\Delta Cu = 0.5$ mm.

From the DMR95 datasheet given by DMEGC, Inc., the core loss coefficients are obtained ($K_c = 8.4683\text{e-}8$, $\alpha = 1.87873$, $\beta = 2.52073$). Therefore, the core loss P_{core} can be evaluated based on Eq. (9.33).

$$P_{core} = K_c f^{\alpha} \left(\frac{\Delta B}{2}\right)^{\beta} V_c = \left(8.4683 \times 10^{-8}\right) \times (16)^{1.87873} \times (325)^{2.52073} \times 52.152 \times 10^{-3} \approx 1.73\,\text{W} \tag{9.39}$$

Actually, due to the skin effect, proximity effect, and fringing effect, the actual winding loss P_{dc} and core loss P_{core} should be larger than the calculated value. Therefore, the FEA tool is used to optimize the inductor design.

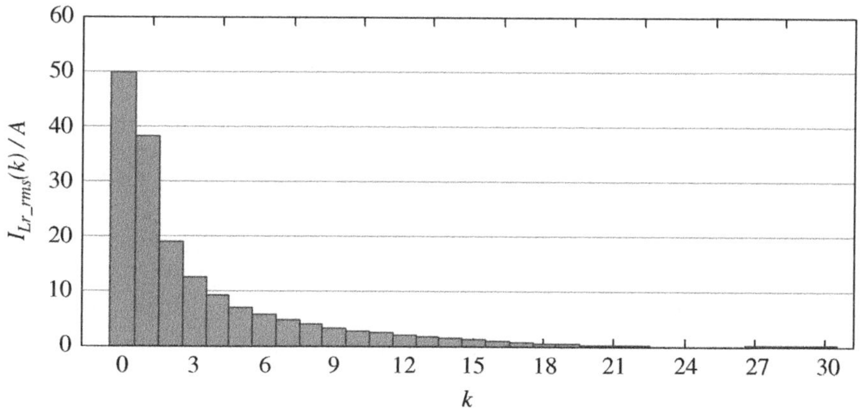

Figure 9.8 Sinusoidal components of different frequencies.

In the FEA simulation, the preliminary design is first verified. By adjusting the length of the air gap l_g and turns of winding N to ensure the inductance $L = 10\ \mu H$ and $B_{max} < 0.3$ T. Finally, $N = 6$ and $l_g = 3$ mm is determined.

In the following, the optimized winding position and winding thickness of barrel winding are discussed. Then, another structure called flat winding is discussed. Finally, the comparison of the barrel winding and flat winding is carried out.

9.4.1 Barrel Winding Discussion

The resonant inductor current i_{Lr} can be decomposed into sinusoidal components of different frequencies by adopting the Fourier analysis as shown in Figure 9.8. In Figure 9.8, k is the order of the current harmonics. When $k = 0$, the Y-axis value represents the amplitude of the DC component of the current. When $k = 1$, the Y-axis value represents the RMS of the switching frequency component of the current. Range from DC to the 30th order AC component is a good approximation to the resonant inductor current. Therefore, these current components ($k = 0, 1, \ldots, 30$) are considered when evaluating winding loss.

In barrel winding design, the winding is arranged vertically in the window area as shown in Figure 9.9 [17]. To satisfy the designed cross-section area of winding $A_w \approx 14.2\ \text{mm}^2$, the copper foil with a width of 35 mm and a thickness of 0.45 mm is selected. The distance between two adjacent copper foils is 0.15 mm to provide reliable isolation. Based on the selected window utilization K_w, the total conducting area is fixed. Theoretically, thinner but larger quantities of the parallel copper foils may result in less skin effect. Nevertheless, when the distance between the air gap and the winding decreases, the fringing effect becomes aggravated. At the same time, the multilayer parallel copper foil also causes the proximity effect. Therefore, many aspects need to be considered and trade-offs are required to realize a minimum winding loss. Accordingly, it is necessary to compare different winding thicknesses and winding positions to find the optimum case.

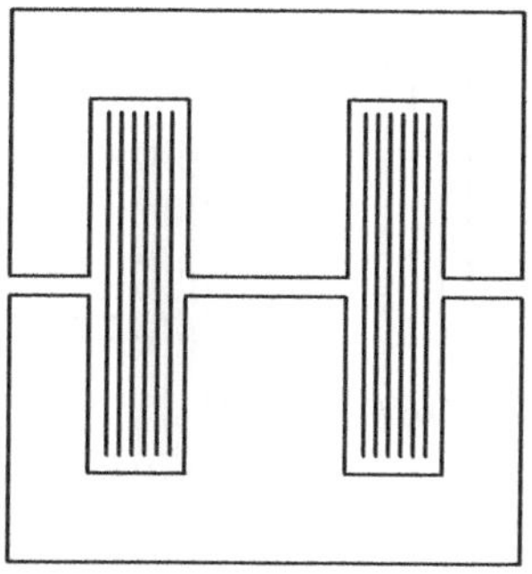

Figure 9.9 The barrel winding structure.

9.4.1.1 Winding Position Discussion

First, it is assumed that the winding is located in the middle of the window to reduce the loss generated by the fringing flux. Through the FEA tool, sinusoidal components of different frequencies are used as current excitation to obtain the winding loss at the corresponding frequency. The simulation results are shown in Figure 9.10. The winding loss caused by the switching frequency component ($k = 1$) is the largest, while the winding loss caused by the DC component is not a dominant part. The high-frequency AC components ($k > 15$) result in less winding loss.

Six turns are marked from A to F as shown in Figure 9.11. The winding loss caused by first order AC component in each turn is calculated by the FEA tool as shown in Figure 9.12. The turn F is the closest to the center pillar. Its winding loss is the largest and accounts for nearly 47%. Then, the winding loss of turn A near the side pillar is the second. It can be concluded that the sum of the winding loss of turn A and turn F accounts for 76% of the entire winding loss. Therefore, it is necessary to reduce the winding loss of turn A and turn F.

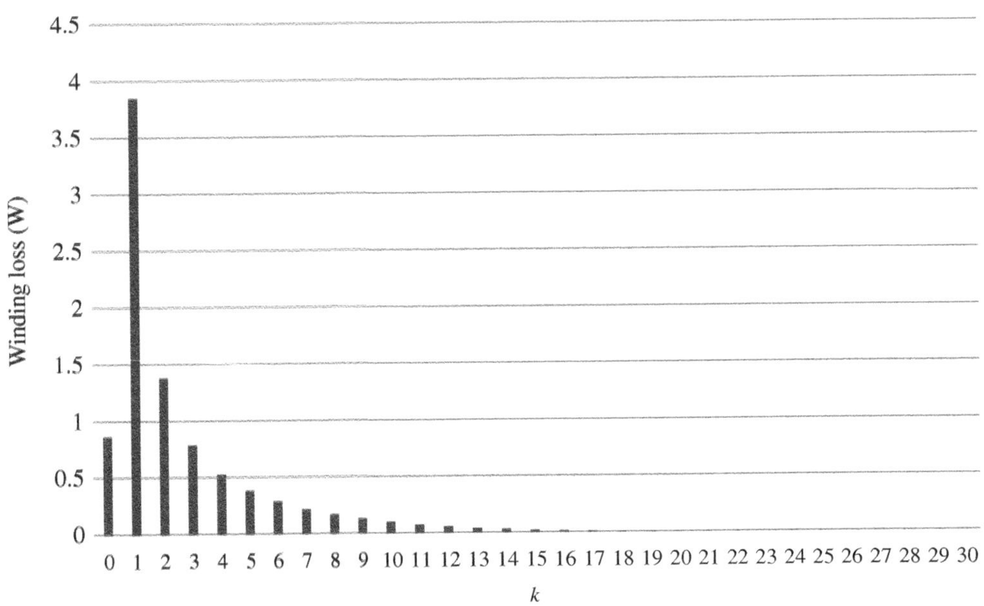

Figure 9.10 Winding loss of different frequencies.

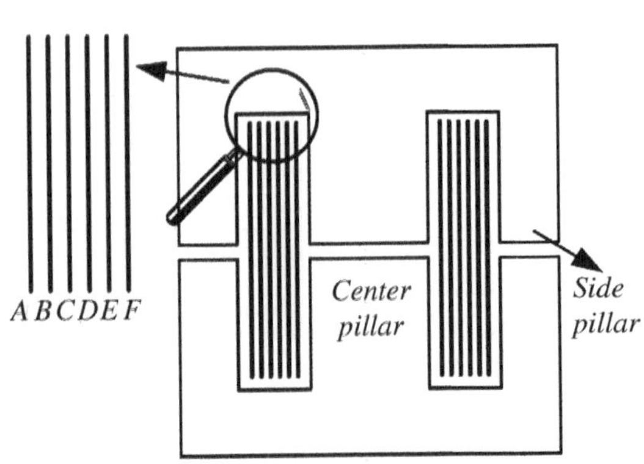

Figure 9.11 The definition of the winding.

9.5 Design Verification

The winding loss of the barrel winding and the flat winding is analyzed and compared by 2D FEA simulation. As shown in Figure 9.29a, the minimum winding loss of the barrel winding is 8.2 W and that of the flat winding is 10.3 W. The inductor volume of the barrel winding is 0.116 L and that of the flat winding is 0.142 L as shown in Figure 9.29b. The winding loss of the barrel winding is 20% smaller than that of the flat winding and its volume is also reduced by 19%. Therefore, the barrel winding is the optimal choice.

9.5.1 Simulation Verification

The winding loss has been estimated before. To understand the total loss of the inductor, the Ansys Maxwell transient mode is used. In this simulation mode, the inductor current in a switching period is used as the current excitation source, and the total loss of the resonant inductor includes the winding loss and core loss is obtained.

The simulation results are shown in Figure 9.30. It can be found that most of the core loss occurs when the inductor current changes sharply. The winding loss is no longer constant, which is varied with the change of the inductor current. The rate of change of the winding loss is large when the inductor current increases dramatically. This is because the significant change of the inductor current induces the magnetic field intensity to change quickly and the fringing flux is relatively strong. The results show that the winding loss is 9.8 W and the core loss is 7.5 W.

According to the previous design, the structure of the resonant inductor is shown in Figure 9.31. The total loss can be uniformly set in the three-dimensional (3D) model, and the temperature rise is estimated. Under natural cooling conditions, the maximum temperature rise is 64.6 °C, which is lower than the design limit of 70 °C and meets the design requirements.

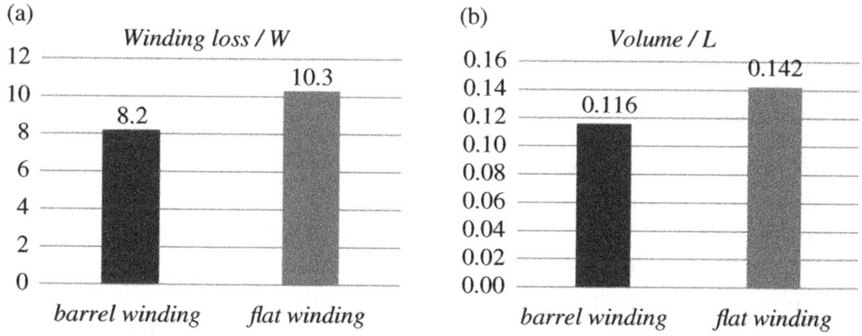

Figure 9.29 Comparison of the different winding structures: (a) winding loss; (b) volume.

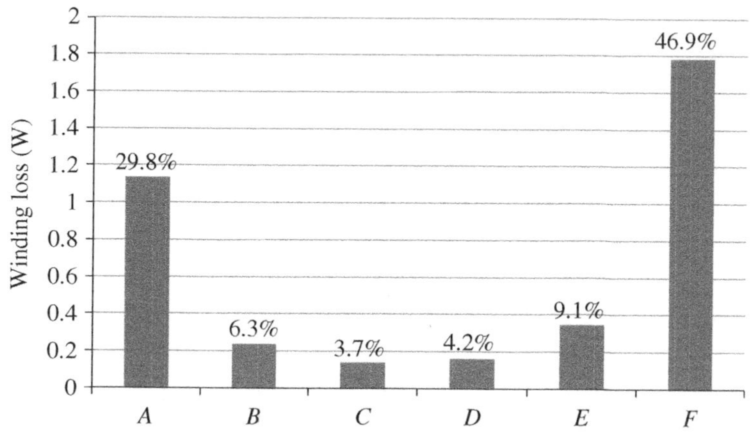

Figure 9.12 Winding loss of each turn in the first order AC component.

In order to illustrate the above results, the specific distribution of the winding loss and magnetic field lines is shown in Figure 9.13. The gray level in the winding area represents the magnitude of winding loss. The light gray area represents high loss distribution and the dark gray area represents low loss distribution. Besides, the line represents the contour of magnetic field intensity. There is high winding loss near the air gap compared with other areas, and the reason for this is that the fringing flux near the air gap causes the eddy current in the winding, which is the fringing effect.

Although turn *A* and turn *F* have the same distance from the air gap, their winding loss is different as shown in Figure 9.12. Therefore, the discussion about the distance between the winding and the air gap is worthwhile.

In the *X*-axis direction, it is assumed that the winding position closest to the center pillar is 0% and that closest to the side pillar is 100% as shown in Figure 9.14. The winding loss caused by the different current components in different positions is shown in Figure 9.15. It should be noted that the total loss I in Figure 9.15 refers to the sum of the winding loss from the DC component to the 30th AC component. The winding loss rises sharply when the winding is close to the center pillar or the side pillar. The winding loss caused by the first order AC component is the dominant winding loss and has the same trend as the total winding loss. The winding loss caused by the second order AC component is second. However, the winding loss caused by the DC component is relatively small.

Only considering the first order AC component, the winding loss of each turn in different positions is shown in Figure 9.16. The total loss II in Figure 9.16 refers to the sum of the winding loss of all turns. The result reveals that the winding loss of turn *F* is the highest when the winding is close to the center pillar, and it is continuously reduced when the winding is far away from the center pillar. Simultaneously, the winding loss of turn *A* is increased and becomes the dominant winding loss. It can be concluded that the winding loss is optimum when the winding loss of turn *A* is almost the same as the winding loss of turn *F*. The distribution of the winding loss can be interpreted that the winding is induced eddy current due to the fringing flux and the winding loss is increased. If the winding position is closer to the air gap, the influence of fringing flux becomes more intense, which results in more winding loss. The winding loss of the middle turns, such as turn *C* and turn *D*, seems to be insensitive to the winding position. It is considered that the induced eddy current in the outer layer winding weakens the influence of the fringing flux. In other words, the outer layer winding has a certain shielding effect.

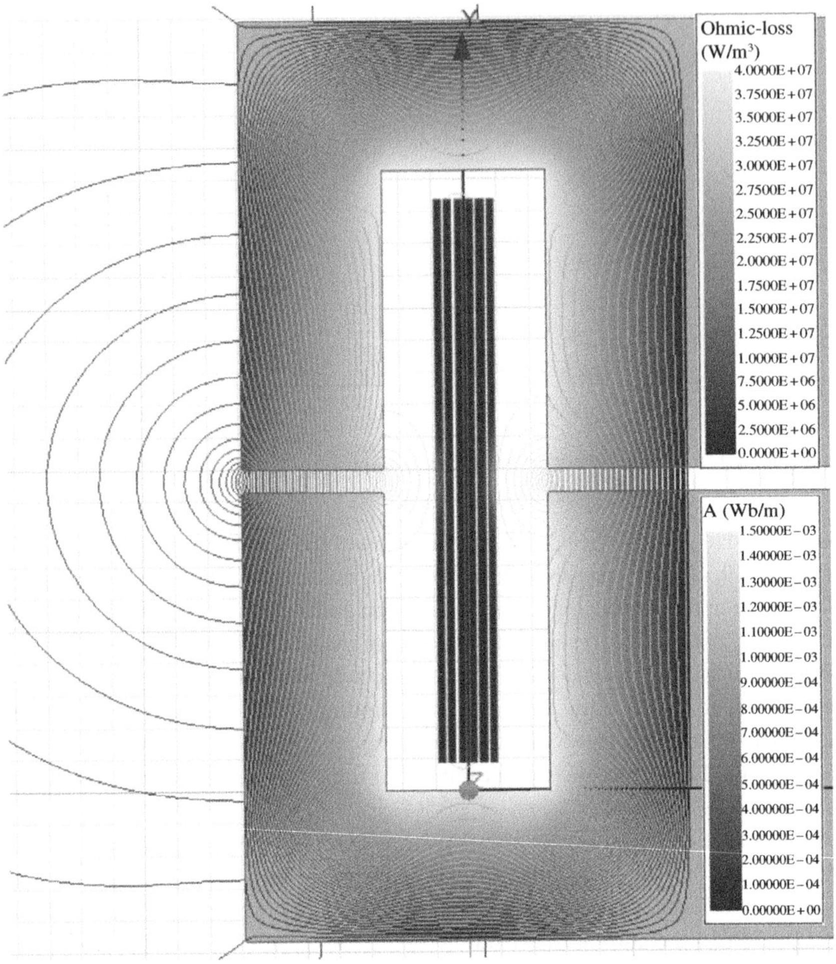

Figure 9.13 Distribution of winding loss and magnetic field lines.

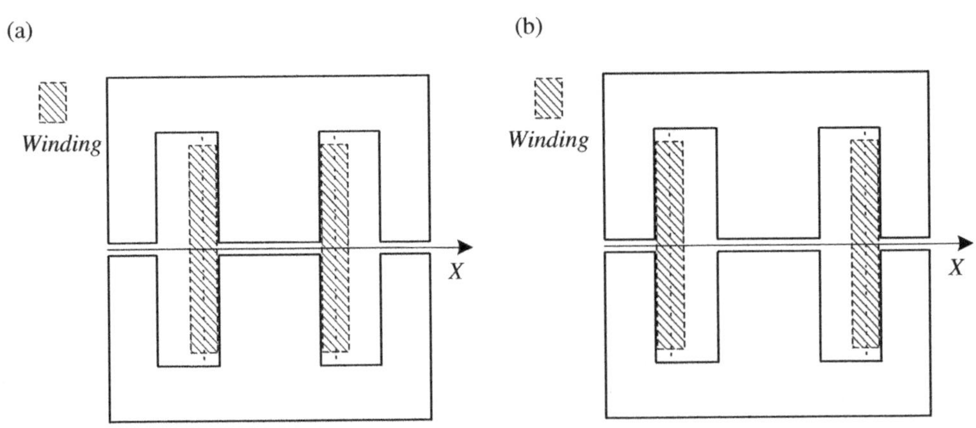

Figure 9.14 The position definition in the core window: (a) 0%; (b) 100%.

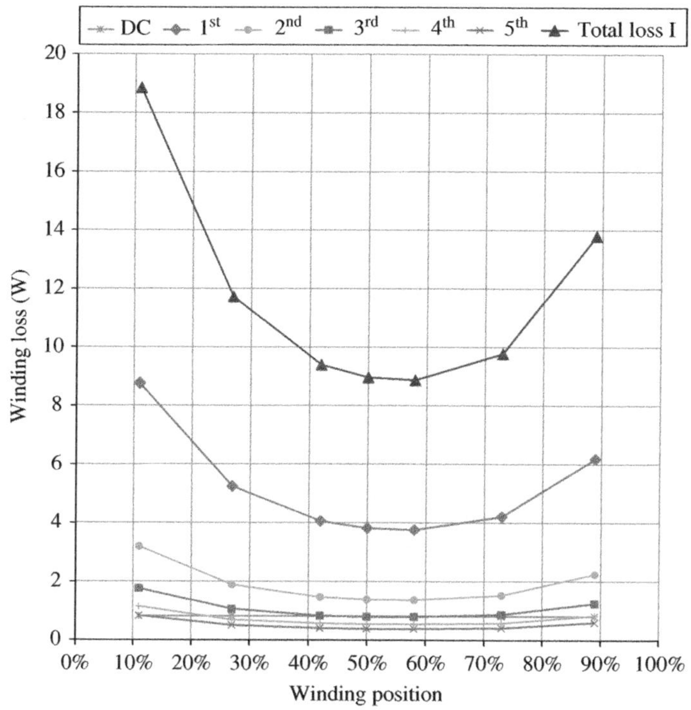

Figure 9.15 Winding loss in different positions.

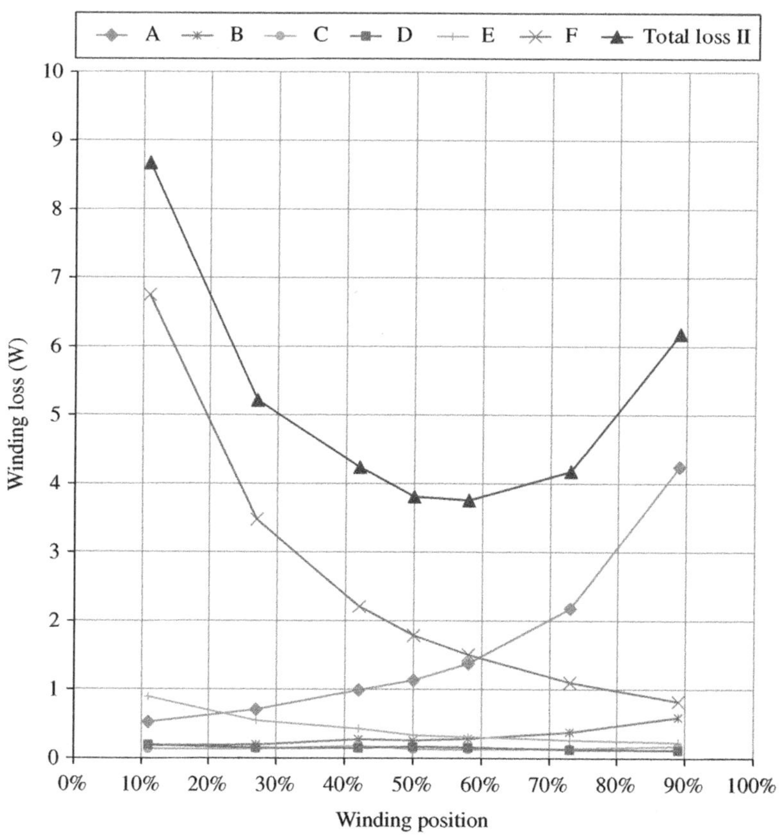

Figure 9.16 Distribution of winding loss in different positions.

9.4.1.2 Winding Thickness Discussion

With a constant insulation distance, a thinner winding thickness will lead to an increase in the distance between the winding and the air gap [12]. As shown in Figure 9.17, the winding position is located at 58% and the winding loss under different winding thicknesses is compared. The results indicate that the winding loss decreases continually when the winding thickness varies from 0.45 to 0.3 mm. When the winding thickness is between 0.3 and 0.35 mm, the winding loss reaches a minimum value. Then, the winding loss begins to increase when the winding thickness decreases continually since the conducting area is reduced. Although the winding loss caused by fringing flux is reduced, the resistance of the winding increases and the conduction loss also increases. The winding loss caused by the DC component increases as the winding thickness decreases. As shown in Figure 9.18, the winding loss of the turn A and F close to the air gap decreases as the winding thickness decreases. However, the turn B to E are less affected by the fringing flux, and their winding loss increases because the conductive area decreases as the winding thickness decreases.

From the above analysis, when the winding position is 58% on *X*-axis, and the winding thickness is between 0.3 and 0.35 mm, the winding loss in the barrel winding structure is optimal, which is 8.2 W.

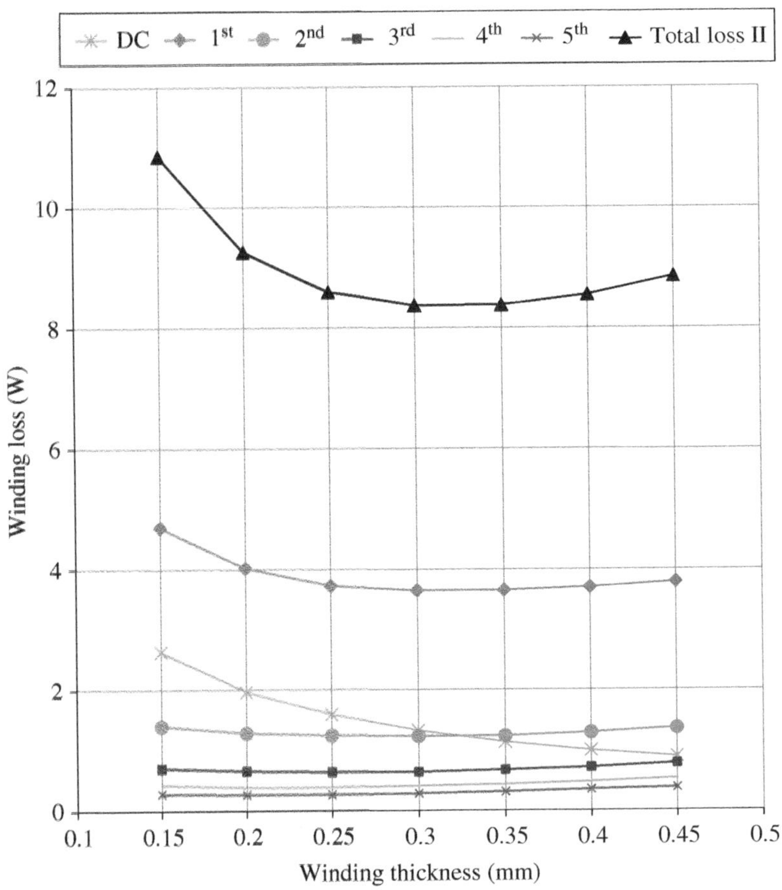

Figure 9.17 Winding loss under different winding thickness.

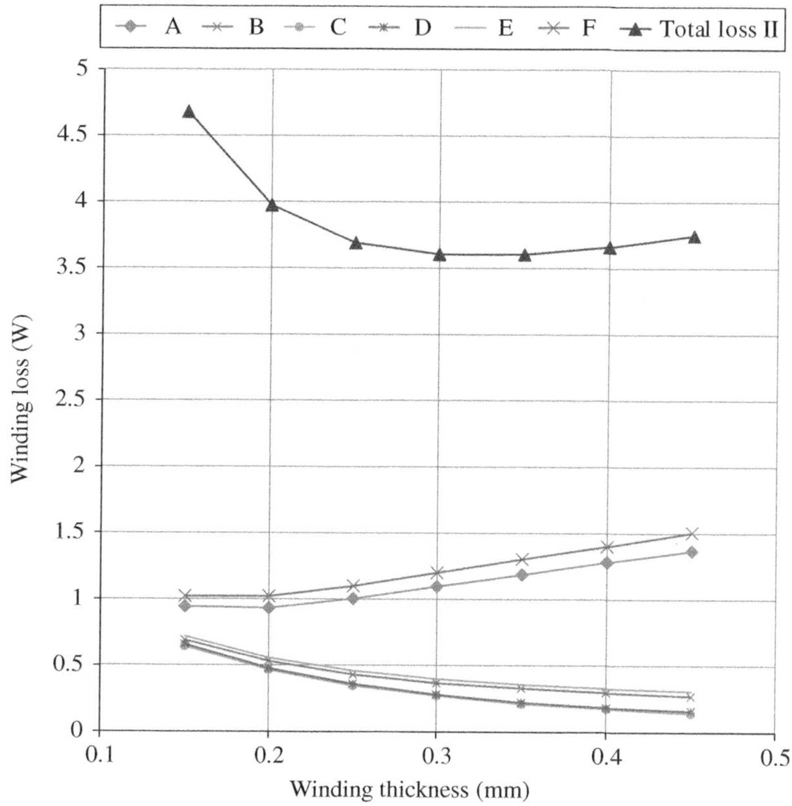

Figure 9.18 Distribution of winding loss under different winding thickness.

9.4.2 Flat Winding Discussion

In the previous section, the optimization process of the barrel winding structure is discussed. Apart from the barrel winding structure, there is another structure called flat winding as shown in Figure 9.19 [17]. In flat winding, the copper foil is placed horizontally. The winding consists of a copper foil with a thickness of 0.45 mm and a width of 8.2 mm. The winding is placed as far away as possible from the air gap to avoid the fringing flux. The cross-section area of the conducting area of two different structures is kept the same.

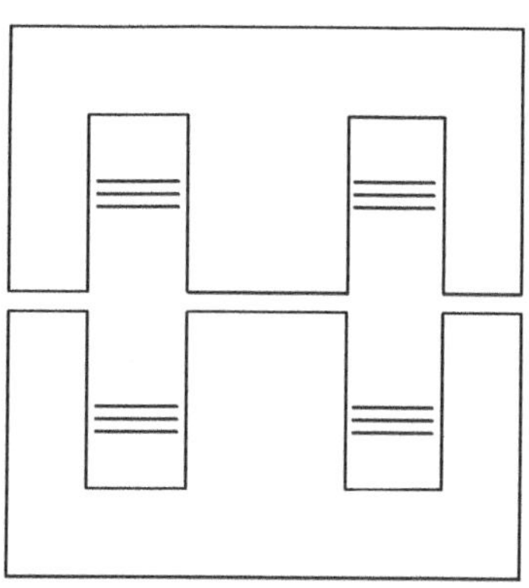

Figure 9.19 Flat winding structure.

9.4.2.1 Different Structures Comparison

To ensure the sufficient conducting area of each turn, four copper foils are connected in parallel to construct one turn. Four paralleled copper foils can be arranged in different methods. Two types of structures are studied as shown in Figure 9.20. One is the laminated structure, and the other is the interleaved structure. For laminated structure, four adjacent copper foils are connected in parallel to form one turn as shown in Figure 9.20a. For interleaved structure, four interleaved copper foils are connected in parallel to form one turn as shown in Figure 9.20b, where five

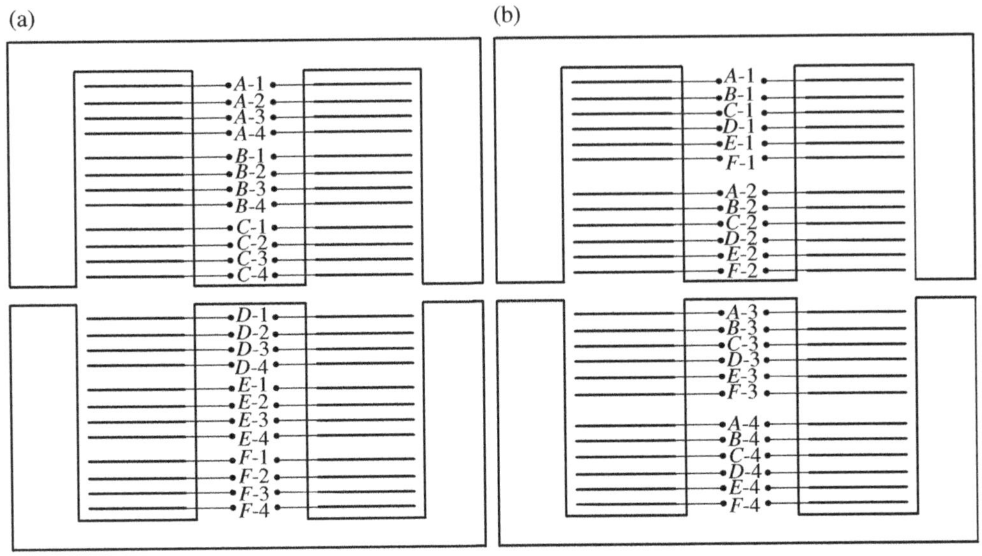

Figure 9.20 Different structures: (a) structure 1: laminated structure; (b) structure 2: interleaved structure.

copper foils are spaced between two interleaved copper foils. The number in Figure 9.20 represents the layer of the parallel winding [13–16].

Utilizing the two-dimensional (2D) FEA to calculate the winding loss of the two different structures, the simulation results are shown in Figure 9.21. Since the cross-section area of the conducting area is the same, the winding loss caused by the DC component in the two structures is the same. However, the winding loss caused by the different AC components in the laminated structure is higher than that in the interleaved structure. The total winding loss of the laminated structure is 21.3 W, which is 10.4 W higher than that in the interleaved structure.

The winding loss caused by the first AC component is taken into account because of its highest amplitude. In the laminated structure, the turn *A*, *B*, *C* and turn *D*, *E*, *F* are symmetrical. Therefore, only the winding loss of the turn *A*, *B*, *C* needs to be extracted and its distribution is shown in Figure 9.22. As shown in Figure 9.22, the horizontal axis is sorted according to the distance between

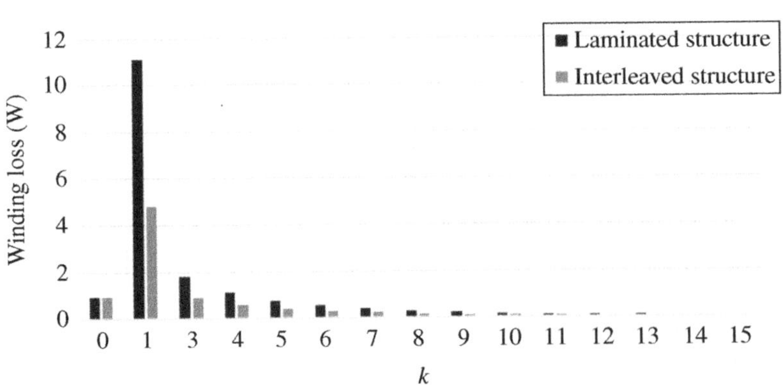

Figure 9.21 Winding loss in different structures.

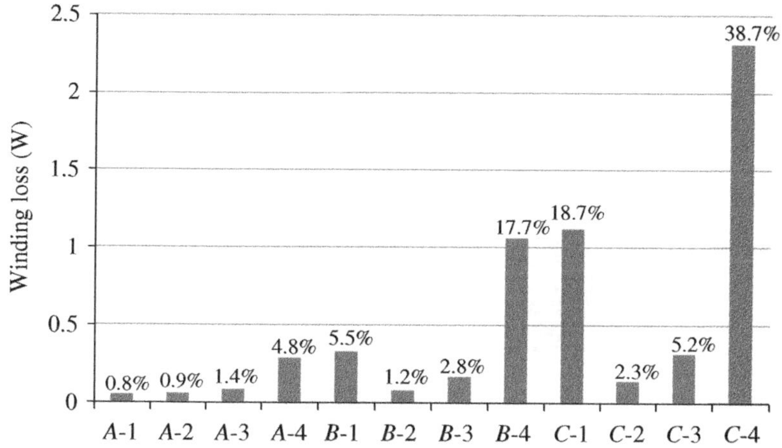

Figure 9.22 Distribution of winding loss in the laminated structure.

the winding and the air gap from far to near. The proportion of the winding loss is marked at the top of the data column.

To compare the winding loss at the same position in the interleaved structure, the distribution of the winding loss in the interleaved structure is extracted as shown in Figure 9.23. It is clear that the distribution of the winding loss in the interleaved structure is much more uniform, which results in smaller winding loss of each turn.

The distribution of the winding loss and magnetic field lines is shown in Figure 9.24. By comparing with the barrel winding in Figure 9.13, the magnetic lines in Figure 9.24 are distributed radially near the air gap. However, it is horizontally distributed inside the window area of the core. It can be found that the distribution density of the magnetic lines inside the window area is relatively uniform. Therefore, it is conceivable that the winding loss should not change significantly when the winding position is closed to the air gap as long as the winding does not reach the radiation area.

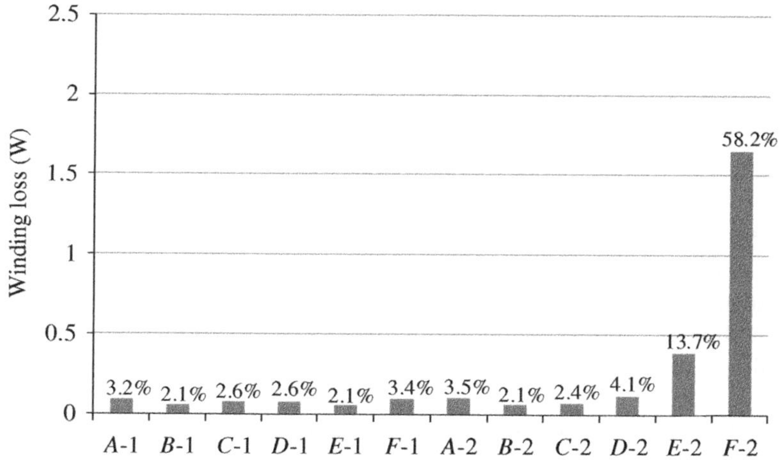

Figure 9.23 Distribution of winding loss in the interleaved structure.

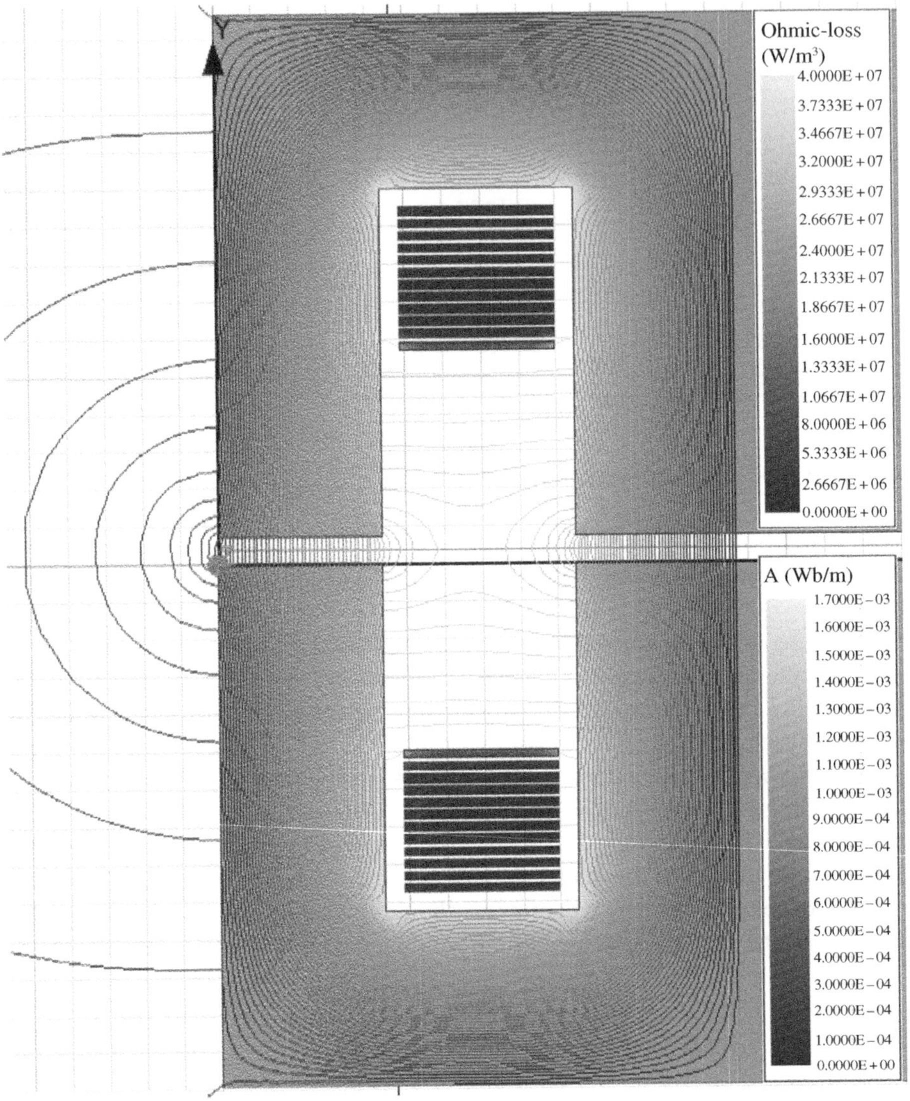

Figure 9.24 Distribution of winding loss and magnetic field lines.

9.4.2.2 Winding Position Discussion

As shown in Figure 9.25, it is assumed that the winding position in the Y-axis direction is 0% when it is far away from the air gap and is 100% when it is closest to the air gap. In the Y-axis direction, the winding loss at different positions is shown in Figure 9.26. After the winding position exceeds 60%, the winding loss increases as the winding approaches the air gap due to the fringing flux.

Inspired by this result, the winding loss can be reduced by increasing the cross-section area of the winding as long as the winding position does not exceed 60%. Therefore, by increasing the winding thickness to increase the cross-section area of the winding, the change in the winding loss is further

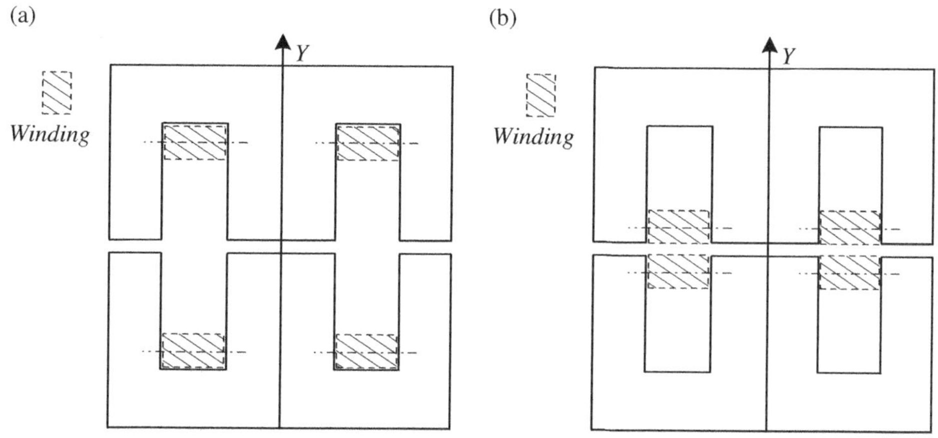

Figure 9.25 The position defined in the core window: (a) 0%; (b) 100%.

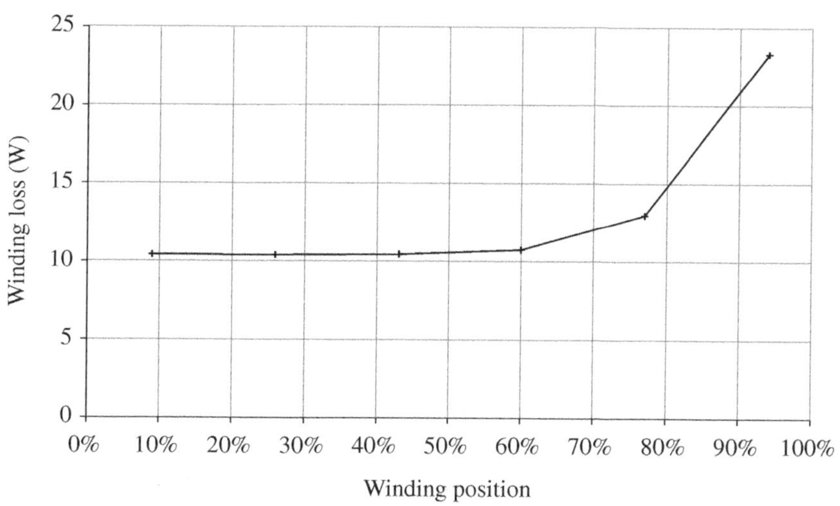

Figure 9.26 Winding loss in different positions.

analyzed. The winding loss under different winding thicknesses is shown in Figure 9.27. From the results, due to the winding resistance reduction, the total winding loss decreases as the winding thickness increases when the winding thickness changes from 0.45 to 0.55 mm. However, the winding loss begins to increase as the winding thickness continues to increase from 0.55 mm. The reason is that the winding loss caused by the fringing flux begins to increase. As a result, the minimum winding loss is 10.3 W when the thickness is 0.55 mm.

The distribution of winding loss caused by the different AC components under different positions is shown in Figure 9.28. The results demonstrate that the winding loss caused by the DC component decreases as the winding thickness increases. Simultaneously, the winding loss caused by the first AC component has a minimum value when the winding thickness is 0.6 mm, while the winding loss caused by other AC components continually increases with increasing winding thickness. The simulation results are consistent with the analysis.

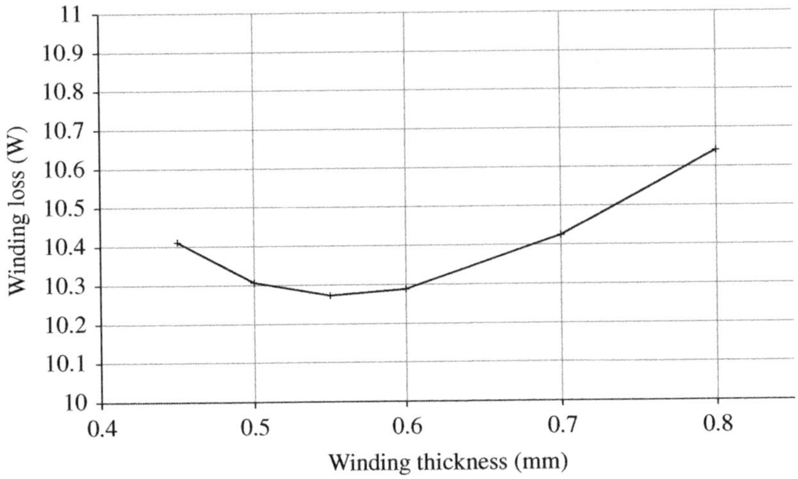

Figure 9.27 Winding loss under different winding thicknesses.

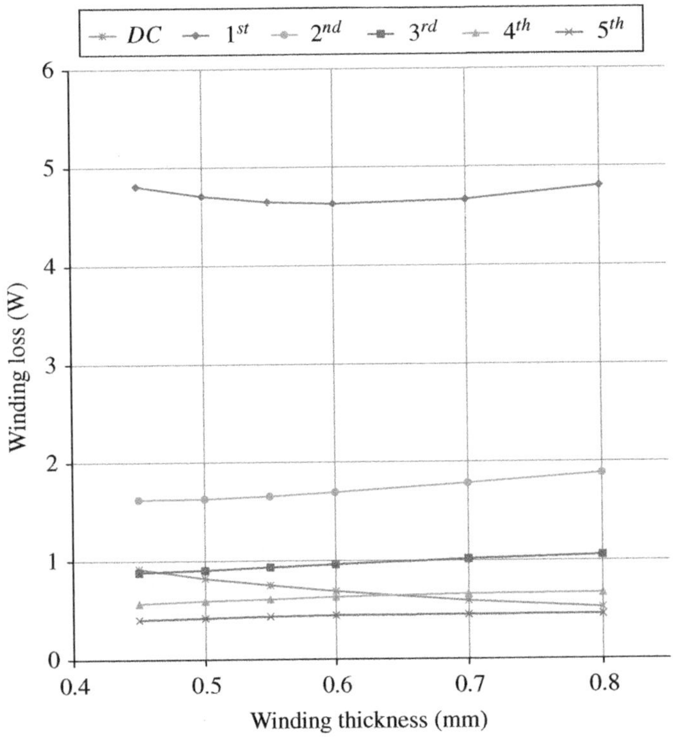

Figure 9.28 Distribution of winding loss under different winding thicknesses.

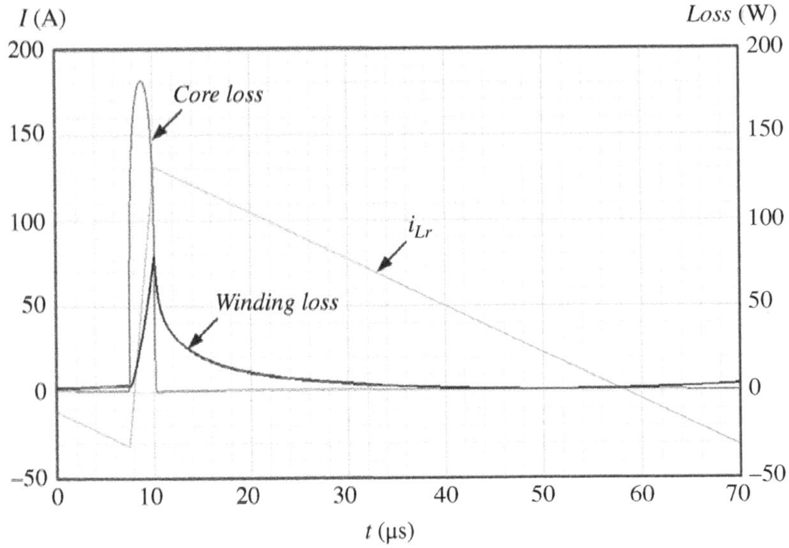

Figure 9.30 Simulation results in a switching period.

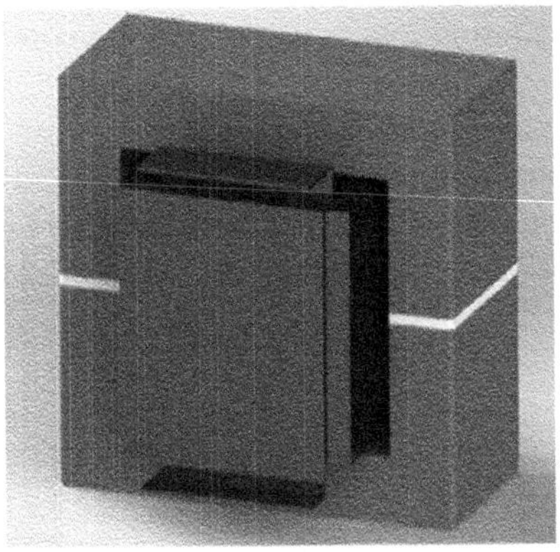

Figure 9.31 Structure of the resonant inductor.

9.5.2 Experimental Verification

Based on the previous design parameters, the resonant inductor is built as shown in Figure 9.32 and applied to the 30 kW ZVS inverter prototype.

When the ZVS inverter works at 30 kW, the voltage and current waveforms of the resonant inductor are shown in Figure 9.33. The peak-to-peak value of the resonant inductor is 150 A.

When the temperature of the resonant inductor is stable, the temperature distribution is shown in Figure 9.34. Under the test environment temperature of 21.0 °C, the maximum winding temperature is 82.6 °C and the maximum core temperature is 55.3 °C. The corresponding maximum temperature rise of the winding is 61.6 °C and that of the core is 34.3 °C.

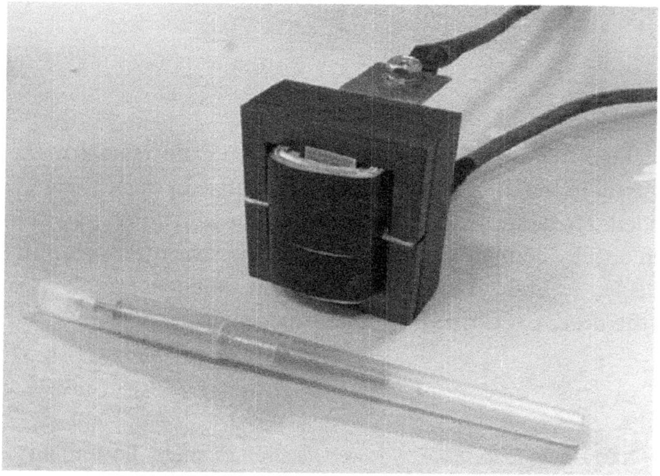

Figure 9.32 Experimental prototype of the resonant inductor.

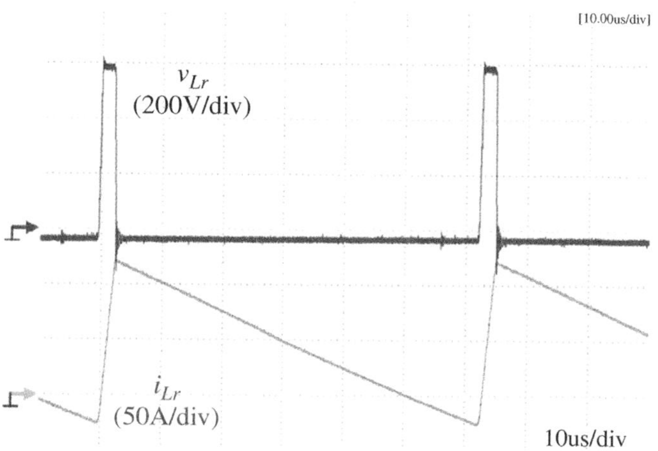

Figure 9.33 Voltage and current of the resonant inductor at 30 kW.

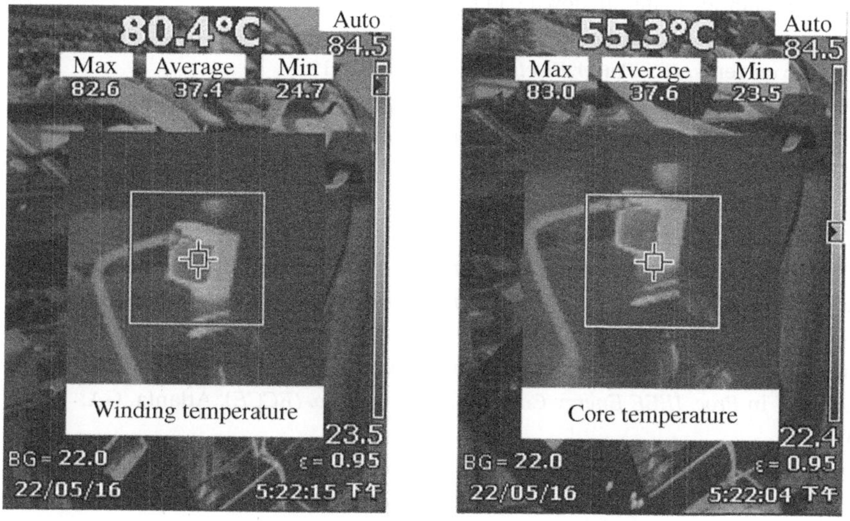

Figure 9.34 Experimental tests of winding and core temperature.

9.6 Summary

The current of the resonant inductor is a high-frequency and large amplitude triangle. How to make it more efficient and miniaturized is critical in the design of a resonant inductor. In this chapter, a design of the resonant inductor is provided. Then, the analysis and simulations with different winding structures and positions are carried out. By comparison, a barrel winding structure with minimum winding loss and volume is selected to build. Finally, the resonant inductor prototype is applied in a 30 kW ZVS inverter, and the design is verified through experiments.

References

1 Erickson, R.W. and Maksimovic, D. (2001). *Fundamentals of Power Electronics*, 2nde. Boston, MA: Springer US.

2 Kazimierczuk, M.K. (2013). *High-Frequency Magnetic Components*, 2nde. Chichester, UK: John Wiley & Sons Ltd.

3 Van den Bossche, A. and Valchev, V.C. (2005). *Inductors and Transformers for Power Electronics*. Boston: CRC Press.

4 Hurley, W.G. and Wölfle, W.H. (2013). *Transformers and Inductors for Power Electronics. Theory, Design and Applications*. Chichester, UK: John Wiley & Sons Ltd.

5 Mohan, N. (2012). Chapter 9: Design of High-frequency Inductors and Transformers, Power Electronics: A First Course. *Hydrogen Gov* 28 (1): 156–159.

6 M. K. Kazimierczuk, H. Sekiya, "Design of ac resonant inductors using area product method", in *Proc. IEEE Energy Conv. Congr. and Expo. (ECCE)*, San Jose, CA, pp. 994–1001, 2009.

7 Reatti, A. and Kazimierczuk, M.K. (2002). Comparison of various methods for calculating the AC resistance of inductors. *IEEE Trans. Magn.* 38 (3): 1512–1518.

8 Wojda, R.P. and Kazimierczuk, M.K. (2013). Analytical optimisation of solid-round-wire windings conducting dc and ac non-sinusoidal periodic currents. *IET Power Electronics* 6 (7): 1462–1474.

9 K. Venkatachalam, C. R. Sullivan, T. Abdallah, and H. Tacca, "Accurate prediction of ferrite core loss with nonsinusoidal waveforms using only Steinmetz parameters," in *Proc. 8th IEEE Workshop on Comput. in Power Electron. (COMPEL)*, pp. 36–41, 2002.

10 C. Du, W. G. Hurley, D. Xu, "Design methodology of resonant inductor in a ZVS inverter," in *Proc. 5th IEEE Int. Symp. Power Electronics for Distributed Generation Systems (PEDG)*, Galway, Ireland, pp. 1–7, 2014.

11 http://www.chinadmegc.com/en/material_download.php?14.

12 Wojda, R.P. and Kazimierczuk, M.K. (2012). Optimum foil thickness of inductors conducting dc and non-sinusoidal periodic currents. *IET Power Electronics* 5 (6): 801–812.

13 R. Prieto, J. A. Cobos, O. Garcia, P. Alou, and J. Uceda, "Using parallel windings in planar magnetic components," in *Proc. 32nd Annual IEEE Power Electron. Specialists Conf. (PESC)*, Vancouver, BC, vol. 4, pp. 2055–2060, 2001.

14 Chen, W., Yan, Y., Hu, Y., and Lu, Q. (2003). Model and design of PCB parallel winding for planar transformer. *IEEE Trans. Magn.* 39 (5): 3202–3204.

15 H. Jiankun, C. R. Sullivan, "Analytical method for generalization of numerically optimized inductor winding shapes", in *Proc. 30th Annu. IEEE Power Electron. Specialists Conf.*, Charleston, SC, vol. 1, pp. 568–573, 1999.

16 R. Prieto, R. Asensi and J. A. Cobos, "Selection of the appropriate winding setup in planar inductors with parallel windings," in *Proc. IEEE Energy Conv. Congr. and Expo. (ECCE)*, Atlanta, GA, pp. 4599–4604, 2010.

17 C. Du, W. G. Hurley and D. Xu, "Design methodology of resonant inductor in a ZVS inverter," in *IEEE Journal of Emerging and Selected Topics in Power Electronics*, vol. 3, no. 4, pp. 1142–1150, Dec. 2015.

Part 4

Impact of SiC Device on Soft-switching Grid Inverter

10

Soft-switching SiC Three-phase Grid Inverter

In this chapter, SiC metal-oxide-semiconductor field-effect transistor (MOSFET) three-phase inverter with zero-voltage-switching space-vector-modulation (ZVS-SVM) technique is introduced to further push the power density of SiC inverter. The loss distributions, conversion efficiency, and volume of passive components between the hard-switching SiC inverter and the soft-switching SiC inverter are compared under wide switching frequency range. In addition, a SiC MOSFET power module with low stray inductance is designed for soft-switching SiC inverter to reduce the voltage overshoots on the switches. Besides, design of low loss resonant inductor with distributed air gap and optimal winding thickness is studied. A 20 kW soft-switching SiC three-phase grid inverter prototype is built to verify the design.

10.1 Introduction

Switching frequency of conventional three-phase grid inverters with Si insulated gate bipolar transistor (IGBT) is generally limited less than 20 kHz due to the rapid rising switching loss with the increase of the switching frequency. SiC MOSFET brings the opportunities of increasing the conversion efficiency, switching frequency, and power density of three-phase grid inverter in comparison with its Si IGBT counterpart. However, the switching loss of SiC MOSFET increases significantly when switching frequency increases to hundreds of kHz, which will restrict the improvement of power density of SiC MOSFET inverter. To further increase the power density of SiC MOSFET applications, soft-switching technique needs to be used.

The soft-switching Si IGBT three-phase inverter with ZVS-SVM has already been investigated and optimally designed in Chapters 7 and 8. Unlike the Si IGBT, SiC MOSFET has no turn-off tail current, which means ZVS is more effective to SiC MOSFET. Thus the ZVS technique can cut down a significant proportion of the switching loss in the SiC MOSFET based inverters. As a result, by applying the ZVS-SVM technique to the three-phase SiC MOSFET inverter, the power density may be raised substantially. The loss distribution, conversion efficiency, and volume of the passive components between hard-switching SiC inverter and soft-switching SiC inverter are compared and the experimental verification is presented. A 7-in-1 SiC MOSFET module with low stray inductance is introduced and evaluated with both simulations and experiments. Finally, the optimal design of resonant inductor with experimental verification is studied in this chapter.

Soft-Switching Technology for Three-phase Power Electronics Converters, First Edition.
Dehong Xu, Rui Li, Ning He, Jinyi Deng, and Yuying Wu.

10.2 Soft-switching Three-phase Inverter

10.2.1 SVM Scheme in Hard-switching Inverter

The topology of traditional hard-switching SiC three-phase grid inverter is shown in Figure 10.1. In order to maintain the fairness of efficiency comparison, a specific discontinuous SVM modulation that is also known as SVM 12 is applied for the hard-switching SiC three-phase inverter [1]. The basic knowledge of SVM is provided in Appendix A.1. By ignoring the unique short-circuit operation of ZVS-SVM in the soft-switching inverter, each main switch in hard-switching inverter with SVM 12 has the same duty cycle in every individual switching period as that of the soft-switching inverter.

The three-phase grid currents and grid voltages waveforms of the hard-switching inverter with unity power factor are shown in Figure 10.2. Since the voltage drop on filter inductor is quite small, the fundamental components of the output voltages of the three-phase bridges can be represented by the three-phase grid voltages. Based on the conventional SVM scheme in Appendix A.1, the line cycle can be divided into six sectors, from sector 1 to sector 6 with 60° phase angle range in each sector. Considering the current polarities of different phases, each sector is further divided into two subsectors as shown in Figure 10.2. In each subsector, phase with the largest absolute value of current has the opposite polarity to the other two phases.

The three-phase bridges' output voltage space vector diagram with 12 subsectors is shown in Figure 10.3. The principle of SVM 12 is that the phase bridge with the largest absolute value of grid

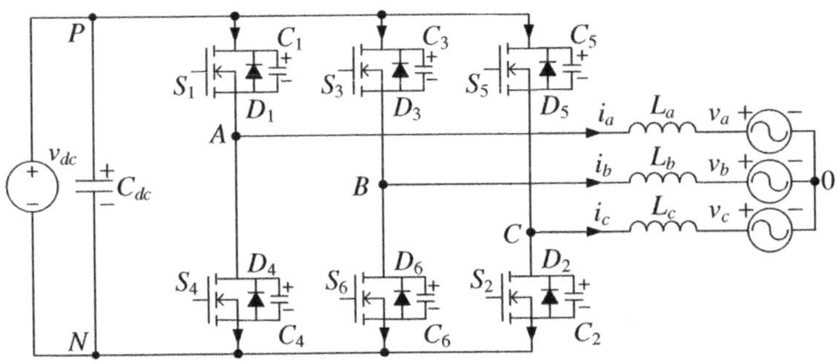

Figure 10.1 Topology of traditional hard-switching SiC three-phase grid inverter.

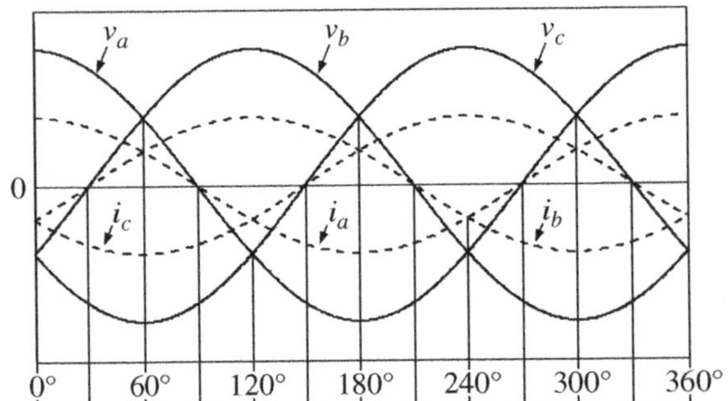

Figure 10.2 Grid voltages and grid currents waveforms of unity power factor hard-switching inverter.

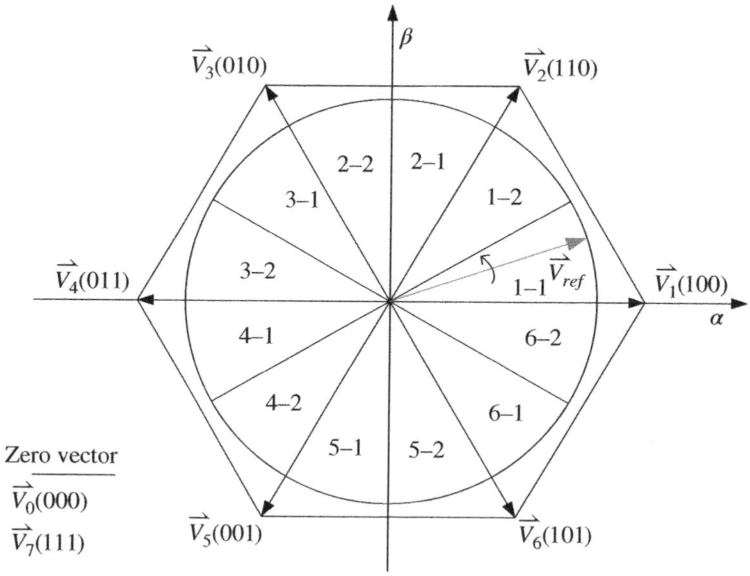

Figure 10.3 Voltage space vector diagram with 12 subsectors.

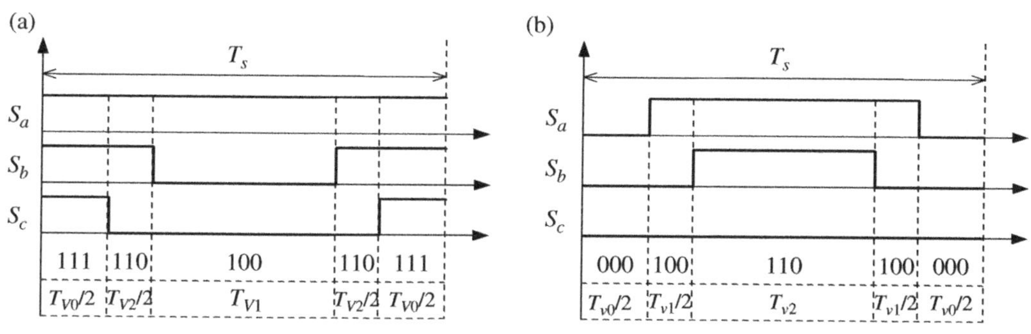

Figure 10.4 Switching patterns of SVM 12: (a) sector 1-1 and (b) sector 1-2.

current has no switching during the entire subsector in order to reduce the switching loss. Taking phase A as an example, main switches S_1 and S_4 will not switch in sector 1-1, sector 3-2, sector 4-1, and sector 6-2. Besides, the closest two basic nonzero vectors are used to compose the reference vector depending on its location. For example, when the reference vector $\vec{V}_{ref}$ is located in sector 1-1 and sector 1-2, basic vectors $\vec{V}_1$ and $\vec{V}_2$ are used to compose the reference vector. Since phase A's main switches have no switching in sector 1-1, zero vector $\vec{V}_7$ is the only choice. Similarly, in sector 1-2 zero vector $\vec{V}_0$ is the only choice. Based on these principles, the switching pattern of SVM 12 in sector 1-1 and sector 1-2 is derived as shown in Figure 10.4. The entire 12 switching patterns are provided in Appendix A.2.

If the dead time is ignored, switching functions S_a, S_b, and S_c represent the switching of upper switches S_1, S_3, and S_5, respectively. In Figure 10.4a the turn-on of lower switches S_6 and S_2 causes Type 2 commutations from the body diodes of the upper switches to their complementary switches due to negative polarity of phase B and phase C grid currents in sector 1-1.

10.2.2 ZVS-SVM Scheme in Soft-switching Inverter

Figure 10.5 shows the topology of the soft-switching SiC three-phase grid inverter [2]. An auxiliary circuit is in series connected in the DC side. The auxiliary circuit is composed of a resonant inductor L_r, a clamping capacitor C_{cl}, and an auxiliary switch S_7. Resonant capacitors, C_1, C_2, C_3, C_4, C_5, C_6, and C_7, are paralleled to the main switches and the auxiliary switch S_7.

The switching pattern of three-phase bridges in ZVS-SVM in sector 1-1 can be obtained by rearranging the sequences of the three basic vectors in SVM 12 as shown in Figure 10.6. The commutations from body diodes to their complementary MOSFETs in the same legs are aligned at the same instant in each switching period. Besides, the switching pattern of auxiliary switch is also provided in Figure 10.6. The DC side voltage across the three-phase bridges v_{bus} will be resonated to zero by turning off the auxiliary switch S_7, which creates the ZVS condition for S_2 and S_6. The entire 12 switching patterns of ZVS-SVM are provided in Appendix A.3.

10.2.3 Operation Stages and ZVS Condition of Soft-switching Inverter

10.2.3.1 Operation Stages Analysis

Sector 1-1 is taken as an example to investigate the operation stages of the ZVS turn-on process. Figure 10.7 shows the waveforms of the key components in the soft-switching three-phase inverter in one switching period. The operation stages can be briefly analyzed as follows.

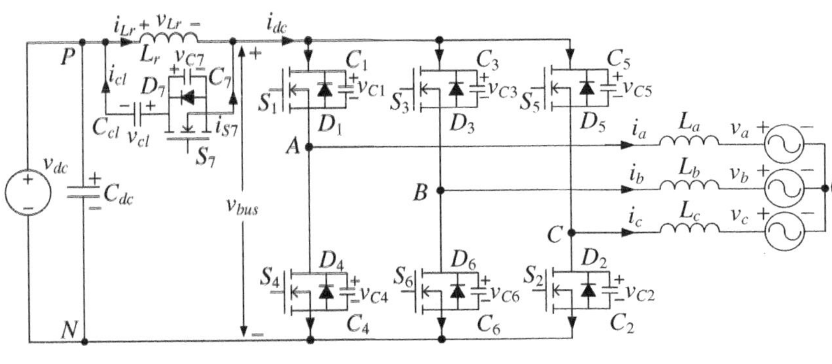

Figure 10.5 Topology of soft-switching SiC three-phase grid inverter.

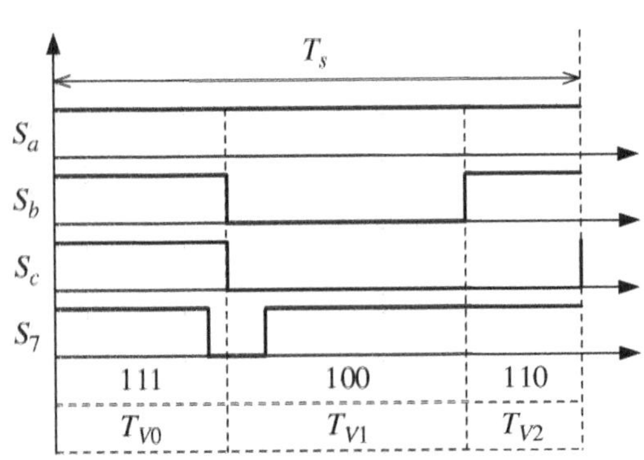

Figure 10.6 Switching patterns of ZVS-SVM in sector 1-1.

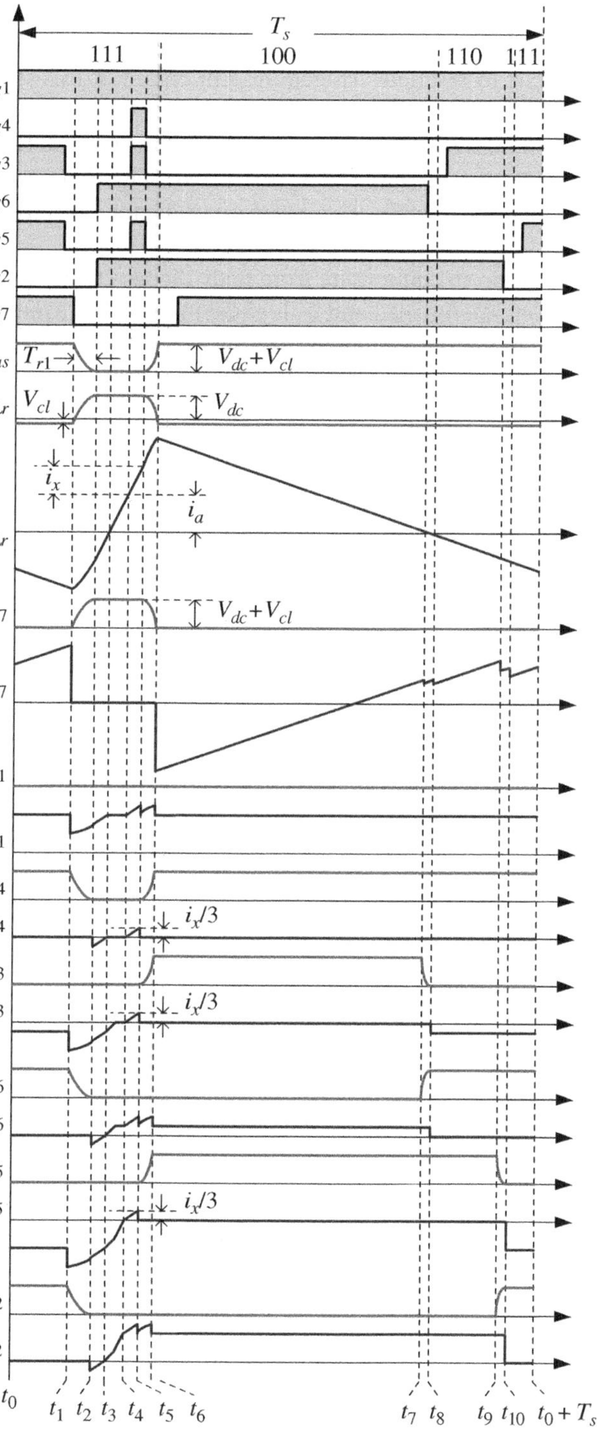

Figure 10.7 Waveforms of the key components in one switching period in sector 1-1.

In stage 1 (t_0–t_1), the inverter output voltage vector is $\vec{V}_7$ (111).

In stage 2 (t_1–t_2), at t_1 the auxiliary switch S_7 is turned off, then the resonant inductor L_r and resonant capacitors C_4, C_6, C_2, and C_7 start to resonate. The equivalent circuit of this stage is shown in Figure 10.8. Stage 2 ends when v_{bus} is resonated to zero.

In stage 3 (t_2–t_3), at t_2 the freewheeling body diodes D_4, D_6, and D_2 begin to conduct and the turn-on signals are added to S_6 and S_2. Thus the ZVS turn-on of S_6 and S_2 is realized. Stage 3 ends at t_3 when the freewheeling diodes are turned off; meanwhile, the resonant inductor current i_{Lr} reaches zero.

In stage 4 (t_3–t_4), the grid currents i_b and i_c begin to commutate from body diodes D_3 and D_5 to main switches S_6 and S_2. At t_4, the commutations are finished and i_{Lr} is equal to the grid current i_a.

In stage 5 (t_4–t_5), at t_4 all six main switches are turned on, which continues clamping the v_{bus} at zero to increase the energy stored in the resonant inductor L_r. The equivalent circuit of this stage is shown in Figure 10.9. The extra added current i_x on resonant current i_{Lr} in this stage is also shown in Figure 10.7.

In stage 6 (t_5–t_6), at t_5 the main switches S_4, S_3, and S_5 are turned off. The resonant inductor L_r and resonant capacitors C_4, C_3, C_5, and C_7 start to resonate as shown in Figure 10.10. Stage 6 ends at t_6 when drain-source voltage of the auxiliary switch v_{C7} is resonated to zero. Meanwhile, the body diode D_7 is conducting and the ZVS turn on of auxiliary switch S_7 is also realized.

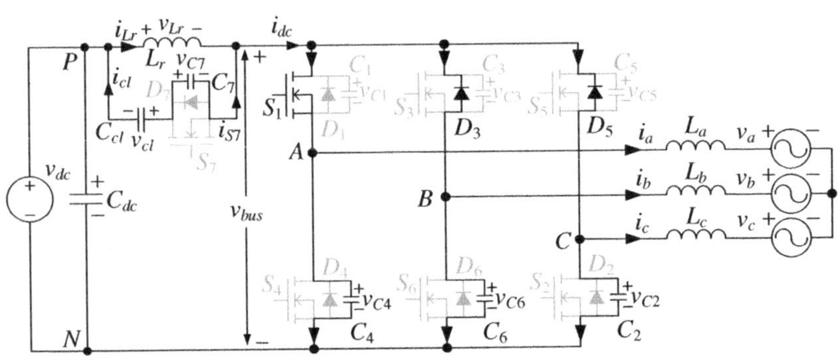

Figure 10.8 Equivalent circuit of the resonant stage for ZVS of the main switches.

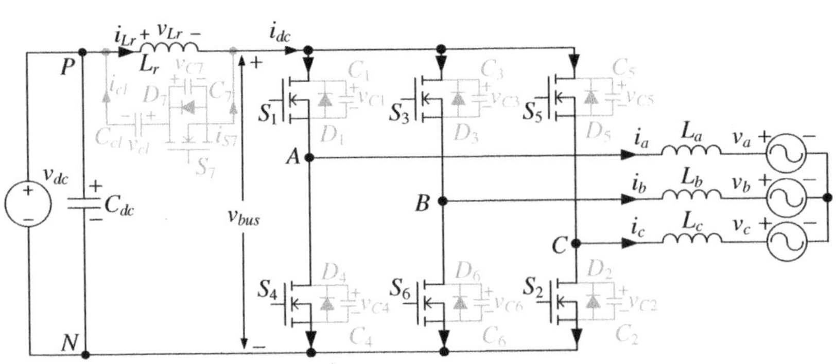

Figure 10.9 Equivalent circuit of the three-phase bridges' short-circuit stage.

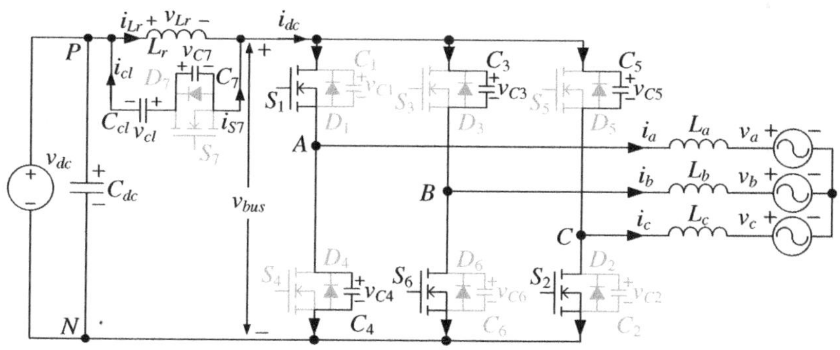

Figure 10.10 Equivalent circuit of the resonant stage for ZVS of the auxiliary switch.

In stage 7 (t_6–t_7), the inverter output voltage vector is $\vec{V}_1(100)$. The ZVS turn-on operations of both main switches and auxiliary switch have been achieved.

In stage 8 (t_7–t_8), stage 9 (t_8–t_9), and stage 10 (t_9–t_{10}), the soft-switching three-phase inverter operates as hard-switching three-phase inverter.

10.2.3.2 ZVS Condition Derivation

10.2.3.2.1 Resonant Stages Analysis

Stage 2 is the resonant stage for the ZVS turn-on of main switches. It is assumed that the resonant capacitors of main switches are equal:

$$C_1 = C_4 = C_3 = C_6 = C_5 = C_2 = C_r \tag{10.1}$$

According to equivalent circuit in Figure 10.8, following equations are derived as:

$$\begin{cases} i_{Lr} + C_7\dfrac{dv_{C7}}{dt} = C_r\dfrac{dv_{C4}}{dt} + i_a + C_r\dfrac{dv_{C6}}{dt} + i_b + C_r\dfrac{dv_{C2}}{dt} + i_c \\ v_{Lr} = L_r\dfrac{di_{Lr}}{dt} \end{cases} \tag{10.2}$$

Besides, following equations are obtained:

$$v_{C4} = v_{C6} = v_{C2} = V_{dc} + V_{cl} - v_{C7} \tag{10.3}$$

$$V_{dc} = v_{Lr} + v_{C4} \tag{10.4}$$

$$i_a + i_b + i_c = 0 \tag{10.5}$$

By substituting (10.3), (10.4), and (10.5) into (10.2), the state equations with resonant inductor current i_{Lr} and resonant capacitor voltage v_{C4} as state variables are derived as:

$$\begin{cases} (3C_r + C_7)\dfrac{dv_{C4}}{dt} = i_{Lr} \\ L_r\dfrac{di_{Lr}}{dt} = -v_{C4} + V_{dc} \end{cases} \tag{10.6}$$

with the initial condition:

$$\begin{cases} v_{C4}(t_1) = V_{cl} + V_{dc} \\ i_{Lr}(t_1) = i_{Lr}(t_1) \end{cases} \tag{10.7}$$

By solving the state equations (10.6) with their initial values, the expressions of resonant capacitor voltage v_{C4} and resonant inductor current i_{Lr} can be obtained as:

$$\begin{cases} v_{C4}(t) = V_{cl}\cos(\omega_r t) + i_{Lr}(t_1)Z_r\sin(\omega_r t) + V_{dc} \\ i_{Lr}(t) = -\dfrac{V_{cl}}{Z_r}\sin(\omega_r t) + i_{Lr}(t_1)\cos(\omega_r t) \end{cases} \tag{10.8}$$

where the characteristic impedance Z_r and the resonant angular frequency ω_r are defined as:

$$Z_r = \sqrt{\frac{L_r}{3C_r + C_7}} \tag{10.9}$$

$$\omega_r = \frac{1}{\sqrt{L_r(3C_r + C_7)}} \tag{10.10}$$

The resonant capacitor voltage v_{C4} can be further rewritten as:

$$v_{C4}(t) = V_{dc} + \sqrt{V_{cl}^2 + i_{Lr}(t_1)^2 Z_r^2}\cos(\omega_r t + \sigma) \tag{10.11}$$

where

$$\sigma = \arctan\left(\frac{|i_{Lr}(t_1)|Z_r}{V_{cl}}\right) \tag{10.12}$$

According to (10.11), the ZVS turn-on can be realized unless the minimum value of resonant capacitor v_{C4} is smaller than zero. Thus the following inequality needs to be satisfied:

$$\sqrt{V_{cl}^2 + i_{Lr}(t_1)^2 Z_r^2} \geq V_{dc} \tag{10.13}$$

Then the ZVS condition for main switches can be further simplified as:

$$i_{Lr}(t_1) \leq -\sqrt{\frac{V_{dc}^2 - V_{cl}^2}{Z_r^2}} \tag{10.14}$$

Stage 6 is the resonant stage for the ZVS turn-on of auxiliary switch. The mathematical model of equivalent circuit in Figure 10.10 can be derived as:

$$\begin{cases} i_{Lr} + C_7\dfrac{dv_{C7}}{dt} = C_r\dfrac{dv_{C4}}{dt} + i_a + C_r\dfrac{dv_{C6}}{dt} + C_r\dfrac{dv_{C2}}{dt} \\ v_{Lr} = L_r\dfrac{di_{Lr}}{dt} \end{cases} \tag{10.15}$$

In addition,

$$v_{C4} = v_{C3} = v_{C5} = V_{dc} + V_{cl} - v_{C7} \tag{10.16}$$

$$V_{cl} + v_{C4} = v_{Lr} \tag{10.17}$$

By substituting (10.16) and (10.17) into (10.15), the state equations are obtained:

$$\begin{cases} (3C_r + C_7)\dfrac{dv_{C7}}{dt} = -i_{Lr} + i_a \\ L_r\dfrac{di_{Lr}}{dt} = v_{C7} - V_{cl} \end{cases} \tag{10.18}$$

with the initial condition:

$$\begin{cases} i_{Lr}(t_5) = i_a + i_x \\ v_{C7}(t_5) = V_{dc} + V_{cl} \end{cases} \tag{10.19}$$

The expressions of resonant capacitor voltage v_{C7} and resonant inductor current i_{Lr} can be obtained by solving the state equation (10.18) with the initial values:

$$\begin{cases} v_{C7}(t) = V_{cl} + V_{dc}\cos(\omega_r t) - i_x Z_r \sin(\omega_r t) \\ i_{Lr}(t) = \dfrac{V_{dc}}{Z_r}\sin(\omega_r t) + i_x \cos(\omega_r t) + i_a \end{cases} \tag{10.20}$$

The resonant capacitor voltage v_{C7} can be further rewritten as:

$$v_{C7}(t) = V_{cl} + \sqrt{{V_{dc}}^2 + {i_x}^2 {Z_r}^2}\cos(\omega_r t + \tau) \tag{10.21}$$

where

$$\tau = \arctan\left(\frac{i_x Z_r}{V_{dc}}\right) \tag{10.22}$$

According to (10.21), since the clamping capacitor voltage V_{cl} is always much smaller than V_{dc}, the minimum value of v_{C7} is always smaller than zero. Therefore, the ZVS condition for auxiliary switch is easy to be satisfied.

At the end of stage 6 at t_6, the resonant capacitor voltage v_{C7} and resonant inductor current i_{Lr} meet:

$$v_{C7}(t_6) = V_{cl} + V_{dc}\cos(\omega_r t_6) - i_x Z_r \sin(\omega_r t_6) = 0 \tag{10.23}$$

$$i_{Lr}(t_6) = \frac{V_{dc}}{Z_r}\sin(\omega_r t_6) + i_x \cos(\omega_r t_6) + i_a \tag{10.24}$$

Equations (10.23) and (10.24) can be modified as:

$$[V_{dc}\cos(\omega_r t_6) - i_x Z_r \sin(\omega_r t_6)]^2 = {V_{cl}}^2 \tag{10.25}$$

$$[V_{dc}\sin(\omega_r t_6) + i_x Z_r \cos(\omega_r t_6)]^2 = {Z_r}^2[i_{Lr}(t_6) - i_a]^2 \tag{10.26}$$

Adding (10.25) to (10.26), the following relationship can be obtained:

$${V_{dc}}^2 + {i_x}^2{Z_r}^2 = {V_{cl}}^2 + {Z_r}^2[i_{Lr}(t_6) - i_a]^2 \tag{10.27}$$

Therefore, the resonant inductor current i_{Lr} at t_6 can be derived as:

$$i_{Lr}(t_6) = \sqrt{\frac{{V_{dc}}^2 - {V_{cl}}^2}{{Z_r}^2} + {i_x}^2} + i_a \tag{10.28}$$

10.2.3.2.2 Steady State Analysis

According to the voltage-second balance principle of resonant inductor, the mean value of voltage across the resonant inductor in a switching period is zero. That is,

$$\frac{1}{T_s}\int_{t_0}^{t_0 + T_s} v_{Lr}\, dt = 0 \tag{10.29}$$

The voltage waveform of the resonant inductor v_{Lr} is shown in Figure 10.7. Since the resonant stages during intervals (t_1–t_2) and (t_5–t_6) are relatively short in a switching period, the integrations of the v_{Lr} during intervals (t_1–t_2) and (t_5–t_6) are ignored to simplify the calculation as:

$$\frac{1}{T_s}\int_{t_0}^{t_0+T_s} v_{Lr}\ dt = \frac{1}{T_s}\left(\int_{t_0}^{t_1} v_{Lr}\ dt + \int_{t_2}^{t_5} v_{Lr}\ dt + \int_{t_6}^{t_0+T_s} v_{Lr}\ dt\right) = 0 \tag{10.30}$$

By substituting the voltage across the resonant inductor at different intervals as shown in Figure 10.7 into (10.30), it can be rewritten as:

$$\frac{1}{T_s}\int_{0}^{T_s} v_{Lr}\ dt = -V_{cl}(1-\lambda_7)T_s + V_{dc}\lambda_7 T_s = 0 \tag{10.31}$$

where λ_7 is the turn-off duty cycle of the auxiliary switch S_7. $\lambda_7 T_s$ can be defined as the time interval from t_2 to t_5 as shown in Figure 10.7 if the two short resonant stages are ignored. The voltage across clamping capacitor V_{cl} can be obtained by simplifying (10.31) as:

$$V_{cl} = \frac{\lambda_7}{1-\lambda_7}V_{dc} \tag{10.32}$$

According to the current-second balance principle of clamping capacitor, the mean value of current through clamping capacitor is also zero during a switching period. That is,

$$\frac{1}{T_s}\int_{t_0}^{t_0+T_s} i_{cl}\ dt = 0 \tag{10.33}$$

By ignoring the two short resonant stages and two Type 1 commutation stages, (10.33) can be expanded as:

$$\begin{aligned}&\int_{t_0}^{t_1} i_{cl\_stage\ 1}(t)dt + \int_{t_2}^{t_5} i_{cl\_stage\ 3,4,5}(t)dt + \int_{t_6}^{t_7} i_{cl\_stage\ 7}(t)\ dt\\ &+ \int_{t_8}^{t_9} i_{cl\_stage\ 9}(t)dt + \int_{t_{10}}^{t_0+T_s} i_{cl\_stage\ 1}(t)\ dt \approx 0\end{aligned} \tag{10.34}$$

The current through clamping capacitor i_{cl} is equal to $-i_{S7}$ as shown in Figure 10.5. Therefore, i_{cl} in stage 3, stage 4, and stage 5 is equal to zero:

$$i_{cl\_stage\ 3,4,5}(t) = 0 \tag{10.35}$$

In other stages, i_{cl} is obtained with the DC bus current i_{dc} and the resonant inductor current i_{Lr} as:

$$i_{cl} = i_{Lr} - i_{dc} \tag{10.36}$$

With the initial value at t_6 in (10.28) and the decreasing rate of V_{cl}/L_r, the resonant inductor current i_{Lr} from t_6 to t_1 of the next switching cycle shown in Figure 10.7 can be expressed as:

$$\begin{aligned}i_{Lr}(t) &= i_{Lr}(t_6) - \frac{V_{cl}}{L_r}(t-t_6)\\ &= \sqrt{\frac{V_{dc}^2 - V_{cl}^2}{Z_r^2} + i_x^2} + i_a - \frac{V_{cl}}{L_r}(t-t_6)\end{aligned} \tag{10.37}$$

Since the output vectors of inverter in stage 7 (t_6–t_7), stage 9 (t_8–t_9), and stage 1 (t_0–t_1, t_{10}–t_0+T_s) are $\vec{V}_7$ (111), $\vec{V}_1$ (100), and $\vec{V}_2$ (110), respectively, the DC bus current i_{dc} in these stages are obtained as:

$$\begin{cases} i_{dc\_stage\ 7}(t) = i_a \\ i_{dc\_stage\ 9}(t) = i_a + i_b \\ i_{dc\_stage\ 1}(t) = 0 \end{cases} \tag{10.38}$$

By substituting Eqs. (10.37) and (10.38) into (10.36), i_{cl} in stage 7, stage 9, and stage 1 can be derived as:

$$\begin{cases} i_{cl\_stage\ 7}(t) = \sqrt{\dfrac{V_{dc}^2 - V_{cl}^2}{Z_r^2} + i_x^2} - \dfrac{V_{cl}}{L_r}(t - t_6) \\ i_{cl\_stage\ 9}(t) = \sqrt{\dfrac{V_{dc}^2 - V_{cl}^2}{Z_r^2} + i_x^2} - \dfrac{V_{cl}}{L_r}(t - t_6) - i_b \\ i_{cl\_stage\ 1}(t) = \sqrt{\dfrac{V_{dc}^2 - V_{cl}^2}{Z_r^2} + i_x^2} - \dfrac{V_{cl}}{L_r}(t - t_6) + i_a \end{cases} \tag{10.39}$$

Since the equivalent vector within the time interval from t_2 to t_5 ($\lambda_7 T_s$) is zero vector and the DC voltage at other stages are $V_{dc} + V_{cl}$, the soft-switching inverter can be viewed as hard-switching inverter with DC voltage at $V_{dc} + V_{cl}$ when calculating the duration of stage 1, stage 7, and stage 9. The basic vectors in stage 1, stage 7, and stage 9 are $\vec{V}_7$, $\vec{V}_1$, and $\vec{V}_2$, respectively. According to Appendix A.1, their durations, T_{V0}, T_{V1}, and T_{V2}, can be obtained as:

$$\begin{cases} T_{V1} = MT_s \sin\left(\dfrac{\pi}{3} - \varphi\right) \\ T_{V2} = MT_s \sin(\varphi) \\ T_{V0} = T_s - \lambda_7 T_s - T_{V1} - T_{V2} \\ \varphi = \omega_l t = 2\pi f_l t \end{cases} \tag{10.40}$$

where M is the equivalent ZVS-SVM modulation ratio of soft-switching inverter, φ is the phase angle of reference vector, ω_l is the angle frequency of grid voltages, and f_l is the line frequency of grid voltages. With equation (10.32), M can be expressed as:

$$M = \frac{\sqrt{3}V_m}{V_{dc} + V_{cl}} = \frac{\sqrt{3}V_m}{V_{dc}}(1 - \lambda_7) = m(1 - \lambda_7) \tag{10.41}$$

where m is the SVM modulation ratio of hard-switching inverter and V_m is the amplitude of grid voltage. If the soft-switching inverter is at unity power factor, then the grid currents meet:

$$\begin{cases} i_a = I_m \cos(\varphi) \\ i_b = I_m \cos(\varphi - 2\pi/3) \\ i_c = I_m \cos(\varphi + 2\pi/3) \end{cases} \tag{10.42}$$

By substituting (10.35), (10.39), (10.40), (10.41), and (10.42) into (10.34), the turn-off duty cycle of the auxiliary switch λ_7 can be derived as:

$$\lambda_7 \approx \frac{\sqrt{\dfrac{V_{dc}^2 - V_{cl}^2}{Z_r^2} + i_x^2} + i_a - \dfrac{\sqrt{3}}{2} m I_m}{\dfrac{V_{dc} T_s}{2L_r}} \tag{10.43}$$

10.2.3.2.3 Soft-Switching Condition

With (10.28), (10.37), and (10.43), the resonant inductor i_{Lr} at t_1 can be calculated as:

$$i_{Lr}(t_1) = i_{Lr}(t_6) - \frac{V_{cl}}{L_r}(1-\lambda_7)T_s = -\sqrt{\frac{{V_{dc}}^2 - {V_{cl}}^2}{{Z_r}^2} + {i_x}^2 - i_a + \sqrt{3}mI_m} \tag{10.44}$$

By substituting (10.44) into (10.14), extra added current i_x meets:

$$-\sqrt{\frac{{V_{dc}}^2 - {V_{cl}}^2}{{Z_r}^2} + {i_x}^2 - i_a + \sqrt{3}mI_m} \le -\sqrt{\frac{{V_{dc}}^2 - {V_{cl}}^2}{{Z_r}^2}} \tag{10.45}$$

The extra added current i_x can be obtained as:

$$i_x \ge \sqrt{\left(\sqrt{\frac{{V_{dc}}^2 - {V_{cl}}^2}{{Z_r}^2}} - i_a + \sqrt{3}mI_m\right)^2 - \frac{{V_{dc}}^2 - {V_{cl}}^2}{{Z_r}^2}} \tag{10.46}$$

Equation (10.46) describes the ZVS condition for ZVS turn-on of main switches. The duration of stage 5 should be controlled to meet (10.46) and store enough energy in resonant inductor for the next first resonant stage. By substituting the minimum value of extra added current i_x as shown in (10.46) into (10.14) and (10.28), the minimum resonant current at t_1 and the maximum resonant current at t_6 can be simplified as:

$$\begin{cases} i_{Lr}(t_1) = -\sqrt{\dfrac{{V_{dc}}^2 - {V_{cl}}^2}{{Z_r}^2}} \approx -\dfrac{V_{dc}}{Z_r} \\ i_{Lr}(t_6) = \sqrt{\dfrac{{V_{dc}}^2 - {V_{cl}}^2}{{Z_r}^2}} + \sqrt{3}mI_m \approx \dfrac{V_{dc}}{Z_r} + \sqrt{3}mI_m \end{cases} \tag{10.47}$$

The turn-off duty cycle of the auxiliary switch λ_7 with the minimum i_x can be further simplified as:

$$\lambda_7 \approx \frac{2L_r}{V_{dc}T_s}\left(\frac{V_{dc}}{Z_r} + \frac{\sqrt{3}}{2}mI_m\right) \tag{10.48}$$

The minimum extra added current i_x is used in the following design.

10.3 Efficiency Comparison of Hard-switching SiC Inverter and Soft-switching SiC Inverter

10.3.1 Parameters Design of Soft-switching SiC Inverter

The input and output parameters of the designed soft-switching SiC inverter are listed in Table 10.1.

Table 10.1 Input and output parameters of designed soft-switching SiC inverter.

Parameters	Descriptions	Value
V_{dc}	DC voltage	600 V
v_a, v_b, v_c	Grid voltage	220 Vrms
P	Rated power	20 kW

10.3.1.1 AC Filter Inductor

According to the derivation in Appendix A.5, the AC filter inductance L can be expressed with the maximum current ripple $\Delta i_{a(max)}$ as:

$$L = \frac{3V_m}{2\Delta i_{a(\max)} V_{dc} f_s} \left[\frac{2}{3} V_{dc} - V_m(1 - \lambda_7)\right] \tag{10.49}$$

If $\Delta i_{a(max)}$ is set to be $0.25 I_m$, the relationship between AC filter inductance L and switching frequency f_s under different turn-off duty cycles of auxiliary switch λ_7 can be drawn as shown in Figure 10.11. It is observed that larger filter inductance L is required with the increment of λ_7 for given current ripple. Besides, larger λ_7 causes larger voltage stress of switching devices, and λ_7 is usually designed to be less than 0.1.

Equation (10.49) can also be used to describe the filter inductance L in hard-switching inverter by setting the turn-off duty cycle of auxiliary λ_7 to zero. Since λ_7 is always smaller than 0.1, its impact on filter inductance is small. In order to simplify the analysis, AC filter inductance in soft-switching inverter is identical to that in hard-switching inverter under the same switching frequency.

10.3.1.2 Resonant Parameters

The resonant parameters in soft-switching SiC inverter includes resonant inductor L_r, resonant capacitor C_r for each three-phase main switch, and resonant capacitor C_7 for auxiliary switch. They should meet the following four requirements [3].

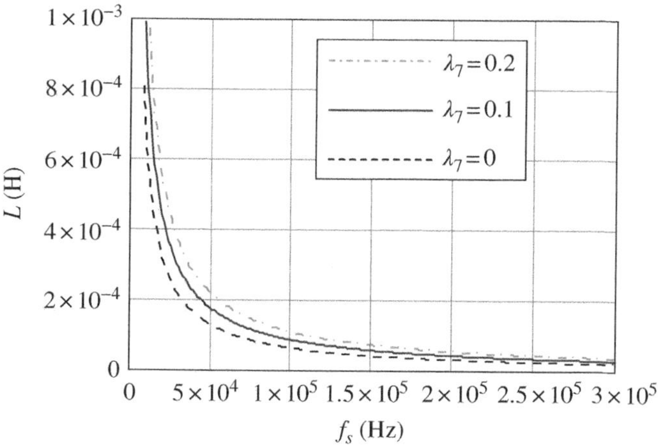

Figure 10.11 Relationship between AC filter inductance L and switching frequency f_s under different turn-off duty cycles of auxiliary λ_7.

10.3.1.2.1 Requirement of Voltage Stress

The turn-off cycle duty of auxiliary switch λ_7 in (10.48) is designed less than 0.1 to avoid high voltage stress on all switches. Thus the resonant parameters are requirement to meet:

$$\frac{2L_r}{V_{dc}T_s}\left(\frac{V_{dc}}{Z_r}+\frac{\sqrt{3}}{2}mI_m\right)\leq 0.1 \tag{10.50}$$

10.3.1.2.2 Requirement of the Current Stress in Main Switches and Auxiliary Switch

As shown in Figure 10.7, the maximum current of auxiliary switch in one switching period is presented by:

$$|i_{S7}(t_6)|=|i_a-i_{Lr}(t_6)|=\left|i_a-\frac{V_{dc}}{Z_r}-\sqrt{3}mI_m\right|=\frac{V_{dc}}{Z_r}+\sqrt{3}mI_m-I_m\cos\varphi \tag{10.51}$$

The current stress of auxiliary switch can be derived when φ is equal to $\pi/6$.

The maximum current of main switch S_1 in one switching period can be expressed as:

$$i_{S1}(t_5)=i_a+\frac{i_x}{3}=I_m\cos\varphi+\frac{1}{3}\sqrt{\left(\frac{V_{dc}}{Z_r}+\sqrt{3}mI_m-I_m\cos\varphi\right)^2-\frac{V_{dc}{}^2}{Z_r{}^2}} \tag{10.52}$$

The current stress of main switch can be derived when φ is equal to 0.

As shown in (10.51) and (10.52), the current stress in both auxiliary switch and main switches are limited by the resonant impedance Z_r. According to the input and output parameters in Table 10.1, SiC MOSFET Wolfspeed C2M0025120D with 1200 V/60 A rating is selected for both main switches and auxiliary switch. The current stress in auxiliary switch and main switches should be limited to 60 A as shown below:

$$\begin{cases}\dfrac{V_{dc}}{Z_r}+\sqrt{3}mI_m-I_m\cos\dfrac{\pi}{6}\leq 60\ \text{A}\\ I_m+\dfrac{1}{3}\sqrt{\left(\dfrac{V_{dc}}{Z_r}-I_m+\sqrt{3}mI_m\right)^2-\dfrac{V_{dc}{}^2}{Z_r{}^2}}\leq 60\ \text{A}\end{cases} \tag{10.53}$$

10.3.1.2.3 Requirement of Resonant Time

In order to maintain the power quality, the first resonant time T_{r1} is selected to be less than 5% of the switching period:

$$T_{r1}\approx\frac{\pi\sqrt{L_r(3C_r+C_7)}}{2}\leq 0.05T_s \tag{10.54}$$

10.3.1.2.4 Requirement of Minimum Resonant Capacitance

The resonant capacitor of each SiC MOSFET in soft-switching inverter includes two parts, the equivalent output capacitance C_o of the selected SiC MOSFET and the external paralleled capacitance C_p. The equivalent output capacitance C_o of the selected SiC MOSFET C2M0025120D is about 0.37 nF, thus the total resonant capacitance $3C_r + C_7$ meets:

$$3C_r+C_7\geq 3\times 0.37\ \text{nF}+0.37\ \text{nF}=1.48\ \text{nF} \tag{10.55}$$

With the parameters in Table 10.1, 300 kHz switching frequency and the aforementioned four requirements, the appropriate region for selecting resonant parameters can be obtained as shown

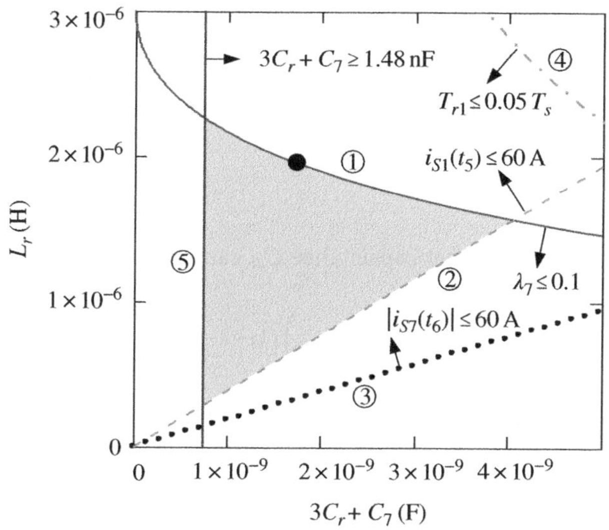

Figure 10.12 Region for selecting resonant parameters.

in Figure 10.12. The curve with number ① is the boundary condition of voltage stress requirement and drawn by using Eq. (10.50). The dashed curves with numbers ② and ③ are the boundary conditions of current stress requirement and drawn by using Eq. (10.53). The dashed curve with number ④ is the boundary condition of resonant time requirement and drawn by using Eq. (10.54), while the vertical line with number ⑤ is the boundary condition of minimum resonant capacitance requirement and drawn by using Eq. (10.55). The dark region shown in Figure 10.12 can be used to select the appropriate resonant inductance and resonant capacitance.

The black point shown in Figure 10.12 is chosen as the resonant parameter at 300 kHz switching frequency. The resonant inductance L_r is 2 μH, the total resonant capacitance is 1.84 nF, the external paralleled capacitance for each main switche C_p is 120 pF, and no external paralleled capacitor is added to auxiliary switch. The resonant impedance Z_r can be calculated as 33 Ω. Similarly, the resonant parameters under other switching frequencies can be derived.

The parameters under different switching frequency are shown in Table 10.2. The values of the resonant inductors L_r and total resonant capacitor are reduced proportionally to the switching

Table 10.2 Parameters under different switching frequency.

Switching frequency f_s/kHz	AC filter inductor (L_a, L_b, L_c)/μH	Resonant inductor L_r/μH	Resonant capacitor ($C_r = C_p + C_o$)/nF	Resonant impedance Z_r/Ω	Turn-off duty cycle of S_7 λ_7
50	200	12	2.2 + 0.37	38.5	0.1
100	100	6	1 + 0.37	36.6	0.1
200	50	3	0.36 + 0.37	34.2	0.1
300	33	2	0.12 + 0.37	33.0	0.1

period to keep a constant turn-off duty cycle of the auxiliary switch. It is clear that the resonant impedances Z_r are almost the same at different switching frequencies in order to keep the same current stress in the auxiliary switch and main switches. The values of the output filter inductors at different switching frequencies can keep the maximum filter inductor ripple current as 25% of the rated peak grid current.

10.3.1.3 DC Filter Capacitor

According to the derivation in Appendix A.6, the DC filter capacitance C_{dc} can be derived as:

$$C_{dc} = \frac{1}{\Delta V_{dc(\max)}}\left[\left(\frac{V_{dc}}{Z_r} + \frac{\sqrt{3}}{2}mI_m\right)\frac{\lambda_7 T_s}{4} + \frac{\sqrt{3}}{2}\left(I_m - \frac{\sqrt{3}}{2}mI_m\right)(1-\lambda_7)mT_s\right] \quad (10.56)$$

where $\Delta V_{dc(max)}$ is the maximum DC filter capacitor voltage ripple. Equation (10.56) can also be used to describe the DC filter capacitance in hard-switching inverter by setting the turn-off duty cycle of auxiliary switch λ_7 to zero. Since λ_7 is always smaller than 0.1, its impact to DC filter capacitance is small. In order to simplify the analysis, DC filter capacitance in soft-switching inverter is identical to that in hard-switching inverter under the same switching frequency.

With the input and output parameters in Table 10.1, resonant impedance Z_r and turn-off duty cycle of auxiliary switch λ_7 in Table 10.2, the relationship between DC filter capacitance C_{dc} and switching frequency f_s can be calculated as shown in Figure 10.13. Resonant impedance Z_r is equal to 35 Ω, λ_7 is equal to 0.1, and the maximum value of DC filter capacitor voltage ripple $\Delta V_{dc(max)}$ is limited to 1% of DC voltage V_{dc}.

10.3.1.4 Clamping Capacitor

With Eqs. (10.38) and (10.39), the waveforms of resonant inductor current i_{Lr}, DC-side current i_{dc}, current of auxiliary switch i_{S7}, and clamping capacitor current i_{cl} in one switching period within sector 1-1 can be drawn from Figure 10.7 as shown in Figure 10.14.

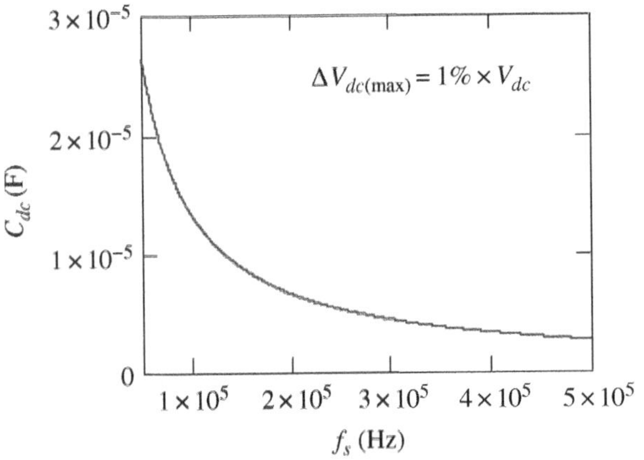

Figure 10.13 Relationship between DC filter capacitance C_{dc} and switching frequency f_s.

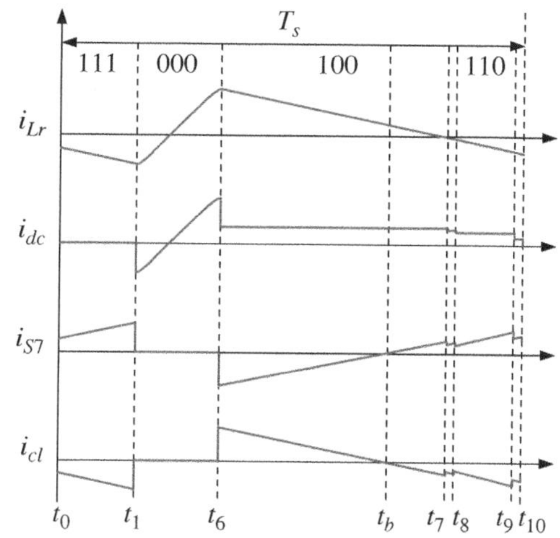

Figure 10.14 Waveform of clamping capacitor current i_{Cc} in one switching period within sector 1-1.

By using (10.47) and (10.48), the resonant current i_{Lr} in (10.37) can be rewritten as:

$$\begin{aligned} i_{Lr}(t) &= i_{Lr}(t_6) - \frac{i_{Lr}(t_6) - i_{Lr}(t_1)}{(1-\lambda_7)T_s}(t - t_6) \\ &= \frac{V_{dc}}{Z_r} + \sqrt{3}mI_m - \frac{2V_{dc}/Z_r + \sqrt{3}mI_m}{(1-\lambda_7)T_s}(t - t_6) \end{aligned} \tag{10.57}$$

With (10.57) and DC-side current i_{dc} in (10.38), clamping capacitor current i_{cl} at different instants in Figure 10.14 can be derived as:

$$\begin{cases} i_{cl}(t_1) = -\dfrac{V_{dc}}{Z_r} \\ i_{cl}(t_6) = \dfrac{V_{dc}}{Z_r} + \sqrt{3}mI_m - i_a \\ i_{cl}(t_7) = \dfrac{V_{dc}}{Z_r} + \sqrt{3}mI_m - \dfrac{2V_{dc}/Z_r + \sqrt{3}mI_m}{(1-\lambda_7)T_s} T_{V1} - i_a \\ i_{cl}(t_8) = \dfrac{V_{dc}}{Z_r} + \sqrt{3}mI_m - \dfrac{2V_{dc}/Z_r + \sqrt{3}mI_m}{(1-\lambda_7)T_s} T_{V1} + i_c \\ i_{cl}(t_9) = \dfrac{V_{dc}}{Z_r} + \sqrt{3}mI_m - \dfrac{2V_{dc}/Z_r + \sqrt{3}mI_m}{(1-\lambda_7)T_s} (T_{V1} + T_{V2}) + i_c \\ i_{cl}(t_{10}) = \dfrac{V_{dc}}{Z_r} + \sqrt{3}mI_m - \dfrac{2V_{dc}/Z_r + \sqrt{3}mI_m}{(1-\lambda_7)T_s} (T_{V1} + T_{V2}) \end{cases} \tag{10.58}$$

With the input and output parameters in Table 10.1, resonant impedance Z_r equals to 35 Ω and turn-off duty cycle of auxiliary switch λ_7 equals to 0.1, the values of clamping capacitor current i_{cl} at different instants within sector 1-1 can be calculated as shown in Figure 10.15. Obviously, clamping capacitor current i_{cl} at t_7, t_8, and t_9 are almost below zero within entire sector 1-1. However, clamping capacitor current i_{cl} at t_{10} is positive within the phase angle range from 0 to φ_2 while it is below zero within the phase angle range from φ_2 to $\pi/6$.

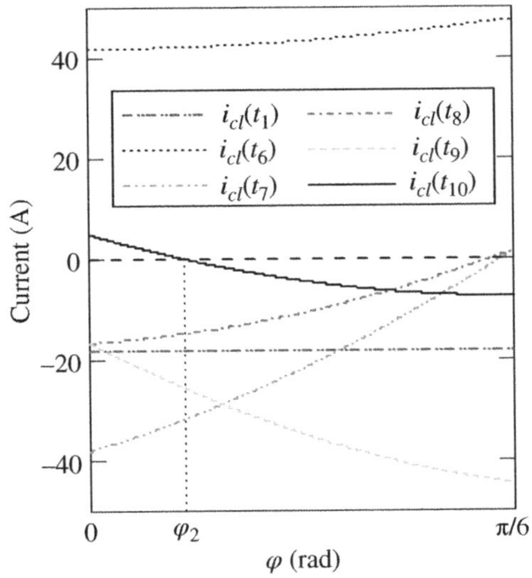

Figure 10.15 Values of clamping capacitor current i_{cl} at different instants within sector 1-1.

Then the instant t_b between t_6 and t_7 as shown in Figure 10.14 can be calculated by the following equation:

$$i_{cl}(t_b) = V_{dc}/Z_r + \sqrt{3}mI_m - \frac{2V_{dc}/Z_r + \sqrt{3}mI_m}{(1-\lambda_7)T_s}(t_b - t_6) - i_a = 0 \tag{10.59}$$

Then t_b can be derived as:

$$t_b = \frac{\left(V_{dc}/Z_r + \sqrt{3}mI_m - i_a\right)(1-\lambda_7)T_s}{2V_{dc}/Z_r + \sqrt{3}mI_m} + t_6 \tag{10.60}$$

According to the current polarity of clamping capacitor current i_{cl} at t_{10}, there are two kinds of waveforms of clamping capacitor current i_{cl} in one switching period as shown in Figure 10.16 as well as the related clamping capacitor voltage ripple.

Obviously, the amplitude of clamping capacitor voltage ripple ΔV_{cl} can be obtained as:

$$\begin{aligned}\Delta V_{cl} &= \frac{1}{C_c}\int_{t_6}^{t_b} i_{cl}dt = \frac{1}{C_{cl}}\int_{t_6}^{t_b}\left(i_{Lr} - i_{dc}\right)\,dt \\ &= \frac{1}{C_{cl}}\int_{t_6}^{t_b}\left(V_{dc}/Z_r + \sqrt{3}mI_m - \frac{2V_{dc}/Z_r + \sqrt{3}mI_m}{(1-\lambda_7)T_s}(t - t_6) - i_a\right)\,dt \\ &= \frac{\left(V_{dc}/Z_r + \sqrt{3}mI_m - I_m\cos\varphi\right)^2(1-\lambda_7)T_s}{C_{cl}\left(4V_{dc}/Z_r + 2\sqrt{3}mI_m\right)}\end{aligned} \tag{10.61}$$

As shown in (10.61), the maximum value of clamping capacitor voltage ripple $\Delta V_{cl(max)}$ can be obtained when phase angle φ is equal to $\pi/6$. With the input and output parameters in Table 10.1, resonant impedance Z_r equals to 35 Ω, turn-off duty cycle of auxiliary switch λ_7 equals to 0.1, switching frequency equals to 300 kHz, and the maximum value of clamping capacitor voltage ripple $\Delta V_{cl(max)}$ limited to 2% of clamping capacitor voltage, the relationship between the maximum value of clamping capacitor voltage ripple $\Delta V_{cl(max)}$ and the clamping capacitance C_{cl} is shown in Figure 10.17.

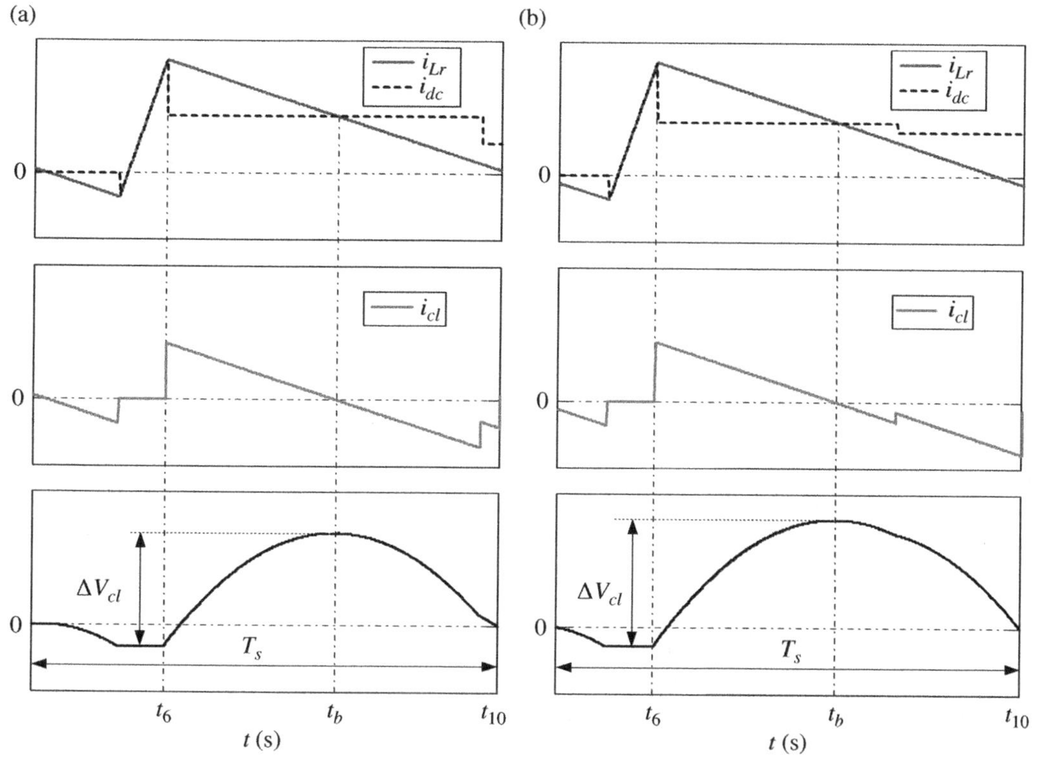

Figure 10.16 Two kinds of waveforms of clamping capacitor current i_{cl} in one switching period: (a) positive polarity and (b) negative polarity of clamping capacitor current i_{cl} at t_{10}.

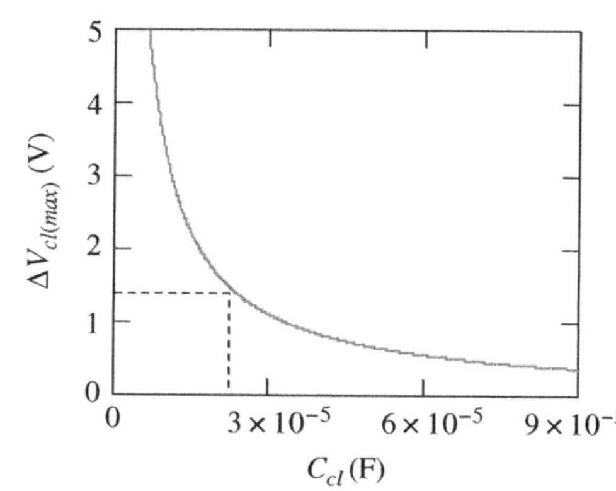

Figure 10.17 The relationship between the maximum value of clamping capacitor voltage ripple $\Delta V_{cl(max)}$ and the clamping capacitance C_{cl}.

10.3.1.5 Cores Selection

AC filter inductors and resonant inductor are designed with Area Product method [4]. A pair of E-shaped (EE) cores made up of DMR95 ferrite material are adopted in the AC filter inductors of both hard-switching SiC inverter and soft-switching SiC inverter with maximum magnetic flux density B_{max} limited to 0.3 T. EE cores made up of DMR51 ferrite material with lower core loss are

Table 10.3 Cores selection and volume evaluation of AC filter inductors and resonant inductor.

Switching frequency	AC filter inductor cores	Three AC filter inductors volume	Resonant inductor cores	Resonant inductor volume
50 kHz	3 × EE70	1.70	2 × EE42*	0.16
100 kHz	2 × EE65	1.01	EE42*	0.10
200 kHz	EE65	0.66	EE42	0.07
300 kHz	EE55A	0.42	2 × EE33	0.04

applied for the resonant inductor of soft-switching SiC inverter due to the large amplitude and high frequency of the resonant inductor current. According to literatures about designing resonant inductor with ferrite material operating at hundreds of kHz or MHz frequency, a maximum magnetic flux density at 0.15 T is selected for resonant inductor.

Physical size and loss of the AC filter inductors and the resonant inductor can be evaluated according to the loss models in Appendix A.4. Cores selection and volume evaluation of AC filter inductors and resonant inductor are summarized in Table 10.3.

10.3.1.6 Switching Loss Measurement

As mentioned in Section 10.3.1.2, SiC MOSFET Wolfspeed C2M0025120D with 1200 V/60 A rating is used in both the hard-switching SiC inverter and the soft-switching SiC inverter. The calculation of conduction loss in the three-phase main bridges of hard-switching SiC inverter uses the parameters in datasheet of C2M0025120D under 100 °C junction temperature, while the calculation of switching loss uses the measured results from double-pulse test (DPT) setup with 4.7 Ω external gate resistor and 100 °C junction temperature provided by the hot plate. The turn-on energy, turn-off energy, and reverse recovery energy of SiC MOSFET are measured from DPT setup, and its scheme is shown in Figure 10.18a with 600 V DC voltage. The equivalent output capacitance of SiC MOSFET is represented as C_{o1} and C_{o4}. Snubber capacitor C_s is placed close to the testing

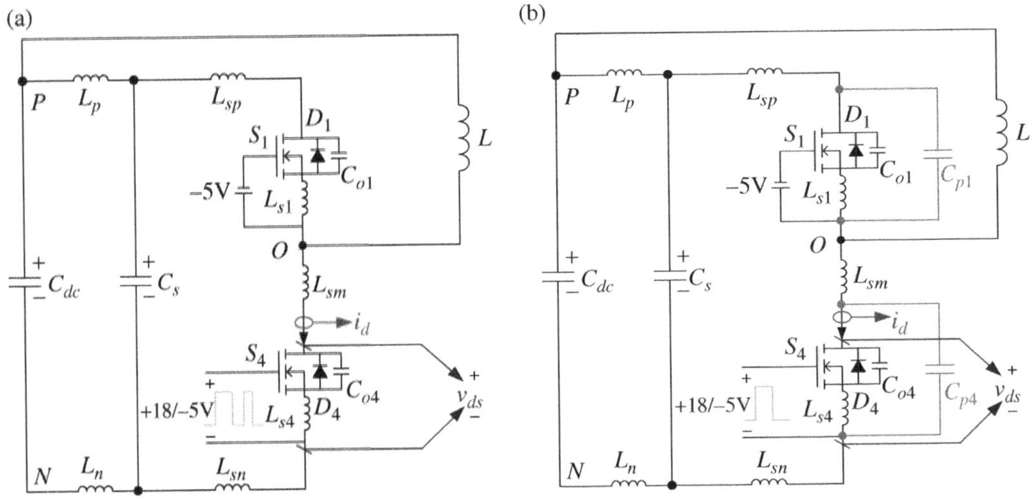

Figure 10.18 Switching loss testing scheme: (a) double-pulse test and (b) single-pulse test.

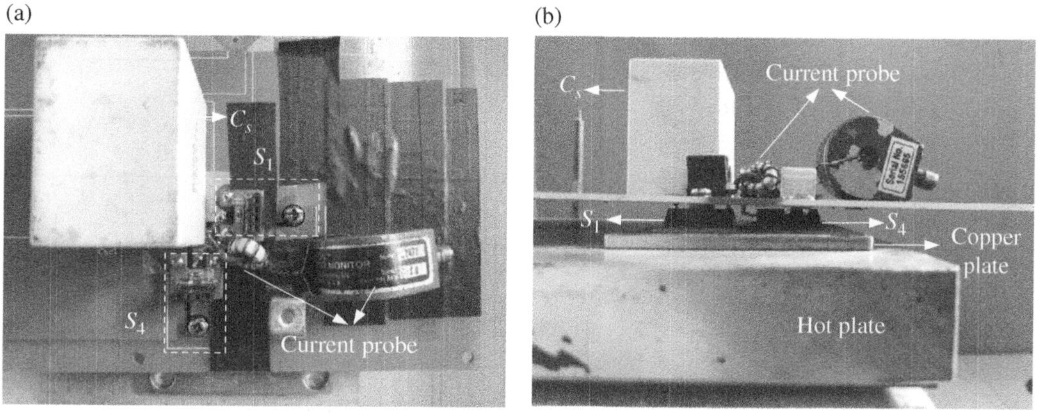

Figure 10.19 Pictures of experimental DPT setup: (a) top view and (b) side view.

SiC MOSFETs half-bridge in order to achieve small commutation loop that consists of stray inductance L_{sp}, L_{sm}, and L_{sn}. L_{s1} and L_{s4} are the shared source stray inductance of both power loop and driving loop introduced by pin length of TO-247 package, which are also hoped to be minimized for achieving low switching loss. In DPT setup, the mounting positions and pin lengths of SiC MOSFETs are similar to those in the inverter setup. The measurement positions of current i_d and voltage v_{ds} are also shown in Figure 10.18a.

The calculation of conduction loss in all seven switches of the soft-switching SiC inverter uses the same parameters as those for hard-switching SiC inverter. The turn-off energy of SiC MOSFET with different external paralleled capacitors is also measured from DPT setup as shown in Figure 10.18b, which will be utilized in calculation of turn-off loss in both main switches and auxiliary switch. Single pulse is used to obtain the turn-off energy instead of the double pulse as the double-pulse brings large and unsafe turn-on current due to the external paralleled capacitors, C_{p1} and C_{p4}.

The pictures of experimental DPT setup are shown in Figure 10.19. A 3 mm thick copper plate is inserted between the SiC MOSFETs and the hot plate to thoroughly heat up the SiC MOSFETs' junction temperature to 100 °C.

CYBERTEK's DP6150B differential voltage probe (200 MHz band width and $\pm 2\%$ accuracy) and a two-stage current transformer including PEARSON 2877 current monitor (200 MHz band width and $+1/-0\%$ accuracy) are used in the current and voltage measurement of tested SiC MOSFET.

Two probes need to be calibrated by measuring the drop voltage and flowing current of a low-inductance resistor as shown in Figure 10.20a. The excitation voltage is supplied by RIGOL's Function Waveform Generator DG4062, which generates an extreme sharp square wave with rise or fall time less than 20 ns. L_s and R_s are the stray inductance and stray resistance in the testing circuit, respectively. The time delay Δt caused by the stray inductance in tested low-inductance resistor L_{sR} is shown in Figure 10.20b and can be derived as:

$$\Delta t = \frac{L_{sR}\frac{di_m}{dt}}{\frac{dv_m}{dt}} = \frac{L_{sR}}{R} \tag{10.62}$$

As shown in (10.62), low size resistor with large resistance is preferred in order to reduce the time delay Δt introduced by the tested low-inductance resistor. The typical stray inductance of the

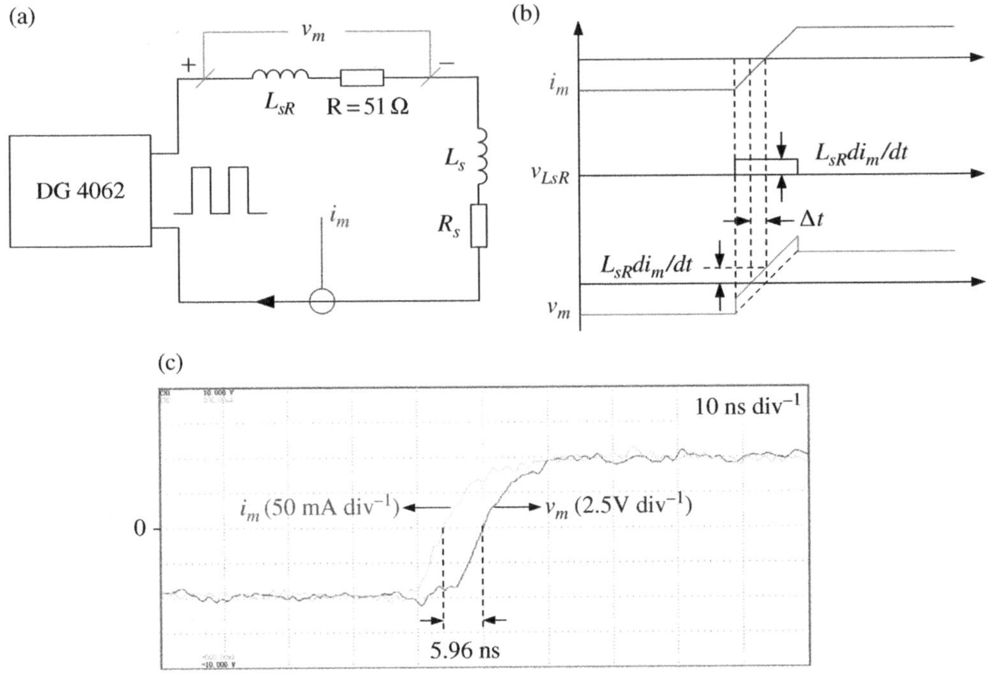

Figure 10.20 Probes calibration: (a) scheme, (b) delay caused by stray inductance in low-inductance resistor, and (c) experimental results.

implemented 51 Ω resistor in 2512 package is 5 nH, which brings only 0.1 ns time delay. The measured voltage and current waveforms of tested low-inductance resistor are shown in Figure 10.20c. The time delay between current probe and voltage probe is 5.96 ns. Thus the time delay Δt caused by the stray inductance in tested low-inductance resistor L_{sR} can be ignored. The time delay of current channel or voltage channel in digital scopes needs to be adjusted with the calibration results.

The tested turn-on energy E_{on}, turn-off energy E_{off}, and reverse recovery energy E_{rr} of SiC MOSFET C2M0025120D at different drain current I_d without any external paralleled capacitor are shown in Figure 10.21a. It is clear that the turn-on energy E_{on} and turn-off energy E_{off} increases with the increment of drain current I_d while the reverse recovery energy E_{rr} keeps almost constant. Besides, the turn-on energy E_{on} is much larger than turn-off energy E_{off} and reverse recovery energy E_{rr}. The tested turn-off energy E_{off} of SiC MOSFET C2M0025120D with different external paralleled capacitors C_p are shown in Figure 10.21b. Obviously, with the increment of external paralleled capacitors C_p from 0.12 to 2.2 nF, the turn-off energy E_{off} decreases significantly when drain current I_d is larger than 15 A. The data in Figure 10.21a and b will be used to calculate the switching loss of SiC MOSFETs in hard-switching inverter and soft-switching inverter, respectively.

10.3.2 Comparison of Two SiC Inverters

By using the designed parameters in Section 10.3.1 and the loss models of inverters in Appendix A.4, the loss distributions of the hard-switching three-phase grid inverter and the soft-switching three-phase grid inverter can be estimated.

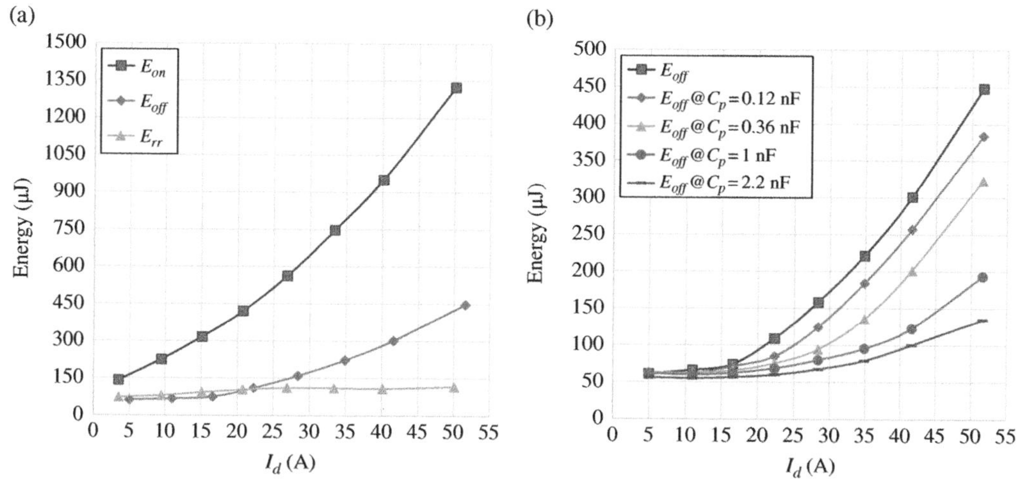

Figure 10.21 Switching energy test results of C2M0025120D from DPT setup: (a) switching energy without external paralleled capacitor and (b) turn-off energy with different external paralleled capacitor.

10.3.2.1 Loss Distributions

The loss distributions of the hard-switching SiC inverter at rated power with the switching frequency f_s changing from 50 to 300 kHz are shown in Figure 10.22. The conduction loss of three-phase main bridges P_{con3} shows small upward trend that is caused by increasing conduction loss due to dead time when switching frequency rises. The turn-on loss P_{on3} increases rapidly and dominates the main part of the total loss P_{total}, while the turn-off loss still occupies a low proportion at high switching frequency. The winding loss P_{L_Cu} takes the majority of AC filter inductor loss and decreases with increasing switching frequency. However, the AC filter inductor loss also occupies a low proportion of the total loss. As a result, the total loss P_{total} increases by 187% from 204 W at 50 kHz switching frequency to 585 W at 300 kHz switching frequency.

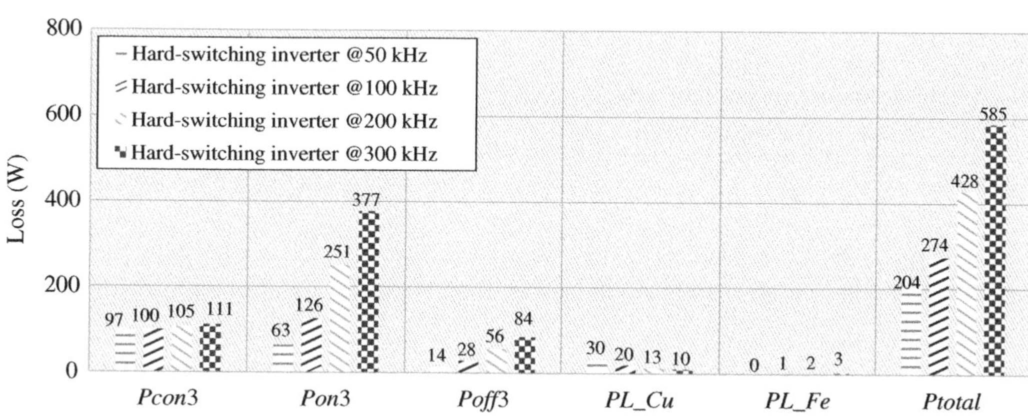

Figure 10.22 Loss distributions of hard-switching SiC inverter with different switching frequencies at rated power (P_{con3}, P_{on3}, and P_{off3} are the conduction loss, turn-on loss, and turn-off loss of three-phase main bridges, respectively; P_{L_Cu} and P_{L_Fe} are the winding loss and core loss of filter inductors; P_{total} is the total loss).

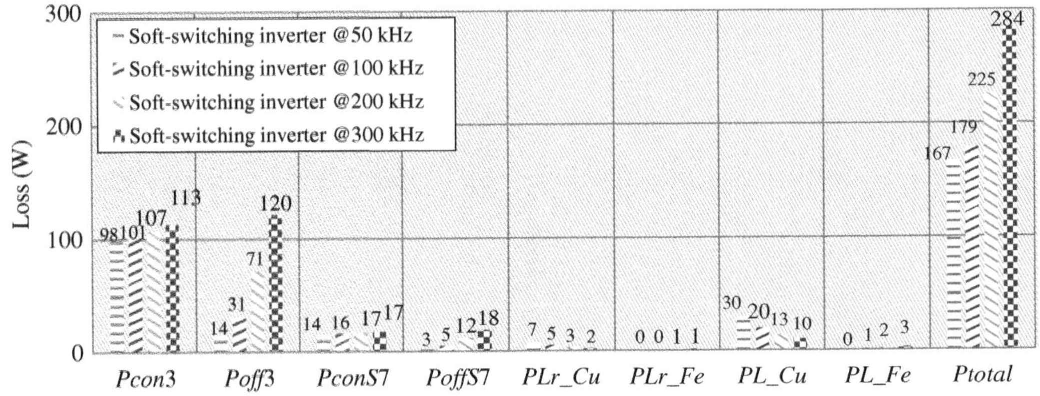

Figure 10.23 Loss distributions of ZVS-SVM inverter with different switching frequencies at rated power (P_{conS7} and P_{offS7} are the conduction loss and turn-off loss of auxiliary switch; P_{Lr_Cu} and P_{Lr_Fe} are the winding loss and core loss of resonant inductor).

The loss distributions of the soft-switching SiC inverter at rated power with the switching frequency f_s changing from 50 to 300 kHz are shown in Figure 10.23. In comparing with the hard-switching SiC inverter, the soft-switching SiC inverter has eliminated the turn-on loss P_{on3} in all SiC MOSFETs. However, it introduces extra turn-off loss in main bridges due to the short-circuit stage and the loss in added components including the auxiliary switch and the resonant inductor. The turn-off loss in three-phase main bridges increases rapidly and dominates the main part of the total loss P_{total} at 300 kHz switching frequency. The loss in auxiliary switch, including the conduction loss P_{conS7} and the switching loss P_{offS7}, and the loss in resonant inductor occupy small portion of the total loss in soft-switching SiC inverter. As a result, the total loss keeps a slow increment when the switching frequency increases from 50 to 300 kHz. The total loss increases by only 70% from 167 W at 50 kHz switching frequency to 284 W at 300 kHz switching frequency.

The loss distributions of the hard-switching SiC inverter with 100 kHz switching frequency and the soft-switching SiC inverter with 300 kHz switching frequency are shown in Figure 10.24. It is observed that by using the ZVS-SVM technique to reassign the loss distribution, the switching

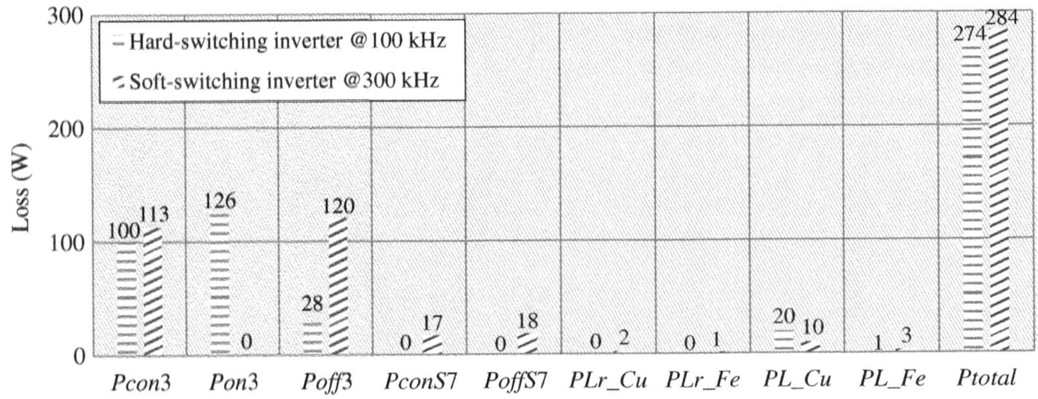

Figure 10.24 Loss distributions comparison of hard-switching inverter (f_s = 100 kHz) and ZVS-SVM inverter (f_s = 300 kHz).

frequency of soft-switching SiC inverter can be raised to 300 kHz, and meanwhile the total loss is kept at almost the same level as that in the hard-switching SiC inverter with 100 kHz switching frequency.

10.3.2.2 Efficiency Stiffness

The theoretical conversion efficiencies of hard-switching SiC inverter and soft-switching SiC inverter at rated power against the switching frequency are shown in Figure 10.25, which reveal significant advantages of conversion efficiency of soft-switching SiC inverter at high switching frequency. The conversion efficiency of hard-switching SiC inverter decreases from 98.99 to 97.16% when switching frequency changing from 50 to 300 kHz, while the conversion efficiency of the soft-switching SiC inverter decreases from 99.17 to 98.60%. The SiC MOSFET device seems not suitable for hard-switching inverter above 100 kHz switching frequency; on the other hand the soft-switching inverter can still keep high conversion efficiency at 98.60% at 300 kHz switching frequency.

The slope of the conversion efficiency curve with the increasing switching frequency as shown in Figure 10.25 can be defined as a new benchmark as efficiency stiffness, which can be calculated as:

$$\text{Efficiency stiffness} = \frac{\Delta\eta}{\Delta f_s} = \frac{\eta(f_{s2}) - \eta(f_{s1})}{f_{s2} - f_{s1}} \tag{10.63}$$

where $\eta(f_{s1})$ is the conversion efficiency at switching frequency f_{s1}, and $\eta(f_{s2})$ is the conversion efficiency at switching frequency f_{s2}. As shown in Figure 10.25, the efficiency stiffness of the hard-switching SiC inverter is about −0.73%/100 kHz while the efficiency stiffness of the soft-switching SiC inverter is about −0.23%/100 kHz. It means the efficiency decreasing rate with the increasing switching frequency in soft-switching SiC inverter is only about one-third of that in hard-switching SiC inverter. It is clear that the ZVS-SVM technique significantly raises the switching frequency of SiC inverter without sacrificing the conversion efficiency, which benefits the miniaturization of passive components in SiC inverters. The efficiency stiffness is used to evaluate not only the

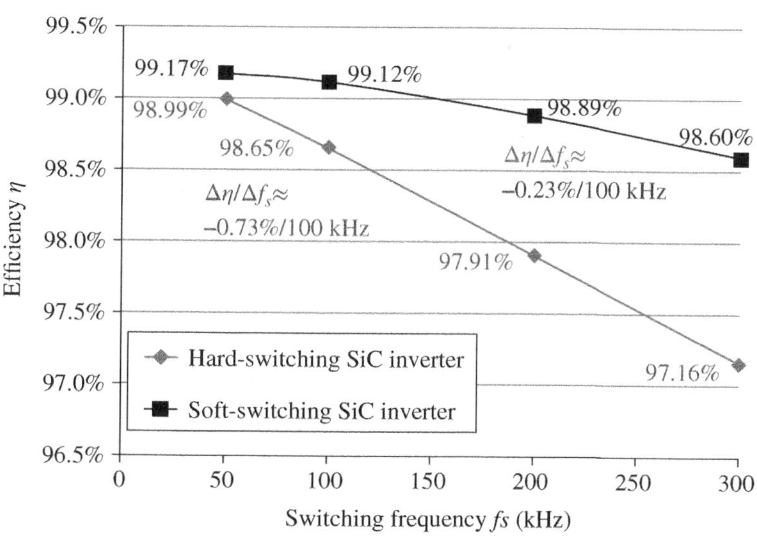

Figure 10.25 Comparison of theoretical conversion efficiencies of hard-switching inverter and soft-switching inverter at 20 kW.

efficiency performance of different circuit topologies with the same devices but also the efficiency performance of different power devices adopted in the same topology.

10.3.2.3 Passive Components Volumes

The physical volumes of the AC filter inductors are identical in both the hard-switching SiC inverter and soft-switching SiC inverter under the same switching frequency. For the soft-switching inverter, the volume of resonant inductor is added. Volumes of the AC filter and the resonant inductor at different switching frequency are shown in Figure 10.26. The volume of the AC filter inductors in the hard-switching SiC inverter with 100 kHz switching frequency is 1.005 L, while the volume of the AC filter inductors and the resonant inductor of the soft-switching SiC inverter with 300 kHz switching frequency is 0.416 and 0.035 L, respectively. A 55% volume reduction has been achieved, and the power density of SiC inverter has been improved.

10.3.3 Experimental Verification

A 20 kW hard-switching SiC inverter under 100 kHz switching frequency and a 20 kW soft-switching SiC inverter under 300 kHz switching frequency both with discrete packaged SiC MOSFET C2M0025120D are built. The main input and output specifications are listed in Table 10.1 and the resonant parameters are listed in Table 10.2. The TMS320F28335 DSP controller is used in the two prototypes. The two inverters are operating at the stand-alone mode with resistive loads.

The comparison of hard turn-on and ZVS turn-on waveforms at 40 A is shown in Figure 10.27. As shown in Figure 10.27a, the overlap of the drain to source voltage v_{S1} and the drain current i_{S1} during switching transition brings turn-on loss in SiC MOSFET. In Figure 10.27b, current commutation in soft-switching SiC inverter starts after the drain to source voltage v_{S1} resonating to zero, which means the ZVS turn-on of switch S_1 is realized and no turn-on loss occurs.

10.3.3.1 Efficiency Test

The measured conversion efficiencies of the 20 kW hard-switching SiC inverter under 100 kHz switching frequency and the 20 kW soft-switching SiC inverter under 300 kHz switching frequency

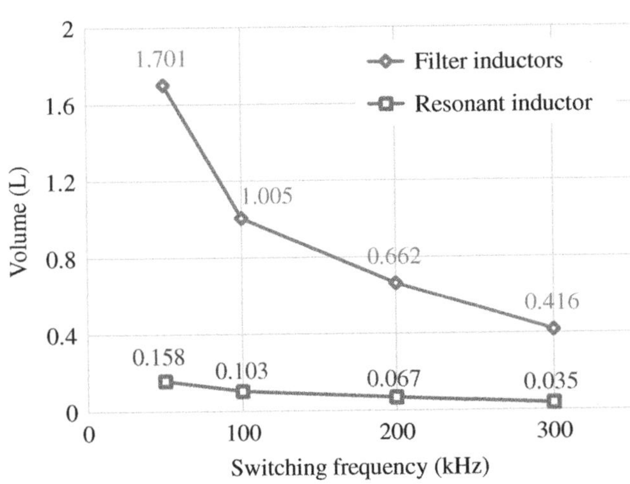

Figure 10.26 Volumes comparison of the key passive components of hard-switching SiC inverter and soft-switching SiC inverter.

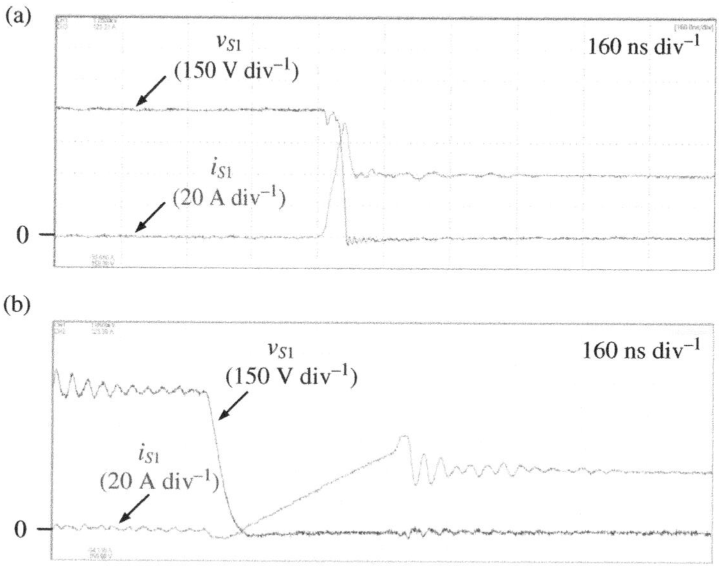

Figure 10.27 Waveforms of hard turn-on and ZVS turn-on at 40 A: (a) hard turn-on and (b) ZVS turn-on.

with resistive load ranging from 3 to 20 kW are shown in Figure 10.28. The peak efficiency of the hard-switching SiC inverter is 98.85 and 0.11% higher than the peak efficiency of the soft-switching SiC inverter. At the light load range, the hard-switching SiC inverter seems to have an advantage. At the heavy load, the soft-switching SiC inverter has a little higher efficiency. At 20 kW load, the soft-switching SiC inverter with 300 kHz switching frequency has almost the same efficiency as that of the hard-switching SiC inverter with 100 kHz switching frequency.

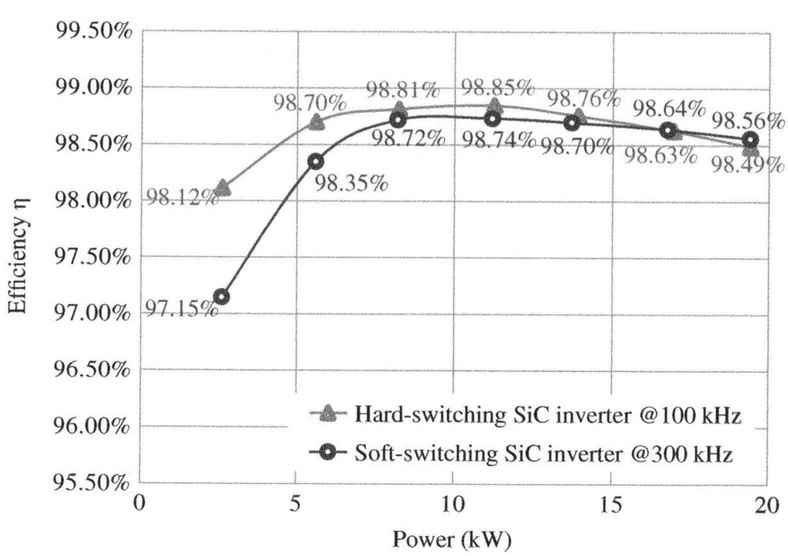

Figure 10.28 Measured conversion efficiency of hard-switching SiC inverter with 100 kHz switching frequency and the soft-switching SiC inverter with 300 kHz switching frequency.

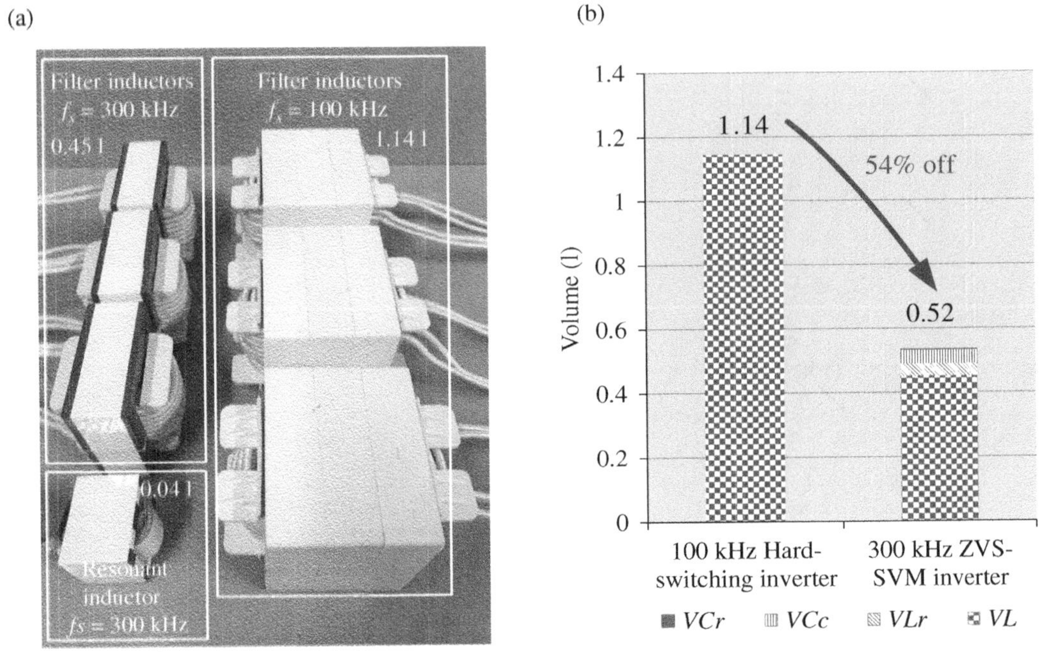

Figure 10.29 (a) Prototypes of inductors and (b) volume comparison of the key passive components of the hard-switching SiC inverter and the soft-switching SiC inverter (V_L is the volume of filter inductors, V_{Lr} is the volume of resonant inductor, V_{cl} is volume of clamping capacitors, and V_{Cr} is the volume of external paralleled capacitors).

10.3.3.2 Passive Components Volumes Comparison

The prototypes of the inductors with dimensions in the hard-switching SiC inverter and the soft-switching SiC inverter are shown in Figure 10.29a. The total volumes of key passive components in the hard-switching inverter under 100 kHz switching frequency including the filter inductors (L_a, L_b, L_c) and the total volumes of key passive components in soft-switching SiC inverter under 300 kHz switching frequency including the filter inductors (L_a, L_b, L_c), resonant inductor (L_r), clamping capacitors (C_{cl}), and the resonant capacitors (C_1, C_2, C_3, C_4, C_5, C_6) are compared in Figure 10.29b.

The volume of the key passive components in soft-switching SiC inverter under 300 kHz switching frequency has been reduced by 54% in comparison with that in hard-switching SiC inverter under 100 kHz switching frequency. As a result, the power density of SiC inverter has been significantly increased.

10.4 Design of Low Stray Inductance Layout in Soft-switching SiC Inverter

10.4.1 Oscillation Model

There are two categories of voltage overshoots in soft-switching SiC inverter. One category is the voltage overshoot during turn-off caused by the current commutation from MOSFET to diode in main bridges. During this instant, the voltage overshoot is produced by the *di*/*dt* at the channel current falling stage and the stray inductance of the single phase's commutation loop including the snubber capacitors, the clamping capacitors, the auxiliary switch, and the single-phase bridge. The other category of voltage overshoot occurring after the second resonant stage is more complicated, which is caused by the commutation loop including the snubber capacitors, the clamping capacitors, the auxiliary switch, and the three-phase bridges.

The equivalent circuit of the oscillation after the second resonant stage in sector 1-1 is shown in Figure 10.30. L_{scs} is the stray inductance in snubber capacitors, L_{sc} is the stray inductance from snubber capacitors to the clamping capacitors, L_{scc} is the stray inductance in clamping capacitors, L_{sc7} is the stray inductance from the clamping capacitors to auxiliary switch, L_{sd7} is the stray inductance in drain electrode of auxiliary switch, L_{ss7} is the stray inductance in source electrode of auxiliary switch, L_{sp} is the stray inductance from auxiliary switch to common point of three upper switches, L_{sap} is the stray inductance from common point to drain electrode of switch S_1 (similar definitions of L_{sbp} and L_{scp}), L_{sd1} is the stray inductance in drain electrode of switch S_1 (similar definitions of L_{sd4}, L_{sd3}, L_{sd6}, L_{sd5}, and L_{sd2}), L_{ss1} is the stray inductance in source electrode of switch S_1 (similar definitions of L_{ss4}, L_{ss3}, L_{ss6}, L_{ss5}, and L_{ss2}), L_{sam} is the stray inductance from source electrode of upper switch S_1 to drain electrode of lower switch S_4 (similar definitions of L_{sbm} and L_{scm}), L_{san} is the stray inductance from source electrode of lower switch S_4 to common point of three lower switches (similar definitions of L_{sbn} and L_{scn}), and L_{sn} is the stray inductance from common point of three lower switches to snubber capacitors. C_{o1} is the equivalent output capacitance (similar definitions of C_{o4}, C_{o3}, C_{o6}, C_{o5}, and C_{o2}), and C_{p1} is the external paralleled capacitance (similar definitions of C_{p4}, C_{p3}, C_{p6}, C_{p5}, and C_{p2}) in upper switch S_1. The stray inductance in external paralleled

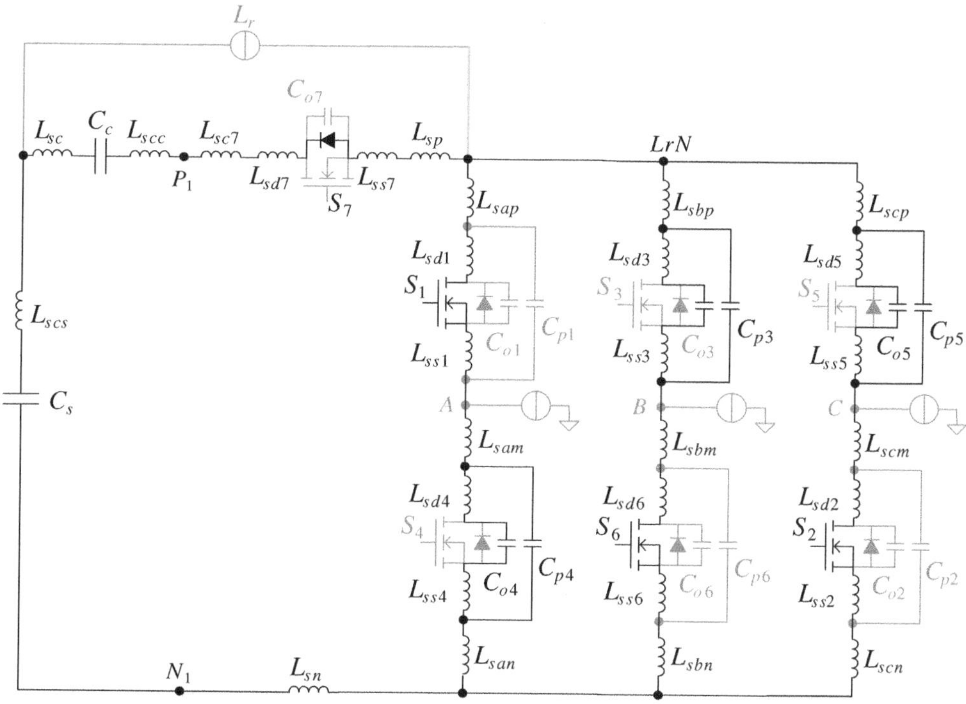

Figure 10.30 Equivalent circuit of the oscillation after the second resonant stage in sector 1-1.

resonant capacitors has not been considered in this equivalent circuit due to their small sizes. These parasitic are assumed as:

$$\begin{cases} L_{sap} = L_{sbp} = L_{scp} \\ L_{sam} = L_{sbm} = L_{scm} \\ L_{san} = L_{sbn} = L_{scn} \\ L_{sd1} = L_{sd4} = L_{sd3} = L_{sd6} = L_{sd5} = L_{sd2} = L_{sd7} \\ L_{ss1} = L_{ss4} = L_{ss3} = L_{ss6} = L_{ss5} = L_{ss2} = L_{ss7} \\ C_{o4} = C_{o3} = C_{o5} \\ C_{p1} = C_{p4} = C_{p3} = C_{p6} = C_{p5} = C_{p2} \end{cases} \tag{10.64}$$

In order to simplify the high frequency equivalent circuit, the large capacitors are assumed as short circuits and the large inductors are assumed as open circuits. Meanwhile, the stray inductance L_{sd4}, L_{ss4}, L_{sd3}, L_{ss3}, L_{sd5}, and L_{ss5} caused by the discrete package of switches S_4, S_3, and S_5 are neglected to simplify the analysis. By using (10.64), the equivalent circuit in Figure 10.30 can be simplified into Figure 10.31. R_s is the total stray resistance of the equivalent circuit.

Obviously, the simplified high frequency equivalent circuit in Figure 10.31 is viewed as a RLC series resonant circuit and expressed as:

$$\begin{cases} R_s i_s + L_s \dfrac{di_s}{dt} + u_s = 0 \\ i_s = C_T \dfrac{du_s}{dt} \end{cases} \tag{10.65}$$

where the stray inductance L_s and resonant capacitance C_T meet:

$$\begin{cases} L_s = L_{scs} + L_{sc} + L_{scc} + L_{sc7} + L_{sd7} + L_{ss7} + L_{sp} + \dfrac{1}{3}\left(L_{sd1} + L_{ss1} + L_{sap} + L_{sam} + L_{san}\right) + L_{sn} \\ C_T = 3C_{o4} + 3C_{p4} \end{cases} \tag{10.66}$$

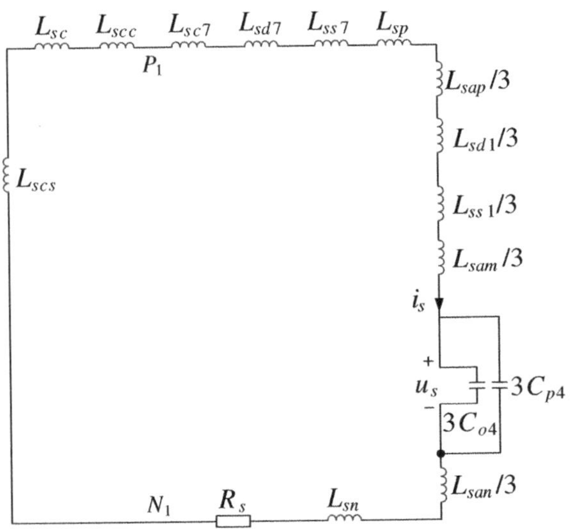

Figure 10.31 Simplified high frequency equivalent circuit.

According to the waveforms in Figure 10.7, there exists a current step at instant t_6 in the high frequency equivalent circuit. With Eq. (10.51), the initial conditions are obtained:

$$\begin{cases} u_s(0) = 0 \\ i_s(0) = \dfrac{V_{dc}}{Z_r} + \sqrt{3}mI_m - i_a \end{cases} \tag{10.67}$$

The voltage across the stray capacitor u_s can be obtained as:

$$u_s(t) = \frac{\dfrac{V_{dc}}{Z_r} + \sqrt{3}mI_m - i_a}{C_T\sqrt{\dfrac{1}{L_sC_T} - \left(\dfrac{R_S}{2L_S}\right)^2}} e^{\frac{-R_S}{2L_S}t} \sin\left(\sqrt{\frac{1}{L_sC_T} - \left(\frac{R_S}{2L_S}\right)^2}\, t\right) \tag{10.68}$$

If the circuit is assumed to be underdamped as:

$$\frac{R_S}{2L_S} \ll \frac{1}{\sqrt{L_sC_T}} \tag{10.69}$$

Then the maximum value U_s of the voltage u_s can be derived as:

$$U_S = \sqrt{\frac{L_S}{C_T}}\left(\frac{V_{dc}}{Z_r} + \sqrt{3}mI_m - i_a\right) \tag{10.70}$$

where voltage U_s is the overshoot voltage on SiC MOSFET. From (10.70), it can be seen that the overshoot voltage U_s is proportional to square root of stray inductance L_s and inversely proportional to stray capacitance C_T. As shown in Table 10.2, the resonant impedance keeps constant under different switching frequency while the external paralleled capacitor C_p is decreased with the increment of switching frequency. Therefore, the overshoot voltage U_s will rise at higher switching frequency if the stray inductance L_s maintains constant. In order to prohibit the overshoot voltage U_s at high switching frequency, method to reduce the stray inductance L_s is required.

10.4.2 Design of Low Stray Inductance 7-in-1 SiC Power Module

As mentioned in the aforementioned oscillation model, the voltage overshoot is not only produced by the stray inductance of the capacitors part including the snubber capacitors and the clamping capacitors but also the power devices part including the seven switches. The stray inductance in the power devices part will be investigated in this section.

In order to reduce the stray inductance of the soft-switching SiC three-phase inverter layout, a 7-in-1 SiC MOSFET power module is designed and its topology is shown in Figure 10.32. No external paralleled capacitors are packaged inside of the power module.

The same bare die CPM2-1200-0025B packaged in discrete SiC MOSFET C2M0025120D is also used in the proposed 7-in-1 power module. Bare die CPM2-1200-0025B has 71 A rated current at 100 °C case temperature and 1200 V rated voltage. The proposed module employing double direct bonding copper (DBC) plates [5] and bond wires is shown in Figure 10.33.

As shown in Figure 10.33, it is clear that the three half-bridges are symmetrically arranged around the auxiliary switch S_7 to achieve an identical stray inductance of three power loops. Besides, a second small DBC plate is soldered in the center of the primary DBC plate. The dotted frame in primary DBC plate reveals the soldering area of the second DBC plate. The second DBC plate enables the three source electrodes of three lower switches, S_4, S_6, and S_2, connecting together to terminal N_1 and staying close to the terminal P_1 to minimize the power loops.

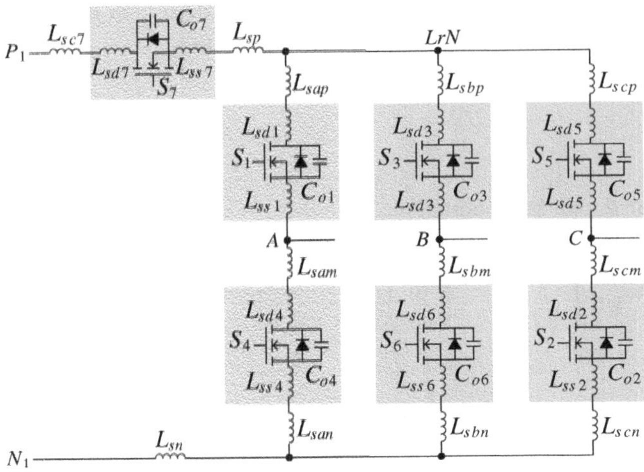

Figure 10.32 Topology of the 7-in-1 SiC MOSFET module.

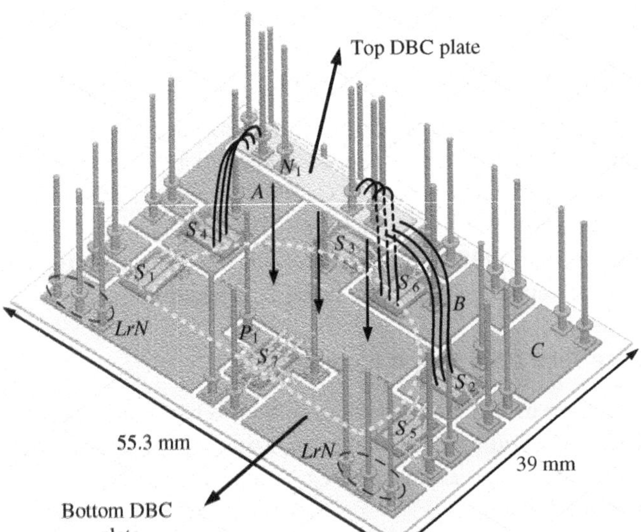

Figure 10.33 3D model of the 7-in-1 SiC MOSFET power module. *Source:* Based on Kicin et al. [5].

Figure 10.34 shows the critical power loop on power devices side from terminal P_1 to terminal N_1 in soft-switching inverter. Four parts of the critical power loop are marked with different curves. The solid line represents the shared power path with auxiliary switch S_7, while the three different dotted lines stand for the power paths of three half-bridges, respectively.

The layout sizes comparison between the original seven discrete devices and the 7-in-1 module is shown in Figure 10.35. The critical power loops in both layouts also have been marked by curves

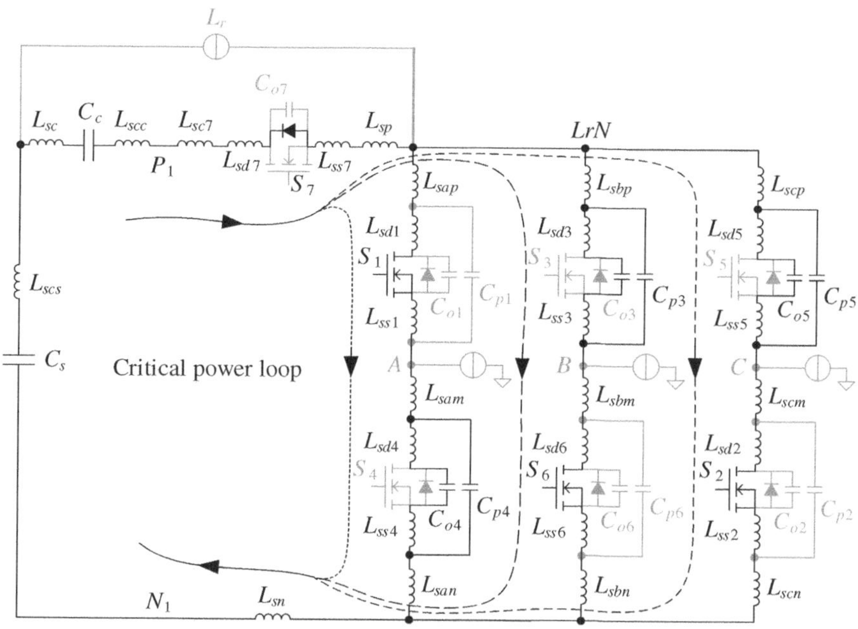

Figure 10.34 Critical power loop in soft-switching inverter.

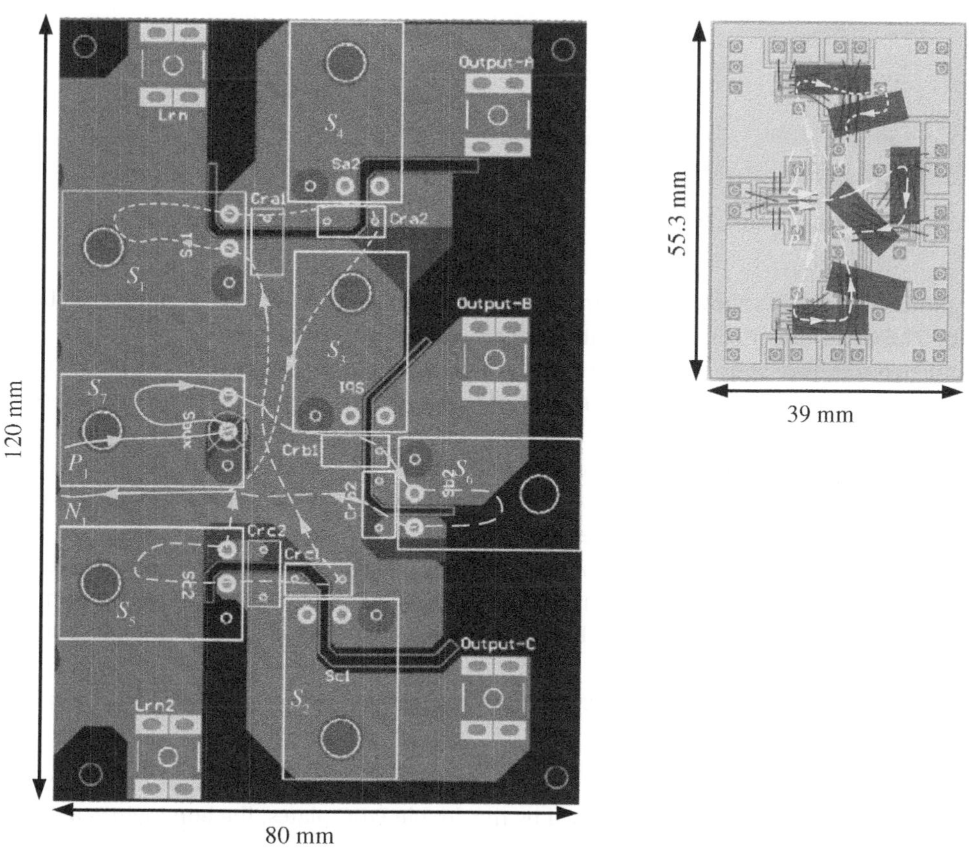

Figure 10.35 Layout sizes comparison of original seven discrete devices and the 7-in-1 module.

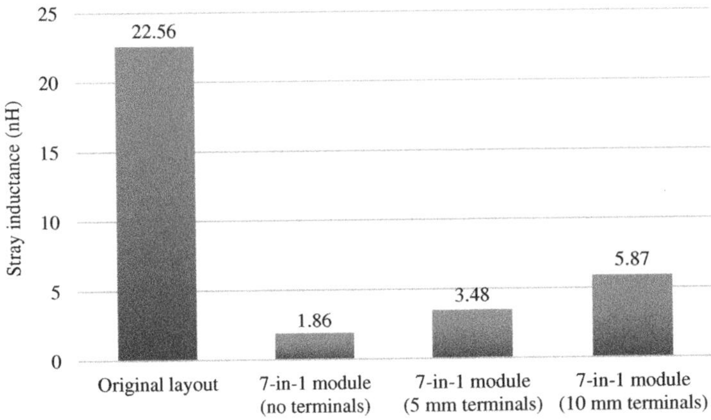

Figure 10.36 Stray inductance comparison of simulation results.

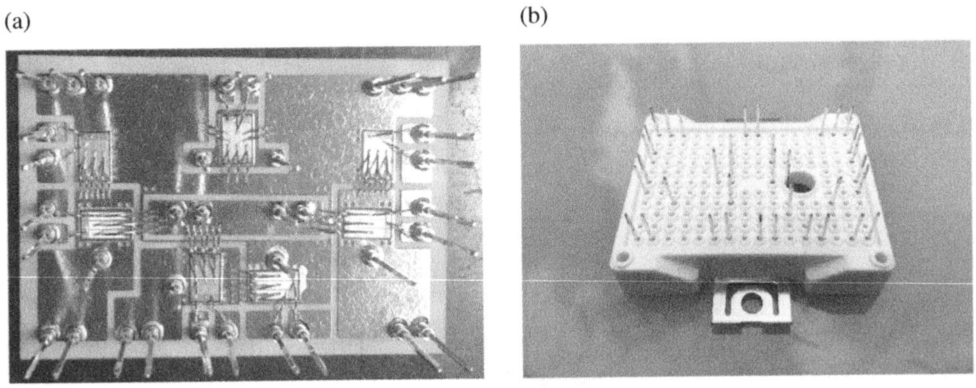

Figure 10.37 Prototype of 7-in-1 SiC MOSFET module: (a) DBC top view and (b) packaged module.

with the similar line style as shown in Figure 10.34. The layout area of the 7-in-1 module has been reduced by 77.5% in comparison to that with original seven discrete devices.

The simulated results of stray inductance of the critical power loop using ANSYS Q3D Extractor under 30 MHz excitation are shown in Figure 10.36. If the length of terminals P_1 and N_1 is not considered, the stray inductance of the critical power loop in 7-in-1 power module is 1.86 μH, which is only 8.2% of the original design with seven discrete devices. If the length of terminals P_1 and N_1 is increased to 10 mm, the stray inductance of the critical power loop in 7-in-1 power module is 5.87 μ H, which is still only 26% of the original design.

Figure 10.37 shows the prototype of 7-in-1 SiC MOSFET power module. The bare dies, bond wires, two-layer structure DBC plate, and the pins are shown in Figure 10.37a. The packaged module is shown in Figure 10.37b, and the pin length inside of the packaged module is about 10 mm.

10.4.3 7-in-1 SiC Power Module Prototype and Testing Results

10.4.3.1 Stray Inductance Measurement

As the stray inductance of bare dies are assumed to be zero, then a power module without bare dies is built to measure the stray inductance caused by the double DBC plates, the bond wires, and the

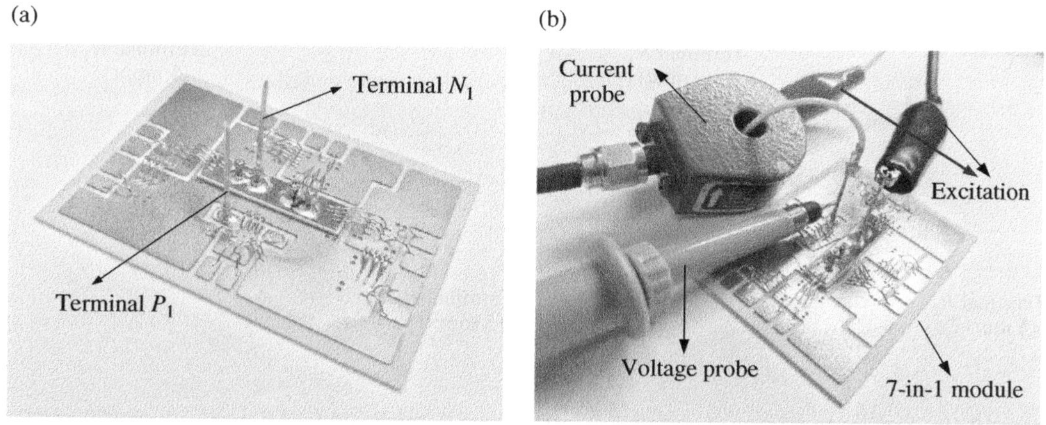

Figure 10.38 Stray inductance measurement of 7-in-1 SiC MOSFET power module: (a) power module without bare dies and (b) measurement setup.

pins as shown in Figure 10.38a. Since most of the impedance under 30 MHz excitation is caused by the stray inductance of the power module, the stray inductance can be obtained with the setup shown in Figure 10.38b. The 30 MHz excitation is generated by RIGOL's Function Waveform Generator DG4062. KEYSIGHT's passive voltage probe N2890A (500 MHz band width) and PEARSON 2877 current monitor (200 MHz band width) are used to measure the voltage across the effective terminals and the current following in the module after deskew calibration. The stray inductance L_s can be calculated by the following equation:

$$L_s = \frac{V_{pp}}{2\pi f_{ex} I_{pp}} \tag{10.71}$$

where V_{pp} is the peak to peak value of the voltage across the effective terminals, I_{pp} is the peak to peak value of the current flowing in the module, and f_{ex} is the excitation frequency equal to 30 MHz in the measurement. The waveforms of the voltage across the effective terminals and the current following in the module are shown in Figure 10.39. The measurement point of the voltage probe in the terminals is 5 mm above the DBC layers. As a result, the stray inductance is obtained as 6.38 nH.

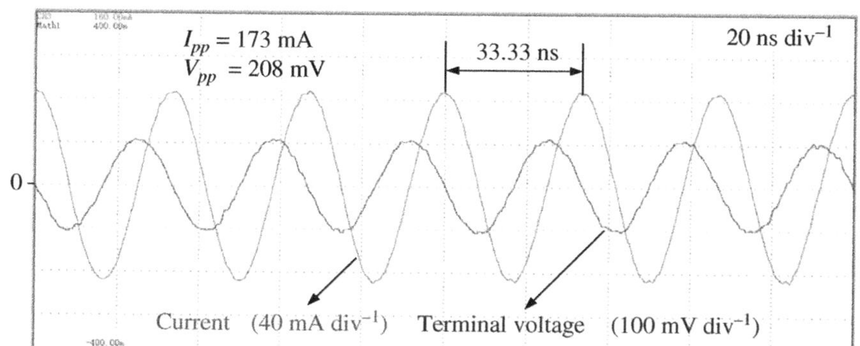

Figure 10.39 Waveforms of the voltage across the effective terminals and the current flowing in the module.

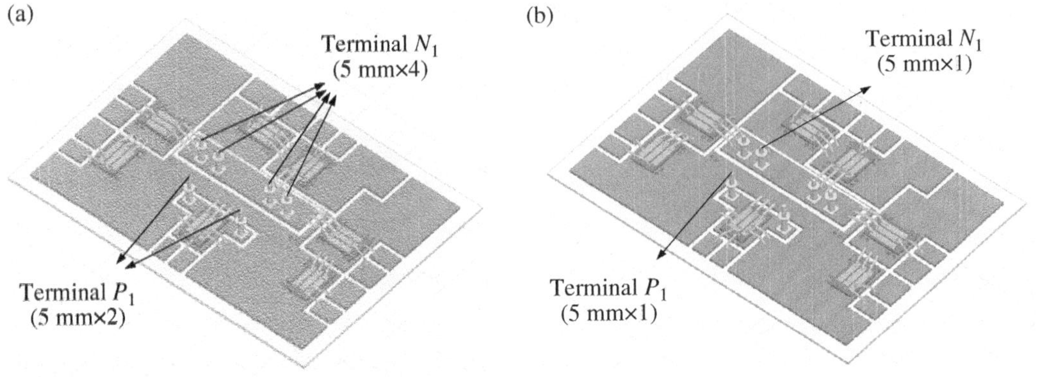

Figure 10.40 Simulation scheme of stray inductance: (a) two P_1 terminals, four N_1 terminals and (b) one P_1 terminal, one N_1 terminal.

Since the numbers of P_1 terminals and N_1 terminals in measurement are different from those in simulation, the measured stray inductance 6.38 nH seems higher than that in simulation as shown in Figure 10.36 with 5 mm terminals. For simulation in Section 10.3.2, there are two P_1 terminals and four N_1 terminals as shown in Figure 10.40a. However, there are only one P_1 terminal and one N_1 terminal in measurement due to the limited excitation terminals of RIGOL's Function Waveform Generator DG4062. The simulation with the same condition of the measurement is investigated as shown in Figure 10.40b. The simulated stray inductance in Figure 10.40b is 6.27 nH, which is very close to the measured result.

10.4.3.2 Voltage Stress Comparison

Figure 10.41 shows the experimental waveforms of the original soft-switching SiC inverter with discrete devices and the soft-switching SiC inverter with 7-in-1 module at rated power after second

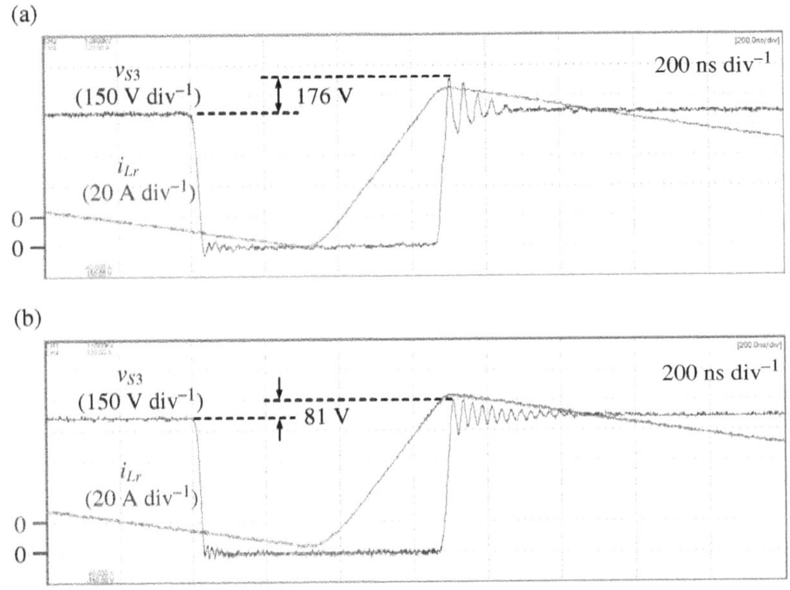

Figure 10.41 Waveforms of the maximum voltage overshoots after the second resonant stage at rated power: (a) discrete devices layout and (b) 7-in-1 module layout.

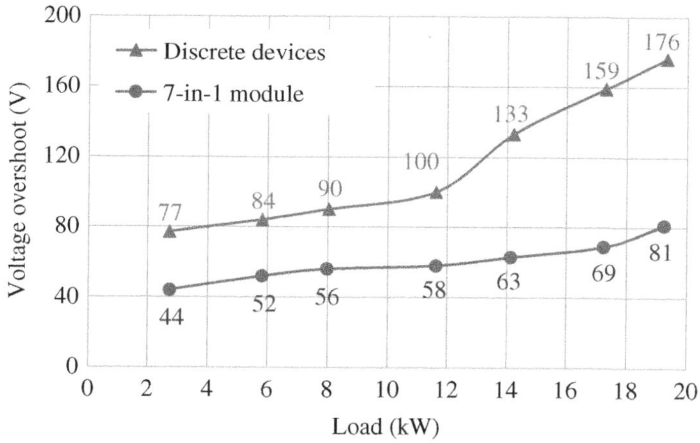

Figure 10.42 Comparison of maximum voltage overshoot after second resonant stage at different loads.

resonant stage. The maximum voltage overshoots of main switch S_3 after the second resonant stage has been reduced by 54% from 176 to 81 V.

Figure 10.42 shows the comparison of maximum voltage overshoot after second resonant stage at different loads. It seems that the voltage overshoots in soft-switching SiC inverter have been significantly reduced by the designed low stray inductance 7-in-1 SiC MOSFET power module.

10.5 Design of Low Loss Resonant Inductor in Soft-switching SiC Inverter

The resonant inductor of soft-switching three-phase SiC inverter operating with hundreds of kHz switching frequency and large current amplitude is a critical component to the whole system. High frequency AC loss induced in inductor windings is due to skin, proximity, fringing, and other AC effects [6]. One feasible solution to eliminate the impact from the fringing fields is using distributed air gap [7]. Besides, both the skin and proximity effects have to be considered to calculate the theoretical AC resistance of the copper foil, which could help to achieve an optimal thickness of copper foil for lowest winding loss [8–10].

10.5.1 Impact of Distributed Air Gap

The original resonant inductor of soft-switching SiC inverter operating at 300 kHz switching frequency is made with centric air gap (2.9 mm for each air gap), two paralleled EE33 cores with ferrite DMR51, and four turns copper foil with 0.4 mm × 18 mm cross section. The distributed air gap is adopted in order to reduce the additional AC winding loss produced by the fringing fields at air gap. Finite Element Analysis (FEA) tool, ANSYS Maxwell 2D, is used to analyze the improvement by using the distributed air gap with the excitation of the time domain resonant inductor current at 20 kW load as shown in Figure 10.43. Two resonant inductors with different number of distributed gaps, six 0.7 mm gaps, and ten 0.4 mm gaps are investigated with the same inductance as the original one. Figure 10.44 shows the comparison of current density in windings with different air gap arrangements at the maximum resonant inductor current (84.9 A). The maximum current density in winding has been significantly reduced from 6.15×10^8 to 2.46×10^8 A/m^2 and 2.79×10^8 A/m^2 with six 0.7 mm gaps and ten 0.4 mm gaps, respectively. The impact from fringing field has been improved.

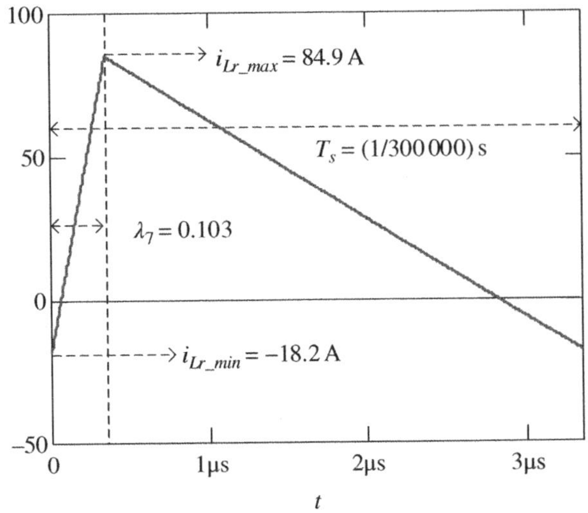

Figure 10.43 Waveform of resonant inductor current in one switching period.

The simulated loss comparison of resonant inductor with different numbers of air gaps is shown in Figure 10.45. With the number of air gaps increasing from 2 to 10, the winding loss is decreased from 21.5 to 10.7 W, while the core loss is only increased from 5.0 to 6.3 W. The winding loss of resonant inductor with six air gaps has been cut by 45.6% in comparison to that of the original one with centric air gaps. The further increment of number of air gaps from 6 to 10 brings another 8.5% reduction of winding loss. By considering both the winding loss and core loss, distributed air gap design with 10 air gaps is chosen for the improvement of the resonant inductor.

10.5.2 Optimal Flux Density Investigation

Since the maximum flux density in the original resonant inductor is only 0.15 T and is far from the saturation flux density of the ferrite material at 0.3 T, the optimal flux density is also investigated with 10 air gaps. Three resonant inductors with the same inductance as the original one and different maximum flux densities are compared in Figure 10.46. Figure 10.46a shows the resonant inductor with three turns winding. Its maximum flux density is 0.22 T, and the length of each air gap is 0.2 mm. Figure 10.46b shows the resonant inductor with four turns winding. Its maximum flux density is 0.17 T, and the length of each air gap is 0.4 mm. Figure 10.46c shows the resonant inductor with five turns winding. Its maximum flux density is 0.14 T, and the length of each air gap is 0.7 mm. With the excitation of the resonant current as shown in Figure 10.43, the simulated loss of three inductors are compared in Figure 10.46d. Obviously, the winding loss increases with the increment of maximum flux density while the core loss decreases. The design with four turns has the optimal flux density with the lowest loss.

10.5.3 Optimal Winding Foil Thickness Investigation

In Appendix A.4, only skin effect is considered in calculating the winding loss of the resonant inductor. However, the proximity effect is severe in multilayer winding if the foil thickness is

Figure 10.44 Current density comparison of different air gap arrangements at the maximum resonant inductor current: (a) centric air gap, (b) distributed air gap design with six gaps, and (c) distributed air gap design with 10 gaps.

(c)

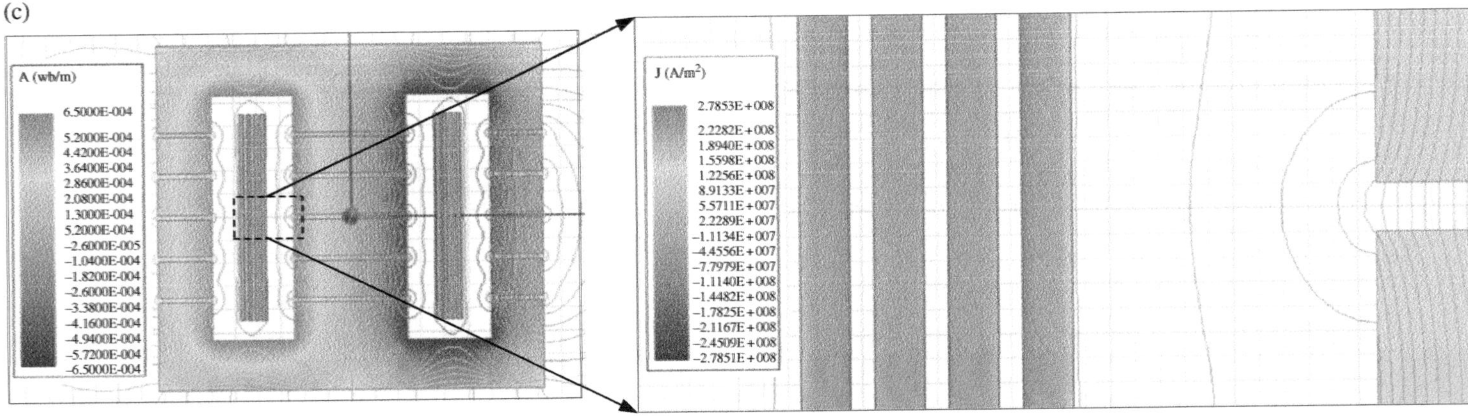

Figure 10.44 (Continued)

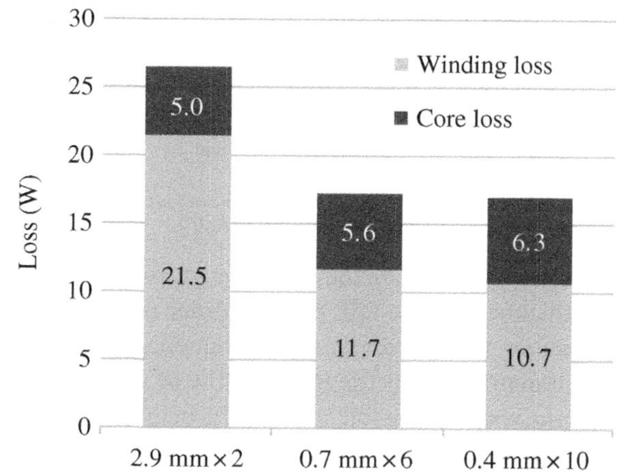

Figure 10.45 Simulated loss comparison of different air gap arrangements.

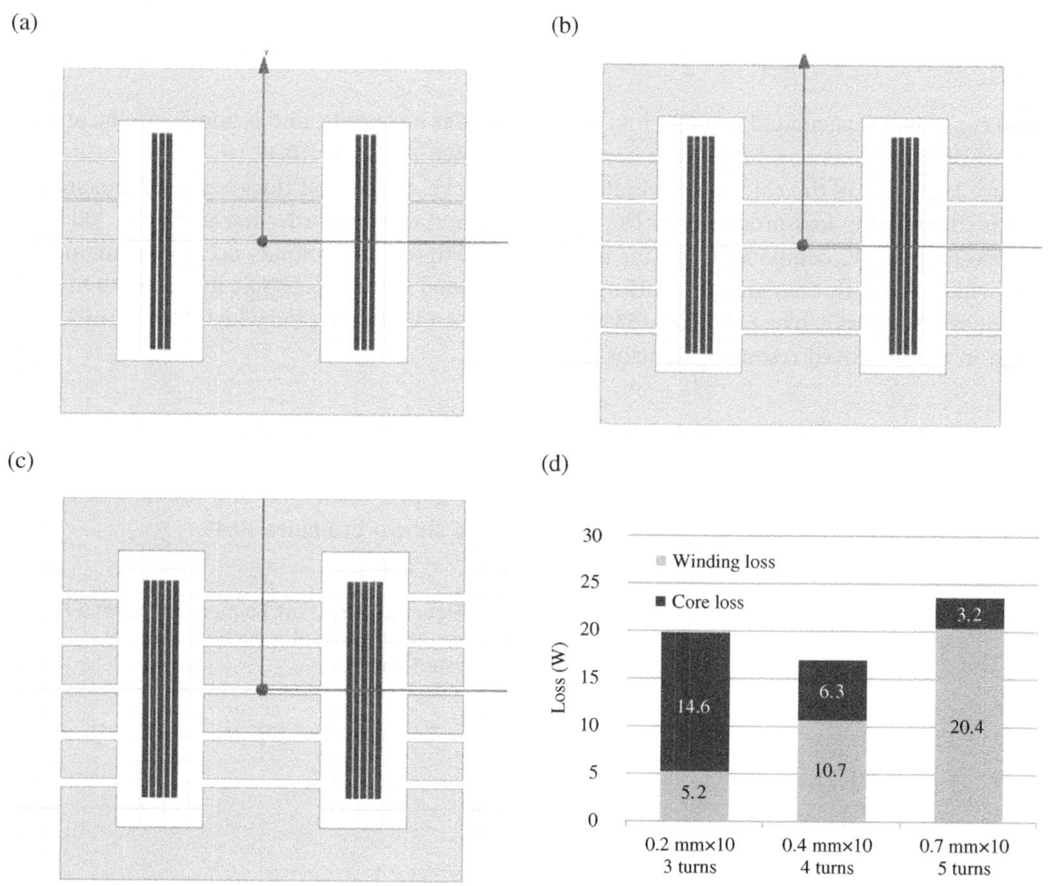

Figure 10.46 Optimal flux density design for resonant inductor: (a) three turns with B_{max} at 0.22T, (b) four turns with B_{max} at 0.17T, (c) five turns with B_{max} at 0.14T, and (d) loss comparison.

not properly designed and leads to large additional AC winding loss [8–10]. The nth ratio of AC resistance to DC resistance of multilayer foil is written as [9]:

$$\begin{cases} \dfrac{R_{acn}}{R_{dc}} = \dfrac{\xi_n'}{2}\left[\dfrac{\sinh \xi_n' + \sin \xi_n'}{\cosh \xi_n' - \cos \xi_n'} + \eta^2\left(\dfrac{4N^2-1}{3}\right)\dfrac{\sinh \xi_n' - \sin \xi_n'}{\cosh \xi_n' + \cos \xi_n'}\right] \\ \xi_n' = \dfrac{\sqrt{\pi}d}{2\delta_n'}, \delta_n' = \dfrac{1}{\sqrt{\pi n f \mu_r \mu_0 \eta \sigma}}, \eta = \dfrac{w_f}{g} \end{cases} \tag{10.72}$$

where R_{acn} and R_{dc} are the nth order AC resistance and the DC resistance of an N-layer winding, d is the thickness of the foil, n is the order of harmonic, δ'_n is the modified skin depth of nth order harmonic, f is the fundamental frequency of the inductor current, μ_r is the relative permeability of foil material, μ_0 is the permeability of vacuum, σ is the conductivity of the conductor material, w_f is the width of the foil, and g is the height of the windows in EE cores as shown in Appendix A.4.

The waveform of the resonant inductor current at rated power is shown in Figure 10.43, and its Fourier analysis result is shown in Figure 10.47.

The winding loss of the resonant inductor without considering the fringing fields can be expressed as:

$$P_{Cu\_m} = R_{dc}I_0{}^2 + \sum_{n=1}^{m} R_{acn}\frac{I_n{}^2}{2} \tag{10.73}$$

where P_{Cu_m} is the calculated winding loss up to mth order harmonic, and I_0 and I_n are the amplitudes of DC component and nth order harmonic as shown in Figure 10.47.

Figure 10.48 shows the calculated winding loss P_{Cu_m} vs. copper foil thickness d. For example, P_{Cu_1} is the winding loss produced by DC component and the first order harmonics at 300 kHz. It is clear that the DC component, and the first to the fourth order harmonics occupy the main part of the winding loss. By changing the thickness of the original four turns copper foil from 0.4 to 0.12 mm, the winding loss can be reduced by 54% from 21.0 to 9.7 W. Copper foil with 0.15 mm thickness is used in the improved resonant inductor prototype.

10.5.4 Resonant Inductor Prototypes and Loss Measurement

Prototypes of the original resonant inductor (centric air gap, $d = 0.4$ mm) and the new designed resonant inductor (distributed air gap, $d = 0.15$ mm) are shown in Figure 10.49.

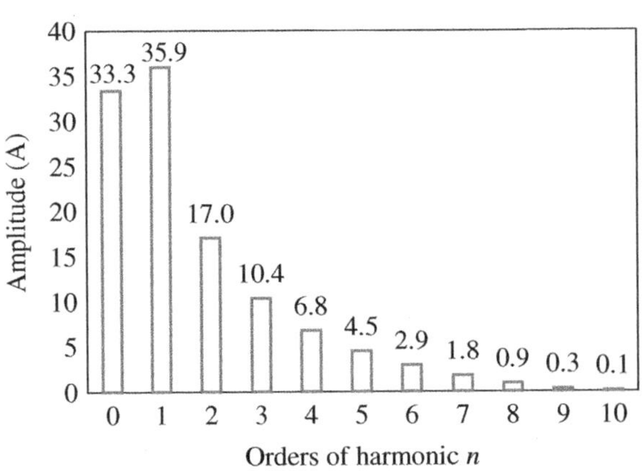

Figure 10.47 Fast Fourier Transform (FFT) result of resonant inductor current at rated power.

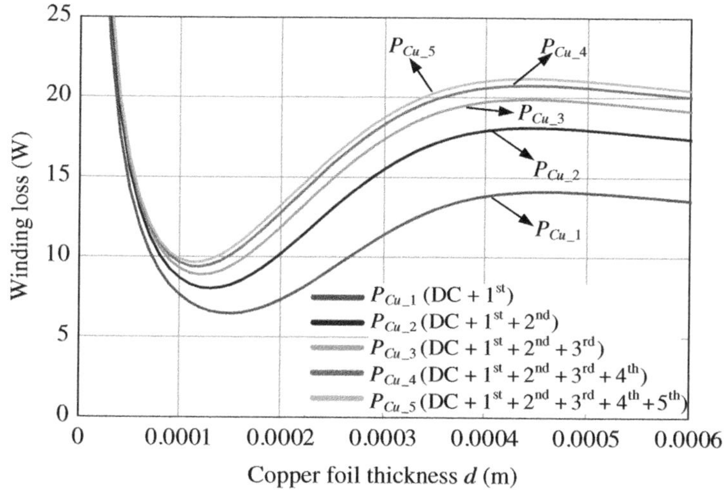

Figure 10.48 Winding loss vs. different copper foil thickness *d*.

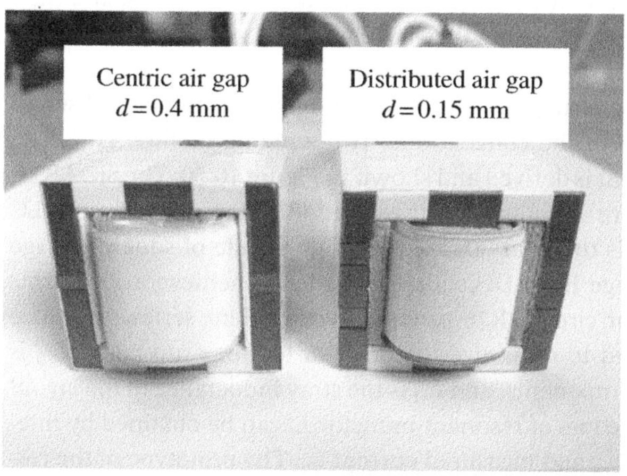

Figure 10.49 Prototypes of original resonant inductor and the new designed resonant inductor.

Precise inductor loss measurement is always difficult [11, 12]. In order to simplify the measurement, we only measure the total loss of the resonant inductor. Besides, instead of the triangular excitation the sinusoidal excitation at 300 kHz will be used in the measurement since it is the main source of the winding loss in the resonant inductor. Two-winding method is the widely used solution to measure the individual winding loss and core loss [13]. The possible error in measurement is caused by the phase discrepancy and the accuracies of probes. The equivalent circuit of an inductor can be assumed as an LR series circuit, then the relative measurement error Δ of the two-winding method caused by the phase discrepancy β can be derived as:

$$\Delta = \left|\frac{UI\cos(\theta+\beta) - UI\cos\theta}{UI\cos\theta}\right| = |-\tan(\theta)\sin(\beta) + \cos(\beta) - 1| \approx \frac{\omega L_r}{R_{Core} + R_{Cu}}\cdot\beta \quad (10.74)$$

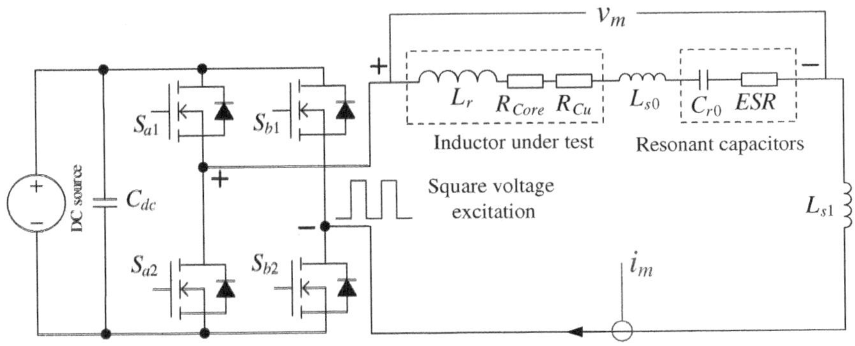

Figure 10.50 Scheme of resonant inductor loss measurement.

where ω is the angular frequency of excitation and $\tan(\theta)$ is the quality factor of the inductor. Phase discrepancy β can be introduced by the probe deskew or the parasitic in the measurement circuit. For example, the parameters in the measurement are assumed as follows: the excitation frequency is 300 kHz, the resonant inductance L_r is 1.7 μH, the equivalent resistance of winding loss R_{Cu} and the equivalent resistance of core loss R_{Core} are 10 mΩ, the probes deskew is about 2.5 ns, and then phase discrepancy β can be calculated as 0.0048. By using the above parameters, the relative measurement error Δ is as large as ±154%.

Capacitive cancellation is used in this chapter to greatly reduce the sensitivity to phase discrepancy, which is the dominating error source in the conventional two-winding method [14]. The scheme of resonant inductor loss measurement is derived and shown in Figure 10.50. The SiC MOSFET H-bridge stage produces a 300 kHz square waveform as the excitation voltage for the series LC resonant circuit whose resonant frequency is tuned at 300 kHz. The amplitude of square voltage will be adjusted by regulating the DC voltage from DC source in order to achieve a 25.4 Arms 300 kHz current flowing in the LC resonant circuit. R76 series low equivalent series resistance (ESR) capacitors from KEMET are employed to minimize the error of loss measurement. L_{s0} is the stray inductance between the resonant components, and L_{s1} is the stray inductance in the layout excluding the resonant components part. The loss of resonant inductor L_r can be obtained by integrating the product of the measured voltage v_m and measured current i_m. The prototype of the resonant inductor loss measurement circuit is shown in Figure 10.51.

The error in measurement of inductor loss results from the phase discrepancy and the accuracies of probes. Errors (ΔL_r and ΔC_{r0}) in resonant inductance and resonant capacitance, stray inductance between the resonant components, and the ESR of resonant capacitors may also cause the incomplete resonance. The relative error Δ caused by phase discrepancy β can be calculated as [14]

$$\Delta \approx \tan(\varphi_{LC})\cdot\beta = \frac{\omega(L_r + \Delta L_r + L_{s0}) - \dfrac{1}{\omega(C_{r0} + \Delta C_{r0})}}{R_{Core} + R_{Cu} + ESR}\cdot\beta \tag{10.75}$$

For the measurement setup in Figure 10.50, $L_r = 1.7$ μH (smaller than the value in Table 10.2 as the connection litz wires are not included), $C_{r0} = 165.7$ nF, maximum $\Delta L_r + L_{s0} \approx 50$ nH, typical $\Delta C_{r0} \approx 2$ nF, minimum $R_{Core} + R_{Cu} + \text{ESR} \approx 10$ mΩ, and typical $\beta = 0.0048$ (2.5 ns time delay between the two probes under 300 kHz excitation). The relative error caused by phase discrepancy is about ±6%. YOKOGAMA's differential voltage probe 700924 (100 MHz band width and ±3% accuracy) and Tek's DC current probe TCP303 (15 MHz band width and ±1% accuracy) are used

Figure 10.51 Prototype of the resonant inductor loss measurement circuit.

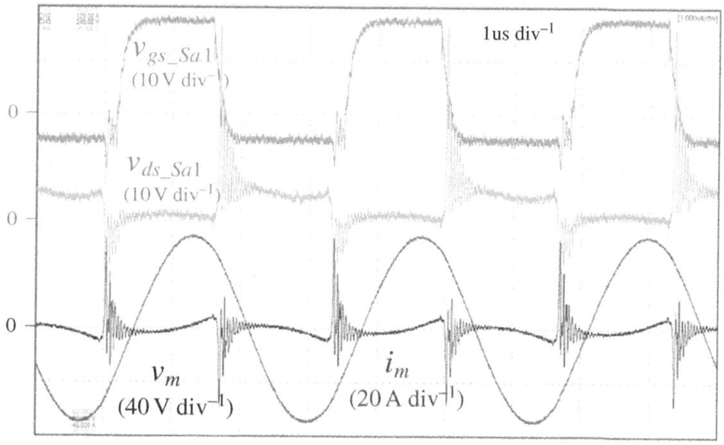

Figure 10.52 Waveforms under 300 kHz and 25.4 Arms current excitation.

in measurement. By considering the accuracy of voltage and current probes, the possible relative error Δ is around ±10%.

The experimental waveforms of the measured voltage v_m and measured current i_m of the resonant inductor with distributed air gaps and copper foil with 0.15 mm × 18 mm cross section are shown in Figure 10.52.

The ESR of resonant capacitors is about 1.33 mΩ at 300 kHz excitation, which is measured by the Precision Impedance Analyzer Wayne kerr 6500B. Therefore, the ESR loss can be compensated. Finally, the measured and simulated loss of resonant inductor with different air gap arrangements is shown in Figure 10.53a. By using distributed air gaps, the loss of resonant inductor has been reduced by 20.9% from 18.2 to 14.4 W. The measured and simulated loss of resonant inductor with different thicknesses of copper foil is shown in Figure 10.53b. By using the optimal thickness of copper foil at 0.15 mm, the loss of the resonant inductor has a further 38.9% reduction from 14.4 to 8.8 W. Finally, a total 51.6% reduction from 18.2 to 8.8 W of resonant inductor loss

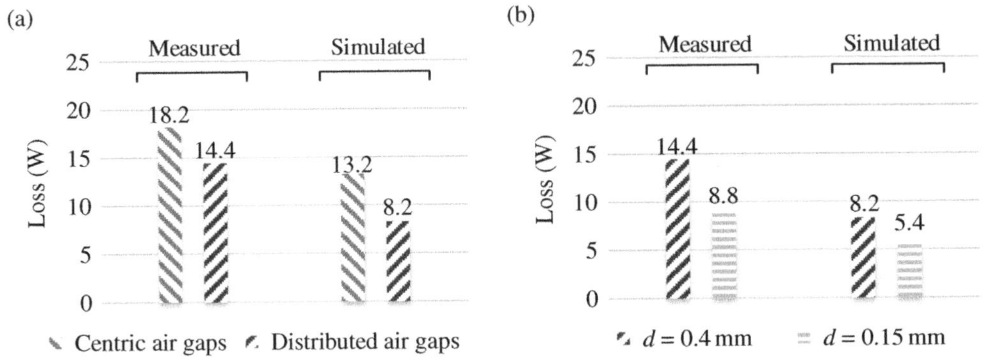

Figure 10.53 (a) Loss reduction with distributed air gaps and (b) loss reduction with different copper foil thickness *d*.

has been achieved. It is observed that the simulated results have similar trend as the measured results, but the measured loss of each case is still larger than the simulated loss.

10.6 Summary

This chapter introduced some potential advantages of ZVS-SVM technique in three-phase SiC inverter at high switching frequency. Soft-switching SiC inverter reveals almost three times stiffness of the efficiency curve than that of the hard-switching inverter, which brings the opportunity for a much higher power density of SiC inverter. By applying the ZVS-SVM technique, the soft-switching SiC inverter at 300 kHz switching frequency almost achieves half of the volume of the passive components with the similar conversion efficiency at 98.5% in comparison with the hard-switching SiC inverter at 100 kHz switching frequency.

A 7-in-1 SiC MOSFET power module is designed for soft-switching SiC inverter with reduced commutation loop stray inductance. The voltage overshoot in soft-switching SiC inverter after the second resonant stage has been reduced by 54%.

Distributed air gaps and the optimal design of the copper foil thickness have also been investigated. By using distributed air gaps, a 20.9% reduction can be obtained in measured loss. The optimal thickness of winding is also calculated by considering both the skin effect and the proximity effect. The resonant inductor with 0.15 mm thickness copper foil achieves another 38.9% measured loss reduction in comparison with the prototype with 0.4 mm thickness copper foil.

References

1 A. M. Hava, R. J. Kerkman, and T. A. Lipo, "A high-performance generalized discontinuous PWM algorithm," *IEEE Transactions on Industry Applications.*, vol. 34, no. 5, pp. 1059–1071, 1998.

2 R. Li and D. Xu, "A zero-voltage switching three-phase inverter," *IEEE Transactions on Power Electronics*, vol.29, no.3, pp.1200–1210, March 2014.

3 N. He, M. Chen, J. Wu, N. Zhu, and D. Xu, "20-kW zero-voltage-switching SiC MOSFET grid inverter with 300 kHz switching frequency," *IEEE Transactions on Power Electronics*, vol. 34, no. 6, pp. 5175–5190, June 2019

4 C. W. T. Mclyman, "*Transformer and Inductor Design Handbook*", fourth edition, CRC Press, 2011.

5 S. Kicin, S. Pettersson, E. Bianda, F. Canales, D. Cottet, G. Cavallini and J. Hamers, "Full SiC half-bridge module for high frequency and high temperature operation," *2015 IEEE 65th Electronic Components and Technology Conference (ECTC)*, San Diego, CA, pp. 950–956, 2015.

6 N. H. Kutkut and D. M. Divan, "Optimal air-gap design in high-frequency foil windings," *IEEE Transactions on Power Electronics*, vol. 13, no. 5, pp. 942–949, September 1998.

7 R. Jez, "Influence of the distributed air gap on the parameters of an industrial inductor," *IEEE Transactions on Magnetics*, vol. 53, no. 11, pp. 1–5, November 2017.

8 M. P. Perry, "Multiple layer series connected winding design for minimum losses," *IEEE Transactions on Power Apparatus and Systems*, vol. PAS-98, no. 1, pp. 116–123, January 1979.

9 J. A. Ferreira, "Appropriate modelling of conductive losses in the design of magnetic components," 21st Annual IEEE Conference on Power Electronics Specialists, San Antonio, TX, USA, pp. 780–785, 1990.

10 M. Kaymak, Z. Shen and R. W. De Doncker, "Comparison of analytical methods for calculating the AC resistance and leakage inductance of medium-frequency transformers," in Proceedings COMPEL, Trondheim, pp. 1–8, 2016.

11 J. Muhlethaler, J. Biela, J. W. Kolar and A. Ecklebe, "Core losses under the DC bias condition based on Steinmetz parameters," *IEEE Transactions on Power Electronics*, vol. 27, no. 2, pp. 953–963, February 2012.

12 T. Shimizu and S. Iyasu, "A practical iron loss calculation for AC filter inductors used in PWM inverters," *IEEE Transactions on Industry Electronics*, vol. 56, no. 7, pp. 2600–2609, July 2009.

13 V. J. Thottuvelil, T. G. Wilson and H. A. Owen, "High-frequency measurement techniques for magnetic cores," *IEEE Transactions on Power Electronics*, vol. 5, no. 1, pp. 41–53, January 1990.

14 M. Mu, Q. Li, D. J. Gilham, F. C. Lee, and K. D. T. Ngo, "New core loss measurement method for high-frequency magnetic materials," *IEEE Transactions on Power Electronics*, vol. 29, no. 8, pp. 4374–4381, August 2014.

11

Soft-switching SiC Single-phase Grid Inverter with Active Power Decoupling

In this chapter, the soft-switching technique is introduced to single-phase grid inverter with active power decoupling (APD). First, commonly used modulation methods for single-phase inverter and the basic concept of APD are introduced. Afterward, the APD technique and the soft-switching technique are combined in a single-phase inverter. Detailed operation principle of the inverter is provided, and soft-switching condition is analyzed. Finally, prototype design of a 1.5-kW silicon carbide (SiC) metal-oxide -semiconductor field effect transistor (MOSFET) soft-switching inverter is presented, and experiments are conducted to verify the analysis.

11.1 Introduction

The zero voltage switching (ZVS) technique for three-phase converters has been introduced in previous chapters. This chapter will extend the technique to single-phase system.

In single-phase inverters, active power decoupling (APD) may be used to cope with the double line frequency power pulsation in the inverter DC side. From the circuit perspective, the single-phase inverter with APD can be seen as three-phase inverter. Therefore, the circuit can be analyzed using similar ways as shown in preceding chapters.

11.1.1 Modulation Methods for Single-phase Inverter

For the conventional hard-switching single-phase inverter shown in Figure 11.1, there are three commonly used modulation strategies: unipolar sinusoidal pulse width modulation (SPWM), bipolar SPWM, and unipolar double-frequency (DF) SPWM [1]. Key waveforms of the three modulation methods are shown in Figure 11.2.

In unipolar SPWM, S_1 and S_4 operate at low frequency, while S_2 and S_3 operate at high frequency.

In bipolar SPWM, all the four switches are operated at high frequency. S_1 and S_3 share the same driving signal, while S_2 and S_4 share the same driving signal.

In DF SPWM, two complementary modulation signals are used. The driving signals for S_1 and S_4 are modulated by v_{m1}, while the driving signals for S_2 and S_3 are modulated by v_{m2}. The four switches are also operated at high frequency.

The position of Type 2 commutation (commutation from diode to its complimentary switch) in a switching cycle depends on the modulation strategies and the polarity of the output current. In Figure 11.2, the position of Type 2 commutation is marked with bold bar.

Soft-Switching Technology for Three-phase Power Electronics Converters, First Edition.
Dehong Xu, Rui Li, Ning He, Jinyi Deng, and Yuying Wu.

It is easy to find that there is only one Type 2 commutation with unipolar SPWM in every carrier period. For bipolar SPWM, since the same driving signal is shared by two switches (S_1 and S_3, S_2 and S_4), Type 2 commutation instants in two legs appear at the same time. For DF SPWM, Type 2 commutation instant of two legs are separated because two modulation signals are used.

Compared with unipolar SPWM, the switching loss of DF SPWM is evenly distributed among the four switches. Besides, it has lower harmonics on the load [1]. Therefore, DF SPWM will be adopted in this chapter for single-phase inverter.

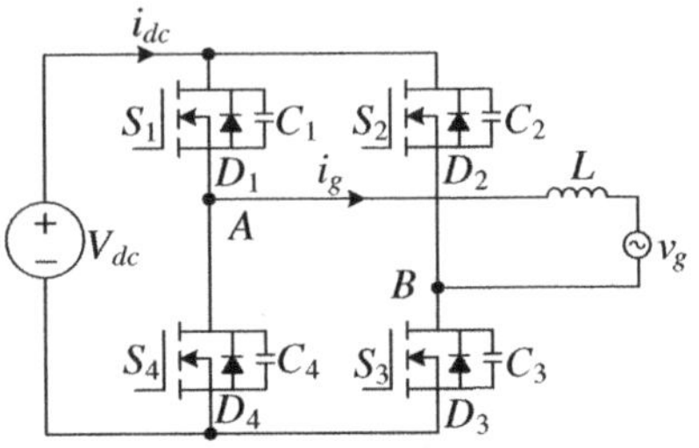

Figure 11.1 Hard-switching single-phase full-bridge inverter.

11.1.2 APD in Single-phase Grid Inverter

It is assumed that the grid voltage $v_g(t)$ and output current $i_g(t)$ of single-phase inverter are expressed as follows:

$$v_g(t) = V_m \sin(\omega t) \tag{11.1}$$

$$i_g(t) = I_m \sin(\omega t + \varphi) \tag{11.2}$$

where V_m is the amplitude of the grid voltage, I_m is the amplitude of the grid current, ω is the angular line frequency, and φ is the power factor angle.

The output power of the inverter $p_{inv}(t)$ is expressed as

$$\begin{aligned} p_{inv}(t) &= v_g(t)i_g(t) = V_m \sin(\omega t) I_m \sin(\omega t + \varphi) \\ &= \frac{1}{2}V_m I_m \cos(\varphi) - \frac{1}{2}V_m I_m \cos(2\omega t + \varphi) \\ &= p_{inv\_dc} + p_{inv\_ac}(t) \end{aligned} \tag{11.3}$$

where

$$p_{inv\_dc} = \frac{1}{2}V_m I_m \cos(\varphi) \tag{11.4}$$

and

$$p_{inv\_ac}(t) = -\frac{1}{2}V_m I_m \cos(2\omega t + \varphi) \tag{11.5}$$

$p_{inv\_dc}(t)$ is the DC component of the output power, while $p_{inv\_ac}(t)$ is the AC component with double line frequency of the output power.

With unity power factor, the output waveforms in a line cycle are shown in Figure 11.3. As can be seen from Figure 11.3, the output power $p_{inv}(t)$ of the inverter fluctuates with double line frequency. This power pulsation would lead to double line frequency current ripples of i_{dc} in DC input side as shown in Figure 11.4. Therefore, bulky electrolytic capacitor C_d is needed in DC side to absorb the low-frequency pulsation current, which will affect the lifetime and power density of the system [2].

APD has been proposed by predecessors for power pulsation suppression in single-phase system [3–7]. Active switches and absorbing capacitors are used in APD circuit as shown in Figure 11.5. The APD is a buck converter. With active control of this APD, the low-frequency current ripple of

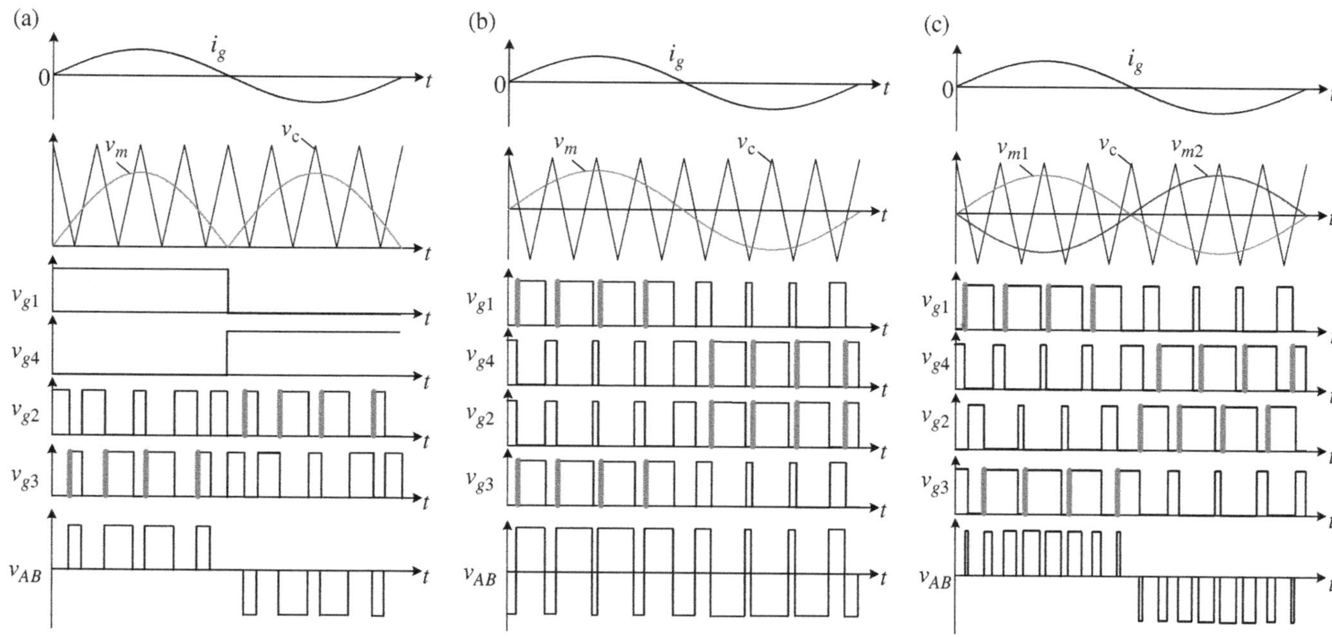

Figure 11.2 Commonly used modulation strategies for single-phase inverter: (a) unipolar SPWM; (b) bipolar SPWM; (c) double-frequency (DF) SPWM.

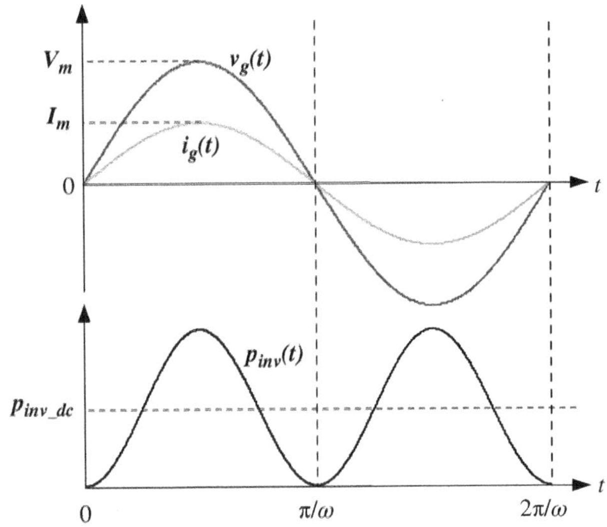

Figure 11.3 Key output waveforms of single-phase inverter.

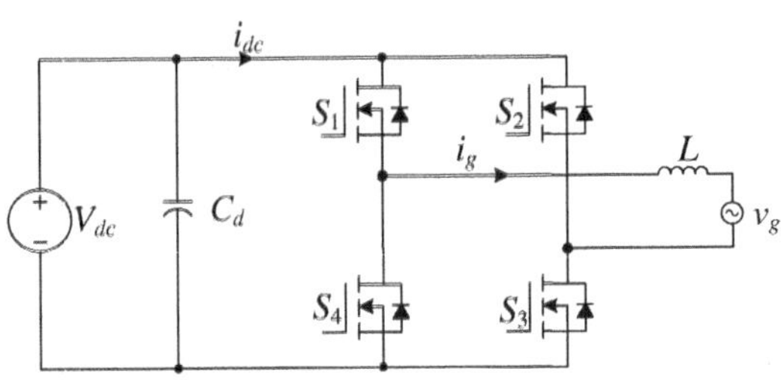

Figure 11.4 Traditional single-phase inverter.

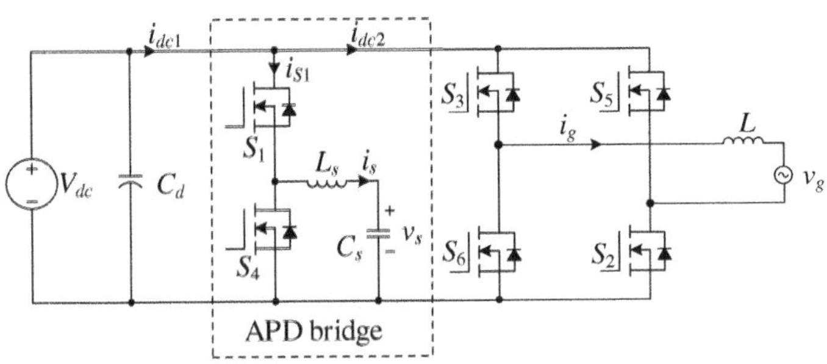

Figure 11.5 Single-phase inverter with APD.

i_{dc1} can be suppressed. As a result, the current ripple flowing into DC-side capacitor C_d can be reduced, and the value of C_d can be decreased.

To mitigate the current ripples of i_{dc1} in DC input side of the inverter, the current i_{S1} provided by the APD branch needs to compensate the low-frequency AC component of i_{dc2}. As mentioned before, the power provided by the APD is expected to be equal to the pulsation power $p_{inv\_ac}(t)$ with opposite direction [7–9]:

$$p_{apd}(t) = -p_{inv\_ac}(t) = \frac{1}{2}V_m I_m \cos(2\omega t + \varphi) \tag{11.6}$$

In principle, both the capacitor and the inductor can be used as energy storage components in an electrical circuit. However, the inductor is not as good as a capacitor in terms of energy density for an application using a few hundred hertz [5]. Therefore, in the following analysis, all the energy in the APD is regarded coming from the APD capacitor C_s.

Since the power in APD comes from the APD capacitor C_s, the energy stored in the APD capacitor C_s can be expressed as

$$E_s(t) = \int_0^t p_{apd}(\tau)d\tau + E_s(0) = \frac{V_m I_m \sin(2\omega t + \varphi)}{4\omega} + E_s(0) \tag{11.7}$$

where $E_s(0)$ is the initial energy stored in the APD capacitor at $t = 0$.

On the other hand, the energy of the APD capacitor C_s can also be expressed by its terminal voltage

$$E_s(t) = \frac{1}{2}C_s v_s^2(t) \tag{11.8}$$

Since the voltage variation of APD capacitor $v_s(t)$ is $0 - V_{dc}$, the maximum energy stored in APD capacitor is

$$E_{s\_max} = \frac{1}{2}C_s V_{dc}^2 \tag{11.9}$$

To fully use the absorbing ability of the capacitor, the initial value $E_s(0)$ is set as half of the maximum value E_{s_max} [7]

$$E_s(0) = \frac{1}{2}E_{s\_max} = \frac{1}{4}C_s V_{dc}^2 \tag{11.10}$$

By substituting (11.10) into (11.7), we obtain the energy stored in the APD capacitor

$$E_s(t) = \frac{V_m I_m \sin(2\omega t + \varphi)}{4\omega} + \frac{1}{4}C_s V_{dc}^2 \tag{11.11}$$

According to (11.8) and (11.11), the voltage between APD capacitor $v_s(t)$ can be derived as

$$v_s(t) = \sqrt{\frac{2E_s(t)}{C_s}} = \sqrt{\frac{V_m I_m \sin(2\omega t + \varphi)}{2\omega C_s} + \frac{V_{dc}^2}{2}} \tag{11.12}$$

Thus, the current through APD capacitor is derived by combining (11.6) and (11.12)

$$i_s(t) = \frac{p_{apd}(t)}{v_s(t)} = \frac{\frac{1}{2}V_m I_m \cos(2\omega t + \varphi)}{\sqrt{\frac{V_m I_m \sin(2\omega t + \varphi)}{2\omega C_s} + \frac{V_{dc}^2}{2}}} \tag{11.13}$$

Typical waveforms of APD capacitor voltage and current are shown in Figure 11.6. The fluctuation amplitudes of the voltage and current are tightly related to output conditions and the value of APD capacitor.

A typical control schematic of the APD bridge is shown in Figure 11.7. An average voltage loop is used to prevent the overcharging or discharging of the APD capacitor. $H_1(s)$ is the regulator of the voltage loop. The inner current loop is used to control the APD inductor current. $H_2(s)$ is the regulator of the current loop. The reference power of the APD bridge p_{apd_ref} is the sum of two parts. The first part p_{ac1} is directly obtained from the inverter output power. The second part p_{ac2} is

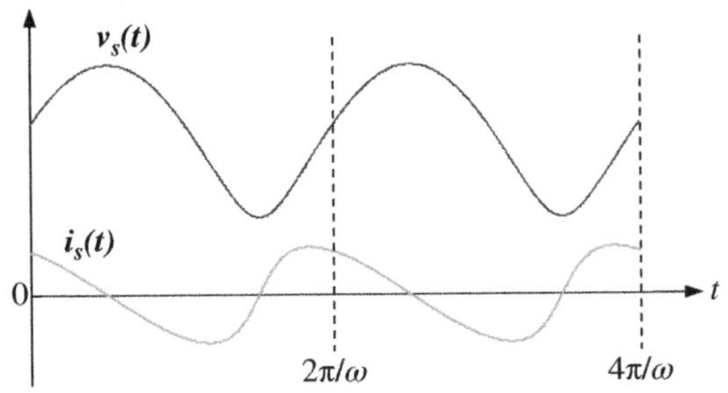

Figure 11.6 Typical waveforms of APD capacitor voltage and current.

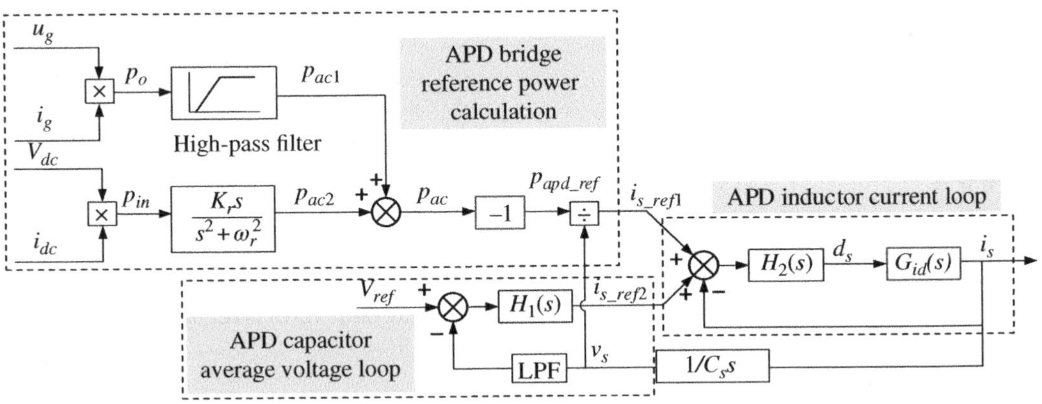

Figure 11.7 Control schematic of the APD bridge.

extracted from the DC input side through a resonant compensator $\frac{K_r s}{s^2 + \omega_r^2}$, where $\omega_r = 2\omega$. The use of the resonant compensator can eliminate the inaccuracies and delays in the signal acquisition and computation of the controller, which makes the compensation effect better [7].

11.2 Operation Principle

In this section, the ZVS technique is extended to the inverter with APD circuit. ZVS operation of all switches could be achieved. The modulation method is redesigned, and the operation principle and stage analysis are introduced.

11.2.1 Topology and Switching Scheme

The single-phase ZVS inverter with APD is shown in Figure 11.8. Compared with traditional inverter with APD as shown in Figure 11.5, an auxiliary branch is inserted in the DC side. The auxiliary branch consists of an auxiliary switch S_7, a clamping capacitor C_{cl}, and a resonant inductor L_r. Meanwhile, resonant capacitors C_1–C_7 are paralleled with switches S_1–S_7, respectively [8, 9].

Figure 11.9 shows the traditional modulation scheme for hard-switching single-phase inverter with APD. Figure 11.9a shows the modulation implementation diagram, and Figure 11.9b shows the key waveforms.

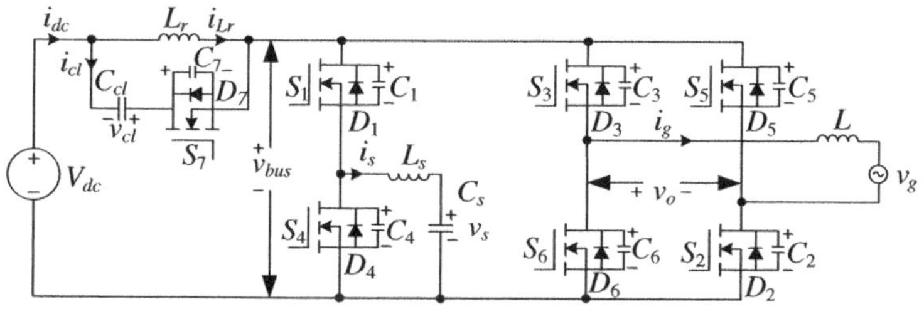

Figure 11.8 Single-phase ZVS inverter with APD.

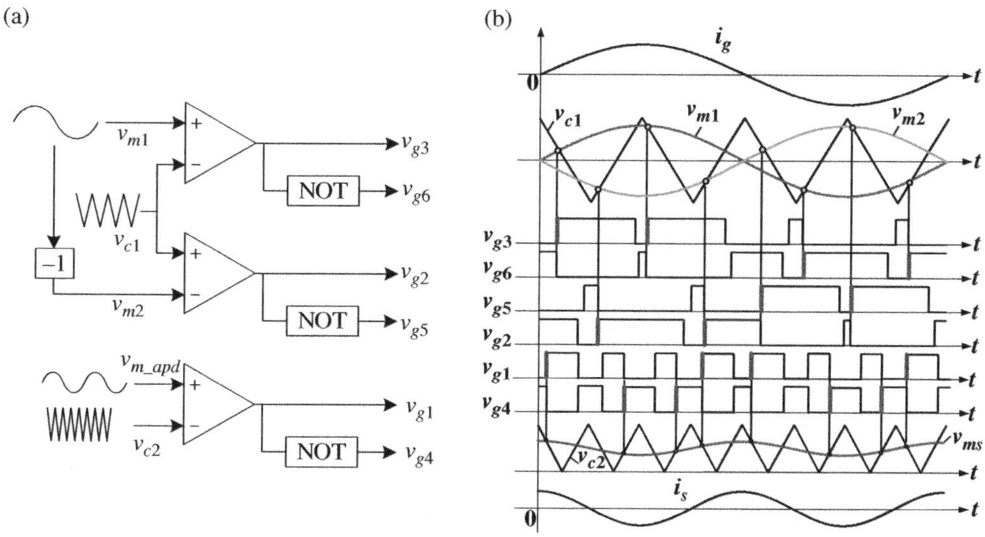

Figure 11.9 Conventional modulation method for single-phase inverter with APD: (a) implementation diagram; (b) key waveforms.

As shown in Figure 11.9b, DF SPWM is used to control the inverter bridges. Two complementary sinusoidal waveforms (v_{m1} and v_{m2}) are used as modulation signals. Modulation signal v_{m1} is used to generate driving signals for switches S_3 and S_6, while modulation signal v_{m2} is used to generate driving signals for switches S_5 and S_2. The APD buck converter is controlled with SPWM. v_{ms} is used as the modulation signal. The frequency of its triangle carrier v_{c2} may be different from that of the inverter bridge carrier v_{c1}. In Figure 11.9b, the frequency of v_{c2} is designed to be twice that of v_{c1}.

Type 2 commutation instants existing in the inverter are decided by the polarity of the inverter output current i_g. Similarly, Type 2 commutation instants in the APD bridge are decided by the polarity of the APD current i_s. In Figure 11.9b, Type 2 instants in inverter bridges and APD bridge are marked in bold. It is clear that Type 2 commutations in inverter bridges and APD bridge do not happen at the same instant. As has been introduced previously, once Type 2 commutation happens, the auxiliary switch needs to be turned off to resonate the bus voltage to zero in order to achieve ZVS. Therefore, if the conventional method is applied, the auxiliary branch has to operate multiple times, which will lead to increased loss and may affect the output current quality.

To synchronize Type 2 commutations in the inverter bridges and APD bridge, edge-aligned PWM (EA-PWM) is introduced for the soft-switching inverter. In EA-PWM, the modulation signals keep

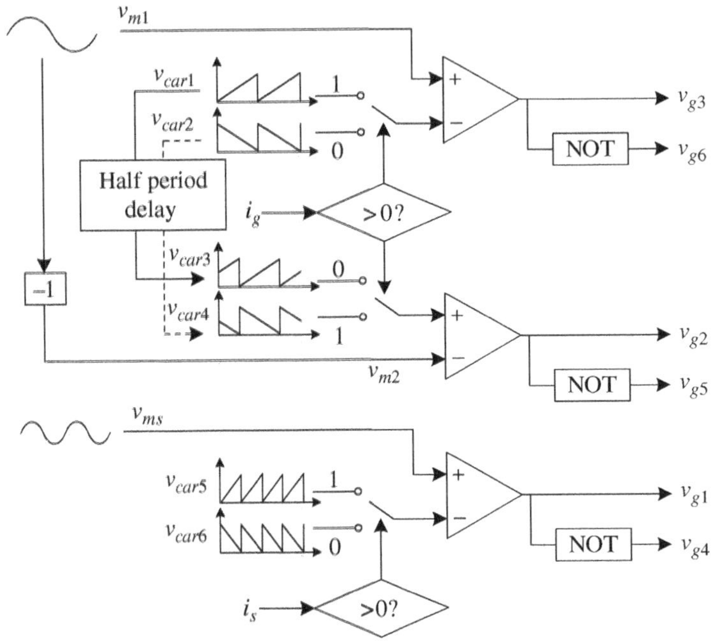

Figure 11.10 Implementation of EA-PWM for soft-switching single-phase inverter with APD.

unchanged, but two symmetrical saw-tooth carriers are used. The polarity of the leg current determines which saw-tooth carrier is to be selected. Ramp-up saw-tooth carrier is selected for positive current, while ramp-down saw-tooth carrier is selected for negative current.

The implementation diagram of EA-PWM is shown in Figure 11.10. The carrier frequency of APD bridge (v_{car5} and v_{car6}) is twice that of inverter bridge carrier (v_{car1}, v_{car2}, v_{car3}, and v_{car4}). For APD bridge, modulation signal v_{ms} is used to generate driving signals for switches S_1 and S_4, and the switching period is defined as T_s. For inverter bridge, two complementary modulation signals v_{m1} and v_{m2} are used. Modulation signal v_{m1} is used to generate driving signals for switches S_3 and S_6, while modulation signal v_{m2} is used to generate driving signals for switches S_2 and S_5. To achieve DF effect in inverter output, the carrier used for v_{m2} (v_{car3} and v_{car4}) is half-period delayed compared with the carrier used for v_{m1} (v_{car1} and v_{car2}).

If $i_g > 0$ and $i_s > 0$, a typical comparison of traditional modulation method and EA-PWM method is shown in Figure 11.11.

For inverter legs, since $i_g > 0$, Type 2 commutation happens when v_{g3} changes from low level to high level and when v_{g2} changes from low level to high level. For APD bridge leg, since $i_s > 0$, Type 2 commutation happens when v_{g1} changes from low level to high level. The Type 2 commutations are marked in bold in Figure 11.11.

As shown in Figure 11.11a, with traditional method, the driving signals are generated by comparing the modulation signal with the triangle carrier. As can be seen, Type 2 commutations in the inverter legs and APD leg do not happen at the same time.

For EA-PWM shown in Figure 11.11b, saw-tooth carriers are adopted. Since $i_g > 0$, ramp-up saw-tooth carrier signal v_{car1} is used to generate driving pulses for the first inverter leg switches S_3 and S_6. The direction of the current in the second inverter leg is reversed; therefore, ramp-down carrier signal v_{car4} is used to generate driving pulses for the second inverter leg switches S_5 and S_2. For the

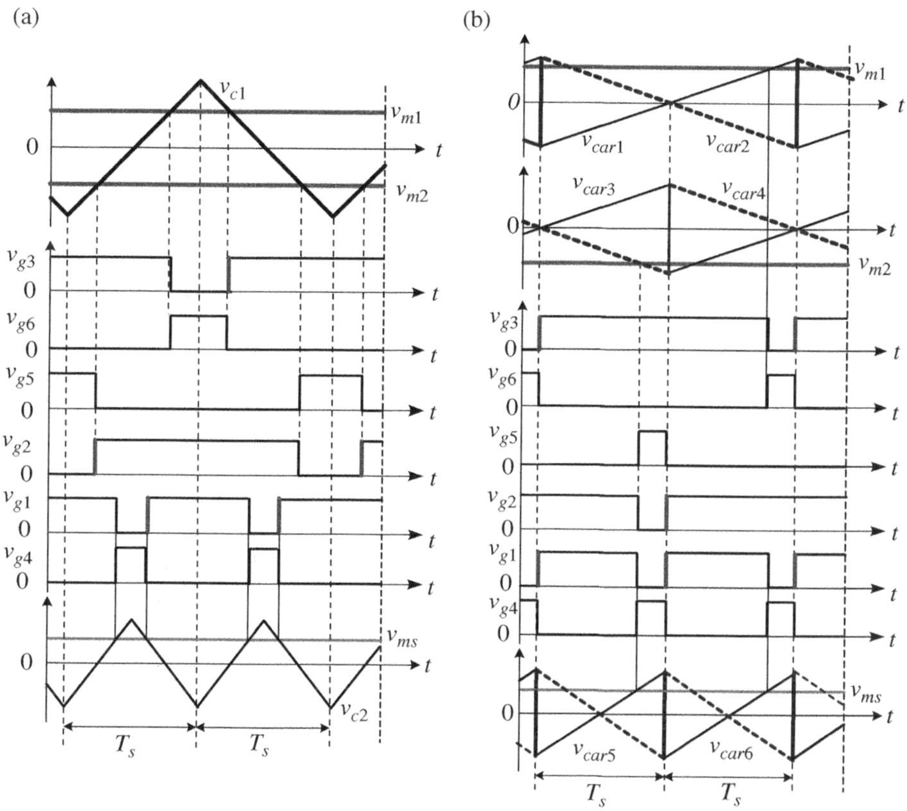

Figure 11.11 Comparison of traditional modulation method and EA-PWM method when $i_g > 0$ and $i_s > 0$: (a) traditional method; (b) EA-PWM method.

APD leg, since $i_s > 0$, ramp-up saw-tooth carrier signal v_{car5} is used to generate driving pulses for S_1 and S_4.

Obviously, in EA-PWM, by introducing saw-tooth carriers, Type 2 commutations in inverter legs and APD leg are shifted to the same instants. Thus, Type 2 commutations existing in inverter legs and APD leg are naturally synchronized, and operation times of the auxiliary branch can be reduced.

11.2.2 Stage Analysis

Stage analysis when inverter output current $i_g > 0$ and APD inductor current $i_s > 0$ are taken as an example.

With EA-PWM introduced in Section 11.2.1, the key waveforms of ZVS inverter are shown in Figure 11.12. Reference directions and variables definitions could be found in Figure 11.8. v_{sc} is the short circuit signal added to all switches through OR logic gates. v_{g1}–v_{g7} are the driving signals for S_1–S_7. v_o is the output voltage of inverter. v_{bus} is the voltage in the inverter DC side. v_{c6} and v_{c7} are voltages across C_6 and C_7, respectively. i_{Lr} is the current through the resonant inductor L_r.

There are 11 stages in a switching period. The equivalent circuits in every stage are drawn from Figures 11.13 to 11.23, respectively.

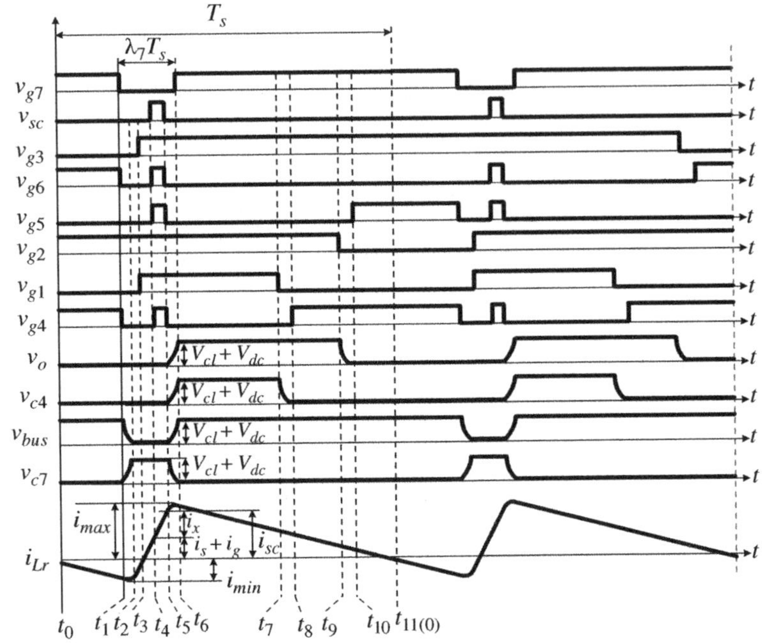

Figure 11.12 Key waveforms of ZVS inverter with APD.

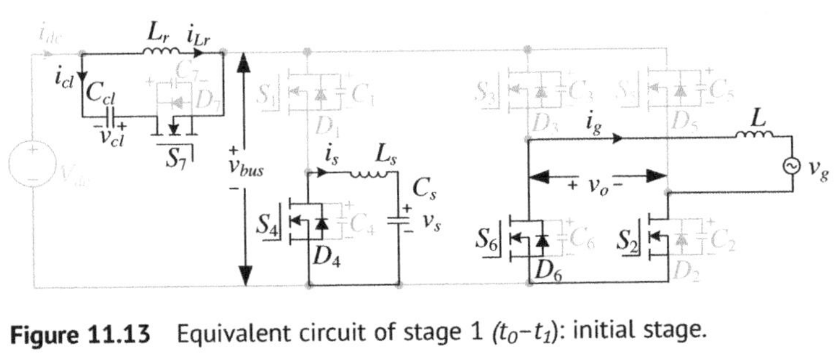

Figure 11.13 Equivalent circuit of stage 1 $(t_0–t_1)$: initial stage.

To simplify the analysis, the following assumptions are made:

1) The circuit is at steady state.
2) The APD inductor current i_s and the inverter output current i_g are constant in a switching period.
3) The clamping capacitor (C_{cl}) is so large that the voltage across the clamping capacitor is kept constant at V_{cl} in a switching period.
4) All switches are ideal switches with their body diodes.
5) Parasitic capacitances and external paralleled capacitors are included in the resonant capacitors (C_1, C_2, C_3, C_4, C_5, C_6, and C_7).
6) $C_1 = C_2 = C_3 = C_4 = C_5 = C_6 = C_r$.

Stage 1 (initial stage, t_0–t_1): The equivalent circuit of this stage is shown in Figure 11.13. The inverter output current i_g is freewheeling through the switches S_2 and S_6. APD inductor current i_s is freewheeling through S_4.

The voltage across the resonant inductor L_r is clamped to $-V_{cl}$, and the slope of its current i_{Lr} is

$$\frac{di_{Lr}}{dt} = -\frac{V_{cl}}{L_r} \tag{11.14}$$

Stage 1 ends when S_7, S_4, and S_6 are turned off by EA-PWM scheme.

Stage 2 (the first resonant stage, t_1–t_2): The equivalent circuit of this stage is shown in Figure 11.14. S_7, S_4, and S_6 are turned off at t_1 by EA-PWM. The currents through switches S_4 and S_6 are shifted to diodes D_4 and D_6. The resonance among the resonant inductor L_r and the resonant capacitors C_1, C_3, C_5, and C_7 begins. The resonant capacitors C_1, C_3, and C_5 are discharged, while C_7 is charged by the resonant inductor. During the resonant process, the current i_{Lr} reaches its minimum value i_{min} as shown in Figure 11.12.

This stage ends when the voltages on C_1, C_3, and C_5 are discharged to zero, while the voltage on C_7 is charged to $V_{dc} + V_{cl}$.

Stage 3 (freewheeling stage, t_2–t_3): The equivalent circuit of this stage is shown in Figure 11.15. After the voltages across the capacitors C_1, C_3, and C_5 are discharged to zero, body diodes D_1, D_3, and D_5 naturally conduct. It creates ZVS turn-on condition for S_1 and S_3. S_1 and S_3 need to be turned on when D_1 and D_3 are conducting. The voltage across the resonant inductor L_r is changed to V_{dc}; the current through it begins to increase and the slope is

$$\frac{di_{Lr}}{dt} = \frac{V_{dc}}{L_r} \tag{11.15}$$

Stage 3 ends when the current through D_1 and D_3 decreases to zero and i_{Lr} becomes zero.

Stage 4 (current commutation stage, t_3–t_4): The equivalent circuit of this stage is shown in Figure 11.16. The switches S_1 and S_3 are ZVS turned on in last stage. The currents in diodes D_4 and D_6 start to commutate to S_1 and S_3. The voltage across the resonant inductor L_r is still clamped to V_{dc}, and the increment slope of its current i_{Lr} is restricted as follows:

$$\frac{di_{Lr}}{dt} = \frac{V_{dc}}{L_r} \tag{11.16}$$

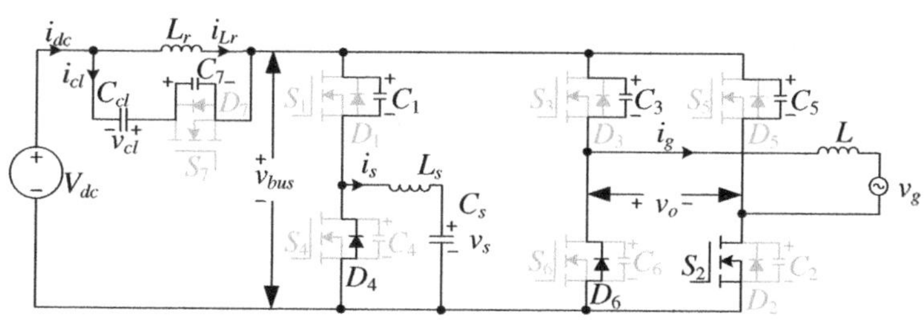

Figure 11.14 Equivalent circuit of stage 2 (t_1-t_2): first resonant stage.

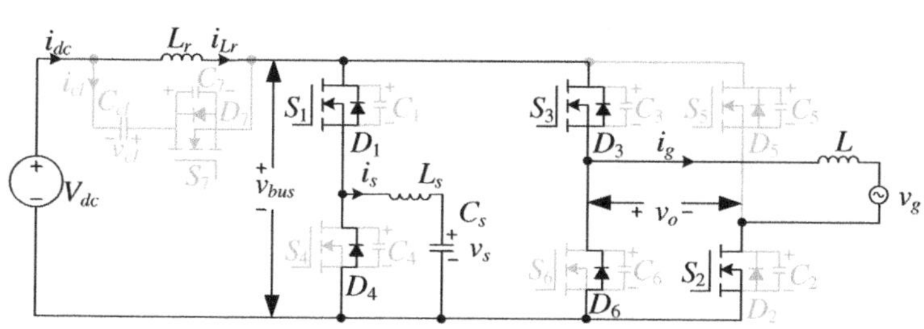

Figure 11.15 Equivalent circuit of stage 3 (t_2-t_3): freewheeling stage.

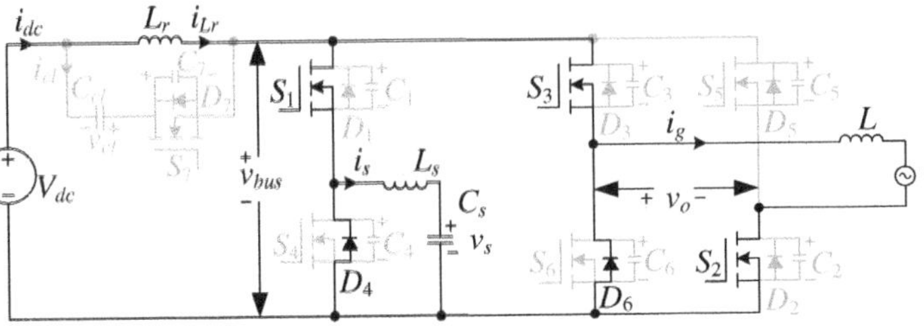

Figure 11.16 Equivalent circuit of stage 4 (t_3–t_4): current commutation stage.

Stage 4 ends when the currents in D_4 and D_6 reach zero. Since *di/dt* of current through both D_4 and D_6 is restricted, the reverse recovery loss of the diodes is suppressed. To realize ZVS turn-on of the first resonance, a short circuit stage is needed by turning on all switches. To add the short circuit stage, driving signals of S_4, S_6, and S_5 should be added before stage 4 ends.

At the end of this stage, the current through the resonant inductor becomes $i_{Lr} = i_s + i_g$.

Stage 5 (current boost stage, t_4–t_5): The equivalent circuit of this stage is shown in Figure 11.17. Switches S_4, S_6, and S_5 are ZVS turned on at t_4. The voltage across the resonant inductor L_r is continuously clamped to V_{dc}, and the increment slope of its current i_{Lr} is

$$\frac{di_{Lr}}{dt} = \frac{V_{dc}}{L_r} \tag{11.17}$$

The resonant inductor store extra energy in this stage by boosting its current by i_x. The duration of this stage must be long enough to satisfy the ZVS condition. Detailed analysis will be given in the following section. At the end of this stage, i_{Lr} reaches $i_{sc} = i_s + i_g + i_x$ as shown in Figure 11.12.

Stage 5 ends when S_4, S_5, and S_6 are turned off according to EA-PWM scheme.

Stage 6 (the second resonant stage, t_5–t_6): The equivalent circuit of this stage is shown in Figure 11.18. S_4, S_5, and S_6 are turned off at t_5 by EA-PWM scheme. The resonance among the resonant inductor L_r and the resonant capacitors C_4, C_5, C_6, and C_7 begins. The resonant capacitors C_4, C_5, and C_6 are charged, while C_7 is discharged by the resonant inductor. During the resonant process the current i_{Lr} reaches its maximum value i_{max} as shown in Figure 11.12.

Stage 6 ends when the voltage across the capacitor C_7 is discharged to zero and the antiparallel diode of S_7, D_7, begins to conduct.

Stage 7 (steady stage 1, t_6–t_7): The equivalent circuit of this stage is shown in Figure 11.19. The conducting D_7 creates ZVS condition for auxiliary switch S_7. Energy in DC side is transferred to APD capacitor and output side.

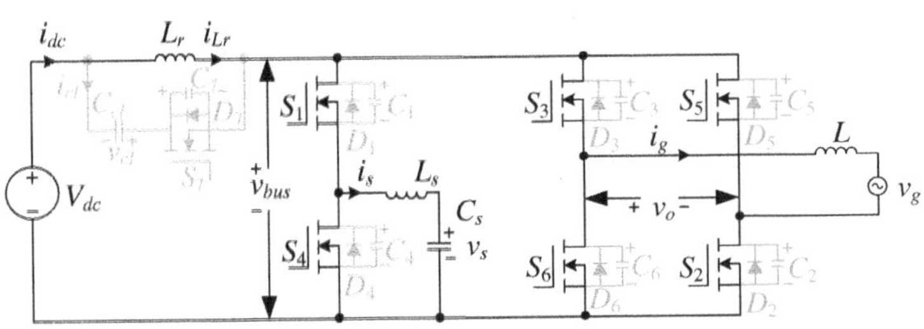

Figure 11.17 Equivalent circuit of stage 5 (t_4–t_5): current boost stage.

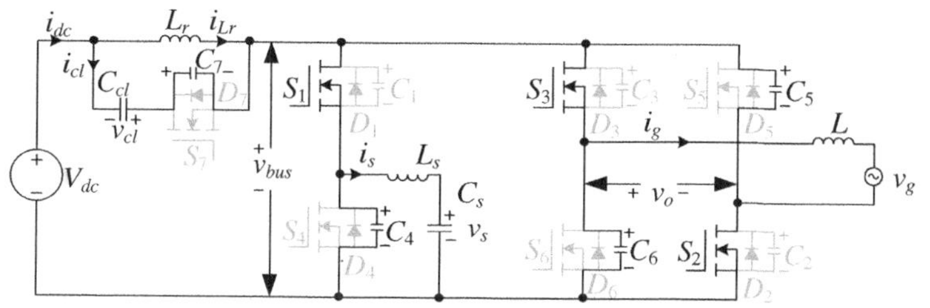

Figure 11.18 Equivalent circuit of stage 6 (t_5-t_6): second resonant stage.

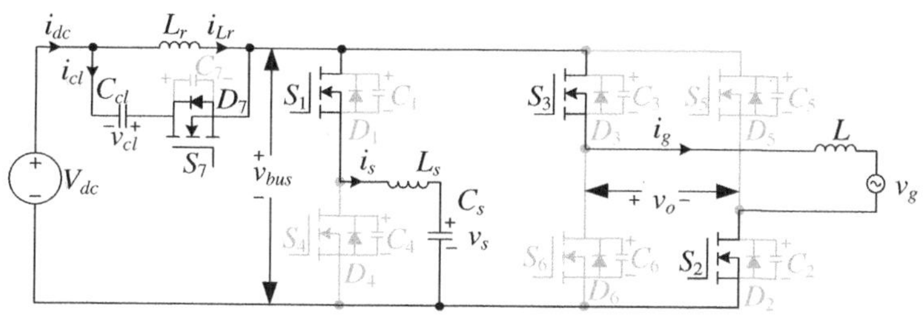

Figure 11.19 Equivalent circuit of stage 7 (t_6-t_7): steady stage 1.

The resonant inductor is clamped by V_{cl}. So the variation slope of the resonant inductor current i_{Lr} in this stage is

$$\frac{di_{Lr}}{dt} = -\frac{V_{cl}}{L_r} \tag{11.18}$$

Starting from this stage, the switch S_7 is kept on till the stage 2 of the next period. The voltage across the resonant inductor L_r is therefore clamped to $-V_{cl}$ continuously. During the process, the variation slope of the resonant inductor current i_{Lr} remains unchanged as (11.18). To realize ZVS for S_7, S_7 needs to be turned on when D_7 is conducting.

Stage 7 ends when S_1 is turned off according to EA-PWM scheme.

Stage 8 (Type 1 commutation stage 1, t_7–t_8): The equivalent circuit of this stage is shown in Figure 11.20. S_1 is turned off at t_7 by EA-PWM scheme, and then C_1 is charged, while C_4 is discharged by APD inductor current i_s. The buffering effect of the capacitor helps to reduce the turn-off loss of S_1. Thus S_1 is ZVS turned off.

Stage 8 ends when the voltage across C_4 reaches zero and D_4 begins to conduct.

Stage 9 (steady stage 2, t_8–t_9): The equivalent circuit of this stage is shown in Figure 11.21. The voltage across C_4 reaches zero at t_8, and D_4 naturally conducts. S_4 is ZVS turned on after the turn-on of D_4.

Energy in DC side is transferred to the output side. The current in the APD inductor i_s starts to freewheel from this stage.

Stage 9 ends when S_2 is turned off according to EA-PWM scheme.

Stage 10 (Type 1 commutation stage 2, t_9–t_{10}): The equivalent circuit of this stage is shown in Figure 11.22. S_2 is turned off at t_9 by EA-PWM scheme, and then C_2 is charged, while C_5 is

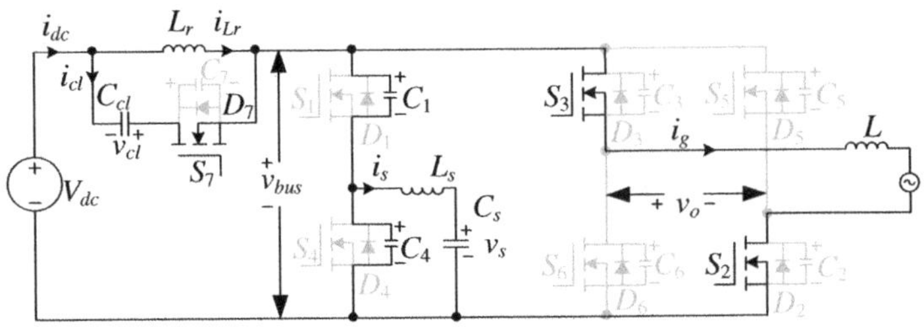

Figure 11.20 Equivalent circuit of stage 8 (t_7–t_8): Type 1 commutation stage 1.

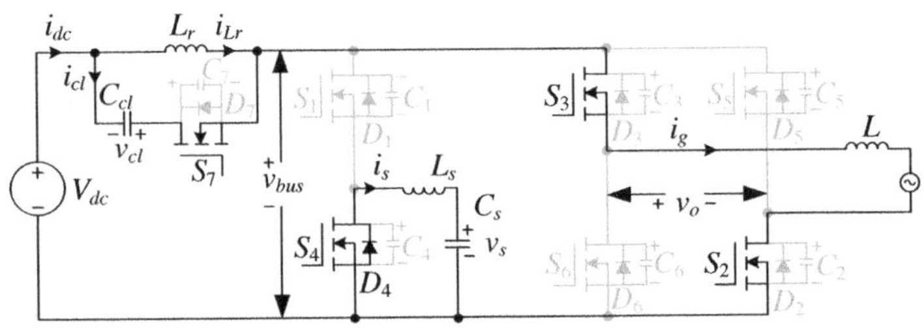

Figure 11.21 Equivalent circuit of stage 9 (t_8–t_9): steady stage 2.

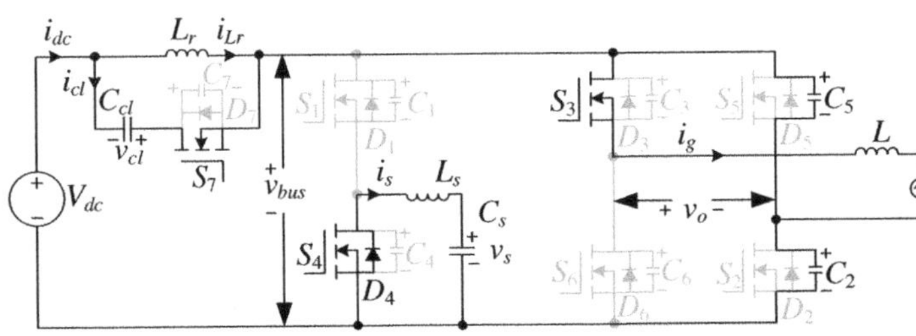

Figure 11.22 Equivalent circuit of stage 10 (t_9–t_{10}): Type 1 commutation stage 2.

discharged by output current i_g. Similarly, the buffering effect of the capacitors helps to reduce the turn-off loss of S_2. Thus S_2 is ZVS turned off.

Stage 10 ends when the voltage across C_5 reaches zero and D_5 starts to conduct.

Stage 11 (freewheeling stage, t_{10}–t_{11}): The equivalent circuit of this stage is shown in Figure 11.23. At t_{10}, D_5 begins to conduct. S_5 is then ZVS turned on after the turn-on of D_5.

In this stage, the DC side is disconnected from the output side. The currents in the APD inductor i_s and the output current i_g are both freewheeling.

After stage 11, the circuit enters a new switching period.

It can be concluded from the previously mentioned analysis that all the seven switches in the circuit can achieve ZVS turn-on. Besides, the auxiliary switch only operates once in a switching period.

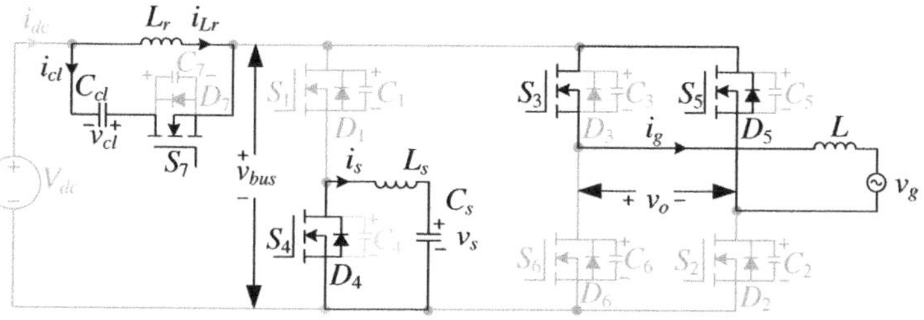

Figure 11.23 Equivalent circuit of stage 11 $(t_{10}–t_{11})$: energy output stage 3.

11.3 Circuit Analysis

As mentioned in Section 11.2.2, there are two resonate stages during a switching period. In this section, ZVS condition of the inverter will be discussed based on state equations.

11.3.1 Resonant Stages Analysis

The original circuit of the first resonance in stage 2 $(t_1–t_2)$ is redrawn in Figure 11.24a. It can be simplified to the circuit shown in Figure 11.24b.

According to the simplified circuit in Figure 11.24b, the state equation can be obtained:

$$\begin{cases} \dfrac{dv_{bus}(t)}{dt} = \dfrac{1}{3C_r + C_7} i_{Lr}(t) \\ \dfrac{di_{Lr}(t)}{dt} = -\dfrac{1}{L_r} v_{bus}(t) + \dfrac{1}{L_r} V_{dc} \end{cases} \tag{11.19}$$

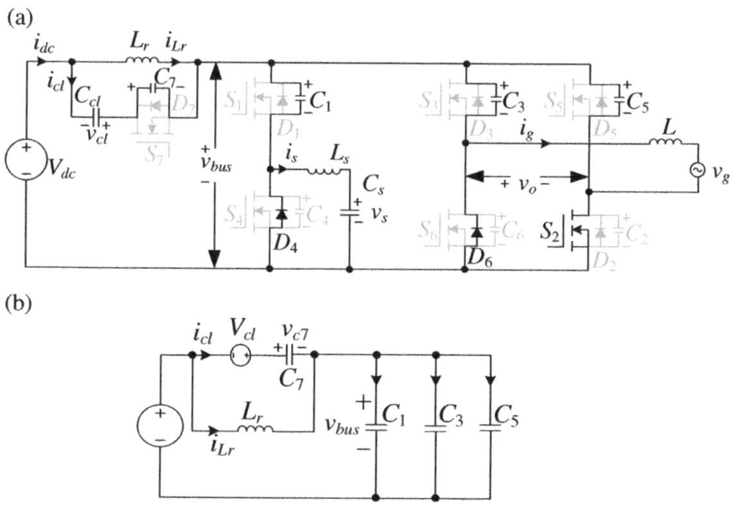

Figure 11.24 Circuits of the first resonance in stage 2: (a) original circuit; (b) simplified circuit.

The initial state of (11.19) is

$$\begin{cases} v_{bus}(t_1) = V_{dc} + V_{cl} \\ i_{Lr}(t_1) = i_{Lr}(t_1) \end{cases} \tag{11.20}$$

Solving state Equation (11.19) with initial state Equation (11.20), the bus voltage $v_{bus}(t)$ and current through the resonant inductor $i_{Lr}(t)$ can be derived

$$\begin{cases} v_{bus}(t) = V_{dc} - \sqrt{V_{cl}{}^2 + [i_{Lr}(t_1)]^2 Z_r^2} \sin[\omega_r(t - t_1) - \varphi_1] \\ i_{Lr}(t) = -\sqrt{\dfrac{V_{cl}{}^2}{Z_r^2} + [i_{Lr}(t_1)]^2} \cos[\omega_r(t - t_1) - \varphi_1] \end{cases} \tag{11.21}$$

where $\omega_r = \dfrac{1}{\sqrt{L_r(3C_r + C_7)}}$, $Z_r = \sqrt{\dfrac{L_r}{3C_r + C_7}}$ and $\varphi_1 = \arctan \dfrac{V_{cl}}{[-i_{Lr}(t_1)] \cdot Z_r}$.

At t_2, the bus voltage v_{bus} is resonated to zero, i.e.

$$v_{bus}(t_2) = V_{dc} - \sqrt{V_{cl}{}^2 + [i_{Lr}(t_1)]^2 Z_r^2} \sin[\omega_r(t_2 - t_1) - \varphi_1] = 0 \tag{11.22}$$

The duration of stage 2 can be calculated from (11.22)

$$t_2 - t_1 = \frac{1}{\omega_r}\left[\arcsin\left(\frac{V_{dc}}{\sqrt{V_{cl}{}^2 + [i_{Lr}(t_1)]^2 Z_r^2}}\right) + \varphi_1\right] \tag{11.23}$$

Then, the current through the resonant inductor at t_2 is derived by substituting (11.23) into (11.21)

$$i_{Lr}(t_2) = -\sqrt{[i_{Lr}(t_1)]^2 + \frac{V_{cl}{}^2 - V_{dc}^2}{Z_r^2}} \tag{11.24}$$

The original circuit of the second resonance in stage 6 (t_5–t_6) is redrawn in Figure 11.25a, and its simplified circuit is shown in Figure 11.25b.

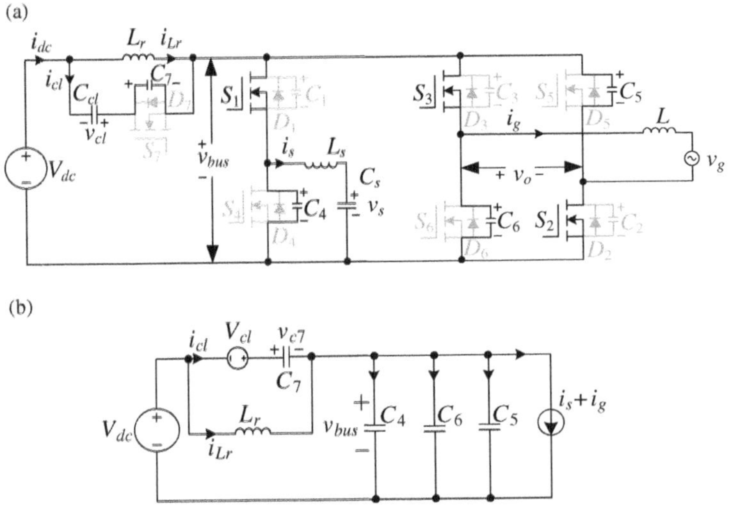

Figure 11.25 Circuits of the second resonance in stage 6: (a) original circuit; (b) simplified circuit.

The state equation can be derived according to the simplified circuit in Figure 11.25b

$$\begin{cases} \dfrac{dv_{bus}(t)}{dt} = \dfrac{1}{3C_r + C_7} i_{Lr}(t) - \dfrac{1}{3C_r + C_7}(i_s + i_g) \\ \dfrac{di_{Lr}(t)}{dt} = -\dfrac{1}{L_r} v_{bus}(t) + \dfrac{1}{L_r} V_{dc} \end{cases} \tag{11.25}$$

According to stage analysis in Section 11.3.1, v_{bus} is 0 at t_5, and the current through the resonant inductor at t_5 is $i_{sc} = i_s + i_g + i_x$. Therefore, the initial state of the circuit is

$$\begin{cases} v_{bus}(t_5) = 0 \\ i_{Lr}(t_5) = i_s + i_g + i_x \end{cases} \tag{11.26}$$

Solving state Equation (11.25) with initial condition (11.26), the following solution is obtained

$$\begin{cases} v_{bus}(t) = V_{dc} + \sqrt{V_{dc}^2 + i_x{}^2 Z_r^2} \sin\left[\omega_r(t - t_5) - \varphi_2\right] \\ i_{Lr}(t) = \sqrt{\dfrac{V_{dc}{}^2}{Z_r^2} + i_x{}^2} \cos\left[\omega_r(t - t_5) - \varphi_2\right] + i_s + i_g \end{cases} \tag{11.27}$$

where $\varphi_2 = \arctan \dfrac{V_{dc}}{i_x \cdot Z_r}$.

11.3.2 Steady-state Analysis

According to the stage analysis, it is observed that the resonant inductor L_r experiences four operation modes: charging by V_{dc} (t_2–t_5), discharging by V_{cl} (t_0–t_1, t_6–t_{11}), and two resonant processes (t_1–t_2, t_5–t_6) in a switching cycle. In steady state, the volt-second balance principle is applicable for the resonant inductor [10]

$$\Delta i_{Lr} = \frac{1}{L_r}\int_{t_0}^{t_{11}} v_{Lr}\ dt = 0 \tag{11.28}$$

Since the resonant processes are quite short, the two resonant stages (t_1–t_2, t_5–t_6) can be ignored. By neglecting these two short duration, (11.28) can be expanded as

$$\begin{aligned} \frac{1}{L_r}\int_{t_0}^{t_{11}} v_{Lr}\ dt &= \frac{1}{L_r}\left(\int_{t_0}^{t_1} v_{Lr}\ dt + \int_{t_2}^{t_5} v_{Lr}\ dt + \int_{t_6}^{t_{11}} v_{Lr}\ dt\right) \\ &= \frac{1}{L_r}\left[\int_{t_0}^{t_1}(-V_{cl})\ dt + \int_{t_2}^{t_5}(V_{dc})\ dt + \int_{t_6}^{t_{11}}(-V_{cl})\ dt\right] \\ &= \frac{1}{L_r}\{V_{dc}(t_5 - t_2) - V_{cl}[T_s - (t_6 - t_1)]\} \\ &= 0 \end{aligned} \tag{11.29}$$

Then, the voltage of the clamping capacitor can be derived from (11.29)

$$V_{cl} = \frac{t_5 - t_2}{T_s - (t_6 - t_1)} V_{dc} \approx \frac{\lambda_7}{1 - \lambda_7} V_{dc} \tag{11.30}$$

where λ_7 is defined as the turn-off time ratio in a switching period of the auxiliary switch S_7

$$\lambda_7 = \frac{t_6 - t_1}{T_s} \tag{11.31}$$

11.3.3 Soft-switching Condition

Both the main switches S_1–S_6 and the auxiliary switch S_7 need to achieve ZVS operation.

The resonance in stage 2 is used to achieve the ZVS of the switches S_1–S_6. The bus voltage v_{bus} has to be resonated to zero in stage 2. The bus voltage v_{bus} during stage 2 is rewritten as

$$v_{bus}(t) = V_{dc} - \sqrt{V_{cl}{}^2 + [i_{Lr}(t_1)]^2 Z_r^2} \sin[\omega_r(t - t_1) - \varphi_1] \tag{11.32}$$

Obviously, the minimum value of $v_{bus}(t)$ has to be equal or lower than zero in order to achieve ZVS. On the right side of (11.32), the first DC item is required to be equal to or lower than the amplitude of the second sinusoidal item

$$V_{dc} \le \sqrt{V_{cl}{}^2 + [i_{Lr}(t_1)]^2 Z_r^2} \tag{11.33}$$

This means that for the sake of ZVS, the current through the resonant inductor at t_1 should satisfy

$$i_{Lr}(t_1) \le -\sqrt{\frac{V_{dc}^2 - V_{cl}{}^2}{Z_r^2}} \tag{11.34}$$

The resonance in stage 6 is used to achieve the ZVS of the auxiliary switch S_7, and the voltage across S_7 has to be resonated to zero in stage 6. Considering (11.27), the voltage across S_7 is derived as

$$\begin{aligned} v_{c7}(t) &= V_{dc} + V_{cl} - v_{bus}(t) \\ &= V_{cl} - \sqrt{V_{dc}^2 + i_x{}^2 Z_r^2} \sin[\omega_r(t - t_5) - \varphi_2] \end{aligned} \tag{11.35}$$

Similarly, following inequality needs to be satisfied to achieve ZVS

$$V_{cl} \le \sqrt{V_{dc}^2 + i_x{}^2 Z_r^2} \tag{11.36}$$

It is easy to find from (11.30) that $V_{cl} < V_{dc}$ is always satisfied. Therefore, (11.36) is always satisfied, and ZVS condition for S_7 is easy to realize.

11.3.4 Short Circuit Current

As illustrated, (11.34) has to be satisfied to achieve ZVS for main switches. This can be done by adjusting the boost current i_x.

According to (11.21), the minimum value of $i_{Lr}(t)$ is derived as

$$i_{min} = -\sqrt{\frac{V_{cl}{}^2}{Z_r^2} + [i_{Lr}(t_1)]^2} \tag{11.37}$$

Thus,

$$i_{Lr}(t_1) = -\sqrt{i_{min}^2 - \frac{V_{cl}{}^2}{Z_r^2}} \tag{11.38}$$

According to (11.27), the maximum value of $i_{Lr}(t)$ is derived as

$$i_{max} = \sqrt{\frac{V_{dc}{}^2}{Z_r^2} + i_x{}^2} + i_s + i_g \tag{11.39}$$

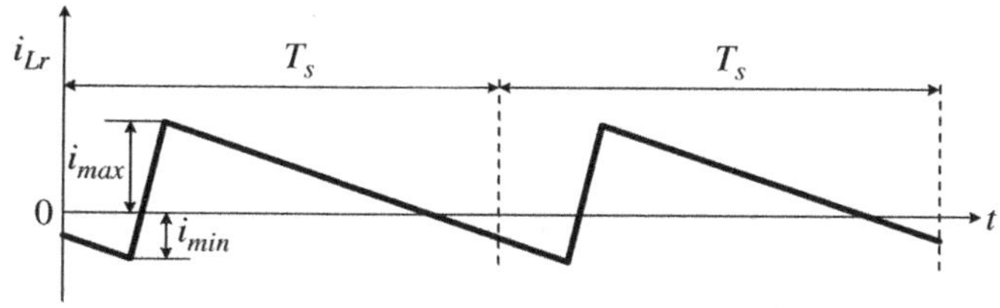

Figure 11.26 Approximate triangular waveforms of i_{Lr}.

In Figure 11.12, if the two short resonant stages are ignored, the current through the resonant inductor can be regarded as periodical triangular waveform as shown in Figure 11.26.

The average current through the resonant inductor can then be deduced as

$$\frac{1}{T_s}\int_0^{T_s} i_{Lr}(t)\ dt \approx \frac{i_{max}+i_{min}}{2} \tag{11.40}$$

As shown in Figure 11.8, the relationship between the DC-side current i_{dc}, the resonant inductor current i_{Lr}, and the clamping capacitor current i_{cl} is

$$i_{dc}(t) = i_{Lr}(t) + i_{cl}(t) \tag{11.41}$$

Considering the charge balance of the clamping capacitor C_{cl}, (11.42) should be satisfied.

$$\Delta v_{cl} = \frac{1}{C_{cl}}\int_0^{T_s} i_{cl}(t)\ dt = 0 \tag{11.42}$$

Combing (11.40), (11.41), and (11.42), the following equation can be obtained

$$\frac{1}{T_s}\int_0^{T_s} i_{dc}(t)\ dt = \frac{1}{T_s}\int_0^{T_s} i_{Lr}(t)\ dt + \frac{1}{T_s}\int_0^{T_s} i_{cl}(t)\ dt \approx \frac{i_{max}+i_{min}}{2} \tag{11.43}$$

The average value of DC-side current i_{dc} in a switching period is

$$\langle i_{dc}\rangle_{T_s} = \frac{1}{T_s}\int_0^{T_s} i_{dc}(t)\ dt \tag{11.44}$$

Combining (11.39), (11.43), and (11.44),

$$i_{min} = 2\langle i_{dc}\rangle_{T_s} - \sqrt{\frac{V_{dc}{}^2}{Z_r^2} + i_x{}^2} - i_s - i_g \tag{11.45}$$

By substituting (11.45) into (11.38), the resonant inductor current at t_1 is obtained,

$$i_{Lr}(t_1) = -\sqrt{\left[2\langle i_{dc}\rangle_{T_s} - \sqrt{\frac{V_{dc}{}^2}{Z_r^2} + i_x{}^2} - i_s - i_g\right]^2 - \frac{V_{cl}{}^2}{Z_r^2}} \tag{11.46}$$

The current increment through the resonant inductor L_r in the boost stage can be solved according to (11.34)

$$i_x \geq \sqrt{\left[2\cdot\langle i_{dc}\rangle_{T_s} - i_s - i_g\right]^2 + 4\cdot\frac{V_{dc}\cdot\langle i_{dc}\rangle_{T_s}}{Z_r} - 2\cdot\frac{V_{dc}}{Z_r}\cdot(i_s + i_g)} \tag{11.47}$$

If the previously mentioned inequality is satisfied, all the switches in Figure 11.8 could achieve ZVS. Since larger i_x leads to more energy stored in the resonant inductor and brings about more loss, the minimum value of i_x is preferred. When the equal sign of the expression in (11.47) is used, we have minimum circulation energy in the circuit.

Once i_x is known, the maximum value of $i_{Lr}(t)$ is obtained from (11.39)

$$i_{max} = \sqrt{\frac{{V_{dc}}^2}{Z_r^2} + \left[2\cdot\langle i_{dc}\rangle_{T_s} - i_s - i_g\right]^2 + 4\cdot\frac{V_{dc}\cdot\langle i_{dc}\rangle_{T_s}}{Z_r} - 2\cdot\frac{V_{dc}}{Z_r}\cdot(i_s + i_g)} + i_s + i_g \tag{11.48}$$

Similarly the minimum value of $i_{Lr}(t)$ is obtained from (11.45)

$$i_{min} = 2\cdot\langle i_{dc}\rangle_{T_s} - \sqrt{\frac{{V_{dc}}^2}{Z_r^2} + \left[2\cdot\langle i_{dc}\rangle_{T_s} - i_s - i_g\right]^2 + 4\cdot\frac{V_{dc}\cdot\langle i_{dc}\rangle_{T_s}}{Z_r} - 2\cdot\frac{V_{dc}}{Z_r}\cdot(i_s + i_g)} - i_s - i_g \tag{11.49}$$

Here, minimum i_x, i.e. i_{x_min} is used.

Theoretically, the short circuit signal should be added in stage 5. However, it is hard to differentiate stage 3, stage 4, and stage 5 in practical implementation. Therefore, the short circuit signal can be added to all the switches from stage 3. According to the stage analysis in Section 11.2.2, it is easy to find that this change does not affect the circuit and the previously discussed mathematical results keep unchanged. Obviously, if the minimum value of i_x in (11.47) is adopted, there will be no extra energy for freewheeling in stage 3, and stage 3 will shrink to a point, i.e. $i_{Lr}(t_2) = i_{Lr}(t_3)$.Thus, the equivalent short circuit current i_{esc} can be defined as

$$i_{esc} = i_{Lr}(t_5) - i_{Lr}(t_3) = i_s + i_g + i_x - i_{Lr}(t_3) \tag{11.50}$$

In (11.50), $i_{Lr}(t_3)$ means the current through the resonant inductor L_r at the end of stage 3 and

$$i_{Lr}(t_3) = 0 \tag{11.51}$$

The equivalent short circuit current i_{esc} can be finally got by substituting (11.47) into (11.50)

$$i_{esc} = \sqrt{\left[2\cdot\langle i_{dc}\rangle_{T_s} - i_s - i_g\right]^2 + 4\cdot\frac{V_{dc}\cdot\langle i_{dc}\rangle_{T_s}}{Z_r} - 2\cdot\frac{V_{dc}}{Z_r}\cdot(i_s + i_g)} + i_s + i_g \tag{11.52}$$

11.4 Design Prototype

11.4.1 Rated Parameters of a 1.5-kW Inverter

A 1.5-kW inverter prototype is designed to verify the theoretical analysis. The circuit of the prototype is shown in Figure 11.27.

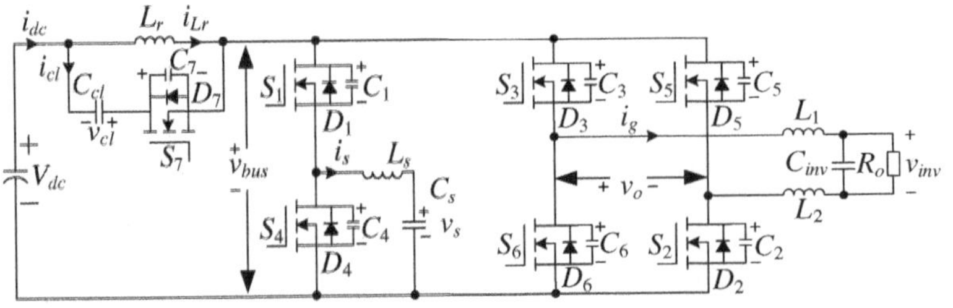

Figure 11.27 Circuit of the prototype.

Table 11.1 Rated parameters of the prototype.

Rated parameters	Value
DC voltage V_{dc}	380 V
Output voltage RMS v_{inv}	220 V
Switching frequency of inverter bridge f_{inv}	250 kHz
Switching frequency of APD bridge f_{apd}	500 kHz
Output frequency f_o	50 Hz
Rated power P_o	1.5 kW
Device type	SiC MOSFET C2M0080120D

The DC input voltage is 380 V_{dc}, and the output voltage is 220 V_{rms}. The switching frequency of inverter bridge is 250 kHz, and the switching frequency of APD bridge is 500 kHz. The rated power is 1.5 kW. Rated parameters of the single-phase inverter with APD are summarized in Table 11.1.

11.4.2 Parameter Design

a) Absorbing capacitor in APD bridge

The energy stored in C_s should satisfy

$$0 \leq E_s(t) \leq E_{S\_max} \tag{11.53}$$

According to (11.11), the minimum value of C_s can be calculated as

$$C_{S\_min} = \frac{V_m I_m}{\omega V_{dc}^2} \tag{11.54}$$

For $V_m = 311\ \text{V}$ and $I_m = 9.64\ \text{A}$,

$$C_{S\_min} = 65\ \mu\text{F} \tag{11.55}$$

The practical value of C_s is selected to be twice the minimum value [7]

$$C_s = 125\ \mu\text{F} \tag{11.56}$$

b) Filter inductor of inverter bridge

Filter inductor is used to reduce output current ripple. At rated power, the output current ripple is designed less than 20% of the amplitude. Output current ripple is expressed as [11]

$$\Delta i_{iL}(\theta) = \frac{V_{dc}\cdot m\cdot \sin\theta\cdot(1-m\cdot\sin\theta)\cdot T_s}{L_1+L_2} \leq \frac{V_{dc}\cdot T_s}{4(L_1+L_2)} \tag{11.57}$$

where θ is the phase angle of output voltage and m is the modulation ratio of the inverter with $m = \sqrt{2}v_{inv}/V_{dc}$ and $T_s = 1/500\ \text{kHz}$.

For the desired 20% current ripple,

$$L_1 + L_2 \geq \frac{V_{dc}\cdot T_s}{4\times 20\%\times 9.64\ \text{A}} = 98\ \mu\text{H} \tag{11.58}$$

In practice, the filter inductors are chosen as

$$L_1 = L_2 = 47\ \mu\text{H} \tag{11.59}$$

c) Filter inductor of APD bridge

Similarly, the current ripple in APD bridge is designed less than 20% of the amplitude. APD inductor current ripple is expressed as [8]

$$\begin{aligned}\Delta i_s(\theta) &= \frac{\frac{v_s(\theta)}{V_{dc}} \cdot T_s[V_{dc} - v_s(\theta)]}{L_s} \\ &= \frac{T_s}{V_{dc}L_s}\left[\frac{1}{4}V_{dc}^2 - \left(v_s(\theta) - \frac{1}{2}V_{dc}\right)^2\right] \leq \frac{V_{dc}T_s}{4L_s}\end{aligned} \tag{11.60}$$

Considering (11.13), the maximum current of APD bridge is 6.6 A at rated power. Then, the filter inductor of APD bridge can be calculated as

$$L_s \geq \frac{V_{dc} \cdot T_s}{4\Delta i_s} = \frac{V_{dc} \cdot T_s}{4 \times 20\% \times 6.6\ \text{A}} = 144\ \mu\text{H} \tag{11.61}$$

The practical filter inductor of APD bridge is chosen as

$$L_s = 143\ \mu\text{H} \tag{11.62}$$

d) Turn-off time ratio of S_7

The turn-off time ratio of S_7 is expressed in (11.31) and can be approximated as [8, 10]

$$\lambda_7 = \frac{t_6 - t_1}{T_s} \approx L_r \frac{i_{max} - i_{min}}{V_{dc}T_s} \tag{11.63}$$

The equivalent DC bus voltage v_{bus} in stage 3, stage 4, and stage 5 is clamped to zero, and the energy cannot be delivered from DC-link to the load. The turn-off time ratio of the auxiliary switch S_7 is expected to be lower than 0.15 to minimize duty cycle loss and to reduce the voltage stress of devices; therefore,

$$\lambda_7 \leq 0.15 \tag{11.64}$$

e) Resonant time

The resonant time of stage 2 (t_1–t_2) and stage 6 (t_5–t_6) can be estimated by one fourth of a resonant period. It is designed to be longer than the turn-off time of SiC MOSFET to reduce the turn-off loss. For the selected SiC MOSFET C2M0080120D, the typical turn-off time is less than 20 ns [12]. The upper limit of the resonant time is set as 5% of the switching period to reduce the duty cycle loss [8, 10].

Considering the limitations mentioned previously, the resonant period is constrained as follows

$$20\ \text{ns} \leq \frac{T_r}{4} \leq 5\% \cdot T_s = 100\ \text{ns} \tag{11.65}$$

where

$$T_r = 2\pi\sqrt{L_r(3C_r + C_7)} \tag{11.66}$$

f) Resonant capacitor

Each device has its intrinsic output capacitance. For the selected SiC MOSFET C2M0080120D, the typical output capacitance is 202 pF [12]. So the lower limit of resonant capacitor is

$$3C_r + C_7 \geq 4C_{oss} = 808\ \text{pF} \tag{11.67}$$

Combining the previously mentioned restrictions (d), (e), and (f), the solution region can be determined as shown in Figure 11.28. Finally, the resonant inductor L_r is chosen as 2.2 μH, and the paralleled capacitor for S_1–S_6 is designed as 100 pF. No external capacitor is paralleled with the auxiliary switch.

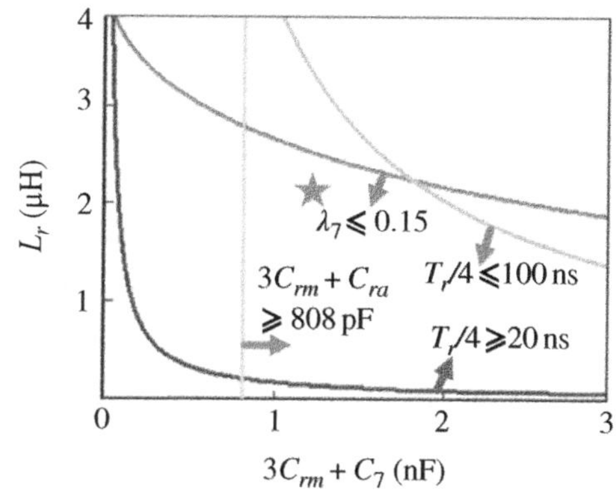

Figure 11.28 Design of resonant parameters.

11.4.3 Experimental Platform and Testing Results

The specifications of the experimental prototype is presented in Table 11.2. The prototype operates with RLC load.

The driving signal-generating diagram is shown in Figure 11.29. Output voltage v_{inv}, output current i_{inv}, DC voltage V_{dc}, DC current i_{dc}, APD capacitor voltage v_s, and APD inductor current i_s are detected to generate the modulation signals (v_{m1},v_{ms}), short circuit duty ratio (d_{sc}), and turn-off duty ratio of the auxiliary switch (λ_7). The generated signals are then sent to the EA-PWM module to produce driving signals for the switches.

Table 11.2 Specifications of the prototype.

Parameter	Value
DC voltage V_{dc}	380 V
Inverter bridge filter inductor L_1–L_2	47 μH
Inverter bridge filter capacitor C_{inv}	470 nF
Load resistance R_o	32 Ω/46 Ω/95 Ω
APD absorbing capacitor C_s	125 μF
APD bridge filter inductor L_s	143 μH
Resonant inductor L_r	2.2 μH
Clamping capacitor C_{cl}	33 μF
Paralleled resonant capacitors C_1–C_6	100 pF
Output voltage RMS v_{inv}	220 V
Output frequency f_o	50 Hz
Switching frequency of inverter bridge f_{inv}	250 kHz
Switching frequency of APD bridge f_{apd}	500 kHz
Device type	SiC MOSFET C2M0080120D

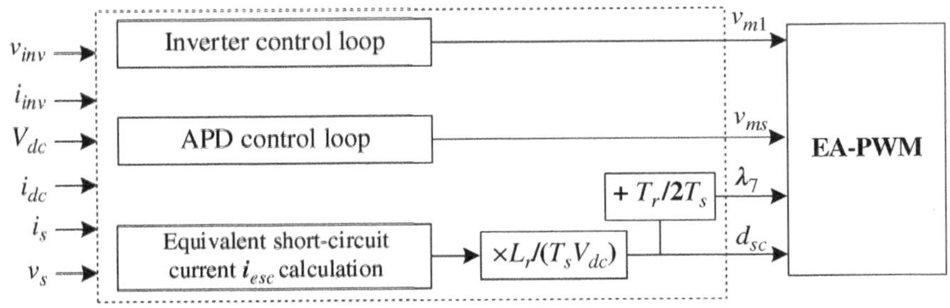

Figure 11.29 Driving signal-producing diagram of ZVS single-phase inverter with APD.

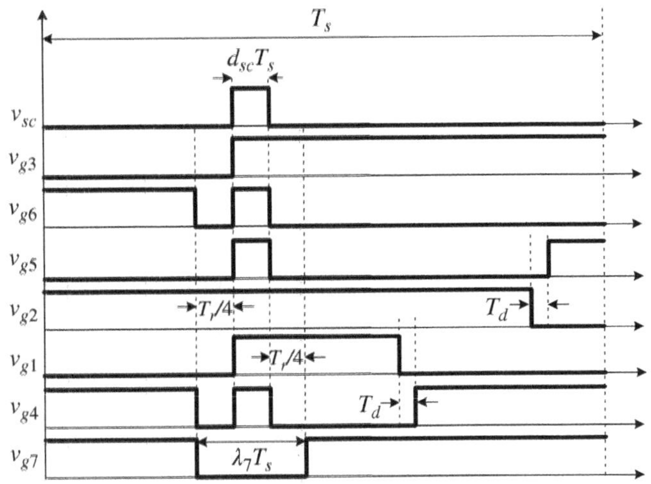

Figure 11.30 Typical driving signal sequence of the ZVS inverter.

The modulation signals for the inverter legs and the APD leg are decided by their control loops, respectively. The short circuit duty ratio d_{sc} is estimated according to Equation (11.52) as follows

$$d_{sc} = \frac{i_{esc}}{V_{dc}/L_r} \tag{11.68}$$

Turn-off duty ratio of the auxiliary switch λ_7 is estimated by

$$\lambda_7 \approx d_{sc}T_s + \frac{T_r}{2} \tag{11.69}$$

A typical gate driver signal sequence of the ZVS inverter is shown in Figure 11.30. The dead time is T_d. The duration of the short circuit state is $d_{sc}T_s$. As shown in Figure 11.30, turn-off operations of the main switches are $T_r/4$ ahead of the short circuit signal. The short circuit signal v_{sc} is superimposed on the normal driving signals of all the main switches. Diagram of the EA-PWM is shown in Figure 11.31.

Figure 11.32 shows the drive signals of S_7, S_1, S_3, and S_5 when $i_s > 0$ and $i_g > 0$. It can be seen that the drive signal of APD bridge (S_1) is synchronized with inverter bridges (S_3, S_5) and the auxiliary switch (S_7).

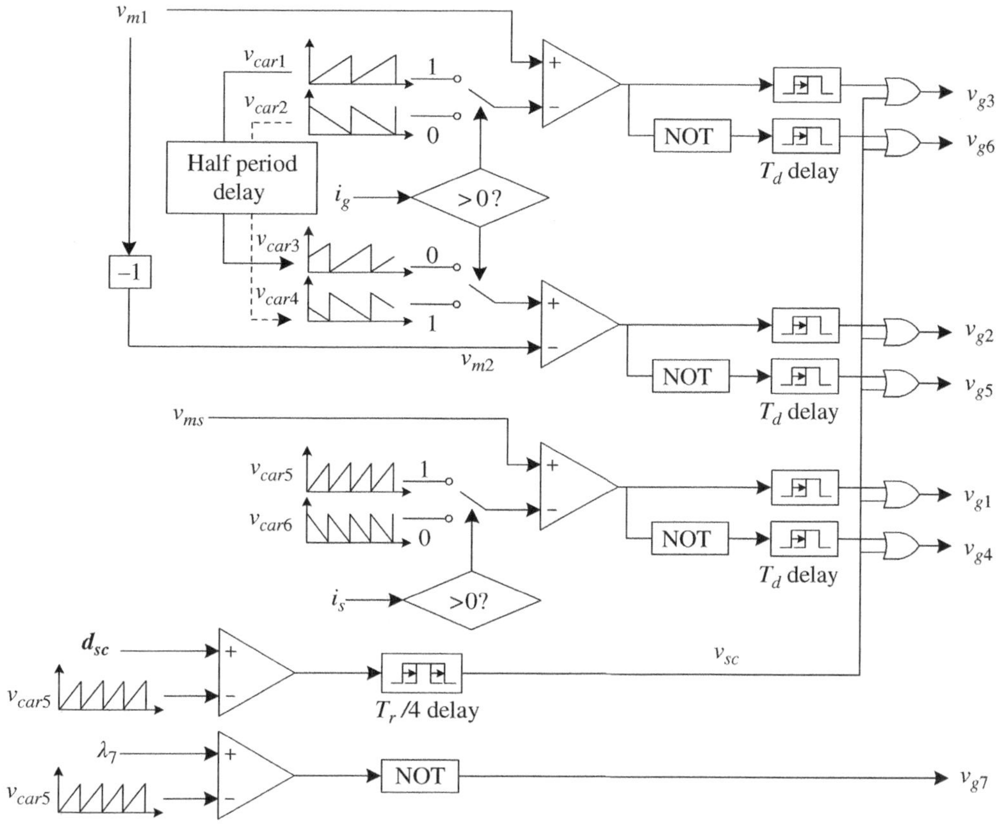

Figure 11.31 Diagram of the EA-PWM module.

Figure 11.32 Experimental EA-PWM driving signals.

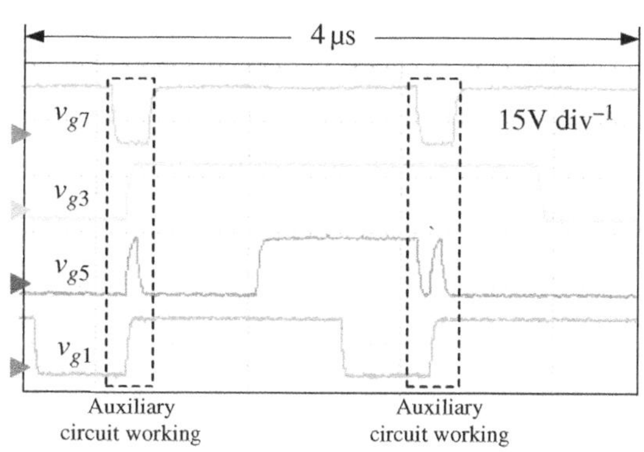

Figure 11.33a shows the theoretical envelopes of the short circuit current i_{sc}, the maximum resonant inductor current i_{max}, and the minimum resonant inductor current i_{min} during one fundamental period under 1.5-kW operation. Figure 11.33b shows the experimental resonant inductor current i_{Lr} during one fundamental period. The experimental waveforms show a good matching with theoretical envelopes.

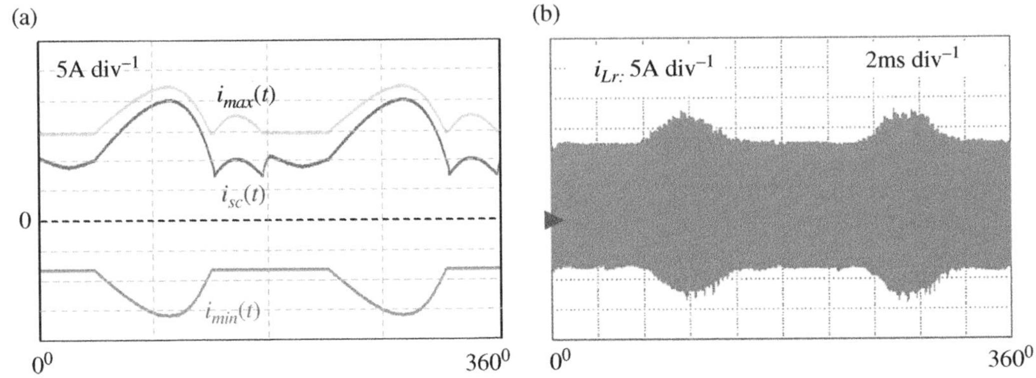

Figure 11.33 Resonant inductor current: (a) theoretical envelopes and (b) experimental waveforms.

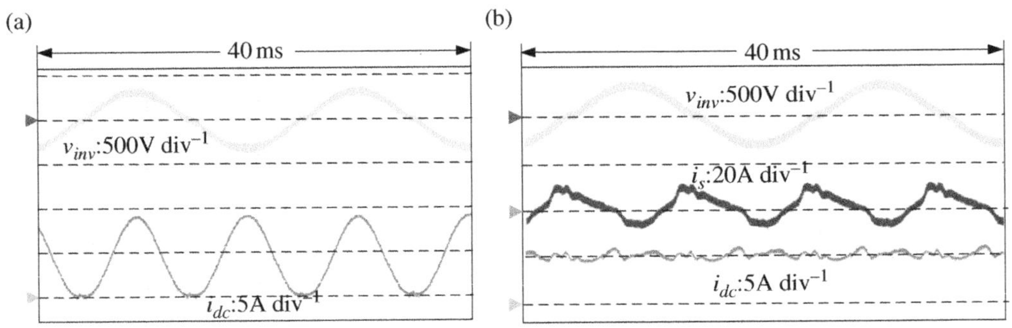

Figure 11.34 Decoupling effect of APD: (a) output voltage v_{inv} and DC input current i_{dc} without APD; (b) output voltage v_{inv}, APD bridge current i_s, and DC input current i_{dc} with APD.

Figure 11.34 shows the output voltage v_{inv} and DC input current i_{dc} without APD and with APD. It can be seen that the double line frequency current ripple is greatly reduced with APD.

Figure 11.35 shows the resonant inductor current i_{Lr}, the DC-link voltage v_{bus}, the drain-source voltage of S_4, and the output voltage of inverter bridge v_o at different low frequency phase angles. The DC-link voltage v_{bus} is resonated to zero in each switching period.

According to previous analysis, if APD inductor current i_s is positive (at 30° and150°), Type 2 commutation that needs ZVS is D_4 to S_1. The rising edge of v_{c4} is aligned with the rising edge of v_{bus} as shown in Figure 11.35a and d. This means Type 2 commutation is accomplished under ZVS condition. If i_s is negative (at 60° and 120°), Type 2 commutation that needs ZVS is D_1 to S_4. The falling edge of v_{c4} is aligned with the falling edge of v_{bus} as shown in Figure 11.35b and c, which also means Type 2 commutation is accomplished under ZVS condition.

The gate drive signal (v_{g1}) of APD switch S_1, its drain-source voltage v_{c1}, and the drain-source voltage (v_{c5}) of S_5 at the phase of 30° is shown in Figure 11.36a. The gate drive signal (v_{g7}) of auxiliary switch S_7 and drain-source voltage v_{c7} is shown in Figure 11.36b. The gate drive signals of S_1 and S_7 rise after their drain-source voltage is resonated to zero, respectively, which means ZVS on is realized. The drive signal (v_{g4}) of APD switch S_4, its drain-source voltage v_{c4}, and the drain-source

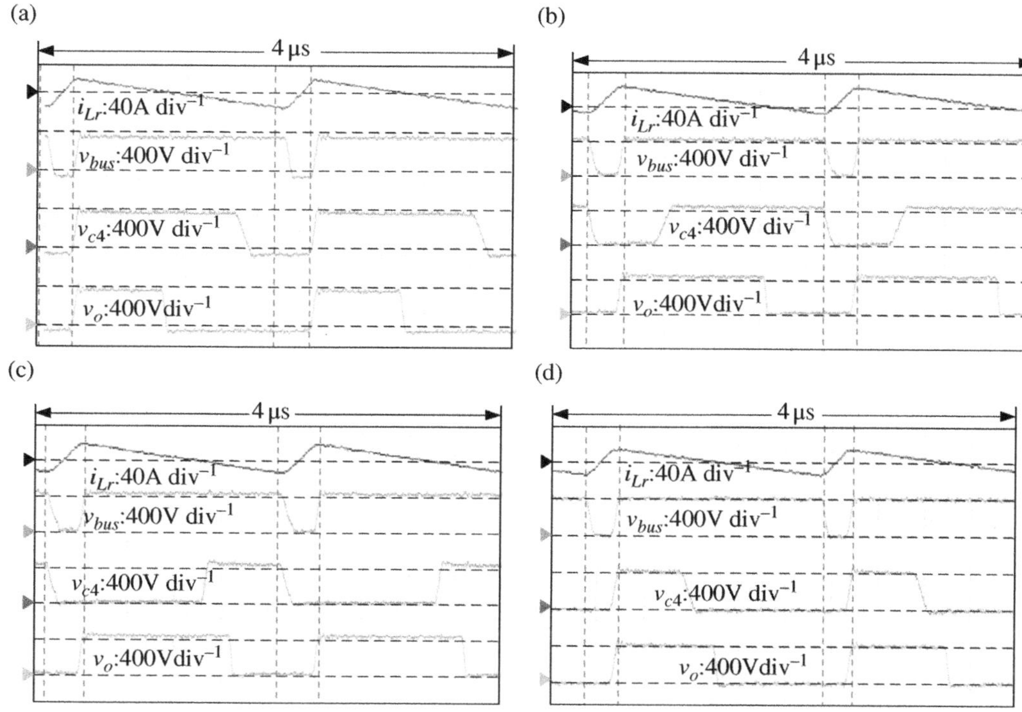

Figure 11.35 Resonant inductor current i_{Lr}, DC-link voltage v_{bus}, drain-source voltage of S_4, and inverter output voltage v_o at different phase angles, (a) 30°, (b) 60°, (c) 120°, and (d) 150°.

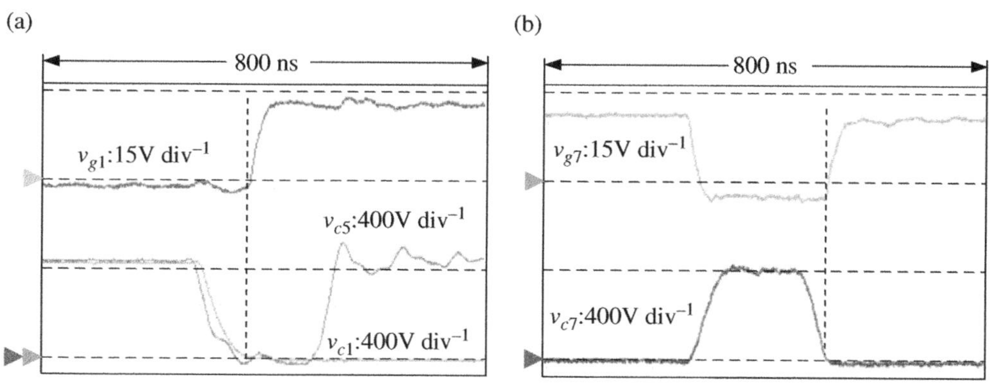

Figure 11.36 ZVS realization at the phase of 30°: (a) main switch and (b) auxiliary switch.

voltage (v_{c5}) of S_5 at the phase of 60° is shown in Figure 11.37a. The gate drive signal (v_{g7}) of auxiliary switch S_7 and drain-source voltage v_{c7} is shown in Figure 11.37b. The gate drive signals of S_4 and S_7 rise after their drain-source voltage is resonated to zero, respectively, which means ZVS turn-on is realized.

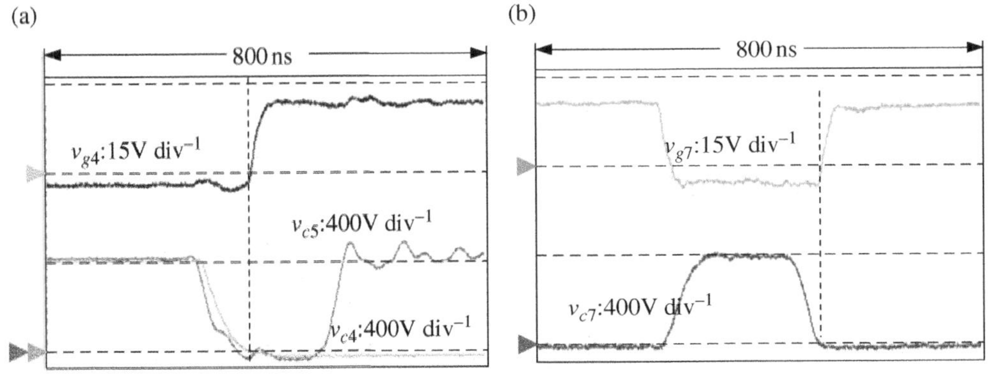

Figure 11.37 ZVS realization at the phase of 60°: (a) main switch and (b) auxiliary switch.

11.5 Summary

In this chapter, the soft-switching technique has been applied to the single-phase inverter.

The traditional modulation methods for single-phase inverter with APD are introduced. Then, the soft-switching technique is extended to single-phase inverter with APD to mitigate double line frequency power pulsation in DC side. By introducing EA-PWM, Type 2 commutations in the inverter bridges and APD bridge are synchronized.

Besides stage analysis for a switch cycle, two resonant stages in a switching cycle are analyzed. Analysis shows that a resonant inductor current boost stage is needed to achieve ZVS of all switches.

Finally, design of a 500 kHz 1.5-kW SiC prototype with RLC load is introduced. Design of inverter parameters and resonant parameters is discussed. Experiment shows that the double line frequency ripple of DC input current is greatly reduced, and ZVS of all switches are realized.

References

1 R.-S. Lai and K. D. T. Ngo, "A PWM method for reduction of switching loss in a full-bridge inverter," *IEEE Transactions on Power Electronics*, vol. 10, no. 3, pp. 326–332, May 1995.

2 H. Hu, S. Harb, N. Kutkut, I. Batarseh and Z. J. Shen, "Power decoupling techniques for micro-inverters in PV systems-a review," in IEEE Energy Conversion Congress and Exposition (ECCE), pp. 3235–3240, 2010.

3 A.C. Kyritsis, N. P. Papanikolaou, and E. C. Tatakis, "A novel parallel active filter for current pulsation smoothing on single stage grid-connected AC-PV modules", in European Conference on Power Electronics and Applications (EPE), pp. 1–10, 2007.

4 Z. Qin, Y. Tang, P. C. Loh, and F. Blaabjerg, "Benchmark of AC and DC active power decoupling circuits for second-order harmonic mitigation in Kilowatt-Scale single-phase inverters," *IEEE Journal of Emerging and Selected Topics in Power Electronics*, vol. 4, no. 1, pp. 15–25, March 2016.

5 R. Wang, F. Wang, D. Boroyevich, R. Burgos, R. Lai, P. Ning, and K. Rajashekara, "A high power density single-phase PWM rectifier with active ripple energy storage," *IEEE Transactions on Power Electronics*, vol. 26, no. 5, pp. 1430–1443, May 2011.

6 S. Qin, Y. Lei, C. Barth, W. Liu, and R. C.N. Pilawa-Podgurski, "Architecture and control of a high energy density buffer for power pulsation decoupling in grid-interfaced applications" in Control and Modeling for Power Electronics(COMPEL), pp. 1–8, 2015.

7 D. Neumayr, D. Bortis, and J. W. Kolar, "Ultra-compact Power Pulsation Buffer for single-phase DC/AC converter systems," in Power Electronics and Motion Control Conference (IPEMC-ECCE Asia), pp. 2732–2741, 2016.

8 M. Chen, Z. Ye, Y. Chen, and D. Xu, "Zero-voltage-switching single-phase full-bridge inverter with active power decoupling," *IEEE Transactions on Power Electronics*, vol. 36, no. 1, pp. 571–582, January 2021.

9 Z. Ye, Y. Chen, and D. Xu, "Zero-voltage-switching single-phase inverter with active power decoupling," in IEEE Applied Power Electronics Conference and Exposition (APEC), pp. 348–355, 2018.

10 Y. Chen, D. Xu, J. Xi, G. Hu, C. Du, Y. Li, and M. Chen, "A ZVS grid-connected full-bridge inverter with a novel ZVS SPWM scheme", *IEEE Transactions on Power Electronics*, vol. 31, no. 5, pp. 3626–3638, May 2016.

11 X. Mao, R. Ayyanar, and H. K. Krishnamurthy, "Optimal variable switching frequency scheme for reducing switching losses in single-phase inverters based on time-domain ripple analysis," *IEEE Transactions on Power Electronics*, vol. 24, no. 4, pp. 991–1001, April 2009.

12 CREE Corporation, C2M0080120D Datasheet [EB/OL]. 2015.10-Rev.C.

12

Soft-switching SiC Three-phase Four-wire Back-to-back Converter

In this chapter, a soft-switching silicon carbide (SiC) three-phase four-wire back-to-back (BTB) converter is introduced. With the edge-aligned pulse-width-modulation (EA-PWM) scheme, all main switches in both the rectifier side and inverter side, as well as the auxiliary switch, can realize zero-voltage-switching (ZVS) turn-on using only one auxiliary circuit. Besides, the auxiliary switch only needs to operate once in each switching period, which can reduce extra loss of the auxiliary circuit. Not only the operation principle and circuit analysis are introduced, but also the ZVS condition is derived. A ZVS BTB converter is studied in both balanced and unbalanced loads. Finally, the operation principle of the ZVS BTB converter is verified on a 9 kW prototype with SiC Metal-oxide-semiconductor field-effect transistor (MOSFET) devices.

12.1 Introduction

A three-phase four-wire BTB converter composed of a front-end PWM rectifier and a load-side PWM inverter is widely used in AC sources, uninterruptible power supply (UPS), motor drives, and micro-grids [1, 2]. The switching frequency of the BTB converter is limited by the switching loss of the power semiconductor device due to its hard-switching operation. It causes a larger size of filters in both rectifier side and inverter side. With the advancement of the power semiconductor device technology, wide-bandgap (WBG) device such as SiC MOSFET has shown superior switching performance compared with traditional Si Insulated-gate bipolar transistor (IGBT). By using the WBG device, power converters can have a considerable efficiency and power density improvement. However, when the switching frequency increases beyond 50 kHz, the switching loss is still a bottleneck for enhancement of the efficiency and power density. In addition, either voltage or current ringing and Electromagnetic interference (EMI) noise due to high dv/dt and di/dt of WBG devices become more and more difficult to deal with when they operate at hard switching. Soft-switching technology seems to be a good way to solve the aforementioned problems [3–6].

In the previous chapters, the soft-switching technique is introduced in the three-phase rectifier or inverter, respectively [7–9]. In this chapter, a soft-switching SiC three-phase four-wire BTB converter is introduced. The rectifier and the inverter in the ZVS BTB converter share one auxiliary circuit to realize the ZVS operation of all the switches [12]. In this way, the components of the auxiliary circuit are reduced. It not only simplifies the circuit but also reduces extra losses caused by the auxiliary circuit. The ZVS operation can eliminate all the turn-on loss and reverse recovery loss of

Soft-Switching Technology for Three-phase Power Electronics Converters, First Edition.
Dehong Xu, Rui Li, Ning He, Jinyi Deng, and Yuying Wu.

the antiparallel diode of the switches. Since the WBG device such as SiC MOSFET or GaN has less turn-off loss than turn-on loss, ZVS operation will result in better efficiency improvement for power converters with WBG devices.

12.2 Operation Principle

Figure 12.1a and b show the conventional three-phase four-wire BTB converter and the ZVS BTB converter, respectively [12]. Comparing to the conventional BTB converter, an auxiliary circuit is installed between the positive DC bus and DC capacitor C_{dc1} in the ZVS BTB converter. The auxiliary circuit is composed of a resonant inductor L_r, a clamping capacitor C_{cl}, and an auxiliary switch S_7. Resonant capacitors C_{i1}–C_{i6}, C_{o1}–C_{o6}, and C_{r7} are paralleled to the main switches of rectifier S_{i1}–S_{i6}, the main switches of the inverter S_{o1}–S_{o6}, and the auxiliary switch S_7, respectively. For most of the time in a switching period, the auxiliary switch S_7 is on-state, and there is circulating current in the auxiliary circuit. Once the auxiliary switch S_7 is turned off, the circulating current will be released to reset the voltage of the main switches to zero so that the ZVS turn-on of the main switches is realized. However, higher circulation energy will cause an extra loss in the circuit. Therefore, it should be accurately controlled according to the load condition to minimize power loss in the circuit. The clamping capacitance C_{cl} is so large that its voltage v_{cl} can be regarded as constant in a switching period.

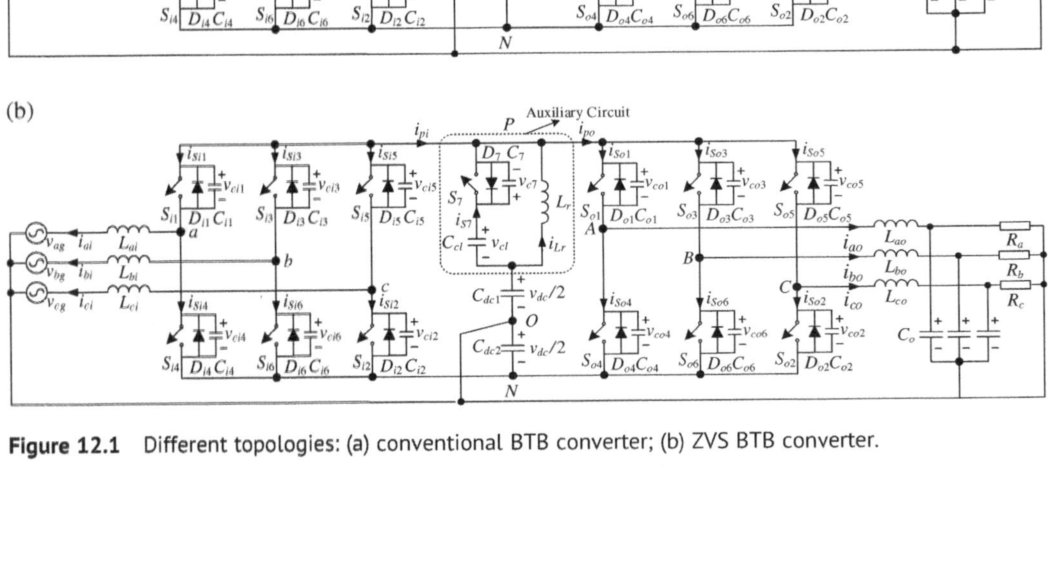

Figure 12.1 Different topologies: (a) conventional BTB converter; (b) ZVS BTB converter.

12.2.1 Commutations Analysis

Although the PWM rectifier and inverter have different control requirements, their switching commutation processes are similar when one switching period is investigated. In the hard-switching converter, there are two types of switching commutation processes in each leg. One type of commutation process is from the switch to the antiparallel diode of its complementary switch, which is defined as Type 1 commutation. Another one is from the antiparallel diode to its complementary switch, which is defined as Type 2 commutation. The detailed description can be found in Chapter 1. For Type 2 commutation, the switch will have turn-on loss and the diode will undergo a severe reverse recovery process. In the ZVS BTB converter, Type 2 commutation is avoided through a resonant process with the help of the auxiliary switch to realize the ZVS turn-on for the main switch. For Type 1 commutation, the switch is ZVS turn-off and the turn-off loss is reduced.

12.2.2 Operation Scheme

In the converter, the current commutation of each leg in one switching period depends on the PWM scheme and the filter inductor current polarity. In order to simplify the analysis, the assumptions are made as follows:

1) The PWM rectifier works with unity power factor, and the inverter works under a balanced three-phase load.
2) The grid voltage, rectifier-side filter current, output voltage, and inverter-side filter current are sinusoidal with the same utility period.

The three-phase grid voltage is represented by

$$\begin{cases} v_{ag} = U_m \cos \omega t \\ v_{bg} = U_m \cos(\omega t - 2\pi/3) \\ v_{cg} = U_m \cos(\omega t + 2\pi/3) \end{cases} \tag{12.1}$$

where U_m is the peak value of the phase voltage.

Ignoring the switching ripple of the rectifier-side filter current, the rectifier-side filter current is expressed as:

$$\begin{cases} i_{ai} = -I_m \cos \omega t \\ i_{bi} = -I_m \cos(\omega t - 2\pi/3) \\ i_{ci} = -I_m \cos(\omega t + 2\pi/3) \end{cases} \tag{12.2}$$

where I_m is the peak value of the phase current.

The output voltage and inverter-side filter current under a balanced three-phase load are expressed as:

$$\begin{cases} v_{ao} = U_m \cos(\omega t + \varphi) \\ v_{bo} = U_m \cos(\omega t + \varphi - 2\pi/3) \\ v_{co} = U_m \cos(\omega t + \varphi + 2\pi/3) \end{cases} \tag{12.3}$$

$$\begin{cases} i_{ao} = I_m \cos(\omega t + \varphi + \theta) \\ i_{bo} = I_m \cos(\omega t + \varphi + \theta - 2\pi/3) \\ i_{co} = I_m \cos(\omega t + \varphi + \theta + 2\pi/3) \end{cases} \tag{12.4}$$

where φ is the phase difference between the grid voltage and the output voltage, and θ is the power factor angle of the load.

When $\varphi = \pi/6$ and $\theta = 0$, the traditional sinusoidal PWM (SPWM) scheme applied in the three-phase four-wire BTB converter is shown in Figure 12.2. The m_{ai}, m_{bi}, m_{ci}, m_{ao}, m_{bo}, and m_{co} are the modulation waveforms of the rectifier and inverter. The modulation waveforms are compared with triangle carrier, and the bridge midpoint voltages v_{aN}, v_{bN}, v_{cN}, v_{AN}, v_{BN}, and v_{CN} are obtained as shown in Figure 12.3. One switching period in the striped band in Figure 12.2 is taken as an example to illustrate the operation principle of the ZVS BTB converter and its modulation scheme. According to the current polarity of the filter inductor in this switching period ($i_{ai} < 0$, $i_{bi} > 0$,

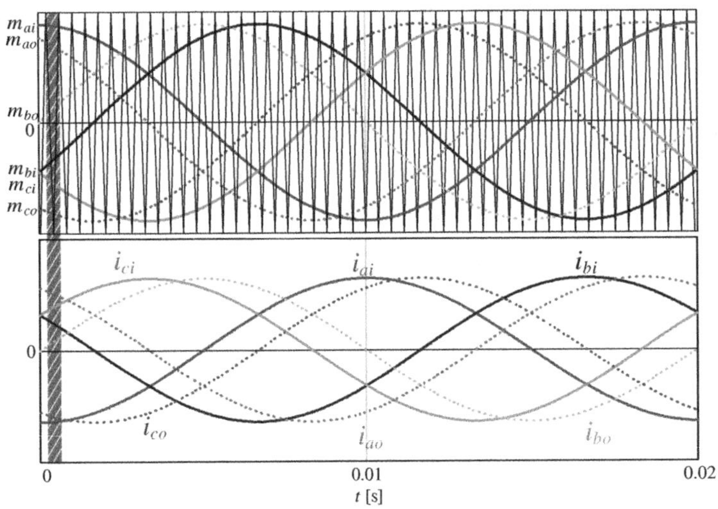

Figure 12.2 Modulation signals of the rectifier and inverter.

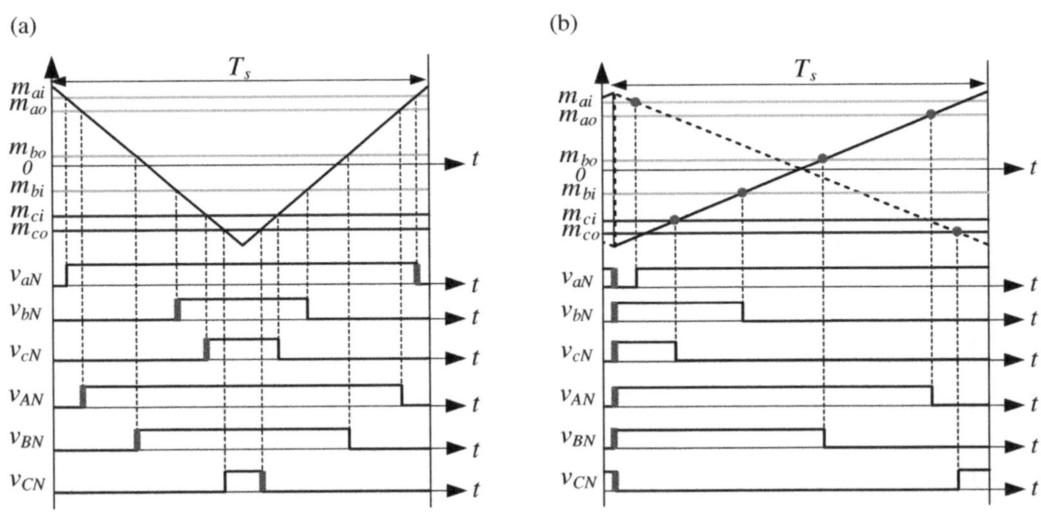

Figure 12.3 Two PWM schemes: (a) traditional SPWM scheme; (b) EA-PWM scheme.

$i_{ci} > 0$, $i_{ao} > 0$, $i_{bo} > 0$, $i_{co} < 0$), six Type 2 commutations from diodes (D_{i1}, D_{i6}, D_{i2}, D_{o4}, D_{o3}, D_{o5}) to MOSFETs (S_{i4}, S_{i3}, S_{i5}, S_{o1}, S_{o6}, S_{o2}) can be easily identified as shown with vertical bold line marked in Figure 12.3a. In order to achieve ZVS operation of all these six Type 2 commutations, the auxiliary circuit may need to operate six times if the conventional SPWM scheme is used, which will cause higher losses in the auxiliary circuit. Instead, an EA-PWM scheme shown in Figure 12.3b is used to replace the conventional SPWM scheme. For the positive filter current, the modulation signal is compared with the ramp-up saw-tooth wave. For the negative filter current, the modulation signal is compared with the ramp-down saw-tooth wave. By changing the triangular carrier into the saw-tooth carrier, the sequences of the driving signals are rearranged and the six Type 2 commutations will occur simultaneously. Therefore, the auxiliary resonant circuit only needs to operate once during each switching period, and ZVS turn-on can be achieved simultaneously in both the rectifier side and inverter side. It should be noted that the rule of equivalent area principle of PWM modulation in each switching period still holds when the triangular carrier is changed to the saw-tooth carrier. Similarly, for the other switching periods, a similar gate signal arrangement style can be deduced according to the filter current polarity of six phases.

12.2.3 Stage Analysis

Before the circuit stage analysis, the following assumptions are made to simplify the analysis.

1) All the switches S_{i1}–S_{i6}, S_{o1}–S_{o6}, and S_7 are regarded as ideal switches and each switch is paralleled to an ideal antiparallel diode.
2) The resonant capacitances (C_{i1}–C_{i6}, C_{o1}–C_{o6}, and C_7) are linear and meet the following equation:

$$C_{i1} = C_{i2} = C_{i3} = C_{i4} = C_{i5} = C_{i6} = C_{o1} = C_{o2} = C_{o3} = C_{o4} = C_{o5} = C_{o6} = C_7 = C_r \tag{12.5}$$

3) The clamping capacitor is so large that its voltage v_{cl} is constant in a switching period.
4) The filter inductor currents (i_{ai}, i_{bi}, i_{ci} and i_{ao}, i_{bo}, i_{co}) have small ripples, which can be regarded as constant in a switching period.

The driving signals of all the main switches and the auxiliary switch can be directly obtained from Figure 12.3b. As shown in Figure 12.4, the driving signals of switches S_{i1}, S_{i3}, and S_{i5} and S_{o1}, S_{o3}, and S_{o5} are complemented to the driving signals of switches S_{i4}, S_{i6}, and S_{i2} and S_{o4}, S_{o6}, and S_{o2}, respectively. The turn-off instant of the auxiliary switch S_7 is designed to be T_r before the turn-on instant of the main switches. In every switching period, there are nineteen operation stages. When the circuit operates in the switching period shown in Figure 12.2, nineteen stages (stage 1–stage 19) are introduced as shown in Figures 12.5–12.23, of which seven stages (stage 1–stage 7) are related to the ZVS operation. For the remaining twelve stages, the ZVS BTB converter operates just like a conventional PWM BTB converter.

Stage 1 (t_0–t_1) is the first steady stage. In this stage, the diodes D_{i1}, D_{i6}, D_{i2} and D_{o4}, D_{o6}, D_{o5} are conducting. The auxiliary switch S_7 is on-state, the voltage across the resonant inductor L_r (v_{Lr}) is clamped to $-V_{cl}$, and its current i_{Lr} decreases at the rate of

$$\frac{di_{Lr}}{dt} = -\frac{V_{cl}}{L_r} \tag{12.6}$$

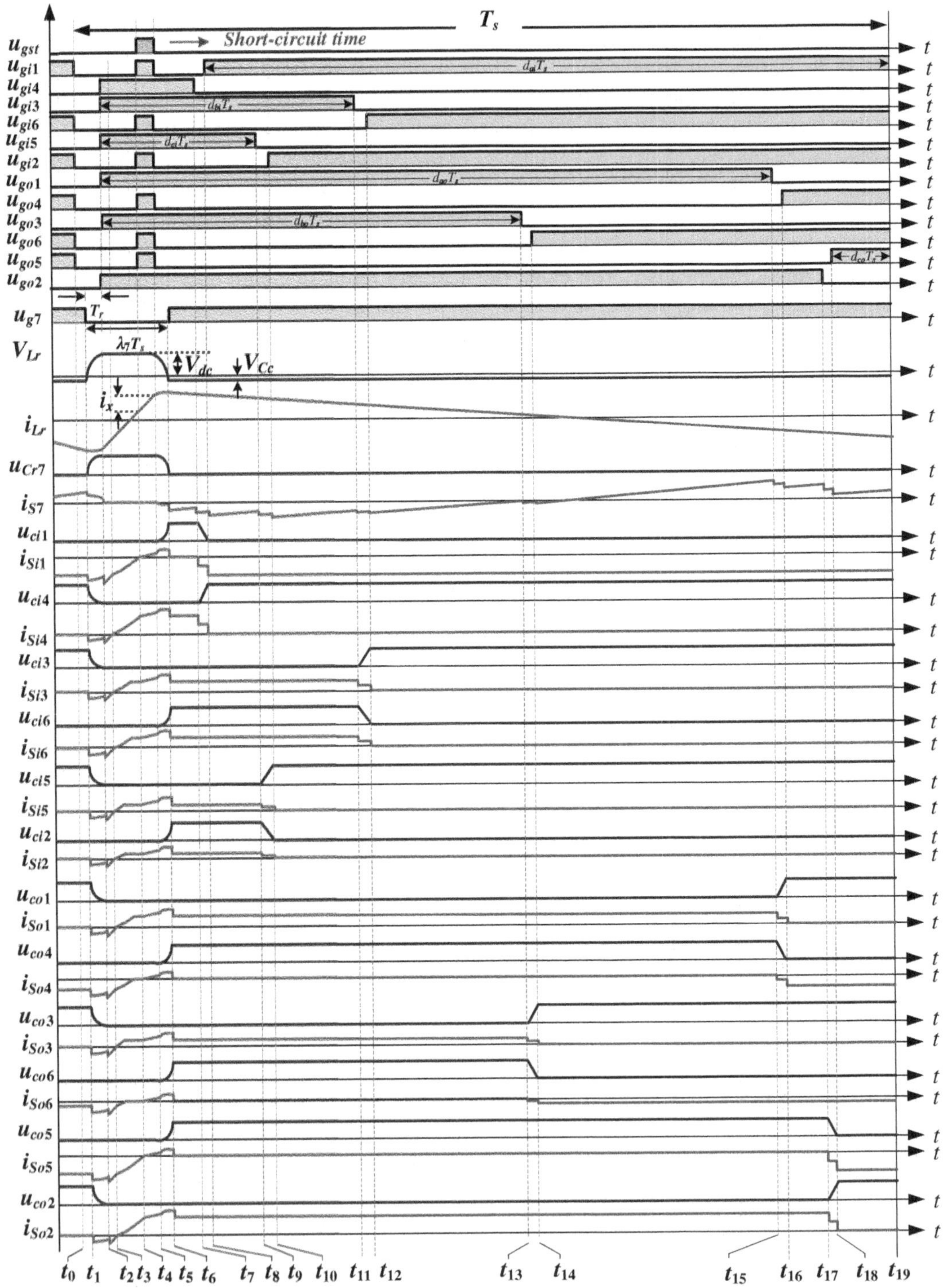

Figure 12.4 Key waveforms in one switching period.

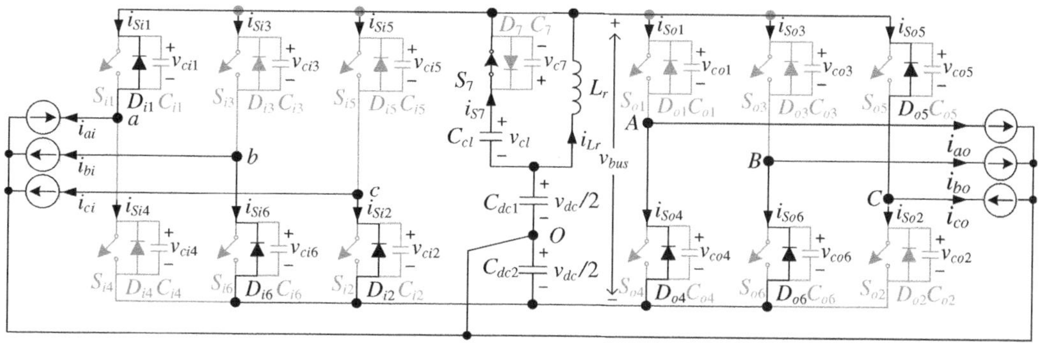

Figure 12.5 Stage 1 (t_0–t_1).

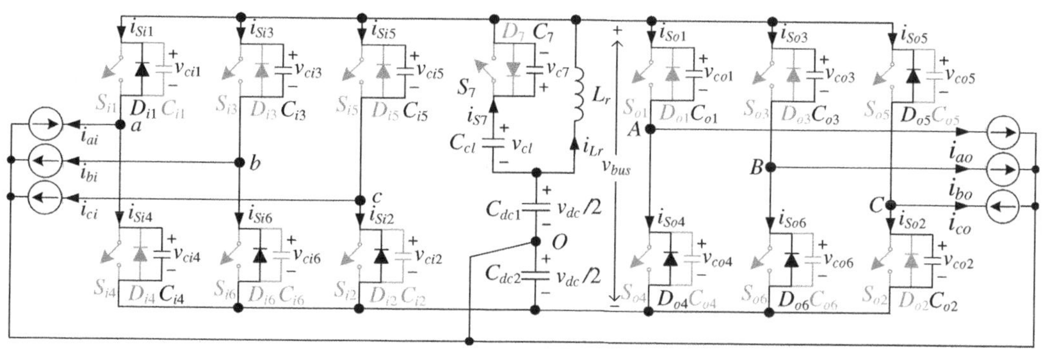

Figure 12.6 Stage 2 (t_1–t_2).

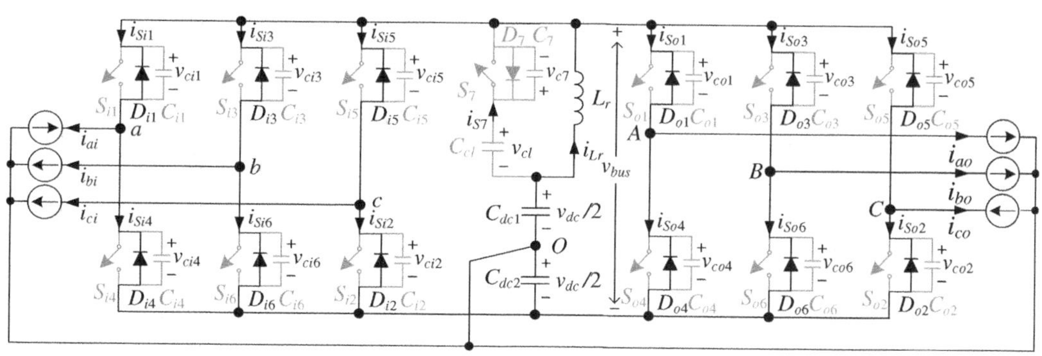

Figure 12.7 Stage 3 (t_2–t_3).

Stage 2 (t_1–t_2) is the first resonant stage by turning off auxiliary switch S_7 at t_1. v_{ci4}, v_{ci3}, v_{ci5} and v_{co1}, v_{co6}, v_{co2} are resonating to zero before adding the turn-on signals for these main switches. At the same time, v_{c7} is resonating to $V_{dc} + V_{cl}$. At t_2, the paralleled diodes D_{i4}, D_{i3}, D_{i5} and D_{o1}, D_{o3}, D_{o2} are conducting, and v_{ci4}, v_{ci3}, v_{ci5}, v_{co1}, v_{co3}, v_{co2} are clamped to zero.

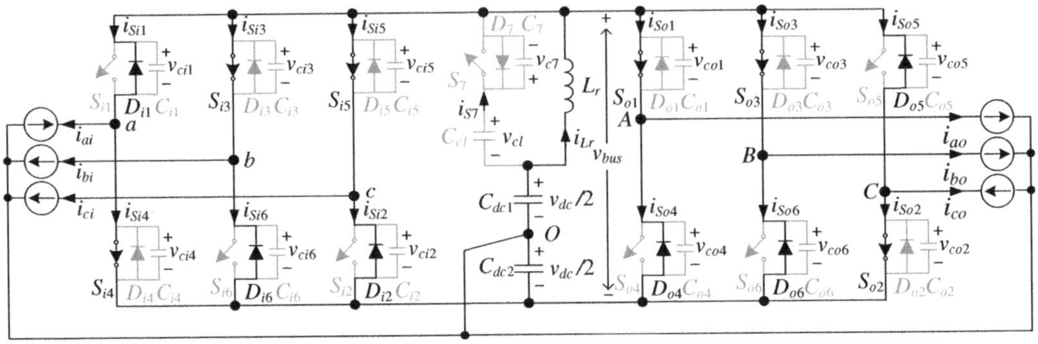

Figure 12.8 Stage 4 (t_3–t_4).

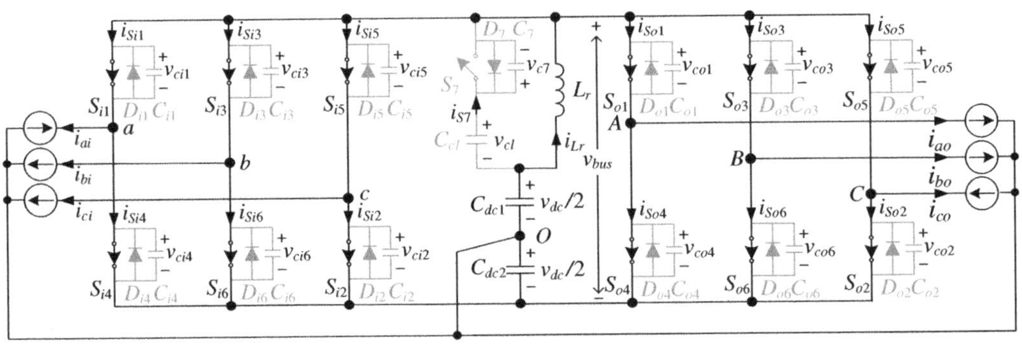

Figure 12.9 Stage 5 (t_4–t_5).

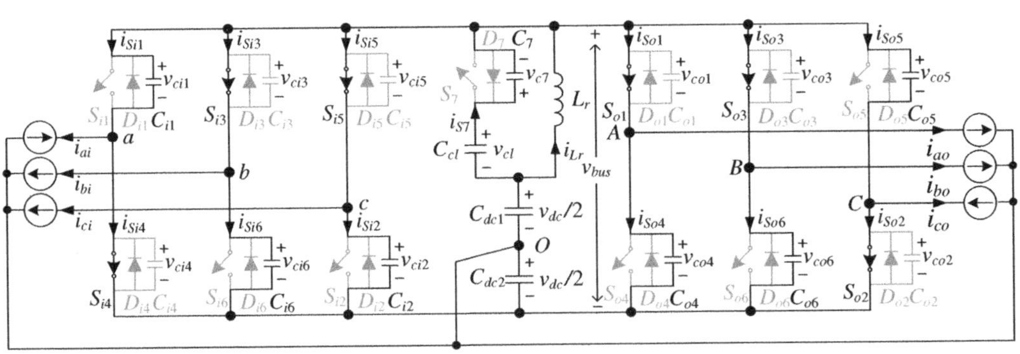

Figure 12.10 Stage 6 (t_5–t_6).

Stage 3 (t_2–t_3) is the diode freewheeling stage. S_{i4}, S_{i3}, S_{i5} and S_{o1}, S_{o3}, S_{o2} are turned on with ZVS condition, and the voltage across the resonant inductor L_r is clamped to V_{dc}. Its current increases at the rate of

$$\frac{di_{Lr}}{dt} = \frac{V_{dc}}{L_r} \tag{12.7}$$

This stage ends when the currents in diodes D_{i4}, D_{i3}, D_{i5} and D_{o1}, D_{o3}, D_{o2} reduce to zero.

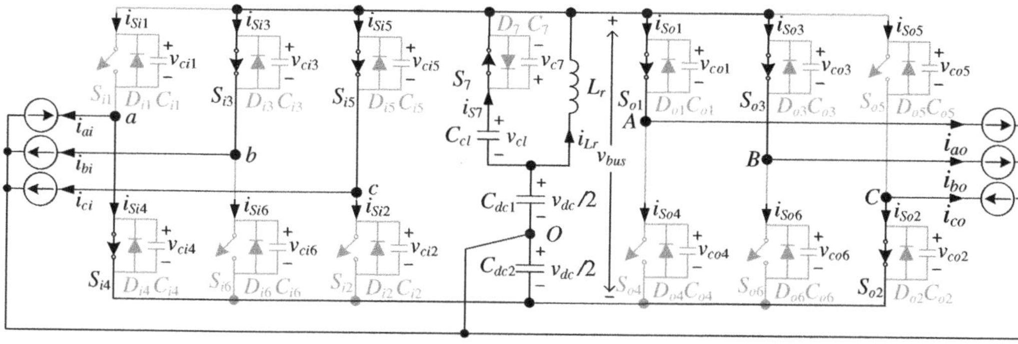

Figure 12.11 Stage 7 (t_6–t_7).

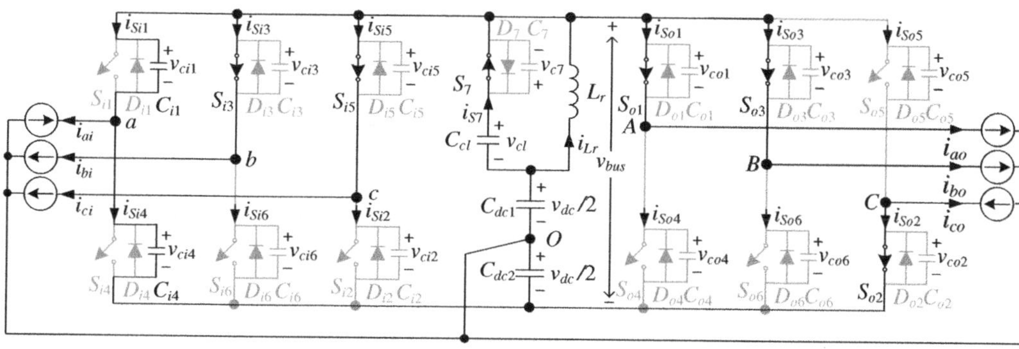

Figure 12.12 Stage 8 (t_7–t_8).

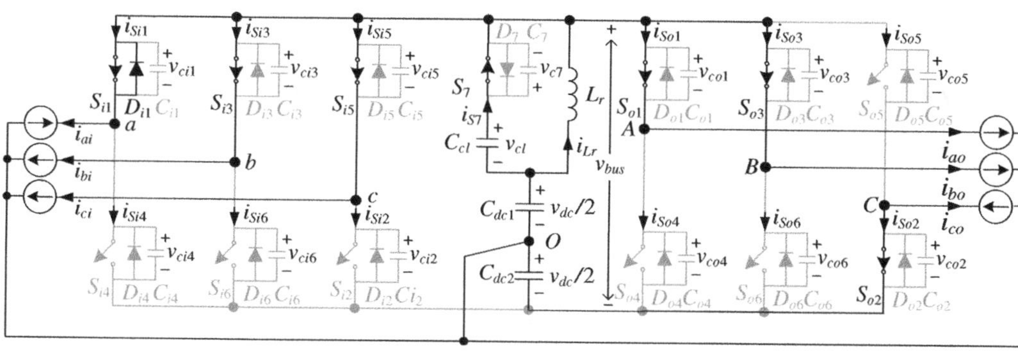

Figure 12.13 Stage 9 (t_8–t_9).

Stage 4 (t_3–t_4) is the Type 2 commutation stage. The currents i_{ai}, i_{bi}, i_{ci}, and i_{ao}, i_{bo}, i_{co} will be commutated to S_{i4}, S_{i3}, S_{i5}, and S_{o1}, S_{o3}, S_{o2} from D_{i1}, D_{i6}, D_{i2} and D_{o4}, D_{o6}, D_{o5} with one phase by one phase. Due to the existence of resonant inductor L_r, the diode reverse recovery processes are suppressed. Meanwhile, v_{Lr} is still clamped to V_{dc}. At the end of this stage, the commutations of the diodes are finished.

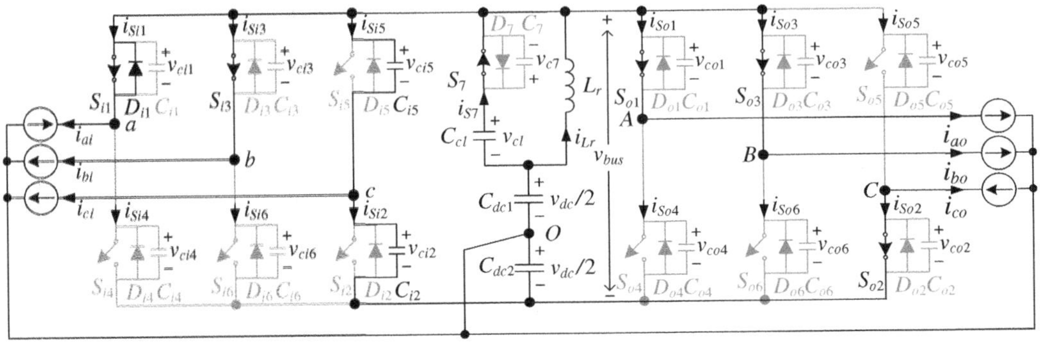

Figure 12.14 Stage 10 (t_9–t_{10}).

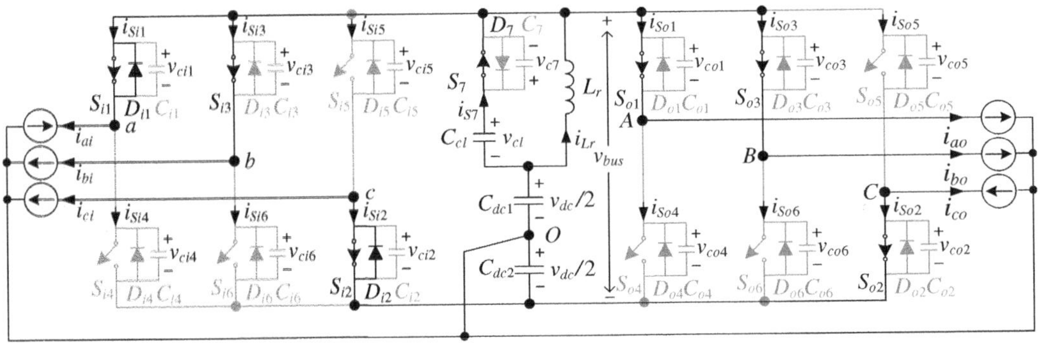

Figure 12.15 Stage 11 (t_{10}–t_{11}).

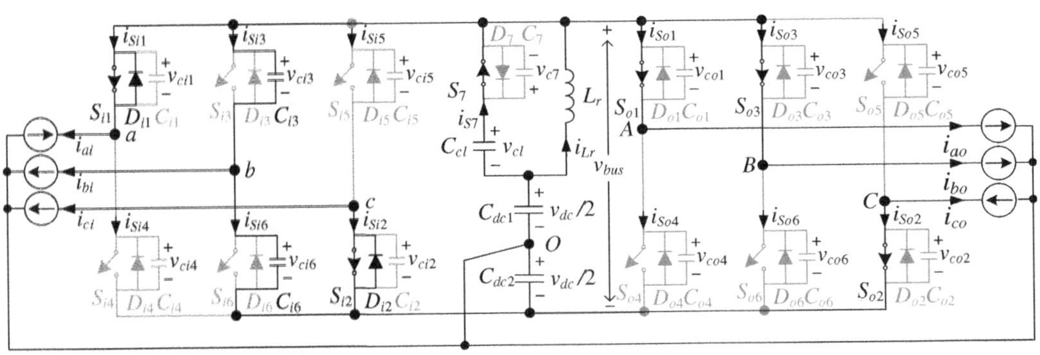

Figure 12.16 Stage 12 (t_{11}–t_{12}).

Stage 5 (t_4–t_5) is the short-circuit stage. After the commutations of the diodes are completed, all main switches will be turned on by a short-circuit signal v_{gst}, and v_{Lr} will continuously be clamped to V_{dc} for energy storage of the resonant inductor. During this stage, an extra current i_x is added to the resonant inductor current i_{Lr}, which is critical to the achievement of the ZVS condition.

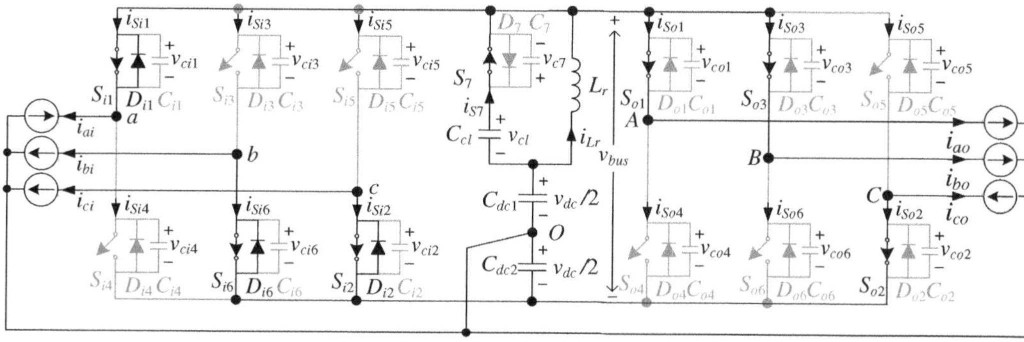

Figure 12.17 Stage 13 (t_{12}–t_{13}).

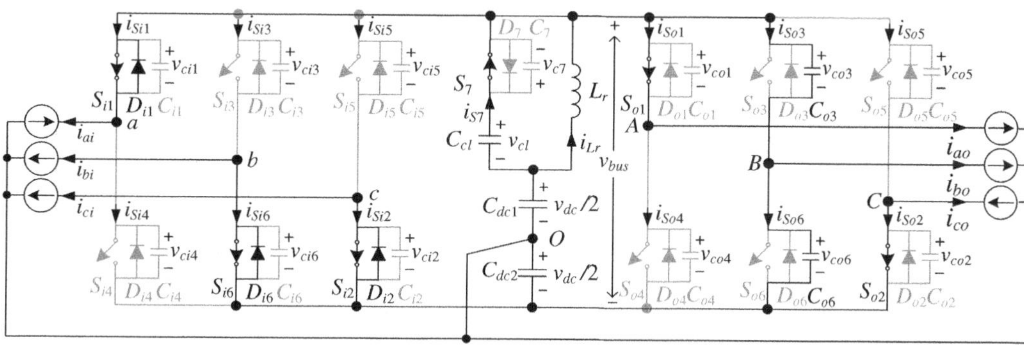

Figure 12.18 Stage 14 (t_{13}–t_{14}).

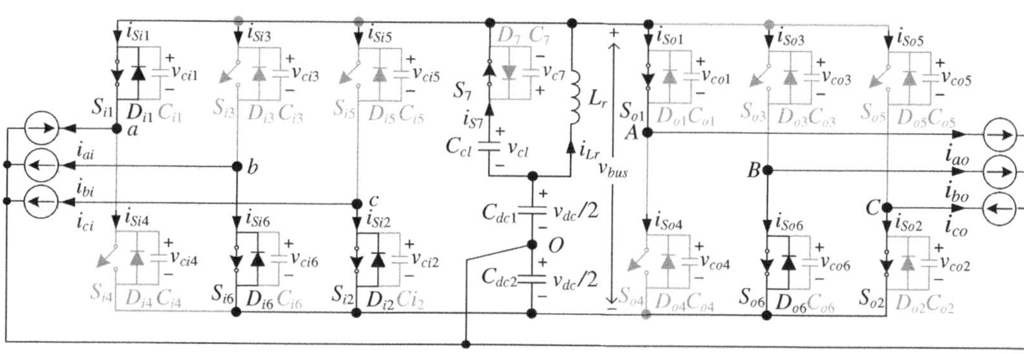

Figure 12.19 Stage 15 (t_{14}–t_{15}).

Stage 6 (t_5–t_6) is the second resonant stage. After S_{i1}, S_{i6}, S_{i2} and S_{o4}, S_{o6}, S_{o5} are turned off, the second resonance happens. At the end of this stage, the voltage across auxiliary switch v_{c7} is resonated to zero. At the same time, the voltages of the main switches (v_{ci1}, v_{ci6}, v_{ci2} and v_{co4}, v_{co6}, v_{co5}) are resonated back to $V_{cl} + V_{dc}$. Then the auxiliary switch S_7 is turned on with ZVS at t_6.

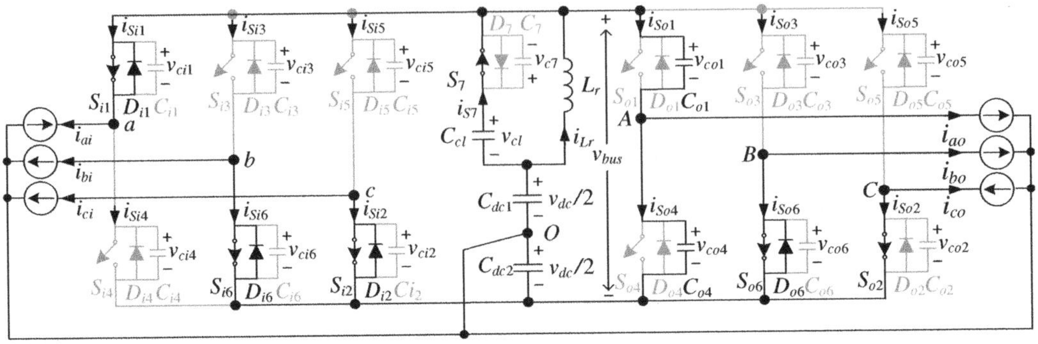

Figure 12.20 Stage 16 (t_{15}–t_{16}).

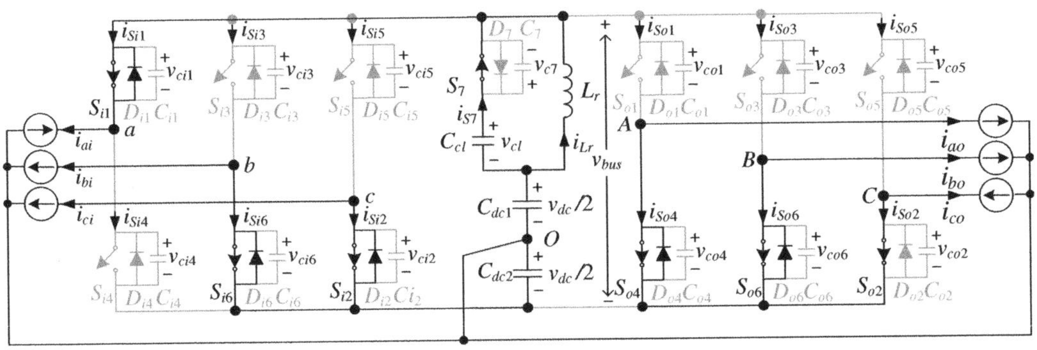

Figure 12.21 Stage 17 (t_{16}–t_{17}).

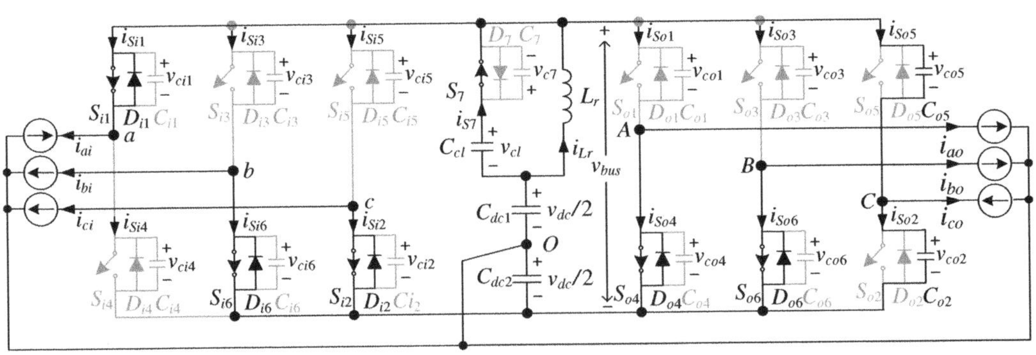

Figure 12.22 Stage 18 (t_{17}–t_{18}).

Stage 7 (t_6–t_7) is the second steady stage. S_{i4}, S_{i3}, S_{i5} and S_{o1}, S_{o3}, S_{o2} are conducting. This stage ends when S_{i4} is turned off at t_7.

Stage 8 (t_7–t_8) is the Type 1 commutation stage in the input A phase. S_{i4} is turned off and the filter inductor current i_{ai} begins to commutate. The resonant capacitor C_{i4} is charging while the resonant

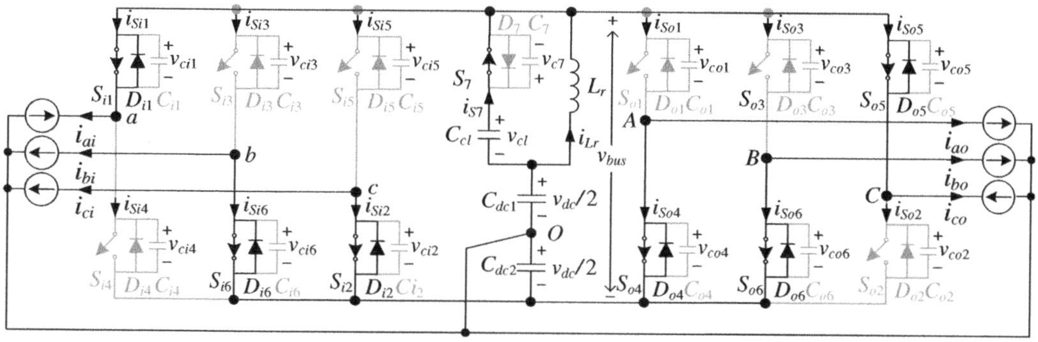

Figure 12.23 Stage 19 (t_{18}–t_{19}).

capacitor C_{i1} is discharging by i_{ai}. At t_8, the resonant capacitor voltage v_{ci4} increases to $V_{cl} + V_{dc}$ and v_{ci1} decreases to zero. When D_{i1} begins to conduct, the Type 1 commutation of the input A phase is over.

Stage 9 (t_8–t_9) is the third steady stage. S_{i1} is turned on and the ZVS turn-on of the main switch S_{i1} is realized because D_{i1} is already conducting.

Stage 10 (t_9–t_{10}) is the Type 1 commutation stage in the input C phase; S_{i5} is turned off, and the filter inductor current i_{ci} begins to commutate. The resonant capacitor C_{i5} is charging while the resonant capacitor C_{i2} is discharging by i_{ci}. At t_{10}, the resonant capacitor voltage v_{ci5} increases to $V_{cl} + V_{dc}$ and v_{ci2} decreases to zero. When D_{i2} begins to conduct, the Type 1 commutation of the input C phase is over.

Stage 11 (t_{10}–t_{11}) is the fourth steady stage. S_{i2} is turned on, and the ZVS turn-on of the main switch S_{i2} is realized because D_{i2} is already conducting.

Stage 12 (t_{11}–t_{12}) is the Type 1 commutation stage in the input B phase, S_{i3} is turned off, and the filter inductor current i_{bi} begins to commutate. The resonant capacitor C_{i3} is charging while the resonant capacitor C_{i6} is discharging by i_{bi}. At t_{12}, the resonant capacitor voltage v_{ci3} increases to $V_{cl} + V_{dc}$ and v_{ci6} decreases to zero. When D_{i6} begins to conduct, the Type 1 commutation of the input B phase is over.

Stage 13 (t_{12}–t_{13}) is the fifth steady stage. S_{i6} is turned on and the ZVS turn-on of the main switch S_{i6} is realized because D_{i6} is already conducting.

Stage 14 (t_{13}–t_{14}) is the Type 1 commutation stage in the output B phase. S_{o3} is turned off and the filter inductor current i_{bo} begins to commute. The resonant capacitor C_{o3} is charging, while the resonant capacitor C_{o6} is discharging by i_{bo}. At t_{14}, the resonant capacitor voltage v_{co3} increases to $V_{cl} + V_{dc}$ and v_{co6} decreases to zero. When D_{o6} begins to conduct, the Type 1 commutation of the output B phase is over.

Stage 15 (t_{14}–t_{15}) is the sixth steady stage, S_{o6} is turned on, and the ZVS turn-on of the main switch S_{o6} is realized because D_{o6} is already conducting.

Stage 16 (t_{15}–t_{16}) is the Type 1 commutation stage in the output A phase. S_{o1} is turned off and the filter inductor current i_{ao} begins to commutate. The resonant capacitor C_{o1} is charging, while the resonant capacitor C_{o4} is discharging by i_{ao}. At t_{16}, the resonant capacitor voltage v_{co1} increases to $V_{cl} + V_{dc}$ and v_{co4} decreases to zero. When D_{o4} begins to conduct, the Type 1 commutation of the output A phase is over.

Stage 17 (t_{16}–t_{17}) is the seventh steady stage. S_{o4} is turned on and the ZVS turn-on of the main switch S_{o4} is realized because D_{o4} is already conducting.

Stage 18 (t_{17}–t_{18}) is the Type 1 commutation stage in the output C phase. S_{o2} is turned off, and the filter inductor current i_{co} begins to commutate. The resonant capacitor C_{o2} is charging, while the resonant capacitor C_{o5} is discharging by i_{co}. At t_{18}, the resonant capacitor voltage v_{co2} increases to $V_{cl} + V_{dc}$ and v_{co5} decreases to zero. When D_{o5} begins to conduct, the Type 1 commutation of the output C phase is over.

Stage 19 (t_{18}–t_{19}) is the eleventh steady stage. S_{o5} is turned on, and the ZVS turn-on of the main switch S_{o5} is realized because the D_{o5} is already conducting. After t_{19}, the circuit state will go back to stage 1.

12.3 Circuit Analysis

Based on the aforementioned stage analysis, the first resonant stage (stage 2: t_1–t_2) will bring down the voltage across the main switches to zero by turning off the auxiliary switch S_7 to realize the ZVS turn-on of the main switches. After $\lambda_7 T_s$, the second resonant stage (stage 6: t_5–t_6) will resonate the voltage across the auxiliary switch to zero to realize the ZVS turn-on of the auxiliary switch. Therefore, the ZVS conditions for the main switches and the auxiliary switch are determined by these two resonant stages.

A short-circuit stage is designed to achieve the ZVS condition. All main switches are turned on, and the resonant inductor voltage v_{Lr} continues to be clamped to V_{dc}. An extra current i_x is added, which is critical under an unbalanced load condition. A larger i_x will cause extra power loss, while a smaller i_x may not be able to realize the ZVS condition.

12.3.1 Resonant Stage Analysis

The first resonant stage shown in Figure 12.24 can be simplified by the following steps:

Step I: In Figure 12.24, all switches and diodes in conducting state can be replaced by conducting wires as shown in Figure 12.25a.

Step II: In Figure 12.25a, the current sources i_{bi}, i_{ci}, i_{ao}, and i_{bo} are eliminated since they are in parallel with a voltage source. Therefore, Figure 12.25a is simplified to Figure 12.25b.

Step III: As shown in Figure 12.25c, the current sources i_{ai} and i_{co} are split into four sources i_{ai1}, i_{ai2}, i_{co1}, and i_{co2} ($i_{ai1} = i_{ai2} = i_{ai}$, $i_{co1} = i_{co2} = i_{co}$). Although the connection of the circuit is changed, Kirchhoff's current law (KCL) and Kirchhoff's voltage law (KVL) are still followed.

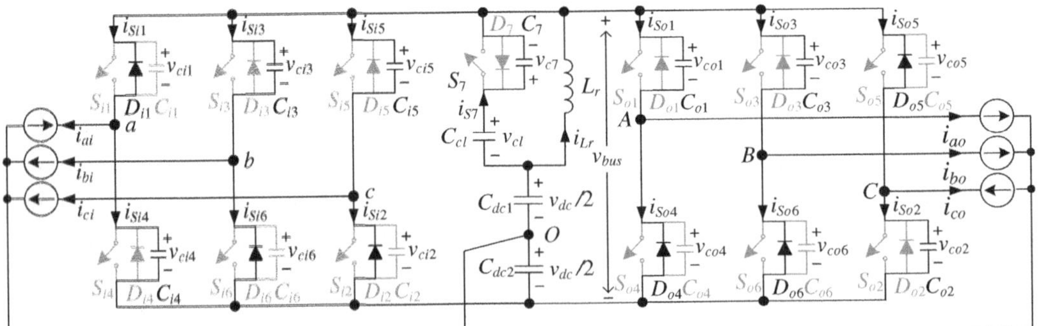

Figure 12.24 First resonant stage.

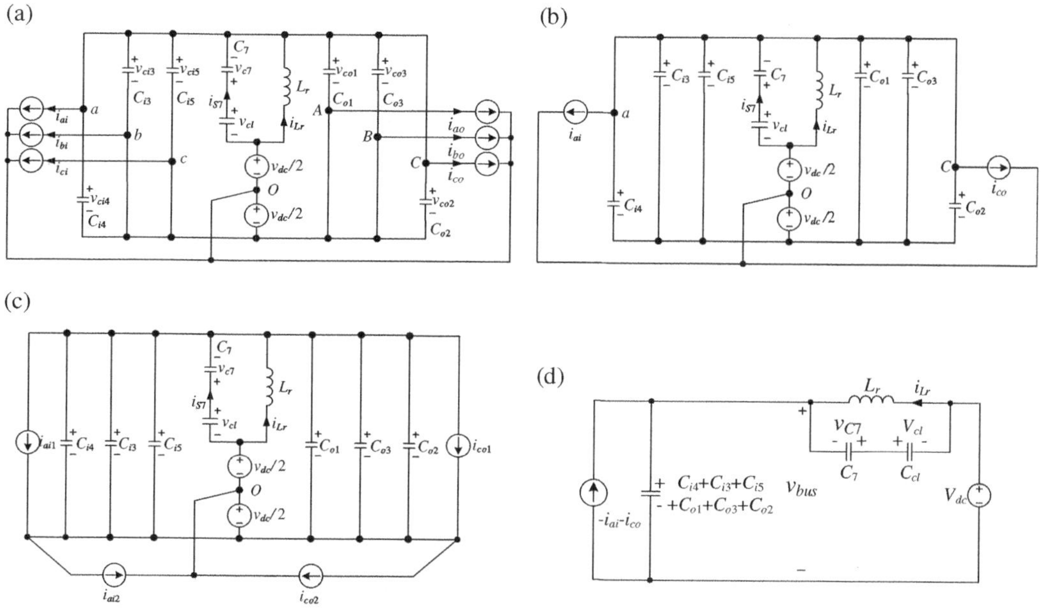

Figure 12.25 Simplified process of the first resonant stage: (a) Step I; (b) Step II; (c) Step III; (d) Step IV.

Step IV: As shown in Figure 12.25d, the current sources i_{ai2} and i_{co2} are eliminated since they are in parallel with a voltage source. The resonant capacitors C_{i4}, C_{i3}, C_{i5}, C_{o1}, C_{o3}, and C_{o2} and current sources i_{ai} and i_{co} are combined into one capacitor and one current source, respectively.

According to the equivalent circuit shown in Figure 12.25d, the following state equation is obtained:

$$\begin{cases} (6C_r + C_7)\dfrac{dv_{bus}(t)}{dt} - i_{Lr}(t) = -i_{ai} - i_{co} \\ v_{bus}(t) + L_r\dfrac{di_{Lr}(t)}{dt} = V_{dc} \end{cases} \tag{12.8}$$

The initial condition for equation (12.8) is:

$$\begin{cases} v_{bus}(t_1) = V_{dc} + V_{cl} \\ i_{Lr}(t_1) = i_{Lr}(t_1) \end{cases} \tag{12.9}$$

Therefore, the expression for v_{bus}(t) and i_{Lr}(t) can be obtained by solving the state equation:

$$\begin{cases} v_{bus}(t) = V_{dc} + \sqrt{V_{cl}^2 + (i_{Lr}(t_1) - i_{ai} - i_{co})^2 Z_r^2}\sin(\omega_r(t-t_1) - \varphi_1) \\ i_{Lr}(t) = \sqrt{(i_{Lr}(t_1) - i_{ai} - i_{co})^2 + \dfrac{V_{cl}^2}{Z_r^2}}\cos(\omega_r(t-t_1) - \varphi_1) + i_{ai} + i_{co} \end{cases} \tag{12.10}$$

where ω_r is the resonance angular frequency, Z_r is the characteristic impedance, φ_1 is the initial angle of the first resonance, which are represented as:

$$\begin{cases} \omega_r = 1/\sqrt{L_r(6C_r + C_{r7})} \\ Z_r = \sqrt{L_r/(6C_r + C_{r7})} \\ \varphi_1 = \arctan\left[V_{cl}/Z_r(-i_{Lr}(t_1) + i_{ai} + i_{co})\right] \end{cases} \tag{12.11}$$

To realize the ZVS condition for main switches, $v_{bus}(t)$ should be resonated to zero. Thus, the following inequality should be satisfied:

$$V_{dc} \leq \sqrt{V_{cl}^2 + (i_{Lr}(t_1) - i_{ai} - i_{co})^2 Z_r^2} \tag{12.12}$$

Then the initial resonant inductor current $i_{Lr}(t_1)$ for ZVS condition of main switches should meet:

$$i_{Lr}(t_1) \leq i_{ai} + i_{co} - \frac{\sqrt{V_{dc}{}^2 - V_{cl}{}^2}}{Z_r} \tag{12.13}$$

The second resonant stage is shown in Figure 12.26, which can be also simplified as shown in Figure 12.27. The following state equation is obtained:

$$\begin{cases} (6C_r + C_7)\dfrac{dv_{c7}(t)}{dt} + i_{Lr}(t) = i_{bi} + i_{ci} + i_{ao} + i_{bo} \\ v_{C7}(t) - L_r\dfrac{di_{Lr}(t)}{dt} = V_{cl} \end{cases} \tag{12.14}$$

The initial condition for equation (12.14) is:

$$\begin{cases} i_{Lr}(t_5) = i_{bi} + i_{ci} + i_{ao} + i_{bo} + i_x \\ v_{c7}(t_5) = V_{dc} + V_{cl} \end{cases} \tag{12.15}$$

where i_x is the extra current added to the resonant inductor current i_{Lr} in stage 5 (t_4–t_5), and i_x is always larger than or equal to zero.

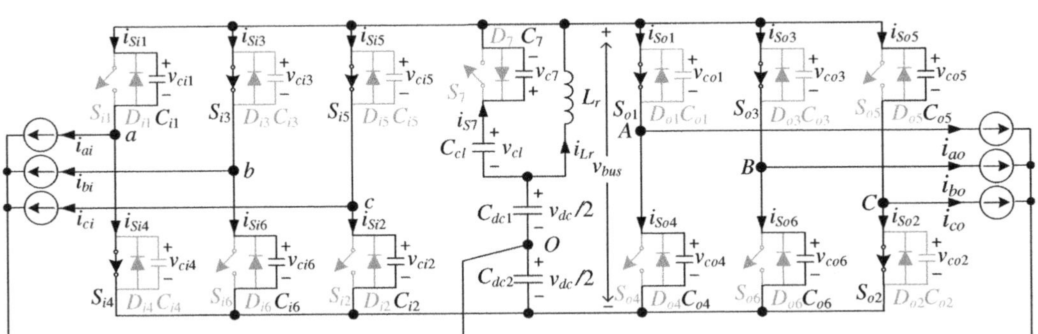

Figure 12.26 Second resonant stage.

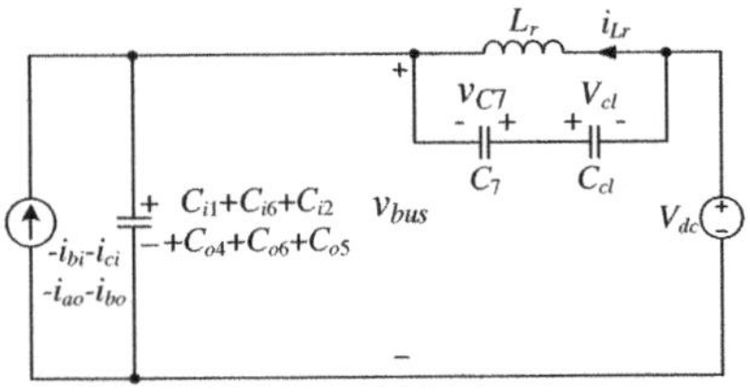

Figure 12.27 Equivalent circuit of the second resonant stage.

By solving equation (12.14), the expression for $v_{c7}(t)$ and $i_{Lr}(t)$ are obtained:

$$\begin{cases} v_{c7}(t) = V_{cl} - \sqrt{V_{dc}^2 + Z_r^2 i_x^2}\sin(\omega_r(t-t_5)-\varphi_2) \\ i_{Lr}(t) = i_{bi} + i_{ci} + i_{ao} + i_{bo} + \sqrt{i_x^2 + \dfrac{V_{dc}^2}{Z_r^2}}\cos(\omega_r(t-t_5)-\varphi_2) \end{cases} \tag{12.16}$$

where φ_2 is the initial angle of the second resonance, it can be expressed as:

$$\varphi_2 = \arctan\frac{V_{dc}}{Z_r i_x} \tag{12.17}$$

To realize the ZVS condition for auxiliary switch S_7, $v_{c7}(t)$ should be resonated to zero. So the following inequality should be satisfied:

$$V_{cl} \le \sqrt{V_{dc}^2 + Z_r^2 i_x^2} \tag{12.18}$$

As clamping capacitor voltage V_{cl} is far less than DC voltage V_{dc} by proper resonance parameter design, the above ZVS condition can always be satisfied. The resonant inductor current $i_{Lr}(t_6)$ can be calculated with $v_{c7}(t_6) = 0$ according to equation (12.16):

$$i_{Lr}(t_6) = i_{bi} + i_{ci} + i_{ao} + i_{bo} + \sqrt{\frac{{V_{dc}}^2 - {V_{cl}}^2}{Z_r^2} + i_x^2} \tag{12.19}$$

The above analysis shows that the second resonance can easily bring down the voltage across the auxiliary switch S_7 to zero and provide the ZVS condition for S_7. However, whether the first resonance can resonate the voltage across the main switches to zero is determined by the condition in inequality (12.13), which depends on the operation condition of the converter.

12.3.2 Steady State Analysis

When the circuit works in a steady state, the resonant inductor L_r satisfies the voltage-second balance principle. The following equation is obtained:

$$\int_0^{T_s} v_{Lr}(t)dt = \int_{t_1}^{t_2} v_{Lr}(t)dt + \int_{t_2}^{t_5} v_{Lr}(t)dt + \int_{t_5}^{t_6} v_{Lr}(t)dt + \int_{t_6}^{t_{19}} v_{Lr}(t)dt = 0 \tag{12.20}$$

Because the resonant time is relatively short, stage 2 (t_1–t_2) and stage 6 (t_5–t_6) can be ignored. The resonant inductor voltage $v_{Lr}(t)$ is equal to $-V_{dc}$ from t_2 to t_5, while it is equal to V_{cl} from t_6 to t_{19}. The equation (12.20) is simplified as:

$$\int_0^{T_s} v_{Lr}(t)dt \approx V_{cl}(t_{19}-t_6) - V_{dc}(t_5-t_2) = 0 \tag{12.21}$$

It is assumed that $t_5 - t_2 = \lambda_7 T$ and $t_{19} - t_6 = (1-\lambda_7)\,T_s$, and the relations between the clamping capacitor voltage V_{cl} and DC voltage V_{dc} is derived:

$$V_{cl} = \frac{\lambda_7}{1-\lambda_7}V_{dc} \tag{12.22}$$

According to the ampere-second balance principle of the clamping capacitor C_{cl}, the clamping capacitor current $i_{cl}(t)$ should obey:

$$\frac{1}{T_s}\int_0^{T_s} i_{cl}(t)dt = 0 \tag{12.23}$$

Since the auxiliary switch S_7 is in series with clamping capacitor C_{cl} and S_7 is turned off during $\lambda_7 T_s$, the $i_{cl}(t)$ should be equal to 0 and the equation (12.23) is rewritten as:

$$\frac{1}{T_s}\int_0^{T_s} i_{cl}(t)dt = \frac{1}{T_s}\int_0^{\lambda_7 T_s} i_{cl}(t)dt + \frac{1}{T_s}\int_{\lambda_7 T_s}^{T_s} i_{cl}(t)dt \approx \frac{1}{T_s}\int_{\lambda_7 T_s}^{T_s} i_{cl}(t)dt \tag{12.24}$$

Combining equations (12.23) and (12.24), the following equation is obtained:

$$\int_{\lambda_7 T_s}^{T_s} i_{cl}(t)dt = 0 \tag{12.25}$$

As shown in Figure 12.1b, the following equation is established according to the KCL at point P:

$$i_{cl}(t) = i_{Lr}(t) + i_{pi}(t) - i_{po}(t) \tag{12.26}$$

where $i_{pi}(t)$ is the current flowing from the rectifier side to the DC side and $i_{po}(t)$ is the current flowing from the DC side to the inverter side.

By substituting equation (12.26) into (12.25), the following equation is obtained:

$$\int_{\lambda_7 T_s}^{T_s} i_{Lr}(t)dt + \int_{\lambda_7 T_s}^{T_s} i_{pi}(t)dt - \int_{\lambda_7 T_s}^{T_s} i_{po}(t)dt = 0 \tag{12.27}$$

When the auxiliary switch S_7 is turned on, the resonant inductor voltage is clamped by the clamping capacitor voltage V_{cl} and the resonant inductor current during $[\lambda_7 T_s, T_s]$ should satisfy:

$$i_{Lr}(t) = i_{Lr}(t_6) - \frac{V_{cl}}{L_r}(t - t_6) \tag{12.28}$$

where $t_6 = \lambda_7 T_s$ is the turn-on instant of the auxiliary switch S_7.

Therefore, the integral of $i_{Lr}(t)$ is derived:

$$\int_{\lambda_7 T_s}^{T_s} i_{Lr}(t)dt = (1-\lambda_7)T_s i_{Lr}(t_6) - \frac{V_{cl}}{2L_r}(1-\lambda_7)^2 T_s^2 \tag{12.29}$$

The bridge currents $i_{pi}(t)$ and $i_{po}(t)$ can be expressed with the state of the switch leg:

$$\begin{cases} i_{pi}(t) = -S_{ai}(t)i_{ai} - S_{bi}(t)i_{bi} - S_{ci}(t)i_{ci} \\ i_{po}(t) = S_{ao}(t)i_{ao} + S_{bo}(t)i_{bo} + S_{co}(t)i_{co} \end{cases} \tag{12.30}$$

where S_{ai}, S_{bi}, and S_{ci} and S_{ao}, S_{bo}, and S_{co} are the states of the corresponding switch leg, which are expressed as follows according to Figure 12.4:

$$S_{ai}(t) = \begin{cases} 1 & \text{if } (1-d_{ai})T_s \le t \le T_s \\ 0 & \text{if } 0 \le t \le (1-d_{ai})T_s \end{cases} \quad S_{ao}(t) = \begin{cases} 1 & \text{if } 0 \le t \le d_{ao}T_s \\ 0 & \text{if } d_{ao}T_s \le t \le T_s \end{cases}$$
$$S_{bi}(t) = \begin{cases} 1 & \text{if } 0 \le t \le d_{bi}T_s \\ 0 & \text{if } d_{bi}T_s \le t \le T_s \end{cases} \quad S_{bo}(t) = \begin{cases} 1 & \text{if } 0 \le t \le d_{bo}T_s \\ 0 & \text{if } d_{bo}T_s \le t \le T_s \end{cases}$$
$$S_{ci}(t) = \begin{cases} 1 & \text{if } 0 \le t \le d_{ci}T_s \\ 0 & \text{if } d_{ci}T_s \le t \le T_s \end{cases} \quad S_{co}(t) = \begin{cases} 1 & \text{if } (1-d_{co})T_s \le t \le T_s \\ 0 & \text{if } 0 \le t \le (1-d_{co})T_s \end{cases} \tag{12.31}$$

Combining equations (12.27)–(12.31), the integrals of $i_{pi}(t)$ and $i_{po}(t)$ are derived as:

$$\begin{cases} \int_{\lambda_7 T_s}^{T_s} i_{pi}(t)dt = [-d_{ai}i_{ai} - (d_{bi} - \lambda_7)i_{bi} - (d_{ci} - \lambda_7)i_{ci}]T_s \\ \int_{\lambda_7 T_s}^{T_s} i_{po}(t)dt = [(d_{ao} - \lambda_7)i_{ao} + (d_{bo} - \lambda_7)i_{bo} + d_{co}(t)i_{co}]T_s \end{cases} \tag{12.32}$$

According to the content of the duty cycle loss described in Section 3.4.2.3 of Chapter 3, there is no duty cycle loss in the input A phase and output C phase, but there is duty cycle loss in the input B phase, input C phase, output A phase, and output B phase. Therefore, the switching cycle average of the midpoint voltages v_{aN}, v_{bN}, v_{cN} and v_{AN}, v_{BN}, v_{CN} are represented by

$$\begin{cases} \langle v_{aN}\rangle_{T_s} = (V_{dc} + V_{cl})d_{ai} = v_{ag}' + \dfrac{V_{dc}}{2} \\ \langle v_{bN}\rangle_{T_s} = (V_{dc} + V_{cl})(d_{bi} - \lambda_7) = v_{bg}' + \dfrac{V_{dc}}{2} \\ \langle v_{cN}\rangle_{T_s} = (V_{dc} + V_{cl})(d_{ci} - \lambda_7) = v_{cg}' + \dfrac{V_{dc}}{2} \end{cases} \quad \begin{cases} \langle v_{AN}\rangle_{T_s} = (V_{dc} + V_{cl})(d_{ao} - \lambda_7) = v_{ao}' + \dfrac{V_{dc}}{2} \\ \langle v_{BN}\rangle_{T_s} = (V_{dc} + V_{cl})(d_{bo} - \lambda_7) = v_{bo}' + \dfrac{V_{dc}}{2} \\ \langle v_{CN}\rangle_{T_s} = (V_{dc} + V_{cl})d_{co} = v_{co}' + \dfrac{V_{dc}}{2} \end{cases} \tag{12.33}$$

where

$$\begin{cases} v_{ag}' = \dfrac{V_{dc}}{2}M\cos\omega t \\ v_{bg}' = \dfrac{V_{dc}}{2}M\cos(\omega t - 2\pi/3) \\ v_{cg}' = \dfrac{V_{dc}}{2}M\cos(\omega t + 2\pi/3) \end{cases} \quad \begin{cases} v_{ao}' = \dfrac{V_{dc}}{2}M\cos(\omega t + \varphi + \theta) \\ v_{bo}' = \dfrac{V_{dc}}{2}M\cos(\omega t + \varphi + \theta - 2\pi/3) \\ v_{co}' = \dfrac{V_{dc}}{2}M\cos(\omega t + \varphi + \theta + 2\pi/3) \end{cases} \tag{12.34}$$

In the above equation, M, the modulation ratio of the rectifier and inverter, is defined as equal to $2U_m/V_{dc}$ for simplicity.

The duty cycles of every switch leg are solved based on the switching cycle average of the midpoint voltage and equation (12.22):

$$\begin{cases} d_{ai} = \dfrac{v_{ag}' + \dfrac{V_{dc}}{2}}{V_{dc}}(1 - \lambda_7) \\ d_{bi} = \dfrac{v_{bg}' + \dfrac{V_{dc}}{2}}{V_{dc}}(1 - \lambda_7) + \lambda_7 \\ d_{ci} = \dfrac{v_{cg}' + \dfrac{V_{dc}}{2}}{V_{dc}}(1 - \lambda_7) + \lambda_7 \end{cases} \quad \begin{cases} d_{ao} = \dfrac{v_{ao}' + \dfrac{V_{dc}}{2}}{V_{dc}}(1 - \lambda_7) + \lambda_7 \\ d_{bo} = \dfrac{v_{bo}' + \dfrac{V_{dc}}{2}}{V_{dc}}(1 - \lambda_7) + \lambda_7 \\ d_{co} = \dfrac{v_{co}' + \dfrac{V_{dc}}{2}}{V_{dc}}(1 - \lambda_7) \end{cases} \tag{12.35}$$

By substituting equation (12.35) into (12.32), the integrals of $i_{pi}(t)$ and $i_{po}(t)$ are rewritten as:

$$\begin{cases} \int_{\lambda_7 T_s}^{T_s} i_{pi}(t)dt = (1 - \lambda_7)T_s\left(-\dfrac{v_{ag}'i_{ai}}{V_{dc}} - \dfrac{v_{bg}'i_{bi}}{V_{dc}} - \dfrac{v_{cg}'i_{ci}}{V_{dc}} - \dfrac{i_{ai}}{2} - \dfrac{i_{bi}}{2} - \dfrac{i_{ci}}{2}\right) \\ \int_{\lambda_7 T_s}^{T_s} i_{po}(t)dt = (1 - \lambda_7)T_s\left(\dfrac{v_{ao}'i_{ao}}{V_{dc}} + \dfrac{v_{bo}'i_{bo}}{V_{dc}} + \dfrac{v_{co}'i_{co}}{V_{dc}} + \dfrac{i_{ao}}{2} + \dfrac{i_{bo}}{2} + \dfrac{i_{co}}{2}\right) \end{cases} \tag{12.36}$$

By combining equations (12.27), (12.29), and (12.36), the following equation is obtained:

$$i_{Lr}(t_6) - \frac{V_{dc}T_s}{2L_r}\lambda_7 = \left(\frac{v_{ag}{}'i_{ai}}{V_{dc}} + \frac{v_{bg}{}'i_{bi}}{V_{dc}} + \frac{v_{cg}{}'i_{ci}}{V_{dc}} + \frac{v_{ao}{}'i_{ao}}{V_{dc}} + \frac{v_{bo}{}'i_{bo}}{V_{dc}} + \frac{v_{co}{}'i_{co}}{V_{dc}} + \frac{i_{ai}}{2} + \frac{i_{bi}}{2} + \frac{i_{ci}}{2} + \frac{i_{ao}}{2} + \frac{i_{bo}}{2} + \frac{i_{co}}{2}\right) \tag{12.37}$$

The turn-off duty cycle λ_7 is solved by substituting equation (12.19) into (12.37):

$$\lambda_7 = \frac{2L_r\left(i_M + \frac{1}{2}(-i_{ai} - i_{co} + i_{ao} + i_{bo} + i_{bi} + i_{ci}) + \sqrt{\frac{V_{dc}{}^2 - V_{cl}{}^2}{Z_r^2} + i_x^2}\right)}{V_{dc}T_s} \tag{12.38}$$

where

$$i_M = -\frac{1}{V_{dc}}\left(v_{ag}{}'i_{ai} + v_{bg}{}'i_{bi} + v_{cg}{}'i_{ci} + v_{ao}{}'i_{ao} + v_{bo}{}'i_{bo} + v_{co}{}'i_{co}\right) \tag{12.39}$$

Before we derive clamping capacitor current i_{cl}. For simplicity of the analysis, i_{cl} during (t_1,t_6) is approximated to be zero by neglecting the resonance process. All Type 1 commutations are ignored. According to Figure 12.11, the clamping capacitor current i_{cl} at t_6 can be derived as:

$$i_{cl}(t_6) = -i_{S7}(t_6) = i_{Lr}(t_6) - i_{bi} - i_{ci} - i_{ao} - i_{bo} \tag{12.40}$$

By substituting equation (12.19) into the above equations, the following equation is obtained:

$$i_{cl}(t_6) = \sqrt{\frac{V_{dc}{}^2 - V_{cl}{}^2}{Z_r^2} + i_x^2} \tag{12.41}$$

During stage 7 (t_6–t_7), the resonant inductor is charged by the clamping capacitor. Therefore, i_{cl} in stage 7 is expressed as:

$$i_{cl}(t) = \sqrt{\frac{V_{dc}{}^2 - V_{cl}{}^2}{Z_r^2} + i_x^2} - \frac{V_{cl}}{L_r}(t - t_6) \qquad t_6 \le t \le t_7 \tag{12.42}$$

After short Type 1 commutation stage 8 (t_7–t_8), the switch S_{i4} is turned off in the input phase-A switch leg and the current source i_{ai} flows into the upper switch in the same switch leg, which causes i_{cl} to suddenly step down by $-i_{ai}$ at t_8. In the same stage, the resonant inductor L_r is charged by the clamping capacitor C_{cl}. Therefore, i_{cl} in stage 9 (t_8–t_9) is expressed as:

$$i_{cl}(t) = \sqrt{\frac{V_{dc}{}^2 - V_{cl}{}^2}{Z_r^2} + i_x^2} - \frac{V_{cl}}{L_r}(t - t_6) - i_{ai} \qquad t_8 \le t \le t_9 \tag{12.43}$$

After short Type 1 commutation stage 10 (t_9–t_{10}), the switch S_{i5} is turned off in the input phase-C switch leg and the current source i_{ci} flows into the lower switch in the same switch leg, which causes i_{cl} to suddenly step up by i_{ci} at t_{10}. During stage 11 (t_{10}–t_{11}), the resonant inductor L_r is still charged by the clamping capacitor C_{cl}. Therefore, i_{cl} in stage 11 (t_{10}–t_{11}) is expressed as:

$$i_{cl}(t) = \sqrt{\frac{V_{dc}{}^2 - V_{cl}{}^2}{Z_r^2} + i_x^2} - \frac{V_{cl}}{L_r}(t - t_6) - i_{ai} + i_{ci} \qquad t_{10} \le t \le t_{11} \tag{12.44}$$

After short Type 1 commutation stage 12 (t_{11}–t_{12}), the switch S_{i3} is turned off in the input phase-B switch leg and the current source i_{bi} flows into the lower switch in the same switch leg. It causes i_{cl} to suddenly step up by i_{bi} at t_{12}. During stage 13 (t_{12}–t_{13}), the resonant inductor L_r is still charged by the clamping capacitor C_{cl}. Therefore, i_{cl} in stage 13 (t_{12}–t_{13}) is expressed as:

$$i_{cl}(t) = \sqrt{\frac{{V_{dc}}^2 - {V_{cl}}^2}{Z_r^2} + i_x^2} - \frac{V_{cl}}{L_r}(t - t_6) - i_{ai} + i_{bi} + i_{ci} \qquad t_{12} \le t \le t_{13} \tag{12.45}$$

After short Type 1 commutation stage 14 (t_{13}–t_{14}), the switch S_{o3} is turned off in the output phase-B switch leg and the current source i_{bo} flows into the lower switch in the same switch leg. It causes i_{cl} to suddenly step up by i_{bo} at t_{14}. During stage 15 (t_{14}–t_{15}), the resonant inductor L_r is still charged by the clamping capacitor C_{cl}. Therefore, i_{cl} in stage 15 (t_{14}–t_{15}) is expressed as:

$$i_{cl}(t) = \sqrt{\frac{{V_{dc}}^2 - {V_{cl}}^2}{Z_r^2} + i_x^2} - \frac{V_{cl}}{L_r}(t - t_6) - i_{ai} + i_{bi} + i_{ci} + i_{bo} \qquad t_{14} \le t \le t_{15} \tag{12.46}$$

After short Type 1 commutation stage 16 (t_{15}–t_{16}), the switch S_{o1} is turned off in the output phase-A switch leg and the current source i_{ao} flows into the lower switch in the same switch leg. It causes i_{cl} to suddenly step up by i_{ao} at t_{16}. During stage 17 (t_{16}–t_{17}), the resonant inductor L_r is still charged by the clamping capacitor C_{cl}. Therefore, i_{cl} in stage 17 (t_{16}–t_{17}) is expressed as:

$$i_{cl}(t) = \sqrt{\frac{{V_{dc}}^2 - {V_{cl}}^2}{Z_r^2} + i_x^2} - \frac{V_{cl}}{L_r}(t - t_6) - i_{ai} + i_{bi} + i_{ci} + i_{ao} + i_{bo} \qquad t_{16} \le t \le t_{17} \tag{12.47}$$

After short Type 1 commutation stage 18 (t_{17}–t_{18}), the switch S_{o2} is turned off in the output phase-C switch leg and the current source i_{co} flows into the upper switch in the same switch leg. It causes i_{cl} to suddenly step down by $-i_{co}$ at $t_{18.}$ During stage 19 (t_{18}–t_{19}), the resonant inductor L_r is still charged by the clamping capacitor C_{cl}. Therefore, i_{cl} in stage 19 (t_{18}–t_{19}) is expressed as:

$$i_{cl}(t) = \sqrt{\frac{{V_{dc}}^2 - {V_{cl}}^2}{Z_r^2} + i_x^2} - \frac{V_{cl}}{L_r}(t - t_6) - i_{ai} + i_{bi} + i_{ci} + i_{ao} + i_{bo} - i_{co} \qquad t_{18} \le t \le t_{19} \tag{12.48}$$

By combining equations (12.42)–(12.48), i_{cl} in different stages is expressed as:

$$i_{cl}(t) = -i_{S7}(t) = \begin{cases} \sqrt{\frac{{V_{dc}}^2 - {V_{cl}}^2}{Z_r^2} + i_x^2} - \frac{V_{cl}}{L_r}(t - t_6) & t_6 \le t \le t_7 \\ \sqrt{\frac{{V_{dc}}^2 - {V_{cl}}^2}{Z_r^2} + i_x^2} - \frac{V_{cl}}{L_r}(t - t_6) - i_{ai} & t_8 \le t \le t_9 \\ \sqrt{\frac{{V_{dc}}^2 - {V_{cl}}^2}{Z_r^2} + i_x^2} - \frac{V_{cl}}{L_r}(t - t_6) - i_{ai} + i_{ci} & t_{10} \le t \le t_{11} \\ \sqrt{\frac{{V_{dc}}^2 - {V_{cl}}^2}{Z_r^2} + i_x^2} - \frac{V_{cl}}{L_r}(t - t_6) - i_{ai} + i_{bi} + i_{ci} & t_{12} \le t \le t_{13} \\ \sqrt{\frac{{V_{dc}}^2 - {V_{cl}}^2}{Z_r^2} + i_x^2} - \frac{V_{cl}}{L_r}(t - t_6) - i_{ai} + i_{bi} + i_{ci} + i_{bo} & t_{14} \le t \le t_{15} \\ \sqrt{\frac{{V_{dc}}^2 - {V_{cl}}^2}{Z_r^2} + i_x^2} - \frac{V_{cl}}{L_r}(t - t_6) - i_{ai} + i_{bi} + i_{ci} + i_{ao} + i_{bo} & t_{16} \le t \le t_{17} \\ \sqrt{\frac{{V_{dc}}^2 - {V_{cl}}^2}{Z_r^2} + i_x^2} - \frac{V_{cl}}{L_r}(t - t_6) - i_{ai} + i_{bi} + i_{ci} + i_{ao} + i_{bo} - i_{co} & t_{18} \le t \le t_{19} \end{cases} \tag{12.49}$$

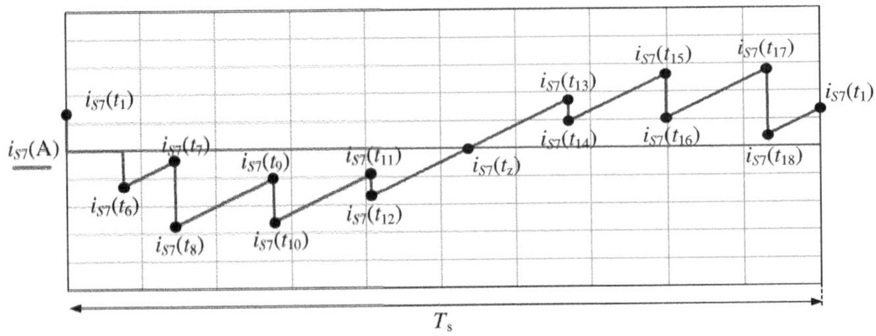

Figure 12.28 Waveform of i_{S7} in a switching period.

Because the clamping capacitor C_{cl} is in series with the auxiliary switch S_7, the current of the clamping capacitor i_{cl} should be equal to $-i_{S7}$. The waveform of the auxiliary switch current i_{S7} in a switching period is shown in Figure 12.28.

12.3.3 ZVS Condition

Since the voltage of resonant inductor $v_{Lr}(t)$ is clamped to V_{dc} from t_1 to t_6, the relationship between $i_{Lr}(t_1)$ and $i_{Lr}(t_6)$ satisfies the following equation.

$$i_{Lr}(t_6) - i_{Lr}(t_1) = \lambda_7 \frac{V_{dc} T_s}{L_r} \tag{12.50}$$

By combining equations (12.19), (12.38), and (12.50), $i_{Lr}(t_1)$ is derived as:

$$i_{Lr}(t_1) = -2i_M + i_{ai} + i_{co} - \sqrt{\frac{{V_{dc}}^2 - {V_{cl}}^2}{Z_r^2} + i_x^2} \tag{12.51}$$

The ZVS condition of the first resonance in equation (12.13) can be rewritten by substituting equation (12.51) as follows:

$$2i_M + \sqrt{\frac{{V_{dc}}^2 - {V_{cl}}^2}{Z_r^2} + i_x^2} \geq \frac{\sqrt{{V_{dc}}^2 - {V_{cl}}^2}}{Z_r} \tag{12.52}$$

From equation (12.52), it is found that the ZVS condition is independent of the filter inductor currents of both input and output. It is suitable for the whole of the utility period. The ZVS condition is always satisfied when $i_M \geq 0$. On the contrary, extra current i_x should be added in stage 5 when $i_M < 0$. Therefore, the minimum i_x can be obtained as:

$$i_{xmin} = \begin{cases} 0 & i_M \geq 0 \\ \sqrt{\left(\frac{\sqrt{{V_{dc}}^2 - {V_{cl}}^2}}{Z_r} - 2i_M\right)^2 - \frac{{V_{dc}}^2 - {V_{cl}}^2}{Z_r^2}} & i_M < 0 \end{cases} \tag{12.53}$$

Equation (12.53) indicates the minimum i_x is related to i_M. By substituting equation (12.53) into (12.38), the turn-off duty cycle λ_7 is represented by:

$$\lambda_7 = \frac{2L_r\left(|i_M| + \frac{1}{2}(-i_{ai} - i_{co} + i_{ao} + i_{bo} + i_{bi} + i_{ci}) + \sqrt{\frac{V_{dc}^2 - V_{cl}^2}{Z_r^2}}\right)}{V_{dc}T_s} \tag{12.54}$$

The above analysis is based on one switching period in the striped band in Figure 12.2. It can extend to other switching periods in a utility period. In different switching periods, only the filter inductor currents and resonant capacitors participating in the resonance are different. A generic λ_7 can be obtained in the whole of the utility period as follows [12]:

$$\lambda_7 = \frac{2L_r\left(|i_M| + \frac{1}{2}(|i_{ai}| + |i_{bi}| + |i_{ci}| + |i_{ao}| + |i_{bo}| + |i_{co}|) + \frac{\sqrt{V_{dc}^2 - V_{cl}^2}}{Z_r}\right)}{V_{dc}T_s} \tag{12.55}$$

12.4 Design Prototype

The resonant parameters are critical for the performance of the ZVS converter. A 9 kW ZVS BTB converter is taken as an example to illustrate the design procedure. The main parameters of the circuit are listed as follows. The rated DC voltage V_{dc} is 700 V. The grid phase voltage and output phase voltage are both 220 V_{rms}/50 Hz. The switching frequency f_s is 150 kHz. The resonant parameters are designed based on the following guideline [12].

12.4.1 Parameters Design

1) Turn-off duty cycle of the auxiliary switch

If the ZVS BTB converter system is balanced, $i_M = 0$. By substituting (12.2) and (12.4) into (12.55), the maximum turn-off duty cycle λ_{7_max} is derived:

$$\lambda_{7\_\max} = \frac{2L_r\left(2I_m + \frac{V_{dc}}{Z_r}\right)}{V_{dc}T_s} \tag{12.56}$$

where V_{cl} is neglected because it is much smaller than V_{dc} in equation (12.55). I_m is the peak value of the phase current.

The voltage stress of both the main switch and auxiliary switch is $V_{cl}+V_{dc}$. According to equation (12.22), the voltage stress of the switches is

$$V_{cl} + V_{dc} = \frac{V_{dc}}{1-\lambda_7} \tag{12.57}$$

Equation (12.57) shows that a larger λ_7 leads to higher voltage stress. To avoid high voltage stress, λ_7 is usually designed to be less than 0.1. Thus

$$\lambda_{7\_\max} = \frac{2L_r\left(2I_m + \frac{V_{dc}}{Z_r}\right)}{V_{dc}T_s} < 0.1 \tag{12.58}$$

2) Current stress

When the inverter of the ZVS BTB converter works under the three-phase unbalanced load, extra current i_x is needed in some operation regions. The maximum current of the main switches in a switching period happens at t_5 when the i_x is added at stage 5.

It is assumed the BTB converter is used as UPS. In this case, the input voltages and output voltages have the same amplitudes. For simplicity, it is assumed the inverter has the resistor load. For the worst case of the unbalanced load, it is assumed that the output C phase is no load and the other two output phases are full load. Thus the amplitude of the three-phase current in the rectifier becomes $2I_m/3$ since the average power of the rectifier is always equal to that of the inverter. Therefore, the equations (12.2) and (12.4) are rewritten as:

$$\begin{cases} i_{ai} = -\frac{2}{3}I_m \cos \omega t \\ i_{bi} = -\frac{2}{3}I_m \cos(\omega t - 2\pi/3) \\ i_{ci} = -\frac{2}{3}I_m \cos(\omega t + 2\pi/3) \end{cases} \quad \begin{cases} i_{ao} = I_m \cos(\omega t + \varphi) \\ i_{bo} = I_m \cos(\omega t + \varphi - 2\pi/3) \\ i_{co} = 0 \end{cases} \tag{12.59}$$

By substituting equations (12.34) and (12.59) into (12.39), i_M is represented as:

$$\begin{aligned} i_M &= -\frac{1}{V_{dc}}(v_{ag}'i_{ai} + v_{bg}'i_{bi} + v_{cg}'i_{ci} + v_{ao}'i_{ao} + v_{bo}'i_{bo} + v_{co}'i_{co}) \\ &= -\frac{1}{V_{dc}}\left(-\frac{V_{dc}}{2}MI_m + \frac{3V_{dc}}{4}MI_m - \frac{V_{dc}}{2}MI_m \cos^2(\omega t + \varphi + 2\pi/3)\right) \\ &= -\frac{1}{4}MI_m + \frac{1}{2}MI_m \cos^2(\omega t + \varphi + 2\pi/3) \\ &= \frac{1}{4}MI_m \cos[2(\omega t + \varphi + 2\pi/3)] \end{aligned} \tag{12.60}$$

According to the above equation, the minimum i_M is derived as:

$$i_{M\_\min} = -\frac{1}{4}MI_m \tag{12.61}$$

By substituting equation (12.61) into (12.53) and ignoring clamping capacitor voltage V_{cl}, the maximum extra current i_x is needed as:

$$i_{x\_\max} \approx \sqrt{\frac{M^2I_m^2}{4} + MI_m\frac{V_{dc}}{Z_r}} \tag{12.62}$$

Since the maximum extra current i_{x_max} is divided by six bridges during stage 5 (t_4–t_5), the maximum current stress i_{Smain_max} of the main switches is obtained:

$$i_{Smain\_\max} \approx I_m + \frac{1}{6}\sqrt{\frac{M^2I_m^2}{4} + MI_m\frac{V_{dc}}{Z_r}} \tag{12.63}$$

The maximum current stress of S_7 can be derived according to equation (12.49), which occurs at t_{15} in three-phase balanced full load condition with $\omega t = \pi/6$:

$$i_{S7\_\max} = -\sqrt{\frac{V_{dc}^2 - V_{cl}^2}{Z_r^2} + i_x^2} + \frac{V_{cl}}{L_r}(t_{15} - t_6) + i_{ai} - i_{bi} - i_{ci} - i_{bo} \tag{12.64}$$

When the ZVS BTB converter works under the three-phase balanced load, both i_M and i_x are equal to 0 according to equations (12.39) and (12.53). As shown in Figure 12.4, the duration from t_6 to t_{15} is derived as:

$$t_{15} - t_6 = (d_{ao} - \lambda_7)T_s \tag{12.65}$$

By combining d_{ao} in equation (12.35) and v_{ao}' in equation (12.34), the above equation is rewritten as:

$$t_{15} - t_6 = \frac{\frac{V_{dc}}{2} M\cos(\omega t + \varphi + \theta) + \frac{V_{dc}}{2}}{V_{dc}}(1 - \lambda_7)T_s \tag{12.66}$$

By substituting equations (12.22), (12.55), and (12.66) into (12.64) and ignoring clamping capacitor voltage V_{cl}, the maximum current stress of S_7 when $\omega t = \pi/6$, $\theta = 0$, $\varphi = 0$ is obtained:

$$i_{S7\_max} \approx \frac{\sqrt{3}}{2} M\left(\sqrt{3}I_m + \frac{V_{dc}}{Z_r}\right) \tag{12.67}$$

Since the turn-off current of the auxiliary switch $i_{S7}(t_1)$ has a direct impact on the turn-off loss, it is worth to be considered. By substituting equation (12.53) into (12.51), $i_{Lr}(t_1)$ is rewritten as:

$$i_{Lr}(t_1) = \begin{cases} -2i_M + i_{ai} + i_{co} - \sqrt{\dfrac{V_{dc}{}^2 - V_{cl}{}^2}{Z_r^2}} & i_M \geq 0 \\ i_{ai} + i_{co} - \sqrt{\dfrac{V_{dc}{}^2 - V_{cl}{}^2}{Z_r^2}} & i_M < 0 \end{cases} \tag{12.68}$$

According to the KCL at point P in Figure 12.5, the relationship between $i_{S7}(t_1)$ and $i_{Lr}(t_1)$ is derived as:

$$i_{S7}(t_1) + i_{Lr}(t_1) = i_{ai} + i_{co} \tag{12.69}$$

By substituting equation (12.68) into (12.69), the turn-off current of the auxiliary switch $i_{S7}(t_1)$ is obtained:

$$i_{S7}(t_1) = \begin{cases} 2i_M + \sqrt{\dfrac{V_{dc}{}^2 - V_{cl}{}^2}{Z_r^2}} & i_M \geq 0 \\ \sqrt{\dfrac{V_{dc}{}^2 - V_{cl}{}^2}{Z_r^2}} & i_M < 0 \end{cases} \tag{12.70}$$

When the ZVS BTB converter works under the three-phase balanced load, i_M is equal to 0. Similarly, the clamping capacitor voltage V_{cl} is ignored. Therefore, $i_{S7}(t_1)$ is expressed as:

$$i_{S7}(t_1) \approx \frac{V_{dc}}{Z_r} \tag{12.71}$$

Equations (12.63), (12.67), and (12.71) show that the current stress of the main switch and auxiliary switch is related to the resonant impedance Z_r. It can be found that smaller resonant impedance leads to larger current stress for both main switch and auxiliary switch. The relationships between i_{Smain_max}, i_{S7_max}, $i_{S7}(t_1)$, and Z_r are shown in Figure 12.29. To reduce the current stress and the switch conduction loss, the resonant impedance Z_r is designed larger than 30 Ω for the given example.

3) Resonant time

To restrict the effect of two resonant durations on the voltage stress of the switches, λ_7 is usually designed to be less than 0.1. Considering that the $\lambda_7 T_s$ is composed of the first resonant

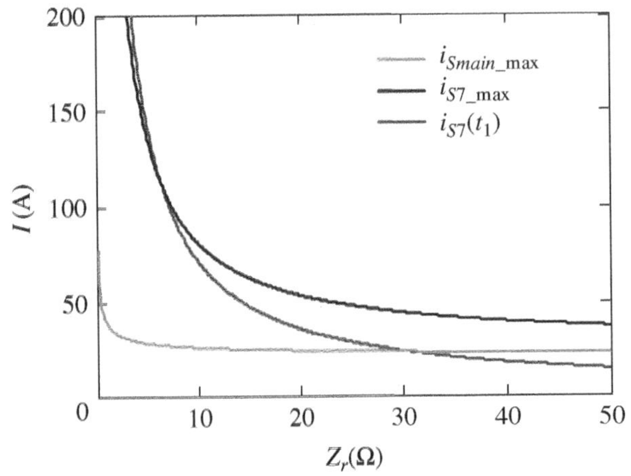

Figure 12.29 Current stress with different Z_r.

time T_{res1}, short-circuit time T_{short}, and the second resonant time T_{res2}, the first resonant time T_{res1} should be limited to a quarter of of $\lambda_7 T_s$.

$$T_{res1} = \frac{\pi\sqrt{L_r(6C_r + C_7)}}{2} < \frac{1}{4} \cdot 0.1T_s \tag{12.72}$$

In this case, the first resonant time T_{res1} is set to 167 ns.

4) Resonant capacitor

When the switch is turned off, the resonant capacitor paralleled to each switch will affect the turn-off loss. The larger the capacitance is, the smaller the turn-off loss will be. However, the larger capacitance will also result in a higher circulation current in the auxiliary circuit, which will cause an extra loss in the circuit. So the capacitance should be selected with a trade-off between the turn-off loss and the auxiliary circuit loss. The resonant capacitor C_r includes the parasitic output capacitance of the SiC MOSFET as well as the external installed capacitor. For the main switches, an external capacitor is installed across each switch to eliminate the inconsistency of the parasitic capacitance. While for the auxiliary switch, there is no external installed capacitor. So the resonant capacitor C_7 is equal to the parasitic capacitor of S_7. The parasitic capacitance for the selected main switches and auxiliary switch are 0.12 nF and 0.35 nF, respectively. So the equivalent resonant capacitance should meet

$$6C_r + C_7 > 6 \times 0.12\text{nF} + 0.35\text{nF} = 1.07\text{nF} \tag{12.73}$$

With the above four restrictions, the design of the resonance parameters will be constrained in a shadow area as shown in Figure 12.30.

Figure 12.31a and b shows the theoretical envelope curve of the clamping capacitor voltage v_{cl} and resonant inductor current i_{Lr} at different load levels when $L_r = 2.7\ \mu\text{H}$ and $6C_r + C_7 = 1.8\text{nF}$. It can be seen that the maximum v_{cl} is about 50 V and the maximum i_{Lr} is 56 A when working under full load.

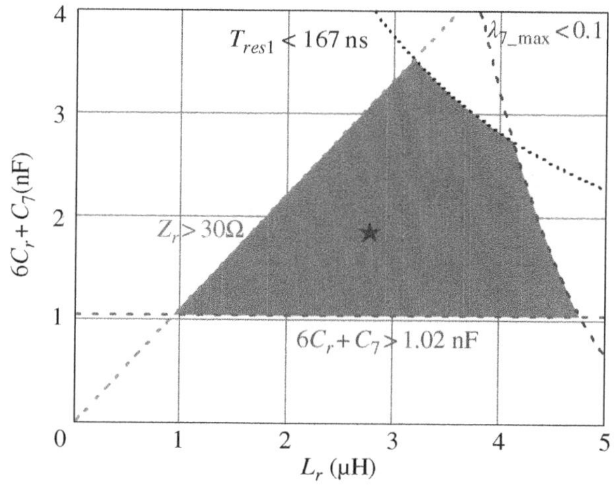

Figure 12.30 Solution region for the resonance parameters.

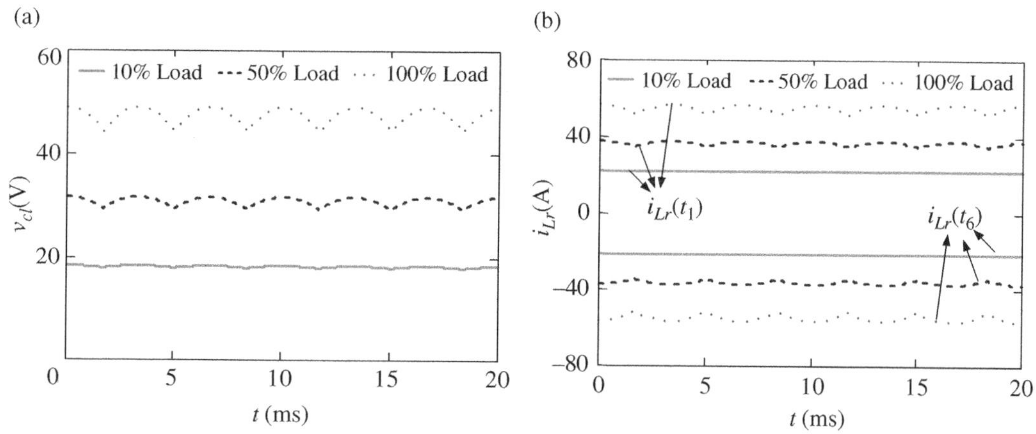

Figure 12.31 Envelope curve of v_{cl} and i_{Lr} at different load levels under a balanced load: (a) v_{cl}; (b) i_{Lr}.

Figure 12.32a and b gives the envelope curve of v_{cl} and i_{Lr} under a three-phase balanced full load and an unbalanced load with no load on phase A and full load on the other two phases. Both v_{cl} and i_{Lr} have smaller values under the unbalanced load than those under the balanced load.

12.4.2 Loss Analysis

The SiC MOSFET SCT2080KE from ROHM is selected as the main switches. Due to the higher current stress on the auxiliary switch, the SiC MOSFET C2M0025120D from CREE is selected as the auxiliary switch. The drain-source on-state resistance R_{dson} of the main switches and auxiliary switch are 108 and 37 mΩ, respectively, which are given in the datasheet when the drive voltage $V_{gs} = 20$ V and junction temperature $T_j = 100$ °C [10, 11]. The switching loss including turn-on loss and turn-off loss with different paralleled capacitors are estimated by the double-pulse test (DPT).

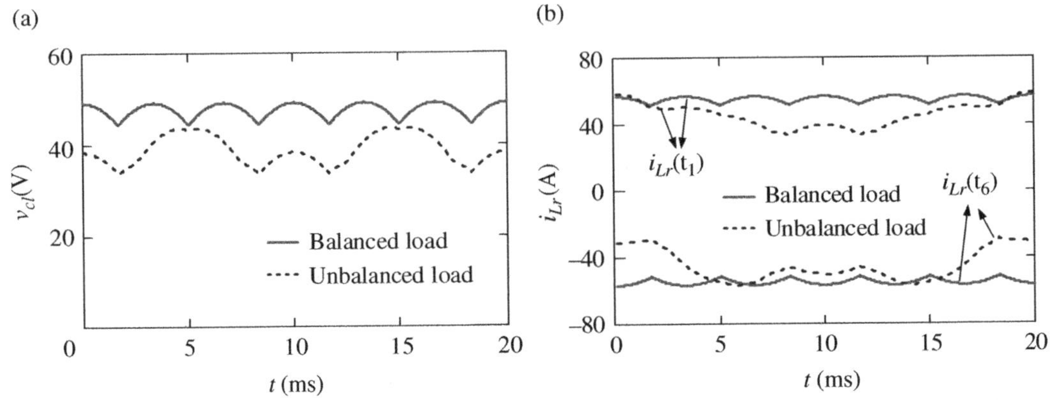

Figure 12.32 Envelope curve of v_{cl} and i_{Lr} under different load conditions: (a) v_{cl}; (b) i_{Lr}.

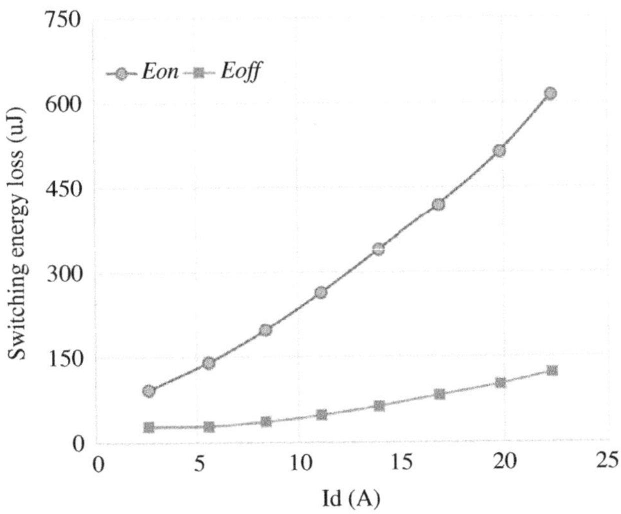

Figure 12.33 Switching loss of SCT2080KE.

The test is carried out under the following conditions: DC voltage $V_{dc} = 600$ V, drive resistance $R_g = 4.7\,\Omega$, drive voltage $V_{gs} = +20$ V/−5 V, and case temperature $T_c = 100\,°\text{C}$. Figure 12.33 shows the switching loss of SCT2080KE in a hard-switching process, which is used to estimate the switching loss of the traditional BTB converter. Figure 12.34 shows the switching loss of SCT2080KE and C2M0025120D with different paralleled resonant capacitors, which is used to estimate the switching loss of the ZVS BTB converter.

To evaluate the improved efficiency, five different switching frequencies such as 20, 50, 100, 150, and 200 kHz are selected. The filter inductance is designed according to 20% of the maximum inductor current ripple at different switching frequencies. The resonant parameters are kept consistent, which the resonant impedance Z_r is 35 Ω and the turn-off duty cycle λ_7 is 0.09. Table 12.1 lists the resonant inductance, resonant capacitance, and filter inductance at different switching frequencies.

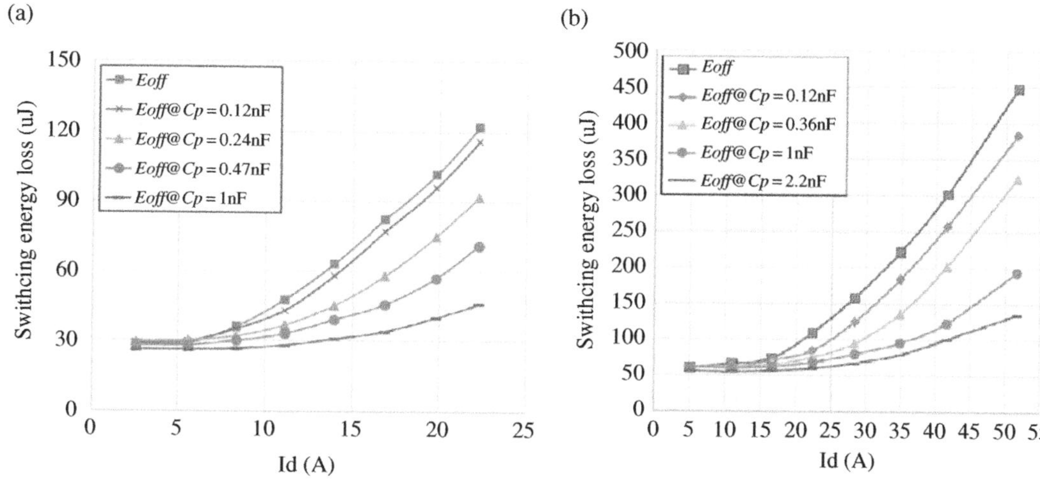

Figure 12.34 Switching loss with different capacitors: (a) SCT2080KE; (b) C2M0025120D.

Table 12.1 Resonant parameter and filter inductance at different switching frequencies.

Switching frequency f_s	20 kHz	50 kHz	100 kHz	150 kHz	200 kHz
Resonant inductor L_r	25 μH	10 μH	5 μH	3.3 μH	2.5 μH
Resonanzt capacitor C_r	2.2 nF	1 nF	0.47 nF	0.24 nF	0.12 nF
Input/output filter inductor	3 mH	1.2 mH	600 μH	400 μH	300 μH

Figures 12.35 and 12.36 show the loss distribution of the traditional BTB converter and ZVS BTB converter at different frequencies in a full load. P_{1L}, P_{1con}, P_{1on}, and P_{1off} are the filter inductor loss, conduction loss, turn-on loss, and turn-off loss of the rectifier side. P_{2L}, P_{2con}, P_{2on}, and P_{2off} are the filter inductor loss, conduction loss, turn-on loss, and turn-off loss of the inverter side. P_{acon}, P_{aoff}, and P_{Lr} are the conduction loss, turn-off loss of the auxiliary switch, and resonant inductor loss. P_{total} is the total loss of the converter.

From the loss distribution of the traditional BTB converter, it can be found that the filter inductor loss decreases with the increase of the switching frequency. The conduction loss is basically unchanged at different switching frequencies. However, the switching loss increases linearly with the increase of the switching frequency and affects the efficiency performance.

The loss distribution of the ZVS BTB converter is different from that of the traditional BTB converter. The total loss of the ZVS BTB converter decreases first and then increases. The minimum loss appears when $f_s = 50$ kHz. The switching loss increase rate of the ZVS BTB converter is obviously slower than that of the traditional BTB converter, which benefits from the realization of ZVS operation.

Figure 12.37 shows the loss comparison between the ZVS BTB converter at 150 kHz and the traditional BTB converter at 20 kHz. It can be concluded that the total loss is similar, which is 290 W of the ZVS BTB converter and 303 W of the traditional BTB converter, respectively. Since the switching frequency of the ZVS BTB converter is higher, the filter inductor loss decreases by 52 W, which

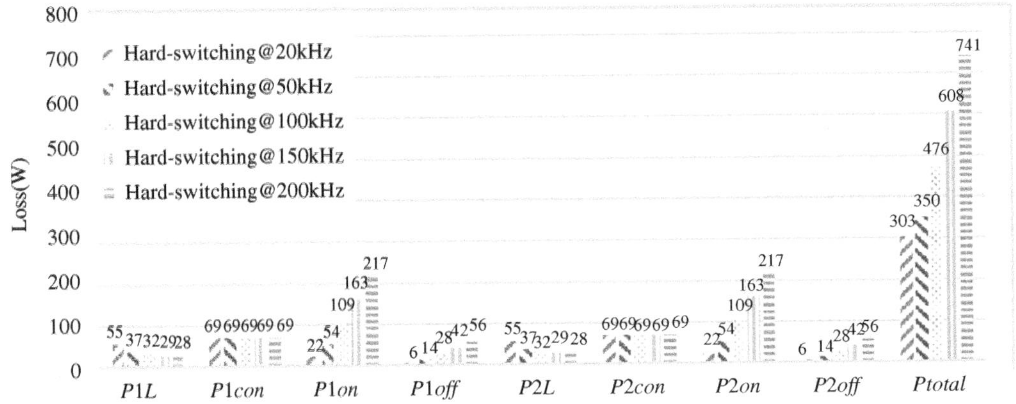

Figure 12.35 Loss distribution of traditional BTB converter at different switching frequencies.

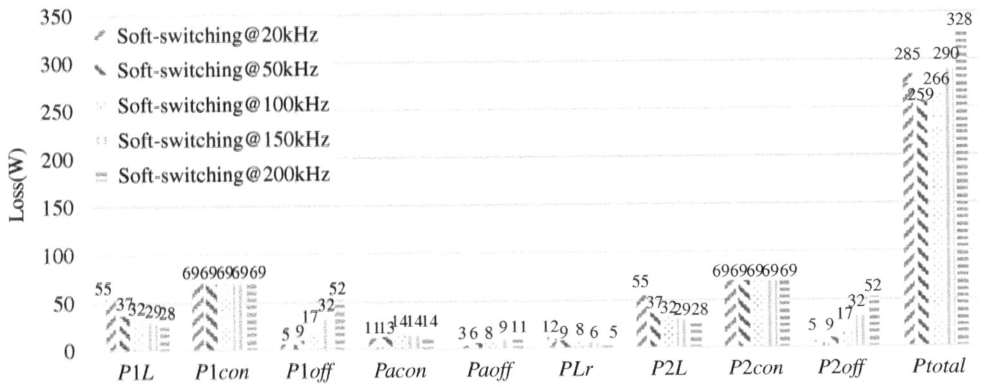

Figure 12.36 Loss distribution of ZVS BTB converter at different switching frequencies.

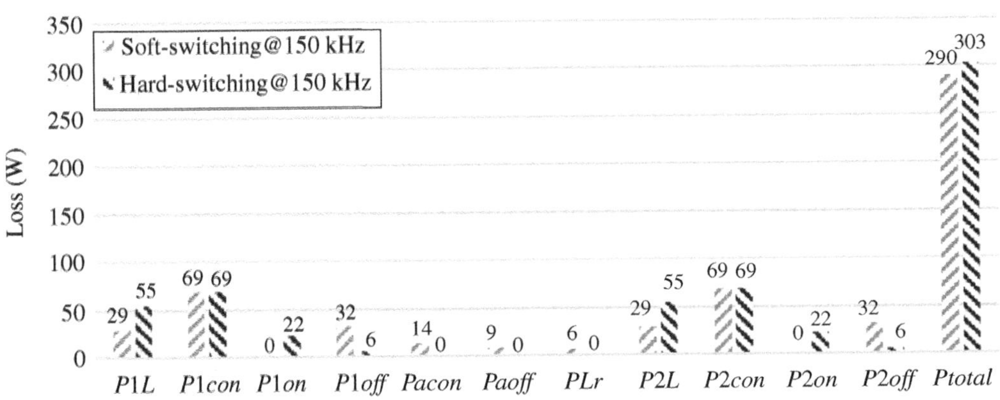

Figure 12.37 Loss comparison – ZVS BTB converter (150 kHz) vs. traditional BTB converter (20 kHz).

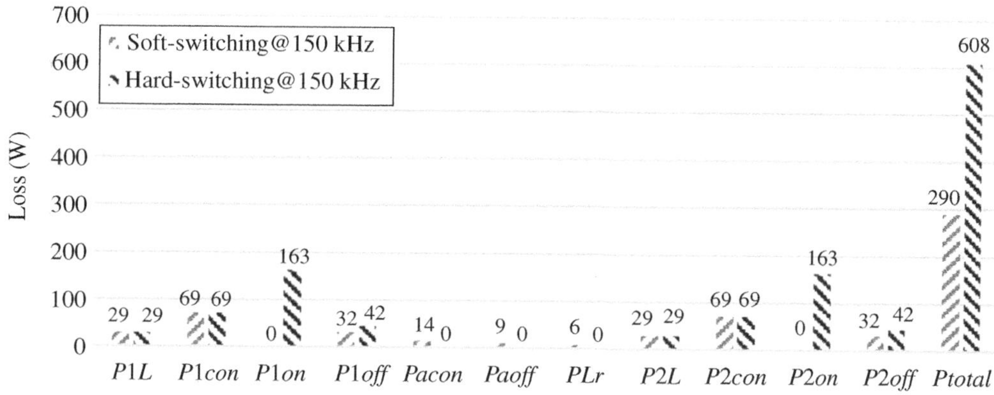

Figure 12.38 Loss comparison – ZVS BTB converter (150 kHz) vs. traditional BTB converter (150 kHz).

accounts for about 0.5% of the rated power. In addition, the ZVS BTB converter introduces an extra loss of about 29 W due to the auxiliary circuit.

Figure 12.38 shows the loss comparison between the ZVS BTB converter and the traditional BTB converter at 150 kHz. The total loss of the ZVS BTB converter decreases by 318 W, which accounts for about 3.2% of the rated power. In the traditional BTB converter, the turn-on loss is 326 W and the turn-off loss is 84 W. In other words, the turn-on loss is about four times the turn-off loss. The ZVS BTB converter can eliminate all turn-on loss and partial turn-off loss. Therefore, the switching loss is reduced by about 346 W in the ZVS BTB converter.

12.4.3 Experimental Results

To verify the operation principle of the ZVS BTB converter, a 9 kW prototype is built with the parameters listed in Table 12.2. Figure 12.39 shows the ZVS BTB prototype. The left picture shows the entire experimental prototype. The right two pictures show the front view and bottom view of the main circuit, in which the main components in the rectifier side, inverter side, as well as auxiliary circuit are demonstrated. The filter inductance is selected according to a 20% switching ripple of the inductor current. Litz wire is used as the winding to reduce the copper loss caused by the skin effect. The magnetic core material is alloy magnetic powder. Besides, the magnetic core material for the resonant inductor is ferrite since its large high frequency current ripple.

1) Balanced three-phase resistive load

   Figure 12.40 shows the waveforms in a three-phase balanced full load. Figure 12.40a shows the steady state waveforms of three-phase inductor currents at the rectifier side. Figure 12.40b shows the three-phase output voltages at the inverter side. It can be seen that both the input currents and output voltages have a low distortion waveform. The Total harmonic distortion (THD) for three-phase input current is 2.6, 2.3, and 2.7%, respectively. The THD for three-phase output voltage is 2.3, 1.8, and 1.8%, respectively. Figure 12.40c shows the waveforms of clamping capacitor voltage v_{cl} and resonant inductor current i_{Lr} in a utility period. It can be seen that both v_{cl} and i_{Lr} have similar low frequency ripple waves as shown in Figure 12.31.

   Figures 12.41–12.43 show the ZVS turn-on waveforms for S_{i4}, $S_{o2,}$ and S_7 with a balanced 3 kW load at different angles. It can be seen that the turn-on gate signals (v_{gi4}, v_{go2}, v_{g7}) are added to MOSFETs after their voltages (v_{Ci4}, v_{Co2}, v_{C7}) are resonated to zero. The first resonant stage

Table 12.2 Circuit parameters for the ZVS BTB prototype.

Parameters	Value
Input/output voltage	$u_{grms} = u_{orms} = 220\text{ V}$
Rated power	$P = 9\text{ kW}$
Input/output filter inductor	$L_i = L_o = 400\ \mu\text{H}$
Resonant inductor	$L_r = 2.7\ \mu\text{H}$
Clamping capacitor	$C_{cl} = 66\ \mu\text{F}/200\text{ V}$
DC voltage	$V_{dc} = 700\text{ V}$
Switching frequency	$f_s = 150\text{ kHz}$
Output filter capacitor	$C_o = 20\ \mu\text{F}$
External resonant capacitor	$C_r = 0.12\text{ nF}$
Main switches	ROHM SCT2080KE
Auxiliary switch	Wolfspeed C2M0025120D

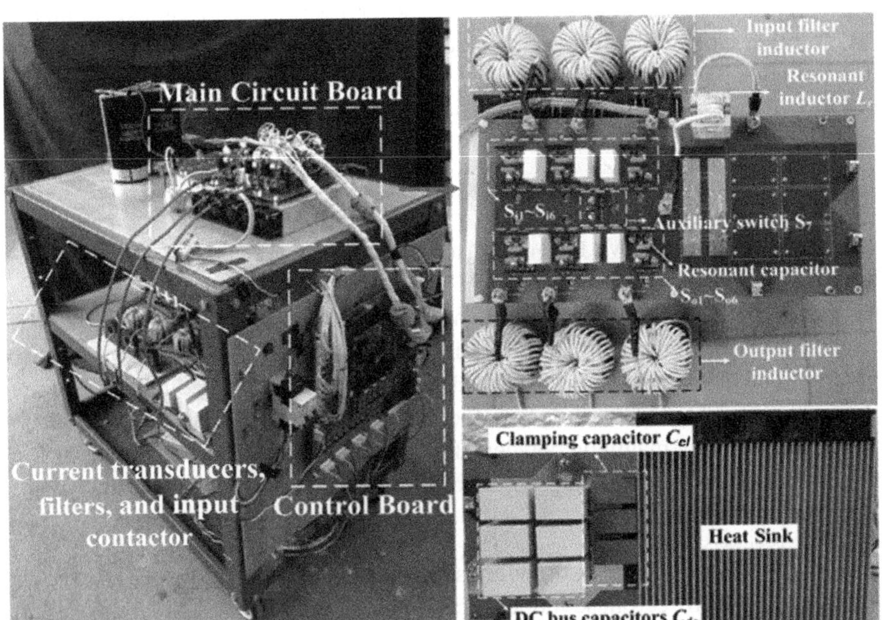

Figure 12.39 9 kW ZVS BTB prototype.

provides the ZVS condition for the main switches, while the second resonant stage provides the ZVS condition for the auxiliary switch S_7.

Figures 12.44–12.46 show the ZVS turn-on waveforms for S_{i4}, S_{o2}, and S_7 with a balanced 9 kW load at different angles. All these switches can also realize ZVS turn-on under full load.

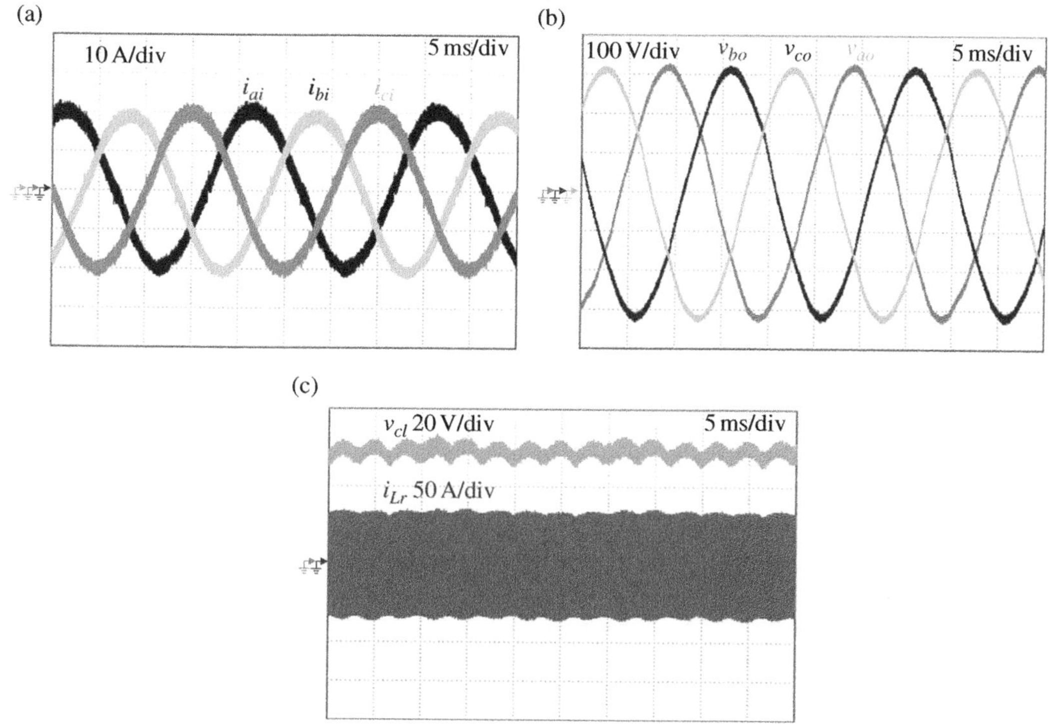

Figure 12.40 Experimental waveforms with a three-phase balanced load. (a) i_{ai}, i_{bi}, i_{ci}; (b) v_{ao}, v_{bo}, v_{co}; (c) v_{cl} and i_{Lr} in a utility period.

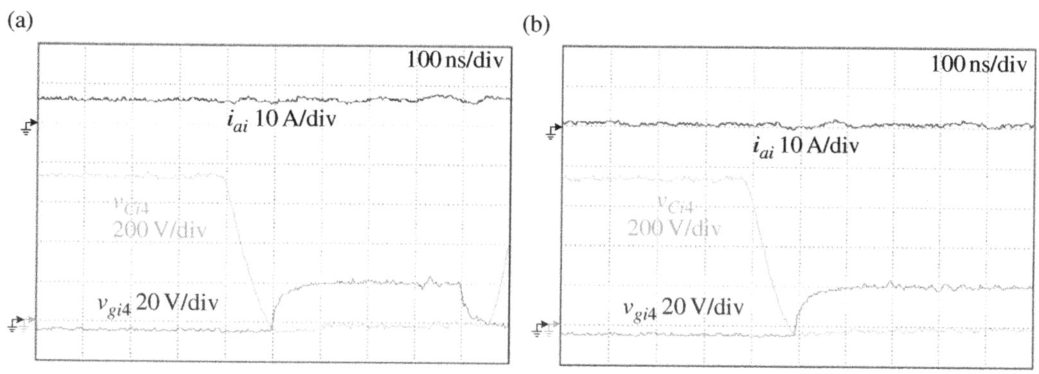

Figure 12.41 Waveforms for ZVS operation of S_{i4} with a balanced 3 kW load at different angles. (a) 0°. (b) 90°.

2) Unbalanced three-phase resistive load

Figure 12.47 shows the waveforms of clamping capacitor voltage v_{cl} and resonant inductor current i_{Lr} under three-phase unbalanced load when phase-B and phase-C are both feeding 3 kW load and phase-A is no load. It can be seen that both v_{cl} and i_{Lr} contain low frequency fluctuation because of unbalanced three-phase power. They have the same low frequency envelope waveform as in Figure 12.32.

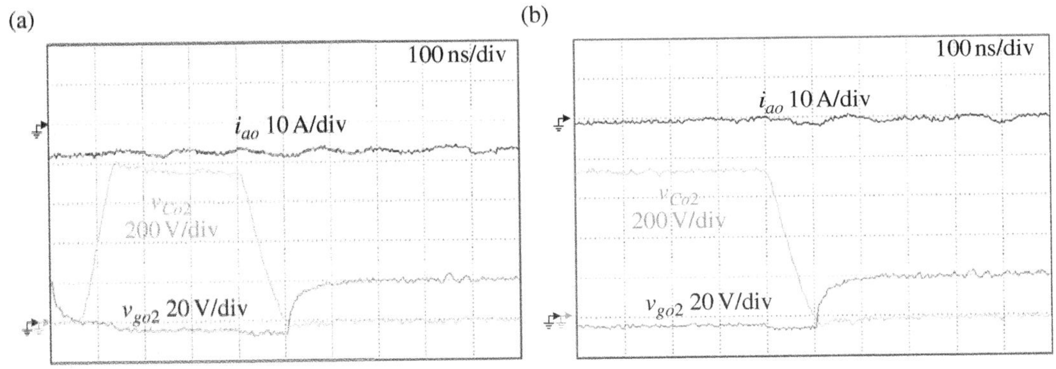

Figure 12.42 Waveforms for ZVS operation of S_{o2} with a balanced 3 kW load at different angles. (a) 0°. (b) 90°.

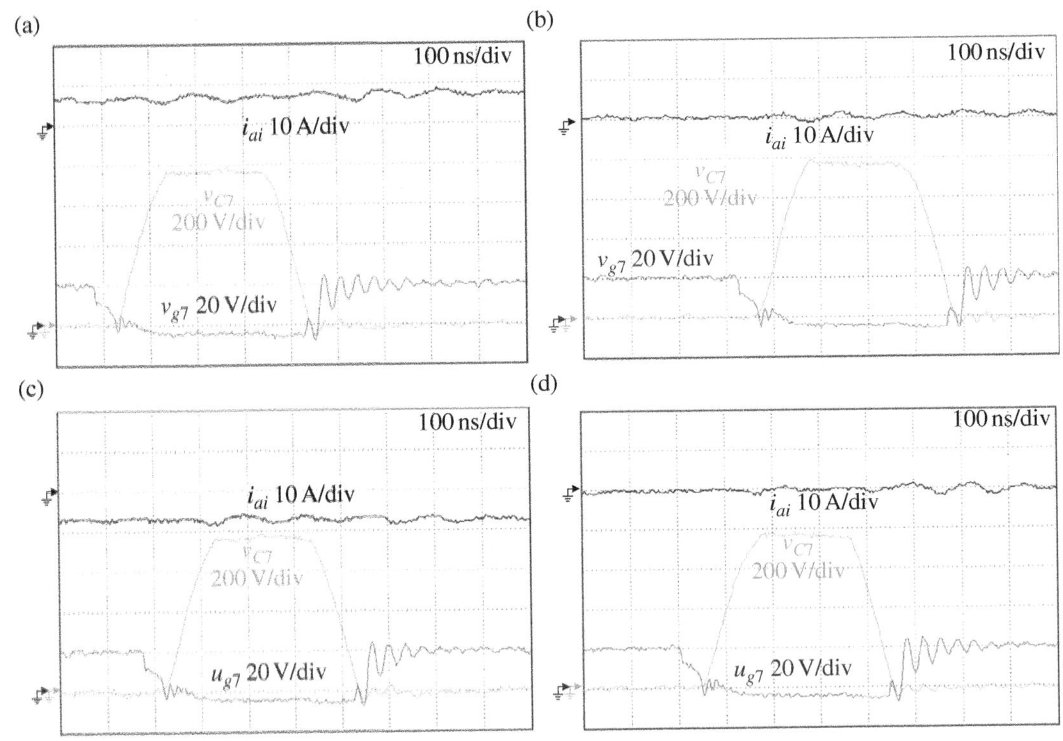

Figure 12.43 Waveforms for ZVS operation of S_7 with a balanced 3 kW load at different angles. (a) 0°. (b) 90°. (c) 180°. (d) 270°.

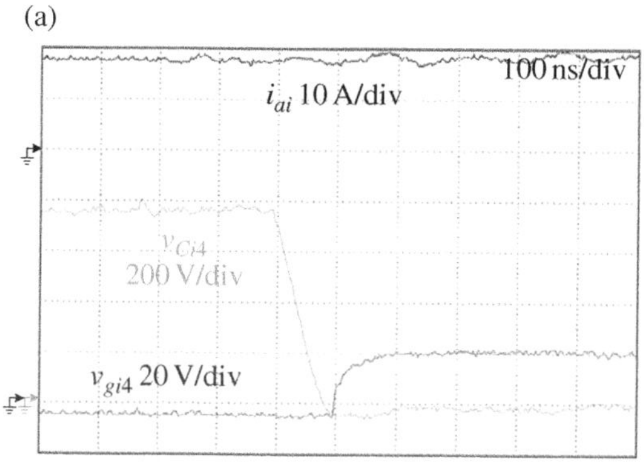

Figure 12.44 Waveforms for ZVS operation of S_{o2} with a balanced 9 kW load at different angles. (a) 0°. (b) 90°.

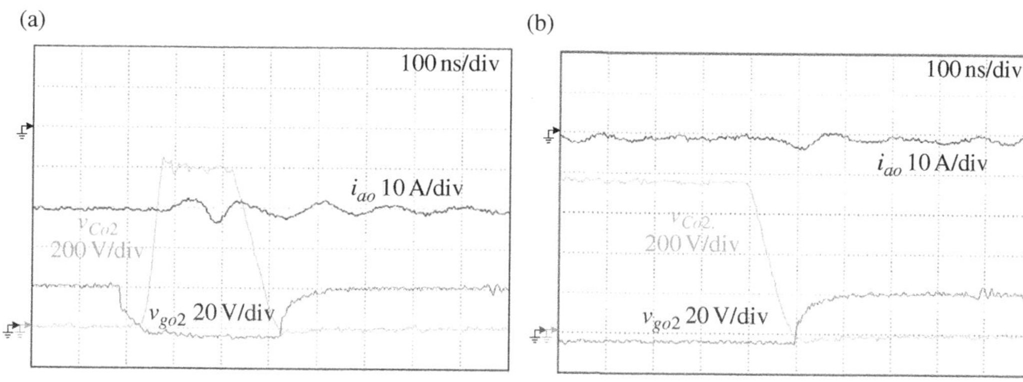

Figure 12.45 Waveforms for ZVS operation of S_{o2} with a balanced 9 kW load at different angles. (a) 0°. (b) 90°.

Figures 12.48–12.50 show the ZVS turn-on waveform for S_{i4}, S_{o2}, and S_7 with an unbalanced load. It can be seen that the turn-on gate signals (v_{gi4}, v_{go2}, v_{g7}) are also added to MOSFETs after their voltages (v_{Ci4}, v_{Co2}, v_{C7}) are resonated to zero. They can also realize ZVS operation in the unbalanced load.

When working under the unbalanced load, a short-circuit stage should be added to control the extra current i_x. During this stage, the bus voltage v_{bus} is clamped to zero and the current of the resonant inductor will decrease with a slope of V_{dc}/L_r. In the theoretical analysis, the short-circuit signal is added after the Type 2 commutation of all six phase legs. However, in practical operation, it is difficult to determine the end time of Type 2 commutations. If the short-circuit signal is added after Type 2 commutations, the second resonation will take place immediately and v_{bus} will be gradually resonated back to $V_{dc} + V_{cl}$. Then the short-circuit stage will cause extra turn-on loss. Actually, the short-circuit signal v_{gst} can be added together with the turn-on signal of the ZVS-on switches. The duration of the short-circuit signal is decided by $\lambda_7 T_s - T_r/2$. In this way, an appropriate i_x will be added to resonant inductor current as long as λ_7 is accurate. The gate signals of all main switches obtained from Figure 12.3b will have a logic OR operation with the short-circuit signal v_{gst} to generate the final gate driving signals. Figure 12.51 shows the diagram of the generation of the gate driving signals for all the main switches and auxiliary switch.

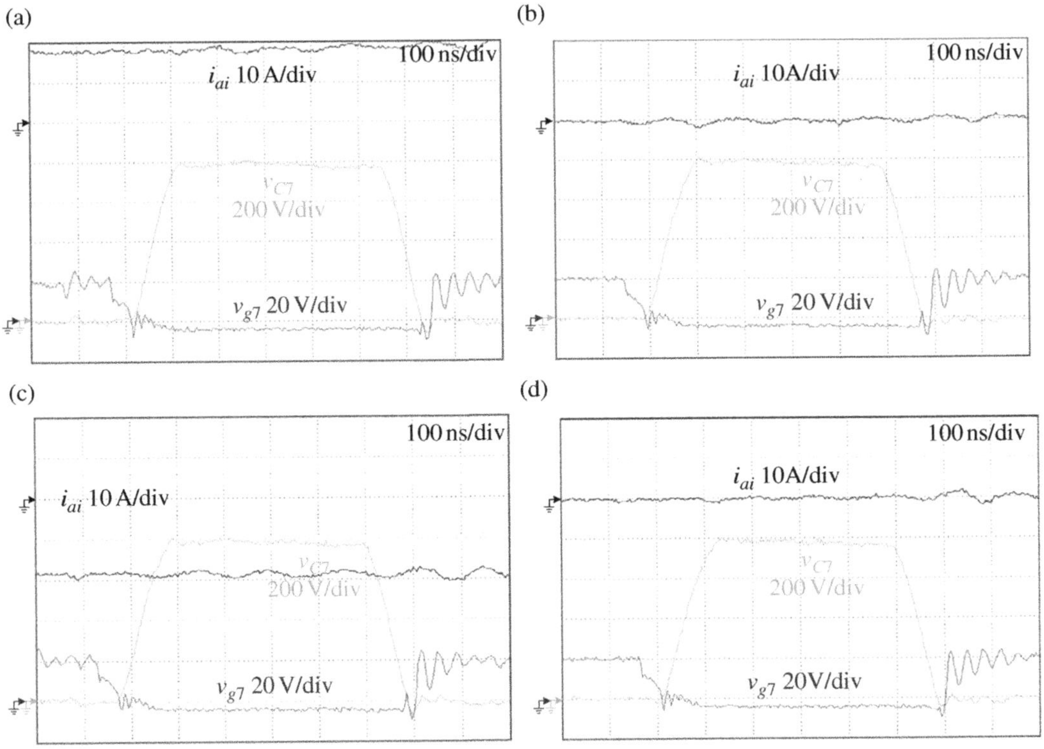

Figure 12.46 Waveforms for ZVS operation of S_7 with a balanced 9 kW load at different angles. (a) 0°. (b) 90°. (c) 180°. (d) 270°.

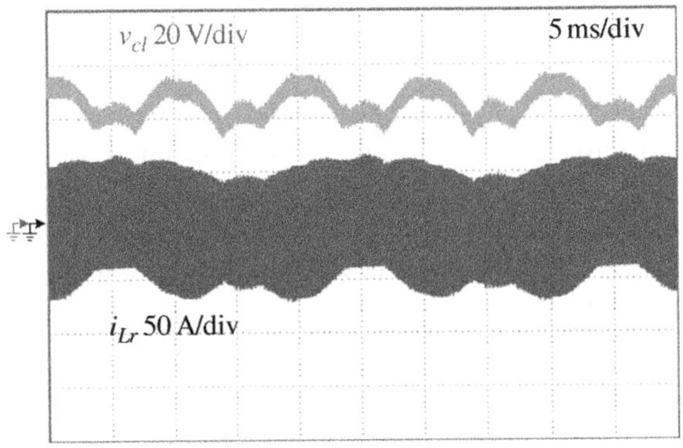

Figure 12.47 Waveforms of v_{cl} and i_{Lr} under three-phase unbalanced load.

Figure 12.52 shows a typical short-circuit signal waveform under an unbalanced load with no load on phase-A and the full load on other two phases. S_{o2} is first turned off at the beginning of a switching period. Then the first resonation stage will resonate v_{bus} to zero. When v_{bus} is resonated to zero, a short-circuit signal is added. S_{o2} is turned on again and v_{bus} is constantly clamped to zero. The current of the resonant inductor is decreasing with a slope of V_{dc}/L_r. When the short-circuit signal is off, the second resonance begins.

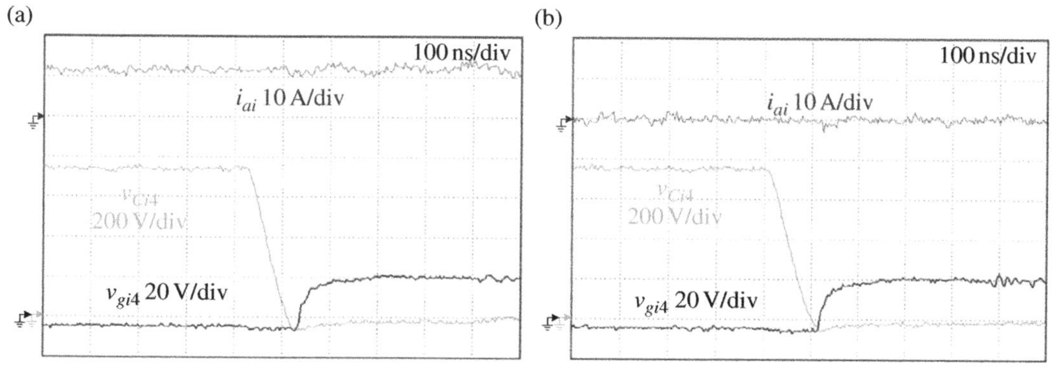

Figure 12.48 Waveforms for ZVS operation of S_{i4} with an unbalanced load at different angles. (a) 0°. (b) 90°.

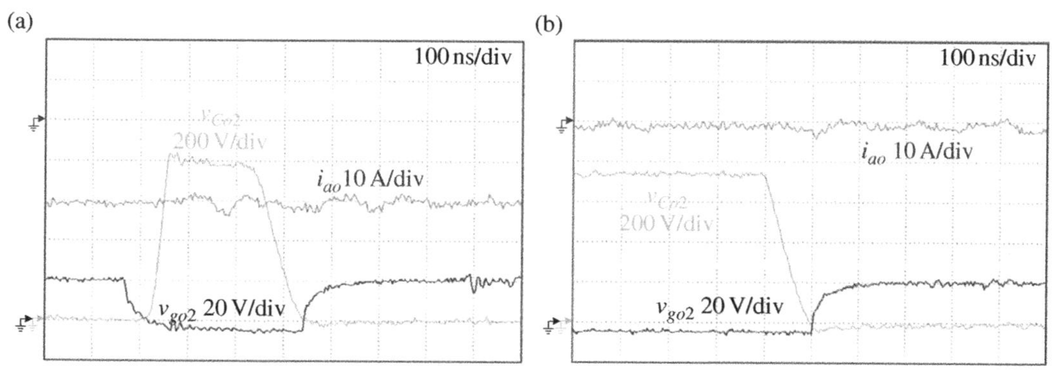

Figure 12.49 Waveforms for ZVS operation of S_{o2} with an unbalanced load at different angles. (a) 0°. (b) 90°.

3) Voltage stress and efficiency comparison

To evaluate the performance of the proposed ZVS BTB converter, a comparison of voltage stress and efficiency is made between the conventional BTB converter and the ZVS BTB converter.

Figure 12.53 shows the voltage stress of the power device for the conventional BTB converter and ZVS BTB converter at both the rectifier side and inverter side. The voltage stress includes the overshooting voltage caused by the parasitic inductor in the current commutation circuit loop. The main switches of the ZVS BTB converter have about 50 V higher voltage stress than those in the conventional BTB converter, which is caused by the voltage of the clamping capacitor v_{cl} according to the analysis in equation (12.57). While the voltage stress for the auxiliary switch is lower than the main switches, this is because when the auxiliary switch S_7 is turned off, its drain-source voltage v_{c7} is resonated to $V_{dc} + V_{cl}$, and theoretically the di/dt of the resonating current is close to zero when v_{c7} reaches $V_{dc} + V_{cl}$.

The experimental efficiency comparison between the conventional BTB converter and the ZVS BTB converter is carried out with the same SiC MOSFET device and the same filter inductors at the same switching frequency of 150 kHz. Figure 12.54 shows the efficiency curve at

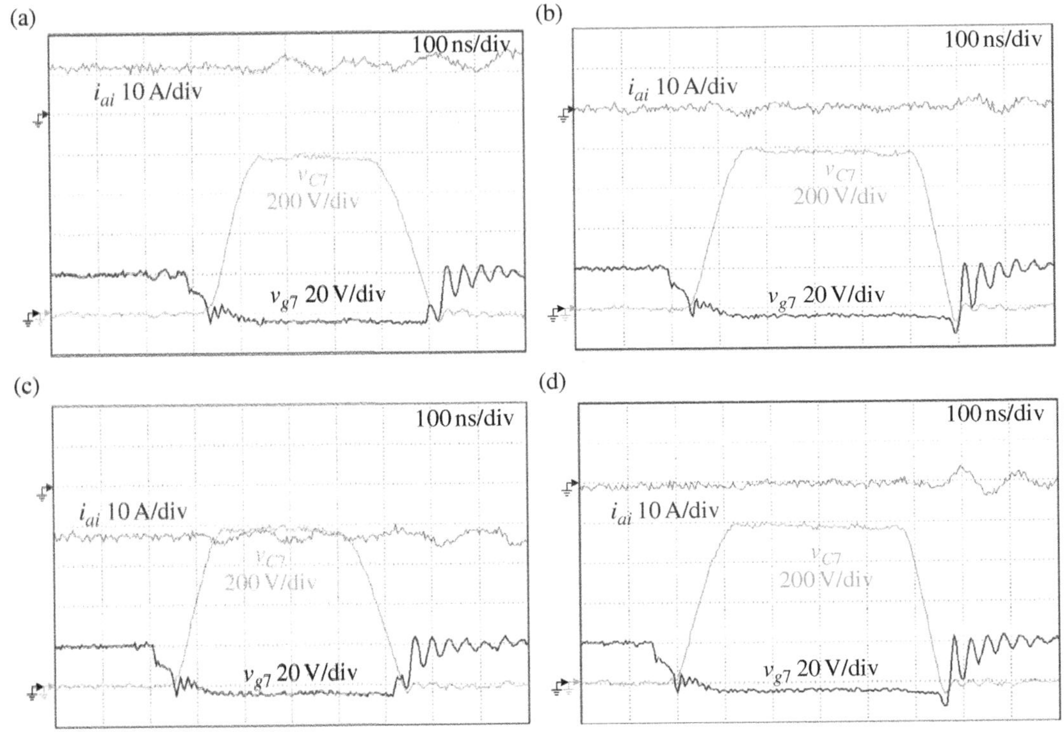

Figure 12.50 Waveforms for ZVS operation of S_7 with an unbalanced load at different angles. (a) 0°. (b) 90°. (c) 180°. (d) 270°.

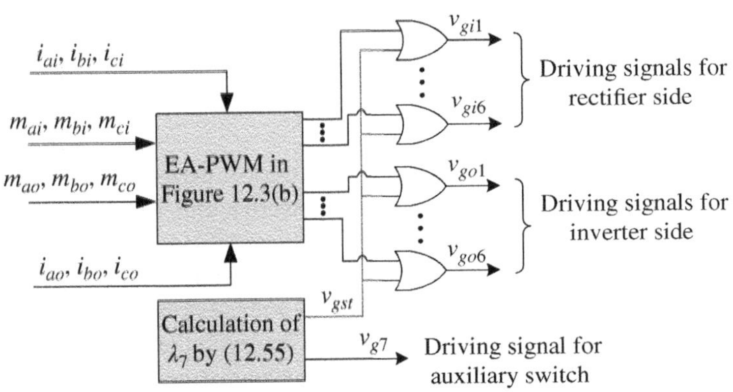

Figure 12.51 Generation of gate driving signals with short-circuit signal.

different load levels. For the proposed ZVS BTB converter, the maximum efficiency point occurs at half rated power, which is 97.5%. And the efficiency at rated power is 97.1%. While the conventional BTB converter has a maximum efficiency of 94.4% at 60% load, its efficiency will decrease to 94.0% at rated power. Comparing to the conventional BTB converter, the proposed

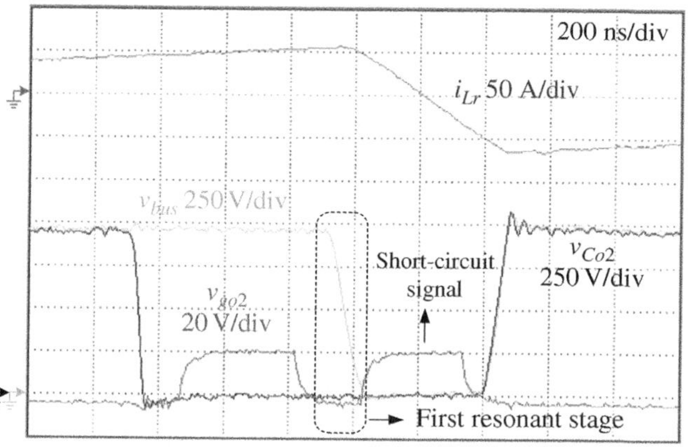

Figure 12.52 Waveforms of short-circuit signal.

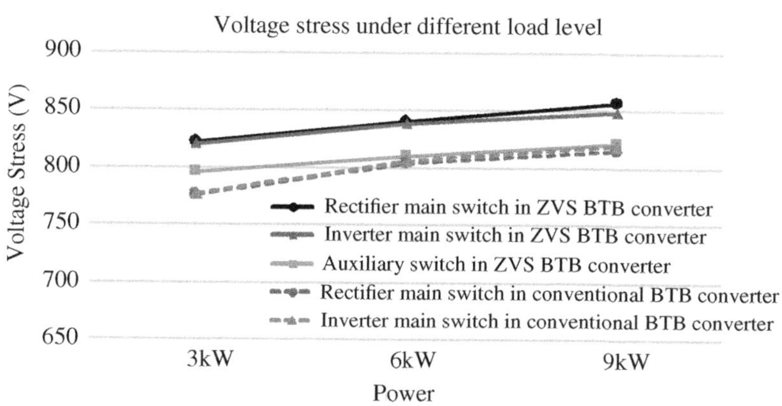

Figure 12.53 Power device voltage stress at different power levels.

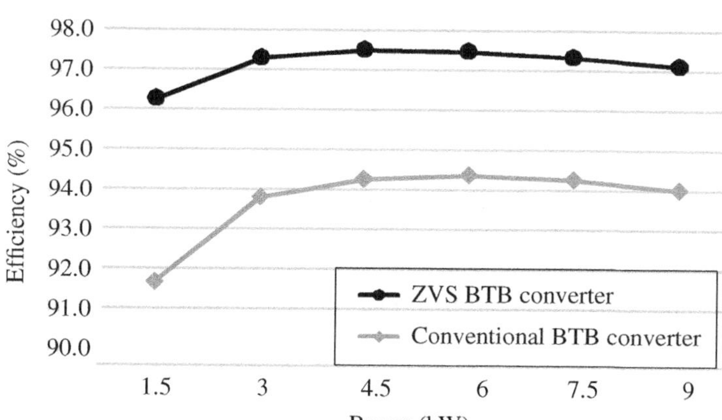

Figure 12.54 Efficiency curve at different power levels.

ZVS BTB converter has an efficiency improvement of over 3% at the entire power range. Especially at 15% load, where the switching loss occupies a larger portion, the ZVS BTB converter has an efficiency improvement of 4.5% due to ZVS operation.

12.5 Summary

The soft-switching SiC three-phase four-wire BTB converter is introduced. When working under the balanced load, all switches can realize ZVS operation at the entire load range. When working under the unbalanced load, an extra current i_x should be added to the resonant inductor in the region where the i_M is negative through a short-circuit stage to fulfill the ZVS condition for main switches. A design guideline of resonant parameters is provided to guarantee a suitable converter performance. Finally, the operation principle of the ZVS three-phase four-wire BTB converter is verified on a 9 kW prototype. The ZVS operation of all switches is realized under the balanced and unbalanced load. The voltage stress for main switches of the ZVS BTB is less than 1.1 times that of the conventional BTB converter. Comparing to conventional BTB converter, the proposed ZVS BTB converter has an efficiency improvement of over 3% at the entire power range at 150 kHz switching frequency.

References

1 M. E. Fraser, C. D. Manning and B. M. Wells, "Transformerless four-wire PWM rectifier and its application in AC-DC-AC converters," *IEE Proceedings-Electric Power Applications*, vol. 142, no. 6, pp. 410–416, 1995.

2 E. H. Kim, J. M. Kwon and B. H. Kwon, "Transformerless three-phase on-line UPS with high performance," *IET Power Electronics*, vol. 2, no. 2, pp. 103–112, 2009.

3 D M. Divan, "The resonant DC link converter-a new concept in static power conversion [J]," *IEEE Transactions on Industry Applications*, vol. 25, no. (2), pp. 317–325, 1989.

4 D. M. Divan and G. Skibinski, "Zero-switching-loss inverters for high-power applications," *IEEE Transactions on Industry Applications*, vol. 25, no. 4, pp. 634–643, 1989.

5 J. J. Jafar and B. G. Fernandes, "A new quasi-resonant DC-link PWM inverter using single switch for soft switching," *IEEE Transactions on Power Electronics*, vol. 17, no. 6, pp. 1010–1016, 2002.

6 H. Chen and D. Divan, "Soft-Switching Solid-State Transformer (S4T)," *IEEE Transactions on Power Electronics*, vol. 33, no. 4, pp. 2933–2947, 2018.

7 R. Li and D. Xu, "A zero-voltage switching three-phase inverter," *IEEE Transactions on Power Electronics*, vol. 29, no. 3, pp. 1200–1210, 2014.

8 Z. Ma, D. Xu, R. Li, C. Du and X. Zhang, "A novel DC-side zero-voltage switching (ZVS) three-phase boost pwm rectifier controlled by an Improved SVM method," *IEEE Transactions on Power Electronics*, vol. 27, no. 11, pp. 4391–4408, 2012.

9 N. He, Y. Zhu and D. Xu, "Zero-voltage-switching SPWM method for three-phase Four-wire inverter," in *2017 IEEE Applied Power Electronics Conference and Exposition (APEC)*, Tampa, FL, pp. 3436–3443, 2017.

10 https://www.wolfspeed.com/downloads/dl/file/id/161/product/643/c2m0025120d.pdf.

11 https://fscdn.rohm.com/en/products/databook/datasheet/discrete/sic/mosfet/sct2080ke-e.pdf.

12 K. Shi, A. Zhao, J. Deng and D. Xu, "Zero-voltage-switching SiC-MOSFET three-phase four-wire back-to-back converter," in *IEEE Journal of Emerging and Selected Topics in Power Electronics*, vol. 7, no. 2, pp. 722–735, June 2019.

Appendix

A.1 Basic of SVM

The symmetrical three-phase voltages v_a, v_b, and v_c can be expressed as:

$$\begin{bmatrix} v_a \\ v_b \\ v_c \end{bmatrix} = V_m \begin{bmatrix} \cos(\omega t) \\ \cos(\omega t - 2\pi/3) \\ \cos(\omega t + 2\pi/3) \end{bmatrix} \tag{A.1.1}$$

It can be transformed by matrix $T_{abc/\alpha\beta\gamma}$ as:

$$\begin{bmatrix} v_\alpha \\ v_\beta \\ v_\gamma \end{bmatrix} = T_{abc/\alpha\beta\gamma} \begin{bmatrix} v_a \\ v_b \\ v_c \end{bmatrix} \tag{A.1.2}$$

where

$$T_{abc/\alpha\beta\gamma} = \frac{2}{3} \begin{bmatrix} 1 & -\frac{1}{2} & -\frac{1}{2} \\ 0 & \frac{\sqrt{3}}{2} & -\frac{\sqrt{3}}{2} \\ \frac{1}{\sqrt{2}} & \frac{1}{\sqrt{2}} & \frac{1}{\sqrt{2}} \end{bmatrix} \tag{A.1.3}$$

Then

$$\begin{bmatrix} v_\alpha \\ v_\beta \\ v_\gamma \end{bmatrix} = V_\mathrm{m} \begin{bmatrix} \cos(\omega t) \\ \sin(\omega t) \\ 0 \end{bmatrix} \tag{A.1.4}$$

If a voltage vector $\vec{V}$ is defined as:

$$\vec{V} = v_\alpha + jv_\beta = V_\mathrm{m}[\cos(\omega t) + j\sin(\omega t)] = V_\mathrm{m}e^{j\omega t} \tag{A.1.5}$$

Soft-Switching Technology for Three-phase Power Electronics Converters, First Edition.
Dehong Xu, Rui Li, Ning He, Jinyi Deng, and Yuying Wu.

Obviously, vector $\vec{V}$ is counterclockwise rotating and the trace of its end point is a circle as shown in Figure A.1.1. The magnitude of vector $\vec{V}$ is the same as the amplitude of phase voltage.

Output of three-phase bridges of the three-phase inverter can be represented by vector $\vec{V}$. If the switching function of an half-bridge S_x, $x = a$, b, and c, is equal to 1 when the upper switch is conducting and equal to 0 when the upper switch is blocked, there will be totally eight combinations, 000, 001, 010, 011, 100, 101, 110, and 111, for three-phase bridges in three-phase inverter. Their equivalent circuits are shown in Figure A.1.2.

The equivalent vectors of the eight combinations can be obtained by using the following equation:

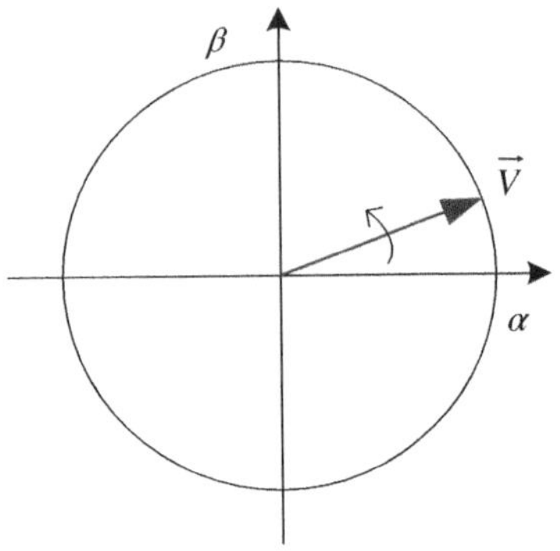

Figure A.1.1 Trace of vector $\vec{V}$ with symmetrical three-phase voltage.

$$\begin{bmatrix} X_\alpha \\ X_\beta \\ X_\gamma \end{bmatrix} = T_{abc/\alpha\beta\gamma} \begin{bmatrix} v_{A0} \\ v_{B0} \\ v_{C0} \end{bmatrix} \tag{A.1.6}$$

where v_{A0}, v_{B0}, and v_{C0} are the output voltages of three-phase bridges and can be expressed as:

$$\begin{cases} v_{A0} = L\dfrac{di_a}{dt} + v_a \\ v_{B0} = L\dfrac{di_b}{dt} + v_b \\ v_{C0} = L\dfrac{di_c}{dt} + v_c \end{cases} \tag{A.1.7}$$

where i_a, i_b, and i_c are the three-phase grid currents. v_a, v_b, and v_c are the three-phase grid voltages and are expressed as:

$$\begin{cases} v_a = V_m \cos(\varphi) \\ v_b = V_m \cos(\varphi - 2\pi/3) \\ v_c = V_m \cos(\varphi + 2\pi/3) \end{cases} \tag{A.1.8}$$

where V_m is the amplitude of phase voltage. Three-phase voltages v_{A0}, v_{B0}, and v_{C0} can also be described as:

$$\begin{cases} v_{A0} = v_{AN} + v_{N0} = S_a V_{dc} + v_{N0} \\ v_{B0} = v_{BN} + v_{N0} = S_b V_{dc} + v_{N0} \\ v_{C0} = v_{CN} + v_{N0} = S_c V_{dc} + v_{N0} \end{cases} \tag{A.1.9}$$

where S_a, S_b, and S_c are switching functions and are defined by:

$$S_a, S_b, S_c = \begin{cases} 1, \text{upper switch is on} \\ 0, \text{upper switch is off} \end{cases} \tag{A.1.10}$$

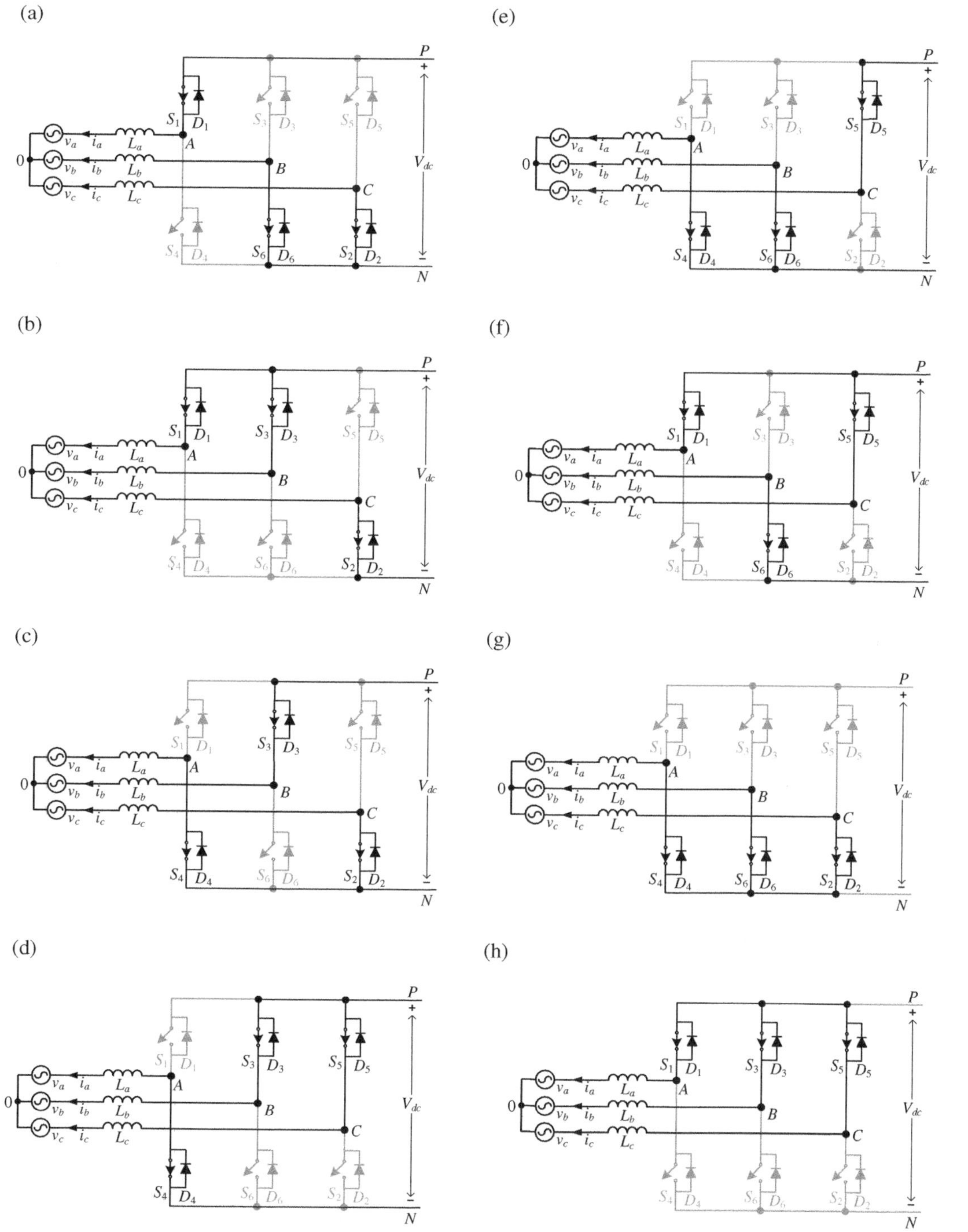

Figure A.1.2 Eight state combinations for three-phase bridges in three-phase inverter: (a) 100, (b) 110, (c) 010, (d) 011, (e) 001, (f) 101, (g) 000, and (h) 111.

When three-phase grid voltages are balanced, $v_a + v_b + v_c = 0$. According to Kirchhoff's current law, $i_a + i_b + i_c = 0$. From (A.1.7), the relationship of three-phase voltages v_{A0}, v_{B0}, and v_{C0} can be obtained as:

$$v_{A0} + v_{B0} + v_{C0} = 0 \tag{A.1.11}$$

By combining (A.1.9) and (A.1.11), voltage v_{N0} can be obtained as:

$$v_{N0} = -\frac{S_a + S_b + S_c}{3} V_{dc} \tag{A.1.12}$$

By substituting (A.1.12) to (A.1.9), three-phase voltages v_{A0}, v_{B0}, and v_{C0} can be modified as:

$$\begin{cases} v_{A0} = \left(S_a - \dfrac{S_a + S_b + S_c}{3}\right) V_{dc} \\ v_{B0} = \left(S_b - \dfrac{S_a + S_b + S_c}{3}\right) V_{dc} \\ v_{C0} = \left(S_c - \dfrac{S_a + S_b + S_c}{3}\right) V_{dc} \end{cases} \tag{A.1.13}$$

Substituting (A.1.13) into equation (A.1.6), following vector is derived:

$$\begin{bmatrix} X_\alpha \\ X_\beta \\ X_\gamma \end{bmatrix} = \frac{2}{3} \begin{bmatrix} 1 & -\frac{1}{2} & -\frac{1}{2} \\ 0 & \frac{\sqrt{3}}{2} & -\frac{\sqrt{3}}{2} \\ \frac{1}{\sqrt{2}} & \frac{1}{\sqrt{2}} & \frac{1}{\sqrt{2}} \end{bmatrix} \begin{bmatrix} \left(S_a - \dfrac{S_a + S_b + S_c}{3}\right) V_{dc} \\ \left(S_b - \dfrac{S_a + S_b + S_c}{3}\right) V_{dc} \\ \left(S_c - \dfrac{S_a + S_b + S_c}{3}\right) V_{dc} \end{bmatrix} = \frac{1}{3} V_{dc} \begin{bmatrix} 2S_a - S_b - S_c \\ \sqrt{3}(S_b - S_c) \\ 0 \end{bmatrix} \tag{A.1.14}$$

Since above vector is located at a plane, it can be rewritten in complex plane:

$$X_\alpha + jX_\beta = \frac{1}{3} V_{dc} \left[(2S_a - S_b - S_c) + \sqrt{3}(S_b - S_c) j\right] \tag{A.1.15}$$

By considering the eight combinations of the switching functions, their equivalent vectors can be derived as:

$$\begin{cases} \vec{V}_1 = \frac{2}{3} V_{dc} & \textit{Combination } 100 \\ \vec{V}_2 = \frac{1}{3} V_{dc}\left(1 + \sqrt{3} j\right) = \frac{2}{3} V_{dc} e^{j\frac{\pi}{3}} & \textit{Combination } 110 \\ \vec{V}_3 = \frac{1}{3} V_{dc}\left(-1 + \sqrt{3} j\right) = \frac{2}{3} V_{dc} e^{j\frac{2\pi}{3}} & \textit{Combination } 010 \\ \vec{V}_4 = -\frac{2}{3} V_{dc} & \textit{Combination } 011 \\ \vec{V}_5 = \frac{1}{3} V_{dc}\left(-1 - \sqrt{3} j\right) = \frac{2}{3} V_{dc} e^{-j\frac{2\pi}{3}} & \textit{Combination } 001 \\ \vec{V}_6 = \frac{1}{3} V_{dc}\left(1 - \sqrt{3} j\right) = \frac{2}{3} V_{dc} e^{-j\frac{\pi}{3}} & \textit{Combination } 101 \\ \vec{V}_0 = \vec{V}_7 = 0 & \textit{Combinations } 000, 111 \end{cases} \tag{A.1.16}$$

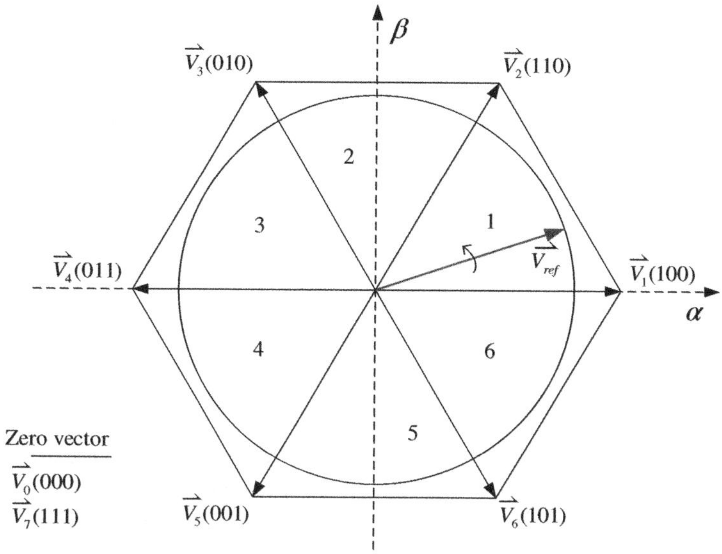

Figure A.1.3 Positions of eight basic vectors of three-phase bridges.

As shown in (A.1.16), six basic vectors with non zero magnitudes, $\vec{V}_1$, $\vec{V}_2$, $\vec{V}_3$, $\vec{V}_4$, $\vec{V}_5$, and $\vec{V}_6$, are called as effective vectors while the other two basic vectors with zero magnitude, $\vec{V}_0$ and $\vec{V}_7$, are called as zero vectors. The positions of eight basic vectors are shown in Figure A.1.3. Obviously, the plane is separated into six regions by the six effective vectors. The regions between each two adjacent effective vectors are called as sectors as shown in Figure A.1.3. Each sector occupies a $\pi/3$ phase angle range.

Since there are only eight separate basic vectors in three-phase bridges, a rotating reference voltage vector $\vec{V}_{ref}$ as shown in Figure A.1.3 is composed by using the area equivalence principle. The reference voltage vector in each switching period can be equivalently composed by selecting basic vectors with certain action durations, and the sum of the all durations is equal to switching period.

For example, by using the two effective basic vectors, $\vec{V}_1$ and $\vec{V}_2$, and zero vector $\vec{V}_0$, the reference voltage vector $\vec{V}_{ref}$ in sector 1 can be composed as shown in Figure A.1.4.

Figure A.1.4 Composition of reference voltage vector in sector 1.

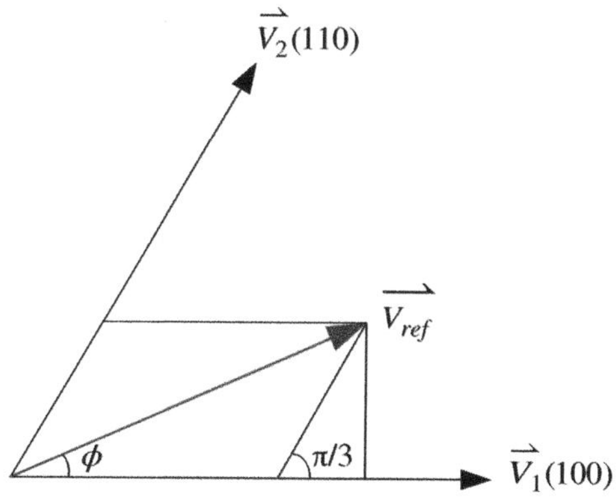

Based on Figure A.1.4, following equation can be obtained as:

$$\vec{V}_1 \cdot T_{V1} + \vec{V}_2 \cdot T_{V2} + \vec{V}_0 \cdot T_{V0} = \vec{V}_{ref} \tag{A.1.17}$$

Above complex number equation can be rewritten as following real number equations:

$$\begin{cases} \left|\vec{V}_{ref}\right| T_s \cos\varphi - \left|\vec{V}_{ref}\right| T_s \sin\varphi \tan\left(\frac{\pi}{2} - \frac{\pi}{3}\right) = \left|\vec{V}_1\right| T_{V1} \\ \left|\vec{V}_{ref}\right| T_s \sin\varphi \tan\left(\frac{\pi}{2} - \frac{\pi}{3}\right) \times 2 = \left|\vec{V}_2\right| T_{V2} \end{cases} \tag{A.1.18}$$

In addition,

$$T_{V1} + T_{V2} + T_{V0} = T_s \tag{A.1.19}$$

where $\left|\vec{V}_{ref}\right|$ is the magnitude of reference vector, $\left|\vec{V}_1\right|$ and $\left|\vec{V}_2\right|$ are the magnitudes of the two effective basic vectors and equal to $2V_{dc}/3$, φ is the phase angle of reference vector if the voltage drop across inductor is quite small, T_s is the switching period, T_{V1} and T_{V2} are the operation durations of the two effective basic vectors, and T_{V0} is the operation duration of zero vector. Thus, the expressions of the durations can be derived as:

$$\begin{cases} T_{V1} = mT_s \sin(\pi/3 - \varphi) \\ T_{V2} = mT_s \sin\varphi \\ T_{V0} = T_s - T_{V1} - T_{V2} \end{cases} \tag{A.1.20}$$

where m is called modulation ratio and can be expressed as:

$$m = \frac{\sqrt{3}\left|\vec{V}_{ref}\right|}{V_{dc}} \approx \frac{\sqrt{3}V_m}{V_{dc}} \tag{A.1.21}$$

where V_m is the amplitude of the phase voltage. It can also be used to represent the amplitude of the vector of inverter output phase voltage if the voltage drop across inductor is quite small.

For given vector composition, there are different ways to generate vector sequence in a switching period. Implementations of different vector sequence will affect the inverter output, total harmonic distortion (THD), and number of switching in a switching period, i.e. switching loss. Thus, many space vector modulation (SVM) schemes have been derived for three-phase converters.

A.2 Switching Patterns of SVM 12

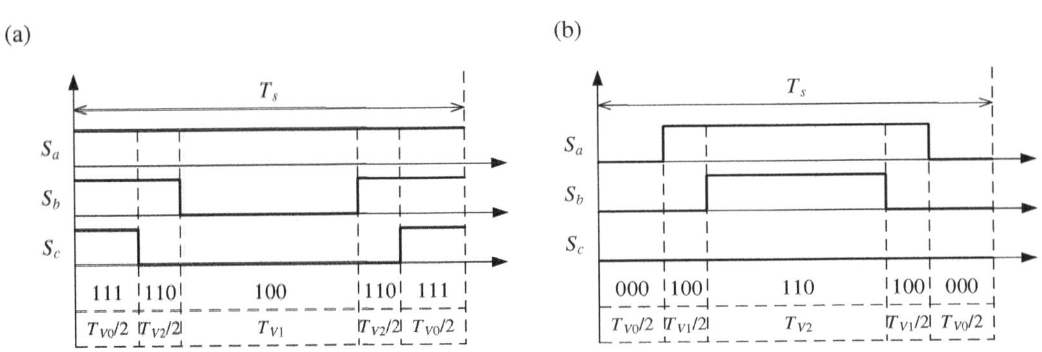

Figure A.2.1 Twelve switching patterns of SVM 12: (a) sector 1-1, (b) sector 1-2, (c) sector 2-1, (d) sector 2-2, (e) sector 3-1, (f) sector 3-2, (g) sector 4-1, (h) sector 4-2, (i) sector 5-1, (j) sector 5-2, (k) sector 6-1, and (l) sector 6-2.

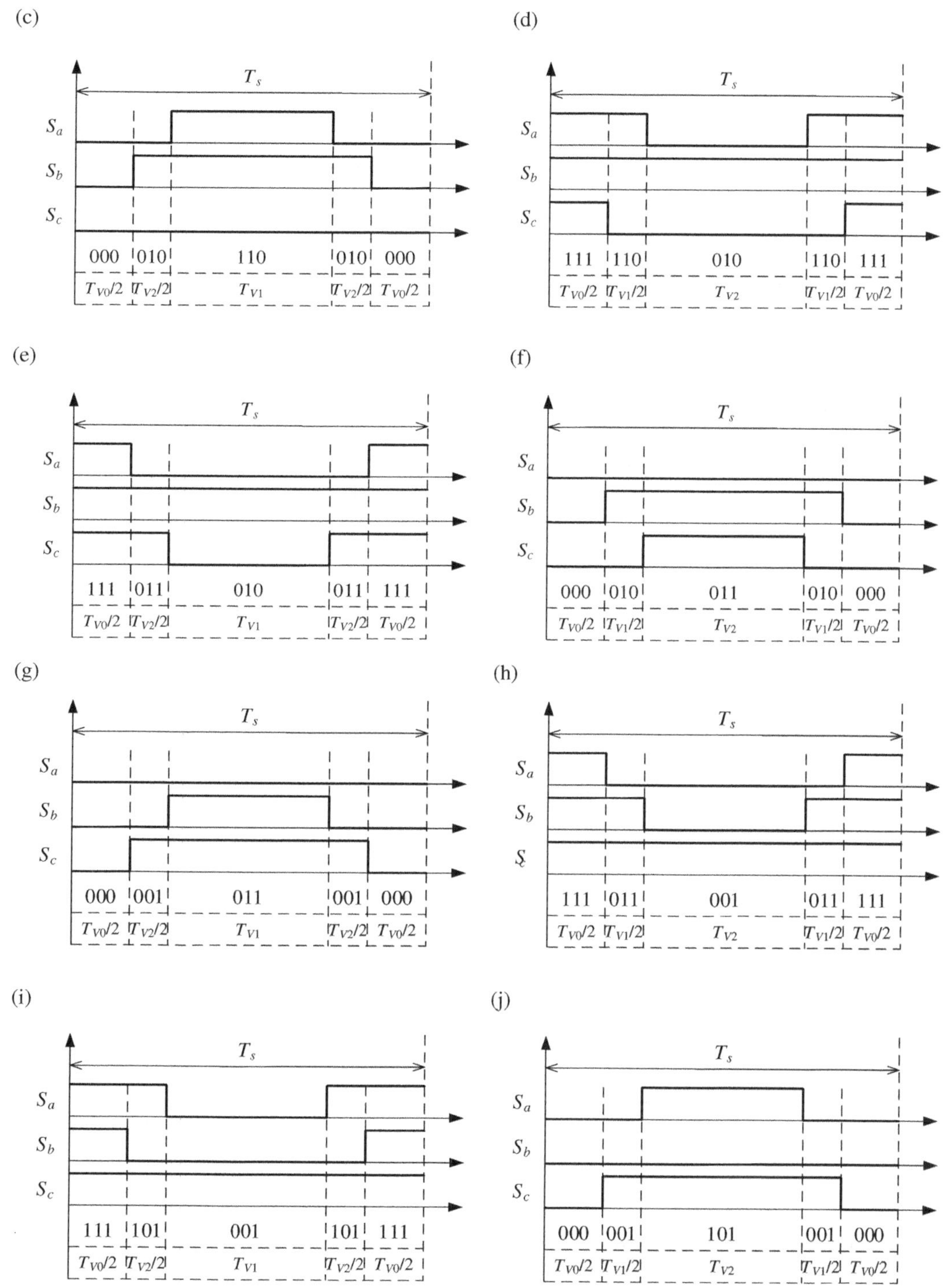

Figure A.2.1 (Continued)

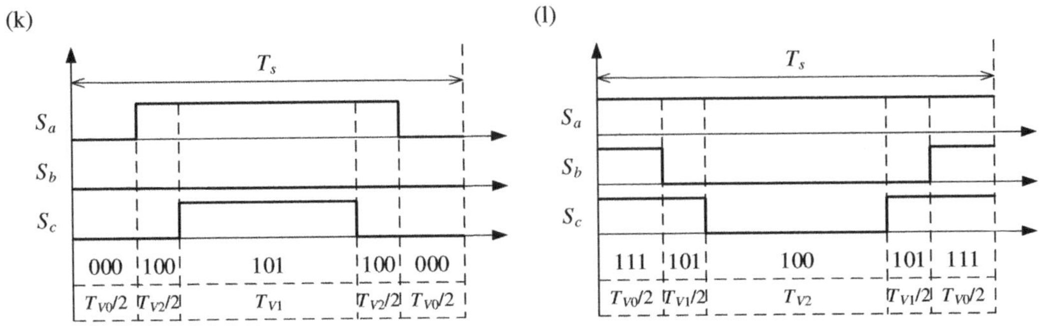

Figure A.2.1 (Continued)

A.3 Switching Patterns of ZVS-SVM

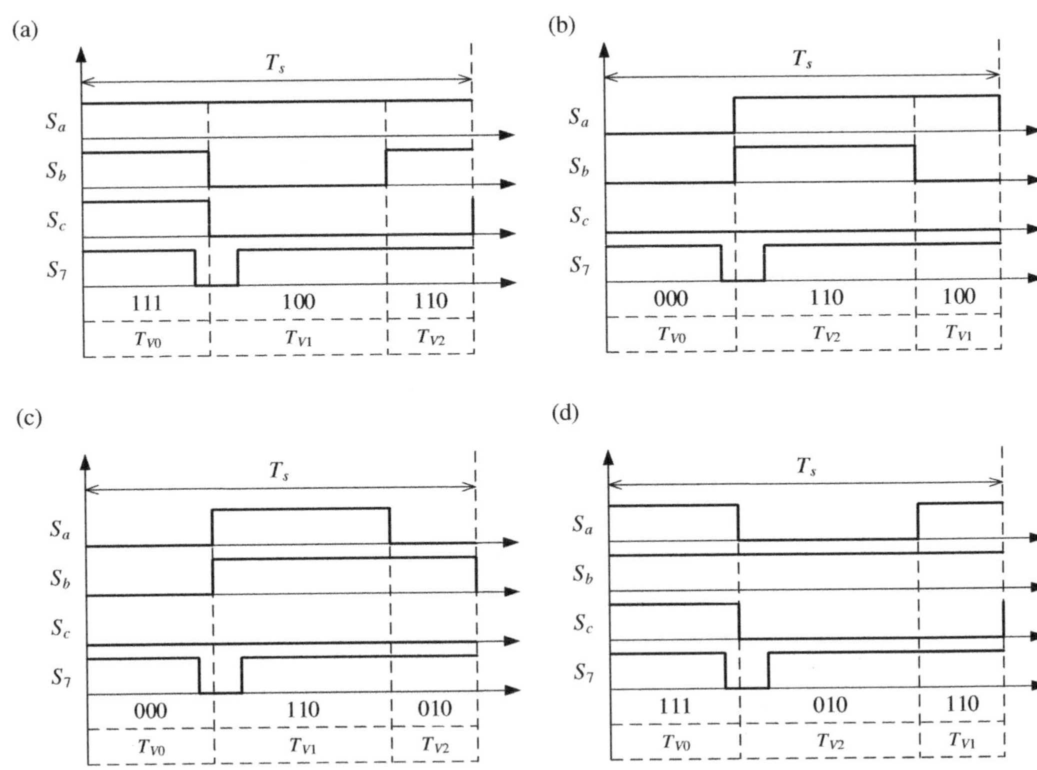

Figure A.3.1 Twelve switching patterns of ZVS-SVM: (a) sector 1-1, (b) sector 1-2, (c) sector 2-1, (d) sector 2-2, (e) sector 3-1, (f) sector 3-2, (g) sector 4-1, (h) sector 4-2, (i) sector 5-1, (j) sector 5-2, (k) sector 6-1, and (l) sector 6-2.

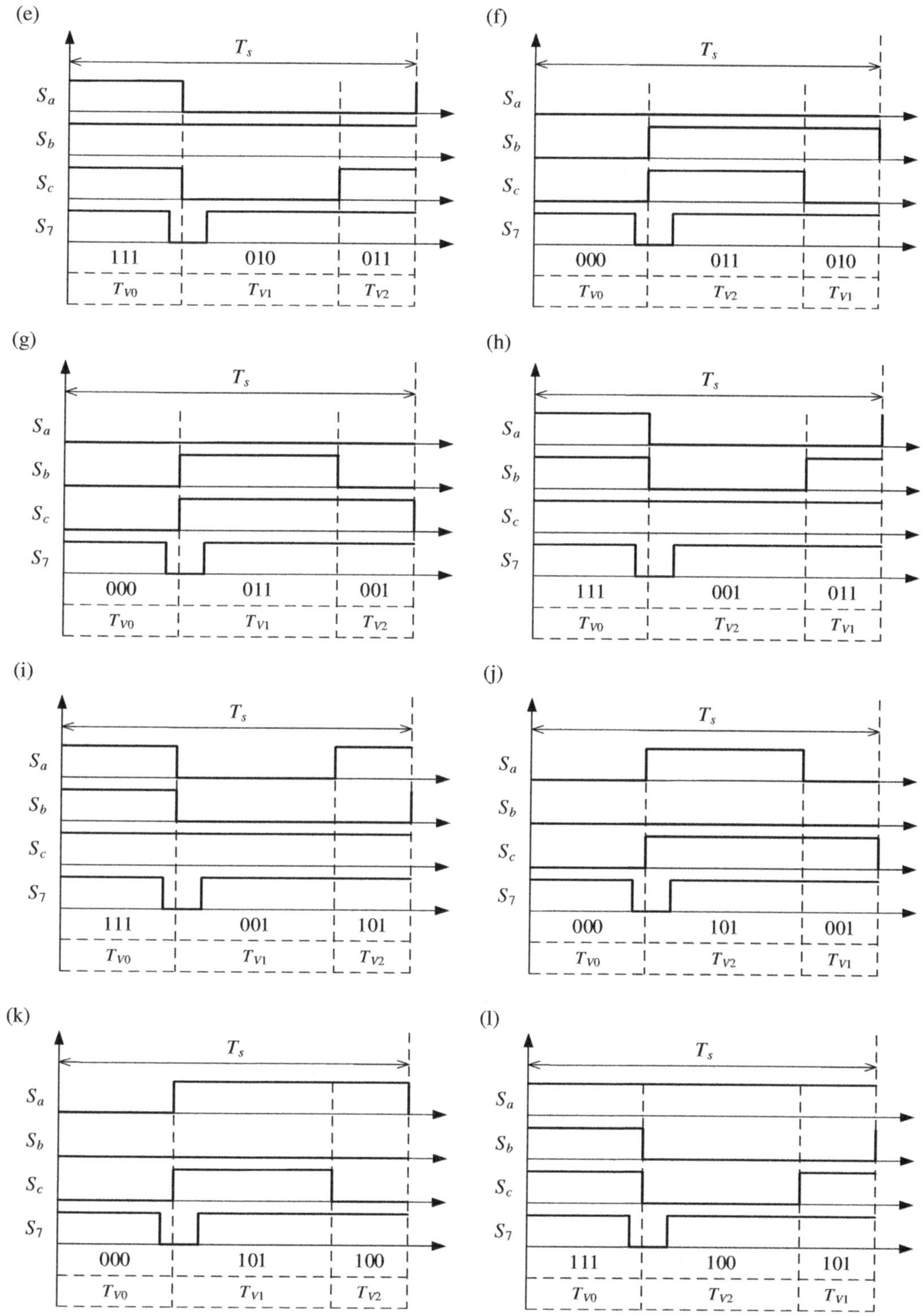

Figure A.3.1 (Continued)

A.4 Inverter Loss Models

The loss model of hard-switching three-phase grid inverter and soft-switching three-phase grid at unity power factor are introduced in the following paragraph. SiC metal-oxide-semiconductor field-effect transistor (MOSFET) device is taken as example to illustrate the loss models.

A.4.1 Loss Model of Hard-switching Three-phase Grid Inverter

Figure A.4.1 shows the topology of hard-switching three-phase grid inverter. The loss of hard-switching inverter can be mainly divided into the loss of switching devices with antiparalleled diodes (S_1, S_4, S_3, S_6, S_5, S_2, D_1, D_4, D_3, D_6, D_5, and D_2) in the three-phase main bridges and the loss in the AC filter inductors (L_a, L_b, and L_c). The loss of switching devices can be further divided into conducting loss and switching loss.

A.4.1.1 Conducting Loss

A line cycle is divided into 12 sectors in SVM 12. The number of switching period p in each sector can be calculated as:

$$p = \frac{f_s}{12 f_l} \tag{A.4.1}$$

where f_s is the switching frequency and f_l is the line frequency.

By considering the dead time and the switching pattern of SVM 12 in section A.2, the gate drive signals of all six main switches in sector 1-1 within one switching period T_s can be derived as shown in Figure A.4.2.

The phase angle φ in (A.1.18) can be discretized as:

$$\varphi(n) = \omega_l t(n) = 2\pi f_l \frac{n}{f_s} \tag{A.4.2}$$

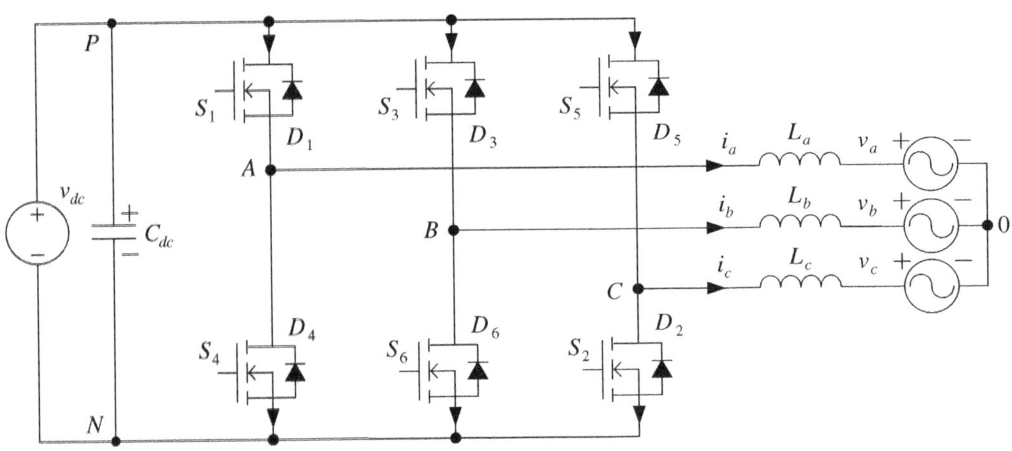

Figure A.4.1 Topology of hard-switching three-phase grid inverter.

Figure A.4.2 Driving sequence of SVM 12 in sector 1-1 within one switching period.

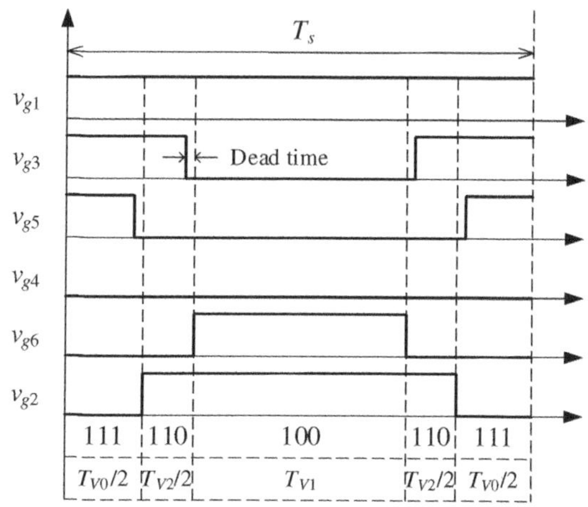

According to (A.1.20) and (A.4.2), the theoretical durations of three basic vectors in the nth switching period, $T_{V1}(n)$, $T_{V2}(n)$, and $T_{V0}(n)$, can be rewritten as:

$$\begin{cases} T_{V1}(n) = mT_s \sin\left[\frac{\pi}{3} - 2\pi f_l\left(\frac{n-k\cdot 2p}{f_s}\right)\right] \\ T_{V2}(n) = mT_s \sin\left[2\pi f_l\left(\frac{n-k\cdot 2p}{f_s}\right)\right] \\ T_{V0}(n) = T_s - T_1(n) - T_2(n) \\ \mathrm{m} = \frac{\sqrt{3}V_m}{V_{dc}} \end{cases} \tag{A.4.3}$$

where n represents the nth switching period, k represents the location of the reference vector. k is defined as:

$$k = \begin{cases} 0 & \text{Sector 1-1, Sector 1-2} \\ 1 & \text{Sector 2-1, Sector 2-2} \\ 2 & \text{Sector 3-1, Sector 3-2} \\ 3 & \text{Sector 4-1, Sector 4-2} \\ 4 & \text{Sector 5-1, Sector 5-2} \\ 5 & \text{Sector 6-1, Sector 6-2} \end{cases} \tag{A.4.4}$$

According to sections A.1 and A.2, the conducting time of phase A's main switches S_1 and S_4 with SVM 12, T_{on_S1} and T_{on_S4}, can be summarized as:

$$T_{on\_S1}(n) = \begin{cases} T_s & \text{Sector 1-1, Sector 6-2} \\ T_{V1}(n) + T_{V2}(n) & \text{Sector 1-2, Sector 6-1} \\ T_{V1}(n) & \text{Sector 2-1} \\ T_{V1}(n) + T_{V0}(n) & \text{Sector 2-2} \\ T_{V0}(n) & \text{Sector 3-1, Sector 4-2} \\ 0 & \text{Sector 3-2, Sector 4-1} \\ T_{V2}(n) + T_{V0}(n) & \text{Sector 5-1} \\ T_{V2}(n) & \text{Sector 5-2} \end{cases} \tag{A.4.5}$$

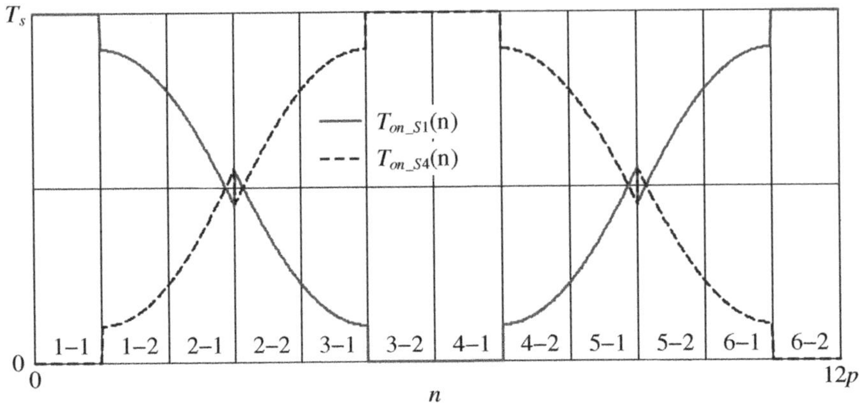

Figure A.4.3 Conduction time of phase A's main switches S_1 and S_4 in one line cycle.

$$T_{on\_S4}(n) = T_s - T_{on\_S1}(n) \tag{A.4.6}$$

Based on (A.4.5) and (A.4.6), the typical conduction time of main switches S_1 and S_4 in one line cycle without considering the dead time is shown in Figure A.4.3.

If the grid current ripple is ignored and the grid current is assumed to be constant in each switching period, the three-phase grid currents at unity power factor can be expressed as:

$$\begin{cases} i_a(n) = I_m \cos(\varphi(n)) = I_m \cos\left(2\pi f_l \frac{n}{f_s}\right) \\ i_b(n) = I_m \cos\left(\varphi(n) - \frac{2\pi}{3}\right) = I_m \cos\left(2\pi f_l \frac{n}{f_s} - \frac{2\pi}{3}\right) \\ i_c(n) = I_m \cos\left(\varphi(n) + \frac{2\pi}{3}\right) = I_m \cos\left(2\pi f_l \frac{n}{f_s} + \frac{2\pi}{3}\right) \end{cases} \tag{A.4.7}$$

If the polarity of grid current is positive, it is flowing in the channel of SiC MOSFET S_1 when S_1 is on or both the channel and antiparalleled diode of SiC MOSFET S_4 in synchronous rectification mode when S_4 is on. If the polarity of grid current is negative, it is flowing in both the channel and antiparalleled diode of SiC MOSFET S_1 in synchronous rectification mode when S_1 is on or the channel of SiC MOSFET S_4 when S_4 is on. By considering the dead time T_{dead}, the actual conducting time $T_{up}(n)$ of SiC MOSFET S_1's channel within the first 1/4 line cycle can be induced as:

$$T_{up}(n) = \begin{cases} T_s & \text{Sector 1-1} \\ T_{V1}(n) + T_{V2}(n) - T_{dead} & \text{Sector 1-2} \\ T_{V1}(n) - T_{dead} & \text{Sector 2-1} \end{cases} \tag{A.4.8}$$

The actual conducting time $T_{down}(n)$ of SiC MOSFET S_4 in synchronous rectification mode within the first 1/4 line cycle can also be obtained as:

$$T_{down}(n) = \begin{cases} 0 & \text{Sector 1-1} \\ T_{V0}(n) - T_{dead} & \text{Sector 1-2} \\ T_{V2}(n) + T_{V0}(n) - T_{dead} & \text{Sector 2-1} \end{cases} \tag{A.4.9}$$

The conducting time of SiC MOSFET S_4's antiparalleled diode has extra conducting duration of $2 \times T_{dead}$ due to the dead time in a switching period of sector 1-2 and sector 2-1. Thus, by considering the symmetry of phase A's grid current in one line cycle and the symmetry of three phase, the conduction loss of three-phase main switches' channel P_{conM3}, the conduction loss of three-phase main switches in synchronous rectification mode P_{conSR3}, and the conduction loss of three-phase main switches' antiparalleled diodes P_{conD3} can be derived as:

$$P_{conM3} = 3 \cdot f_l \cdot 4 \sum_{n=1}^{3p} \left[|i_a(n)|^2 \cdot R_{ds(on)} \cdot T_{up}(n) \right] \tag{A.4.10}$$

$$P_{conSR3} = 3 \cdot f_l \cdot 4 \sum_{n=1}^{3p} \left[v_{sd}(n) \cdot i_a(n) \cdot T_{down}(n) \right] \tag{A.4.11}$$

$$P_{conD3} = 3 \cdot f_l \cdot 4 \sum_{n=1}^{3p} \left[v_f(n) \cdot i_a(n) \cdot 2T_{dead} \right] \tag{A.4.12}$$

where $R_{ds(on)}$ is the drain-source on-state resistance of SiC MOSFET, $v_{sd}(n)$ is the source-drain voltage drop in synchronous rectification mode with grid current $i_a(n)$, and $v_f(n)$ is the forward voltage of antiparalleled diode with grid current $i_a(n)$.

Thus the total conducting loss of three-phase main switches can be obtained as:

$$P_{con3} = P_{conM3} + P_{conSR3} + P_{conD3} \tag{A.4.13}$$

A.4.1.2 Switching Loss

As shown in Figure A.4.2, there are no switching actions for S_1 and S_4 of phase A in sector 1-1. Meanwhile, each of the other phase has one Type 2 switch commutation from diode to SiC MOSFET and another Type 1 current commutation from SiC MOSFET to diode in each switching period. By considering the polarities of grid currents of phase B and phase C, the commutation from diode to SiC MOSFET brings reverse recovery loss in their antiparalleled diodes and turn-on loss in SiC MOSFET. Besides, the commutation from SiC MOSFET to diode brings turn-off loss in SiC MOSFET and no turn-on loss in their antiparalleled diode.

As shown in Figure A.4.3, the switching actions of phase A exist in sector 1-2 and sector 2-1 within the first 1/4 line cycle. Thus, by considering the symmetry of phase A's grid current in one line cycle and the symmetry of three-phase grid currents, the turn-on loss P_{on3}, turn-off loss P_{off3}, and the reverse recovery loss P_{rr3} of three-phase main switches can be derived as:

$$P_{on3} = 3f_l \left[4 \sum_{n=p+1}^{3p} E_{on}(|i_a(n)|, V_{dc}) \right] \tag{A.4.14}$$

$$P_{off3} = 3f_l \left[4 \sum_{n=p+1}^{3p} E_{off}(|i_a(n)|, V_{dc}) \right] \tag{A.4.15}$$

$$P_{rr3} = 3f_l \left[4 \sum_{n=p+1}^{3p} E_{rr}(|i_a(n)|, V_{dc}) \right] \tag{A.4.16}$$

where E_{on3}, E_{off3}, and E_{rr3} are the turn-on energy, turn-off energy, and reverse recovery energy for once switching process, which are related to the conducting current i_a and the DC-side voltage V_{dc} during the switching process. E_{on3}, E_{off3}, and E_{rr3} can be obtained from the datasheet of SiC MOSFET or experimental results of double-pulse test (DPT). The total switching loss of three-phase main bridges P_{sw3} can be obtained as:

$$P_{sw3} = P_{on3} + P_{off3} + P_{rr3} \tag{A.4.17}$$

A.4.1.3 AC Filter Inductor Loss and Volume Estimations

As the switching frequency of SiC inverter is always set much higher than Si insulated gate bipolar transistor (IGBT) inverter, a pair of E-shaped (EE) cores made up of ferrite material with lower core loss is adopted in the AC filter inductor for SiC inverter. The AC filter inductor of the hard-switching inverter is designed based on the widely used Area Product method. With the required inductance L and the maximum inductor current I_{max}, the energy-handling capability of the required core E_L can be expressed as:

$$E_L = \frac{1}{2} L I_{max}{}^2, \quad [\mathrm{J}] \tag{A.4.18}$$

and E_L of the required core is related to core's area product, AP, by the equation:

$$AP = \frac{2E_L \times 10^4}{B_{max} J K_u}, \quad [\mathrm{cm}^4] \tag{A.4.19}$$

where B_{max} is the maximum flux density, J is the current density, and K_u is the window utilization factor. The EE cores with the closest AP can be selected according to (A.4.18) and (A.4.19). The typical EE cores with dimensions are shown in Figure A.4.4.

If the number of turns of winding is defined as N_w, the air gap length is defined as l_g, the effective magnetic path length is defined as MPL, the material permeability is defined as μ_m, and the iron area is defined as A_c, the inductance L and the maximum flux density B_m meet:

$$L = \frac{0.4\pi N_w^2 A_c \times 10^{-8}}{l_g + \frac{MPL}{\mu_m}}, \quad [\mathrm{H}] \tag{A.4.20}$$

$$B_{max} = \frac{0.4\pi N_w I_{max} \times 10^{-4}}{l_g + \frac{MPL}{\mu_m}}, \quad [\mathrm{T}] \tag{A.4.21}$$

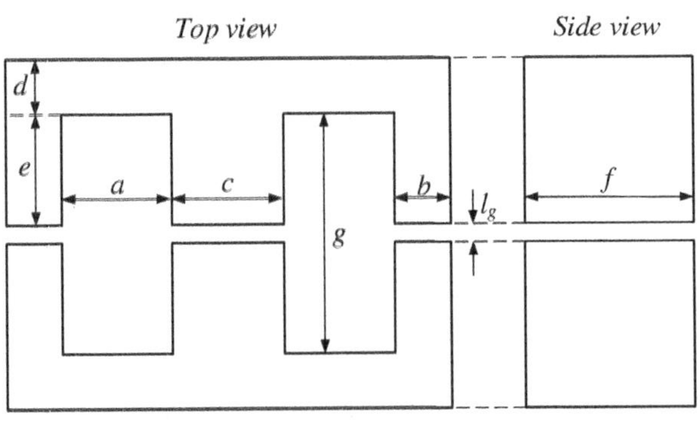

Figure A.4.4 Typical EE cores with dimensions.

where the iron area A_c, the effective magnetic path length MPL, and the material permeability μ_m can be obtained from the datasheet of selected cores. According to the relationship between (A.4.20) and (A.4.21), the number of winding turns N_w can be derived as:

$$N_w = \frac{LI_{max}}{B_{max}A_c \times 10^{-4}} \tag{A.4.22}$$

By considering the fringing flux effect caused by the air gap, the fringing flux factor F is:

$$F = 1 + \frac{l_g}{\sqrt{A_c}} \ln \frac{2g}{l_g} \tag{A.4.23}$$

where g is the window height of EE cores as shown in Figure A.4.4. Thus the inductance L can be modified as:

$$L = \frac{0.4\pi N_w^2 F A_c \times 10^{-8}}{l_g + \frac{MPL}{\mu_m}}, \ [\text{H}] \tag{A.4.24}$$

By substituting (A.4.23) into (A.4.24), with the required L, B_{max}, and winding turns N_w in (A.4.22), the air gap length l_g can be derived. Litz wire is selected as the winding material due to its low AC resistance at high frequency inductor current. With the designed parameters, the inductor loss and the physical size of inductor can be evaluated.

The loss in AC filter inductor can be divided into winding loss and core loss. The winding loss of three-phase filter inductors P_{L_Cu} can be calculated as:

$$P_{L\_Cu} = 3\left(R_0 I_0{}^2 + \sum R_h I_h{}^2\right) \tag{A.4.25}$$

where R_0 is the AC resistance of winding at fundamental frequency, I_0 is the root-mean-square (RMS) value of fundamental frequency component of inductor current, R_h is the AC resistance of winding at harmonic frequency, and I_h is the RMS value of harmonic frequency component of inductor current. The core loss can also be evaluated by using Superposition principle. The flux density B_0 of fundamental frequency component of inductor current and the flux density of harmonic frequency component of inductor current B_h can be calculated based on (A.4.21) as:

$$B_0 = \frac{0.4\pi N_w \sqrt{2} I_0 \times 10^{-4}}{l_g + \frac{MPL}{\mu_m}}, \ [\text{T}] \tag{A.4.26}$$

$$B_h = \frac{0.4\pi N_w \sqrt{2} I_h \times 10^{-4}}{l_g + \frac{MPL}{\mu_m}}, \ [\text{T}] \tag{A.4.27}$$

Then the core loss of per unit volume, P_{cv0} and P_{cvh}, caused by fundamental frequency component and harmonic frequency components can be obtained with datasheet look-up.

Finally, the core loss of three-phase filter inductors P_{L_Fe} can be calculated as:

$$P_{L\_Fe} = 3\left(P_{cv0} + \sum P_{cvh}\right) V_{core} \tag{A.4.28}$$

where V_{core} is the volume of EE cores of each phase's filter inductor and can also be obtained from the datasheet of selected EE cores.

If the window is assumed to be full filled with winding and the winding thickness is a, by ignoring the narrow air gap the volume of three-phase filter inductors can be estimated with the dimensions shown in Figure A.4.4 as:

$$V_L = 3(2a + c + 2b)(2d + 2e)(2a + f) \tag{A.4.29}$$

A.4.2 Loss Model of Soft-switching Three-phase Grid Inverter

Zero-voltage-switching (ZVS) SVM scheme in sector 1-1 is shown in Figure A.4.5. In comparison to the SVM 12 scheme in Figure A.4.2, the two effective vectors of three-phase output voltage in ZVS-SVM are still 100 and 110. However, the two commutations from diodes to MOSFETs of the main bridges are aligned at the same instant in each switching period. As an auxiliary switch S_7 is added in the DC side as well as the resonant inductor and clamping capacitor, the DC-side voltage across the three-phase main bridges will be resonated to zero by turning off the auxiliary switch S_7, which creates the ZVS turn-on condition for S_2 and S_6. A short-circuit stage is also added to S_3, S_5, and S_4 to increase the stored energy in resonant inductor.

In comparing with the hard-switching SiC inverter, although the soft-switching SiC inverter has eliminated the turn-on loss and reverse recovery loss in all SiC MOSFET, it introduces extra turn-off loss in main bridges due to the short-circuit stage and the loss in added components including the auxiliary switch and the resonant inductor. Besides, resonant capacitors are paralleled to switches and their impact on turn-off process of main switches and auxiliary switch has to be considered.

Figure A.4.6 shows the waveforms of the key components in the soft-switching three-phase inverter in one switching period.

A.4.2.1 Loss in Main Switches

As the operation time of auxiliary circuit is very short, typically below 10% of switching period, its impact on conduction loss of main switches can be ignored. Therefore, the conduction loss of main switches in soft-switching inverter can also be calculated with equations (A.4.10), (A.4.11), and (A.4.12). As shown in Figure A.4.6, the three-phase output voltage vector when auxiliary switch is off is equivalent to 000, which will not impact the output voltage. Besides, due to the existence

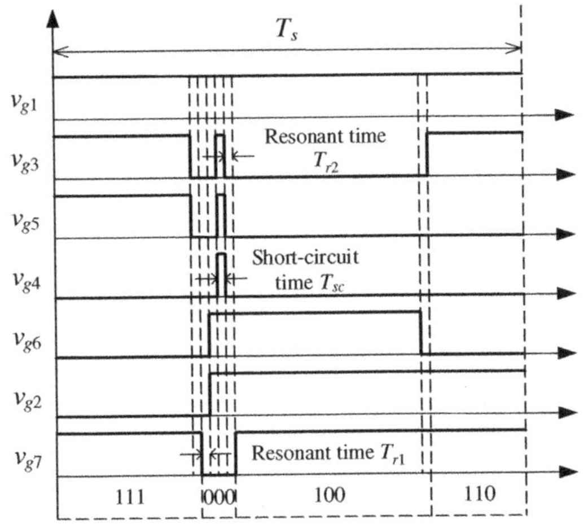

Figure A.4.5 Driving sequence of ZVS-SVM in sector 1-1 within one switching period.

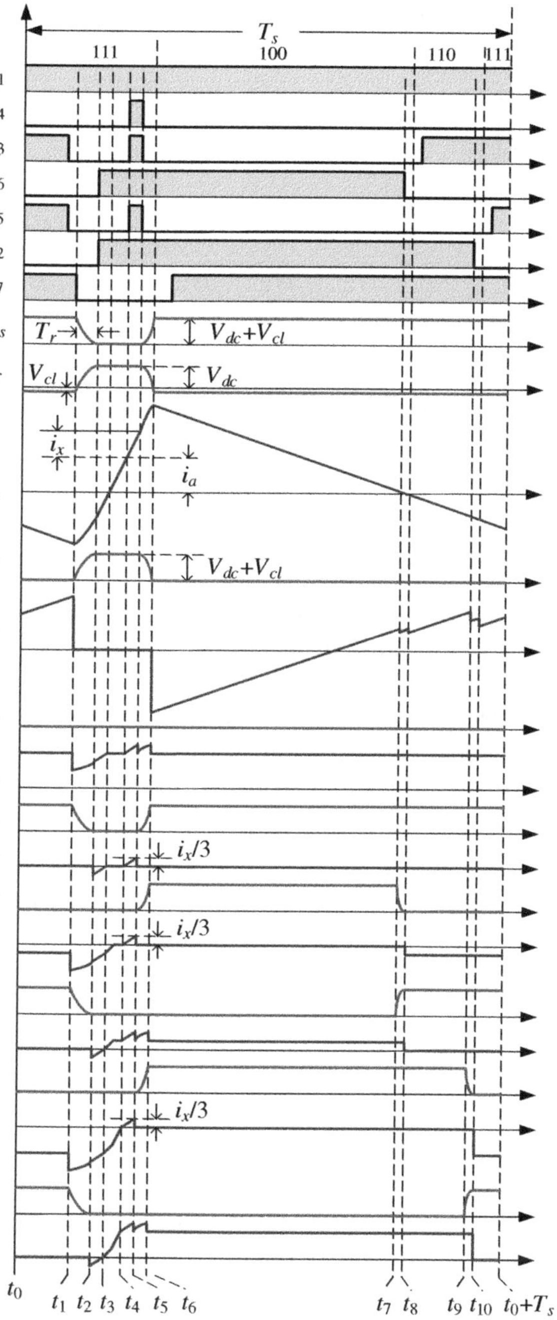

Figure A.4.6 Key waveforms in sector 1-1.

of clamping capacitor the equivalent DC-side voltage has been increased to $V_{cl} + V_{dc}$ when auxiliary switch is on. The operation time of three vectors should be modified as:

$$\begin{cases} T_{V1}(n) = MT_s \sin\left[\frac{\pi}{3} - 2\pi f_l\left(\frac{n-k\cdot 2p}{f_s}\right)\right] \\ T_{V2}(n) = MT_s \sin\left[2\pi f_l\left(\frac{n-k\cdot 2p}{f_s}\right)\right] \\ T_{V0}(n) = T_s - T_{V1}(n) - T_{V2}(n) \\ M = \frac{\sqrt{3}V_m}{V_{dc}+V_{cl}} \end{cases} \tag{A.4.30}$$

The turn-off loss P_{offA} of phase A's main switches can be obtained by modifying equation (A.4.15) as:

$$P_{offA} = f_l\left[4\cdot\sum_{n=p+1}^{3p} E_{off}(|i_a(n)|, (V_{dc}+V_{cl}), 2C_r)\right] \tag{A.4.31}$$

where $2C_r$ is the equivalent paralleled capacitor for each main switch, $V_{cl} + V_{dc}$ is the equivalent DC-side voltage for main switches, E_{off} is the turn-off energy and obtained by DPT experiment.

According to the ZVS-SVM scheme, the extra added current $i_x(n)$ on resonant current is repetitive in 12 subsectors and equally distributed in three-phase bridges as shown in Figure A.4.6. Each switch shares one-third of the current i_x. The turn-off loss P_{offA_sc} of phase A's main switches after short-circuit stage can be derived as:

$$P_{offA\_sc} = f_l\left[12\cdot\sum_{n=1}^{p} E_{off}\left(\frac{i_x(n)}{3}, (V_{dc}+V_{cl}), \frac{(3C_r+C_7)}{3}\right)\right] \tag{A.4.32}$$

Thus the total switching loss of three-phase main switches in soft-switching inverter can be obtained as:

$$P_{off3} = 3(P_{offA} + P_{offA\_sc}) \tag{A.4.33}$$

A.4.2.2 Loss in Auxiliary Switch

According to the ZVS-SVM scheme, the operation of auxiliary switch S_7 is repetitive in 12 subsectors. In order to simplify the loss analysis, the same drain-source on-state resistance $R_{ds(on)}$ will be used to calculate the conduction loss in auxiliary switch since the major part of the current in auxiliary switch is also flowing through the SiC MOSFET's channel in synchronous rectification mode. Then the conduction loss in auxiliary switch P_{consS7} can be derived as:

$$P_{consS7} = f_l\cdot 12\cdot\sum_{n=1}^{p}\left[\int_0^{T_s} R_{ds(on)}\cdot i_{S7}(n,t)^2 dt\right] \tag{A.4.34}$$

where $i_{S7}(n,t)$ is the time-domain expression of current in auxiliary switch in nth switching period as shown in Figure A.4.6. The turn-off current in auxiliary switch $i_{S7}(t_1)$ is equal to $-i_{Lr}(t_1)$ as shown in Figure 7.9. According to equation (7.49), following equation is obtained:

$$i_{S7}(t_1) = \frac{V_{dc}}{Z_r} = \frac{V_{dc}}{\sqrt{L_r/(3C_r + C_7)}} \tag{A.4.35}$$

Then the turn-off loss in auxiliary switch P_{offS7} can be obtained as:

$$P_{offS7} = f_l \sum_{n=1}^{12p} E_{off}(i_{S7}(t_1), (V_{dc} + V_{cl}), (3C_r + C_7)) \tag{A.4.36}$$

As shown in (A.4.36), total equivalent resonant capacitor during turn-off of the auxiliary switch is $3C_r + C_7$.

A.4.2.3 Loss and Volume of Filter Inductor and Resonant Inductor

The design procedures of AC filter inductor and resonant inductor of soft-switching inverter are similar to that of filter inductor of hard-switching inverter. However, copper foil is used instead of Litz wire in order to increase core's window utilization in resonant inductor. The loss of AC filter inductor and resonant inductor can also be calculated by using equations (A.4.25) and (A.4.28). Their volumes can also be estimated by equation (A.4.29).

A.5 AC Filter Inductance Calculation

Figure A.5.1 shows the topology of soft-switching SiC three-phase grid inverter. If the inverter operates at unity power factor, the three-phase grid currents can be expressed as:

$$\begin{cases} i_a = I_m \cos(\varphi) \\ i_b = I_m \cos(\varphi - 2\pi/3) \\ i_c = I_m \cos(\varphi + 2\pi/3) \end{cases} \tag{A.5.1}$$

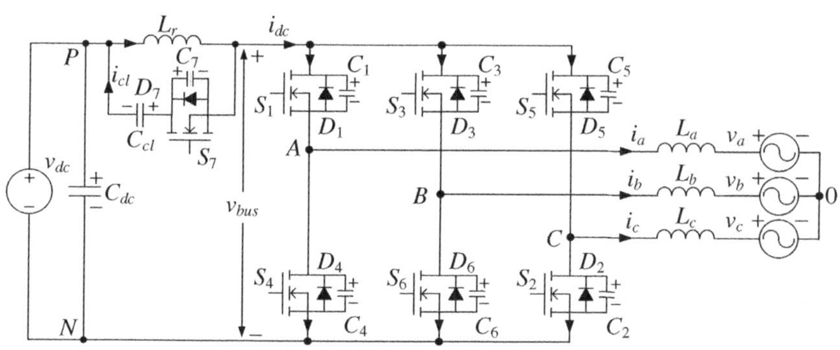

Figure A.5.1 Topology of soft-switching three-phase grid inverter.

Based on (A.1.13), three-phase voltages v_{A0}, v_{B0}, and v_{C0} can be obtained as:

$$\begin{cases} v_{A0} = L\dfrac{di_a}{dt} + v_a = \left(S_a - \dfrac{S_a + S_b + S_c}{3}\right)(V_{dc} + V_{cl}) \\ v_{B0} = L\dfrac{di_b}{dt} + v_b = \left(S_b - \dfrac{S_a + S_b + S_c}{3}\right)(V_{dc} + V_{cl}) \\ v_{C0} = L\dfrac{di_c}{dt} + v_c = \left(S_c - \dfrac{S_a + S_b + S_c}{3}\right)(V_{dc} + V_{cl}) \end{cases} \tag{A.5.2}$$

Taking phase A as an example, voltage v_{A0} at different vectors can be calculated as:

$$v_{A0} = \begin{cases} -\dfrac{2}{3}(V_{dc} + V_{cl}) & \text{Vector 011} \\ -\dfrac{1}{3}(V_{dc} + V_{cl}) & \text{Vector 010, Vector 001} \\ 0 & \text{Vector 000, Vector 111} \\ \dfrac{1}{3}(V_{dc} + V_{cl}) & \text{Vector 110, Vector 101} \\ \dfrac{2}{3}(V_{dc} + V_{cl}) & \text{Vector 100} \end{cases} \tag{A.5.3}$$

According to the symmetry of phase A's AC filter inductor current, the relationship between voltage v_{A0} and grid voltage v_a in three subsectors within the first 1/4 line cycle will be analyzed as follows. In sectors 1-1 and 1-2 effective vectors 100 and 110 are used while in sector 2-1 effective vectors 110 and 010 are used. Then the relationship between voltage v_{A0} and grid voltage v_a can be drawn as shown in Figure A.5.2.

As shown in Figure A.5.2 (b), the phase angle α in sector 1-2 meets:

$$\frac{V_{dc} + V_{cl}}{3} = u_a(\alpha) = V_m \cos\alpha \tag{A.5.4}$$

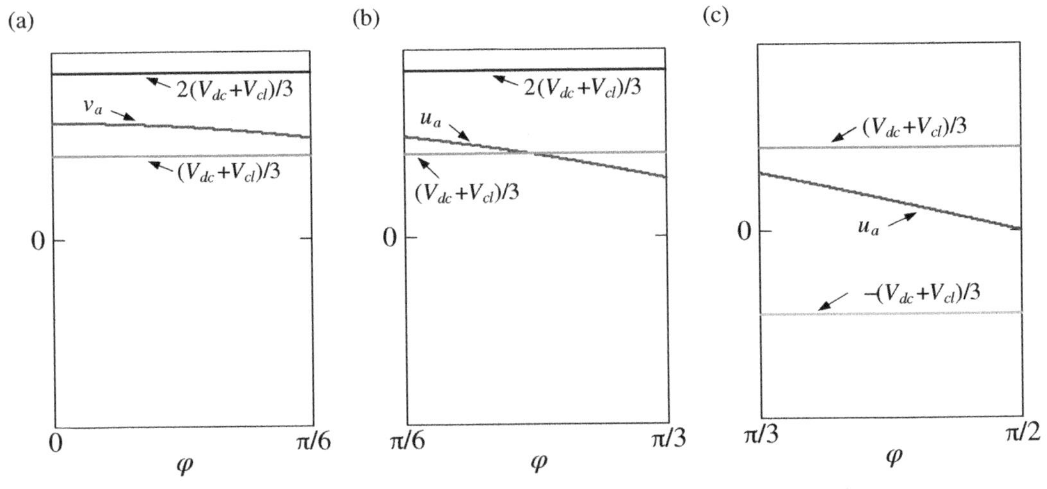

Figure A.5.2 Relationship between voltage v_{A0} and grid voltage v_a within 1/4 the first line cycle: (a) sector 1-1, (b) sector 1-2, and (c) sector 2-1.

Then the expression of phase angle α can be obtained as:

$$\alpha = \arccos\left(\frac{V_{dc} + V_{cl}}{3V_m}\right) \tag{A.5.5}$$

Based on the relationship between voltage v_{A0} and grid voltage v_a, the waveforms of switching functions S_a, S_b, and S_c; voltage v_{A0}; grid voltage v_a; and phase A' AC filter inductor current ripple within the first 1/4 line cycle can be derived as shown in Figure A.5.3. Then the expression of AC filter inductor current ripple Δi_a can be derived as:

$$\Delta i_a = \begin{cases} \frac{1}{L}\left[\frac{2}{3}(V_{dc} + V_{cl}) - v_a\right] T_{V1}(\varphi) & 0 \le \varphi < \alpha \\ \frac{1}{L} v_a T_{V0}(\varphi) & \alpha \le \varphi < \frac{\pi}{3} \\ \frac{1}{L}\left[\frac{1}{3}(V_{dc} + V_{cl}) - v_a\right] T_{V1}\left(\varphi - \frac{\pi}{3}\right) & \frac{\pi}{3} \le \varphi \le \frac{\pi}{2} \end{cases} \tag{A.5.6}$$

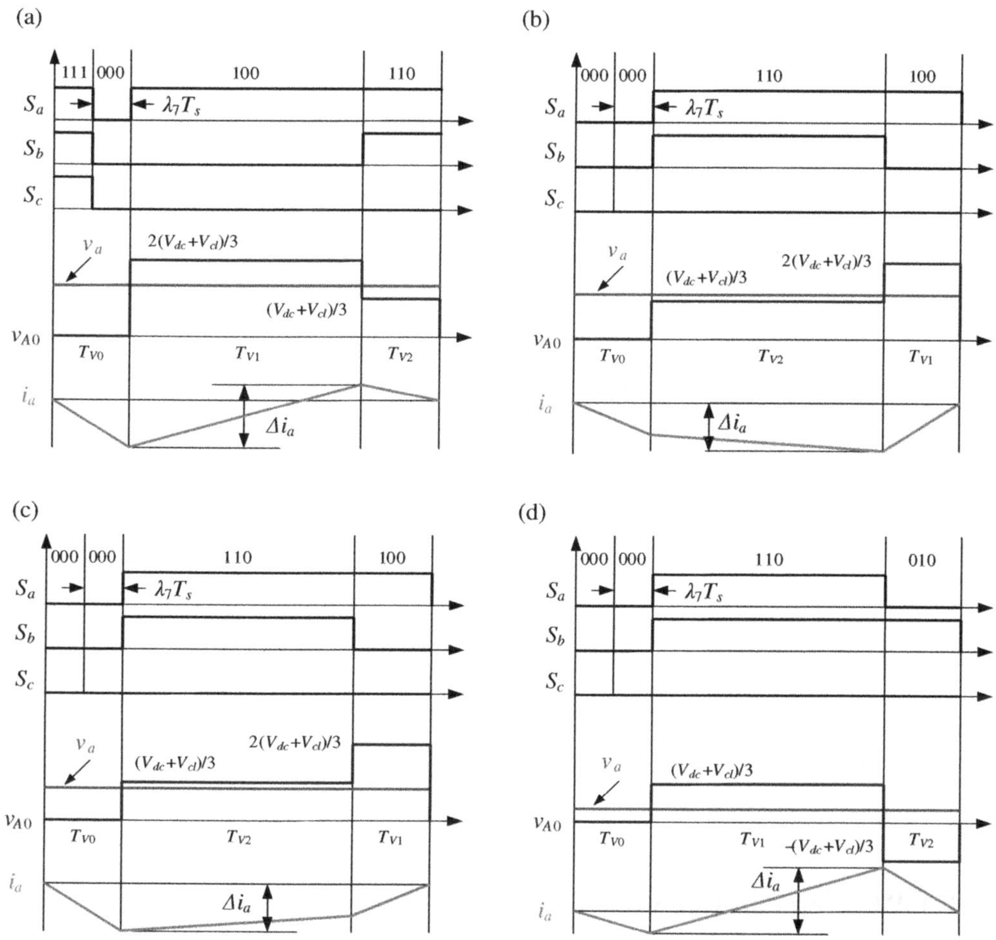

Figure A.5.3 Waveforms derivation of AC filter inductor current ripple with the first 1/4 line cycle: (a) sector 1-1, (b) sector 1-2 within phase angle from $\pi/6$ to α, (c) sector 1-2 within phase angle from α to $\pi/3$, and (d) sector 2-1.

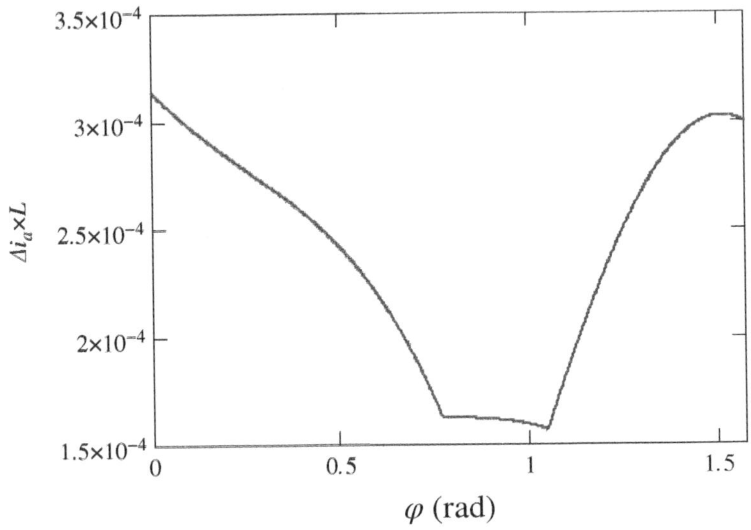

Figure A.5.4 $\Delta i_a \times L$ within the first 1/4 line cycle.

where voltage $V_{dc} + V_{cl}$ and operation time of effective vectors and zero vectors, $T_{V1}(\varphi)$, $T_{V2}(\varphi)$, and $T_{V0}(\varphi)$, meet:

$$V_{dc} + V_{cl} = \frac{V_{dc}}{1-\lambda_7} \tag{A.5.7}$$

$$\begin{cases} T_{V1}(\varphi) = MT_s \sin(\pi/3 - \varphi) \\ T_{V2}(\varphi) = MT_s \sin\varphi \\ T_{V0}(\varphi) = T_s - T_{V1}(\varphi) - T_{V2}(\varphi) \end{cases} \tag{A.5.8}$$

where λ_7 is the turn-off duty cycle of auxiliary switch.

With the parameters in Table 10.1 and turn-off duty cycle of auxiliary switch λ_7 is equal to 0.1, the value of $\Delta i_a \times L$ within the first 1/4 line cycle is calculated by using equation (A.5.6) as shown in Figure A.5.4.

Obviously, the maximum value is obtained at phase angle φ equal to 0 and its expression can be simplified as:

$$\Delta i_{a(\max)} = \frac{1}{L}\left[\frac{2}{3}(V_{dc} + V_{cl}) - V_m \cos(0)\right] T_{V1}(0) = \frac{3V_m}{2V_{dc} f_s L}\left[\frac{2}{3}V_{dc} - V_m(1-\lambda_7)\right] \tag{A.5.9}$$

Then the AC filter inductance L can be expressed with the maximum current ripple $\Delta i_{a(\max)}$ as:

$$L = \frac{3V_m}{2\Delta i_{a(\max)} V_{dc} f_s}\left[\frac{2}{3}V_{dc} - V_m(1-\lambda_7)\right] \tag{A.5.10}$$

A.6 DC Filter Capacitance Calculation

DC filter capacitor is used to suppress the DC-side voltage ripple due to the injection of the switched AC filter inductor current. The DC filter capacitor voltage ripple in soft-switching inverter will be analyzed first. Three-phase soft-switching grid inverter is assumed to be operating at unity power

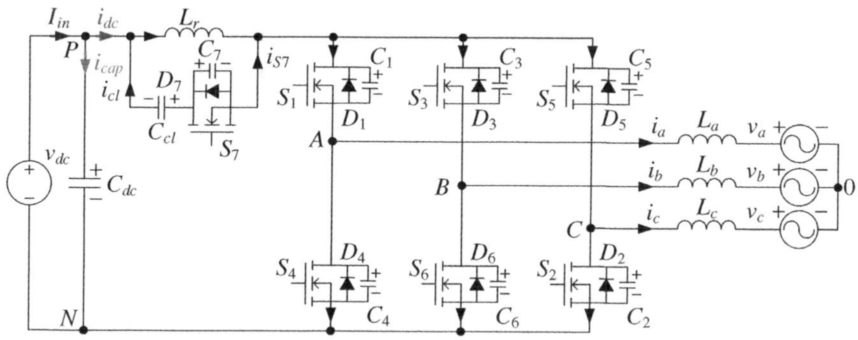

Figure A.6.1 Definitions of DC-side currents.

factor with balanced and ripple-free three-phase grid currents. The definitions of DC bus current i_{dc}, DC filter capacitor current i_{cap}, and DC input current I_{in} are shown in Figure A.6.1. Their relationship meets:

$$i_{cap} = I_{in} - i_{dc} \tag{A.6.1}$$

The waveforms of resonant inductor i_{Lr} and DC bus current i_{dc} in one switching period within sector 1-1 are shown in Figure A.6.2. It is clear that there are four stages with different output voltage vectors. When auxiliary is turned off, DC bus current i_{dc} is equal to resonant inductor i_{Lr}, the equivalent output voltage vector is 000, and the duration of this stage is $\lambda_7 T_s$. The duration of stage with output voltage vector 111 is $T_{V0}-\lambda_7 T_s$, in which DC bus current i_{dc} is equal to zero. The duration of stage with output voltage vector 100 is T_{V1}, in which DC bus current i_{dc} is equal to phase A's AC filter inductor current i_a. The duration of the last stage with output voltage vector 110 is T_{V2}, in

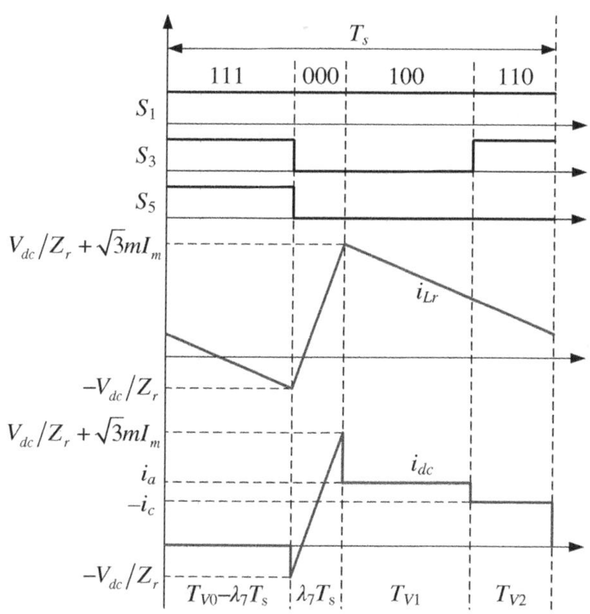

Figure A.6.2 Waveforms of resonant inductor i_{Lr} and DC bus current i_{dc} in one switching period.

which DC bus current i_{dc} is equal to sum of phase A's AC filter inductor current i_a and phase B's AC filter inductor current i_b, that is $-i_c$.

The resonant inductor i_{Lr} in one switching period can be expressed as:

$$i_{Lr} = \begin{cases} \dfrac{V_{dc}}{Z_r} + \sqrt{3}mI_m - \dfrac{2V_{dc}/Z_r + \sqrt{3}mI_m}{(1-\lambda_7)T_s}(t + T_s - T_{V0}) & 0 < t \le (T_{V0} - \lambda_7 T_s) \\ -\dfrac{V_{dc}}{Z_r} + \dfrac{2V_{dc}/Z_r + \sqrt{3}mI_m}{\lambda_7 T_s}[t - (T_{V0} - \lambda_7 T_s)] & (T_{V0} - \lambda_7 T_s) < t \le T_{V0} \\ \dfrac{V_{dc}}{Z_r} + \sqrt{3}mI_m - \dfrac{2V_{dc}/Z_r + \sqrt{3}mI_m}{(1-\lambda_7)T_s}(t - T_{V0}) & T_{V0} < t \le T_s \end{cases} \tag{A.6.2}$$

Then the DC bus current i_{dc} in one switching period within sector 1-1 can be derived from Figure A.6.2 as:

$$i_{dc} = \begin{cases} 0 & 0 < t \le (T_{V0} - \lambda_7 T_s) \\ -\dfrac{V_{dc}}{Z_r} + \dfrac{2V_{dc}/Z_r + \sqrt{3}mI_m}{\lambda_7 T_s}(t - T_{V0} + \lambda_7 T_s) & (T_{V0} - \lambda_7 T_s) < t \le T_{V0} \\ i_a & T_{V0} < t \le (T_{V0} + T_{V1}) \\ -i_c & (T_{V0} + T_{V1}) < t \le T_s \end{cases} \tag{A.6.3}$$

Due to the high efficiency of soft-switching inverter, the DC-side input power is almost equal to the AC-side output power. Thus the DC input current I_{in} can be derived as:

$$I_{in} = \frac{v_a i_a + v_b i_b + v_c i_c}{V_{dc}} \tag{A.6.4}$$

By substituting (A.1.8) and (A.5.1) into (A.6.4), the DC input current I_{in} can be simplified as:

$$I_{in} = \frac{3V_m I_m}{2V_{dc}} = \frac{\sqrt{3}}{2} mI_m \tag{A.6.5}$$

With the input and output parameters in Table 10.1, the waveforms of DC bus current i_{dc}, DC filter capacitor current i_{cap}, and DC input current I_{in} within sector 1-1 can be drawn as shown in Figure A.6.3.

As shown in Figure A.6.3, phase angle φ_1 meets:

$$I_{in} = \frac{\sqrt{3}}{2} mI_m = -i_c(\varphi_1) = -I_m \cos\left(\varphi_1 + \frac{2\pi}{3}\right) \tag{A.6.6}$$

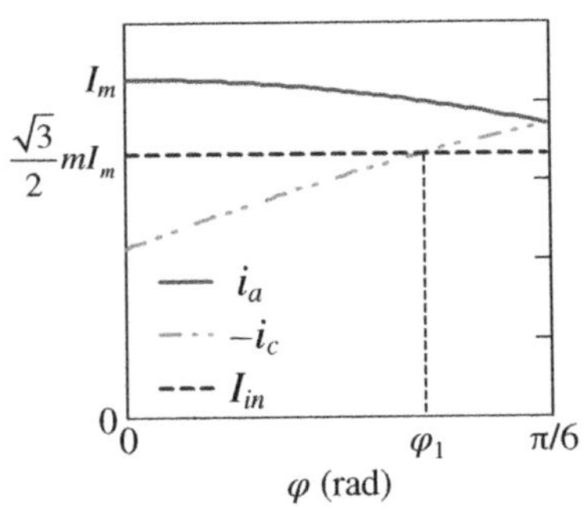

Figure A.6.3 Waveforms of i_a, $-i_c$, and DC input current I_{in} within sector 1-1.

Then phase angle φ_1 can be expressed as:

$$\varphi_1 = \arccos\left(-\frac{\sqrt{3}}{2}m\right) - \frac{2\pi}{3} \tag{A.6.7}$$

It is clear that the value range of φ_1 is from 0 to $\pi/6$. Within phase angle from 0 to φ_1, DC input current I_{in}, phase A's AC filter inductor current i_a, and phase C's AC filter inductor current i_c meet:

$$i_a > I_{in} > -i_c \tag{A.6.8}$$

Within phase angle from φ_1 to $\pi/6$, DC input current I_{in}, phase A's AC filter inductor current i_a, and phase C's AC filter inductor current i_c meet:

$$i_a > -i_c > I_{in} \tag{A.6.9}$$

The waveforms of time-domain DC bus current i_{dc}, DC filter capacitor current i_{cap}, and DC input current I_{in} in one switching period can be detained as show in Figure A.6.4.

The value of DC bus current i_{dc} at instant t_a meets:

$$i_{dc}(t_a) = -\frac{V_{dc}}{Z_r} + \frac{2V_{dc}/Z_r + \sqrt{3}mI_m}{\lambda_7 T_s}(t_a - T_{V0} + \lambda_7 T_s) = \frac{\sqrt{3}}{2}mI_m \tag{A.6.10}$$

Then the expression of instant t_a can be derived as:

$$t_a = T_{V0} - \frac{\lambda_7 T_s}{2} \tag{A.6.11}$$

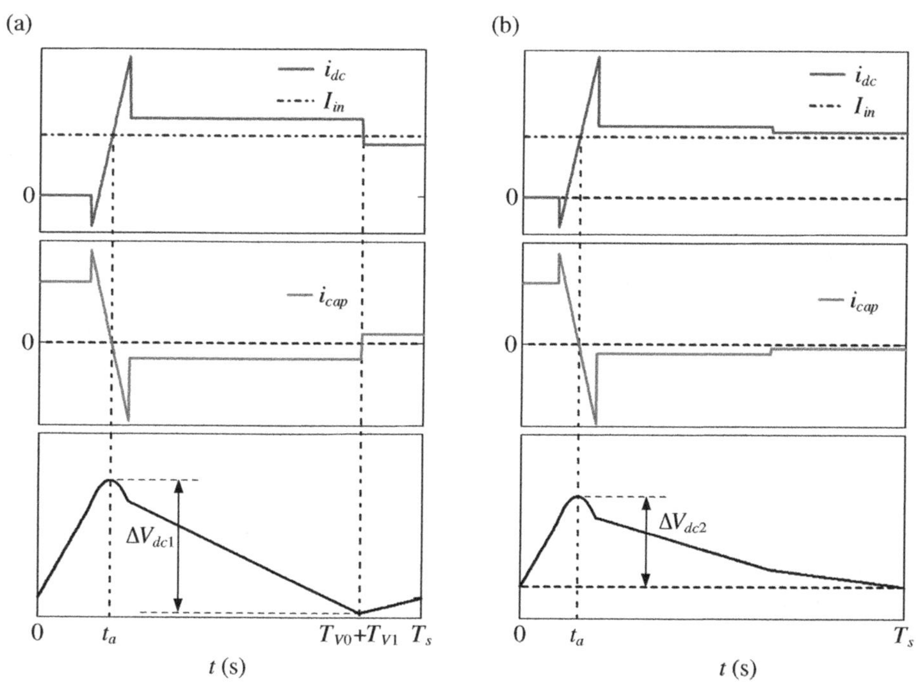

Figure A.6.4 Waveforms of time-domain DC bus current i_{dc}, DC filter capacitor current i_{cap}, and DC input current I_{in} in one switching period: (a) phase angle from 0 to φ_1 and (b) phase angle from φ_1 to $\pi/6$.

According to Figure A.6.4a, the DC filter capacitor voltage ripple ΔV_{dc1} within phase angle from 0 to φ_1 can be derived as:

$$\Delta V_{dc1} = \frac{1}{C_{dc}}\int_{t_a}^{T_{V0}+T_{V1}}(-i_{cap})dt = \frac{1}{C_{dc}}\int_{t_a}^{T_{V0}+T_{V1}}(i_{dc}-I_{in})dt$$
$$= \frac{1}{C_{dc}}\left\{\begin{array}{l}\int_{T_{V0}-\frac{\lambda_7 T_s}{2}}^{T_{V0}}\left(-\frac{V_{dc}}{Z_r}-\frac{\sqrt{3}}{2}mI_m+\frac{2V_{dc}/Z_r+\sqrt{3}mI_m}{\lambda_7 T_s}(t-T_{V0}+\lambda_7 T_s)\right)dt\\ +\int_{T_{V0}}^{T_{V0}+T_{V1}}\left(i_a-\frac{\sqrt{3}}{2}mI_m\right)dt\end{array}\right\}$$
$$= \frac{1}{C_{dc}}\left\{\left(\frac{V_{dc}}{Z_r}+\frac{\sqrt{3}}{2}mI_m\right)\frac{\lambda_7 T_s}{4}+\left[\cos(\varphi)-\frac{\sqrt{3}}{2}m\right]\sin\left(\frac{\pi}{3}-\varphi\right)mI_m(1-\lambda_7)T_s\right\} \quad (A.6.12)$$

As shown in (A.6.12), the DC filter capacitor voltage ripple ΔV_{dc1} decreases with the increment of phase angle from 0 to φ_1. Therefore, the minimum value of the DC filter capacitor voltage ripple ΔV_{dc1} can be obtained with phase angle equal to φ_1 as:

$$\Delta V_{dc1\_min} = \frac{1}{C_{dc}}\left\{\begin{array}{l}\left(\frac{V_{dc}}{Z_r}+\frac{\sqrt{3}}{2}mI_m\right)\frac{\lambda_7 T_s}{4}\\ +\left[\cos(\varphi_1)-\frac{\sqrt{3}}{2}m\right]\sin\left(\frac{\pi}{3}-\varphi_1\right)mI_m(1-\lambda_7)T_s\end{array}\right\} \quad (A.6.13)$$

The maximum value of the DC filter capacitor voltage ripple ΔV_{dc1} can be obtained with phase angle equal to zero as:

$$\Delta V_{dc1\_max} = \frac{1}{C_{dc}}\left[\left(\frac{V_{dc}}{Z_r}+\frac{\sqrt{3}}{2}mI_m\right)\frac{\lambda_7 T_s}{4}+\frac{\sqrt{3}}{2}\left(I_m-\frac{\sqrt{3}}{2}mI_m\right)(1-\lambda_7)mT_s\right] \quad (A.6.14)$$

According to Figure A.6.4b, the DC filter capacitor voltage ripple ΔV_{dc2} within phase angle from φ_1 to $\pi/6$ can be derived as:

$$\Delta V_{dc2} = \frac{1}{C_{dc}}\int_0^{t_a} i_{cap}dt$$
$$= \frac{1}{C_{dc}}\left\{\begin{array}{l}\int_0^{T_{V0}-\lambda_7 T_s}\frac{\sqrt{3}}{2}mI_m dt\\ +\int_{T_{V0}-\lambda_7 T_s}^{T_{V0}-\frac{\lambda_7 T_s}{2}}\left(\frac{\sqrt{3}}{2}mI_m+\frac{V_{dc}}{Z_r}-\frac{2V_{dc}/Z_r+\sqrt{3}mI_m}{\lambda_7 T_s}(t-T_{V0}+\lambda_7 T_s)\right)dt\end{array}\right\}$$
$$= \frac{1}{C_{dc}}\left\{\left[\frac{\sqrt{3}}{2}-\frac{\sqrt{3}}{2}m\sin\left(\frac{\pi}{3}+\varphi\right)\right]mI_m(1-\lambda_7)T_s+\left(\frac{\sqrt{3}}{2}mI_m+\frac{V_{dc}}{Z_r}\right)\frac{\lambda_7 T_s}{4}\right\} \quad (A.6.15)$$

As shown in (A.6.15), the DC filter capacitor voltage ripple ΔV_{dc2} decreases with the increment of phase angle from φ_1 to $\pi/6$. Therefore, the maximum value of the DC filter capacitor voltage ripple ΔV_{dc2} can be obtained with phase angle equal to φ_1 as:

$$\Delta V_{dc2\_max} = \frac{1}{C_{dc}}\left\{\begin{array}{l}\left(\frac{\sqrt{3}}{2}mI_m+\frac{V_{dc}}{Z_r}\right)\frac{\lambda_7 T_s}{4}\\ +\left[\frac{\sqrt{3}}{2}-\frac{\sqrt{3}}{2}m\sin\left(\frac{\pi}{3}+\varphi_1\right)\right]mI_m(1-\lambda_7)T_s\end{array}\right\} \quad (A.6.16)$$

By substituting the expression of φ_1 in (A.6.7) to (A.6.13) and (A.6.16), it is found that ΔV_{dc1_min} is equal to ΔV_{dc2_amx}. Obviously, the DC filter capacitor voltage ripple decreases with the increment of phase angle from 0 to $\pi/6$. Thus the maximum value of DC filter capacitor voltage ripple $\Delta V_{dc(\max)}$ in sector 1-1 can be expressed as (A.6.14). Then the DC filter capacitance C_{dc} can be derived as:

$$C_{dc} = \frac{1}{\Delta V_{dc(\max)}}\left[\left(\frac{V_{dc}}{Z_r} + \frac{\sqrt{3}}{2}mI_m\right)\frac{\lambda_7 T_s}{4} + \frac{\sqrt{3}}{2}\left(I_m - \frac{\sqrt{3}}{2}mI_m\right)(1-\lambda_7)mT_s\right] \quad \text{(A.6.17)}$$

Equation (A.6.17) can also be used to calculate the DC filter capacitance in hard-switching inverter by setting the turn-off duty cycle of auxiliary λ_7 to zero. Since λ_7 is always smaller than 0.1, its impact to DC filter capacitance is small. In order to simplify the analysis, DC filter capacitance in soft-switching inverter is identical to that in hard-switching inverter under the same switching frequency.

Index

Soft-Switching Technology for Three-phase Power Electronics Converters, First Edition.
Dehong Xu, Rui Li, Ning He, Jinyi Deng, and Yuying Wu.

c

d

t

u

Books in the IEEE Press Series on Power and Energy Systems

Series Editor: Ganesh Kumar Venayagamoorthy, Clemson University, Clemson, South Carolina, USA.

The mission of the IEEE Press Series on Power and Energy Systems is to publish leading-edge books that cover a broad spectrum of current and forward-looking technologies in the fast-moving area of power and energy systems including smart grid, renewable energy systems, electric vehicles and related areas. Our target audience includes power and energy systems professionals from academia, industry and government who are interested in enhancing their knowledge and perspectives in their areas of interest.

1. *Electric Power Systems: Design and Analysis, Revised Printing*
Mohamed E. El-Hawary

2. *Power System Stability*
Edward W. Kimbark

3. *Analysis of Faulted Power Systems*
Paul M. Anderson

4. *Inspection of Large Synchronous Machines: Checklists, Failure Identification, and Troubleshooting*
Isidor Kerszenbaum

5. *Electric Power Applications of Fuzzy Systems*
Mohamed E. El-Hawary

6. *Power System Protection*
Paul M. Anderson

7. *Subsynchronous Resonance in Power Systems*
Paul M. Anderson, B.L. Agrawal, J.E. Van Ness

8. *Understanding Power Quality Problems: Voltage Sags and Interruptions*
Math H. Bollen

9. *Analysis of Electric Machinery*
Paul C. Krause, Oleg Wasynczuk, and S.D. Sudhoff

10. *Power System Control and Stability, Revised Printing*
Paul M. Anderson, A.A. Fouad

11. *Principles of Electric Machines with Power Electronic Applications*, Second Edition
Mohamed E. El-Hawary

12. *Pulse Width Modulation for Power Converters: Principles and Practice*
D. Grahame Holmes and Thomas Lipo

13. *Analysis of Electric Machinery and Drive Systems*, Second Edition
Paul C. Krause, Oleg Wasynczuk, and S.D. Sudhoff

14. *Risk Assessment for Power Systems: Models, Methods, and Applications*
Wenyuan Li

15. *Optimization Principles: Practical Applications to the Operations of Markets of the Electric Power Industry*
Narayan S. Rau

16. *Electric Economics: Regulation and Deregulation*
Geoffrey Rothwell and Tomas Gomez

17. *Electric Power Systems: Analysis and Control*
Fabio Saccomanno

18. *Electrical Insulation for Rotating Machines: Design, Evaluation, Aging, Testing, and Repair*
Greg C. Stone, Edward A. Boulter, Ian Culbert, and Hussein Dhirani

19. *Signal Processing of Power Quality Disturbances*
Math H. J. Bollen and Irene Y. H. Gu

20. *Instantaneous Power Theory and Applications to Power Conditioning*
Hirofumi Akagi, Edson H. Watanabe and Mauricio Aredes

21. *Maintaining Mission Critical Systems in a 24/7 Environment*
Peter M. Curtis

22. *Elements of Tidal-Electric Engineering*
Robert H. Clark

23. *Handbook of Large Turbo-Generator Operation and Maintenance,* Second Edition
Geoff Klempner and Isidor Kerszenbaum

24. *Introduction to Electrical Power Systems*
Mohamed E. El-Hawary

25. *Modeling and Control of Fuel Cells: Distributed Generation Applications*
M. Hashem Nehrir and Caisheng Wang

26. *Power Distribution System Reliability: Practical Methods and Applications*
Ali A. Chowdhury and Don O. Koval

27. *Introduction to FACTS Controllers: Theory, Modeling, and Applications*
Kalyan K. Sen, Mey Ling Sen

28. *Economic Market Design and Planning for Electric Power Systems*
James Momoh and Lamine Mili

29. *Operation and Control of Electric Energy Processing Systems*
James Momoh and Lamine Mili

30. *Restructured Electric Power Systems: Analysis of Electricity Markets with Equilibrium Models*
Xiao-Ping Zhang

31. *An Introduction to Wavelet Modulated Inverters*
S.A. Saleh and M.A. Rahman

32. *Control of Electric Machine Drive Systems*
Seung-Ki Sul

33. *Probabilistic Transmission System Planning*
Wenyuan Li

34. *Electricity Power Generation: The Changing Dimensions*
Digambar M. Tagare

35. *Electric Distribution Systems*
Abdelhay A. Sallam and Om P. Malik

36. *Practical Lighting Design with LEDs*
Ron Lenk, Carol Lenk

37. *High Voltage and Electrical Insulation Engineering*
Ravindra Arora and Wolfgang Mosch

38. *Maintaining Mission Critical Systems in a 24/7 Environment, Second Edition*
Peter Curtis

39. *Power Conversion and Control of Wind Energy Systems*
Bin Wu, Yongqiang Lang, Navid Zargari, Samir Kouro

40. *Integration of Distributed Generation in the Power System*
Math H. Bollen, Fainan Hassan

41. *Doubly Fed Induction Machine: Modeling and Control for Wind Energy Generation Applications*
Gonzalo Abad, Jesús López, Miguel Rodrigues, Luis Marroyo, and Grzegorz Iwanski

42. *High Voltage Protection for Telecommunications*
Steven W. Blume

43. *Smart Grid: Fundamentals of Design and Analysis*
James Momoh

44. *Electromechanical Motion Devices, Second Edition*
Paul Krause, Oleg Wasynczuk, Steven Pekarek

45. *Electrical Energy Conversion and Transport: An Interactive Computer-Based Approach, Second Edition*
George G. Karady, Keith E. Holbert

46. *ARC Flash Hazard and Analysis and Mitigation*
J.C. Das

47. *Handbook of Electrical Power System Dynamics: Modeling, Stability, and Control*
Mircea Eremia, Mohammad Shahidehpour

48. *Analysis of Electric Machinery and Drive Systems, Third Edition*
Paul C. Krause, Oleg Wasynczuk, S.D. Sudhoff, Steven D. Pekarek

49. *Extruded Cables for High-Voltage Direct-Current Transmission: Advances in Research and Development*
Giovanni Mazzanti, Massimo Marzinotto

50. *Power Magnetic Devices: A Multi-Objective Design Approach*
S.D. Sudhoff

51. *Risk Assessment of Power Systems: Models, Methods, and Applications, Second Edition*
Wenyuan Li

52. *Practical Power System Operation*
Ebrahim Vaahedi

53. *The Selection Process of Biomass Materials for the Production of Bio-Fuels and Co-Firing*
Najib Altawell

54. *Electrical Insulation for Rotating Machines: Design, Evaluation, Aging, Testing, and Repair, Second Edition*
Greg C. Stone, Ian Culbert, Edward A. Boulter, and Hussein Dhirani

55. *Principles of Electrical Safety*
Peter E. Sutherland

56. *Advanced Power Electronics Converters: PWM Converters Processing AC Voltages*
Euzeli Cipriano dos Santos Jr., Edison Roberto Cabral da Silva

57. *Optimization of Power System Operation, Second Edition*
Jizhong Zhu

58. *Power System Harmonics and Passive Filter Designs*
J.C. Das

59. *Digital Control of High-Frequency Switched-Mode Power Converters*
Luca Corradini, Dragan Maksimoviæ, Paolo Mattavelli, Regan Zane

60. *Industrial Power Distribution, Second Edition*
Ralph E. Fehr, III

61. *HVDC Grids: For Offshore and Supergrid of the Future*
Dirk Van Hertem, Oriol Gomis-Bellmunt, Jun Liang

62. *Advanced Solutions in Power Systems: HVDC, FACTS, and Artificial Intelligence*
Mircea Eremia, Chen-Ching Liu, Abdel-Aty Edris

63. *Operation and Maintenance of Large Turbo-Generators*
Geoff Klempner, Isidor Kerszenbaum

64. *Electrical Energy Conversion and Transport: An Interactive Computer-Based Approach*
George G. Karady, Keith E. Holbert

65. *Modeling and High-Performance Control of Electric Machines*
John Chiasson

66. *Rating of Electric Power Cables in Unfavorable Thermal Environment*
George J. Anders

67. *Electric Power System Basics for the Nonelectrical Professional*
Steven W. Blume

68. *Modern Heuristic Optimization Techniques: Theory and Applications to Power Systems*
Kwang Y. Lee, Mohamed A. El-Sharkawi

69. *Real-Time Stability Assessment in Modern Power System Control Centers*
Savu C. Savulescu

70. *Optimization of Power System Operation*
Jizhong Zhu

71. *Insulators for Icing and Polluted Environments*
Masoud Farzaneh, William A. Chisholm

72. *PID and Predictive Control of Electric Devices and Power Converters Using MATLAB®/Simulink®*
Liuping Wang, Shan Chai, Dae Yoo, Lu Gan, Ki Ng

73. *Power Grid Operation in a Market Environment: Economic Efficiency and Risk Mitigation*
Hong Chen

74. *Electric Power System Basics for Nonelectrical Professional, Second Edition*
Steven W. Blume

75. *Energy Production Systems Engineering*
Thomas Howard Blair

76. *Model Predictive Control of Wind Energy Conversion Systems*
Venkata Yaramasu, Bin Wu

77. *Understanding Symmetrical Components for Power System Modeling*
J.C. Das

78. *High-Power Converters and AC Drives, Second Edition*
Bin Wu, Mehdi Narimani

79. *Current Signature Analysis for Condition Monitoring of Cage Induction Motors: Industrial Application and Case Histories*
William T. Thomson, Ian Culbert

80. *Introduction to Electric Power and Drive Systems*
Paul Krause, Oleg Wasynczuk, Timothy O'Connell, Maher Hasan

81. *Instantaneous Power Theory and Applications to Power Conditioning, Second Edition*
Hirofumi, Edson Hirokazu Watanabe, Mauricio Aredes

82. *Practical Lighting Design with LEDs, Second Edition*
Ron Lenk, Carol Lenk

83. *Introduction to AC Machine Design*
Thomas A. Lipo

84. *Advances in Electric Power and Energy Systems: Load and Price Forecasting*
Mohamed E. El-Hawary

85. *Electricity Markets: Theories and Applications*
Jeremy Lin, Jernando H. Magnago

86. *Multiphysics Simulation by Design for Electrical Machines, Power Electronics and Drives*
Marius Rosu, Ping Zhou, Dingsheng Lin, Dan M. Ionel, Mircea Popescu, Frede Blaabjerg, Vandana Rallabandi, David Staton

87. *Modular Multilevel Converters: Analysis, Control, and Applications*
Sixing Du, Apparao Dekka, Bin Wu, Navid Zargari

88. *Electrical Railway Transportation Systems*
Morris Brenna, Federica Foiadelli, Dario Zaninelli

89. *Energy Processing and Smart Grid*
James A. Momoh

90. *Handbook of Large Turbo-Generator Operation and Maintenance, 3rd Edition*
Geoff Klempner, Isidor Kerszenbaum

91. *Advanced Control of Doubly Fed Induction Generator for Wind Power Systems*
Dehong Xu, Dr. Frede Blaabjerg, Wenjie Chen, Nan Zhu

92. *Electric Distribution Systems, 2nd Edition*
Abdelhay A. Sallam, Om P. Malik

93. *Power Electronics in Renewable Energy Systems and Smart Grid: Technology and Applications*
Bimal K. Bose

94. *Distributed Fiber Optic Sensing and Dynamic Rating of Power Cables*
Sudhakar Cherukupalli, and George J Anders

95. *Power System and Control and Stability,* Third Edition
Vijay Vittal, James D. McCalley, Paul M. Anderson, and A. A. Fouad

96. *Electromechanical Motion Devices: Rotating Magnetic Field-Based Analysis and Online Animations*, Third Edition
Paul Krause, Oleg Wasynczuk, Steven D. Pekarek, and Timothy O'Connell

97. *Applications of Modern Heuristic Optimization Methods in Power and Energy Systems*
Kwang Y. Lee and Zita A. Vale

98. *Handbook of Large Hydro Generators: Operation and Maintenance*
Glenn Mottershead, Stefano Bomben, Isidor Kerszenbaum, and Geoff Klempner

99. *Advances in Electric Power and Energy: Static State Estimation*
Mohamed E. El-hawary

100. *Arc Flash Hazard Analysis and Mitigation,* Second Edition
J.C. Das

101. *Maintaining Mission Critical Systems in a 24/7 Environment,* Third Edition
Peter M. Curtis

102. *Real-Time Electromagnetic Transient Simulation of AC-DC Networks*
Venkata Dinavahi and Ning Lin

103. *Probabilistic Power System Expansion Planning with Renewable Energy Resources and Energy Storage Systems*
Jaeseok Choi and Kwang Y. Lee

104. *Power Magnetic Devices: A Multi-Objective Design Approach,* Second Edition
Scott D. Sudhoff

105. *Optimal Coordination of Power Protective Devices with Illustrative Examples*
Ali R. Al-Roomi

106. *Resilient Control Architectures and Power Systems*
Craig Rieger, Ronald Boring, Brian Johnson, and Timothy McJunkin

107. *Alternative Liquid Dielectrics for High Voltage Transformer Insulation Systems: Performance Analysis and Applications*
Edited by U. Mohan Rao, I. Fofana, and R. Sarathi

108. *Introduction to the Analysis of Electromechanical Systems*
Paul C. Krause, Oleg Wasynczuk, and Timothy O'Connell

109. *Power Flow Control Solutions for a Modern Grid using SMART Power Flow Controllers*
Kalyan K. Sen and Mey Ling Sen

110. *Power System Protection: Fundamentals and Applications*
John Ciufo and Aaron Cooperberg

111. *Soft-Switching Technology for Three-phase Power Electronics Converters*
Dehong Xu, Rui Li, Ning He, Jinyi Deng, Yuying Wu

Printed and bound by CPI Group (UK) Ltd, Croydon, CR0 4YY
17/09/2026
14964322-0007